Plant MicroRNAs and Stress Response

Editors
Deepu Pandita
Senior Lecturer, Government Department of School Education
Jammu, Union Territory of Jammu and Kashmir, India

Anu Pandita
Senior Dietician, Vatsalya Clinic
New Delhi, India

CRC Press is an imprint of the
Taylor & Francis Group, an **informa** business

Cover Credit: Cover illustration courtesy of Viki-2505 - Freepik.com

First edition published 2024
by CRC Press
6000 Broken Sound Parkway NW, Suite 300, Boca Raton, FL 33487-2742

and by CRC Press
4 Park Square, Milton Park, Abingdon, Oxon, OX14 4RN

© 2024 Deepu Pandita and Anu Pandita

CRC Press is an imprint of Taylor & Francis Group, LLC

Reasonable efforts have been made to publish reliable data and information, but the author and publisher cannot assume responsibility for the validity of all materials or the consequences of their use. The authors and publishers have attempted to trace the copyright holders of all material reproduced in this publication and apologize to copyright holders if permission to publish in this form has not been obtained. If any copyright material has not been acknowledged please write and let us know so we may rectify in any future reprint.

Except as permitted under U.S. Copyright Law, no part of this book may be reprinted, reproduced, transmitted, or utilized in any form by any electronic, mechanical, or other means, now known or hereafter invented, including photocopying, microfilming, and recording, or in any information storage or retrieval system, without written permission from the publishers.

For permission to photocopy or use material electronically from this work, access www.copyright.com or contact the Copyright Clearance Center, Inc. (CCC), 222 Rosewood Drive, Danvers, MA 01923, 978- 750-8400. For works that are not available on CCC please contact mpkbookspermissions@tandf.co.uk

Trademark notice: Product or corporate names may be trademarks or registered trademarks and are used only for identification and explanation without intent to infringe.

ISBN: 978-1-032-34449-2 (hbk)
ISBN: 978-1-032-34450-8 (pbk)
ISBN: 978-1-003-32221-4 (ebk)
DOI: 10.1201/9781003322214

Typeset in Times New Roman
by Innovative Processors

Deepu Pandita dedicates this book to

Dr. Saurabh Raghuvanshi

Professor & Project Lead, Department of Plant Molecular Biology,
University of Delhi South Campus, New Delhi, India

and

Indian Biological Data Centre (IBDC), Regional Centre for Biotechnology,
NCR Biotech Science Cluster, Faridabad, India

who introduced her to the world of Plant microRNAs

Preface

MicroRNAs (miRNAs) are small (20–24 nucleotides), single stranded, non-coding, regulatory RNA molecules or gene regulators. These molecules show critical transcriptional or post-transcriptional gene regulation in plants and respond to numerous biotic and abiotic stress factors. The miRNAs are evolutionarily conserved and widely distributed throughout the plant kingdom. Plant stresses like drought, salinity, heat, cold, UV radiation, heavy metals, and microbial infections affect survival, growth and development, quality and quantity and produce of plants. The miRNAs are master regulators of plant growth and development. The stress-responsive miRNAs attenuate growth and development under various environmental stress cues. Stress induced miRNAs down regulate their target mRNA levels. The downregulation of mRNA levels leads to accumulation and function of positive miRNA regulators, recognizing their roles in crop stress response and tolerance. Plant miRNA mediated modifications include overexpression or repression of stress-responsive miRNAs and/ or their target complementary or partially complementary gene products, miRNA-resistant target genes, target-mimics and artificial miRNAs. Thus, miRNAs serve as novel and potent "genomic gold mines", and potential targets in plant genetic manipulations and miRNA-based biotechnology that aid in plant genetic engineering for crop improvement and plant tolerance to different environmental stress scenarios.

This book reviews recent knowledge on plant miRNAs, Biogenesis, Physiological Significance, Converging and Diverging Insights, Regulation of Plant Growth and Development, Expression Profiling of miRNAs, miRNA encoded peptides, Crosstalk of MicroRNAs with Phytohormone Signalling Pathways, role of miRNAs as Molecular Markers in Plants, Plant-Microbe Interactions, Crop Improvement and Epigenetics, and miRNA based biotic and abiotic stress-response regulations in green kingdom. This book on Plant MicroRNAs and Stress Response holds tremendous value in terms of:

1. With the increase in global climate change and challenges to agriculture, this book will be valuable gem for understanding the regulatory networks of stress responsive genes and miRNAome.
2. The companies working on genetically engineered plants can take a cue from this book for designing of climate resilient crops.

3. The book will prove as an important chaperon for the post graduate students, university and college teachers, and plant science researchers, biotechnologists, bioinformaticians, scientists in academia, private sectors and industries.

No such book with compiled knowledge on stress associated microRNAs is available currently. This will be first of its kind.

Deepu Pandita
Anu Pandita

About the Editors

Deepu Pandita is working as a Senior Lecturer in Government Department of School Education, Jammu, Union Territory of Jammu and Kashmir, India. She has 20 years of teaching experience and has done her Masters in Botany (M.Sc. from University of Kashmir) and Master of Philosophy (M.Phil.) in Biotechnology from University of Jammu, Jammu & Kashmir, India. She has a number of international and national courses to her credit and qualified fellowships like JRF NET and SRF from Council of Scientific & Industrial Research (CSIR), New Delhi, India; Biotechnology Fellowship, Government Department of Science and Technology, Jammu & Kashmir, India and IAS-INSA-NASI Summer Research Teacher Fellowship, India. Deepu Pandita has presented her research papers at both the national and international conferences and is recipient of the Women Researcher Award and Research Excellence Award from various professional associations in India. She is a life-time member of various scientific societies and is reviewer (of 18 journals), associate editor and editor of a number of international journals. She has published a number of editorials, book chapters (Springer, Elsevier, CRC, etc.), reviews and research articles in various journals of national and international repute like *Cells*, *Frontiers in Plant Sciences*, *Journal of Fungi* and *Frontiers in Physiology* and currently seven of her books are under production.

Anu Pandita is working as a Senior Dietician at Vatsalya Clinic, New Delhi, India. Previously she worked as a Lecturer at Bee Enn College of Nursing, Talab Tillo, Jammu, and as Dietician in Ahinsa Dham Bhagwan Mahavir Charitable Health Centre, New Delhi. She has done her MSC internship and a course at the Dietetics Department of PGI, Chandigarh, India. She has completed various trainings, refresher courses and workshops and is a life-time member of Indian Dietetic Association and Indian Science Congress Association, Kolkata, India. Anu Pandita has published various book chapters for Springer, CRC, etc. and written research articles in various journals of national and international repute, like *Cells*, *Journal of Fungi* and *Frontiers in Physiology* besides having seven books under production.

Contents

Preface		v
About the Editors		vii
List of Contributors		xi

1. MicroRNA-induced Silencing Complex Assembly
   and MicroRNA Turnover   1
   Lekha Bhagtaney and Priya Sundarrajan

2. MicroRNAs in Plants and Animals: Converging and Diverging Insights   15
   *Humaira Shah, Auqib Manzoor, Tabasum Ashraf, Rouf Maqbool
   and Ashraf Dar*

3. Regulatory Roles of Plant MicroRNAs   50
   *Sehrish Ijaz, Vajiha Sahar Khan, Ayesha Ghazanfar
   and Zulqarnain Khan*

4. MicroRNA-mediated Regulation of Plant Growth and Development   65
   Seyed Alireza Salami and Shirin Moradi

5. Silencing of Stress-regulated miRNAs in Plants   74
   *Abeer Hashem, Hesham Ali El Enshasy, Roshanida Rahmat,
   Ghazala Muteeb and Elsayed Fathi Abd_Allah*

6. MicroRNA-mediated Regulation of Heat Stress Response   90
   Vincent Ezin and Rachael C. Symonds

7. MicroRNA-mediated Regulation of Drought Stress Response   120
   Seyed Alireza Salami, Neda Arad

8. MicroRNA-mediated Regulation of UV Radiation Stress Response   144
   Sonam Dwivedi, Elhan Khan, Iffat Zareen Ahmad

9. MicroRNA-mediated Regulation of Salinity Stress   167
   Seyed Alireza Salami, Shirin Moradi

10. MicroRNA-mediated Regulation of Cold and Chilling Stress Response   175
    Seyed Alireza Salami, Shirin Moradi

11. MicroRNA-mediated Regulation of Heavy Metal Stress in Plants   183
    Swarnavo Chakraborty, Aryadeep Roychoudhury

12.	MicroRNA-mediated Regulation of Osmotic and Oxidative Stress	198
	Ramachandra Reddy Pamuru, T. Chandrasekhar and Arifullah Mohammed	
13.	MicroRNA-mediated Regulation of Herbicide Resistance	220
	Rafiq Lone, Parvaiz Yousuf, Shahid Razzak, Semran Parvaiz	
14.	MicroRNA-mediated Regulation of Plant Viral Disease Development	242
	Saurabh Pandey, Suresh H. Antre, Saumya Kumari, and Ashutosh Singh	
15.	Crosstalk of MicroRNAs with Phytohormone Signalling Pathways	257
	Shilpy Singh, Ruth Assumi, Pooja Bhadrecha	
16.	Role of MicroRNAs in Plant-Microbe Interactions	277
	Bushra Hafeez Kiani	
17.	Micro-RNA: A versatile tool as Molecular Markers in Plants	302
	Parthasarathy Seethapathy, Reena Sellamuthu, Dhivyapriya Dharmaraj, Harish Sankarasubramanian, Anandhi Krishnan, Anu Pandita and Deepu Pandita	
18.	MicroRNAs and other Non-coding RNAs in Plant Epigenetics	329
	Auqib Manzoor, Tabasum Ashraf, Humaira Shah, Rouf Maqbool, Rachna Kaul and Ashraf Dar	
19.	MicroRNA-based Plant Genetic Engineering for Crop Improvement	360
	Heena Tabassum, Iffat Zareen Ahmad	
20.	The miRNA-encoded Peptides	379
	Pooja Bhadrecha, Shilpy Singh, Arun Kumar	
21.	Plant MicroRNAs: Physiological Significance in Plants and Animals	394
	Idris Ali Dar, Masarat Bashir, Ashraf Dar	
Index		424

List of Contributors

Abeer Hashem
Botany and Microbiology Department, College of Science, King Saud University, P.O. Box 2460, Riyadh, Saudi Arabia
Mycology and Plant Disease Survey Department, Plant Pathology Research Institute, ARC, Giza, Egypt

Hesham Ali El Enshasy
Institute of Bioproduct Development (IBD), Universiti Teknologi Malaysia (UTM), Skudai, Johor Bahru, Malaysia
School of Chemical and Energy Engineering, Faculty of Engineering, Universiti Teknologi Malaysia (UTM), Skudai, Johor Bahru, Malaysia
City of Scientific Research and Technology Applications, New Burg Al Arab, Alexandria, Egypt

Roshanida Rahmat
Institute of Bioproduct Development (IBD), Universiti Teknologi Malaysia (UTM), Skudai, Johor Bahru, Malaysia
School of Chemical and Energy Engineering, Faculty of Engineering, Universiti Teknologi Malaysia (UTM), Skudai, Johor Bahru, Malaysia

Ghazala Muteeb
College of Applied Medical Sciences, King Faisal University, Al-Hasa, Saudi Arabia

Elsayed Fathi Abd_Allah
Plant Production Department, College of Food and Agricultural Sciences, King Saud University, P.O. Box 2460, Riyadh, Saudi Arabia

Seyed Alireza Salami
Department of Horticultural Sciences, Faculty of Agricultural Science and Engineering, University of Tehran, Iran

Shirin Moradi
Department of Horticultural Sciences, Faculty of Agricultural Science and Engineering, University of Tehran, Iran

Neda Arad
School of Plant Sciences, University of Arizona, Tucson, AZ, USA

Ruth Assumi
Scientist, ARS, Division of System Research and Engineering, ICAR Research Complex for NEH Region, Umiam, Meghalaya, India

Zulqurnain Khan
Department of Biotechnology, Institute of Plant Breeding and Biotechnology (IPBB), MNS University of Agriculture, Old Shujaabad Road, Multan, Pakistan

Sehrish Ijaz
Department of Biotechnology, Institute of Plant Breeding and Biotechnology (IPBB), MNS University of Agriculture, Old Shujaabad Road, Multan, Pakistan

Vajiha Sahar Khan
Department of Biotechnology, Institute of Plant Breeding and Biotechnology (IPBB), MNS University of Agriculture, Old Shujaabad Road, Multan, Pakistan

Ayesha Ghazanfar
Department of Biotechnology, Institute of Plant Breeding and Biotechnology (IPBB), MNS University of Agriculture, Old Shujaabad Road, Multan, Pakistan

Vincent Ezin
Department of Crop Production, Faculty of Agricultural Science, University of Abomey-Calavi, 01 BP 526 Cotonou, Benin

Ramachandra Reddy Pamuru
Department of Biochemistry, Yogi Vemana University, Vemanapuram, Kadapa, A.P., India

Lekha Bhagtaney
Caius Research Laboratory, St. Xavier's College (Autonomous), Mumbai

Priya Sundarrajan
Caius Research Laboratory, St. Xavier's College (Autonomous), Mumbai
Department of Life Science and Biochemistry, St. Xavier's (Autonomous), Mumbai

Humaira Shah
Department of Biochemistry, University of Kashmir, Srinagar, Jammu & Kashmir, India

Auqib Manzoor
Department of Biochemistry, University of Kashmir, Srinagar, Jammu & Kashmir, India

Tabasum Ashraf
Department of Biochemistry, University of Kashmir, Srinagar, Jammu & Kashmir, India

Rouf Maqbool
Department of Biochemistry, University of Kashmir, Srinagar, Jammu & Kashmir, India

Ashraf Dar
Department of Biochemistry, University of Kashmir, Srinagar, Jammu & Kashmir, India

List of Contributors

Rachna Kaul
Bombay College of Pharmacy, Kalina, Mumbai, Maharashtra, India

Idris Ali Dar
Department of Biochemistry, University of Kashmir, Srinagar, Jammu & Kashmir, India

Masarat Bashir
College of Temparate Sericulture, SKUAST-K, Shalimar, Srinagar, India

Rachael C. Symonds
School of Biological and Environmental Sciences, Liverpool John Moores University, UK

Anu Pandita
Vatsalya Clinic, Krishna Nagar, New Delhi, India

Sonam Dwivedi
Natural Products Laboratory, Department of Bioengineering and Biosciences, Integral University, Dasauli, Kursi Road, Lucknow, Uttar Pradesh, India

Heena Tabassum
Dr. D. Y. Patil Biotechnology and Bioinformatics Institute, Dr. D. Y. Patil Vidyapeeth, Pune, India

Elhan Khan
Natural Products Laboratory, Department of Bioengineering and Biosciences, Integral University, Dasauli, Kursi Road, Lucknow, Uttar Pradesh, India

Iffat Zareen Ahmad
Natural Products Laboratory, Department of Bioengineering and Biosciences, Integral University, Dasauli, Kursi Road, Lucknow, Uttar Pradesh, India

Swarnavo Chakraborty
Post Graduate Department of Biotechnology, St. Xavier's College (Autonomous), 30, Mother Teresa Sarani, Kolkata, West Bengal, India

Aryadeep Roychoudhury
Post Graduate Department of Biotechnology, St. Xavier's College (Autonomous), 30, Mother Teresa Sarani, Kolkata, West Bengal, India

T. Chandrasekhar
Department of Environmental Science, Yogi Vemana University, Kadapa, A.P., India

Arifullah Mohammed
Department of Agrotechnology, Faculty of Agro Based Industry (FIAT), Universiti Malaysia Kelantan, Jeli, Kelantan, Malaysia

Saurabh Pandey
Department of Agriculture, Guru Nanak Dev University, Amritsar, Punjab, India

Suresh H. Antre
University of Agricultural Sciences, GKVK Campus, Bangalore, India

Saumya Kumari
University of Agricultural Sciences, GKVK Campus, Bangalore, India

Ashutosh Singh
Centre for Advanced Studies on Climate Change, RPCAU, Pusa, Samastipur, India

Pooja Bhadrecha
University Institute of Biotechnology, Chandigarh University, Punjab, India

Shilpy Singh
Department of Biotechnology, Noida International University, Uttar Pradesh, India

Arun Kumar
Department of Agriculture Biotech, Sardar Vallabhbhai Patel University of Agriculture and Technology, Uttar Pradesh, India

Bushra Hafeez Kiani
Department of Biological Sciences (Female Campus) International Islamic University, Islamabad, Pakistan

Parthasarathy Seethapathy
Department of Plant Pathology, Amrita School of Agricultural Sciences, Amrita Vishwa Vidyapeetham, Coimbatore, India

Reena Sellamuthu
Department of Plant Biotechnology, Amrita School of Agricultural Sciences, Amrita Vishwa Vidyapeetham, Coimbatore, India

Dhivyapriya Dharmaraj
Department of Plant Breeding and Genetics, Amrita School of Agricultural Sciences, Amrita Vishwa Vidyapeetham, Coimbatore, India

Harish Sankarasubramanian
Department of Plant Pathology, Tamil Nadu Agricultural University, Coimbatore, India

Anandhi Krishnan
Department of Plant Breeding and Genetics, Tamil Nadu Agricultural University, Coimbatore, India

Deepu Pandita
Government Department of School Education, Jammu, Jammu and Kashmir, India

Rafiq Lone
Department of Botany, Central University of Kashmir, Ganderbal, Jammu and Kashmir, India

Parvaiz Yousuf
Department of Zoology, Central University of Kashmir, Ganderbal, Jammu and Kashmir, India

Shahid Razzak
Department of Zoology, Central University of Kashmir, Ganderbal, Jammu and Kashmir, India

Semran Parvaiz
Department of Zoology, Central University of Kashmir, Ganderbal, Jammu and Kashmir, India

CHAPTER

1

MicroRNA-induced Silencing Complex Assembly and MicroRNA Turnover

Lekha Bhagtaney[1] and Priya Sundarrajan[1,2*]

[1] Caius Research Laboratory, St. Xavier's College (Autonomous), Mumbai
[2] Department of Life Science and Biochemistry, St. Xavier's College (Autonomous), Mumbai

1. Introduction

Exploring the transcriptome of the eukaryotic genome revealed that only 2% of the 90% transcribed RNA yield protein products. The remaining transcribed RNAs are non-coding RNAs that are essential at many levels of gene expression regulation (Waititu et al., 2020). Non-coding RNAs operate by interacting with other proteins and forming ribonucleotide complexes. These complexes recognise their target RNAs via base pairing and facilitate transcriptional or post-transcriptional silencing. As a result, these assemblies are extensively studied. Essential factors participating in gene regulation are small, non-coding RNAs. These RNAs are typically 20-35 nucleotides long and can be classified mainly into four groups: transfer RNA-derived small RNAs (tsRNAs), small interfering RNA (siRNA), PIWI-interacting RNA (piRNA) and microRNAs (miRNAs). This classification is based on their origin, how they are processed and the associated downstream proteins. The effect of siRNAs and piRNAs is exercised either at the transcriptional or post-transcriptional level whereas miRNAs are responsible for gene regulation, principally at the post-transcriptional level (Wang et al., 2019).

The generation of microRNA (miRNA), a conserved regulator of post-transcriptional gene expression, involves several steps: transcription, processing of precursor, methylation and ultimately, the miRNA-induced silencing complex (miRISC) ensemble. Regulation by miRNAs is indispensable in many processes involving development and metabolism of an organism. miRNA transcription occurs in a way that is similar to that of genes coding for proteins; however, it is highly

---

*Corresponding author: priya.s@xaviers.edu

regulated as the expression is specific to the tissue or developmental stage. Drosha and dicer enzymes of the RNase III family are responsible for the conversion of precursor miRNAs to mature miRNAs. Proteins regulating this step can interact with the miRNA precursor or associate with Drosha and Dicer. The primary core of miRISC is generated by miRNA association with the Argonaute (AGO) protein family and Glycine-tryptophan protein of 182 kDa (GW182). miRISC acts by stimulating target degradation of mRNA repression of translation. The RNA binding proteins incident on the target mRNA may influence the activity of the miRISC (Krol *et al.*, 2010; Pandita 2019; Pandita and Wani, 2019; Pandita, 2021; 2022a; 2022b; 2022c).

Since long, miRNAs have been perceived as highly stable biomolecules owing to observations that they remain for several hours even after a halt in their production (Rüegger and Großhans, 2012). It has also been indicated that miRNA turnover is rapid or slow, depending on the cell type. The regulation of this miRNA stability is facilitated by the presence of uracil or adenosine residues at the 3′ end of the miRNA. Cellular components, such as multivesicular bodies, may be involved in miRNA inactivation or degradation (Krol *et al.*, 2010).

2. miRNA Biogenesis

Gene duplication in several species has given rise to numerous miRNA loci with associated sequences. The mature miRNAs containing the same sequence from position two to eight are considered to belong to the same family. Fourteen paralogous foci, belonging to the let-7 family, have been observed in the human genome (Ha and Kim, 2014).

RNA polymerase II or III produces the precursor miRNA which is processed co- or post-transcriptionally, marking the beginning of miRNA biogenesis. Most of the known miRNAs arise from introns and are intragenic. Some miRNAs are intergenic, transcribed autonomously and are under the control of their own promoters. Occasionally, a lengthy transcript, containing several miRNAs with common seed regions, will be generated and these transcripts are called 'clusters' (O'Brien *et al.*, 2018). Co-transcription occurs in the same miRNA cluster whilst regulation of individual miRNAs occurs at the level of post-transcription. The cluster involving miR-100~let-7~miR-125 is evolutionary conserved and is essential for bilaterian animal development (Ha and Kim, 2014). Biogenesis of miRNA is categorised into canonical and non-canonical pathways.

2.1 Canonical Pathway

Processing of miRNAs through canonical pathway is a predominant way of miRNA biogenesis. The miRNA transcribed from the gene, known as pri-miRNA, is modified with the help of a microprocessor complex. The complex consists of the proteins, namely Drosha, a ribonuclease III enzyme and DiGeorge Syndrome Critical Region 8 (DGCR8) which is a RNA binding protein (RBP). N6-methyladenylated GGAC motif containing pri-miRNA that is identified by DGCR8 and the pri-miRNA duplex is cleaved at the hairpin base by the protein Drosha, leading to the production of a pre-miRNA. The pre-miRNA 3′ end consists of a two-nucleotide overhang. After the production, the pre-miRNA is bound by exportin 5 (XPO5)/RanGTP complex and exported to the cytoplasm. Cleavage of the terminal loop by Dicer, an RNase III endonuclease, induces the generation of a mature miRNA from the pre-miRNA.

The mature miRNA is labelled according to the orientation of the miRNA strand. The 5p strand arises from the 5′ end of the hairpin of the pre-miRNA whereas the 3′ end gives the 3p strand. The Argonaute (AGO) family of proteins can accommodate any one of the two strands in an ATP-dependent way. The cell type or the environment of the cell determines the proportion of 5p or 3p strand that is accommodated on AGO protein. It can range from preponderantly one or the other to approximately equivalent proportions. Partially, the selection of the strands is dictated by a 5′ uracil being present at position 1 or the 5′ end of the miRNA duplex in terms of thermodynamic stability. Loading of the less stable 5′ end or 5′ uracil is preferential on to AGO and is regarded as the guide strand. Through processes related to the degree of complementarity, the additional strand, called the 'passenger strand', is uncoiled from the guide strand. AGO2 cleaves the passenger strand which is later cleaved by the cellular machinery (O'Brien *et al.*, 2018).

2.2 Non-canonical Pathway

Although certain components of the canonical pathway participate in the biogenesis of these miRNAs, they differ from the classic miRNA by genomic origin. Mirtrons are a specific type of miRNA derived from the intragenic segments of the genome. Mirtrons had been first identified in *Caenorhabditis elegans* and *Drosophila melanogaster* and since then, it has been observed in various entities from plants to mammals (García-López *et al.*, 2013; Abdelfattah *et al.*, 2014).

Like canonically processed miRNAs, mirtrons arising from the mRNA intronic regions create structures like the double-stranded loops. However, the spliceosome machinery converts them to pre-miRNA. After intron lariat formation, the spliced mRNA product acquires a pre-miRNA-like structure which further continues its processing in the canonical pathway. This type of miRNA biogenesis is Drosha/DGCR8 independent and has been understood through knockout studies of DGCR8 and mutation studies of Drosha. A study utilised embryonic stem cells (ESCs) with knockout DGCR8 and found that even though miR-484 and miR-320 are produced non-canonically, they are produced in low concentrations. Such models of DGCR8 knockout can help in understanding molecular mechanisms that cause phenotypes on canonical miRNA loss (Guo and Wang, 2019). In most mirtrons, the double-stranded regions overlap with intron ends and their prime ends, derived by excision through cleavage at donor as well as acceptor sites, are produced by the spliceosome. While in some pre-mitrons, an additional processing step is required as it may maintain a tail that is single-stranded at the 5′ or the 3′ end. This step is accomplished by the exosomal elements of RNA, like the exonuclease of the nucleus, Rrp6 which trims at the 3′ tail and halts at the pre-mirtron double-stranded stem (García-López *et al.*, 2013). Although the 5′ tailed mirtron biogenesis is not well understood, the configuration studies have revealed the possible involvement of XRN family of 5′-3′ exoribonucleases (Yang and Lai, 2011). After this, the mirtron can proceed to be exported to the cytoplasm and cleaved by Dicer.

In another non-canonical pathway, Drosha/DCGR8 cleave the pri-miR-451 in the nucleus and produce a short pre-miRNA. Dicer substrates are required to be 20-27 bp in length in order to be processed by it. Being 18 bp long, this pre-miRNA can be cleaved by an endonuclease at the 3′ end and loaded on AGO2 or is directly loaded on AGO1 which is the non-slicing form. This is the Dicer-independent pathway. It was demonstrated that AGO2 generated from a loss of function mutation significantly reduces the miR-451 produced by the Dicer-independent pathway. This suggests that AGO2 is an important regulator of the Dicer-independent pathway (Havens *et al.*, 2012). A study reported that E-cadherin, a junctional complex protein, can induce AGO2 accumulation in the cell membrane. This can lead to an enhanced biogenesis of miR-451a facilitated by AGO2 and Dicer-independent pathway. The study indicates the implication of miRNA biogenesis in metastasis of cancers (Li *et al.*, 2018).

The variant of a mirtron, known as a simtron, has been reported. The features of simtron are that it is processed independently of spliceosomes, DGCR8, Dicer, XPO5 or AGO2. These variants are processed by Drosha. The knockout or Drosha's expression in the dominant negative form reduced the biogenesis of simtrons (Havens *et al.*, 2012; García-López *et al.*, 2013; Abdelfattah *et al.*, 2014).

2.3 sno-derived miRNAs

An abundant non-coding RNA, found in the nucleolus of several organisms, is the small nucleolar RNA (snoRNAs). It has been demonstrated that these RNAs modulate translational level gene expression by AGO binding. It has been believed that these RNAs can modify snRNA, tRNA and rRNA. When the deep sequencing data of snoRNAs related to AGO1 and AGO2 was analysed, it was found that these RNAs are an alternative source of miRNAs following the canonical pathway occasionally. ACA45 snoRNA found in humans, however, underwent processing in the Dicer-dependent, Drosha/ DGCR8-independent fashion. The structure of precursor ACA45 is a hinge connecting hairpins of two pre-miRNA. Translational repression through GlsR17 snoRNA was observed in *Giardia lamblia* which is processed in a Dicer-dependent way. These studies have implied that snoRNA plays a role in RNA silencing. However, more studies in this field are required for a better understanding of their operation (Abdelfattah *et al.*, 2014).

2.4 miRNA from tRNA

It was also discovered that the tRNA maturation pathway generates by-products in the form of miRNA-like species. The non-stoichiometric accumulation of miRNA-like species in relation to the mature tRNA produced suggests that it may exhibit some functional activity. Even though their activity is not facilitated by AGO proteins, 3′ miRNA-like species are tractable to AGO levels. They participate in RNA silencing but are different from the classical miRNAs. These RNAs are produced by tRNA-processing enzymes, like tRNase Z. The pre-miRNAs are then converted to their mature form by Dicer.

Murine gamma herpes virus 68 (MHV68) has many different miRNAs. The pri-miRNA stage is connected to a tRNA which is processed to a mature miRNA from one arm of the hairpin of a miRNA. The 5′ end of the hairpins is defined by the tRNase Z, whereas the mature miRNA is generated from pre-miRNA by Dicer (Abdelfattah *et al.*, 2014; Yang and Lai, 2011).

2.5 shRNA

When non-mirtronic regions in the genome were analysed for microprocessor independent processing, mir-320 and mir-484 were discovered to be DGCR8-independent and Dicer-independent. These RNAs were labelled as short hairpin RNAs (shRNA) as they could also form short hairpin structures. In order to be accurately identified by the microprocessor complex, the canonically processed pre-miRNAs have conserved flanked regions. Since shRNAs are processed independently of the microprocessor complex, they lack the conserved flanked regions (Abdelfattah *et al.*, 2014).

2.6 siRNA

Processing of long stem-loop structures or double-stranded RNAs by Dicer gives rise to siRNAs that are ~21 nucleotides in length. Their main function is regulation at the post-transcriptional regulation of mRNAs and transposons. The RNAi system facilitates antiviral defence in several plants and invertebrates by producing siRNAs from dsRNA during viral life cycles. The protein from the Dicer family that is involved in biogenesis of siRNA is Dicer-2 to form RISC siRNAs interact with AGO2 through the cytosine residue at 5′ end (Ha and Kim, 2014).

AGO1 has seldom been shown to bind to siRNAs. The amount of miRNA is modulated by a 5′ to 3′ RNA degrading enzyme Fiery, in *Arabidopsis*. It functions by repression of biogenesis of siRNAs from ribosomal RNAs. If such siRNAs are loaded on to AGO1, it causes decreased miRNA loading (You *et al.*, 2019).

2.7 Regulation of Biogenesis

Discrimination between the pri-miRNA transcript and other transcripts containing hairpin for recognition by the microprocessor complex needs identifiers. The identifiers include SR protein (SRSF3)-binding CNNC motif present downstream of pri-miRNA hairpins, UGUG motif at the apex, basal UG motif and a stem ~35 bp long which has a mismatched motif with GHG. Another way by which the pri-miRNA is recognised is by a N6-methyladenosine (m6 A) mark near the stem loop region. A2/B1, a ribonucleoprotein, reads the mark, binds to DGCR8 and activates the processing of miRNA. The heterotrimeric complex of Drosha and two DGCR8 molecules comprises the microprocessor complex. The efficiency of miRNA processing is imparted by the RNA binding heme region and dsRNA-binding domains of the DGCR8 dimer interacting with the stem and apical components of the pri-miRNA (Michlewski and Caceres, 2019).

The quantity of miRNA produced is dependent on the Drosha-mediated processing efficiency. Drosha is regulated at three levels – specificity, activity and expression. The homeostasis of the microprocessor complex is maintained via an autoregulatory mechanism between DGCR8 and Drosha. The binding of DGCR8 leads to the stabilisation of Drosha through protein-protein interaction. Cleavage of DGCR8 mRNA at the second exon hairpin by Drosha destabilises it.

Microprocessor activity, nuclear localisation and protein stabilisation can be controlled by post-translational modification. Acetylation and phosphorylation of Drosha leads to different outcomes (Ha and Kim, 2014). Drosha is acetylated to prevent its proteasomal degradation, hence stabilises it. P300, GCN5 or CBP are thought to participate in the acetylation of Drosha at the N'-terminal region (Pradhan *et al.*, 2017). Glycogen synthase kinase 3 β (GSK3 β) phosphorylates Drosha for its localisation to the nucleus. Similar to regulation of Drosha, DGCR8 is regulated by phosphorylation, deacetylation and MECP2. The stability of DGCR8 is enhanced when it is phosphorylated by ERK. Histone deacetylase 1 (HDAC1) is responsible for the deacetylation of DGCR8 and brings about an increased specificity for pri-miRNAs. Methyl-CpG-binding protein 2 (MECP2), when phosphorylated, binds to DGCR8. In an instance of neuronal activity, there is rapid dephosphorylation of MECP2 which releases DGCR8 to allow for growth of dendrites and production of miRNA (Ha and Kim, 2014).

Apart from miRNAs, transposon-derived transcripts, non-coding RNAs and mRNAs are also regulated by the microprocessor complex. The cellular RNA turnover is affected by the microprocessor's non-canonical activity. The expression of neurogenin, for instance, is negatively regulated by Drosha as the hairpin in neurogenin is analogous to the one in pri-miRNAs. A population of cellular RNAs has been shown to be controlled by alternative DGCR8 complexes, which are DGCR8 interacting with other nucleases. For instance, Drosha-independent snoRNAs are cleaved by these DGCR8 complexes.

Dicer and chaperone Hsp90 bind to pre-miRNA in the cytoplasm, forming a complex. HIV-1 TAR RNA RBP (TRBP) and protein activator of PKR (PACT), in conjunction with Dicer, cleave the pre-miRNA. The site where the cleavage occurs is next to the apical loop and Dicer assesses the distance from the basal end of the pre-miRNA to the site. After cleavage, the apical loop is released and the RNA duplex is generated. This duplex further goes on to interact with AGO2.

Cleavage facilitated by Dicer is frequently inaccurate because the structurally pre-miRNAs are variable. The products of the cleavage are miRNA duplex variants that will produce mature miRNAs. RISC complex is formed by loading of miRNA duplex on to AGO2 mediated by HSC70/HSP90 chaperones – a process that is ATP-dependent. On loading, the duplex is unwound by AGO2, the passenger strand is expelled and a mature RISC complex is generated (Michlewski and Caceres, 2019).

3. RISC Assembly

RISC assembly includes two phases: loading miRNA duplexes on to AGO proteins, also regarded as RISC loading, and unwinding the duplex to dissociate the miRNA strands. Pre-RISC is referred to the stage where the miRNA duplex is associated with AGO and mature RISC or holo-RISC is the stage where only the guide miRNA remains on the AGO protein (Fig. 1) (Kawamata and Tomari, 2010).

3.1 RISC Loading

Associated RISC loading proteins are essential for loading duplex miRNAs on to AGO. In Drosophila, R2D2, a dsRNA binding protein along with Dicer-2 form the RISC loading complex for AGO2. This heterodimer mediates RISC loading based on the thermodynamic disparity of the duplexes, resulting in Dicer-2 binding to the end with a lesser stability and R2D2 binding to the end with higher stability. Hence, the RNA duplex polarisation is determined on loading (Kawamata and Tomari, 2010). Four dsRNA binding protein isoforms are coded by the Loquacious (LOQS) R3D1 gene. miRNA production is carried out by Dicer where Loqs-PB acts as a cofactor (Liang *et al.*, 2015). LOQS may play a role in loading of miRNA on to AGO1 mediating RISC assembly (Liang *et al.*, 2015). Analogous to the Dicer-2-R2D2 mechanism, LOQS may also partner Dicer-1 to load miRNA on to AGO-1 (Kawamata *et al.*, 2009).

RNA features, such as the thermodynamic disparity of the duplex, the nucleotide present on the 5′ end and the structure of the duplex in Drosophila are judged by the loading complex before preferentially loading small RNA on to AGO1 or AGO2. Therefore, the miRNA-miRNA* orientation, with the 5′ nucleotide U on the miRNA,

MicroRNA-induced Silencing Complex Assembly and MicroRNA Turnover

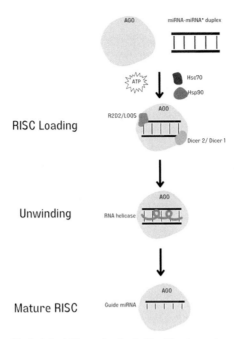

Fig. 1: RISC assembly (original illustration by Lekha Bhagtaney (created using Canva))

is loaded on AGO1 and the miRNA*-miRNA orientation, with the miRNA* nucleotide C, is loaded on AGO2. Additionally, the miRNA of the miRNA-miRNA* possess mismatches at the 9-10 positions.

Sorting of small RNA in plants, like *Arabidopsis*, show the affinity of AGO proteins to a specific 5' nucleotide - AGO1 is biased to U while AGO2 and AGO4 to A, whereas AGO5 to C. This suggests that the sorting in plants is highly dependent on the identity of the 5' nucleotide. The identification of domains that are responsible for 5' nucleotide recognition was conducted by switching the domains conserved in AGO proteins. These domains include PIWI, MID, PAZ and N-terminal. When PIWI and MID were exchanged in AGO1 and AGO2 proteins of *Arabidopsis*, their nucleotide preferences also changed. As in the case of mammals, the RISC loading associated small RNA sorting is not complicated. Central mismatches at positions 8-11 in miRNA duplexes can help in the integration of the duplex in any of the four human AGO proteins. Human AGO proteins do not display differences in preference for the 5' nucleotide (Kawamata and Tomari, 2010). The event of RISC loading is ATP dependent. The duplexes are large enough to fit comfortably into AGO proteins as demonstrated by structural studies on eukaryotic as well as prokaryotic AGO proteins. A conformational change of AGO proteins, facilitated by Hsc70/Hsp90 chaperones along with ATP hydrolysis, allows for acceptance of the miRNA duplex in flies and plants. RISC loading in humans is attenuated by inhibition of Hsc70/Hsp90 chaperones, thereby demonstrating their importance (Betancur, 2012).

3.2 Unwinding of siRNA

Unwinding as a process can be dependent on slicer or independent of slicer. siRNA unwinding involves the cleavage of the passenger strand by AGO proteins. The presence of central mismatches prevents miRNA-miRNA* duplexes from employing the unwinding mechanism involving passenger strand cleavage. Helicases have been demonstrated to be important in assembly of RISC (Ambrus and Frolov, 2009). The helicase domain DExD/H-box is an important feature of Dicer (Kawamata *et al.*, 2009). The role of helicases for unwinding is essential at two steps – one at RISC assembly, where the guide strand is separated from the passenger strand post Dicer-processing and the other, the target mRNA gets separated from the guide miRNA. Apart from these, RNA helicases have been implicated in several steps in the entire RNA interference process. The generation of pre-miRNA from pri-RNAs is mediated by Drosha/DGCR8 complex which may include p68, a helicase important for its function. Dicer complex processed pre-miRNAs into mature miRNA duplexes. The loading of the duplex on RISC may be facilitated by RNA helicase A and the unwinding of the duplex occurs with the help of a putative RNA helicase. The guide strand which is a part of the mature RISC can now target its mRNA. Conversely, the guide strand may load on another RISC complex after dissociating from the previous one, aided by an RNA helicase Armi. The target mRNA undergoes translational repression following binding with miRNA. The subsequent separation of miRNA from RISC is mediated by another RNA helicase (Ambrus and Frolov, 2009).

This type of unwinding is inherently slower as compared to slicer-dependent unwinding since absence of passenger strand cleavage promotes stability. However, internal mismatches in the 3′ mid region (12-16 position) or the seed region (2-8 position) encourage slicer-independent unwinding. True miRNA-miRNA* complexes containing 3′ mid, seed and central mismatches advance RISC generation along with any of the AGO proteins, mammalian in origin (Betancur, 2012). Additionally, the PAZ domain thermal dynamics regulates unwinding. When the PAZ domain was truncated, the passenger release decreased (Nakanishi, 2016). The event of unwinding is ATP independent as opposed to RISC loading. This suggests that RISC loading ATP-dependent step pre-charges the duplex for separation of the strands (Betancur, 2012).

4. Translational Repression and Target Degradation

The ~22 nt long miRNA within the miRISC binds to its target mRNA to carry out translational repression. The target sequence in the mRNA where miRNA binds is known as miRNA responsive element (MRE). In plants and seldom in animals, the miRNA and the target mRNA exhibit precise base pairing, resulting in target mRNA endonucleolytic cleavage by AGO proteins. A partial complementarity is observed more often in animals where cleavage is prevented and the target RNA is consequently destabilised and undergoes translational repression (Chipman and Pasquinelli, 2019). Here the endonuclease activity of AGO proteins is impaired due to central mismatches in the guide miRNA (O'Brien *et al.*, 2018). Seed is the region in a miRNA (2-7 nucleotides long) where the target site of the mRNA is engaged. Binding of the target mRNA to miRISC results in a change in the conformation of AGO protein that uncovers the 3′ region of the miRNA so that it can form more contacts with the target (Chipman and Pasquinelli, 2019).

GW182 proteins are recruited to miRISC, allowing for silencing miRISC complex to be generated. The framework needed for recruitment of other proteins, like CCR4-NOT, PAN2-PAN3 and poly(A)-deadenylase complexes, is provided by GW182. PAN2/3 commences target mRNA poly(A)-deadenylation and CCR4-NOT complex concludes it. Deadenylation is the most efficient when poly(A)-binding protein C (PABPC) interacts with tryptophan repeats in GW182. Following this, decapping protein 2 (DCP2), along with other proteins, mediates de-capping. Finally, exoribonuclease 1 (XRN1) facilitates degradation of the target mRNA in the 5'-3' direction (O'Brien *et al.*, 2018).

The downregulation of gene expression is facilitated by miRNA through mRNA cleavage or translational repression. Once RISC translocates to the cytoplasm, the type of mechanism is chosen based on complementarity of the mRNA to the miRNA. Cleavage occurs on adequate complementarity whereas inadequate complementarity will result in repressed translation. There exist differences in sequence composition at 12 and 13 positions of the metazoan miRNAs and siRNAs, indicating their preference towards target mRNA translational repression or cleavage. miR172, a plant miRNA regulates the Apetala2 gene through translational repression instead of mRNA cleavage even though Apetala2 ORF has a complementary site for the miRNA. The cleavage occurs at the site complementary to the 10 or 11 position in the miRNA. The cleavage position remains consistent when the 5' end pairing of the target to the miRNA is not precise. This implies that the relative miRNA residues and not the miRNA-target mRNA sequence pairing is the determinant of the cleavage site.

A classic example of translational repression is lin-4 miRNA of *Caenorhabditis elegans* which causes the repression of lin-14 mRNA. This was observed as a correlation between an increase in lin-4 expression leading to a decrease in lin-14 protein without a change in the lin-14 mRNA. A similar observation was made with lin-28 mRNA which is also a target of lin-4 miRNA. To understand the mechanism behind these results, two explanations were put forth. One indicates the repression of translation post initiation by stalling of ribosomes on the mRNA; the other explanation says that the rate of translation does not change, however the polypeptide that is synthesised is selectively degraded (Bartel, 2004).

The understanding of the complete mechanism of translational repression remains incomplete whereas there has been a significant study of mRNA destabilisation by miRNA and the associated molecular networks.

Translational repression and mRNA degradation and their comparative contribution have been analysed by high throughput methods at a global scale, though ribosome protection assays have revealed that translational repression occurs more frequently than hypothesised earlier. This assay involves ribosome which is immobilised on mRNAs and the RNA content, which is protected by the following RNase treatment, is sequenced via high-throughput techniques. A comparison was made of the data obtained on sequencing and the proteomics data and total mRNA levels. Several researchers have also demonstrated that translational repression occurs, irrespective of the mRNA going for further degradation. The effect of miRNA on translation has been demonstrated *in vitro*. Reporter mRNAs with a non-poly(A) sequence following the poly(A) sequence were resistant to deadenylation, validating that there is no need for removal of the poly(A) tail for translational repression. Another way to show that the degradation of mRNA is not necessarily following translational repression was the reactivation of translationally repressed mRNA (Wilczynska and Bushell, 2015).

4.1 miRNA Turnover

The consensus on miRNA has been that it is inherently stable. However, recent discoveries indicate that the stability of a miRNA is highly dependent on its environment. In some cell types or environments, there has been an observation of accelerated decay and hence the activity of miRNA is regulated. The factors influencing the stability of miRNA are discussed below.

miRNAs are essential in cell cycle regulation as they modulate cyclin/CDK complexes. Conversely, there are some stages of the cell cycle that observe the accumulation of specific RNAs. For instance, miR-29b is transcribed in HeLa cells along with miR29a as a polycistronic transcript. These miRNAs differ from each other by six nucleotides at the 3′ end and a nucleotide at the 10th position. The concentration levels of miR-29a remain more or less constant throughout the cell cycle; however, miR-29b levels are enhanced in cells undergoing mitosis. Transfection experiments carried out with synthetic miR-29b demonstrated that the accumulation of this miRNA occurs in cells arrested at mitosis and its regulation takes place after it has been converted to its mature form. The half-life of miR-29b was found to be four hours in cells undergoing division and more than 12 hours in mitosis-arrested cells, whereas that of miR-29a was more than 12 hours in both cases, as observed by pulse-chase experiments. The relatively quick degradation of miR-29b has been attributed, although not entirely, to uracil nucleotides at 9-11 positions.

A decrease in levels of miRNA have been observed upon addition of growth factors in *in vitro* studies. For example, when MCF10A cells (immortalised human breast epithelial cells) were subjected to EGF after starvation of serum to halt their proliferation, there was a decrease of 23 miRNAs by more than 50% in an hour. These experiments indicated that the miRNA downregulation by addition of growth factors can induce a physiological response by a cell. In case of MCF10A cells, the miRNA targets that are activated by presence of EGF were downregulated. Nevertheless, it is unsure whether there is alteration of unstable miRNA transcription or processing or whether the miRNA is degraded on induction by EGF (Rüegger and Großhans, 2012).

The major participants in biogenesis and function of miRNA are AGO proteins. They also help in the miRNA stabilisation and prevent their degradation in animals and plants. Presence of loss-of-function mutations in AGO2 causes drastic reduction in the stability of miRNA, whereas if AGO2 is overexpressed, the half-life of miRNA increases and the miRNA is protected from degradation. Similarly, mutations in AGO1 can significantly decrease the abundance of miRNAs. The function of AGO involved in stability of miRNAs is unrelated to its slicing function as demonstrated by a slicing-deficient mutant which showed no reduction in miRNA abundance.

The AGO protein PAZ domain binds the 3′ end of miRNA whereas the phosphate of the 5′ end is bound by the MID domain. This protects both ends from the action of nucleases, thereby preventing degradation. The significant difference in levels of the guide and the passenger strand itself demonstrates the role of AGO in stabilisation of miRNA and presence of a system for decay of small RNAs that are unprotected (Sanei and Chen, 2015).

Mimicking the target mRNA has a regulatory effect on the activity of miRNA. In *Arabidopsis*, miR-399 binds to a site on the non-protein coding IPS1 transcript. The cleavage activity of miR-399 is prevented as IPS1 mRNA, which has a three-nucleotide long bulge between position 10 and 11, acts as a target mimic. This causes sequestration of miR-399 and therefore, miR-399 cannot act on its endogenous target. Such target mimics have been commonly observed in plants and could regulate the activity of several miRNAs. Apart from non-coding regions, certain intragenic and annotated genes also code for target mimic RNAs. There also exists a tissue-specific variation in expression of target mimics which impart a spatial and temporal regulation of miRNA. Expression of target mimics leads to degradation of miRNA. This was seen in a study that used artificial target mimic RNAs in transgenic lines which lead to the reduction in related miRNAs.

It has been proposed that target mimics might affect the stability of miRNA by stimulating changes in the structure of miRISC. On analysing the structure of the AGO of *Thermus thermophilus*, its target RNA and guide DNA, it was found that the guide is released from the 3′ end attached to the PAZ domain on binding of a highly complementary target RNA. However, the same is not observed with a target RNA having lower complementarity to the guide. If applied to the eukaryotic context with target RNA interaction with miRISC, it would imply that an extensive complementarity of the target RNA to its miRNA could allow for the exonuclease to come in contact with the 3′ end of the miRNA, thus, contributing to its stability (Sanei and Chen, 2015).

Another factor imparting stability to miRNAs is adenylation, which can be accomplished by a number of enzymes. In the liver, miR-122 3′ end adenylation is facilitated by GLD2, a poly(A) RNA polymerase. The neutralisation of this process is mediated by PARN, a poly(A) ribonuclease. This was observed when an increased level of miR-122 with enhanced stability was observed in cell lines of human liver cancer with diminished PARN. miRNA 3′ end containing uracil residues curbs the activity of the miRNA. The addition of residues is mediated by the enzyme terminal uridylyltransferase 4 (TUT4). The miR-26b uridylation by TUT4 resulted in a decrease in repression of target in human adenocarcinoma cell line. It has also been observed that different species have preferences for uridylation and adenylation, for instance, in humans and mice, adenylation is preferred whereas C. elegans prefers uridylation (Gebert and MacRae, 2019).

The sequence of miRNA dictates the turnover of miRNAs as indicated in the case of the presence of the nucleotide at the 5′ end and a uracil prevents quick degradation whereas a cytosine or guanine accelerates it. The turnover is also known to be miRNA-specific. In 3T3 mouse fibroblast cells, miR-16 is inherently unstable whereas all the other miRNAs are regarded as stable molecules. The case with miR-16 instability in fact helps in facilitating cell cycle progression. Tissue-specific turnover rates have also been observed. For instance, neuronal miRNAs have a higher turnover rate compared to other tissues (Gebert and MacRae, 2019).

Degradation of miRNAs is facilitated by a number of exonucleases. In plants, such as *Arabidopsis*, the degradation of miRNA in 3′ to 5′ direction is carried out by small RNA degrading nuclease (SND). The levels of miRNA had significantly increased on knockout of SDN proteins and it also led to certain developmental defects. Methylation of miRNA leads to partial suppression of SDN1. SDN1 acts on targets which do not have a uracil residue on the 3′ end. XRN-2 is an exonuclease found in *C. elegans* that degrades miRNAs in the 5′ to 3′ direction. *Arabidopsis* also codes for

XRN protein which mainly acts on de-capped miRNAs. Certain XRN proteins cause the degradation of precursor miRNA loops during biogenesis. Mutations in these proteins lead to increase miRNA loop levels.

The 3′ to 5′ exonuclease DIS3L2, in mammals, induces pre-let-7 degradation due to presence of oligouridylation. The targets that are uridylated are more tightly bound by DIS3L2 compared to the non-uridylated ones. Uridylated miRNAs are also degraded by enzymes, like Suppressor of Varicose (SOV) in *Arabidopsis* (Sanei and Chen, 2015).

5. miRNA Editing

RNA sequence modification, as a result of particular changes to the RNA molecule, is referred to as RNA editing. This process includes several different enzymes present in different species of plants and animals. A process that includes substitution of nucleotides, known as substitutional RNA editing, occurs across metazoans and is facilitated by deaminases. Adenosine deaminases causing A-to-I editing are known as adenosine deaminases acting on RNA (ADARs) and cytidine deaminases mediating C-to-U editing are known as activation-induced cytidine deaminases/apolipoprotein B mRNA editing enzyme (AID/APOBEC). A-to-I editing is regarded as a process where inosine (I) is produced from adenosine (A) deamination, whereas C-to-U editing is where uridine (U) is produced from cytidine (C) deamination. Despite there being evidence of RNA editing on coding RNAs, it has also been observed that RNA editing is more predominant in non-coding RNAs. Instances of similarity in the editing mechanisms of both bird and mammal miRNAs suggest that it is a conserved mechanism occurring in the miRNA seed region.

In mice and *C. elegans*, the knockout of ADAR results in altered miRNA levels due to improper pri- and pre-miRNA processing. miRNA editing can lead to either an increased turnover or an enhanced stability. Certain mismatches are created on adenosine deaminations, which are identified by RNA-binding proteins specific for inosine with endonuclease activity, such as endonuclease V (ENDOV) and Tudor staphylococcal nuclease (Tudor-SN). For instance, in HEK293 cells pri-miR-142 editing led to a halt in the cleavage induced by DROSHA and edited miRNA degradation by Tudor-SN. Another study demonstrated that the maturation of oncogenic miRNAs in the brain, like miR-21, miR-222 and miR-221, is disabled by editing their precursors mediated by ADAR2. An example of increased miRNA stability by editing is that of miR-376a-1. Editing miR-376a-1 at position 4 enhances its stability (Correia de Sousa *et al.*, 2019).

miRISC formation and interaction with target mRNA can also be influenced by editing of miRNA. A study was conducted on a white spot syndrome virus (WSSV)-affected *Marsupenaeus japonicus* to understand how ADAR of the host edits the viral miRNA. The miR-N12 of WSSV undergoes A-to-I editing at position 16. Since the edit was made in a non-seed region, the interaction of the miRNA with the AGO protein was spared. However, a disruption between the association of miR-N12 with its target wsv399 was observed. This ultimately resulted in the demonstration of virus latency and the WSSV infection being inhibited (Cui *et al.*, 2015).

The variants of miRNA – edited and unedited and their existence – could impact biological systems in the healthy as well as diseased conditions.

6. Conclusion

Scientists have made huge strides in understanding the mechanisms right to the molecular level of synthesis and activity of miRNAs. Specificity for substrate and mediating translational repression and target degradation are the two fundamental characteristics of miRISC. The components on which these characteristics are dependent include target mRNA and AGO protein. Several organisms exhibit miRNA turnover, demonstrating its importance in the regulation of miRNA. It is essential that regulation of gene expression is understood spatio-temporally. This is due to the fact that the activity exhibited miRISC is reversible. Till now, many studies have focused on individual miRNA for understanding the regulation of gene expression by miRISC. The future of these studies is to focus on miRNAs and RBPs in combination.

References

Abdelfattah, A.M., Park, C. and Choi, M.Y. (2014). Update on non-canonical microRNAs, *Biomol. Concepts*, 5(4): 275-287.

Ambrus, A.M. and Frolov, M.V. (2009). The diverse roles of RNA helicases in RNAi, *Cell Cycle*, 8(21): 3500-3505.

Bartel, D.P. (2004). MicroRNAs: Genomics, biogenesis, mechanism, and function, *Cell*, 116(2): 281-297.

Betancur, J.G., Yoda, M. and Tomari, Y. (2012). miRNA-like duplexes as RNAi triggers with improved specificity, *Front. Genet.*, 3: 127.

Chipman, L.B. and Pasquinelli, A.E. (2019). miRNA targeting: Growing beyond the seed, *Trends Genet.*, 35(3): 215-222.

Correia de Sousa, M., Gjorgjieva, M., Dolicka, D., Sobolewski, C. and Foti, M. (2019). Deciphering miRNAs' action through Arg editing, *Int. J. Mol. Sci.*, 20(24): 1-22.

Cui, Y., Huang, T. and Zhang, X. (2015). RNA editing of microRNA prevents RNA-induced silencing complex recognition of target mRNA, *Open Biol.*, 5(12): 150126.

García-López, J., Brieño-Enríquez, M.A. and Del Mazo, J. (2013). MicroRNA biogenesis and variability, *Biomol. Concepts*, 4(4): 367-380.

Gebert, L.F.R. and MacRae, I.J. (2019). Regulation of microRNA function in animals, *Nat. Rev. Mol. Cell Biol.*, 20(1): 21-37.

Guo, W.T. and Wang, Y. (2019). Dgcr8 knockout approaches to understand microRNA functions *in vitro* and *in vivo*, *Cell. Mol. Life Sci.*, 76(9): 1697-1711.

Ha, M. and Kim, V.N. (2014). Regulation of microRNA biogenesis, *Nat. Rev. Mol. Cell Biol.*, 15(8): 509-524.

Havens, M.A., Reich, A.A., Duelli, D.M. and Hastings, M.L. (2012). Biogenesis of mammalian microRNAs by a non-canonical processing pathway, *Nucleic Acids Res.*, 40(10): 4626-4640.

Kawamata, T., Seitz, H. and Tomari, Y. (2009). Structural determinants of miRNAs for RISC loading and slicer-independent unwinding, *Nat. Struct. Mol. Biol.*, 16(9): 953-960.

Kawamata, T. and Tomari, Y. (2010). Making RISC, *Trends Biochem. Sci.*, 35(7): 368-376.

Krol, J., Loedige, I. and Filipowicz, W. (2010). The widespread regulation of microRNA biogenesis, function and decay, *Nat. Rev. Genet.*, 11(9): 597-610.

Li, J.N., Kuo, Y.L., Ku, W.H., Lu, Y.-J., Wang, M.Y. and Chen, P.S. (2018). Membrane-anchored E-cadherin/AGO2 complex promote non-canonical miRNA biogenesis of miR-451a, *FASEB J.*, 32(S1).

Liang, C., Wang, Y., Murota, Y., Liu, X., Smith, D., Siomi, M.C. and Liu, Q. (2015). TAF11 assembles the RISC loading complex to enhance RNAi efficiency, *Mol. Cell.*, 59(5): 807-818.

Michlewski, G. and Cáceres, J.F. (2019). Post-transcriptional control of miRNA biogenesis, *RNA*, 25(1): 1-16.

Nakanishi, K. (2016). Anatomy of RISC: How do small RNAs and chaperones activate Argonaute proteins? Wiley Interdiscip., *Rev. RNA*, **7**(5): 637-660.

O'Brien, J., Hayder, H., Zayed, Y. and Peng, C. (2018). Overview of microRNA biogenesis, mechanisms of actions, and circulation, *Front. Endocrinol.* (Lausanne), 9: 402.

Pandita, D. (2019). Plant MIRnome: miRNA biogenesis and abiotic stress response. pp. 449-474. *In:* Hasanuzzaman, M., Hakeem, K., Nahar, K. and Alharby, H. (Eds.). *Plant Abiotic Stress Tolerance.* Springer, Cham. Doi https://doi.org/10.1007/978-3-030-06118-0_18

Pandita, D. and Wani, S.H. (2019). MicroRNA as a tool for mitigating abiotic stress in rice (*Oryza sativa* L.). pp. 109-133. *In:* Wani, S. (Ed.). *Recent Approaches in Omics for Plant Resilience to Climate Change.* Springer, Chamz. Doi https://Doi.org/10.1007/978-3-030-21687-0_6

Pandita, D. (2021). Role of miRNAi technology and miRNAs in abiotic and biotic stress resilience. pp. 303-330. *In:* Aftab, T. and Roychoudhury, A. (Eds.). *Plant Perspectives to Global Climate Changes.* Academic Press, Elsevier. https://Doi.org/10.1016/B978-0-323-85665-2.00015-7

Pandita, D. (2022a). How microRNAs regulate abiotic stress tolerance in wheat? A snapshot. pp. 447-464. *In:* Roychoudhury, A., Aftab, T. and Acharya, K. (Eds.). *Omics Approach to Manage Abiotic Stress in Cereals.* Springer, Singapore. https://Doi.org/10.1007/978-981-19-0140-9_17

Pandita, D. (2022b). MicroRNAs shape the tolerance mechanisms against abiotic stress in maize. pp. 479-493. *In:* Roychoudhury, A., Aftab, T. and Acharya, K. (Eds.). *Omics Approach to Manage Abiotic Stress in Cereals.* Springer, Singapore. https://Doi.org/10.1007/978-981-19-0140-9_19

Pandita, D. (2022c). miRNA- and RNAi-mediated metabolic engineering in plants. pp. 171-186. *In:* Aftab, T. and Hakeem, K.R. (Eds.). *Metabolic Engineering in Plants.* Springer, Singapore. https://Doi.org/10.1007/978-981-16-7262-0_7

Pradhan, A.K., Emdad, L., Das, S.K., Sarkar, D. and Fisher, P.B. (2017). The enigma of miRNA regulation in cancer, *Adv. Cancer Res.*, 135: 25-52.

Rüegger, S. and Großhans, H. (2012). MicroRNA turnover: When, how, and why, *Trends Biochem. Sci.*, 37(10): 436-446.

Sanei, M. and Chen, X. (2015). Mechanisms of microRNA turnover, *Curr. Opin. Plant Biol.*, 27: 199-206.

Waititu, J.K., Zhang, C., Liu, J. and Wang, H. (2020). Plant non-coding RNAs: Origin, biogenesis, mode of action and their roles in abiotic stress, *Int. J. Mol. Sci.*, 21(21): 8401.

Wang, J., Mei, J. and Ren, G. (2019). Plant microRNAs: Biogenesis, homeostasis, and degradation, *Front. Plant Sci.*, 10: 360.

Wilczynska, A. and Bushell, M. (2015). The complexity of miRNA-mediated repression, *Cell Death Differ.*, 22: 22-33.

Yang, J.-S. and Lai, E.C. (2011). Alternative miRNA biogenesis pathways and the interpretation of core miRNA pathway mutants, *Mol. Cell.*, 43(6): 892-903.

You, C., He, W., Hang, R., Zhang, C., Cao, X., Guo, H., Chen, X., Cui, J. and Mo, B. (2019). FIERY1 promotes microRNA accumulation by suppressing rRNA-derived small interfering RNAs in *Arabidopsis*, *Nat. Commun.*, 10(1): 4424.

CHAPTER
2

MicroRNAs in Plants and Animals: Converging and Diverging Insights

Humaira Shah, Auqib Manzoor, Tabasum Ashraf, Rouf Maqbool and Ashraf Dar*

Department of Biochemistry, University of Kashmir, Srinagar - 190006, Jammu & Kashmir, India

1. Introduction

miRNAs are a diverse family of single-stranded, generally non-coding endogenous RNA molecules, varying in length of 19- 25 nucleotides. These short RNA molecules are instrumental in regulating gene expression post-transcriptionally (Miller and Waterhouse, 2015). Lin-4 was the first miRNA to be discovered in 1993 in *Caenorhabditis elegans* (Wightman *et al.*, 1993). Since then, thousands of miRNAs have been discovered in eukaryotes (plants and animals) which have emerged as critical post-transcriptional regulators of mRNA stability and expression. miRNAs have been demonstrated to play important roles in timing of development, health, disease and ageing (Zhao *et al.*, 2018). miRNAs suppress expression of genes by associating with their complementary target mRNA, either by inhibiting translation or by degrading the target mRNAs. A single miRNA may bind hundreds of different mRNA transcripts and similarly, one mRNA transcript can have several binding sites for different miRNAs with synergistic function (Perge *et al.*, 2013).

In both plant and animal systems, the majority of miRNAs are transcribed as lengthy stretches of double-stranded RNA, called primary miRNAs (pri-miRNAs) by RNA Polymerase II or RNA Polymerase III. These pri-miRNAs are cleaved and processed into precursor miRNAs (pre-miRNAs). The pre-miRNAs are further processed into 19-25 nt long, mature miRNAs by RNaseIII in animals or RNaseIII-like enzymes in plants (Zhao *et al.*, 2018; Pandita and Wani, 2019; Pandita 2019; 2021; 2022a; 2022b; 2022c). miRNAs, in the majority of cases, decrease gene expression by binding to the 3′ UTR of target mRNAs. Nevertheless, interaction of miRNAs with additional sequences like 5′ UTR, exons, and gene promoters has been elucidated by several studies (Ha *et al.*, 2014).

---

*Corresponding author: ashrafdar@kashmiruniversity.ac.in

There are clear parallels in miRNAs from plants and animals. For example, (a) the length of mature miRNAs are usually between 19 and 25 bases and (b) they modulate gene expression by interacting with target mRNAs that have a role in regulating crucial physiological processes, like growth, development and stress response (Carrington *et al.*, 2003; Miller and Waterhouse, 2015). Nevertheless, plant miRNAs show certain dissimilarities from their animal counterparts (Table 1). The initial step of miRNA biogenesis in animals requires a nuclease, known as Drosha, whereas a different nuclease, DCL1 plays an equivalent role in plants. A small number of animal miRNAs are produced from polycistronic transcripts that are present in chromosomal intergenic regions, whereas most of them are produced from intragenic regions. In comparison, most miRNAs in plants are created from single main transcripts found in intergenic regions. Furthermore, animal miRNAs mostly supress translation by binding at 3'-UTR of target mRNAs, whereas plant miRNAs primarily suppress gene expression by cleaving mRNAs at coding regions (Zhu *et al.*, 2016). The evolution of plant and animal miRNAs has taken diverse routes as is evident from their differences in biogenesis and regulation of target gene expression. Plants miRNAs have a high rate of target: miRNA suggesting they might have co-evolved with their target genes. On the other hand, animal miRNAs appear to have arisen first and subsequently acquired target genes. This chapter discusses biogenesis, mechanism of action and regulation of miRNAs in plant and animal systems.

Table 1: Similarities and differences in animal and plant miRNA systems

miRNA Feature	Animal System	Plant System
miRNA genes present within the genome	Less than 500 genes	Less than 200
Position within the genome	Mostly intergenic regions and introns, rarely clustered	Mostly intergenic regions, rarely clustered
Polymerase involved in biosynthesis	RNA Polymerase II, III	RNA polymerase II
Completion of biosynthesis	Nucleus and cytoplasm	Nucleus
Regulation mechanism	Repression of translation	mRNA degradation
Complementarity	Mostly perfect complementarity	No perfect complementarity, presence of seed region and flanking region
Location of miRNA binding motifs within target genes	Mostly 3'-UTR of multiple targets	Mostly open frame
miRNA binding sites within the targets	More than one	Mostly single

2. miRNA Biogenesis

miRNAs are categorised as either intragenic (also called 'intronic') or intergenic, depending on the location of their synthesis within the genome. Intronic miRNAs are generated by processing intronic regions of parent transcripts, whereas intergenic miRNAs are transcribed from coding regions as independent transcription units. The

MicroRNAs in Plants and Animals: Converging and Diverging Insights **17**

intergenic miRNAs have their own intrinsic promoters from where transcription is initiated by RNA Polymerase II (Budak *et al.*, 2015; Millar, 2005; de Rie *et al.*, 2017; Kim and Kim, 2007). Most of the plant and animal gene templates for miRNA synthesis are located in intergenic regions; however, a significant number of animal miRNAs are present in the introns of pre-miRNAs sequence (Millar, 2005).

2.1 Biogenesis of miRNA in Plants

Most of the plant genomes house hundreds and thousands of miRNA genes (MiR), many of which are organised into different families (Budak *et al.*, 2015; Nozawa *et al.*, 2012). In plants, the primary transcripts of MiRs, called as pri-miRNAs, are products of RNA polymerase II. The pri-miRNA transcripts attain a hairpin-like secondary structures, get capped and polyadenylated at 5′ and 3′ ends respectively. These post-transcriptional modifications are followed by splicing (Rogers and Chen, 2013; Xie *et al.*, 2005). The pri-miRNA differs in size from about 70 to >500 bases and consists of a terminal loop region, an upper stem, the miRNA/miRNA region, a lower stem, and two arms (Wang *et al.*, 2019). The plant pri-miRNAs are identified and cleaved by Dicer-like RNase III ribonucleases (DCLs), which vary in number in different plant species. The biogenesis of plant miRNAs has been comprehensively studied in *Arabidopsis thaliana*, which has four DCLs. The DCL1 is complex with other proteins, such as dsRNA-binding protein, hyponastic leaves 1 (HYL1) and the zinc-finger serrate protein (SE) is involved in production of the most of the miRNAs (Fang and Spector, 2007; Dong *et al.*, 2008). The other DCLs, such as AtDCL4, mediate the production of miR822 and miR839. The OsDCL3a is linked to production of a class of miRNAs (24-nt) in rice (Rajagopalan *et al.*, 2006; Wu *et al.*, 2007).

Generally, the processing of plant primary miRNAs (pri-miRNAs) occurs from the loop-proximal site to the loop-distal site or vice versa. DCL1 mediates the processing of pri-miRNA mainly in two steps – first, it cleaves the misfolded end of the pri-miRNA and generates the stem-loop hairpin secondary structure containing pre-miRNA; next, the pre-miRNA generated is again cleaved by DCL1 to form a miRNA/miRNA* duplex of (Fig. 1) (Addo-Quaye *et al.*, 2009; Bologna *et al.*, 2009; Mateos *et al.*, 2010; Werner *et al.*, 2010; Voinnet, 2009). The miRNA/miRNA* duplex generated possess 2-nt overhangs at 3′ ends with hydroxyl groups. The 5′ ends of both strands in duplex have phosphate groups. The initially released nascent plant miRNAs are 2′-O-methylated at 3′-terminal by 2′-O-methyl transferase, Hua Enhancer (HEN)1. This methylation is commonly present in all plant miRNAs which protects them from uridylation and subsequent degradation (Yu *et al.*, 2005; Li *et al.*, 2005; Yang *et al.*, 2006). The methylated miRNA/miRNA* duplexes are released into cytoplasm for loading on to protein Argonaute (AGO) with the assistance of Hasty (HST) (B24), a homolog of Exportin 5 (EXPO5) protein in animals. From the miRNA/miRNA* duplex, the guide strand (which is miRNA) is preferentially retained on to AGO. The second strand is released and degraded and is called 'passenger strand' (miRNA*) (Zhang *et al.*, 2015). In plants, 10 different AGO family members have been discovered which possess RNA cleavage activities (Baumberger and Baulcombe, 2005; Qi *et al.*, 2005; Vaucheret, 2006). However, AGO1 is known as a major player in miRNA-mediated gene-silencing reactions. Most of the plant pri-miRNA hairpins generate a single miRNA/miRNA* duplex with the exception of certain loci, such as MIR159 and MIR319 which produce more than one duplexes (Addo-Quaye *et al.*, 2009; Bologna *et al.*, 2009). The biogenesis of miRNA/miRNA* duplex is completed at specific sites in the nucleus, called as 'nuclear processing centres' or D-bodies (dicing bodies) (Park et al., 2005; Fang and Spector, 2007; Song et al., 2007).

2.2 Biogenesis of miRNA in Animals

Various studies in mammals, *C. elegans* and *Drosophila*, have suggested a common mechanism for miRNA biogenesis that is similar but not identical to miRNA generation in plants. In animals, miRNAs are generated by canonical pathways where long pri-miRNAs are synthesised by RNA polymerase II-mediated transcription. However, a sub-class of animal miRNAs are also transcribed by RNA pol III (Fig. 1) (Cai et al., 2004; Faller and Guo, 2008; Lee et al., 2004). The pri-miRNAs have one or more than one stable stem loop hairpin structures. These transcripts are capped with 7-methylguanosine (m7G) at their 5′ end and are polyadenylated at the 3′ end (Li et al., 2018). The miRNA biogenesis in animals differs from that in plants, mostly in the cleavage steps of miRNA precursors by nuclear and cytoplasmic RNase III enzymes (Axtell et al., 2011). The animal pri-miRNA hairpins are identified by the nuclear microprocessor complex and cleaved to produce pre-miRNA hairpin precursor (~ 55–70 nucleotides long). The microprocessor complex contains nuclear RNase III, Drosha and a dsRNA binding protein DGCR8 (DiGeorge critical region 8). The equivalent protein of DGCR8 in invertebrates is called 'Pasha' (Partner of Drosha) (Denli et al., 2004; Gregory et al., 2004; Lee at al., 2003; Landthaler et al., 2004). The pre-miRNA precursor has a characteristic 5′ monophosphate group and a 2-nt overhang at 3′-end, a characteristic of RNase III enzyme cleavage product. The precursor miRNA is recognised by a different RNase III, Exportin-5 (XPO5)

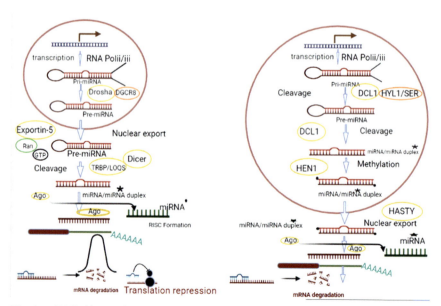

Fig. 1: miRNA biogenesis pathways in animals and plants. In animals, precursor miRNAs (pre-miRNAs) are converted into mature miRNAs in the cytoplasm, whereas in plants, miRNAs are generated and matured in the nucleus. Plants miRNAs are methylated to avoid degradation, whereas animal miRNAs do not require methylation for stability (This figure has been generated in Biorender)

and carried to the cytoplasm through nuclear pore complex with the help of its Ran-GTP co-factor (Lund et al., 2004; Yi et al., 2003). In cytoplasm, DICER (which also possesses RNase III activity) in association with TRBP in mammals (dsRNA-binding partner transactivation-response RNA-binding protein) or Loqs (in *Drosophila*) cuts off pre-miRNA stem loop to release ~25 nt mature miRNA/miRNA* duplex with a 2-nt overhangs at 3′ end in both strands (Bernstein et al., 2001; Haase et al., 2005). To ensure correct and efficient cleavage, DICER recognises structural features of pre-miRNA hairpin and counts ~22 nucleotides from the 5′ monophosphate as well as the 3′ overhang. The miRNA/miRNA* duplex is then loaded on to miRNA effector, Argonaute or AGO. The AGO protein is a constituent protein of larger RISC complex. One strand in miRNA/miRNA* duplex is chosen to generate an active miRNA–RISC complex (miRISC). The second strand (miRNA*) is unwound, released and decayed further (Liu et al., 2004).

3. Target Recognition

Understanding the recognition of corresponding target sequence in gene transcripts is necessary to get an insight into the functional roles played by miRNAs in cellular physiology. According to the seed-pairing rule, base pairing with 5′ end nucleotides of miRNA (from 2–8) governs the recognition of miRNA targets (Fig. 2) (Lewis et al., 2005). Earlier studies on miRNAs in *C. elegans* suggested that the *lin-4* RNA contains sequence elements that are complementary to various conserved sites in lin-14 mRNA (Lee et al., 1993). Lin-14 plays a major role in *C. elegans* development. Further molecular and genetic studies highlight that the sequence elements at 3′-UTR of lin-14 were important for lin-4 mediated lin-14 repression (Wightman et al., 1993). The 5′ ends are highly conserved regions in animal miRNAs whereas in some *Drosophila* miRNAs, the 5′ portions are completely identical to 3′-UTR regions of target transcripts that play a role in miRNA-mediated gene silencing. Later studies, which focussed on responses of whole transcriptome and proteome to miRNAs,

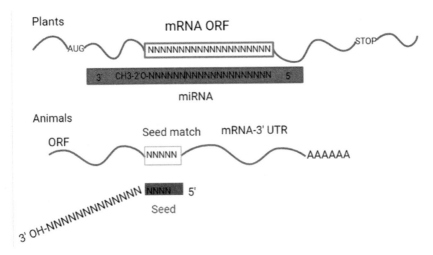

Fig. 2: Schematics of target recognition by plant and animal miRNAs (This figure has been generated in Biorender)

further confirmed the seed paring rule (Selbach *et al.*, 2008). However, a small number of miRNA target sites are not recognised on the basis of seed-pairing complementarity rule (Lu *et al.*, 2010). Several studies have shown that perfect complementary miRNA seeds are not required absolutely, nor are they enough for identifying all active target miRNA sites. Newer and advanced techniques in transcriptomics studies have helped to decipher that a great number of miRNA-mRNA interactions not only depend on seed regions, but also on non-canonical regions (Fig. 3) (Didiano and Hobert, 2006; Kim *et al.*, 2016). Additional factors that affect the miRNA target recognition include seed-like, motifs nucleation bulge and central pairing miRNA-binding sites (Li and Li, 2018).

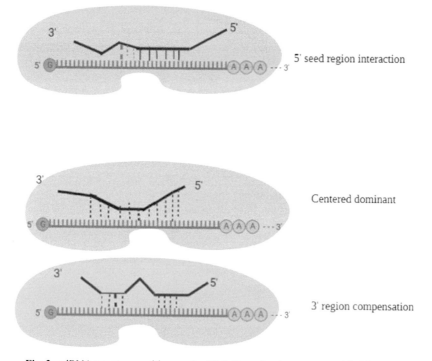

Fig. 3: miRNA target recognition modes (This figure has been generated in Biorender)

3.1 Target Recognition in Plants

Identification of targets by plant miRNAs was earlier considered to be based on perfect or near-perfect complementarity. However, more recent studies indicate that plant miRNAs are involved in translational repression and therefore indicate that there may be variations in target recognition rules (Brodersen *et al.*, 2008). There are studies that suggest that for a functional miRNA-target interaction, there should be complete matching at positions 2-13 nt, particularly at position 9-11 nt, and mismatches at positions at 1 and 14-21 are tolerated to a great degree (Wang *et al.*, 2015). However, there are some exceptional studies that do not corroborate this rule, for example, miR398a, in spite of a GU wobble and a bulge at the essential 12-13 nucleotide position of target represses (CSD)2 copper/zinc superoxide dismutase in *Arabidopsis*

MicroRNAs in Plants and Animals: Converging and Diverging Insights

(Jones *et al.*, 2004). Similarly, *A. thaliana* miRNA398 regulates, blue copper-binding protein post transcriptionally (BCB), despite the presence of 6-nucleotide bulge at locations 6 and 7 of the miRNA (Brousse *et al.*, 2014). These studies suggest that abundance of secondary structures in target mRNA, miRNA: mRNA stoichiometric ratio and availability of target mRNA may influence the target recognition in plants (Li and Li, 2018).

3.2 Target Recognition in Animals

Experimental and computational studies have suggested the seed-matching rule of miRNA targets in animals (Brennecke *et al.*, 2005). Secondary features that affect miRNA target recognition include the presence of target site in the 3'-UTR, an Ado base after seed region, and secondary structure in target mRNA. Many studies have concluded that miRNA binding sites are not highly conserved in animals. Several extensive transcriptomic and proteomic studies suggest that most active miRNA: mRNA interactions are not highly conserved (Bartel, D., 2009). miRNAs in animals have a wider potential to exert combinatorial effect on gene regulation due to the reason that various mRNAs are regulated by a common miRNA. This indicates that animal miRNAs have evolved to effect different regulatory processes across the transcriptome (Schnall-Levin *et al.*, 2010).

miRNAs in animals were previously assumed to bind target transcripts generally on the 3' UTRs, whereas plant miRNAs target coding regions (CDS). Later studies have revealed that in both plants and animals, miRNAs can regulate mRNAs by targeting UTRs ends and coding regions, that hints miRNA target sites in both plant and animal kingdoms are similar but diverse (Schnall-Levin *et al.*, 2010). The exogenous let-7 mRNAs injected into zebra fish and containing target sites for let-7 miRNA in either coding region or 5' UTR were silenced significantly (Kloosterman *et al.*, 2004). A study carried out in *Arabidopsis* revealed that members of the miR398 family are complementary to the 5' UTR of At3g15640 (cytochrome c oxidase) and miR399 possesses several complimentary sites within the 5' UTR of At2g337709 (E2, ubiquitin conjugative enzyme) (Sunkar and Zhu, 2004).

4. Mechanism of miRNA Action

MiRNA-mediated gene regulation (silencing) has been shown to work by two main mechanisms in plants and animals: translational repression and miRNA mediated mRNA decay (Ameres and Zamore, 2013). These different protein machineries involved in the mechanistic regulation of gene regulation are listed in Table 2. All miRNAs regulate gene expression at post-transcriptional level, regardless of which mechanism is involved. The mode of regulation has been shown to depend greatly upon the level of complementarity between the miRNA and the target mRNA binding site. miRNAs with perfect or near-perfect complementarity to the target mRNAs promote mRNA decay. This mode of miRNA action is prevalent in plant gene regulation. The low or partial complementary of miRNAs, called as 'seed matching' promotes translational repression and is considered as main mechanism of miRNAs action in animals (Huntzinger and Izaurralde, 2011). However, most recent studies in animals suggest that miRNAs not only mediate translational inhibition but also promote decay of partially complementary targets. As an example, miR196 completely degrades HOXB8 mRNA, which play an essential function in animal development

Table 2: Proteins components involved in miRNA-mediated mechanisms of gene expression

Protein	Role	Effect
TNRC6A-C	Binding of miRNA.	Transcript deadenylation and decapping, as well as translation inhibition or up-regulation.
TNR6A	navigator protein AGO2.	Silencing of genes in the nucleus.
PABPA-C	Interaction with initiation factors and binding to the poly(A) tail of the transcript.	Deadenylation of transcripts and suppression of translation.
PAN2-PAN3 complex:		Deadenylation of the first transcript.
PAN2	Nuclease activity.	
PAN3	Protein binding activity.	
CCR4-NOT complex:		Deadenylation of transcripts.
CNOT1	CCR4-NOT complex scaffolding.	
Caf1a/b	Deadenylase subunit.	
Ccr4a/b	Deadenylase subunit.	
DCP1-DCP2 complex:		Decapping of transcript.
DCP1	Activator of DCP2.	
DCP2	Decapping subunit.	
RCK/p54	DCP2 activator, which is responsible for DCP1-DCP2 P-body localisation.	Cap-dependent translational suppression, decapping of transcript.
EDC4	Linking of DCP1 and DCP2, decapping stimulation.	Decapping of transcript.
XRN1	5′ 3′ exonuclease activity.	Degradation of transcript.
FMRP	Component of stress granules.	Regulation of translation.
PNRC2	Stimulation of decapping, synergy with DCP1.	Decapping of transcript.
EDD	E3 ubiquitin ligase.	Deadenylation of transcript, inhibition of translation.
Importin-8	Nuclear import.	Inhibition of translation.
Exportin-1	Export of TNRC6A from nucleus.	Gene silencing.

(Zhang *et al.*, 2007). Similarly, in plants, in spite of differences in recognition of targets, miRNAs can result in target mRNA translational repression, e.g. regulation of AP2 by miR 172 is a consequence of translational repression, despite the fact that there is perfect complementarity between miR 172 and target AP2 mRNA (Aukerman and Sakai, 2003; Carthew and Sontheimer, 2009; Iwakawa Tomari, 2015).

During the post-transcriptional regulatory silencing (PTGS) in plants and animals, miRNAs need miRNA-induced silencing complex (miRISC). The miRISC contains a

MicroRNAs in Plants and Animals: Converging and Diverging Insights **23**

small RNA (sRNA) as guide strand and a member of Argonaute family protein (Ago), which provides a special platform for identification and silencing of target (Ameres and Zamore, 2013; Kawamata and Tomari, 2010; Wilson and Doudna, 2013). The interaction of miRISC with the complementary sequences on target transcripts, known as miRNA response elements (MREs), is essential for target specificity. The degree of complementarity between MRE and miRNA determines the mode of action of miRISC complex in deciding whether there is AGO2-mediated slicing of target mRNA or miRISC-dependent inhibition of translation or there is target mRNA decay (Jo _et al._, 2015). The differences in plant and animal miRNAs regarding their mechanisms of regulation of target have now become irrelevant with recent findings indicating that translational inhibition and miRNA-mediated mRNA decay can occur in both the systems.

4.1 miRNA-mediated Decay of MRNA in Animals

miRNAs in animals were earlier believed to act by suppressing protein synthesis carried by target mRNAs with minor or no effect on mRNA levels. However, research in zebrafish and _C. elegans_ revealed that miRNAs can also mediate the decay of their target mRNAs by recruiting deadenylases (Olsen and Ambros, 1999; Bagga _et al._, 2005). The animal miRNAs usually mediate the mRNA degradation in three phases (Fig. 4). The first step is deadenylation in which miRNAs initiate the shortening of polyadenylic acid tail, poly(A) by recruitment of various proteins which include two deadenylases complexes, poly(A)-nuclease (PAN)2-PAN3 and carbon catabolite repressor (CCR)4-negative on TATA (NOT) and poly(A)-binding protein (PABP) (Huntzinger and Izaurralde, 2011; Chekulaeva _et al._, 2011; Moretti _et al._, 2012). The PABP increases the availability of the poly(A) tail to deadenylating exoribonucleases, thus enhancing the efficacy of removal of adenyl groups (Moretti _et al._, 2012). The recruitment of these proteins on target mRNAs is mediated by TNRC6A–C in mammals or GW182 (glycine-tryptophan protein of 182 kDa) in _Drosophila_. The second step is decapping, which is mediated by decapping protein (DCP)2, which needs accessory proteins DCP1, DEAD box helicase 6, EDC and EDC4 for stability and to perform its functions (Jonas and Izaurralde, 2013). In the third stage, after deadenylation and decapping, target mRNA transcripts are decayed by 5′-to-3′ exoribonuclease, (XRN)1 in the cytoplasm (Huntzinger and Izaurralde, 2011).

4.2 miRNA-mediated Decay of MRNA in Plants

miRNAs in plants, unlike their animal counterparts, do not promote deadenylation of target mRNAs; rather, they promote slicing of their targets at particular locations (Iwakawa and Tomari, 2013). The P-element induced wimpy testes (PIWI) domain of Ago proteins, which has an RNase H-like motif and possess endonuclease activity, is required for slicing the targets (Qi _et al._, 2006; Bologna and Voinnet, 2014; Carbonell _et al._, 2012). The slicing of mRNA generates two fragments: 5′ fragment which has 5′ end capped structure and an exposed 3′ end and a 3′ fragment with unprotected 5′ end and a poly(A)-attached 3′ end (Fig. 5). The 3′ ends generated are degraded by a 5′-to-3′ exoribonuclease, XRN4. The loss-of-function mutation in XRN4 has been found to cause accumulation of 3′ fragments. The 5′ cleavage fragments are also known to be degraded by XRN4-like degradation of 3′ fragments (Souret _et al._, 2004). Moreover, the cytoplasmic exosome may also be involved in this degradation process

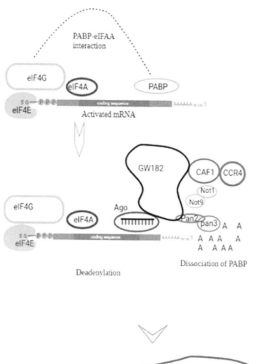
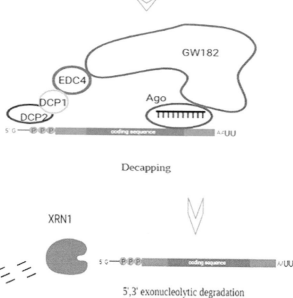

Fig. 4: In animals, miRNA-mediated mRNA degradation occurs in three stages. Deadenylation is the initial stage. By assembling protein complexes, CCR4–NOT and PAN2–PAN3 deadenylase to target micro RNAs via the protein GW182, miRNAs cause poly (A) shortening. Decapping is the next phase. Decapping activators facilitate the removal of the 5′ m7 G-cap by the DCP2 (decapping enzyme). The exonucleolytic mRNA degradation (5′-to-3′) by XRN1 is the third and last step. (This figure has been generated in Biorender)

MicroRNAs in Plants and Animals: Converging and Diverging Insights   **25**

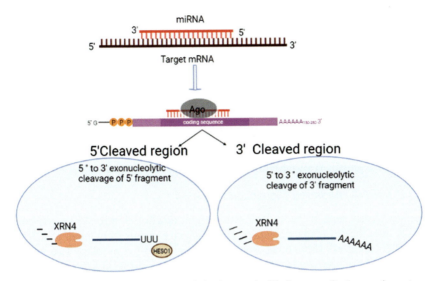

Fig. 5: Plant miRNAs promote endonucleolytic cleavage by binding to perfectly complementary target sites, which are mostly found in the open reading frame. In *Arabidopsis*, the 5' fragment is uridylated by HEN1 suppressor 1 (HESO1) and is decayed by XRN4(5'-to-3' direction). XRN4 also break down the 3' left fragments (This figure has been generated in Biorender).

because its various subunits, SKI2, SKI3 and SKI8, are indispensable for the decay of RISC-generated 5' fragments (Yu *et al.*, 2017). Further, degradation of 5' fragment (in *Arabidopsis*) is accelerated by HUA ENHANCER 1 (HEN1) suppressor 1 (HESO1) catalysed 3'end uridylation (Ren *et al.*, 2014).

4.3 miRNA-mediated Repression of Translation in Animals

Different studies involving various methods and different organisms have revealed that miRNAs block the first step of translation (Braun *et al.*, 2012; Fabian *et al.*, 2010; Huntzinger and Izaurralde, 2011). Genome-wide studies of endogenous miRNA targets have shown that miRNAs mediate translational regulation at the initiation step (Eichhorn *et al.*, 2014; Guo *et al.*, 2010). Three primary mechanisms have been postulated in recent years: (i) GW182-mediated PABP displacement, (ii) GW182-mediated recruitment of translational repressors, (iii) dissociation of eIF4A from the cap-binding complex eIF4F (Li *et al.*, 2018). These three processes of inhibition are not totally exclusive; they may occur simultaneously or may overlap with one another with distinct kinetic effects on gene silencing.

4.3.1 W182-mediated PABP Displacement

In addition to deadenylation activity and target mRNA cleavage, GW182 proteins promote translation repression (Huntzinger *et al.*, 2013). Various studies have shown that GW182 proteins directly associate with C-terminal MLLE domain of PABP through PABP-interacting motif (PAM)2 and block PABP. However, there are evidences which suggest that miRNAs induce PABP dissociation from mRNA targets, independent of deadenylation (Moretti *et al.*, 2012). This phenomenon is consistent with the existing model according to which GW182-carried dissociation of PABP from the poly(A) tail disrupts the closed-loop architecture created by binding of PABP and eIF4G, which in turn inhibits initiation of protein synthesis (Zekri *et al.*, 2009).

4.3.2 GW182-mediated Recruitment of Translational Repressors

GW182 inhibit protein synthesis from nonpolyadenylated mRNA targets by interacting with protein complex, CCR4–NOT. GW182 binds to target mRNAs and subsequently recruits translational repressors, DDX6 or the eIF4E transporter (Presnyak Coller, 2013). Recent studies have indicated that an eIF4E-binding protein, 4E-T, is recruited on to CCR4-NOT complex to inhibit protein synthesis via LSM14, DDX6 or PATL 1(PAT1-like protein 1) (Waghray *et al.*, 2015; Nishimura *et al.*, 2015).

4.3.3 miRNA-mediated Dissociation of eIF4A

Various studies suggest that miRNAs can inhibit function of mRNAs by displacing eIF4A from the cap-binding complex, eIF4F, which prevents recruitment of 43S pre-initiation complexes (PIC) or ribosomal scanning in cell-free systems. Two separate approaches have been used to test the effect of miRNAs on the dissociation of EIF4A from target mRNAs and formation of initiation complex. In the first instance, it was shown *in vitro* that from mammalian mRNA the addition of miRNA preferentially liberates eIF4AI and eIF4AII from eIF4F. In the second approach, UV crosslinking of cap-labelled target mRNAs helped to identify the association of eIF4A and eIF4E near the 5′end. However, in the presence of cognate miRNAs, eIF4A did not crosslink with mRNA (Meijer *et al.*, 2013; Fukao *et al.*, 2014; Fukaya *et al.*, 2014).

4.4 miRNA-mediated Repression of Translation in Plants

Initially it was thought that plant miRNAs inhibit the gene expression only by endonucleolytic cleavage of mRNA by Ago proteins. Plants lack GW182 homologues, which are essential proteins for miRNA-mediated translational repression or mRNA degradation in animals. Nonetheless, it has been discovered that miRNAs have a differential influence on expression at the mRNA and protein levels, indicating that miRNAs may potentially inhibit the translational process in plants. It has been observed that miRNAs exert unequal effect on the target gene expression at the transcription and protein levels, suggested that miRNAs may also suppress translational process in plants (Fig. 6) (Chen, 2004; Todesco *et al.*, 2010). As an example, miR172 and miR156/7, regulate Apetala and squamosa promoter binding protein-like, respectively by translational inhibition (Chen, 2004). In *Arabidopsis* mutant screen (of RNA silencing defectives), Brodersen *et al.* discovered that KATANIN 1 (microtuble severing enzyme) is required for translation repression. They further found that AGO1, AGO10, VARICOSE (VCS) and decapping activator Ge-1 homolog play roles in translational repression. Other proteins involved in translation repression include AMP1, like AMP (LAMP) 1 and GW-repeat protein, SUO (Brodersen *et al.*, 2008; Li *et al.*, 2013; Yang *et al.*, 2012). The concentration of SUO and VCS in processing bodies, where various mRNA decaying factors are localised, and interaction of KTN1 with microtubules suggests that a variety of dynamic cellular mechanisms may be involved in miRNA-mediated translational repression (Li *et al.*, 2013; Yang *et al.*, 2012). The mechanistic details regarding the molecular action of these proteins and their interconnecting roles in plant miRNA repression are not clear.

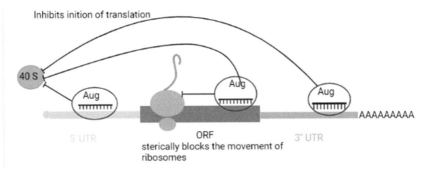

Fig. 6: Mechanisms of miRNA-regulated translational suppression in plants. An *in vitro* system revealed that an *Arabidopsis* AGO1–RISC catalytic mutant may limit translation initiation without causing deadenylation or mRNA degradation. Plant AGO1–RISC can bind to the ORF and prevent ribosome translocation (This figure has been generated in Biorender)

5. Regulation of miRNA Biogenesis in Plants

miRNAs regulate many aspects of cell biology including, metabolism, differentiation, proliferation, development, tumorigenesis, apoptosis and immune responses, by their roles in gene expression (Lee and Dutta, 2009). The biogenesis of miRNAs is strictly regulated, and discrepancies in their biogenesis are frequently linked to various physiological disorders in plants and animals. miRNAs are regulated at several stages, which include transcription, Dicer and Drosha processing, loading on to AGO proteins and miRNA stability. The cells employ a number of strategies to either impede or promote the effect of miRNAs at each stage by recruiting transcription factors, protein-modifying enzymes, RNA-binding proteins, RNA-modifying enzymes, endoribonucleases and exoribonucleases (Kim *et al.*, 2009; Siomi and Siomi, 2010).

5.1 Regulation of miRNA Transcription

So far, a few hundred miRNAs have been found in plants. These miRNAs regulate various developmental and physiological processes, like maturation of different plant parts (root, shoot, flowers, seeds), phase change and responsiveness to various biological and chemical stresses. Deregulation in miRNA levels has been implicated in numerous developmental defects or pathologies, suggesting that biogenesis of miRNAs is a precisely regulated phenomenon (Li and Yu, 2021).

The transcription of Pol II-dependent pri-miRNAs is regulated by RNA Pol II-associated transcription factors and several epigenetic regulators. In addition, the transcription of a small number of primary miRNAs is spatiotemporally regulated to make certain that the miRNAs function properly (Mcgraw *et al.*, 2006). The mediator co-activator complex acts as a universal transcription factor which recruits RNA polymerase II to miRNA (MIR) gene promoters at onset of transcription (Kim *et al.*, 2012). The other proteins found to promote MIR transcription include, protein phosphatase 4 (PP4), short valve 1 (STV1), elongator complex ELP2 and ELP5, negative on TATA less 2 (NOT2), the splicing factor small (SMA1), cell division cycle 5 (CDC5), hasty (HST), homologue of exportin 5, the ATPase CHR2, the pore-associated proteins THP1 and the TREX-2 complex suppressor of actin 3A (SAC3A) (Fang *et al.*, 2015; Wang *et al.*, 2019; Lang *et al.*, 2018). Inactivation or mutation (loss-of-function) in these proteins are associated with lower occupancy of polymerase II at

MIR promoters, lower activities at MIR promoters and decreased pri-miRNA levels, implying that the above proteins are involved in regulating MIR transcription directly or indirectly by recruiting pol II to MIR gene promoters (Li and Yu, 2018).

Different proteins have been shown to suppress miRNA transcription, e.g. disturbance of the CPR1 (constitutive expresser of PR gene 1) F-Box protein, raises the mRNA levels of SNC1 (suppressor of npr1-1, constitutive 1), a disease-resistance protein shown to decrease transcription of pri-miRNA. Furthermore, increased expression of SNC1 and TPR1 (transcriptional corepressor) inhibit MIR transcription. These findings indicate that SNC1 acts as a repressor of MIR transcription (Cai *et al.*, 2018).

There are protein factors that function as both activators and repressors of MIR transcription. For example, mutation in CDF2 (cyclin DOF transcription factor 2) promotes the transcription of some, while suppressing the transcription of other MIRs (Sun *et al.*, 2015). The changes in Pol II CTD serine phosphorylation levels by CDKDs and CDKF have been shown to regulate MIR transcription, capping at 5′ ends, addition of adenyl groups at 3′ ends and splicing (Hajheidari *et al.*, 2012). Histone modifications, such as acetylation and methylation, also have a profound effect on MIR gene expression levels. PRC2 (polycomb repressive complex 2), for example, inhibits MIR156A and C transcription by increasing H3K27me3 accumulation at their promoters, whereas the acetylation of H3K14 by GCN5 up-regulates subset of MIRs (Hajheidari *et al.*, 2012). Some transcription factors regulate the transcription of MIRs at specific developmental stages. For example, during flower formation, MIR172 transcription is co-ordinately regulated by several proteins. Powerdress (PWR) can help in the positioning of Pol II at MIR172a, b and c promoters, while showing no influence on MIR172d or MIR172e (Yumul *et al.*, 2013). Likewise, ABI3 (Abscisic Acid Insensetive3), a B3 transcription factor increases the miR156 expression levels during the initial seed development, while supressing their expression during later stages of development. ABI3 has also been involved in repression the miR160B transcription (Tian *et al.*, 2020).

5.2 Regulation of the miRNA Core Processing Machinery

The pri-miRNA processing mediated by DCL1 at subnuclear regions, known as D-bodies (dicing bodies), requires two additional proteins, SE and HYL1. To facilitate accurate and effective cleavage of pri-miRNAs, SE and HYL1 form an assembly with DCL1 and form a heterocomplex. This protein assembly is called 'core processing machinery'. HYL1 is critical for recruitment and proper placing of pri-miRNAs into DCL1. The absence of this protein causes improper DCL1 mediated primary miRNA cleavage, resulting in a reduction in miRNA levels (Li and Yu, 2021). Sequestration of HYL1 from miRNA precursors by overexpressed SINE transposons results in decreased miRNA levels (Pouch-Pélissier *et al.*, 2008).

The miRNA core processing machinery is regulated at transcriptional, post-transcriptional and post-translational levels (Wang *et al.*, 2019). DCL1 transcription is generally inhibited by GCN5 histone acetyltransferase, whereas XCT and STA1 positively regulate its transcription (Fang *et al.*, 2015; Kim *et al.*, 2009). Furthermore, there are two negative feedback mechanisms involved in regulation of DCL1 levels. The first mechanism involves miR162, a miRNA produced by DCL1. MiR162 targets DCL1 mRNA for degradation. The second negative feedback mechanism results from

MicroRNAs in Plants and Animals: Converging and Diverging Insights **29**

DCL1-mediated processing of MIR838 (an intragenic miRNA gene that is located in the 14th intron of DCL1 precursor mRNAs). The processing of MIR838 negatively affects transcription of DCL1 (Rajagopalan *et al.*, 2006; Xie *et al.*, 2003).

The activity and stability of HYL1 is regulated by post-translational phosphorylation. The phosphorylated HYL1 present in the nucleus is inactive and is protected from proteolysis by darkness. Light-mediated de-phosphorylation results in activation of HYL1 (Achkar *et al.*, 2020). Inactive SE interacts with 20S proteasome and undergoes degradation in a ubiquitination, independent manner. The degradation of dysfunctional SE is critical for the stabilisation of the active DCL1 complex (Li *et al.*, 2020).

Studies have shown that SnRK2 and MPK3 interact with HYL1 in cells and in cell-free conditions phosphorylate HYL1 (Su *et al.*, 2017). What is the physiological significance of this HYL1 phosphorylation remains to be seen (Wang *et al.*, 2019). PP4/Suppressor of MEK 1 (SMEK1) complex and CPL1, on the other hand, dephosphorylate HYL1. Moreover, HYL1 is a cytoplasmic protein and is transported to nucleus by protein Ketchi to enhance the production of miRNAs.

The intrinsically disordered regions (IDRs) present in SE are critical for phase separation and effective miRNA generation. Any disruption in SE production causes a decrease in miRNA levels. During later stages of bacterial infection, SE levels are regulated by a negative feedback mechanism by miR863-3p.

5.3 Regulation of miRNA Biogenesis by Structure of Pri-miRNA and RNA Splicing

Pri-miRNAs influence processing efficiency of DCL1 complex in a variety of ways. The structure of pri-miRNA is not only an essential factor for the selection of cleavage site, but also for processing efficiency of core machinery protein, DCL1. Furthermore, base modifications, splicing and alternative splicing of pri-miRNAs also affect the processing of pri-miRNAs.

Majority of pri-miRNAs have a 15-bp imperfect lower stem that is frequently proceeded by a bulge-like region beneath the miRNA/miRNA* duplex. This arrangement is needed for the precise initial base-to-loop cleavage. Some pri-miRNAs, however, such as pri-miR159a and pri-miR319a, have an extended upper stem that results in loop-to-base cleavage. Multi-branched terminal loops containing pri-miRNAs are processed bidirectionaly; however, miRNAs are produced effectively by base-to-loop processing (Zhu *et al.*, 2013). Furthermore, biogenesis of mi-RNAs not only requires presence of specific base pairs at the cleavage site of miRNA, but also at mismatched positions on the stem (Rojas *et al.*, 2020). Moreover, various plant species have conserved structures and sequences next to the miRNA/miRNA* duplex, indicating the importance of these sequences in plant miRNA biogenesis (Chorostecki *et al.*, 2017).

RNA (intron) splicing plays a critical role in the biogenesis of some pri-miRNAs. Intron splicing can alter the structure of some pri-miRNAs, thereby affecting processing, e.g. the splicing of introns in rice pri-miR842 and pri-miR846 is important for the formation of stem-loop structure and ultimately for processing of these pri-miRNAs (Li *et al.*, 2020). The stem-loop of pri-miR162a, on the other hand, is made up of unspliced intron and exon. As a result, intron splicing inhibits miR162 production (Hirsch *et al.*, 2006). Furthermore, the introns following the stem loop from pri-miR161, pri-miR163 and pri-miR172b are required for efficient miRNA production. Studies have also shown that 5′ splice site introns of these pri-miRNAs influences the selection of polyadenylation site which consequently increases efficiency of biogenesis (Bielewicz *et al.*, 2013).

5.4 Control of Pri-miRNA Stability

Studies have shown that the 5′-to-3′ exoribonucleases (XRNs) and the 3′-to-5′ exoribonuclease complex exosomes are involved in removal of pri-miRNA processing intermediates (Kurihara *et al.*, 2012). Furthermore, majority of pri-miRNAs are possible targets of RNAase machinery in nucleus because of the presence of premature stop codon. Thus, protecting pri-miRNAs from destruction during co-transcriptional processing is very important for maintaining the pri-miRNA levels. Indeed, it has been demonstrated that accelerating the degradation of pri-miRNAs reduces build-up of miRNA levels (Yu *et al.*, 2008). Studies have shown that RNA-binding protein DDL stabilise miRNA levels in plants and deficiency of DDL decreases the pri-miRNA and therefore, miRNA levels despite having no effect on promoter efficiency or transcription of MIR, imply that DDL is involved in the protection of pri-miRNAs from degradation (Yu *et al.*, 2008).

The other protein factors of the MAC (MOS4-associated complex), e.g. PRL1, MAC3 are involved in stabilising pri-miRNAs without affecting their transcription status (Li *et al.*, 2018). Over-expression XRN inhibitor, PAP, increases the pri-miRNA levels. Furthermore, loss of function mutation in XRN2 or XRN3, another component of MAC, decreases the stability of pri-miRNA and increased their half-lives. These findings indicate that XRN2/XRN3 mediates the degradation of pri-miRNA in the nucleus (Li *et al.*, 2020). Some studies have also suggested that exosome is involved in the degradation of pri-miRNAs. Increased levels of pri-miRNAs have been observed in HEN2 (HUA1 enhancer 2) mutated Next (nuclear exosome targeting complex) which mediates RNA degradation by nuclear exosome (Li *et al.*, 2020). Experimental evidences suggest that the interaction (crosstalk) between proteins involved in miRNA biogenesis with ribonucleases results in pri-miRNA degradation and stabilisation of mRNAs in plants. SE, an important component of the DCL1 complex, appears to be involved in the recruitment of ribonucleases to pri-miRNAs. On the other hand, the unprocessed pri-miRNAs are protected from these exoribonucleases by coordinated action of other DCL1 complex components, such as DDL, HYL1 and MAC (Bielewicz *et al.*, 2013).

6. Regulation of miRNA Biogenesis in Animals

MiRNA processing pathway in animals is regulated at multiple levels for controlled generation of miRNAs in distinct cell types under varying conditions or different developmental stages. Transcriptional regulation of miRNAs appears to be the primary mechanism responsible for their biogenesis (Alberti *et al.*, 2018). However, pri-miRNA maturation may also be controlled by the microprocessor in the nucleus and by the Dicer in the cytoplasm, thereby adding to the abundance and accumulation of various miRNAs (Conrad *et al.*, 2014). Furthermore, regulation of miRNA decay is also crucial in regulating the optimal levels of miRNAs in animals (Reichholf *et al.*, 2019; Zhou *et al.*, 2018).

6.1 Transcriptional Control

Animal miRNA biogenesis is primarily regulated at the transcription level (Fig. 7). The transcriptional regulation of miRNA genes is mediated by a number of transcription factors and epigenetic regulators associated with Pol II. For instance, the transcription of miR1 and miR133 during myogenesis takes place by the binding of myogenic transcription factors, like myoblast determination 1 (MyoD1) and myogenin, upstream of miR133 and miR1 loci (Chen *et al.*, 2006; Rao *et al.*, 2006). Tumour suppressor, p53, regulates and activates miR-34 family of miRNAs. Furthermore, oncogenic factors, like MYC increases or supresses a number of miRNAs that play a role in cell cycle progression and cell death (He *et al.*, 2007; Chang *et al.*, 2007). Studies have found that c-Myc over-expression results in decreased expression of tumour suppressor miRNA gene families, like miR-29, miR-15, miR-34 and Let-7 (Frenzel

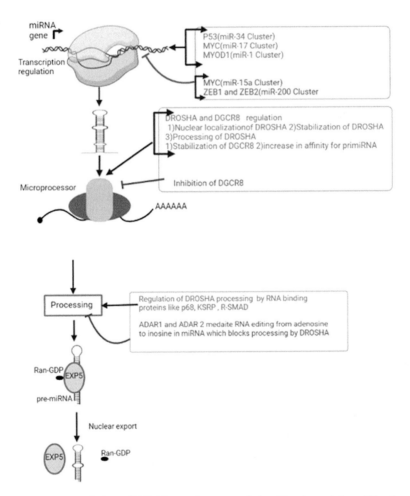

Fig. 7: Regulation of Plant miRNA biogenesis at transcriptional level, regulation of Drosha and DGCR8 proteins by different modifications; regulation of microprocessor complex by other proteins (This figure has been generated in Biorender)

et al., 2010). Moreover, expression of c-Myc-repressed miRNAs exogenously decreased the growth of lymphoma cells, indicating that repression of a class of miRNAs is an essential mechanism of c-Myc-induced tumorigenesis. miRNA gene regulation is also influenced by epigenetic factors, such as DNA methylation and histone modifications, e.g. in T-cell lymphoma, miR-203 locus is commonly methylated but not in normal T lymphocytes (Bueno *et al.*, 2008).

6.2 Regulation of the Microprocessor

DGCR8 and Drosha of microprocessor interact with pri-miRNAs by recognising a hairpin secondary structure followed by single-stranded RNA. Drosha cleaves and releases a precursor hairpin. The efficiency and precision of microprocessor interaction with pri-mRNA are critical for determining miRNA abundance and targeting specificity. The expression level, target specificity and cleavage activity of Drosha are controlled by a variety of mechanisms (Lee *et al.*, 2003; Sohn *et al.*, 2007). The autoregulation of Drosha and DGCR8 by cross-regulatory loop is important for the homeostatic levels of microprocessor activity throughout the animal world (Han *et al.*, 2009).

Drosha-catalysed processing of pri-mRNA is frequently influenced by RNA-binding proteins that either interact with Drosha or pri-miRNAs. MiR21 and mir-199a are stimulated by bone morphogenetic protein (BMP) and TGFβ without effecting their transcription. It has been postulated that BMP and TGF-activated R-SMAD proteins interact with Drosha, p68 and the stem of pri-miRNAs to accelerate processing of these pri-miRNAs by Drosha (Davis *et al.*, 2008). Another example of Drosha-mediated processing is miR-18a by hnRNPA1 (a miR-17-92 cluster member), which directly binds to the terminal loop of pri-miRNA18a hairpin and generates accessible cleavage site for Drosha (Guil *et al.*, 2007). The Drosha-mediated processing of miRNAs is also negatively regulated by several mechanisms. For example, let-7 miRNA processing is supressed by RNA binding protein LIN-28. It is understood that LIN28 interferes with pre-let-7 processing in cytoplasm by initiating uridylation of pre-let-7 terminals. The uridylated 3′ end of pre-let-7 consequently suppresses Dicer-mediated processing and promotes the degradation of pre-let-7 miRNA by recruiting 3′ to 5′ exonucleases (Heo *et al.*, 2008). Lin-28A mediated uridylation has also been found in other pre-miRNAs e.g. miR-143, -107 and -200c. These miRNAs contain GGAG motifs in their terminal loops (Ha and Kim, 2014).

6.3 Regulation of miRNA Stability

Stability of miRNAs contributes to their abundance. The miRNA stability varies from a few minutes to several days. Studies in different model organisms suggest that controlled decay of individual miRNAs is a way to regulate their expression or abundance (Ha and Kim, 2014).

It has been postulated that various nucleases mediate the cleavage and degradation of miRNAs, but little is known about their specificity. It is also less clear whether the miRNA decay machinery is conserved across the species. XRN-1 and XRN-2 (5′-3′ exoribonucleases) are involved in the decay of mature miRNAs in *C. elegans* (Chatterjee *et al.*, 2011; Krol *et al.*, 2010).

In mammals, various sensory neural-specific miRNAs, miR-183/-182/-96/-192 cluster and miR-204 and miR-211 are subjected to fast decay during dark adaptation by yet to be identified nucleases. Moreover, many neuronal miRNAs decay very

MicroRNAs in Plants and Animals: Converging and Diverging Insights   **33**

quickly, both in primary mammalian cell cultures as well as in dead brain tissues. It has been found that neuronal miRNAs in *Aplysia*, a marine snail were quickly degraded upon exposure to the neurotransmitter serotonin. Furthermore, in human melanoma cells, an interferon inducible enzyme, PNPT1 ($3'$–$5'$ exonuclease) decays some mature miRNAs, like miR-106, miR-222 and miR-221(Das *et al.*, 2010; Krol *et al.*, 2010). Likewise, absence of ERI1, $3'$-$5'$ exonuclease increased miRNA levels in mouse immune cells (Thomas *et al.*, 2012). In addition to mature miRNAs, pri-miRNAs and pre-miRNAs might also be subjected to the action of RNA decay enzymes. The mammalian endoribonuclease, MCP-induced protein 1, cleaves pre-miRNAs like miR-135b and miR-146a at their terminal loops (Suzuki *et al.*, 2011).

7. Evolution of miRNA in Plants and Animals

Since their discovery more than 20 years ago in *C. elegans*, miRNAs are now known to be essential for a wide range of biological processes across the species and also in plant and animal development. It is believed that miRNAs in plant and animal kingdoms have evolved separately from a common ancestral siRNA mechanism. The absence of homologous sequences among animal and plant miRNA families and dissimilarities in processing machinery and their working mechanism suggest that plant and animal miRNAs have evolved separately (Table 3) (Moran *et al.*, 2017).

Table 3: Comparison of evolutionary features in plant and animal miRNAs

Feature	Animals	Plants
MIR gene location	80% are located on introns.	90% are exonic.
Enzyme for transcription	RNA polymerase II and III.	RNA polymerase II.
pri-miRNA processing	Two step process and takes place in nucleus and completed in cytoplasm.	Completed in nucleus.
Processing enzymes	Drosha/DGCR8 (nucleus) Dicer (cytoplasm).	Micro-processor complex with components DCL1/HYL1/SE.
Methylation at $3'$OH end	NA	Prior to export of miRNA/miRNA* duplex into cytoplasm, HEN1 methylates its $3'$OH ends.
Addition of Uridine	Terminal uridylyl transferases are more commonly used (TUTases).	HEN1 suppressor 1 (HESO1) uridylates unmethylated miRNAs.
Addition of adenine	Single stranded miRNA is adenylated by GLD2, adenylation increases its stability.	Unknown enzyme for miRNA adenylation, miRNA stability is increased.
RNA editing (A-»I)	ADAR converts A-»I, may impede pri-micro RNA processing or affect miRNA targeting specificity.	NA
N6A methylation	Pri-micro RNA is methylated by METTL3 and is assigned to be processed further.	Methyl transferase (MTA) methylates pri-miRNA, its effect is unknown.

7.1 Origin of MIR Genes in Plants

Plant miRNAs are widely dispersed across the genome, with only a few members grouped together within a few kilobases. This miRNA gene dispersion may be explained by substantial shuffling the plant genomes has experienced since the amplification of earlier miRNA families. miRNA genes in plants are often generated from intergenic or, to a lesser degree, intragenic sections of the genome (Merchan et al., 2009). Three models have been proposed to understand the *de novo* origin of MIR genes in plants (Fig. 8). (1) *Inverted duplication of the target gene*: It has been shown that plant MIR genes were formed by inverted duplication of target gene sequences. Sequences surrounding mature miRNAs in MIR genes were shown to be similar to the sequences flanking miRNA-binding sites in target genes in *Arabidopsis*. Many examples indicate sequence similarity between MIR genes and target genes and this supports this approach. Furthermore, studies of miRNA precursors has shown that 16 miRNA loci have originated by inverted duplication of protein encoding regions (Allen et al., 2004; Fahlgren et al., 2010). (2) *Spontaneous origin of miRNAs*: Some immature miRNA genes arise from sequences that have few MIR gene properties, such as the capacity to generate a hairpin-like RNA. These genes were not subjected to severe selection pressure, which resulted in the emergence of new MIR genes that underwent advantageous mutations (Felippes et al., 2008). (3) *Miniature transposable elements, with inverted repeats (MITE)*: MITEs, which are short non-autonomous DNA-type TEs, are another source of immature miRNA genes. According to this model, some MITEs may be able to fold into stem loops which may be then processed into precursor miRNAs. It has been found that some rice miRNAs, one wheat miRNA and many miRNAs in *Arabidopsis* are derived from MITEs (Yu et al., 2014; Li et al., 2011).

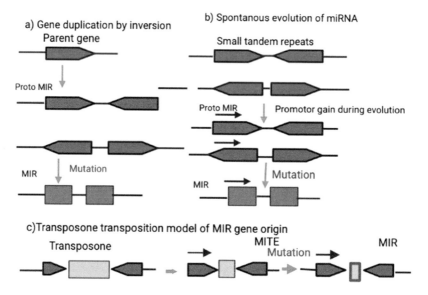

Fig. 8: Models for *de novo* origination and development of plant MIRs include inverted gene duplication, spontaneous evolution and transposon transposition (This figure has been generated in Biorender)

7.2 miRNA Divergence in Plants

Several families of miRNA, including miR156, miR159/319, miR156, miR166, miR160, miR408, miR390/391, and miR395 are conserved in embryophytes. Also, miR396 family has been discovered in all vascular plants and miR397/miR398 family in most of the seeded plants (Cuperus *et al.*, 2011).

miRNA gene duplication produces paralogous genes. This increases miR gene families and perhaps thereby expanding their function. Changes in sequences, expression patterns and functions can cause paralogous miRNA genes to diverge. In *A. thaliana*, most of the duplicated genes have divergent miRNA-binding sites between paralogs (Wang and Adams, 2015). Also, miR159/319 miRNAs have same sequence at 17 of 21 nucleotides but have evolved to target two different target gene families. This is an example of functional divergence of paralogous gene families. Another example of functional divergence of paralogous miRNAs is miR156 and miR529 families in land plants. These two miR gene families encode similar miRNAs which vary in their distribution and expression patterns in different species (Palatnik *et al.*, 2007). Furthermore, in *Arabidopsis,* miR156 and miR529 miRNAs have shown partial functional redundancy in which few miR156 family members compensate the loss of function by miR529. Different expression patterns of paralogous miRNAs may also have played a role in plant adaption to crop domestication patterns and unfavourable environmental conditions (Morea *et al.*, 2016; Zhang *et al.*, 2018).

7.3 miRNA and Target Gene Co-evolution in Plant

Mutual selection between miRNAs and their target genes is most likely to occur during evolution. Strong complementarity between miRNAs and their targets and the evolutionary conservation of these interactions among rice and *Arabidopsis*, as well as the existence of miRNA-binding motifs in many members of a gene family has allowed a reliable prediction about miRNA target genes. Recent studies have shown that MIR gene and the corresponding target gene evolution may have been influenced by factors like interaction of miRNAs with their targets and their expression levels (Liu *et al.*, 2016).

In rare circumstances, conserved miRNA-target pairings might be lost in specific species of plants, which necessitates the reorganisation of target gene networks for plant survival. For example, in *Salvia miltiorrhiza*, miR397 substitutes the function of miR162. Likewise, target site for miR162 in SmDCL1 is lost and the function is performed by miR397 to maintain its homeostatic feedback loop (Shao *et al.*, 2015). Furthermore, miRNA-triggered RNA cleavage of some miRNA targets, including as PPR, NBSLRR, MYB and other non-coding TAS loci, produce secondary phased siRNAs (phasiRNAs) (Fei *et al.*, 2013). This plant-specific sRNA amplification is most likely to be an efficient and cost-effective method of suppressing a large number of comparable genes with only one or a few starting miRNAs (Zhang *et al.*, 2018).

7.4 Origin of MIR Genes in Animals

The inverted duplication model for origin of MIR genes from target does not appear to be widely applicable in animals because miRNAs in animals very rarely show complementarity to their target genes. Nevertheless, new miRNA genes with different functions are generated from duplication of existing miRNA genes. This kind of duplication might result in the formation of larger miRNA clusters, which are subsequently classified as miRNA families (Zhang *et al.*, 2018). On the contrary, it has been believed that the key source of *de novo* miRNA origin is the appearance of RNA hairpins that gain capability for biogenesis of miRNAs (Axtell *et al.*, 2011). Transposable elements, also, are thought to be a possible mechanism for the generation of novel miRNA genes in animals (Piriyapongsa and Jordan, 2007; Sinzelle *et al.*, 2009). Mirtron, which is derived from short hairpin shaped introns, is another source of animal miRNA genes. They have evolved the ability for one RNase III cleavage by Dicer, as opposed to normal animal miRNAs, which must be cleaved by Drosha and Dicer separately (Zhang *et al.*, 2018).

8. Conservation of Expression Pattern in Animal miRNAs

The expression patterns of conserved miRNAs are often preserved in orthologous tissues and organ systems in animals over huge evolutionary gaps (Christodoulou *et al.*, 2010). Nevertheless, it has been found that patterns of expression vary in closely related species, like medaka and zebrafish, depending on the time and position of multiple orthologous miRNAs (Ason *et al.*, 2006; Liu *et al.*, 2008). Furthermore, in both vertebrates and invertebrates, there is variation in location and timing of expression pattern in duplicated miRNAs within miRNA families (Zhang *et al.*, 2018).

9. Functions of miRNA in Animals

Despite limited complementarity of mammalian miRNAs with their target mRNAs, they still regulate a variety of physiological functions in the host. Loss-of-function mutations in the two initially discovered miRNAs in *C. elegans*, lin-4 (abnormal cell lineage-4) and let-7 (lethal-7), caused developmental abnormalities in larvae. Lin-4 is thought to govern early embryonic phases, whereas let-7 is thought to play a key role in late development in *C. elegans* and other species (Lin *et al.*, 2003; Boehm *et al.*, 2005).

The lsy-6 (laterally symmetric-6) miRNA influences the cell fate of two anatomically different neurons, the ASE left (ASEL) and ASE right (ASER), respectively. The ASEL neurons release lsy-6, which supresses the expression of its target gene, cog-1 (connection of gonad defective-1 and causes asymmetry loss. MiR-273 suppresses the translation of die-1 in ASER neurons when it is triggered by the lsy-6 target cog-1 (dorsal intercalation and elongation defect-1). This causes lsy-6 to be down-regulated and the GCY-5 (guanylyl cyclase-5) receptor to be expressed in the ASER (Johnston *et al.*, 2003).

The over-expression of miRNA, bantam in *D. melanogaster* promotes growth and suppresses apoptosis (Nolo *et al.*, 2006). By acting on *D. melanogaster* IL1-beta convertase (DRICE), which is increased in the absence of miR-14, miR-14 decreases the rate of cell death. The GY-box motif is regulated by miR-7 and a reduction in miR-7 expression causes downstream notch targets, such as Cut, to be down-regulated and as a result causes reduced vein spacing and thickening.

The significance of miRNAs in mammalian developmental processes has been studied, using knockout gene method in several mammals (Stark *et al.*, 2003). Recent findings have revealed that miRNAs regulate late-stage mouse development, which is corroborated with modulation of Hox genes by miR-196. This miRNA is expressed

MicroRNAs in Plants and Animals: Converging and Diverging Insights   **37**

in the hind limb, where it cleaves Hoxa B8 and inhibits Hoxc8, Hoxd8 and Hoxa7 translation (Yekta *et al.*, 2004). Heart and neural crest derivatives-expressed protein 2 (HAND2) is targeted by the muscle-specific miRNA miR-1, resulting in muscle atrophy and premature cardiomyocyte differentiation (Zhao *et al.*, 2005). MiR181, which is expressed in the B lymphocytes of mice's bone marrow and thymus, promotes an increase in B lymphocytes and controls hematopoietic lineage differentiation in mice. Similarly, miR-143 expression in human fat tissues influences fat differentiation by increasing the level of extracellular signal-regulated kinase-5 (ERK5). MiR-375 and miR-16 are two miRNAs that influence a variety of physiological functions. MiR-375 is found in the pancreatic islet and suppresses glucose-stimulated insulin secretion by regulating the myotrophin gene, demonstrating that miR-375 is a glucose-stimulated insulin secretion inhibitor. The negative regulation of IgG1 and T-cell lineage by differentiation of T helper type 1 and type 2 cells is caused by miR-155, which targets the transcription factors PU.1 and c-Maf.

Some endogenous miRNAs have recently been discovered and these are involved in antiviral defence mechanisms, like the retrovirus type 1 (PFV-1) is inhibited by miR-32, which protects human cells against infection (Lecellier *et al.*, 2005). Furthermore, carcinogenesis is triggered by miR-372 and miR-373, which are expressed in primary human fibroblasts and target the tumour suppressor gene, LATS2. MiR-372 and miR-373 are specifically expressed in germ cell cancers of the testis. Based on the information presented above, it is reasonable to conclude that miRNAs regulate a variety of physiological processes in humans and other animals via a variety of targets. Table 4 lists a number of animal miRNAs and their corresponding biological roles.

10. Functions of miRNA in Plants

The initial evidence regarding the role of miRNAs in plant development came from mutants with different growth patterns due to defects in short RNA production or their function. The decreased miRNA activity caused a slew of developmental problems. Target gene expression pattern expansion in the absence of miRNA regulation indicates the involvement of miRNAs in target accumulation. This restriction was hypothesised, based on miR-165/166 prohibiting (PHB) regulation in *Arabidopsis* and maize rolled leaf 1 (RLD1) regulation in maize (Kim *et al.*, 2005). The miR-165/166 genes play a crucial role in the establishment and maintenance of abaxial polarity. Similarly, mutations in the miR-165/166 complementary region of the maize homeodomain leucine zipper (HD-ZIP) gene RLD1 induce an overabundance of RLD1-mRNA, which adaxialises leaf primordia (Juarez *et al.*, 2004). When the mutations in miR-165/166 were discovered and mapped to the miRNA complementary site, it was assumed that the changed phenotypes were due to the lack of miRNA-directed control.

The regulatory role of miR-167 in plant reproductive development has been discovered by two recent studies (Ru *et al.*, 2006). In immature flowers, the ARF6 and ARF8 genes control stamen formation. The ARF6/ARF8 double-mutant phenotypes are recapitulated in miR-167-over-expressing *Arabidopsis* plant, in which the plant produces flowers with small stamens and anthers that are unable to discharge pollen. In both ovules and anthers, where miR-167 is typically present, mutations of the miR-167 target sites for ARF6 or ARF8 cause aberrant expression of these genes (Wu *et al.*, 2006). Transgenic plants producing the miRNA-resistant form of MYB33 and the double-mutant miR-159ab showed developmental abnormalities, such as hyponastic

Table 4: Biological role of different miRNAs in animal

Micro RNAs	Gene Target	Biological Functions	Species
Bantam	hid	Cell proliferation and apoptosis.	*Drosophila melanogaster*
let-7	HBL-1, lin-41	Developmental transition.	*Caenorhabditis elegans*
MicroRNA-1	HAND 2	Proliferation and differentiation of Cardiomyocyte.	Mammals
MicroRNA-7	Notch target genes	Notch signalling.	*Drosophila melanogaster*
MicroRNA-14	Caspase	Cell death and lipid metabolism.	*Drosophila melanogaster*
MicroRNA-16	Various genes	Instability of mRNA.	*Homo sapiens*
MicroRNA-17-92	E2F1, C-Myc	Lymphoma of B-cells.	*Homo sapiens*
MicroRNA-32	Retrovirus PFV1	Defence against viruses.	*Homo sapiens*
Micro-RNA-143	ERK5	Adenocarcinoma formation of colon.	*Homo sapiens*
MicroRNA-146	ROCK1, c-Myc	Immune system development and functioning.	*Homo sapiens*
MicroRNA-155	c-Maf, PU-1	T-cell formation and in immunity (innate).	Mouse
MicroRNA-181	Unknown	Hematopoietic cell fate regulation.	*Mus. musculus*
MicroRNA-223	Mef2c, NFI-A	Granulocytic maturation.	*Homo sapiens*
MicroRNA-375	Myotrophin	Secretion of insulin.	*Mus. musculus*

leaves (Schwab *et al.*, 2005). These deficits were reduced in the quadruple mutant of miR-159ab, MYB33 and MYB65, concluding the function of miRNA-based MYB gene regulation in these abnormalities. Furthermore, over-expression of miR-164 promotes down-regulation of cup-shaped cotyledon (CUC1), (CUC2) and the NAC family genes, resulting in lateral leafing and rooting induction (Guo *et al.*, 2005).

Studies have found that miR-156 plays an important role in the process of leaf initiation by regulating members of Squamosa promotor, like gene family (SPL). For example, reduced miR-156 levels lead to over-expression of SPL9, promoting delayed leaf initiation in *Arabidopsis* (Guo *et al.*, 2005). Several miRNAs, such as miR-393, which targets transport inhibitor response 1 (TIR1) and three F-box proteins, are linked to signalling mediators that respond to plant hormones (Kepinski *et al.*, 2005). Table 5 summarises some of the known plant miRNAs and their corresponding biological roles. In conclusion, miRNAs clearly are involved in the regulation of numerous developmental and other physiological processes in plants.

MicroRNAs in Plants and Animals: Converging and Diverging Insights

Table 5: Biological role of different miRNAs in plants

Micro RNAs	Target Gene	Biological Functions	Species
Micro RNA 156	SPL	Transition in developmental stages.	*Arabidopsis. thaliana*
Micro RNA 157	SPL	Developmental timing.	*Gossypium hirsutum*
Micro RNA 159	MYBTFS	Development of flowers.	*Arabidopsis thaliana*
Micro RNA 160	ARF	Development of roots and leaf, auxin response to auxins, floral organ identity.	*Glycine max*
Micro RNA 164	NAC-TF	Development of roots and shoots.	*Arabidopsis thaliana*
Micro RNA 164a	NAC-TF	Leaf development, leaf patter formation and leaf polarity.	*Arabidopsis thaliana*
Micro RNA 164c	NAC-TF	Floral identity and flower development.	*Arabidopsis thaliana*
Micro RNA 165/166	PHB, HD-ZIP	Maintenance of meristem, development of vascular system and organ polarity.	*Arabidopsis thaliana*
Micro RNA 167	ARF6, ARF 8	Response to auxin.	*Arabidopsis thaliana*
Micro RNA 170/171	SCL	Development of roots.	*Populus trichocarpa*
Micro RNA 172	AP2	Developmental timing and floral organ identity.	*Oryza sativa, Arabidopsis thaliana*
Micro RNA 319	TCP	Leaf development.	*Arabidopsis thaliana*
Micro RNA 390	TAS3	Auxin response, developmental timing, lateral organ polarity.	*Zea mays*

References

Achkar, N.P., Cho, S.K., Poulsen, C., Arce, A.L., Re, D.A., Giudicatti, A.J., Karayekov, E., Ryu, M.Y., Choi, S.W., Harholt, J., Casal, J.J., Yang, S.W. and Manavella, P.A. (2020). A quick HYL1-dependent reactivation of microRNA production is required for a proper developmental response after extended periods of light deprivation, *Developmental Cell*, 46(2): 236-247 e6.

Addo-Quaye, C., Snyder, J.A., Park, Y.B., Li, Y.F., Sunkar, R. and Axtell, M.J. (2009). Sliced microRNA targets and precise loop-first processing of MIR319 hairpins revealed by analysis of the *Physcomitrella patens* degradome, *RNA*, New York, N.Y., 15(12): 2112-2121.

Alberti, C., Manzenreither, R.A., Sowemimo, I., Burkard, T.R., Wang, J. and Mahofsky, K. (2018). Cell-type specific sequencing of microRNAs from complex animal tissues, *Nat. Methods*, 15: 283-289.

Allen, E., Xie, Z., Gustafson, A.M., Sung, G.H., Spatafora, J.W. and Carrington, J.C. (2004). Evolution of microRNA genes by inverted duplication of target gene sequences in *Arabidopsis thaliana*, *Nature Genetics*, 36(12): 1282-1290.

Ameres, S.L. and Zamore, P.D. (2013). Diversifying microRNA sequence and function, *Nature Reviews: Molecular Cell Biology*, 14(8): 475-488.

Ason, B., Darnell, D.K., Wittbrodt, B., Berezikov, E., Kloosterman, W.P., Wittbrodt, J., Antin, P.B. and Plasterk, R.H. (2006). Differences in vertebrate microRNA expression, *Proceedings of the National Academy of Sciences of the United States of America*, 103(39): 14385-14389.

Aukerman, M.J. and Sakai, H. (2003). Regulation of flowering time and floral organ identity by a microRNA and its APETALA2-like target genes, *The Plant Cell*, 15(11): 2730-2741.

Axtell, M.J., Westholm, J.O. and Lai, E.C. (2011). *Vive la différence*: Biogenesis and evolution of microRNAs in plants and animals, *Genome Biology*, 12(4): 221.

Bagga, S., Bracht, J., Hunter, S., Massirer, K., Holtz, J., Eachus, R. and Pasquinelli, A.E. (2005). Regulation by let-7 and lin-4 miRNAs results in target mRNA degradation, *Cell*, 122(4): 553-563.

Bartel, D.P. (2009). MicroRNAs: Target re cognition and regulatory functions, *Cell*, 136(2): 215-233.

Baumberger, N. and Baulcombe, D.C. (2005). *Arabidopsis* Argonaute1 is an RNA slicer that selectively recruits microRNAs and short interfering RNAs, *Proceedings of the National Academy of Sciences of the United States of America*. 102(33): 11928-11933.

Bernstein, E., Caudy, A.A., Hammond, S.M. and Hannon, G.J. (2001). Role for a Bidentate ribonuclease in the initiation step of RNA interference, *Nature*, 409(6818): 363-366.

Bielewicz, D., Kalak, M., Kalyna, M., Windels, D., Barta, A., Vazquez, F., Szweykowska-Kulinska, Z. and Jarmolowski, A. (2013). Introns of plant pri-miRNAs enhance miRNA biogenesis, *EMBO Reports*, 14(7): 622-628.

Boehm, M. and Slack, F. (2005). A developmental timing microRNA and its target regulate life span in *C. elegans, Science* (New York, N.Y.), 310(5756): 1954-1957.

Bologna, N.G., Mateos, J.L., Bresso, E.G. and Palatnik, J.F. (2009). A loop-to-base processing mechanism underlies the biogenesis of plant microRNAs miR319 and miR159, *The EMBO Journal*, 28(23): 3646-3656.

Bologna, N.G. and Voinnet, O. (2014). The diversity, biogenesis and activities of endogenous silencing small RNAs in *Arabidopsis, Annual Review of Plant Biology*, 65: 473-503.

Braun, J.E., Huntzinger, E. and Izaurralde, E. (2012). A molecular link between miRISCs and deadenylases provides new insight into the mechanism of gene silencing by microRNAs, Cold Spring Harbour, *Perspectives in Biology*, 4(12): a012328.

Brennecke, J., Stark, A., Russell, R.B. and Cohen, S.M. (2005). Principles of microRNA-target recognition, *PLoS Biology*, 3(3): e85.

Brodersen, P., Sakvarelidze-Achard, L., Bruun-Rasmussen, M., Dunoyer, P., Yamamoto, Y.Y., Sieburth, L. and Voinnet, O. (2008). Widespread translational inhibition by plant miRNAs and siRNAs, *Science*, New York, N.Y., 320(5880): 1185-1190.

Brousse, C., Liu, Q., Beauclair, L., Deremetz, A., Axtell, M.J. and Bouché, N. (2014). A non-canonical plant microRNA target site, *Nucleic Acids Research*, 42(8): 5270-5279.

Budak, H., Kantar, M., Bulut, R. and Akpinar, B.A. (2015). Stress responsive miRNAs and isomiRs in cereals, *Plant Sci.*, 235: 1-13.

Bueno, M.J., Pérez de Castro, I., Gómez de Cedrón, M., Santos, J., Calin, G.A., Cigudosa, J.C., Croce, C.M., Fernández-Piqueras, J. and Malumbres, M. (2008). Genetic and epigenetic silencing of microRNA-203 enhances ABL1 and BCR-ABL1 oncogene expression, *Cancer Cell*, 13(6): 496-506.

Cai, Q., Liang, C., Wang, S., Hou, Y., Gao, L., Liu, L., He, W., Ma, W., Mo, B. and Chen, X. (2018). The disease resistance protein SNC1 represses the biogenesis of microRNAs and phased siRNAs, *Nature Communications*, 9(1): 5080.

Cai, X., Hagedorn, C.H. and Cullen, B.R. (2004). Human microRNAs are processed from capped, polyadenylated transcripts that can also function as mRNAs. *RNA*, New York, N.Y., 10(12): 1957-1966.

Carbonell, A., Fahlgren, N., Garcia-Ruiz, H., Gilbert, K.B., Montgomery, T.A., Nguyen, T., Cuperus, J.T. and Carrington, J.C. (2012). Functional analysis of three *Arabidopsis* Argonautes using slicer-defective mutants, *The Plant Cell*, 24(9): 3613-3629.

Carrington, J.C. and Ambros, V. (2003). Role of microRNAs in plant and animal development, *Science*, 301(5631): 336-338.

Carthew, R.W. and Sontheimer, E.J. (2009). Origins and mechanisms of miRNAs and siRNAs, *Cell*, 136(4): 642-655.

Chang, T.C., Wentzel, T.A., Kent, O.A., Ramachandran, K., Mullendore, M., Lee, K.H., Feldmann, G., Yamakuchi, M., Ferlito, M., Lowenstein, C.J., Arking, D.E., Beer, M.A., Maitra, A. and Mendell, J.T. (2007). Transactivation of miR-34a bt p53 broadly influences gene expression and promotes apoptosis, *Molecular Cell*, 26(5): 745-752.

Chatterjee, S., Fasler, M., Büssing, I. and Grosshans, H. (2011). Target-mediated protection of endogenous microRNAs in *C. elegans*, *Developmental Cell*, 20(3): 388-396.

Chekulaeva, M., Mathys, H., Zipprich, J.T., Attig, J., Colic, M., Parker, R. and Filipowicz, W. (2011). miRNA repression involves GW182-mediated recruitment of CCR4-NOT through conserved W-containing motifs, *Nature: Structural & Molecular Biology*, 18(11): 1218-1226.

Chen, J.F., Mandel, E.M., Thomson, J.M., Wu, Q., Callis, T.E., Hammond, S.M., Conlon, F.L. and Wang, D.Z. (2006). The role of microRNA-1 and microRNA-133 in skeletal muscle proliferation and differentiation, *Nature Genetics*, 38(2): 228-233.

Chen, X. (2004). A microRNA as a translational repressor of APETALA2 in *Arabidopsis* flower development, *Science*, New York, N.Y., 303(5666): 2022-2025.

Chorostecki, U., Moro, B., Rojas, A., Debernardi, J.M., Schapire, A.L., Notredame, C. and Palatnik, J.F. (2017). Evolutionary footprints reveal insights into plant microRNA biogenesis, *The Plant Cell*, 29(6): 1248-1261.

Christodoulou, F., Raible, F., Tomer, R., Simakov, O., Trachana, K., Klaus, S., Snyman, H., Hannon, G.J., Bork, P. and Arendt, D. (2010). Ancient animal microRNAs and the evolution of tissue identity, *Nature*, 463(7284): 1084-1088.

Conrad, T., Marsico, A., Gehre, M. and Orom, U.A. (2014). Microprocessor activity controls differential miRNA biogenesis *in vivo*, *Cell Rep.*, 9: 542-554.

Cuperus, J.T., Fahlgren, N. and Carrington, J.C. (2011). Evolution and functional diversification of miRNA genes, *The Plant Cell*, 23(2): 431-442.

Das, S.K., Sokhi, U.K., Bhutia, S.K., Azab, B., Su, Z.Z., Sarkar, D. and Fisher, P.B. (2010). Human polynucleotide phosphorylase selectively and preferentially degrades microRNA-221 in human melanoma cells, *Proceedings of the National Academy of Sciences of the United States of America*, 107(26): 11948-11953.

Davis, B.N., Hilyard, A.C., Lagna, G. and Hata, A. (2008). SMAD proteins control DROSHA-mediated microRNA maturation, *Nature*, 454(7200): 56-61.

Denli, A.M., Tops, B.B., Plasterk, R.H., Ketting, R.F. and Hannon, G.J. (2004). Processing of primary microRNAs by the Microprocessor complex, *Nature*, 432(7014): 231-235.

de Rie, D, Abugessaisa, I. and Alam, T. (2017). An integrated expression atlas of miRNAs and their promoters in human and mouse, *Nat. Biotechnol.*, 35(9): 872-878.

Didiano, D. and Hobert, O. (2006). Perfect seed pairing is not a generally reliable predictor for miRNA-target interactions, *Nature: Structural & Molecular Biology*, 13(9): 849-851.

Dong, Z., Han, M.H. and Fedoroff, N. (2008). The RNA-binding proteins HYL1 and SE promote accurate *in vitro* processing of pri-miRNA by DCL1, *Proceedings of the National Academy of Sciences of the United States of America*, 105(29): 9970-9975.

Eichhorn, S.W., Guo, H., McGeary, S.E., Rodriguez-Mias, R.A., Shin, C., Baek, D., Hsu, S.H., Ghoshal, K., Villén, J. and Bartel, D.P. (2014). MRNA destabilisation is the dominant effect of mammalian microRNAs by the time substantial repression ensues, *Molecular Cell*, 56(1): 104-115.

Fabian, M.R., Sonenberg, N. and Filipowicz, W. (2010). Regulation of mRNA translation and stability by microRNAs, *Annual Review of Biochemistry*, 79: 351-379.

Fahlgren, N., Jogdeo, S., Kasschau, K.D., Sullivan, C.M., Chapman, E.J., Laubinger, S., Smith, L.M., Dasenko, M., Givan, S.A., Weigel, D. and Carrington, J.C. (2010). MicroRNA gene evolution in *Arabidopsis lyrata* and *Arabidopsis thaliana*, *The Plant Cell*, 22(4): 1074-1089.

Faller, M. and Guo, F. (2008). MicroRNA biogenesis: There's more than one way to skin a cat, *Biochimica et Biophysica Acta*, 1779(11): 663-667.

Fang, X., Cui, Y., Li, Y. and Qi, Y. (2015). Transcription and processing of primary microRNAs are coupled by Elongator complex in *Arabidopsis*, *Nature Plants*, 1: 15075.

Fang, X., Shi, Y., Lu, X., Chen, Z. and Qi, Y. (2015). CMA33/XCT Regulates small RNA production through modulating the transcription of dicer-like genes in *Arabidopsis*, *Molecular Plant*, 8(8): 1227-1236.

Fang, Y. and Spector, D.L. (2007). Identification of nuclear dicing bodies containing proteins for microRNA biogenesis in living *Arabidopsis* plants, *Current Biology: CB*, 17(9): 818-823.

Fei, Q., Xia, R. and Meyers, B.C. (2013). Phased, secondary, small interfering RNAs in posttranscriptional regulatory networks, *The Plant Cell*, 25(7): 2400-2415.

Felippes, F.F., Schneeberger, K., Dezulian, T., Huson, D.H. and Weigel, D. (2008). Evolution of *Arabidopsis thaliana* microRNAs from random sequences, *RNA*, New York, N.Y., 14(12): 2455-2459.

Frenzel, A., Lovén, J. and Henriksson, M.A. (2010). Targeting MYC-regulated miRNAs to combat cancer, *Genes & Cancer*, 1(6): 660-667.

Fukao, A., Mishima, Y., Takizawa, N., Oka, S., Imataka, H., Pelletier, J., Sonenberg, N., Thoma, C. and Fujiwara, T. (2014). MicroRNAs trigger dissociation of eIF4AI and eIF4AII from target mRNAs in humans, *Molecular Cell*, 56(1): 79-89.

Fukaya, T., Iwakawa, H.O. and Tomari, Y. (2014). MicroRNAs block assembly of eIF4F translation initiation complex in *Drosophila*, *Molecular Cell*, 56(1): 67-78.

Gregory, R.I., Yan, K.P., Amuthan, G., Chendrimada, T., Doratotaj, B., Cooch, N. and Shiekhattar, R. (2004). The microprocessor complex mediates the genesis of microRNAs, *Nature*, 432(7014): 235-240.

Guil, S. and Cáceres, J.F. (2007). The multifunctional RNA-binding protein hnRNP A1 is required for processing of miR-18a, *Nature: Structural & Molecular Biology*, 14(7): 591-596.

Guo, H., Ingolia, N.T., Weissman, J.S. and Bartel, D.P. (2010). Mammalian microRNAs predominantly act to decrease target mRNA levels, *Nature*, 466(7308): 835-840.

Guo, H.S., Xie, Q., Fei, J.F. and Chua, N.H. (2005). MicroRNA directs mRNA cleavage of transcriptional factor NAC1 to downregulate auxin signals for arabidopsis lateral root development. *The Plant Cell*, 17(5): 1276-1386.

Ha, M. and Kim, V.N. (2014). Regulation of microRNA biogenesis, nature reviews, *Molecular Cell Biology*, 15(8): 509-524.

Haase, A.D., Jaskiewicz, L., Zhang, H., Lainé, S., Sack, R., Gatignol, A. and Filipowicz, W. (2005). TRBP, a regulator of cellular PKR and HIV-1 virus expression, interacts with Dicer and functions in RNA silencing, *EMBO Reports*, 6(10): 961-967.

Hajheidari, M., Farrona, S., Huettel, B., Koncz, Z. and Koncz, C. (2012). CDKF; 1 and CDKD protein kinases regulate phosphorylation of serine residues in the C-terminal domain of *Arabidopsis* RNA polymerase II, *The Plant Cell*, 24(4): 1626-1642.

Han, J., Pedersen, J.S., Kwon, S.C., Belair, C.D., Kim, Y.K., Yeom, K.H., Yang, W.Y., Haussler, D., Blelloch, R. and Kim, V.N. (2009). Post-transcriptional cross-regulation between Drosha and DGCR8, *Cell*, 136(1): 75-84.

He, L., He, X., Lim, L.P., de Stanchina, E., Xuan, Z., Liang, Y., Xue, W., Zender, L., Magnus, J., Ridzon, D., Jackson, A.L., Linsley, P.S., Chen, C., Lowe, S.W., Cleary, M.A. and Hannon, G.J. (2007). A microRNA component of the p53 tumor suppressor network. *Nature*, 447(7148): 1130-1134.

Heo, I., Joo, C., Cho, J., Ha, M., Han, J. and Kim, V.N. (2008). Lin28 mediates the terminal uridylation of let-7 precursor MicroRNA, *Molecular Cell*, 32(2): 276-284.

MicroRNAs in Plants and Animals: Converging and Diverging Insights **43**

Hirsch, J., Lefort, V., Vankersschaver, M., Boualem, A., Lucas, A., Thermes, C., d'Aubenton-Carafa, Y. and Crespi, M. (2006). Characterisation of 43 non-protein-coding mRNA genes in *Arabidopsis*, including the MIR162a-derived transcripts, *Plant Physiology*, 140(4): 1192-1204.

Huntzinger, E. and Izaurralde, E. (2011). Gene silencing by microRNAs: Contributions of translational repression and mRNA decay: Nature reviews, *Genetics*, 12(2): 99-110.

Huntzinger, E., Kuzuoglu-Öztürk, D., Braun, J.E., Eulalio, A., Wohlbold, L. and Izaurralde, E. (2013). The interactions of GW182 proteins with PABP and deadenylases are required for both translational repression and degradation of miRNA targets, *Nucleic Acids Research*, 41(2): 978-994.

Iwakawa, H.O. and Tomari, Y. (2015). The functions of MicroRNAs: mRNA decay and translational repression, *Trends in Cell Biology*, 25(11): 651-665.

Iwakawa, H.O. and Tomari, Y. (2013). Molecular insights into microRNA-mediated translational repression in plants, *Molecular Cell*, 52(4): 591-601.

Jo, M.H., Shin, S., Jung, S.R., Kim, E., Song, J.J. and Hohng, S. (2015). Human Argonaute2 has diverse reaction pathways on target RNAs, *Molecular Cell*, 59(1): 117-124.

Johnston, R.J. and Hobert, O. (2003). A microRNA controlling left/right neuronal asymmetry in *Caenorhabditis elegans, Nature*, 426(6968): 845-849.

Jonas, S. and Izaurralde, E. (2013). The role of disordered protein regions in the assembly of decapping complexes and RNP granules, *Genes & Development*, 27(24): 2628-2641.

Jones-Rhoades, M.W. and Bartel, D.P. (2004). Computational identification of plant microRNAs and their targets, including a stress-induced miRNA, *Molecular Cell*, 14(6): 787-799.

Juarez, M., Kui, J. and Thomas, J. (2004). microRNA-mediated repression of rolled leaf1 specifies maize leaf polarity, *Nature*, 428: 84-88.

Kawamata, T. and Tomari, Y. (2010). Making RISC, *Trends in Biochemical Sciences*, 35(7): 368-376.

Kepinski, S. and Leyser, O. (2005). The Arabidopsis F-box protein TIR1 is an auxin receptor, *Nature*, 435(7041): 446-451.

Kim, D., Sung, Y.M., Park, J., Kim, S., Kim, J., Park, J., Ha, H., Bae, J.Y., Kim, S. and Baek, D. (2016). General rules for functional microRNA targeting, *Nature Genetics*, 48(12): 1517-1526.

Kim, V.N., Han, J. and Siomi, M.C. (2009). Biogenesis of small RNAs in animals: Nature reviews, *Molecular Cell Biology*, 10(2): 126-139.

Kim, W., Benhamed, M., Servet, C., Latrasse, D., Zhang, W., Delarue, M. and Zhou, D.X. (2009). Histone acetyltransferase GCN5 interferes with the miRNA pathway in *Arabidopsis, Cell Research*, 19(7): 899-909.

Kim, Y.K. and Kim, V.N. (2007). Processing of intronic microRNAs, *The EMBO Journal*, 26(3): 775-783.

Kim, J., Jung, J.H., Reyes, J.H., Kim, Y.S., Kim, S.Y., Chung, K.S., Kim, J.A., Lee, N., Lee, Y., Narry Kim, V., Chua, N.H. and Park, C.M. (2005). microRNA-directed cleavage of ATHB15 mRNA regulates vascular development in *Arabidopsis inflorescence* stems, *The Plant Journal: For Cell and Molecular Biology*, 41(1): 84-94.

Kim, V.N. (2005). microRNA biogenesis: Coordinated cropping and dicing, *Nature Reviews: Molecular Cell Biology*, 6(5): 375-385.

Kloosterman, W.P., Wienholds, E., Ketting, R.F. and Plasterk, R.H. (2004). Substrate requirements for let-7 function in the developing zebrafish embryo, *Nucleic Acids Research*, 32(21): 6284-6291.

Krol, J., Busskamp, V., Markiewicz, I., Stadler, M.B., Ribi, S., Richter, J., Duebel, J., Bicker, S., Fehling, H.J., Schübeler, D., Oertner, T.G., Schratt, G., Bibel, M., Roska, B. and Filipowicz, W. (2010). Characterising light-regulated retinal microRNAs reveals rapid turnover as a common property of neuronal microRNAs, *Cell*, 141(4): 618-631.

Kurihara, Y., Schmitz, R.J., Nery, J.R., Schultz, M.D., Okubo-Kurihara, E., Morosawa, T., Tanaka, M., Toyada, T., Seki, M. and Ecker, J.R. (2012). Surveillance of 3' noncoding transcripts requires FEIRY1 and XRN3 in *Arabidopsis. G3(Bethesda, Md)*, 2(4): 487-498.

Landthaler, M., Yalcin, A. and Tuschl, T. (2004). The human DiGeorge syndrome critical region gene 8 and Its D. melanogaster homolog are required for miRNA biogenesis, *Current Biology: CB*, 14(23): 2162-2167.

Lang, P.L.M., Christie, M.D., Dogan, E.S., Schwab, R., Hagmann, J. and Van de Weyer, A.L. (2018). A role for the F-Box Protein HAWAIIAN SKIRT in plant microRNA function, Plant Physiol., 176: 730-741.

Lecellier, C.H., Dunoyer, P., Arar, K., Lehmann-Che, J., Eyquem, S., Himber, C., Saib, A. and Voinnet, O. (2005). A cellular microRNA mediates antiviral defence in human cells. *Science* (New YORK, N.Y.), 308(5721): 557-560.

Lee, R.C., Feinbaum, R.L. and Ambros, V. (1993). The *C. elegans* heterochronic gene lin-4 encodes small RNAs with antisense complementarity to lin-14, *Cell*, 75(5): 843-854.

Lee, Y., Ahn, C., Han, J., Choi, H., Kim, J., Yim, J., Lee, J., Provost, P., Rådmark, O., Kim, S. and Kim, V.N. (2003). The nuclear RNase III Drosha initiates microRNA processing, *Nature*, 425(6956): 415-419.

Lee, Y., Kim, M., Han, J., Yeom, K.H., Lee, S., Baek, S.H. and Kim, V.N. (2004). MicroRNA genes are transcribed by RNA polymerase II, *The EMBO Journal*, 23(20): 4051-4060.

Lee, Y.S. and Dutta, A. (2009). MicroRNAs in cancer, *Annual Review of Pathology*, 4: 199-227.

Lewis, B.P., Burge, C.B. and Bartel, D.P. (2005). Conserved seed pairing, often flanked by adenosines, indicates that thousands of human genes are microRNA targets, *Cell*, 120(1): 15-20.

Li, J., Yang, Z., Yu, B., Liu, J. and Chen, X. (2005). Methylation protects miRNAs and siRNAs from a 3'-end uridylation activity in *Arabidopsis*, *Current Biology: CB*, 15(16): 1501-1507.

Li, M. and Yu, B. (2021). Recent advances in the regulation of plant miRNA biogenesis, *RNA Biology*, 18(12): 2087-2096.

Li, S., Li, M., Liu, K., Zhang, H., Zhang, S., Zhang, C. and Yu, B. (2020). MAC5, an RNA-binding protein, protects pri-miRNAs from Serrate-dependent exoribonuclease activities, *Proceedings of the National Academy of Sciences of the United States of America*, 117(38): 23982-23990.

Li, S., Liu, K., Zhou, B., Li, M., Zhang, S., Zeng, L., Zhang, C. and Yu, B. (2018). MAC3A and MAC3B, two core subunits of the MOS4-associated complex, positively influence miRNA biogenesis, *The Plant Cell*, 30(2): 481-494.

Li, S., Xu, R., Li, A., Liu, K., Gu, L., Li, M., Zhang, H., Zhang, Y., Zhuang, S., Wang, Q., Gao, G., Li, N., Zhang, C., Li, Y. and Yu, B. (2018). SMA1, a homolog of the splicing factor Prp28, has a multifaceted role in miRNA biogenesis in *Arabidopsis*, *Nucleic Acids Research*, 46(17): 9148-9159.

Li, S., Liu, L., Zhuang, X., Yu, Y., Liu, X., Cui, X., Ji, L., Pan, Z., Cao, X., Mo, B., Zhang, F., Raikhel, N., Jiang, L. and Chen, X. (2013). MicroRNAs inhibit the translation of target mRNAs on thendoplasmic reticulum in *Arabidopsis*, *Cell*, 153(3): 562-574.

Li, Y., Sun, D., Ma, Z., Yamaguchi, K., Wang, L., Zhong, S., Yan, X., Shang, B., Nagashima, Y., Koiwa, H., Han, J., Xie, Q., Zhou, M., Wang, Z. and Zhang, X. (2020). Degradation of Serrate via ubiquitin-independent 20S proteasome to survey RNA metabolism, *Nature Plants*, 6(8): 970-982.

Li, Y., Li, C., Xia, J. and Jin, Y. (2011). Domestication of transposable elements into MicroRNA genes in plants, *PLoS ONE*, 6(5): e19212.

Li, Z., Xu, R. and Li, N. (2018). MicroRNAs from plants to animals, do they define a new messenger for communication? *Nutrition & Metabolism*, 15: 68.

Lin, S.Y., Johnson, S.M., Abraham, M., Vella, M.C., Pasquinelli, A., Gamberi, C., Gottlieb, E. and Slack, F.J. (2003). The Celegans hunchback homolog, HBL-1, controls temporal patterning and is a orobable microRNA target, *Developmental Cell*, 4(5): 639-650.

Liu, J., Carmell, M.A., Rivas, F.V., Marsden, C.G., Thomson, J.M., Song, J.J., Hammond, S.M., Joshua-Tor, L. and Hannon, G.J. (2004). Argonaute2 is the catalytic engine of mammalian RNAi, *Science*, New York, N.Y., 305(5689): 1437-1441.

Liu, N., Okamura, K., Tyler, D.M., Phillips, M.D., Chung, W.J. and Lai, E.C. (2008). The evolution and functional diversification of animal microRNA genes, *Cell Research*, 18(10): 985-996.

Liu, T., Fang, C., Ma, Y., Shen, Y., Li, C., Li, Q., Wang, M., Liu, S., Zhang, J., Zhou, Z., Yang, R., Wang, Z. and Tian, Z. (2016). Global investigation of the co-evolution of miRNA genes and microRNA targets during soybean domestication, *The Plant Journal: For Cell and Molecular Biology*, 85(3): 396-409.

Lu, L.F., Boldin, M.P., Chaudhry, A., Lin, L.L., Taganov, K.D., Hanada, T., Yoshimura, A., Baltimore, D. and Rudensky, A.Y. (2010). Function of miR-146a in controlling Treg cell-mediated regulation of Th1 responses, *Cell*, 142(6): 914-929.

Lund, E., Guttinger, S., Calado, A., Dahlberg, J.E. and Kutay, U. (2004). Nuclear export of microRNA precursors, *Science* (New York, N.Y.), 303(5654): 95-98.

Mateos, J.L., Bologna, N.G., Chorostecki, U. and Palatnik, J.F. (2010). Identification of microRNA processing determinants by random mutagenesis of *Arabidopsis* MIR172a precursor, *Current Biology: CB*, 20(1): 49-54.

Mcgraw, M., Baev, V., Rusinov, V., Jensen, S.T., Kalantidis, K. and Hatzigeorgiou, A.G. (2006). MicroRNA promoter element discovery in *Arabidopsis*, *RNA*, New York, N.Y., 12(9): 1612-1619.

Meijer, H.A., Kong, Y.W., Lu, W.T., Wilczynska, A., Spriggs, R.V., Robinson, S.W., Godfrey, J.D., Willis, A.E. and Bushell, M. (2013). Translational repression and eIF4A2 activity are critical for microRNA-mediated gene regulation, *Science*, New York, N.Y., 340(6128): 82-85.

Merchan, F., Boualem, A., Crespi, M. and Frugier, F. (2009). Plant polycistronic precursors containing non-homologous microRNAs target transcripts encoding functionally related proteins, *Genome Biology*, 10(12): R136.

Miller, A.A. and Waterhouse, P.M. (2015). Plant and animal microRNAs: Similarities and differences, *Functional & Integrative Genomics*, 5(3): 129-135.

Moran, Y., Agron, M., Praher, D. and Technau, U. (2017). The evolutionary origin of plant and animal microRNAs, *Nature Ecology & Evolution*, 1(3): 27.

Morea, E.G., da Silva, E.M., e Silva, G.F., Valente, G.T., Barrera Rojas, C.H., Vincentz, M. and Nogueira, F.T. (2016). Functional and evolutionary analyses of the miR156 and miR529 families in land plants, *BMC Plant Biology*, 16: 40.

Moretti, F., Kaiser, C., Zdanowicz-Specht, A. and Hentze, M.W. (2012). PABP and the poly(A) tail augment microRNA repression by facilitated miRISC binding, *Nature: Structural and Molecular Biology*, 19(6): 603-608.

Nishimura, T., Padamsi, Z., Fakim, H., Milette, S., Dunham, W.H., Gingras, A.C. and Fabian, M.R. (2015). The eIF4E-binding protein 4E-T is a component of the mRNA decay machinery that bridges the 5′ and 3′ termini of target mRNAs, *Cell Reports*, 11(9): 1425-1436.

Nolo, R., Morrison, C.M., Tao, C., Zhang, X. and Halder, G. (2006). The bantam microRNA is a target of the hippo tumor-suppressor pathway, *Current Biology: CB*, 16(19): 1895-1904.

Nozawa, M., Miura, S. and Nei, M. (2012). Origins and evolution of microRNA genes in plant species, *Genome Biol Evol.*, 4(3): 230-239.

Olsen, P.H. and Ambros, V. (1999). The lin-4 regulatory RNA controls developmental timing in *Caenorhabditis elegans* by blocking LIN-14 protein synthesis after the initiation of translation, *Developmental Biology*, 216(2): 671-680.

Palatnik, J.F., Wollmann, H., Schommer, C., Schwab, R., Boisbouvier, J., Rodriguez, R., Warthmann, N., Allen, E., Dezulian, T., Huson, D., Carrington, J.C. and Weigel, D. (2007). Sequence and expression differences underlie functional specialisation of *Arabidopsis* microRNAs miR159 and miR319, *Developmental Cell*, 13(1): 115-125.

Pandita, D. and Wani S.H. (2019). MicroRNA as a tool for mitigating abiotic stress in rice (*Oryza sativa* L.). *In:* Wani, S. (Ed.). *Recent Approaches in Omics for Plant Resilience to Climate Change*, Springer, Chamz., 109-133. https://Doi.org/10.1007/978-3-030-21687-0_6

Pandita, D. (2019). Plant MIRnome: miRNA biogenesis and abiotic stress response. *In:* Hasanuzzaman, M., Hakeem, K., Nahar, K., Alharby, H. (Eds.). *Plant Abiotic Stress Tolerance*, Springer, Cham, 449-474. https://Doi.org/10.1007/978-3-030-06118-0_18

Pandita, D. (2021). Role of miRNAi technology and miRNAs in abiotic and biotic stress resilience. *In:* Aftab, T. and Roychoudhury, A. (Eds.). *Plant Perspectives to Global Climate Changes*, Academic Press, Elsevier, 303-330. https://Doi.org/10.1016/B978-0-323-85665-2.00015-7

Pandita, D. (2022a). How microRNAs regulate abiotic stress tolerance in wheat? A Snapshot. *In:* Roychoudhury, A., Aftab, T. and Acharya, K. (Eds.). *Omics Approach to Manage Abiotic Stress in Cereals*, Springer, Singapore, 447-464. https://Doi.org/10.1007/978-981-19-0140-9_17

Pandita, D. (2022b). MicroRNAs shape the tolerance mechanisms against abiotic stress in maize. *In:* Roychoudhury, A., Aftab, T. and Acharya, K. (Eds.). *Omics Approach to Manage Abiotic Stress in Cereals*, Springer, Singapore, 479-493. https://Doi.org/10.1007/978-981-19-0140-9_19

Pandita, D. (2022c). miRNA- and RNAi-mediated metabolic engineering in plants. *In:* Aftab, T. and Hakeem, K.R. (Eds.). *Metabolic Engineering in Plants*, Springer, Singapore, 171-186. https://Doi.org/10.1007/978-981-16-7262-0_7

Park, M.Y., Wu, G., Gonzalez-Sulser, A., Vaucheret, H. and Poethig, R.S. (2005). Nuclear processing and export of microRNAs in *Arabidopsis*, *Proceedings of the National Academy of Sciences of the United States of America*, 102(10): 3691-3696.

Perge, P., Nagy, Z., Decmann, Á., Igaz, I. and Igaz, P. (2017). Potential relevance of microRNAs in inter-species epigenetic communication, and implications for disease pathogenesis, *RNA Biology*, 14(4): 391-401.

Piriyapongsa, J. and Jordan, I.K. (2007). A family of human microRNA genes from miniature inverted-repeat transposable elements, *PLoS ONE*, 2(2): e203.

Pouch-Pélissier, M.N., Pélissier, T., Elmayan, T., Vaucheret, H., Boko, D., Jantsch, M.F. and Deragon, J.M. (2008). Sine RNA induces severe developmental defects in *Arabidopsis thaliana* and interacts with HYL1 (DRB1), a key member of the DCL1 complex, *PLoS Genetics*, 4(6): e1000096.

Presnyak, V. and Coller, J. (2013). The DHH1/RCKp54 family of helicases: An ancient family of proteins that promote translational silencing, *Biochimica et Biophysica Acta*, 1829(8): 817-823.

Qi, Y., Denli, A.M. and Hannon, G.J. (2005). Biochemical specialisation within *Arabidopsis* RNA silencing pathways, *Molecular Cell*, 19(3): 421-428.

Qi, Y., He, X., Wang, X.J., Kohany, O., Jurka, J. and Hannon, G.J. (2006). Distinct catalytic and non-catalytic roles of Argonaute4 in RNA-directed DNA methylation, *Nature*, 443(7114): 1008-1012.

Rajagopalan, R., Vaucheret, H., Trejo, J. and Bartel, D.P. (2006). A diverse and evolutionarily fluid set of microRNAs in *Arabidopsis thaliana*, *Gene Dev.*, 20: 3407-3425.

Rao, P.K., Kumar, R.M., Farkhondeh, M., Baskerville, S. and Lodish, H.F. (2006). Myogenic factors that regulate expression of muscle-specific microRNAs, *PNAS*, 103(23): 8721-8726.

Reichholf, B., Herzog, V.A., Fasching, N., Manzenreither, R.A., Sowemimo, I. and Ameres, S.L. (2019). Time-resolved small RNA sequencing unravels the molecular principles of microRNA homeostasis, *Mol. Cell*, 75: 756-768

Ren, G., Xie, M., Zhang, S., Vinovskis, C., Chen, X. and Yu, B. (2014). Methylation protects microRNAs from an AGO1-associated activity that uridylates 5′ RNA fragments generated by AGO1 cleavage, *Proceedings of the National Academy of Sciences of the United States of America*, 111(17): 6365-6370.

Rogers, K. and Chen, X. (2013). Biogenesis, turnover, and mode of action of plant microRNAs, *The Plant Cell*, 25(7): 2383-2399.

Rojas, A., Drusin, S.I., Chorostecki, U., Mateos, J.L., Moro, B., Bologna, N.G., Bresso, E.G., Schapire, A., Rasia, R.M., Moreno, D.M. and Palatnik, J.F. (2020). Identification of key sequence features required for microRNA biogenesis in plants, *Nature Communications*, 11(1): 5320.

MicroRNAs in Plants and Animals: Converging and Diverging Insights   **47**

Ru, P., Xu, L., Ma, H. and Huang, H. (2006). Plant fertility defects induced by the enhanced expression of microRNA167, *Cell Research*, 16(5): 457-465.

Schnall-Levin, M., Zhao, Y., Perrimon, N. and Berger, B. (2010). Conserved microRNA targeting in *Drosophila* is as widespread in coding regions as in 3′UTRs, *Proceedings of the National Academy of Sciences of the United States of America*, 107(36): 15751-15756.

Schwab, R., Palatnik, J.F., Riester, M., Schommer, C., Schmid, M. and Weigel, D. (2005). Specific effects of microRNAs on the plant transcriptome, *Developmental Cell*, 8(4): 517-527.

Selbach, M., Schwanhäusser, B., Thierfelder, N., Fang, Z., Khanin, R. and Rajewsky, N. 2008. Widespread changes in protein synthesis induced by microRNAs, *Nature*, 455(7209): 58-63.

Shao, F., Qiu, D. and Lu, S. (2015). Comparative analysis of the Dicer-like gene family reveals loss of miR162 target site in SmDCL1 from *Salvia miltiorrhiza*, *Scientific Reports*, 5: 9891.

Siomi, H. and Siomi, M.C. (2010). Post-transcriptional regulation of microRNA biogenesis in animals, *Molecular Cell*, 38(3): 323-332.

Sinzelle, L., Izsvák, Z. and Ivics, Z. (2009). Molecular domestication of transposable elements: From detrimental parasites to useful host genes. Cellular and molecular life sciences, *CMLS*, 66(6): 1073-1093.

Sohn, S.Y., Bae, W.J., Kim, J.J., Yeom, K.H., Kim, V.N. and Cho, Y. (2007). Crystal structure of human DGCR8 core, *Nature Structural & Molecular Biology*, 14(9): 847-853.

Song, L., Han, M.H., Lesicka, J. and Fedoroff, N. (2007). *Arabidopsis* primary microRNA processing proteins HYL1 and DCL1 define a nuclear body distinct from the Cajal body, *Proceedings of the National Academy of Sciences of the United States of America*, 104(13): 5437-5442.

Souret, F.F., Kastenmayer, J.P. and Green, P.J. (2004). AtXRN4 degrades mRNA in *Arabidopsis* and its substrates include selected miRNA targets, *Molecular Cell*, 15(2): 173-183.

Stark, A., Brennecke, J., Russell, R.B. and Cohen, S.M. (2003). Identification of drosophila microRNA targets, *PLoS Biology*, 1(3): E60.

Su, C., Li, Z., Cheng, J., Li, L., Zhong, S., Liu, L., Zheng, Y. and Zheng, B. (2017). The Protein phosphatase 4 and SMEK1 complex dephosphorylates HYL1 to promote miRNA biogenesis by antagonising the MAPK cascade in *Arabidopsis*, *Developmental Cell*, 41(5): 527-539. e5.

Sun, Z., Guo, T., Liu, Y., Liu, Q. and Fang, Y. (2015). The roles of *Arabidopsis* CDF2 in transcriptional and posttranscriptional regulation of primary microRNAs, *PLoS Genetics*, 11(10): e1005598.

Sunkar, R. and Zhu, J.K. (2004). Novel and stress-regulated microRNAs and other small RNAs from *Arabidopsis*, *The Plant Cell*, 16(8): 2001-2019.

Suzuki, H.I., Arase, M., Matsuyama, H., Choi, Y.L., Ueno, T., Mano, H., Sugimoto, K. and Miyazono, K. (2011). MCPIP1 ribonuclease antagonizes dicer and terminates microRNA biogenesis through precursor microRNA degradation, *Molecular Cell*, 44(3): 424-436.

Thomas, M.F., Abdul-Wajid, S., Panduro, M., Babiarz, J.E., Rajaram, M., Woodruff, P., Lanier, L.L., Heissmeyer, V. and Ansel, K.M. (2012). Eri1 regulates microRNA homeostasis and mouse lymphocyte development and antiviral function, *Blood*, 120(1): 130-142.

Tian, R., Wang, F., Zheng, Q., Niza, V., Downie, A.B. and Perry, S.E. (2020). Direct and indirect targets of the *Arabidopsis* seed transcription factor ABSCISIC acid insensitive3, *The Plant Journal: For Cell and Molecular Biology*, 103(5): 1679-1694.

Todesco, M., Rubio-Somoza, I., Paz-Ares, J. and Weigel, D. (2010). A collection of target mimics for comprehensive analysis of microRNA function in *Arabidopsis thaliana*, *PLoS Genetics*, 6(7): e1001031.

Vaucheret, H. (2006). Post-transcriptional small RNA pathways in plants: Mechanisms and regulations, *Genes & Development*, 20(7): 759-771.

Voinnet, O. (2009). Origin, biogenesis, and activity of plant microRNAs, *Cell*, 136(4): 669-687.

Waghray, S., Williams, C., Coon, J.J. and Wickens, M. (2015). Xenopus CAF1 requires NOT1-mediated interaction with 4E-T to repress translation *in vivo*. *RNA*, New York, N.Y., 21(7): 1335-1345.

Wang, F., Polydore, S. and Axtell, M.J. (2015). More than meets the eye? Factors that affect target selection by plant miRNAs and heterochromatic siRNAs, *Current Opinion in Plant Biology*, 27: 118-124.

Wang, S., Quan, L., Li, S., You, C., Zhang, Y., Gao, L., Zeng, L., Liu, L., Qi, Y., Mo, B. and Chen, X. (2019). The Protein Phosphatase4 Complex Promotes Transcription and Processing of Primary microRNAs in *Arabidopsis*, *The Plant Cell*, 31(2): 486-501.

Wang, J., Mei, J. and Ren, G. (2019). Plant microRNAs: Biogenesis, Homeostasis, and Degradation, *Frontiers in Plant Science*, 10: 360.

Wang, S. and Adams, K.L. (2015). Duplicate gene divergence by changes in microRNA binding sites in *Arabidopsis* and Brassica, *Genome Biology and Evolution*, 7(3): 646-655.

Werner, S., Wollmann, H., Schneeberger, K. and Weigel, D. (2010). Structure determinants for accurate processing of miR172a in *Arabidopsis thaliana*, *Current Biology: CB*, 20(1): 42-48.

Wightman, B., Ha, I. and Ruvkun, G. (1993). Post-transcriptional regulation of the heterochronic gene lin-14 by lin-4 mediates temporal pattern formation in *C. elegans*, *Cell*, 75(5): 855-862.

Wilson, R.C. and Doudna, J.A. (2013). Molecular mechanisms of RNA interference, *Annual Review of Biophysics*, 42: 217-239.

Wu, M.F., Tian, Q. and Reed, J.W. (2006). Arabidopsis microRNA167 controls patterns of ARF6 and ARF8 expression, and regulates both female and male reproduction, *Development* (Cambridge, England), 133(21): 4211-4218.

Wu, F., Yu, L., Cao, W., Mao, Y., Liu, Z. and He, Y.L. (2007). The N-terminal double-stranded RNA binding domains of Arabidopsis HYPONASTIC LEAVES1 are sufficient for pre-microRNA processing, *The Plant Cell*, 18(3): 914-925.

Xie, Z., Allen, E., Fahlgren, N., Calamar, A., Givan, S.A. and Carrington, J.C. (2005). Expression of *Arabidopsis* miRNA genes, *Plant Physiology*, 138(4): 2145-2154.

Xie, Z., Kasschau, K.D. and Carrington, J.C. (2003). Negative feedback regulation of Dicer-like1 in *Arabidopsis* by microRNA-guided mRNA degradation, *Current Biology: CB*, 13(9): 784-789.

Yang, L., Wu, G. and Poethig, R.S. (2012). Mutations in the GW-repeat protein SUO reveal a developmental function for microRNA-mediated translational repression in *Arabidopsis*, *Proceedings of the National Academy of Sciences of the United States of America*, 109(1): 315-320.

Yang, Z., Ebright, Y.W., Yu, B. and Chen, X. (2006). HEN1 recognizes 21-24 nt small RNA duplexes and deposits a methyl group onto the 2' OH of the 3' terminal nucleotide, *Nucleic Acids Research*, 34(2): 667-675.

Yekta, S., Shih, I.H. and Bartel, D.P. (2004). MicroRNA-directed cleavage of HOXB8 mRNA, *Science* (New York, N.Y.), 304(5670): 594-596.

Yi, R., Qin, Y., Macara, I.G. and Cullen, B.R. (2003). Exportin-5 mediates the nuclear export of pre-microRNAs and short hairpin RNAs, *Genes & Development*, 17(24): 3011-3016.

Yu, B., Bi, L., Zheng, B., Ji, L., Chevalier, D., Agarwal, M., Ramachandran, V., Li, W., Lagrange, T., Walker, J.C. and Chen, X. (2008). The FHA Domain Proteins DAWDLE in Arabidopsis and SNIP1 in Humans Act in Small RNA Biogenesis, *Proceedings of the National Academy of Sciences of the United States of America*, 105(29): 10073-10078.

Yu, B., Yang, Z., Li, J., Minakhina, S., Yang, M., Padgett, R.W., Steward, R. and Chen, X. (2005). Methylation as a crucial step in plant microRNA biogenesis, *Science*, New York, N.Y, 307(5711): 932-935.

Yu, M., Carver, B.F. and Yan, L. (2014). TamiR1123 originated from a family of miniature inverted-repeat transposable elements (MITE) including one inserted in the Vrn-A1a promoter in wheat, *Plant Science: An International Journal of Experimental Plant Biology*, 215-216: 117-123.

MicroRNAs in Plants and Animals: Converging and Diverging Insights

Yu, Y., Jia, T. and Chen, X. (2017). The 'how' and 'where' of plant microRNAs, *The New Phytologist*, 216(4): 1002-1017.

Yumul, R.E., Kim, Y.J., Liu, X., Wang, R., Ding, J., Xiao, L. and Chen, X. (2013). Powerdress and diversified expression of the MIR172 gene family bolster the floral stem cell network, *PLoS Genetics*, 9(1). e1003218.

Zekri, L., Huntzinger, E., Heimstädt, S. and Izaurralde, E. (2009). The silencing domain of GW182 interacts with PABPC1 to promote translational repression and degradation of microRNA targets and is required for target release, *Molecular and Cellular Biology*, 29(23): 6220-6231.

Zhang, B., Wang, Q. and Pan, X. (2007). MicroRNAs and their regulatory roles in animals and plants, *Journal of Cellular Physiology*, 210(2): 279-289.

Zhang, H., Xia, R., Meyers, B.C. and Walbot, V. (2015). Evolution, functions, and mysteries of plant Argonaute proteins, *Current Opinion in Plant Biology*, 27: 84-90.

Zhang, Y., Yun, Z., Gong, L., Qu, H., Duan, X., Jiang, Y. and Zhu, H. (2018). Comparison of miRNA evolution and function in plants and animals, *MicroRNA*, Shariqah, United Arab Emirates, 7(1): 4-10.

Zhang, Z., Guo, X., Ge, C., Ma, Z., Jiang, M., Li, T., Koiwa, H., Yang, S.W. and Zhang, X. (2017). Ketchi Imports HYL1 to Nucleus for miRNA Biogenesis in Arabidopsis, *Proceedings of the National Academy of Sciences of the United States of America*, 114(15): 4011-4016.

Zhao, Y., Samal, E. and Srivastava, D. (2005). Serun response factor regulates a muscle-specific microRNA that target *Hand2* durin cardiogenesis, *Nature*, 436: 4011-4016.

Zhao, Y., Cong, L. and Lukiw, W.J. (2018). Plant and animal microRNAs (miRNAs) and their potential for inter-kingdom communication, *Cellular and Molecular Neurobiology*, 38(1): 133-140.

Zhou, L., Lim, M.Y.T., Kaur, P., Saj, A., Bortolamiol-Becet, D. and Gopal, V. (2018). Importance of miRNA stability and alternative primary miRNA isoforms in gene regulation during *Drosophila* development, *eLife*, 7: e38389.

Zhu, H., Zhou, Y., Castillo-González, C., Lu, A., Ge, C., Zhao, Y.T., Duan, L., Li, Z., Axtell, M.J., Wang, X.J. and Zhang, X. (2013). Bidirectional processing of pri-miRNAs with branched terminal loops by *Arabidopsis* Dicer-like1, *Nature: Structural & Molecular Biology*, 20(9): 1106-1115.

Zhu, R., Zhang, Z., Li, Y., Hu, Z., Xin, D., Qi, Z. and Chen, Q. (2016). Discovering numerical differences between animal and plant microRNA, *PLoS ONE*, 11(10): e0165152.

CHAPTER

3

Regulatory Roles of Plant MicroRNAs

Sehrish Ijaz, Vajiha Sahar Khan, Ayesha Ghazanfar and Zulqarnain Khan*

Institute of Plant Breeding and Biotechnology, MNS University of Agriculture, Multan

1. Introduction

Plant growth and development are controlled by precise temporal and spatial gene expression. However, it seems complicated as to how it is done by the plant to control the gene expression in different tissues. The discovery of miRNA was made possible when scientists tried to explore the reasons for the regulatory mechanism of gene expression. MiRNAs are non-coding, regulatory, small RNAs produced endogenously to regulate the gene expression at post-transcriptional level (Bartel, 2004). This class of RNAs is studied well in flowering plants, animals and some viruses. The first ever reported miRNA was found in a soil nematode *Caenorhabditis elegans* in 1993 (Lee *et al.*, 1993). Currently, it's far widely known that miRNAs are extensively present in flowering plants, animals and in a few viruses. miRNAs are found to play an important role in defining different developmental phases by regulation metabolic process, organ differentiation and signalling pathways. These are particularly important in controlling the gene expression in all biotic and environmental stresses (Bushati and Cohen, 2007). Previously a lot of research was conducted to identify the synthesis phenomenon of miRNAs and how these tiny molecules searched their targets and controlled their gene expression. It was reported in Drosophila, that small pieces of RNAs (21-22 nts) generated from dsRNA, were responsible for the initiation of gene silencing in cell extracts (Zamore *at al.*, 2004). It was found that in silencing machinery, there is a ribonuclease that is responsible for the generation of these small fragments of RNA. Then another important member of RNA interference family was identified and was named DICER, that chops down the dsRNA into small fragments (Ketting *et al.*, 2001) It was found laterally that a ribonuclease concerned within the RNA-caused silencing complex (RISC) generates these tiny RNAs of 21-23 nts (Hammond *et al.*, 2001). It is interesting to know how the miRNA is made and how it works. MIR genes usually located in intergenic regions are transcribed first into primary miRNAs. This primary microRNA forms imperfect stem loop structures, named as precursor miRNA, which

---

*Corresponding author: zulqurnain.khan@mnsuam.edu.pk

Regulatory Roles of Plant MicroRNAs

are further processed by DICER and methylated by Hua protein. These miRNAs are loaded on the RISC where they go systemically for gene silencing. Complementarity of miRNA with target genes directs the RISC to probe and silence thet genes, showing complementarity with the guided strand with the help of Argonaute proteins (Rogers and Chen, 2013). Majority of plants have more than hundred miRNA genes, mostly present in intergenic regions of genome (Nozawa *et al.*, 2012).

2. Role in Plant Regulation

MicroRNA is a master regulator that is involved in developmental transitions in plants. A variety of miRNAs controls the transcription factors that regulate crucial pathways. In many plants, miR172 regulates the flowering time, targeting APETALA2 (AP2) in *Arabidopsis* (Aukerman and Sakai, 2003; Khraiwesh *et al.*, 2012). Mature miRNA inhibits the translation of the target genes by complementarily pairing with the target genes, thus altering the expression of plant genes. Their mode of action is via binding, either at 3'UTR or 5'UTR of target genes or repression of the target gene expression post transcriptionally.

It is established that miRNAs are importantly involved in developmental and hormonal pathways in plants enabling them to cope with outer environmental stress and nutrient homeostasis (Khraiwesh *et al.*, 2012). This gene repression through mature miRNAs controls the plant's organ development; regulates hormonal pathways and the capacity of the plant to respond to extreme conditions and control signalling transduction (Naqvi *et al.*, 2012).

2.1 Gene Expression Regulation

Plant miRNAs use the conserved mechanism of transcript cleavage and translation repression to regulate gene expression (Rogers and Chen, 2013) and this is explained below.

2.1.1 Messenger RNA Cleavage

The Argonaute (AGO) protein has a particular slicer endonuclease activity, thus it is capable of cleaving RNA targets that show complementarity to loaded guided strand (Liu *et al.*, 2004). Slicer activity of several AGOs, like AGO1, AGO2, AGO4, AGO7 and AGO10 in *Arabidopsis* has been proved (Mi *et al.*, 2008; Maunoury and Vaucheret, 2011) By knowing this fact that small RNA is present in plants (Llave *et al.*, 2002), since then mRNA degradome sequencing has revealed a huge number of miRNAs targets via splicing (Addo-Quaye *et al.*, 2008). Good complementarity in miRNA/target is the basic necessity for notable target splicing via AGO protein and this complementarity is exhibited by miRNA throughout its life. AGO splicing may also pass the general requirements for 3' de-adenylation or 5' de-capping before mRNA degradation by exonucleases (Fahlgren and Carrington, 2010).

2.1.2 Translational Repression

The mRNA degradation leading towards translational repression is also another mechanism for miRNA-mediated gene regulation. As explained previously, that complementarity among miRNAs and the targeted mRNA is necessary. RNA cleavage is most frequently observed in the case of animals when there is a sufficient amount of complementarity in miRNA/target. However, in case of compromised complementarity, the chances of translation repression increase. The degree of repression is variable according to the available binding sites of miRNA in targeted mRNA (Cuellar and McManus, 2005). In this mechanism, some miRNAs bind to the 5' UTR and/or the ORF (Zeng *et al.*, 2002; Doench and Sharp, 2004). Although an appropriate mechanism of translational repression has not been established yet, studies propose that miRNA may also prevent ribosome machinery to work along the mRNAs and suppress protein formation (Carrington and Ambros, 2003). However, not all miRNAs comply with this role always. Like in the case of plants, usually the miRNA shows enough complementarity, but translation repression is most frequently observed among the common ones. For instance, even though miR172 perfectly binds the APETALA2 (AP2) mRNA, but it regulates gene expression, using repression translation mechanism (Aukerman and Sakai, 2003; Chen, 2004).

However, another phenomenon of miRNA-mediated gene expression has been observed. A poly(A) tail is attached to the 3' end of the mRNA to stabilise it and avoid the mRNA decay after transcription (Coller and Parker, 2004). It was observed that miRNAs also target mRNA by de-adenylation of their poly-A tails actively. In this way, miRNA affects the integrity of mRNA and possibly it would be the initial step in mRNA decay (Giraldez *et al.*, 2006). For example, in zebrafish, it was observed that miR430 was involved in the continuous de-adenylation of a particular mRNA and that ultimately wiped out all maternal mRNA and hence embryo development was affected badly. This kind of activity is exhibited by several miRNAs both *in vivo* and *in vitro* conditions (Bagga *et al.*, 2005; Jing *et al.*, 2005; Wu and Belasco, 2005).

2.2 Regulation by miRNAs in Abiotic Stresses

Plant miRNAs are actively involved in regulation of every developmental phase, such as root formation, plant biomass development, shifting of plant from vegetative stage to flowering and ultimately seed development. Moreover, miRNAs become active in controlling the plant response in severe stress conditions (Xie *at al.*, 2015; Pandita and Wani, 2019; Pandita, 2019; 2021; 2022a; 2022b; 2022c).

Abiotic stresses, including cold, drought, salinity and many others, can cause severe restrictions in plants, ultimately affecting their vigour and development. Plants, due to their immobile nature, have adopted various strategies to combat these stresses. Stomatal regulation is an easy approach to deal with drought and water pressure and to decrease water loss. Plant hormonal pathways, like abscisic acid (ABA), are activated to combat the stress. To efficiently address these harsh circumstances, the plant adopts various strategies. These strategies involve the down/up regulation of diverse genes; hence controlling of the specific pathways. The function of non-coding RNAs, which include miRNAs, is frequently observed in gene regulation via hindering the translation of mRNAs of genes (Großhans and Filipowicz, 2008). Usually, transcription factors are targeted by plant miRNAs that smartly regulate the diverse developmental components even in extreme conditions. For example, miR399 and miR395 functionally regulate the sulphate transporter and phosphate transporter, respectively (Jones-Rhoades and Bartel, 2004). Normally, miR160, miR167 and miR393 are down-regulated and transcriptional elements involved in auxin-mediated signalling remain active. While in unfavourable conditions, this situation is vice versa due to the activation of miRNAs. These miRNAs target the transcription factors involved in developmental pathways and ultimately save the energy of plants

Regulatory Roles of Plant MicroRNAs **53**

to combat the stress. However, miRNA conservation in particular is not associated with their function as many miRNAs behave differently in different species. Some miRNAs, found in *Arabidopsis*, are considered to regulate HD-Zip gene family and are found in all plants, ranging from simple ferns to flowering plants (Floyd and Bowman, 2004). *Arabidopsis* miR159 expression is regulated by gibberellin activity (Achard *et al.*, 2004; Wang, Zhou *et al.*, 2004), while in rice, its expression was not changed in response to gibberellin application (Tsuji *et al.*, 2006).

2.2.1 Role of MicroRNA in Drought Stress

Drought stress is an unfavourable environmental situation, when several plant metabolic activities, such as formation of photosynthetic components, stomatal conductance, nutrient uptake and balance are affected and ultimately leading to crop loss (Jaleel *et al.*, 2009). Drought avoidance is commonly carried out via adoption of physical changes in plants consisting of decreased stomatal conductance, narrowing of leaf area and extended root system (Pollock, 1990). Several miRNAs have been identified primarily by sequencing of small RNAs library in *Arabidopsis* seedlings exposed to numerous stresses (Sunkar and Zhu, 2004). A variety of pathways and genes are regulated by drought stress (Nezhadahmadi *et al.*, 2013). The miRNA uses the same strategy of gene silencing to provide drought tolerance in plants, however, sometimes it seems that the response of miRNAs to drought is species-specific (Arenas-Huertero, 2009). For Example, miR156 was found to be upregulated during drought in *Arabidopsis*, *Prunus persica*, wheat, barley and *Panicum virgatum* (Eldem *et al.*, 2012). However, the same miRNA is down-regulated in rice and maize (Lv *et al.*, 2010). Until now, a large number of miRNAs conferring drought tolerance have been identified in sugarcane (Gentile *et al.*, 2015), *Arabidopsis* (Liu *et al.*, 2008) and rice (Zhou *et al.*, 2010) and the count is ever-growing. Fourteen different stress-inducible miRNAs in *Arabidopsis* were observed; 10 out of them had been involved in NaCl pathways-regulation, four were drought-regulated and 10 have been low temperature (4°C)-regulated and miR168, miR171 and miR396 (with TFs as their anticipated goals) spoke back to all the stresses. Several miRNA families were confirmed to be involved in controlling auxin signalling to regulate plant growth and improvement, during drought. TIR1 (transport inhibitor reaction 1) is targeted by miR393, an auxin receptor in *Arabidopsis*. The TIR1 enzyme is involved in the degradation of Aux/IAA proteins through ubiquitination (Dharmasiri and Estelle, 2002). It was suggested that the overexpression of miR393 in control conditions changed and it was suppressed when subjected to drought stress. Overexpression of miR393 in rice seedlings was also validated, in response to artificial auxin analogue treatments (Bian *et al.*, 2012). Thus, accelerated stages of miR393 might down-regulate auxin signalling and may lessen plant development in drought stress. In some plant species, miR319 family members showed a variable response to drought, e.g. some miRNA (mir319 family) were up-regulated while the other ones were down-regulated during drought (Zhou *et al.*, 2010). Even the same miRNA in other plant species can show variable responses to drought, depending on the precise conditions. For example, in Medicago, expression degree of miR398a/b was accelerated in drought stress (Trindade *et al.*, 2010), while the expression of the same miRNA declined in the same plant species (Wang *et al.*, 2011). Such variations may additionally reflect different stages of drought stress and the high sensitivity of some miRNAs to different situations. Like in response to PEG-mediated drought stress, the miRNAs (miR167, miR172, miR393, miR395, miR396, miR398 and miR399) confirmed different phases of up/down regulation in tobacco plants (Frazier *et al.*, 2011).

The miR474 is found to be targeting proline dehydrogenase (PDH) and abundantly found during the duration of drought stress in maize. With much less PDH, proline accumulates inside the plant and enables to guard against drought pressure (Rayapati and Stewart, 1991). Similarly, both miR160 and miR167 had been observed as important regulators in drought and ABA reaction in flora. Microarray analysis showed that miR167 is triggered by drought in *Arabidopsis* (Liu *et al.*, 2008).

In *Arabidopsis* and rice, down-regulation of miR65 and 66 resulted in a high drought resistance. These miRNAs are particularly regulated plant development normally. Nuclear Transcription Factor Y Subunit Alpha 5 (NFYA5) is another protein that initiates the signalling cascade during stress. It was found in *Arabidopsis* that when miRNA expression is down, it will induce NYFA5. Hence, NFYA5 high expression mediates the signalling response against drought (Song *et al.*, 2019).

2.2.2 Role of Micro RNA in Salinity Stress

Salinity creates osmotic and ionic stress in vegetation which reduces plant growth. In saline conditions, higher uptake of Na^+ and Cl^- by cells reaches toxic levels that badly affect the plants by chlorosis or leaf senescence (James *et al.*, 2002). To cope with the excessive salt stress in their immobile lifestyle, flowering plants have evolved a massive diploma of developmental regulation through cascades of molecular networks. Obvious observation in response to salt tolerance is expression modification of genes.

Expression profiles for genes involved in metabolism, signal transduction, transcription, protein biosynthesis and membrane trafficking and photosynthesis are regulated in stress (Vinocur and Altman, 2005). The statement that a few plant miRNAs respond to stressful situations and their targets are stress-related genes indicates that miRNAs may play vital roles in plant stress reactions (Phillips *et al.*, 2007). Several stress responsive miRNAs have also been identified in *planta* under various biotic and abiotic stress conditions, including nutrient deficiency (Fujii *et al.*, 2005; Aung *et al.*, 2015). MiR156 delays flowering and root improvement in alfalfa, enhances biomass accumulation in addition to improving heat stress tolerance in *Arabidopsis* (Stief *et al.*, 2014). Salinity induces the expression of several genes to facilitate adaptive and protection responses in tolerant versus susceptible ones (Arshad *et al.*, 2017).

In maize and alfalfa, the function of miR156 in salinity stress was studied and it showed miR156 is actively involved in modulating the physiological and molecular mechanisms of salt tolerance and vitamins pathways. Other research has shown that miR398 expression was suppressed by both ABA and salinity (Jia *et al.*, 2009). MicroRNA and short interfering RNA are the main regulators of eukaryotic gene expression playing interactively to control plant expression profiling (Brant and Budak, 2018). In cotton, altered expression of five miRNAs in response to stress (miR156, miR162, miR159, miR395 and miR396) was studied. MiR156 is conserved across plant life and its position has widely been investigated in all plant species (Xie *et al.*, 2015).

2.2.3 Role of MicroRNA in Temperature Stress

Extreme temperatures damage nearly all components of plant growth and yield (Wahid *et al.*, 2012). The reproductive stage during plant cycle is the most sensitive degree to temperature stress or even a small alteration in temperature at seed formation stage will cause devastating losses for grain crops (Mittler *et al.*, 2012).

World climate is continuously changing and annual increase in environmental temperature is reported due to global warming (Neilson *et al.*, 2010). Three diverse miRNAs (miR156, miR160 and miR168) have been recognised in response to high temperature stress in seven plant species. Moreover, these temperature stress-regulated miRNAs had been also shown to be controlling the developmental activities (Liu and Chen, 2010). These findings suggest that the role of miRNAs in development and regulation of abiotic stress may be tightly connected and probably employs the same mechanism in all plant species. Though many temperature and stress-regulated miRNAs had been shared among specific plants, their expression patterns during heat stress seem to be species-specific. For example, miR172 is prompted by cold in *Brachypodium* and *Prunus persica* (Barakat *et al.*, 2012), while it turned to be repressed at some stage in response to cold in grapevine and wheat (Tang *et al.*, 2012). Similarly, miR156 was up-regulated in response to heat stress in *Arabidopsis*, wheat and Chinese cabbage. However, it was observed to be down-regulated in rice and *Saccharinum japonica* (Stief *et al.*, 2014). Eighty-five temperature responsive miRNAs were observed in tolerant rice genotypes and 26 were identified in sensitive ones. Additionally, out of the nine (commonly identified in both types) miRNAs, three (miR159a.1, miR159b and miR528-3p) were up-regulated in sensitive genotypes and the same set of miRNAs turned down-regulated in the tolerant genotype (Liu *et al.*, 2017). As one of the highly conserved miRNAs in flowers, miR394 has been tested to regulate leaf morphological development and stem cell identification using targeting leaf curling response (LCR) which encodes a putative F-box protein (Song *et al.*, 2012; Knauer *et al.*, 2013). Recently, miR394 was additionally identified as being involved in the modification of plant responses to low temperature, salt, and drought stresses. The transcription of miR394 and LCR was both significantly induced via low temperatures. The over-expressed miR394 and lcr mutants were found to confer higher cold tolerance. In comparison, the LCR over-expressing flowers exhibited a hypersensitive phenotype when compared to the wild ones in cold (Song *et al.*, 2016). MiR397 is another miRNA that has been also involved in the regulation of CBF-based pathway to control the tolerance mechanism against low temperature (Dong and Pie, 2014). Over-expression of miR397a in *Arabidopsis* significantly contributed tolerance to chilling and freezing stresses, accompanied with decrease leaf electrolyte leakage as compared to wild ones. It was also found that the CBF2 and downstream cold-responsive gene transcripts were abundant in miR397a over-expressing population in comparison to wild plants (Dong and Pie, 2014). MiR398 is affected by high temperatures and makes the plant more sensitive to HT stress by repressing the expression level of copper superoxide dismutase (a class of essential scavengers of ROS) (Lu *et al.*, 2013) and miR156 is found to be up-regulated by high temperature in *Arabidopsis* (Hwan Lee *et al.*, 2012; Stief *et al.*, 2014). Similarly different reports show that in plants miR160, miR166, miR167, miR172 and miR393, among others, are regulated in HT temperature stress (Kruszka *et al.*, 2014; Xu *et al.*, 2014). Such differential expression may be due to the species-specific function of miRNAs in cold reactions (Jeong and Green, 2013). In another study, many miRNAs have been identified in response to cold stress (Zhang *et al.*, 2014). High-throughput sequencing analyses showed the up-regulation of 31 and down-regulation of 43 miRNAs in tea (*Camellia sinensis*) plants. Moreover, their targeted genes have also been identified via degradome sequencing (Liu *et al.*, 2014).

2.2.4 Role of Micro RNA in Metals Stress

The major metallic contaminants are arsenic (As), cadmium (Cd), lead (Pb) and mercury (Hg). Recent research has diagnosed conserved and non-conserved miRNAs in response to heavy metals (Xie *et al.*, 2007). Twenty-one miRNAs have been analysed computationally in Brassica napus, where four miRNAs (miR156, miR171, miR393 and miR396a) turned suppressed upon exposure to Cd (Sun *et al.*, 2008). *A. thaliana*, having mutant gene for Pi transporter, showed a twofold increase in As accumulation as compared to control, while the overexpression of same gene resulted in more efficient translocation and accumulation of As (Catarecha *et al.*, 2007).

miR159 and miR167 have been found to target metallic transporter genes that are responsible for metal uptake, movement and vacuolar accumulation (Talke *et al.*, 2006; Xu *et al.*, 2013). An oligopeptide transporter (OPT1) has been studied as a putative target of miR159. OPTs regulate the movement of heavy metals in cells (Srivastava *et al.*, 2013). In rice, overexpression of miR166 conferred higher cadmium tolerance by actively reducing oxidative stress. It was found that Cd movement in plants was improved and hence cadmium was stored in grain (Ding *et al.*, 2018). Mir395 was shown to be up-regulated in response to arsenic-induced stress in *Arabidopsis* (Jagadeeswaran *et al.*, 2014). In other findings, however, miR395 overexpression in rice resulted in starvation response against sulphur (Liang *et al.*, 2010; Yuan *et al.*, 2016). So it suggests that this miRNA may be involved in the homeostasis process (Song *et al.*, 2019).

2.3 Role of miRNAs in Biotic Stresses

Bacteria, fungi, viruses, insects and nematodes are the predominant living entities taking the plant into biotic stress phase. In response, plants regulate their gene expression to cope with such situations. Several conserved and novel miRNAs are found in tomatoes that are regulating several genes in case of biotic stresses (Zuo *et al.*, 2011). Several studies have proven that small RNAs play an important role to initiate disease resistance responses (Zhang *et al.*, 2011). Until far, many mutants for genes of miRNA pathways show low resistance/enhanced susceptibility to pathogens which clearly demonstrates the role of miRNA in biotic stresses.

2.3.1 Response in Bacterial Infection

Bacterial pathogens take control of plant vital genes to escape immunity and making successful infection. Plants combat this kind of negative situations by allowing the expression of miRNAs (Sunkar *et al.*, 2012). The expression of miR2911 and miR1030 was induced in response to exposure to *Xanthomonas axonopodis* in Populas species. MiR393 was found to trigger PAMP immunity in response to bacterial elicitor flflg22. It resulted in blockage of auxin signalling due to suppressed gene expression of auxin receptors (Navarro, Dunoyer *et al.*, 2006). Here, the role of auxin can be explained as it promotes plant growth, so indirectly it provide biomass and foliage nutrition. In other aspects, auxin also suppressed the salicylic acid-mediated response against biotrophs that can be prompted by and contribute to the pathogen-associated molecular sample (PAMP)-prompted immunity (PTI) by silencing auxin receptors and eventually suppressing auxin signalling (Grant and Jones, 2009). Small RNA profiling in tomato showed that miR393, miR167 and miR160 were additionally prompted with the aid of a non-pathogenic pseudomonas syringae PV (Liu *et al.*, 2007). Upon induction

Regulatory Roles of Plant MicroRNAs **57**

of bacterial elicitor, PAMP triggered immunity which is also triggered by miR160a, miR773 and miR398b (Li *et al.*, 2010). Similarly, miR398 was found to regulate stress responses (Jagadeeswaran, Saini *et al.*, 2009). It was suppressed in response to an infection by pst DC3000. T-DNA genes of agrobacterium can induce RNA silencing. miR393 and miR167 accumulation significantly reduced the tumour formation by C58 agro-strain (Dunoyer *et al.*, 2006). Furthermore, pathogenesis-related (PR) gene expression was provoked by overexpression of miR393 (Mlotshwa *et al.*, 2008).

2.3.2 Response in Fungal Infection

The first acknowledged miRNA lin-4 was observed in *Caenorhabditis elegans*. The 2d miRNA was no longer identified till 2000 (Reinhart, Slack *et al.*, 2000). Subsequently, thousands of miRNAs were identified by using bioinformatic approaches and molecular cloning. The most current release of the miRNA database, miR Base launch 18.0, carries over 18,000 miRNAs in all lineages of life (Wang *et al.*, 2009). The unicellular organisms had also been shown to encode miRNAs (Zhao and Srivastava, 2007). Recently, it was stated that miRNA-like small RNAs additionally exist within the filamentous fungus *Neurospora crassa*. It was found that miR166a can also target osmotin-like protein (OLP) precursor. This protein is not found usually in healthy plants. In a transgenic potato, two OLP genes were activated by using a fungal pathogen. Infection with *Phytophthora infestans* ended in robust OLP expression. Over-expression of a plant miRNA (miR393) resulted in improved bacterial resistance (Navarro, Dunoyer *et al.*, 2006). Previous reports show that many fungi encode Dicer-like proteins and Argonaute proteins. It supported the idea that the silencing mechanism is active in fungus also to regulate gene expression (Segers *et al.*, 2007).

2.3.3 Role of MicroRNA in Viral Stress

Viruses are the devastating pathogens causing severe losses in plants. Flowering plants have developed various complex mechanisms during evolution to cope with the viruses. Gene silencing is one of these conserved mechanisms. Recent studies support that miRNA are actively involved against plant viruses. Interestingly, miRNA discovery has opened new horizons to study complex gene expression, and plant-pathogen molecular interactions (Dunoyer *et al.*, 2004; Simón-Mateo and García, 2006). The discovery of miRNAs has opened up a brand-new avenue for understanding gene expression, plant genetic engineering, and plant pathogenesis molecular investigations. Silencing mechanisms via miRNA are considered to be an ancient phenomenon developed at unicellular level; so these theories also support that miRNA may a have a common ancestry (Zhang *et al.*, 2005); that ptr-miR473a, ptrmiR478a and ptr-miR482 play variable roles in Populus in comparison to their activities in rice (Lu, Sun *et al.*, 2005). Many plant miRNA sites were identified in chimeric regions of plum pox virus (PPV) chimeras bearing plant miRNA target sequences, that have been reported to be playing an important role in *Arabidopsis*, (Simón-Mateo and García, 2006). Viruses are pathogenic in nature, having DNA/RNA as genetic material, which cause infection either by recruiting the cell machinery for their own replication or integrating into host genome. Virus are very smart players as they have evolved their protein into silencing suppresses that interact with the host silencing machinery to stop. So, to target a viral genome, several miRNAs work in coordination. Viral genome can be detected either by complete complementarity in case of compromised complementarity by miRNAs (Simón-Mateo and García, 2006). Based on the diversity of virus families, it is reasonable to expect that there can be numerous categories of virally encoded miRNAs. Many RNA viruses has been studied to find the miRNA via cDNA cloning, but the actual information is lacking. It is may be due to the lack of DNA-dependent RNA polymerase which is required for the pri-mRNA formation (Pfeffer, Zavolan *et al.*, 2004; Sullivan and Ganem, 2005).

Rice plants, infected by rice stripe virus were found to have several miRNAs, like miRNA171, miRNA535 and miR390. Similarly, during RSV infection, miR444 was found to interact with MIKCc MADS Box proteins that block RNA-dependent RNA polymerase to promote RNA silencing against it (224). Similarly, miR168 is induced by RSV bind with AGO18. Thus miRNA relieves its target AGO1 that is involved in multiple steps of RNAi to create viral resistance. In another report, miR393 was known to be induced by rice dwarf virus. TMV resistance in *A. tumefaciens* was correlated with a multiplexed expression of miR393 (Mlotshwa, Pruss *et al.*, 2008).

Numerous miRNAs were demonstrated to be involved in plant protection. For instance, in a previous study, nine out of 48 miRNAs were identified to be active during defence in a primitive moss (Physcomitrella). Proteolysis is a major step in pathogen invasion and it is found that miR139 targets a pathogen gene that does coding for a mucin-like protein to have a dense sugar coating in opposition to proteolysis. MiR163 targets the pathogenesis-related genes present intracellularly. MiR-408 plays in defence via interacting with various genes, e.g. a copper ion-binding protein and with electron transporter genes or phytocyanin homolog (Fattash *et al.*, 2007). In another study, 476 ESTs were identified as probable miRNAs against biotic stress (Zhang *et al.*, 2005). Similarly, 130 miRNAs were identified in Populus; majority of them were particularly regulating plant genes in defence (Lu *et al.*, 2005).

3. Conclusion

After a first report of the miRNA in 2000, several research studies were conducted to investigate the role of miRNA in various living organisms. These studies indicate that these tiny molecules play smartly to regulate plant developmental activities as well as regulation of plant genes in unfavourable conditions. Previous reports indicate a huge variety of conserved as well as non-conserved miRNAs, which either regulate gene during development or in stress conditions or work in both situations. The mechanism exploited by these molecules is gene regulation by suppression. Majority of miRNAs are particularly conserved and present in intergenic regions, hence, are particularly involved in regulation. Thus number of miRNAs are increasing day by day with the development of high-throughput sequencing techniques which help to find the diverse miRNAs in different tissues under different conditions. These techniques enable the researchers to identify the miRNA genome widely and their expression analysis in variable tissues. Many computational tools and web resources are available which facilitate in the identification, expression analysis and recognition of probable targets of miRNAs and stress response studies of miRNAs in unfavourable conditions. In the changing climatic conditions, it is necessary to characterise miRNAs which are responsible for regulating multiple stress responses in plants. Likewise, miRNAs also take part in modulation of gene expression via translational repression and in a large variety of organic alterations, promoting up- and down-regulation of genes. However, the available information is inadequate to fulfil all the gaps to completely understand

Regulatory Roles of Plant MicroRNAs

the capacity of miRNAs. Therefore, it is necessary to elucidate their unpredictable roles as principal components in development to derive their potential benefit.

References

Achard, P., Herr, A., Baulcombe, D.C. and Harberd, N.P. (2004). Modulation of floral development by a gibberellin-regulated microRNA, *Development*, 131(14): 3357-3365.

Addo-Quaye, C., Eshoo, T.W., Bartel, D.P. and Axtell, M.J. (2008). Endogenous siRNA and miRNA targets identified by sequencing of the *Arabidopsis* degradome, *Current Biology*, 18(10): 758-762.

Arshad, M., Guber, M.Y., Wall, K. and Hannoufa, A. (2017). An insight into microRNA156 role in salinity stress responses of alfalfa, *Frontiers in Plant Science*, 8: 356.

Aukerman, M.J. and Sakai, H. (2003). Regulation of flowering time and floral organ identity by a microRNA and its APETALA2-like target genes, *The Plant Cell*, 15(11): 2730-2741.

Aung, B., Gruber, M.Y. and Hannoufa, A. (2015). The microRNA156 system: A tool in plant biotechnology, *Biocatalysis and Agricultural Biotechnology*, 4(4): 432-442.

Bagga, S., Brach, J., Hunter, S., Massirer, K., Holtz, J., Eachus, R. and Pasquinellia, A.E. (2005). Regulation by let-7 and lin-4 miRNAs results in target mRNA degradation, *Cell*, 122(4): 553-563.

Barakat, A., Sriram, A., Park, J., Zhebentyayeva, T., Mian, D. and Abbott, A. (2012). Genome wide identification of chilling responsive microRNAs in *Prunus persica*, *BMC Genomics*, 13(1): 1-11.

Bartel, D.P. (2004). MicroRNAs: Genomics, biogenesis, mechanism, and function, *Cell*, 116(2): 281-297.

Bian, H., Xie, Y., Guo, F., Han, N., Ma, S., Zeng, Z., Wang, J., Yang, Y. and Zhu, M. (2012). Distinctive expression patterns and roles of the miRNA393/TIR1 homolog module in regulating flag leaf inclination and primary and crown root growth in rice (*Oryza sativa*), *New Phytologist*, 196(1): 149-161.

Brant, E.J. and Budak, H. (2018). Plant small non-coding RNAs and their roles in biotic stresses, *Frontiers in Plant Science*, 9: 1038.

Bushati, N. and Cohen, S.M. (2007). MicroRNA functions, *Annu. Rev. Cell Dev. Biol.*, 23: 175-205.

Carrington, J.C. and Ambros, V. (2003). Role of microRNAs in plant and animal development, *Science*, 301(5631): 336-338.

Catarecha, P., Segura, M.D., Zorrilla, J.M., Ponce, B.G., Lanza, M., Solano, R., Ares, J.P. and Leyva, A. (2007). A mutant of the *Arabidopsis* phosphate transporter PHT1 displays enhanced arsenic accumulation, *The Plant Cell*, 19(3): 1123-1133.

Chen, X. (2004). A microRNA as a translational repressor of APETALA2 in *Arabidopsis* flower development, *Science*, 303(5666): 2022-2025.

Coller, J. and Parker, R. (2004). Eukaryotic mRNA decapping, *Annual Review of Biochemistry*, 73(1): 861-890.

Cuellar, T.L. and McManus, M.T. (2005). MicroRNAs and endocrine biology, *Journal of Endocrinology*, 187(3): 327-332.

Dharmasiri, S. and Estelle, M. (2002). The role of regulated protein degradation in auxin response, *Plant Molecular Biology*, 49: 401-408.

Ding, Y., Gong, S., Wang, Y., Wang, F., Bao, H., Sun, J., Cai, C., Yi, K., Chen, Z. and Zhu, C. (2018). MicroRNA166 modulates cadmium tolerance and accumulation in rice, *Plant Physiology*, 177(4): 1691-1703.

Doench, J.G. and Sharp, P.A. (2004). Specificity of microRNA target selection in translational repression, *Genes & Development*, 18(5): 504-511.

Dong, C.-H. and Pei, H. (2014). Overexpression of miR397 improves plant tolerance to cold stress in *Arabidopsis thaliana*, *Journal of Plant Biology*, 57(4): 209-217.

Dunoyer, P., Himber, C. and Voinnet, O. (2006). Induction, suppression and requirement of RNA silencing pathways in virulent *Agrobacterium tumefaciens* infections, *Nature Genetics*, 38(2): 258-263.

Dunoyer, P., Himber, C. and Voinnet, O. (2006). Induction, suppression and requirement of RNA silencing pathways in virulent *Agrobacterium tumefaciens* infections, *Nature Genetics*, 38(2): 258-263.

Dunoyer, P., Lecellier, C.-H., Parizotto, E.A., Himber, C. and Voinnet, O. (2004). Retracted: Probing the MicroRNA and small interfering RNA pathways with virus-encoded suppressors of RNA silencing, *The Plant Cell*, 16(5): 1235-1250.

Eldem, V., Akcay, U.C., Ozhuner, E., Bakir, Y., Uranbey, S. and Unver, T. (2012). Genome-wide identification of miRNAs responsive to drought in peach (*Prunus persica*) by high-throughput deep sequencing, *PloS ONE*, 7(12): e50298.

Fahlgren, N. and Carrington, J.C. (2010). miRNA target prediction in plants. *In: Plant MicroRNAs*, Springer, 51-57.

Fattash, I., Voß, B., Reski, R., Hess, W.R. and Frank, W. (2007). Evidence for the rapid expansion of microRNA-mediated regulation in early land plant evolution, *BMC Plant Biology*, 7(1): 1-19.

Floyd, S.K. and Bowman, J.L. (2004). Ancient microRNA target sequences in plants, *Nature*, 428(6982): 485-486.

Frazier, T.P., Sun, G., Burklew, C.E. and Zhang, B. (2011). Salt and drought stresses induce the aberrant expression of microRNA genes in tobacco, *Molecular Biotechnology*, 49(2): 159-165.

Fujii, H., Chiou, T.J., Lin, S.I., Aung, K. and Zhu, J.K. (2005). A miRNA involved in phosphate-starvation response in *Arabidopsis*, *Current Biology*, 15(22): 2038-2043.

Gentile, A., Dias, L.I., Mattos, R.S., Ferreira, T.H and Menossi, M. (2015). MicroRNAs and drought responses in sugarcane, *Frontiers in Plant Science*, 6: 58.

Giraldez, A.J., Mishima, Y., Rihel, J., Grocock, R.J., Dongen, S.V., Inoue, K., Enright, A.J. and Schier, A.F. (2006). Zebrafish MiR-430 promotes deadenylation and clearance of maternal mRNAs, *Science*, 312(5770): 75-79.

Grant, M.R. and Jones, J.D. (2009). Hormone (dis)harmony moulds plant health and disease, *Science*, 324(5928): 750-752.

Großhans, H. and Filipowicz, W. (2008). The expanding world of small RNAs, *Nature*, 451(7177): 414-416.

Hammond, S.M., Bernstein, E., Beach, D. and Hannon, G.J. (2001). An RNA-directed nuclease mediates post-transcriptional gene silencing in Drosophila cells, *Nature*, 404(6775): 293-296.

Huertero, C.A, Perez, B., Rabanal, F., Melo, D.B., Rosa, C.D., Navarrete, G.E., Sancher, F., Covarrubias, A.A. and Reyes, J.L. (2009). Conserved and novel miRNAs in the legume *Phaseolus vulgaris* in response to stress, *Plant Molecular Biology*, 70(4): 385-401.

Hwan Lee, J., Joon Kim, J. and Ahn, J.H. (2012). Role of SEPALLATA3 (SEP3) as a downstream gene of miR156-SPL3-FT circuitry in ambient temperature-responsive flowering, *Plant Signalling & Behaviour*, 7(9): 1151-1154.

Jagadeeswaran, G., Saini, A. and Sunkar, R. (2009). Biotic and abiotic stress down-regulate miR398 expression in *Arabidopsis*, *Planta*, 229(4): 1009-1014.

Jagadeeswaran, G., Li, Y.F. and Sunkar, R. (2014). Redox signalling mediates the expression of a sulphate-deprivation-inducible micro RNA 395 in *Arabidopsis*, *The Plant Journal*, 77(1): 85-96.

Jaleel, C.A., Manvannan, P., Wahid, A., Farooq, M., Juburi, H.J.A., Somasundaram, R. and Vam, R.P. (2009). Drought stress in plants: A review on morphological characteristics and pigments composition, *Int. J. Agric. Biol.*, 11(1): 100-105.

James, R.A., Rivelli,A. R., Munns, R. and Caemmerer, S.V. (2002). Factors affecting CO_2 assimilation, leaf injury and growth in salt-stressed durum wheat, *Functional Plant Biology*, 29(12): 1393-1403.

Jeong, D.-H. and Green, P.J. (2013). The role of rice microRNAs in abiotic stress responses, *Journal of Plant Biology*, 56(4): 187-197.

Regulatory Roles of Plant MicroRNAs

Jia, X., Wang, W.X., Rwn, L., Chen, Q.J., Mendu, V., Willcut, B., Dinkins, R., Tang, X. and Tang, G. (2009). Differential and dynamic regulation of miR398 in response to ABA and salt stress in *Populus tremula* and *Arabidopsis thaliana*, *Plant Molecular Biology*, 71(1): 51-59.

Jing, Q., Huang, S., Guth, S., Zarubin, T., Motoyama, A., Chen, J., Padova, F.D., Lin, S.C., Gram, H. and Han, J. (2005). Involvement of microRNA in AU-rich element-mediated mRNA instability, *Cell*, 120(5): 623-634.

Ketting, R.F., Fischer, S.E.J., Bernstein, E., Sijen, T., Hannon, G.J. and Plasterk, R.H.A. (2001). Dicer functions in RNA interference and in synthesis of small RNA involved in developmental timing in *C. elegans*, *Genes & Development*, 15(20): 2654-2659.

Khraiwesh, B., Zhu, J.-K. and Zhu, J. (2012). Role of miRNAs and siRNAs in biotic and abiotic stress responses of plants, *Biochimica et Biophysica Acta (BBA)-Gene Regulatory Mechanisms*, 1819(2): 137-148.

Knauer, S., Holt, A.L., Somoza, I.R., Tucker, E.J., Hinze, A., Pisch, M., Tavelle, M., Timmwemans, M.C., Tucker, M.R. and Laux, J. (2013). A protodermal miR394 signal defines a region of stem cell competence in the *Arabidopsis* shoot meristem, *Developmental Cell*, 24(2): 125-132.

Kruszka, K., Pacak, A., Barteczka, A.S., Nuc, P., Alaba, S., Wroblewska, Z., Karlowski,W., Jarmolowski, Z. and Kulinska, Z.S. (2014). Transcriptionally and post-transcriptionally regulated microRNAs in heat stress response in barley, *Journal of Experimental Botany*, 65(20): 6123-6135.

Lee, R.C., Feinbaum, R.L. and Ambros, V. (1993). The *C. elegans* heterochronic gene lin-4 encodes small RNAs with antisense complementarity to lin-14, *Cell*, 75(5): 843-854.

Li, Y., Zhang, Q., Zhang, J., Wu, L., Qi, Y. and Zhou, J.-M. (2010). Identification of microRNAs involved in pathogen-associated molecular pattern-triggered plant innate immunity, *Plant Physiology*, 152(4): 2222-2231.

Liang, G., Yang, F. and Yu, D. (2010). MicroRNA395 mediates regulation of sulfate accumulation and allocation in *Arabidopsis thaliana*, *The Plant Journal*, 62(6): 1046-1057.

Liu, H., Qin, C., Chen, Z., Zuo, T., Yang, X., Zhou, H., Xu, M., Cao, S., Shen, Y., Lin, H., He, X., Zhang, Y., Li, L., Ding, H., Libbersted, T., Zhnag, Z. and Pan, G. (2014). Identification of miRNAs and their target genes in developing maize ears by combined small RNA and degradome sequencing, *BMC Genomics*, 15(1): 1-18.

Liu, H.H., Tian, X., Li, X.J., Wu, C.A. and Zheng, C.C. (2008). Microarray-based analysis of stress-regulated microRNAs in *Arabidopsis thaliana*, *RNA*, 14(5): 836-843.

Liu, J., Carmell, M.A., Rivas, F.V., Marsden, C.G., Thomson, M., Song, J.J., Hammond, S.M., Tor, L.J. and Hannon, G.J. (2004). Argonaute2 is the catalytic engine of mammalian RNAi, *Science*, 305(5689): 1437-1441.

Liu, P.P., Montgomery, T.A., Fahlgren, N., Kasschau, K.D., Nonogaki, H. and Carrington, J.C. (2007). Repression of auxin response factor 10 by microRNA160 is critical for seed germination and post-germination stages, *The Plant Journal*, 52(1): 133-146.

Liu, Q. and Chen, Y.-Q. (2010). A new mechanism in plant engineering: The potential roles of microRNAs in molecular breeding for crop improvement, *Biotechnology Advances*, 28(3): 301-307.

Liu, Q., Yan, S., Yang, T., Zhang, S., Chen, Y.Q. and Liu, B. (2017). Small RNAs in regulating temperature stress response in plants, *Journal of Integrative Plant Biology*, 59(11): 774-791.

Llave, C., Xie, Z., Kasschau, K.D. and Carrington, J.C. (2002). Cleavage of Scarecrow-like mRNA targets directed by a class of *Arabidopsis* miRNA, *Science*, 297(5589): 2053-2056.

Lu, S., Sun, Y.-H., Shi, R., Clark, C., Li, L. and Chiang, V.L. (2005). Novel and mechanical stress–responsive microRNAs in *Populus trichocarpa* that are absent from *Arabidopsis*, *The Plant Cell*, 17(8): 2186-2203.

Lu, X., Guan, Q. and Zhu, J. (2013). Down-regulation of CSD2 by a heat-inducible miR398 is required for thermotolerance in *Arabidopsis*, *Plant Signalling & Behaviour*, 8(8): e24952.

Lv, D.-K., Bai, X., Li, Y., Ding, X.D., Ge, Y., Cai, H., Ji, W., Wu, N. and Zhu, Y.M. (2010). Profiling of cold-stress-responsive miRNAs in rice by microarrays, *Gene*, 459(1-2): 39-47.

Maunoury, N. and Vaucheret, H. (2011). AGO1 and AGO2 act redundantly in miR408-mediated Plantacyanin regulation, *PLoS ONE*, 6(12): e28729.

Mi, S., Cai, T., Hu, Y., Chen, Y., Hodges, E., Ni, F., Wu, L., Li, S., Zhou, H., Long, C., Chen, S., Hannon, G.J. and Qi, Y. (2008). Sorting of small RNAs into *Arabidopsis* Argonaute complexes is directed by the 5′ terminal nucleotide, *Cell*, 133(1): 116-127.

Mittler, R., Finka, A. and Goloubinoff, P. (2012). How do plants feel the heat? *Trends in Biochemical Sciences*, 37(3): 118-125.

Mlotshwa, S., Pruss, G.J., Peragine, A., Endres, M.W., Li, J., Chen, X., Poethig, R.S., Bowman, L.H. and Vance, V. (2008). Dicer-like2 plays a primary role in transitive silencing of transgenes in *Arabidopsis*, *PLoS ONE*, 3(3): e1755.

Naqvi, A.R., Sarwar, M., Hasan, S. and Roychodhury, N. (2012). Biogenesis functions and fate of plant microRNAs, *Journal of Cellular Physiology*, 227(9): 3163-3168.

Navarro, L., Dounoyer, P., Jay, F., Arnold, B., Dharmasiri, N., Estell, M., Voinnet, O. and Jones, D.G. (2006). A plant miRNA contributes to antibacterial resistance by repressing auxin signalling, *Science*, 312(5772): 436-439.

Neilson, K.A., Gammulla, C.G., Mirzaei, M., Imin, N. and Haynes, P.A. (2010). Proteomic analysis of temperature stress in plants, *Proteomics*, 10(4): 828-845.

Nezhadahmadi, A., Prodhan, Z.H. and Faruq, G. (2013). Drought tolerance in wheat, *The Scientific World Journal*, 2013.

Nozawa, M., Miura, S. and Nei, M. (2012). Origins and evolution of microRNA genes in plant species, *Genome Biology and Evolution*, 4(3): 230-239.

Pandita, D. (2019). Plant MIRnome: miRNA biogenesis and abiotic stress response. *In:* Hasanuzzaman, M., Hakeem, K., Nahar, K. and Alharby, H. (Eds.). *Plant Abiotic Stress Tolerance*. Springer, Cham., 449-474. Doi https://Doi.org/10.1007/978-3-030-06118-0_18

Pandita, D. (2021). Role of miRNAi technology and miRNAs in abiotic and biotic stress resilience. *In:* Aftab, T. and Roychoudhury, A. (Eds.). *Plant Perspectives to Global Climate Changes*. Academic Press, Elsevier, 303-330. https://Doi.org/10.1016/B978-0-323-85665-2.00015-7

Pandita, D. and Wani, S.H. (2019). MicroRNA as a tool for mitigating abiotic stress in rice (*Oryza sativa* L.). *In:* Wani, S. (Ed.). *Recent Approaches in Omics for Plant Resilience to Climate Change*, Springer, Chamz, 109-133. Doi https://Doi.org/10.1007/978-3-030-21687-0_6

Pandita, D. (2022a). How microRNAs regulate abiotic stress tolerance in wheat? A snapshot. *In:* Roychoudhury, A., Aftab, T. and Acharya, K. (Eds.). *Omics Approach to Manage Abiotic Stress in Cereals*, Springer, Singapore, 447-464. https://Doi.org/10.1007/978-981-19-0140-9_17

Pandita, D. (2022b). MicroRNAs shape the tolerance mechanisms against abiotic stress in maize. *In:* Roychoudhury, A., Aftab, T. and Acharya, K. (Eds.). *Omics Approach to Manage Abiotic Stress in Cereals*, Springer, Singapore, 479-493. https://Doi.org/10.1007/978-981-19-0140-9_19

Pandita, D. (2022c). miRNA- and RNAi-mediated metabolic engineering in plants. *In:* Aftab, T. and Hakeem, K.R. (Eds.). *Metabolic Engineering in Plants*, Springer, Singapore, 171-186. https://Doi.org/10.1007/978-981-16-7262-0_7

Pfeffer, S., Zavolan, M., Grasser, F.A., Chien, M., Russo, J.J., Ju, J., John, B., Enright, A.J., Marks, D. and Sander, C. (2004). Identification of virus-encoded microRNAs, *Science*, 304(5671): 734-736.

Phillips, J.R., Dalmay, T. and Bartel, D. (2007). The role of small RNAs in abiotic stress, *FEBS Letters*, 581(19): 3592-3597.

Pollock, C. (1990). The response of plants to temperature change, *The Journal of Agricultural Science*, 115(1): 1-5.

Rayapati, P.J. and Stewart, C.R. (1991). Solubilisation of a proline dehydrogenase from maize (*Zea mays* L.) mitochondria, *Plant Physiology*, 95(3): 787-791.

Reinhart, B.J., Slack, F.J., Basson, M., Pasquinelli, A.E., Bettinger, J.C., Rougvie, A.E., Horvitz, H.R. and Ruvkun, G. (2000). The 21-nucleotide let-7 RNA regulates developmental timing in *Caenorhabditis elegans*, *Nature*, 403(6772): 901-906.

Regulatory Roles of Plant MicroRNAs

Rhoades, M.W.J. and Bartel, D.P. (2004). Computational identification of plant microRNA and their target, including a stress induced miRNA, *Molecular Cell*, 14: 787-799.

Rogers, K. and Chen, X. (2013). Biogenesis, turnover, and mode of action of plant microRNAs, *The Plant Cell*, 25(7): 2383-2399.

Segers, G.C., Zhang, X., Deng, F., Sun, Q. and Nuss, D.L. (2007). Evidence that RNA silencing functions as an antiviral defense mechanism in fungi. *Proceedings of the National Academy of Sciences*, 104(31): 12902-12906.

Simón-Mateo, C. and García, J.A. (2006). MicroRNA-guided processing impairs *Plum pox* virus replication, but the virus readily evolves to escape this silencing mechanism, *Journal of Virology*, 80(5): 2429-2436.

Song, J.B., Huang, S.Q., Dalmay, T. and Yang, Z.M. (2012). Regulation of leaf morphology by microRNA394 and its target leaf curling responsiveness, *Plant and Cell Physiology*, 53(7): 1283-1294.

Song, J.B., Gao, S., Wang, Y., Li, B.W., Zhang, Y.L. and Yang, Z.M. (2016). MiR394 and its target gene LCR are involved in cold stress response in *Arabidopsis*, *Plant Gene*, 5: 56-64.

Song, X., Li, Y., Cao, X. and Qi, Y. (2019). MicroRNAs and their regulatory roles in plant–environment interactions, *Annual Review of Plant Biology*, 70: 489-525.

Srivastava, S., Srivastava, A.K., Suprasanna, P. and Souza, S.F.D. (2013). Identification and profiling of arsenic stress-induced microRNAs in *Brassica juncea*, *Journal of Experimental Botany*, 64(1): 303-315.

Stief, A., Atmann, S., Hoffmann, K., Pant, B.D., Scheible, W.R. and Baurle, I.S. (2014). *Arabidopsis* miR156 regulates tolerance to recurring environmental stress through SPL transcription factors, *The Plant Cell*, 26(4): 1792-1807.

Sullivan, C.S. and Ganem, D. (2005). MicroRNAs and viral infection, *Molecular Cell*, 20(1): 3-7.

Sun, Y., Zhou, Q. and Diao, C. (2008). Effects of cadmium and arsenic on growth and metal accumulation of Cd-hyperaccumulator *Solanum nigrum* L., *Bioresource Technology*, 99(5): 1103-1110.

Sunkar, R. and Zhu, J.-K. (2004). Novel and stress-regulated microRNAs and other small RNAs from *Arabidopsis*, *The Plant Cell*, 16(8): 2001-2019.

Sunkar, R., Li, Y.-F. and Jagadeeswaran, G. (2012). Functions of microRNAs in plant stress responses, *Trends in Plant Science*, 17(4): 196-203.

Talke, I.N., Hanikenne, M. and Krämer, U. (2006). Zinc-dependent global transcriptional control, transcriptional deregulation, and higher gene copy number for genes in metal homeostasis of the hyperaccumulator *Arabidopsis halleri*, *Plant Physiology*, 142(1): 148-167.

Tang, Z., Zhong, L., Xu, C., Yuan, S., Zhang, F., Zheng, Y. and Zhao, C. (2012). Uncovering small RNA-mediated responses to cold stress in a wheat thermosensitive genic male-sterile line by deep sequencing, *Plant Physiology*, 159(2): 721-738.

Trindade, I., Capita, C., Dalmay, I., Fevereiro, M.P. and Santos, D.M.D. (2010). MiR398 and miR408 are up-regulated in response to water deficit in *Medicago truncatula*, *Planta*, 231(3): 705-716.

Tsuji, H., Aya, K., Ueguchi-Tanaka, M., Shimada, Y., Nakazono, M., Watanabe, R., Nishizawa, N.K., Gomi, K., Shimada, A. and Kitano, H. (2006). GAMYB controls different sets of genes and is differentially regulated by microRNA in aleurone cells and anthers, *The Plant Journal*, 47(3): 427-444.

Vinocur, B. and Altman, A. (2005). Recent advances in engineering plant tolerance to abiotic stress: Achievements and limitations, *Current Opinion in Biotechnology*, 16(2): 123-132.

Wahid, A., Farooq, M., Hussain, I., Rasheed, R. and Galani, S. (2012). Responses and management of heat stress in plants. *In: Environmental Adaptations and Stress Tolerance of Plants in the Era of Climate Change*, Springer, 135-157.

Wang, J.F., Zhou, H., Chen, Y.Q., Luo, Q.J. and Qu, L.H. (2004). Identification of 20 microRNAs from *Oryza sativa*, *Nucleic Acids Research*, 32(5): 1688-1695.

Wang, W., Ye, R., Xin, Y., Fang, X., Li, C., Shi, H., Zhou, X. and Qi, Y. (2011). An importin β protein negatively regulates microRNA activity in *Arabidopsis*, *The Plant Cell*, 23(10): 3565-3576.

Wang, W.-C., Lin, F.-M., Chang, W.-C., Lin, K.-Y., Huang, H.-D. and Lin, N.-S. (2009). MiRExpress: Analysing high-throughput sequencing data for profiling microRNA expression, *BMC Bioinformatics*, 10(1): 1-13.

Wu, L. and Belasco, J.G. (2005). Micro-RNA regulation of the mammalian lin-28 gene during neuronal differentiation of embryonal carcinoma cells, *Molecular and Cellular Biology*, 25(21): 9198-9208.

Xie, F., Wang, Q., Sun, R. and Zhang, B. (2015). Deep sequencing reveals important roles of microRNAs in response to drought and salinity stress in cotton, *Journal of Experimental Botany*, 66(3): 789-804.

Xie, F., Wang, Q., Sun, R. and Zhang, B. (2015). Small RNA sequencing identifies miRNA roles in ovule and fibre development, *Plant Biotechnology Journal*, 13(3): 355-369.

Xie, F.L., Huang, S.Q., Guo, K., Xiang, A.L., Zhu, Y.Y., Nie, L. and Yang, Z.M. (2007). Computational identification of novel microRNAs and targets in *Brassica napus*, *Febs Letters*, 581(7): 1464-1474.

Xu, F., Liu, Q., Chen, L., Kuang, J., Walk, T., Wang, J. and Liao, H. (2013). Genome-wide identification of soybean microRNAs and their targets reveals their organ-specificity and responses to phosphate starvation, *BMC Genomics*, 14(1): 1-30.

Xu, M.Y., Zhang, L., Li, K., Wang, J., Hu, X.L., Wang, M.B., Fan, Y.L., Zhang, C.Y. and Wang, L. (2014). Stress-induced early flowering is mediated by miR169 in *Arabidopsis thaliana*, *Journal of Experimental Botany*, 65(1): 89-101.

Yuan, N., Yuan, S., Li, Z., Li, D., Hu, Q. and Luo, H. (2016). Heterologous expression of a rice miR395 gene in *Nicotiana tabacum* impairs sulfate homeostasis, *Scientific Reports*, 6(1): 1-14.

Zamore, P.D., Tuschl, T., Sharp, P.A. and Bartel, D.P. (2004). RNAi: Double-stranded RNA directs the ATP-dependent cleavage of mRNA at 21 to 23 nucleotide intervals, *Cell*, 101(1): 25-33.

Zeng, Y., Wagner, E.J. and Cullen, B.R. (2002). Both natural and designed micro RNAs can inhibit the expression of cognate mRNAs when expressed in human cells, *Molecular Cell*, 9(6): 1327-1333.

Zhang, B.H., Pan, X.P., Wang, Q.L., Cobb, G.P. and Anderson, T.A. (2005). Identification and characterisation of new plant microRNAs using EST analysis, *Cell Research*, 15(5): 336-360.

Zhang, W., Gao, S., Zhou, X., Chellappan, P., Chen, Z., Zhou, X., Zhang, X., Fromuth, N., Coutino, G. and Coffey, M. (2011). Bacteria-responsive microRNAs regulate plant innate immunity by modulating plant hormone networks, *Plant Molecular Biology*, 75(1): 93-105.

Zhang, Y., Zhu, X., Chen, X., Song, C., Zou, Z., Wang, Y., Wang, M., Fang, W. and Li, X. (2014). Identification and characterisation of cold-responsive microRNAs in tea plant (*Camellia sinensis*) and their targets using high-throughput sequencing and degradome analysis, *BMC Plant Biology*, 14(1): 1-18.

Zhao, Y. and Srivastava, D. (2007). A developmental view of microRNA function, *Trends in Biochemical Sciences*, 32(4): 189-197.

Zhou, L., Liu, Y., Liu, Z., Kong, D., Duan, M. and Luo, L. (2010). Genome-wide identification and analysis of drought-responsive microRNAs in *Oryza sativa*, *Journal of Experimental Botany*, 61(15): 4157-4168.

Zhou, M., Gu, L., Li, P., Song, X., Wei, L., Chen, Z. and Cao, X. (2010). Degradome sequencing reveals endogenous small RNA targets in rice (*Oryza sativa* L. ssp. *indica*), *Frontiers in Biology*, 5(1): 67-90.

Zuo, J., Wang, Y., Liu, H., Ma, Y., Ju, Z., Zhai, B., Fu, D., Zhu, Y., Luo, Y. and Zhu, B. (2011). MicroRNAs in tomato plants, *Science China Life Sciences*, 54(7): 599-605.

CHAPTER

4

MicroRNA-mediated Regulation of Plant Growth and Development

Seyed Alireza Salami* and Shirin Moradi

Department of Horticultural Sciences, Faculty of Agricultural Science and Engineering,
University of Tehran, Iran

1. Introduction

Plant growth means enlargement of plant's mass, organs, tissues, cells or cell organelles. However, the definitions of growth differ from different points of view. At cellular level, growth is known as cell division, cell enlargement, vacuole increment and cell wall deposition. At the organ level, growth is known as enlargement or expedition of different organs (leaves, stems and roots). At the plant level, on its own, growth is defined as biomass accumulation. While considering the plant cycle, it refers to the different stages of germination, leaves, branches, flowers, fruits and seeds production (Brukhin and Morozov, 2011; Hilty *et al.*, 2021). Development is defined as the sum of growth and differentiation process. These stages involve all changes in plants organs, from seed germination to senescence.

Eukaryotes contain large populations of small non-coding RNAs, ranging in size from 21 to 24 nucleotides and are known as microRNAs (miRNAs). For biosynthesis of miRNAs, larger RNAs are required as precursors which should have the complementary part with miRNA to make double strand structures. miRNAs then link to the target mRNAs which contain specific sequence motif to the complementary part of miRNA to regulate the expression (Garcia, 2008).

MiRNAs are a class of important cellular, organic molecules that finely modulate gene expression in response to developmental and environmental conditions (Taheri-Dehkordi *et al.*, 2021). They are also involved in many vital processes, such as plant defence against biotic and abiotic stresses, growth and development (Pandita, 2019; Pandita and Wani, 2019; Pandita, 2021; 2022a; 2022b; 2022c). miRNAs have various roles in plant growth and development, such as regulation of terpenoids biosynthesis (Chen *et al.*, 2020), alkaloids biosynthesis (Verma *et al.*, 2020), shoot apical meristem

---

*Corresponding author: asalami@ut.ac.ir

66   *Plant MicroRNAs and Stress Response*

(SAM), root apical meristem (RAM), leaf development and flowering transition (D'Ario *et al.*, 2017). These crucial roles of miRNAs in plant development were discovered in mutant plants (Moss and Poethig, 2002; Carrington and Ambros, 2003).

Various miRNAs control the expression of certain genes in specific tissues and different developmental stages. This chapter focuses on the roles and functions of micro-RNAs in growth and development. This review compiles the current understanding of growth and development-responsive miRNAs in plants towards better management of crop production.

2. Seed Development, Embryogenesis and Germination

Seed production is a crucial stage in flowering plants' life cycle that insures plant survival. Successful pollination and fertilisation guarantee the seed development of flowering plants. In the double fertilisation process, two male gametes reach and fertilise two female gametes with the help of a pollen tube, and as a result, two structures of embryo and endosperm are made (Bleckmann *et al.*, 2014). Embryogenesis, known is the first step of development as the embryo structure is derived from non-embryo cells, sexually or asexually, followed by dormancy and germination. Germination is the first step in plant growth. This critical process is defined as sporting of seeds which usually occurs after dormancy and can be initiated by absorbing water, light or chilling (Srivastava, 2002). For germination, the first step is water absorption; then the embryo develops to the radicals and hypocotyl and after that, the apical meristems appear in the first visible stage of germination. Understanding these stages from different points of view, especially at the molecular level, is necessary for improving plant yield and quality.

Several genes, transcription factors and miRNAs are involved in seed development, embryogenesis and germination. miRNAs regulate seed development (Fig. 1). Activation or suppression of miRNAs in different tissues and developmental stages regulates the expression of related genes in the context of comprehensive gene network, eventually resulting in an expected phenotype which is not only the consequent of mRNA expression.

The first step of embryo development is known as heart-shape which is under miRNAs regulation. It has been reported that mutants that don't have the embryo defective76 (*emb76*) cannot be further developed to the heart shape. Not only *emb76*, but also *sus1,* which controls suspensor growth after heart shape stage and *SIN1* (Short-Integuments1), which control embryo viability are another target genes during embryogenesis (Bleckmann *et al.*, 2014). The gene, named *DICER-Like1* (*DCL1*), encoding enzymes of miRNA biosynthesis, seems to regulate the expression and function of the above genes (Sunkar, 2012).

DCL1 also controls the target genes involved in seed development, such as *LEC2* (Leafy-Cotyledon2) and *FUS3* (*FUSCA3*), which are necessary for early maturation of embryo and seed. Study of *DCL1* mutants indicated that this miRNA represses the function of these target genes during early development of seeds (Sunkar, 2012). Hyponastic leave 1(*hyl1*) and serrate (*se*) were found to be essential for making miRNA processing complex. The *se* mutant showed abnormal cell division in heart-shape stage and the *hyl1* mutant showed over-sensitivity to abscisic acid (ABA) during seed germination. Argonaute protein (AGO) family is involved in gene silencing. AGO proteins can moderate expression of target RNAs via siRNAs or miRNAs (Höck

MicroRNA-mediated Regulation of Plant Growth and Development **67**

and Meister, 2008). In *Arabidopsis*, *Ago1* gene negatively regulates the viability of embryo and the mutants of this gene don't show embryo lethal. The miR168 controls embryo development by affecting the expression of *hyl1*, *se* and *ago1* as target genes (Sunkar, 2012).

It has been reported that miR156, miR159, miR160, miR166, miR319, miR824 (more abundant at eight cell stage), miR168, miR393 and miR400 (embryo global stage) are involved in embryogenesis and embryo development (Sunkar, 2012). MiR156 seems to have a crucial role in early and late stages of embryogenesis. MiR156 mediates the expression of squamosa promoter-binding protein-like*10* (*SPL10*) as target gene which results in embryo maturation. It was reported that miR156 prevents the maturation of embryo by repression of *SPL10* (Nodine and Bartel, 2010). The auxin response factor10 (*ARF10*) gene, which is responsible for biosynthesis of auxin, is the target of miR160. The amount of auxin is affected during early and late stages of embryogenesis due to miR160-*ARF10* interaction (Nodine and Bartel, 2010; Sunkar, 2012). The Cup-Shaped Cotyledon1 (*CUC1*) gene has a crucial role in shoot apical meristem formation. It has been reported that miR164 affects the embryogenesis development by regulating *CUC1* expression (Jin *et al*., 2013).

The regulation of seed-size and seed-shape pattern seems to be under miRNA159 function by modulating the multiple GAMYB-like genes (*MYB33/65*) as target genes. Similarly, there is a group of transcription factors, named APETALA2 (*AP2*), that have an important role in seed weight and size. It was reported that miR172, by suppression of *AP2*, has a positive effect of seed weight increment and seed oil accumulation (Sunkar, 2012). The *Arabidopsis* auxin response factor 6 and 8 (*ARF6/8*) gene is targeted by miR167, which is necessary for seed dispersal. The distraction in the miR167 function resulted in lower seed production (Jin *et al*., 2013). It has been found that L-ascorbate oxidases (*AO*) are target genes for both miR397 and miR528 in rice seed development (Jin *et al*., 2013).

2.1 Leaf Development

The establishment of primary meristem occurs during embryo development and then germination ensues and leaf primordia appear. Leaves are the crucial part of vascular plant as they are the factory of food production. A better understanding of the leaf development process at the molecular level paves the way for increasing plant quality and yield by enhancing photosynthesis. The miRNAs function as post-transcriptional regulators of target genes that affect various stages of leaf development. Some miRNAs involved in leaf development process are shown in Fig. 1.

Leaf primordia formation and other related changes, such as organ boundaries and leaf polarity, are regulated by *CUC1* gene which is the target of miR164 (Jin *et al*., 2013; Li and Zhang, 2016). The Homeodomain-leucine Zipper (HD-Zip) transcription factors are a group of genes, unique in plant kingdom. It has been reported that the abundance of miRNA165/miR166 increased in leaf primordial affected patterning of leaves radial orientation by mediating HD zip genes *PHV* and *PHB*. These genes have high expression in leaf primordia and detect the abaxial and adaxial leaf fate (Carrington and Ambros, 2003; Liu *et al*., 2018). The cell proliferation is an important part of leaf development which is under the control of Growth Regulating Factors (*GRFs*) genes. MiR396 family were abundantly found in young leaves, perhaps due to their roles in modulating *GRF* genes. It was reported that when miR396 overexpressed in plants, the number of leaves decreased (Jin *et al*., 2013). Leaf curling is the morphological process which is under control of *leaf curling responsiveness (LCR)* and miR172 (Jin *et al.*, 2013). TCP is the gene family that encodes plant-specific genes protein and their member, especially in class 1, promotes the cell proliferation and expedition by regulating plant hormones. The TCP-miR319 interaction regulates the leaf development. Mutant of miR310 developed lesser number of leaves (Garcia, 2008; Jin *et al.*, 2013). TCP is also the target of miR159 which is involved in leaf development. Abnormal shape with flat and big flag leaves was observed overexpressed in miR393 mutant. This miRNA modulates leaves shape development by targeting group of genes that have role as auxin receptors, including Transport Inhibitor Response1 (*TIR1*), Auxin Signalling F-Box (*AFB1*) (Jin *et al.*, 2013). The transition from juvenille to adult leaf is under the control of glossy15 (*gl15*). It was found that miR172 negatively affects the juvenal to adult leaf transition by regulating *gl15* (Jin *et al.*, 2013).

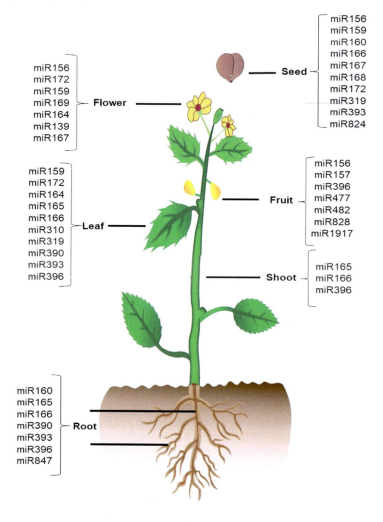

Fig. 1: Important miRNAs involved in different plant growth and development stages

Leaf senescence is the last stage of leaf and plant development. This process is controlled by miRNAs. It has been reported that miR166 targets the *WRKY53* gene which has a positive effect on expression of senescence-associated genes (SAG). The ORESARA1 (*ORE1*) gene was known as a leaf aging gene, which functions by promoting activity of SAG29, Bifunctional Nuclease1 (*BFN1*), Non-Yellow Coloring 1 (*NYC1*) and Pheporbide A Oxygenase (*PAO*) genes. MiR164 delays leaf senescence by repressing ORE1 (Liu *et al.*, 2018). The Auxin Response Factor2 (*ARF2*) regulates auxin transcriptional activation/repression. It was found that *ARF2* is the target gene of miR390 and their interaction promotes leaf senescence by modulating auxin signalling in plant leaves (Liu *et al.*, 2018).

2.2 Apical Meristem, Shoot and Vascular Development

During embryogenesis, the shoot apical meristem (SAM) and root apical meristem (RAM) are established in the shoot apex and root apex, respectively. Establishment of the SAM is the first step of shoot development. Shoot organs, such as leaves, branches and flowers are generated from the SAM. The next step is development of procambium to xylem and phloem; after these, other developmental changes happen towards development of complete shoot. As mentioned earlier, miRNAs regulate different stages of development, so it is expected that there are microRNAs committed to controlling branch development (Fig. 1). MiR165/166 seems to target Class III Homeodomain-Leucine Zipper (HD-ZIP III). These transcription factors are necessary for development of SAM to vascular organ (Miguel *et al.*, 2020). It was found that HD-ZIP III with high level of auxin, stops proliferation and differentiation of vascular cell and promotes xylem cell production (Miguel *et al.*, 2020). Until now, two members of miR165 and seven members of miR166 were found which related to shoot and vascular development. This is consistent with the nature of auxin as auxin induces dedifferentiation at high concentrations. However, the miR165/166 restricts the HD-ZIP III frequency and increases xylem cell production. AGO1 positively interacts with miR165/166, whereas, AGO10 promotes making inactive complex of AGO1-miR165/166, enhances HD-ZIP III frequency and SAM differentiation process by controlling the AGO1 (Liu *et al.*, 2018; Miguel *et al.*, 2020). The miR396 suppresses the *GRFs* as target genes; on the other hand, *GRFs* were critical for cell division and transition of meristem from SAM to RAM. It was reported that miR396s regulate SAM to RAM transition by moderating *GRFs* expression (Miguel *et al.*, 2020).

2.3 Root Development

Roots absorb nutrients and water from soil or any media surrounding it. Development of roots starts at the end of embryo development when RAM derives from embryo cells. Several miRNAs have crucial roles during root development (Fig. 1). MiR160 regulates root development by moderating *ARF10*, *ARF16* and *ARF17* transcription factors as their target genes. These transcription factors are involved in auxin signalling pathway and control root process, including root cap formation, primary root growth and lateral root development. It was reported that miR160 negatively regulates *ARFs* (Liu *et al.*, 2018). Root cell fate was determined by down-regulation of *HD-ZIP III* genes as target of miR165/166 during the early stage of root meristem establishment (de Lima *et al.*, 2012; Meng *et al.*, 2010; Wu, 2013).

The Plethora (*PLT*) is a gene which was found in root meristem at early developmental stage. *PLT* specialises stem cell in root meristem. It was suggested that miR396-GFRs' interaction negatively regulates *PLT* expression. On the other hand, in meristem niche *PLT* increases activation of miR396 to decrease *GRFs* function (Liu *et al.*, 2018). Another miRNA that interacts with *ARFs* is miR167 which regulates expression of *ARF6* and *ARF8*, both of which react with Gretchen Hagen3 (*GH3*). *GH3* gene encodes IAA-conjugating enzyme. *MiR167* regulates lateral root development (Li and Zhang, 2016; Liu *et al.*, 2018). *MiR390* also has a role in auxin signalling pathway and inhibits lateral root formation by targeting *ARF2* and *ARF3* (Meng *et al.*, 2010). MiR393-*NAC1* interaction modulates lateral root emergence (Meng *et al.*, 2010). MiR847 targets Indole-3-Acetic Acid 28 (*AA28*) gene that has the function in auxin pathway. It was reported both over-expressed and silenced mutant of miR847 cussed early lateral root formation. It means that miR847 controls lateral root formation (Meng *et al.*, 2010; Wu, 2013).

Most of the miRNAs involved in root development are tightly connected with auxin signalling pathway, which reveals the crucial role of auxin in root development.

2.4 Reproductive Phase and Flowering

Reproductive phase starts by developing floral organs during transition from vegetative to reproductive phase. Like other developmental stages during plant life cycle, several miRNAs seem to be involved in reproductive phase and flowering. During transition from juvenile to adult phase, two miRNAs, that work opposite, together play a crucial role (Fig. 1). The miR156 promotes juvenile growth by interacting with a group of genes, known as Squamosa Promoter Binding-Likes (*SPLs*). It positively enhances expression of *SPL3*, *SPL4*, *SPL5*, *SPL9*, *SPL10* and *SPL15* genes which promote juvenility (Li and Zhang, 2016). On the other hand, miR172 completely reacts opposite of miR156 and promotes enhanced flowering. This miRNA negatively interacts with a specific member of *APG2* transcription factors that are specialists in expressing flowering time, including Schlafmutze (*SMZ*), Schnarchzapfen (*SNZ*), Target Of EAT1 (*TOE1*) and *TOE2*. These genes promote juvenility in plants and delay flowering. It was observed that over-exertion of miR172 induced early flowering in both monocotyledonous and dicotyledonous plants; on the other side, silencing this miRNA delay flowering in mutant plant. It can be explained that miR172 down-regulates its target genes from *APG2* family (Liu *et al.*, 2018; Teotia and Tang, 2015; Wu, 2013).

MiR159 down-regulates *GAMYB-like* genes and represses flowering, although, it's very dependent on gibberellic acid (GA) signalling (Liu *et al.*, 2018; Teotia and Tang, 2015). The miR284 is specific to Brassicaceae, works similar to miR159, and regulates Agamous-Like 6 (*AGL16*) as target gene.

Flowering is also affected by different environmental factors and some miRNAs seem to control the genes related to the environment factor signalling. For example, both miR156 and miR172 are temperature-dependent miRNAs. It was observed that expression of miR156 increased at 16°C but the expression of miR172 increased at temperature higher than 23°C. The target gene of miR172 has a Short Vegetative Phase (*SVP*) which is a member of MADS-BOX genes that function in thermos-sensory pathway of flowering (Li and Zhang, 2016; Liu *et al.*, 2018; Teotia and Tang, 2015). The miR172 controls flowering, dependent on photoperiod pathway by regulating the expression of GIGANTEA (*GI*) as the target gene with a crucial role

MicroRNA-mediated Regulation of Plant Growth and Development **71**

in circadian clock pathway (Teotia and Tang, 2015). Another miRNA which reacts to environmental factor, is miR169. Its target gene is Nuclear Transcription Factor Y Subunit Alpha (_NF-YA_). MiR169 promotes flowering under stress conditions by supressing the expression of _NF-YA_ (Ferdous _et al._, 2015; Liu _et al._, 2018).

There are a number of miRNAs that specialise in flower shape development, such as miR164, which produces abnormal flowers with rotten petals and fused sepals by modulating the expression of the _CUC1_ gene (Liu _et al._, 2018). In contrast to miR164, miR319 increases flower petals via decreasing expression of TCP Family Transcription Factor 4 (_TCP4_) in tomato plants (Liu _et al._, 2018; Teotia and Tang, 2015). In _Arabidopsis_, miR167 also plays an important role in flower shape development. It was suggested that miR167 affects the flower shape by moderating genes in auxin pathway. The narrow and long sepal and petal phenotype, with irregular flower organs, was seen while miR167 was silenced and the small flower organs were reported while miR167 was over-expressed (Garcia, 2008b; Jin _et al._, 2013; Liu _et al._, 2018).

Fruit Development

Fruit development is very important in terms of commercial food production as well as the survival strategy to produce seeds. Finding the miRNAs which regulate fruit development is momentous scientifically and economically. The miRNA156/ miRNA157 control the development of carpel and fruit in tomato and _Arabidopsis_ plants by modulating the Squamosa Promoter-Binding Protein-Like (_SBP/SPL_) transcription factors (Miguel _et al._, 2020).

Also, miR396 has a key role in gynoecium development by regulating the expression of _GRF_ genes (Jin _et al._, 2013; Miguel _et al._, 2020). In apple, anthocyanin biosynthesis is under control of miR828 and miR858 (Miguel _et al._, 2020). It was reported that in the fleshy fruit development, miR477 and mirR482 affect cell expanding, cell wall enzymes production and gibberellins level in different stages by regulating DELLA protein GAI1-like (Garcia, 2008; Miguel _et al._, 2020).

MiR1917 has a crucial role in plant fruit development by moderating CTR1-Like Protein Kinase; it directly affects ethylene synthesis and response (Llave, 2004).

The F-box protein family has a role in signal transduction of ethylene. It was known that ethylene has a critical role in fruit ripening during its development stage. Both miR394 and miR828 are regulating expressions of different members of F-box family as target genes. Accordingly, for fruit ripening, high expression of miR394 and miR828 is necessary (Miguel _et al._, 2020).

At an early stage of berry fruit development, abundant of miR156 level was reported. It was suggested that this miRNA regulates light signalling in fruit development (Jin _et al._, 2013; Miguel _et al._, 2020).

3. Conclusion

Plant growth and development consist of complex various processes which are under the control of several mRNAs, transcription factors and regulatory RNAs, such as miRNAs. These tiny non-coding RNAs regulate their expression at a certain time and in a specific tissue. In this chapter, microRNAs which are involved in different stages of plant development, including embryogenesis, seed development, germination, leaf, shoot, root development, flowering, and fruit development were reviewed. The data showed that some miRNAs are tissue- or time-specific and others have different roles in several processes at different times. In general, it can be concluded that microRNAs cause plants to react properly in response to various environmental and developmental conditions and turn the plant into an automatic high-quality factory. In other words, miRNAs can participate in the regulation of multiple mRNAs in a wide variety of plant growth and developmental stages that make them a versatile tool for metabolic engineering and crop improvement.

References

Bleckmann, A., Alter, S. and Dresselhaus, T. (2014). The beginning of a seed: Regulatory mechanisms of double fertilisation, *Front. Plant Sci.*, 5: 452.

Brukhin, V. and Morozova, N. (2011). Plant growth and development – Basic knowledge and current views, *Math. Model. Nat. Phenom.*, 6(2): 1-53.

Carrington, J.C. and Ambros, V. (2003). Role of microRNAs in plant and animal development, *Science*, 301(5631): 336-338.

Chen, C., Zhong, Y., Yu, F. and Xu, M. (2020). Deep sequencing identifies miRNAs and their target genes involved in the biosynthesis of terpenoids in *Cinnamomum camphora*, *Ind. Crops Prod.*, 145: 111853.

D'Ario, M., Grifths-Jones, S. and Kim, M. (2017). Small RNAs: Big impact on plant development, *Trends in Plant. Sci.*, 22: 1056-1068.

de Lima, J.C., Loss-Morais, G. and Margis, R. (2012). MicroRNAs play critical roles during plant development and in response to abiotic stresses, *Genet. Mol. Biol.*, 35(4): 1069-1077.

Ferdous, J., Hussain, S.S. and Shi, B.J. (2015). Role of microRNAs in plant drought tolerance, *Plant Biotechnol. J.*, 13(3): 293.

Garcia, D. (2008). A miRacle in plant development: Role of microRNAs in cell differentiation and patterning, *Semin. Cell Dev. Biol.*, 19(6): 586-595.

Hilty, J., Muller, B., Pantin, F. and Leuzinger, S. (2021). Plant growth: The what, the how and the why, *New Phytol.*, 232(1): 25-41.

Höck, J. and Meister, G. (2008). The Argonaute protein family, *Genome Biol.*, 9(2): 210.

Jin, D., Wang, Y. and Zhao, Y. (2013). MicroRNAs and their cross-talks in plant development. *JGG*, 40(4): 161-170.

Li, C. and Zhang, B. (2016). MicroRNAs in control of plant development, *J. Cell. Physiol.*, 231(2): 303-313.

Liu, H., Yu, H., Tang, G. and Huang, T. (2018). Small but powerful: Function of microRNAs in plant development, *Plant Cell Rep.*, 37(3): 515-528.

Llave, C. (2004). MicroRNAs: More than a role in plant development? *Mol. Plant Pathol.*, 5(4): 361-366.

Meng, Y., Ma, X., Chen, D., Wu, P. and Chen, M. (2010). MicroRNA-mediated signalling involved in plant root development, *Biochem. Biophys. Res. Commun.*, 393(3): 345-349.

Miguel, C., Dalmay, T. and Chaves, I. (2020). *Plant MicroRNAs*, Springer, Amsterdam.

Moss, E.G. and Poethig, R.S. (2002). MicroRNAs: Something new under the sun, *Curr. Biol.*, 12(20): R688-R690.

Nodine, M.D. and Bartel, D.P. (2010). MicroRNAs prevent precocious gene expression and enable pattern formation during plant embryogenesis, *Genes Dev.*, 24(23): 2678-2692.

Pandita, D. (2019). Plant miRnome: miRNA biogenesis and abiotic stress response. *In:* Hasanuzzaman, M., Hakeem, K., Nahar, K. and Alharby, H. (Eds.). *Plant Abiotic Stress Tolerance*, Springer, Cham., 449-474. Doi https://Doi.org/10.1007/978-3-030-06118-0_18

Pandita, D. and Wani, S.H. (2019). MicroRNA as a tool for mitigating abiotic stress in rice (*Oryza sativa* L.). *In:* Wani, S. (Eds.). *Recent Approaches in Omics for Plant Resilience to Climate Change*. Springer, Chamz., 109-133. Doi https://Doi.org/10.1007/978-3-030-21687-0_6

Pandita, D. (2021). Role of miRNAi technology and miRNAs in abiotic and biotic stress resilience. *In:* Aftab, T. and Roychoudhury, A. (Eds.). *Plant Perspectives to Global Climate Changes*. Academic Press, Elsevier, 303-330. https://Doi.org/10.1016/B978-0-323-85665-2.00015-7

Pandita, D. (2022a). How microRNAs regulate abiotic stress tolerance in wheat? A snapshot. *In:* Roychoudhury, A., Aftab, T. and Acharya, K. (Eds.). *Omics Approach to Manage Abiotic Stress in Cereals*. Springer, Singapore, 447-464. https://Doi.org/10.1007/978-981-19-0140-9_17

Pandita, D. (2022b). MicroRNAs shape the tolerance mechanisms against abiotic stress in maize. *In:* Roychoudhury, A., Aftab, T. and Acharya, K. (Eds.). *Omics Approach to Manage Abiotic Stress in Cereals*. Springer, Singapore, 479-493. https://Doi.org/10.1007/978-981-19-0140-9_19

Pandita, D. (2022c). miRNA- and RNAi-mediated metabolic engineering in plants. *In:* Aftab, T. and Hakeem, K.R. (Eds.). *Metabolic Engineering in Plants*. Springer, Singapore, 171-186. https://Doi.org/10.1007/978-981-16-7262-0_7

Srivastava, L.M. (2002). Seed germination, mobilisation of food reserves, and seed dormancy. *In:* Srivastava, L.M. (Eds.). *Plant Growth and Development*. Elsevier, New York, USA, 447-471.

Sunkar, R. (2012). *MicroRNAs in Plant Development and Stress Responses*. Springer, Amsterdam.

Taheri-Dehkordi, A., Naderi, R., Martinelli, F. and Salami, S.A. (2021). Computational screening of miRNAs and their targets in saffron (*Crocus sativus* L.) by transcriptome mining, *Planta*, 254: 117.

Teotia, S. and Tang, G. (2015). To bloom or not to bloom: Role of microRNAs in plant flowering, *Mol. Plant Pathol.*, 8(3): 359-377.

Verma, P., Singh, N., Khan, S.A., Mathur, A.K., Sharma, A. and Jamal, F. (2020). TIAs pathway genes and associated miRNA identification in *Vinca minor*: Supporting Aspidosperma and Eburnamine alkaloids linkage via transcriptomic analysis, *Physiol. Mol. Biol. Plants*, 26: 1695-1711.

Wu, G. (2013). Plant microRNAs and development, *J. Genet. Genom.*, 40(5): 217-230.

CHAPTER

5

Silencing of Stress-regulated miRNAs in Plants

Abeer Hashem[1,2], Hesham Ali El Enshasy[3,4,5], Roshanida Rahmat[3,4], Ghazala Muteeb[6] and Elsayed Fathi Abd_Allah[7]*

[1] Botany and Microbiology Department, College of Science, King Saud University, P.O. Box. 2460, Riyadh 11451, Saudi Arabia

[2] Mycology and Plant Disease Survey Department, Plant Pathology Research Institute, ARC, Giza 12511, Egypt

[3] Institute of Bioproduct Development (IBD), Universiti Teknologi Malaysia (UTM), Skudai, Johor Bahru 81310, Malaysia

[4] School of Chemical and Energy Engineering, Faculty of Engineering, Universiti Teknologi Malaysia (UTM), Skudai, Johor Bahru 81310, Malaysia

[5] City of Scientific Research and Technology Applications, New Burg Al Arab, Alexandria 21934, Egypt

[6] College of Applied Medical Sciences, King Faisal University, Al-Hasa, Saudi Arabia

[7] Plant Production Department, College of Food and Agricultural Sciences, King Saud University, P.O. Box. 2460, Riyadh 11451, Saudi Arabia

1. Introduction

MicroRNAs (miRNAs) are small RNAs that have a significant function in controlling endogenous target gene expression. The miRNAs, with a length of 20-22 nucleotides (nt), perform this job by attaching to the target genes near perfect complementary sites. This binding cause mRNA degradation and/or translational suppression in target mRNA. The control of target mRNA by miRNAs is critical for plant development and function. Overexpression or silencing of miRNA expression, resulting in suppressed or enhanced expression of their specific genes, are the two basic techniques for studying miRNA functions (Teotia and Tang, 2017). Different types of approaches mimic (Franco-Zorrilla *et al.*, 2007) short tandem target mimics (STTMs) (Yan *et al.*, 2012) and molecular sponges (Reichel *et al.*, 2015) were used to down-regulate the miRNA in plants and control the regulation of certain functions. Each method has a different level of success in down-regulating a specific miRNA. STTM was developed

---

*Corresponding author: efabdallah@gmail.com

Silencing of Stress-regulated miRNAs in Plants **75**

lately and used efficiently to down-regulate a variety of miRNAs produced in different plant species. STTM is a noncoding RNA-expressing short (96 nt) DNA segment that can be delivered to plants through transient expression or by transformation process (Sha *et al.*, 2014). STTM is made up of two opposite (complimentary) binding sites that bind specific miRNA into the plants. For that purpose, empirically (48-88 nt) spacer sequence separates the binding sites. The binding sites contain trinucleotide, usually (11th-13th) positions, to the target miRNAs. These incompatible bases will not connect to the specific miRNA, forming a knob that will allow the fake position (binding site) to evade the target miRNA's cleavage (Yan *et al.*, 2012). STTM will be able to degrade/sequester target miRNAs, thanks to its cleavage resistance. The spacer produces a weak stem-loop that aids in preventing RISC collision between two miRNA binding sites (Teotia *et al.*, 2016). A variety of miRNAs were reported under different stress conditions, triggered due to abiotic or biotic stress response in plants (Khraiwesh *et al.*, 2012; Pandita 2019; Pandita and Wani, 2019; Pandita 2021; 2022a; 2022b; 2022c). The miR159 is one of the leading miRNAs reported under different abiotic stress functions in crop plants. Initial research found that *Arabidopsis* miR159 levels rise in response to drought or exogenous ABA regulated during seed germination, implying drought resistance function (Reyes and Chua, 2007; Sunkar and Zhu, 2004). Similarly, miR159 is regulated by a variety of stressors in various plant species. Drought and ABA, both up-regulate miR159 in *Phaseolus vulgaris* (Arenas-Huertero *et al.*, 2009), Cowpea (Barrera-Figueroa *et al.*, 2011), *Triticum aestivum* (Gupta *et al.*, 2014), *Oryza sativa* (Khraiwesh *et al.*, 2012), maize (Wang *et al.*, 2014) under drought conditions. Similarly, under salinity conditions, crop plants, such as *Triticum aestivum* (Gupta *et al.*, 2014), *Arabidopsis* (Khraiwesh *et al.*, 2012), banana (Lee *et al.*, 2015), phasleous (Arenas-Huertero *et al.*, 2009) and potato (Kitazumi *et al.*, 2015), in wheat by heat (Xin *et al.*, 2010), *Arabidopsis* due to hypoxia (Khraiwesh *et al.*, 2012), cold in wheat (Gupta *et al.*, 2014), by ultraviolet (UV) radiation response in wheat (Wang *et al.*, 2013). Drought has been discovered to down-regulate miR159 in potatoes (Yang *et al.*, 2014) and in rice (Barrera-Figueroa *et al.*, 2012) is reported due to heavy metals in plant species (Min Yang and Chen, 2013). Knocking down the expression of these miRNAs in stress responses using techniques like STTM reveals their exact functions in stress responses. STTM has been shown to effectively down-regulate several miRNAs (Teotia *et al.*, 2016). STTM was previously mentioned as having a limited ability to down-regulate miR159 in model plants, i.e. *A. thaliana* and several other crop plants (Reichel *et al.*, 2015). The current study highlights the detailed knowledge of miRNA-mediated modulation on significant genes suppression and regulatory pathways under stress. The current research was designed to overview the status, application, advantages and future concerns of the miRNA application to suppress the stress responsive genes in crop plants.

2. Types of miRNAs

Small RNAs (sRNAs) are non-coding molecules with 21-24 nt that influence plant development and physiological functions (Bologna and Voinnet, 2014). Based on their biosynthesis and method of action, sRNAs are divided into two classes. Micro-RNAs (miRNAs) are members of the first group, which consists of sRNAs containing specific genes that produce a single-stranded, self-complementary, non-coding RNA which form a hairpin structure and processed by Dicer-like proteins (DCL) which form a mature miRNA (21-22 nt) in length. Small interfering RNAs (siRNAs) were formed due to breakdown of double-stranded RNA (dsRNA) molecule by RNA-dependent RNA polymerase-6 (RDR6) controlled by DCL. Single-stranded (RNAs) are linked with Argonaute (AGO) proteins to form RNA-induced silencing complexes (RISC). The AGO1 protein is primarily used to load miRNAs. Different methods, such as biotin-based pull down test, gene expression profiling and RNA immune precipitation were developed to figure the intercellular miRNA targets significant for understanding the numerous cellular functions linked with miRNAs (Begum, 2022).

The RISC uses sRNA to detect the specific mRNA, cleaved and finally destroyed, based on sequence homology (Borges and Martienssen, 2015). The sRNAs act locally within a plant; however, they also move between plasmodesmata cells and are delivered systemically across large distances towards the vasculature (Li *et al.*, 2021; Molnar *et al.*, 2011). sRNA-induced gene silencing has the potential to spread throughout an entire plant. It's still unclear how these sRNAs migrate around within the plant. MiR399 is a phloem-mobile microRNA formed due to the response phosphate (PO_4^{-2}) deficiency. The miR399 is noticed in phloem sap of numerous plants (Pant *et al.*, 2008). Here we can infer that the miR399 expression was determined under PO_4-deficient soils as well as plants, as a result of miR399 translocation confirmed in the root area, where it regulates the expression. Similarly, the phosphate transporter enzyme (PHO_2) translates ubiquitin conjugating E2 and have the function of negative regulator of phosphorous shortage in signal transduction pathways (Bari *et al.*, 2006). However, miR399 translocation in the root system causes PHO1 activation and enhanced phosphate uptake in PO_4-deficient plants. MiR156 is another miRNA that has been known as a mobile element in the phloem (Bhogale *et al.*, 2014). Similarly, squamosa-promoter binding protein-like (SPL) transcription factors (TFs) are repressed by miR156, which influences various developmental features. However, the reproductive stage in *Arabidopsis* is one of the processes mediated by the miR156/SPL module. As a result of a longer juvenile period, over-expresses of miR156 blossom exceedingly late (Wu *et al.*, 2009). sRNAs are mobile signalling and monitoring molecules among the plant and migrate between plants as well as in other species, including pathogens used to silence the biotic and abiotic stress genes. Cross-kingdom/organism RNA interference (RNAi) is the term used for this phenomenon (Cai, He, *et al.*, 2018). Cotton plants produce more miR166 and miR159 in response to *Verticillium dahliae* infection, and both are exported from fungal hyphae. However two verticillium genes are required for fungal virulence (Zhang *et al.*, 2016). *Botrytis cinerea* secretes sRNAs in *Arabidopsis* cells to suppress the immunity gene in the host plant (Weiberg *et al.*, 2013). Similarly, sRNAs are transferred to *B. cinerea* plant. *Arabidopsis* secretes exosome-like extracellular vesicles (Cai, Qiao *et al.*, 2018). Exogenous RNAs can be absorbed from leaf surfaces by the fungal hyphae. The absorbance is taken up first by the plant cells and then moved to the hyphae to the other cells of the plants (Koch *et al.*, 2016). The results also suggest that artificial application of dsRNAs on the leaf surface systematically reach the plants (Mitter *et al.*, 2017). Plants also exchange small RNA (sRNAs) molecules. Dodders (*Cuscuta* spp.) is an obligatory parasite plant that obtains nutrients and water through the host plant called 'haustoria'. During parasitism, *Cuscuta campestris* haustoria accumulates a large number of 22-nucleotide miRNAs that target *Arabidopsis* mRNAs. As a result cleavage is formed in secondary siRNA and there is reduced mRNA accumulation (Shahid *et al.*, 2018). These findings reveal that dodder miRNAs served as modulators in trans-species host and gene expression,

Silencing of Stress-regulated miRNAs in Plants 77

implying that they have an impact on parasitic plant's pathogenicity. Exogenous RNAs from the outer source can cause RNAi (interference). Environmental RNAi are detected in insects and worms as well (Whangbo and Hunter, 2008), but not in plants. Topical laboratory-made dsRNAs target the insect at different developmental genes and inhibit the insect growth at specific stages of growth and development (Dubrovina and Kiselev, 2019; Wang *et al.*, 2017). The results also suggest that ingestion of either dsRNAs or sRNAs suppresses gene expression in insect pests; but cannot be sure about the RNA picked by insects on the treated plant. Some miRNAs exhibit systemic effects in plants, representing their function as itinerant signalling molecules (Li *et al.*, 2021). However, it is unknown if they're accessible in the surroundings, or taken up by the plants, or these result due to miRNA target genes receiving plants or being silenced post-transcriptionally.

2.1 miRNA-mediated Stress Regulation in Crop Plants

Plants are exposed to different types of stresses, i.e. heat, drought, salinity, radiation, insect pests and pesticides. As a result, the growth attributes are severely affected and so is the ultimate yield. Plants acquire systematic mechanisms to tolerate environmental pressures induced by diverse biotic or abiotic reasons as plants are sessile in nature. At different growth stages, plants respond differently, based on their susceptibility and resistant types of genes present. Therefore, response towards specific stress was different at morphological, physiological and cellular levels. Diverse mechanisms of stress response contribute to stress tolerance or the avoidance mechanism (Covarrubias and Reyes, 2010). Plants respond to stress at different levels, such as growth, protein level, mRNA as well as post-transcriptional level (Phillips *et al.*, 2007). The plant stress response is regulated by short RNAs (sRNAs), regulated at post-transcriptional or transcriptional silencing stage (Khraiwesh *et al.*, 2012). Different processes, such as RNA slicing (Baumberger and Baulcombe, 2005; Ma *et al.*, 2013), translational repression (Lanet *et al.*, 2009), histone modification and DNA methylation processes are used to silence genes (Khraiwesh *et al.*, 2010; Schramke and Allshire, 2004). Plant sRNA processes are complicated in nature; however, their biosynthesis and genomic locus structure are used to classify them into different processes (Phillips *et al.*, 2007). This includes small interfering RNAs (siRNAs), small temporal RNAs (stRNAs), piwi interacting RNAs (piRNAs), tiny non-coding RNAs (tncRNAs) and small modular RNAs (smRNAs) which are all examples of sRNAs. The miRNAs are short, in a sequence range of 20-22 nt in length, endogenic and single-stranded non-coding regulatory RNAs linked with various MIR genes. Lin-4 is a microRNA that was identified in *Caenorhabditis elegans* in 1993 (Lee *et al.*, 1993; Reinhart *et al.*, 2002). However, the miRNA term was first used in 2001. Application of miRNA respond in different physiological developments. miRNA recorded different roles in physiological and developmental processes (He and Hannon, 2004). It is also involved in different metabolic processes, development, hormonal expression, shape formation stress responses and miRNA biogenesis self-regulation (Jones-Rhoades *et al.*, 2006). The miRNAs are briefly described in the research and suggest that the miRNA discovery is mainly linked with diverse stress responses as well as complex diversity of sRNAs synthesised during the process under specific stress conditions. Plant miRNA is important for growth, physiological activities and stress responses. The miRNAs are shown to regulate many stress-regulated genes during the stress condition. Furthermore, it was also suggested that these are the essential molecules used in gene regulation networks and are encoded by different MIR genes. Agronomic traits can be influenced by miRNA manipulation due to the key role that miRNAs play in plant development and their potential role in stress response. Previously, Zhang et al. (2016) provided an overview of the various gene silencing pathways and the mechanisms responsible for them in maize. They describe how miRNA regulate plant traits and how they may be used to improve agronomic characteristics.

2.2 Biogenesis

More than 100 MIR genes were reported in plant genome, spread over at various genomic positions. These genomic regions could be intergenic, intronic or cluster (Budak and Akpinar, 2015) and modulated by different transcription factors (TFs) (Samad et al., 2017). The RNA polymerase II enzyme (PolII) transcribes primary miRNAs (pri-miRNAs) with hundred nucleotides (NDs) sequence from MIR genes-encoded miRNA biogenesis pathway (Lee et al., 2004). The hairpin structure of pri-miRNA is well-known. RNase-III-like enzymes, also named as Dicer-like (DCL1), hyponastic leaves (HYL1) and serrate (SE) catalyse the processing of pri-miRNAs into mature miRNAs (Park et al., 2005). HUA enhancer 1 (HEN1) methylates the 3' terminus of the pre-miRNA hairpin precursor, converting it into a (20-22nt) miRNA/miRNA*duplex (HEN1). HASTY (HST1), an exportin protein, transports it into the cytoplasm (Park et al., 2005). One of the duplex strands of miRNA is turned towards exosome and destroyed by small RNA degrading nuclease (SDN) when it enters the cytoplasm (Lima et al., 2012). The mature miRNA is exposed to the complex RNA-induced silencing complexes (RISCs). It's integrated with protein Argonaute (AGO), used in complementarity sequence to guide it towards the specific transcripts.

As a result, the mature miRNA unwinds in the cytoplasm area and is transferred to the gene silencing complex, formed the miR-induced gene silencing complex and controls the expression. AGO protein maintains the RISC complex and mature miRNA strand is driven to exosomes for disintegration as the miRNA duplex predominantly. Furthermore, it unwinds the duplex with the aid of the AGO1 protein, while the other strand is left attached to the RISC (Baumberger and Baulcombe, 2005). When it comes down to it, the mature miRNA directs the AGO1-containing RISC complex either direct to site-specific cleavage of complementary mRNA with high homology or impedes the translation of the targeted mRNA by poor base combination. The presumption on miRNA-mediated gene regulation claimed that only mature miRNA block mRNAs and translation. However, the recent study has shown that a different miRNA strand also controls the expression of the corresponding genes and has its own targeted mRNA (Zhang and Unver, 2018).

2.3 Modulation and Regulation

MicroRNAs normalise the expression of the gene by cleaving specific mRNA or repressing translation (Guleria et al., 2011; Lu and Huang, 2008). The target mRNA cleavage occurs when mRNA and miRNA are perfectly paired; however, the repression occurs only when they are not present in pair form. Plant miRNA is involved in the cleavage of target mRNA. Researchers recently discovered a new miRNA-driven regulator called miRNA-mediated mRNA degradation. When target mRNA and miRNA have just a partial or no resemblance, the decay of the target mRNA occurs due to deletion of poly A tail, making erratic degradation (Guleria et al., 2011; Jing

Silencing of Stress-regulated miRNAs in Plants **79**

et al., 2005; Wu *et al.*, 2006). Some important examples of silencing of stress-regulated miRNAs in plants are in Table 1.

Table 1: Some important miRNAs silencing stress response under different crop plants

miRNAs	Methods	Functions	Name	References
miR159	STTMs	Abiotic stresses, like ABA	*Arabidopsis*	Teotia and Tang, 2017
miRNA165/166	STTM	ABA stress response	*Arabidopsis*	Teotia and Tang, 2017
miR-399	Target mimicry	phosphate (Pi) starvation	*Arabidopsis*	Franco-Zorrilla *et al.*, 2007
STTM165/166	MIMICs; STTMs; Molecular Sponge	Loss of function phenotypes	*Arabidopsis*	Reichel *et al.*, 2015
miR165/166	virus-based miRNA silencing/ suppression (VbMS)		*Nicotiana benthamiana*	Zhao *et al.*, 2016
STTM and CRISPR	–	Plant traits	Rice	Chen *et al.*, 2021
miR156; multiple miRNAs	–	Drought and heat stress response	Alfalfa	Li *et al.*, 2017; Zhao *et al.*, 2019
Hv-miR827; multiple miRNA	–	Drought and salinity	Barley	Cui *et al.*, 2020
miR1885	BraTNL1 expression	Immune response	Brassica	Ramírez *et al.*, 2013
miR399	PvPHR1 transcription factor	Phosphorous deficiency	Bean	
miR2111 and miR2118	Fungal infected salt treated	Biotic stress response	Chickpea	Kohli *et al.*, 2014

3. Silencing of Biotic Stress-regulated miRNAs in Plants

MiR393 was the first miRNA whose role in plant antibacterial inhibits the auxins signalling process by pattern-triggered immunity (PTI) (Jin, 2008). The mode of action was stimulated, using bacteria-based molecular pattern. There was a report that suggested that three miR160, miR167 and miR393 were recorded showing increase when infected with *Pseudomonas syringae* PV. In tomato (DC3000hrcC), however, miR825 was shown to be down-regulated (Fahlgren *et al.*, 2007). On bacterial infection, miR398 was also reported to be down-regulated (Jagadeeswaran *et al.*, 2009). The RNA silencing repressor P1/HC-Pro, which is encoded by the turnip mosaic virus (TuMV), has been linked to miR171. P1/HC-Pro was found to promote miR156 and miR164 in *Arabidopsis* (Kasschau *et al.*, 2003; Zhou and Luo, 2013). TuMV infection was reported to generate a new miRNA – brassica-miR1885 (He *et al.*, 2008). The expression of miR159 was discovered to be higher in the plants infected with tomato leaf curl virus (ToLCV), while miR164 and miR171 were shown to be down-regulated (Naqvi *et al.*, 2008). The genes, miR (159/319) and miR172, identification was reported in tomato leaf curl in Delhi's virus infections (Naqvi *et al.*, 2010). However, the single miRNAs have not yet been studied for viral infection.

Ten miRNAs recorded differential gene expressions relative to control when infected with *Cronartium quercuum* f. sp. Fusiforme (Lu *et al.*, 2007). Another study suggested different miRNA genes expression in wheat powdery mildew disease, infected with *Blumeria graminis* f. *sp. tritici* (Bgt). The study further suggested that miR (156, 159, 164, 171 and 396) were recorded to be down-regulated. However, miR (393, 444 and 827) was up-regulated (Xin *et al.*, 2010). Similarly, another study suggested that overexpression of osa-miR (7696) was recorded in rice blast infection (Campo *et al.*, 2013). However, many miR (161, 164, 167a, 172c, 396a-b and 398a) were down-regulated due to infection of *Heterodera schachtii* in *Arabidopsis* (Hewezi *et al.*, 2008). However, the cyst nematode infection triggered to activate the more than 100 miRNAs from different families, as identified in soybean (Li *et al.*, 2012).

4. Silencing of Abiotic Stress-regulated miRNAs in Plants

The abiotic stress response produced hydroxyl radicals, hydrogen peroxide and superoxide radicals, generating reactive oxygen species (ROS) due to different abiotic stresses (Bartels and Sunkar, 2005; Mittler, 2002). SODs are enzymes that convert extremely harmful superoxide radicals (O_2) into less hazardous hydrogen peroxide (H_2O_2) (Fridovich, 1995). MiR398 levels were shown to be required for up-regulation of the two Cu-Zn superoxide dismutase (CSD) genes. The CSD1 (systolic) and CSD2 (plastidic) are targeted by the miR (398). The overexpression of miR-398 and down-regulation of these genes reduced the stress by limiting the free radical ions (Jagadeeswaran *et al.*, 2009; Kruszka *et al.*, 2012; Sunkar *et al.*, 2006). The miRNA (398) recorded that three loci (miRNA a, b, c) have different functions (Bonnet *et al.*, 2004; Jones-Rhoades and Bartel, 2004; Sunkar and Zhu, 2004). Another study proposed that seven miRNAs (169, 397, 528, 827, 1425, 319a.2 and 408-5p) were present in the rice seedling (Li *et al.*, 2011), recording different abiotic stress functions in crops.

Water scarcity is an important stress in which lack of water or precipitation occurs in the root zone due to excessive evaporation from the soil surface (Shukla *et al.*, 2008). Drought tolerance is achieved through increased water absorption, limiting the water loss and modifications in different metabolic pathways (Bartels and Sunkar, 2005). miRNAs differential expression was recorded under drought stress in wheat (*Triticum dicoccoides*) (Kantar *et al.*, 2011), *Vigna unguiculata* (Barrera-Figueroa *et al.*, 2011), *Glycine max* (Kulcheski *et al.*, 2011), *Phaseolus vulgaris* (Arenas-Huertero *et al.*, 2009) and *Nicotiana tabacum* (Frazier *et al.*, 2011). Other different miRNAs (156, 158, 159, 165, 167, 168, 169, 171, 393, 394 and 396) were reported up-regulated under water scarcity conditions (Liu *et al.*, 2008). Mi393 function enhanced the growth and development processes under drought conditions by lowering the TIR1 level. The miR (169g, 171a and 393) genes were recorded to control the dehydration processes and regulate the water supply in the root zone area in rice (Jian *et al.*, 2010; Zhao *et al.*, 2007; Zhou *et al.*, 2010). The transcription factor (TF) CBF/DREBs was discovered to be potentially used for regulating the photosynthate accumulation under drought

Silencing of Stress-regulated miRNAs in Plants **81**

conditions due to miR169g mRNA (Kruszka *et al.*, 2012). The other miRNAs (156, 159, 168, 170, 171, 172, 319, 396, 397, 408, 529, 896, 1030, 1035, 1050, 1088 and miR1126) recorded down-regulation under water deficit conditions in *Oryza sativa* L. However, other 14 miRNAs were up-regulated under water deficit conditions in rice. Lu *et al.* (2008) recorded different miRNAs (171l-n, 1445, 1446a-e, 1444a, 1450, 482.2, 530a, 827, 1448 and 1447) showing miRNA gene expression under drought conditions in Populus. Furthermore, another study suggested that three miRNAs (miRS1, 1514a and miR2119) recorded moderately regulation relative to control; however, three miRNAs (159.2, 393 and 2118) were up-regulated in *P. vulgaris* (Arenas-Huertero *et al.*, 2009). The miR169 was shown to be down-regulated solely in the roots of *Medicago truncatula*, while miR (398 a, b and 408) were recorded to express higher manifold (Trindade *et al.*, 2010). Thirteen differently regulated miRNAs were identified in a study on drought-resistant wild emmer wheat (*Triticum turgidums* sp. *Dicoccoides*) (Kantar *et al.*, 2011).

Salinity had the most destructive effect on plant growth and development relative to other abiotic stresses in the crop plants. High salinity inhibits plants' ability to absorb water, resulting in drought (Munns, 2005). There were different types of miRNA found in rice belonging to miR169 faimily, significantly cleaving the NF-Ya gene transcript (Gao *et al.*, 2011; Zhao *et al.*, 2009). The other microRNA (156, 158, 159, 165, 167, 168, 169, 171, 319, 393, 394, 396) and miR397 were up-regulated in *Arabidopsis*, whilst miR398 was shown to be down-regulated (Liu *et al.*, 2008). MiRS1 and miR159.2 were shown to be substantially elevated in *P. vulgaris* (Arenas-Huertero *et al.*, 2009). The miRNA belongs to MiR (530a, 445, 1446a-e, 1447 and 171l-n) from different families and were down-regulated in *P. trichocarpa*, while miR482.2 and miR1450 were up-regulated (Lu and Huang, 2008). Similarly, another study identified 98 miRNA belonging to 27 different families in salt-sensitive and tolerant maize genotypes. The results showed that miRNA (156, 164, 167 and 396) and RNA were down-regulated; however, members (162, 168, 395 and 474) belonging to miRNA families were up-regulated when maize crop was exposed to salt-stress condition (Ding *et al.*, 2009).

Chilling injury severely influences the plant's growth and development. Identification of miRNA for chilling stress in plants helps the researcher to transform the miRNA, TFs while chilling response genes help to withstand the plants under harsh conditions. For instance, miR319c recorded as biomarkers the expression of cold stress in plants. Furthermore, the miR397 (LAC2, LAC4 and LAC17) specifies the Laccase family members, which are significant in the lignification process (Abdel-Ghany and Pilon, 2008; Bosch *et al.*, 2011). It also regulates the mRNA level with TCP TFs (Palatnik *et al.*, 2007). Sunkar and Zhu (2004) reported the miRNA related to cold stress and suggested four miRNAs (393, 397b, 402 and 319c) were up-regulated during the study. Other studies also reported the up-regulation of different cold stress related to miRNA which were moderately up-regulated in *Arabidopsis* (Liu *et al.*, 2008; Sunkar and Zhu, 2004), in Populus (Lu and Huang, 2008) and in *Brachypodium* (Zhang *et al.*, 2009); but miR156g-j, miR475a, b and miR476a were down-regulated (Lu and Huang, 2008). Eighteen miRNAs found in rice: miR (156k, 166k, 166m, R167a/b/c, 168b, 169e, 169f, 169h, 171a, 535, 319a/b, 1884b, 444a.1, 1850, 1868, 1320, 1435 and 876 (Lv *et al.*, 2010). Potential indicators in response to nutritional stress, such as phosphate-miR399, sulfate-miR395, and copper-miR398 have been found in *Arabidopsis* (Khraiwesh *et al.*, 2012). Phosphate encoding genes (PHO$_2$)

controlled by miR-399 (Khraiwesh *et al.*, 2012) down-regulated under prosperous deficient conditions (Bari *et al.*, 2006; Chiou *et al.*, 2006; Fujii *et al.*, 2005). Phosphate starvation response (PHR1) transcription factors, which bind to GNATATNC cis regions for regulation of phosphate-responsive genes, stimulate the production of miR-399 (Chiou, 2007; Franco-Zorrilla *et al.*, 2004; Rubio *et al.*, 2001). MiR399 matures in shoots and then travels to roots via the phloem, where it cleaves the targeted PHO2 transcripts (Pant *et al.*, 2008), inducing phosphate absorption by up-regulating the phosphate transporters Pht1;8 and Pht1;9. IPS1 (induced by phosphate starvation) works as a target mimic of miR399 to prevent PHO2 transcript degradation as the phosphate balance improves. The miR399-PHO2-IPS1 cycle is thus critical for maintaining phosphate homeostasis in the Pi-deficient signalling pathway (Kruszka *et al.*, 2012). MiR395 was involved in controlling the low affinity sulphate transporter and internal translocation from roots to shoot area. Furthermore, it also regulates the ATP sulfurylases involved in sulphur assimilation pathway (Allen *et al.*, 2005; Jones-Rhoades and Bartel, 2004; Liang and Yu, 2010), though the mechanism for miR395 that modulates sulphate homeostasis is still unknown. Copper homeostasis is dependent on miR398. MiR398 is up-regulated in Cu-stressed condition (Beauclair *et al.*, 2010). Under copper deficiency, three conserved miRNA family members: miR (397, 408 and 857) were up-regulated and targets the Cu-containing plantacyanin to regulate the nutrients under Cu-deficient conditions (Abdel-Ghany and Pilon, 2008). However, the miR (408) targets the plantacyanin transcripts to overcome the nutrient deficiency stress under field conditions.

Heat shock alters the proteins (enzymes) and their structure with the ability of cellular machinery to restore proteins and membranes. Up till now, many miRNAs were discovered that respond better under heat stress condition. A study on wheat showed differential expression of heat miRNAs under heat stress conditions. There were nine conserved heat-sensitive miRNAs from the total 32 miRNAs members in wheat. The miRNAs (156, 159, 160, 166, 168, 169, 393 and 827) were recorded up-regulated; however, miR172 was only down-regulated under heat stress conditions (Xin *et al.*, 2010). Mechanical stress regulates miRNAs, which play a role in plant structure and mechanical fitness. Another study suggested that six different miRNAs were recorded to be down-regulated in *P. trichocarpa*; however, miR408 was found to be up-regulated under heat stress conditions. Only compressed-stressed tissue showed down-regulation in miR160 and miR172, whereas tension-stressed tissue showed up-regulation in miR168 (Khraiwesh *et al.*, 2012; Lu *et al.*, 2005).

5.   Summary and Future Concerns

Ten years after miRNA was discovered in *C. elegans,* it was discovered in *Arabidopsis*. Since then, a great deal of work has gone into figuring out what miRNAs are and what they do. Plants and mammals have ribo-regulators that control gene expression. The miR base serves as the central source and database for all published miRNA data. The current release-20 contains 24,521 precursor miRNA entries from 206 species, demonstrating the rapid increase in the number of miRNAs discovered in recent years. The miRNA is an important component of stress regulation networks. For the identification of unique stress-response miRNA, several approaches were used to enhance the miRNA research and their unique sequences were engineered to silence the target genes and get the desired genes suppressed for the desired function. Therefore, more research into miRNA-mediated gene regulation can help to understand the complicated attributes of plant stress tolerance. The identification and characterisation of new miRNA biomarkers can be aided by miRNA profiling of diverse plant species exposed to various environmental stressors. Researchers have used high-throughput sequencing tools to perform genome-wide miRNA expression profiling under abiotic stress. The information gained from analysing the new miRNA biomarkers and developing strategies to improve plant tolerance to associated stress will be aided by regulatory networks stress resistance or tolerance to a variety of conditions. Degradome sequencing is one of the technological advancements that has allowed the effective and quick identification of multiple-miRNA targets. However, it has limitations in identifying miRNA targets, particularly those that target gene expression via translation repression, necessitating the development of better alternatives. However, most miRNA functions remain unknown, leaving a wide gap between discovered miRNAs and their functions. Furthermore, the miRNA for plant stress response that has been explored so far has only been limited to a few distinct species. miRNAs are promising targets for engineering abiotic stress tolerance in major crops using transgenic technologies because of their critical roles in post-transcriptional regulation of gene expression in response to abiotic stresses and resulting growth attenuation. As a result, greater discovery and classification of plant miRNA biomarkers from diverse key plant species is required for a better understanding of the protocol.

References

Abdel-Ghany, S.E. and Pilon, M. (2008). MicroRNA-mediated systemic down-regulation of copper protein expression in response to low copper availability in *Arabidopsis*, *Journal of Biological Chemistry*, 283(23): 15932-15945.

Allen, E., Xie, Z., Gustafson, A.M. and Carrington, J.C. (2005). MicroRNA-directed phasing during trans-acting siRNA biogenesis in plants, *Cell*, 121(2): 207-221.

Arenas-Huertero, C., Pérez, B., Rabanal, F., Blanco-Melo, D., De la Rosa, C., Estrada-Navarrete, G., Sanchez, F., Covarrubias, A.A. and Reyes, J.L. (2009). Conserved and novel miRNAs in the legume *Phaseolus vulgaris* in response to stress, *Plant Molecular Biology*, 70(4): 385-401.

Bari, R., Datt Pant, B., Stitt, M. and Scheible, W.-R. (2006). PHO2, microRNA399 and PHR1 define a phosphate-signalling pathway in plants, *Plant Physiology*, 141(3): 988-999.

Barrera-Figueroa, B.E., Gao, L., Diop, N.N., Wu, Z., Ehlers, J.D., Roberts, P.A., Close, T.J., Zhu, J.-K. and Liu, R. (2011). Identification and comparative analysis of drought-associated microRNAs in two cowpea genotypes, *BMC Plant Biology*, 11(1): 1-11.

Barrera-Figueroa, B.E., Gao, L., Wu, Z., Zhou, X., Zhu, J., Jin, H., Liu, R. and Zhu, J.-K. (2012). High throughput sequencing reveals novel and abiotic stress-regulated microRNAs in the inflorescences of rice, *BMC Plant Biology*, 12(1): 1-11.

Bartels, D. and Sunkar, R. (2005). Drought and salt tolerance in plants, *Critical Reviews in Plant Sciences*, 24(1): 23-58.

Baumberger, N. and Baulcombe, D. (2005). *Arabidopsis* Argnaute1 is an RNA slicer that selectively recruits microRNAs and short interfering RNAs, *Proceedings of the National Academy of Sciences*, 102(33): 11928-11933.

Beauclair, L., Yu, A. and Bouché, N. (2010). microRNA-directed cleavage and translational repression of the copper chaperone for superoxide dismutase mRNA in *Arabidopsis*, *The Plant Journal*, 62(3): 454-462.

Begum, Y. (2022). Regulatory role of microRNAs (miRNAs) in the recent development of abiotic stress tolerance of plants, *Gene*, 821: 146283.

Bhogale, S., Mahajan, A.S., Natarajan, B., Rajabhoj, M., Thulasiram, H.V. and Banerjee, A.K. (2014). MicroRNA156: A potential graft-transmissible microRNA that modulates plant architecture and tuberization in *Solanum tuberosum* ssp. *Andigena*, *Plant Physiology*, 164(2): 1011-1027.

Bologna, N.G. and Voinnet, O. (2014). The diversity, biogenesis and activities of endogenous silencing small RNAs in *Arabidopsis*, *Annual Review of Plant Biology*, 65: 473-503.

Bonnet, E., Wuyts, J., Rouzé, P. and Van de Peer, Y. (2004). Detection of 91 potential conserved plant microRNAs in *Arabidopsis thaliana* and *Oryza sativa* identifies important target genes, *Proceedings of the National Academy of Sciences*, 101(31): 11511-11516.

Borges, F. and Martienssen, R.A. (2015). The expanding world of small RNAs in plants, *Nature Reviews Molecular Cell Biology*, 16(12): 727-741.

Bosch, M., Mayer, C.-D., Cookson, A. and Donnison, I.S. (2011). Identification of genes involved in cell wall biogenesis in grasses by differential gene expression profiling of elongating and non-elongating maize internodes, *Journal of Experimental Botany*, 62(10): 3545-3561.

Budak, H. and Akpinar, B.A. (2015). Plant miRNAs: Biogenesis, organisation and origins, *Functional and Integrative Genomics*, 15(5): 523-531.

Cai, Q., He, B., Kogel, K.-H. and Jin, H. (2018). Cross-kingdom RNA trafficking and environmental RNAi—Nature's blueprint for modern crop protection strategies, *Current Opinion in Microbiology*, 46: 58-64.

Cai, Q., Qiao, L., Wang, M., He, B., Lin, F.-M., Palmquist, J., Huang, S.-D. and Jin, H. (2018). Plants send small RNAs in extracellular vesicles to fungal pathogen to silence virulence genes, *Science*, 360(6393): 1126-1129.

Campo, S., Peris-Peris, C., Siré, C., Moreno, A.B., Donaire, L., Zytnicki, M., Notredame, C., Llave, C. and San Segundo, B. (2013). Identification of a novel micro RNA (miRNA) from rice that targets an alternatively spliced transcript of the N ramp6 (Natural resistance-associated macrophage protein 6) gene involved in pathogen resistance, *New Phytologist*, 199(1): 212-227.

Chen, J., Teotia, S., Lan, T. and Tang, G. (2021). MicroRNA techniques: Valuable tools for agronomic trait analyses and breeding in rice, *Frontiers in Plant Science*, 1929.

Chiou, T.-J., Aung, K., Lin, S.-I., Wu, C.-C., Chiang, S.-F. and Su, C.-L. (2006). Regulation of phosphate homeostasis by microRNA in *Arabidopsis*, *The Plant Cell*, 18(2): 412-421.

Chiou, T.J. (2007). The role of microRNAs in sensing nutrient stress, *Plant, Cell and Environment*, 30(3): 323-332.

Covarrubias, A.A. and Reyes, J.L. (2010). Post-transcriptional gene regulation of salinity and drought responses by plant microRNAs, *Plant, Cell & Environment*, 33(4): 481-489.

Cui, C., Wang, J.-J., Zhao, J.-H., Fang, Y.-Y., He, X.-F., Guo, H.-S. and Duan, C.-G. (2020). A Brassica miRNA regulates plant growth and immunity through distinct modes of action, *Molecular Plant*, 13(2): 231-245.

Ding, D., Zhang, L., Wang, H., Liu, Z., Zhang, Z. and Zheng, Y. (2009). Differential expression of miRNAs in response to salt stress in maize roots, *Annals of Botany*, 103(1): 29-38.

Dubrovina, A.S. and Kiselev, K.V. (2019). Exogenous RNAs for gene regulation and plant resistance, *International Journal of Molecular Sciences*, 20(9): 2282.

Fahlgren, N., Howell, M.D., Kasschau, K.D., Chapman, E.J., Sullivan, C.M., Cumbie, J.S., Givan, S.A., Law, T.F., Grant, S.R. and Dangl, J.L. (2007). High-throughput sequencing of *Arabidopsis* microRNAs: Evidence for frequent birth and death of miRNA genes, *PLoS ONE*, 2(2): e219.

Franco-Zorrilla, J.M., Valli, A., Todesco, M., Mateos, I., Puga, M.I., Rubio-Somoza, I., Leyva, A., Weigel, D., García, J.A. and Paz-Ares, J. (2007). Target mimicry provides a new mechanism for regulation of microRNA activity, *Nature Genetics*, 39(8): 1033-1037.

Silencing of Stress-regulated miRNAs in Plants

Franco-Zorrilla, J.M., González, E., Bustos, R., Linhares, F., Leyva, A. and Paz-Ares, J. (2004). The transcriptional control of plant responses to phosphate limitation, *Journal of Experimental Botany*, 55(396): 285-293.

Frazier, T.P., Sun, G., Burklew, C.E. and Zhang, B. (2011). Salt and drought stresses induce the aberrant expression of microRNA genes in tobacco, *Molecular Biotechnology*, 49(2): 159-165.

Fridovich, I. (1995). Superoxide radical and superoxide dismutases, *Annual Review of Biochemistry*, 64(1): 97-112.

Fujii, H., Chiou, T.-J., Lin, S.-I., Aung, K. and Zhu, J.-K. (2005). A miRNA involved in phosphate-starvation response in *Arabidopsis*, *Current Biology*, 15(22): 2038-2043.

Gao, P., Bai, X., Yang, L., Lv, D., Pan, X., Li, Y., Cai, H., Ji, W., Chen, Q. and Zhu, Y. (2011). osa-MIR393: A salinity- and alkaline stress-related microRNA gene, *Molecular Biology Reports*, 38(1): 237-242.

Guleria, P., Mahajan, M., Bhardwaj, J. and Yadav, S.K. (2011). Plant small RNAs: Biogenesis, mode of action and their roles in abiotic stresses, *Genomics, Proteomics and Bioinformatics*, 9(6): 183-199.

Gupta, O.P., Meena, N.L., Sharma, I. and Sharma, P. (2014). Differential regulation of microRNAs in response to osmotic, salt and cold stresses in wheat, *Molecular Biology Reports*, 41(7): 4623-4629.

He, L. and Hannon, G.J. (2004). MicroRNAs: Small RNAs with a big role in gene regulation, *Nature Reviews Genetics*, 5(7): 522-531.

He, X.-F., Fang, Y.-Y., Feng, L. and Guo, H.-S. (2008). Characterisation of conserved and novel microRNAs and their targets, including a TuMV-induced TIR–NBS–LRR class R gene-derived novel miRNA in Brassica, *FEBS Letters*, 582(16): 2445-2452.

Hewezi, T., Howe, P., Maier, T.R. and Baum, T.J. (2008). *Arabidopsis*, small RNAs and their targets during cyst nematode parasitism, *Molecular Plant-Microbe Interactions*, 21(12): 1622-1634.

Jagadeeswaran, G., Saini, A. and Sunkar, R. (2009). Biotic and abiotic stress down-regulate miR398 expression in *Arabidopsis*, *Planta*, 229(4): 1009-1014.

Jian, X., Zhang, L., Li, G., Zhang, L., Wang, X., Cao, X., Fang, X. and Chen, F. (2010). Identification of novel stress-regulated microRNAs from *Oryza sativa* L., *Genomics*, 95(1): 47-55.

Jin, H. (2008). Endogenous small RNAs and antibacterial immunity in plants, *FEBS Letters*, 582(18): 2679-2684.

Jing, Q., Huang, S., Guth, S., Zarubin, T., Motoyama, A., Chen, J., Di Padova, F., Lin, S.-C., Gram, H. and Han, J. (2005). Involvement of microRNA in AU-rich element-mediated mRNA instability, *Cell*, 120(5): 623-634.

Jones-Rhoades, M.W. and Bartel, D.P. (2004). Computational identification of plant microRNAs and their targets, including a stress-induced miRNA, *Molecular Cell*, 14(6): 787-799.

Jones-Rhoades, M.W., Bartel, D.P. and Bartel, B. (2006). MicroRNAs and their regulatory roles in plants, *Annu. Rev. Plant Biol.*, 57: 19-53.

Kantar, M., Lucas, S.J. and Budak, H. (2011). miRNA expression patterns of *Triticum dicoccoides* in response to shock drought stress, *Planta*, 233(3): 471-484.

Kasschau, K.D., Xie, Z., Allen, E., Llave, C., Chapman, E.J., Krizan, K.A. and Carrington, J.C. (2003). P1/HC-Pro, a viral suppressor of RNA silencing, interferes with *Arabidopsis* development and miRNA function, *Developmental Cell*, 4(2): 205-217.

Khraiwesh, B., Arif, M.A., Seumel, G.I., Ossowski, S., Weigel, D., Reski, R. and Frank, W. (2010). Transcriptional control of gene expression by microRNAs, *Cell*, 140(1): 111-122.

Khraiwesh, B., Zhu, J.-K. and Zhu, J. (2012). Role of miRNAs and siRNAs in biotic and abiotic stress responses of plants, *Biochimica et Biophysica Acta (BBA)-Gene Regulatory Mechanisms*, 1819(2): 137-148.

Kitazumi, A., Kawahara, Y., Onda, T.S., De Koeyer, D. and de los Reyes, B.G. (2015). Implications of miR166 and miR159 induction to the basal response mechanisms of an andigena potato (*Solanum tuberosum* subsp. *andigena*) to salinity stress, predicted from network models in *Arabidopsis*, *Genome*, 58(1): 13-24.

Koch, A., Biedenkopf, D., Furch, A., Weber, L., Rossbach, O., Abdellatef, E., Linicus, L., Johannsmeier, J., Jelonek, L. and Goesmann, A. (2016). An RNAi-based control of Fusarium graminearum infections through spraying of long dsRNAs involves a plant passage and is controlled by the fungal silencing machinery, *PLoS Pathogens*, 12(10): e1005901.

Kohli, D., Joshi, G., Deokar, A.A., Bhardwaj, A.R., Agarwal, M., Katiyar-Agarwal, S., Srinivasan, R. and Jain, P.K. (2014). Identification and characterisation of wilt and salt stress-responsive microRNAs in chickpea through high-throughput sequencing, *PLoS ONE*, 9(10): e108851.

Kruszka, K., Pieczynski, M., Windels, D., Bielewicz, D., Jarmolowski, A., Szweykowska-Kulinska, Z. and Vazquez, F. (2012). Role of microRNAs and other sRNAs of plants in their changing environments, *Journal of Plant Physiology*, 169(16): 1664-1672.

Kulcheski, F.R., de Oliveira, L.F., Molina, L.G., Almerão, M.P., Rodrigues, F.A., Marcolino, J., Barbosa, J.F., Stolf-Moreira, R., Nepomuceno, A.L. and Marcelino-Guimarães, F.C. (2011). Identification of novel soybean microRNAs involved in abiotic and biotic stresses, *BMC Genomics*, 12(1): 1-17.

Lanet, E., Delannoy, E., Sormani, R., Floris, M., Brodersen, P., Crété, P., Voinnet, O. and Robaglia, C. (2009). Biochemical evidence for translational repression by *Arabidopsis* microRNAs, *The Plant Cell*, 21(6): 1762-1768.

Lee, R.C., Feinbaum, R.L. and Ambros, V. (1993). The *C. elegans* heterochronic gene lin-4 encodes small RNAs with antisense complementarity to lin-14, *Cell*, 75(5): 843-854.

Lee, W.S., Gudimella, R., Wong, G.R., Tammi, M.T., Khalid, N. and Harikrishna, J.A. (2015). Transcripts and microRNAs responding to salt stress in *Musa acuminata* Colla (AAA Group) cv. Berangan roots, *PLoS ONE*, 10(5): e0127526.

Lee, Y., Kim, M., Han, J., Yeom, K.H., Lee, S., Baek, S.H. and Kim, V.N. (2004). MicroRNA genes are transcribed by RNA polymerase II, *The EMBO Journal*, 23(20): 4051-4060.

Li, S., Wang, X., Xu, W., Liu, T., Cai, C., Chen, L., Clark, C.B. and Ma, J. (2021). Unidirectional movement of small RNAs from shoots to roots in interspecific heterografts, *Nature Plants*, 7(1): 50-59.

Li, T., Li, H., Zhang, Y.-X. and Liu, J.-Y. (2011). Identification and analysis of seven H2O2-responsive miRNAs and 32 new miRNAs in the seedlings of rice (*Oryza sativa* L. ssp. *indica*), *Nucleic Acids Research*, 39(7): 2821-2833.

Li, X., Wang, X., Zhang, S., Liu, D., Duan, Y. and Dong, W. (2012). Identification of soybean microRNAs involved in soybean cyst nematode infection by deep sequencing, *PLoS ONE*, 7(6): e39650.

Li, Y., Wan, L., Bi, S., Wan, X., Li, Z., Cao, J., Tong, Z., Xu, H., He, F. and Li, X. (2017). Identification of drought-responsive microRNAs from roots and leaves of alfalfa by high-throughput sequencing, *Genes*, 8(4): 119.

Liang, G. and Yu, D. (2010). Reciprocal regulation among miR395, APS and SULTR2; 1 in *Arabidopsis thaliana*, *Plant Signalling & Behaviour*, 5(10). 1257-1259.

Lima, J.C.D., Loss-Morais, G. and Margis, R. (2012). MicroRNAs play critical roles during plant development and in response to abiotic stresses, *Genetics and Molecular Biology*, 35(4): 1069-1077.

Liu, H.-H., Tian, X., Li, Y.-J., Wu, C.-A. and Zheng, C.-C. (2008). Microarray-based analysis of stress-regulated microRNAs in *Arabidopsis thaliana*, *Rna*, 14(5): 836-843.

Lu, S., Sun, Y.-H., Shi, R., Clark, C., Li, L. and Chiang, V.L. (2005). Novel and mechanical stress-responsive microRNAs in Populus trichocarpa that are absent from *Arabidopsis*, *The Plant Cell*, 17(8): 2186-2203.

Lu, S., Sun, Y.H., Amerson, H. and Chiang, V.L. (2007). MicroRNAs in loblolly pine (*Pinus taeda* L.) and their association with fusiform rust gall development, *The Plant Journal*, 51(6): 1077-1098.

Lu, S., Sun, Y.H. and Chiang, V.L. (2008). Stress-responsive microRNAs in Populus, *The Plant Journal*, 55(1): 131-151.

Lu, X.-Y. and Huang, X.-L. (2008). Plant miRNAs and abiotic stress responses, *Biochemical and Biophysical Research Communications*, 368(3): 458-462.

Lv, D.-K., Bai, X., Li, Y., Ding, X.-D., Ge, Y., Cai, H., Ji, W., Wu, N. and Zhu, Y.-M. (2010). Profiling of cold-stress-responsive miRNAs in rice by microarrays, *Gene*, 459(1-2): 39-47.

Ma, H., Williams, P.L. and Diamond, S.A. (2013). Ecotoxicity of manufactured ZnO nanoparticles – A review, *Environmental Pollution*, 172: 76-85.

Min Yang, Z. and Chen, J. (2013). A potential role of microRNAs in plant response to metal toxicity, *Metallomics*, 5(9): 1184-1190.

Mitter, N., Worrall, E.A., Robinson, K.E., Li, P., Jain, R.G., Taochy, C., Fletcher, S.J., Carroll, B.J., Lu, G. and Xu, Z.P. (2017). Clay nanosheets for topical delivery of RNAi for sustained protection against plant viruses, *Nature Plants*, 3(2): 1-10.

Mittler, R. (2002). Oxidative stress, antioxidants and stress tolerance, *Trends in Plant Science*, 7(9): 405-410.

Molnar, A., Melnyk, C. and Baulcombe, D.C. (2011). Silencing signals in plants: A long journey for small RNAs, *Genome Biology*, 12(1): 1-8.

Munns, R. (2005). Genes and salt tolerance: Bringing them together, *New Phytologist*, 167(3): 645-663.

Naqvi, A.R., Choudhury, N.R., Rizwanul Haq, Q.M. and Mukherjee, S.K. (2008). *MicroRNAs as Biomarkers in Tomato Leaf Curl Virus (ToLCV) Disease*, Nucleic Acids Symposium Series.

Naqvi, A.R., Haq, Q.M. and Mukherjee, S.K. (2010). MicroRNA profiling of tomato leaf curl new delhi virus (tolcndv) infected tomato leaves indicates that deregulation of mir159/319 and mir172 might be linked with leaf curl disease, *Virology Journal*, 7(1): 1-16.

Palatnik, J.F., Wollmann, H., Schommer, C., Schwab, R., Boisbouvier, J., Rodriguez, R., Warthmann, N., Allen, E., Dezulian, T. and Huson, D. (2007). Sequence and expression differences underlie functional specialisation of *Arabidopsis* microRNAs miR159 and miR319, *Developmental Cell*, 13(1): 115-125.

Pandita, D. (2019). Plant MIRnome: miRNA biogenesis and abiotic stress response. *In:* Hasanuzzaman, M., Hakeem, K., Nahar, K. and Alharby, H. (Eds.). *Plant Abiotic Stress Tolerance*. Springer, Cham., 449-474. Doi https://Doi.org/10.1007/978-3-030-06118-0_18

Pandita, D. and Wani S.H. (2019). MicroRNA as a tool for mitigating abiotic stress in rice (*Oryza sativa* L.). *In:* Wani, S. (Ed.). *Recent Approaches in Omics for Plant Resilience to Climate Change*. Springer, Chamz., 109-133. Doi https://Doi.org/10.1007/978-3-030-21687-0_6

Pandita, D. (2021). Role of miRNAi technology and miRNAs in abiotic and biotic stress resilience. *In:* Aftab, T. and Roychoudhury, A. (Eds.). *Plant Perspectives to Global Climate Changes*. Academic Press, Elsevier, 303-330. https://Doi.org/10.1016/B978-0-323-85665-2.00015-7

Pandita, D. (2022a). How microRNAs regulate abiotic stress tolerance in wheat? A snapshot. *In:* Roychoudhury, A., Aftab, T. and Acharya, K. (Eds.). *Omics Approach to Manage Abiotic Stress in Cereals*. Springer, Singapore, 447-464. https://Doi.org/10.1007/978-981-19-0140-9_17

Pandita, D. (2022b). MicroRNAs shape the tolerance mechanisms against abiotic stress in maize. *In:* Roychoudhury, A., Aftab, T. and Acharya, K. (Eds.). *Omics Approach to Manage Abiotic Stress in Cereals*. Springer, Singapore, 479-493. https://Doi.org/10.1007/978-981-19-0140-9_19

Pandita, D. (2022c). miRNA- and RNAi-mediated metabolic engineering in plants. *In:* Aftab, T. and Hakeem, K.R. (Eds.). *Metabolic Engineering in Plants*. Springer, Singapore, 171-186. https://Doi.org/10.1007/978-981-16-7262-0_7

Pant, B.D., Buhtz, A., Kehr, J. and Scheible, W.R. (2008). MicroRNA399 is a long-distance signal for the regulation of plant phosphate homeostasis, *The Plant Journal*, 53(5): 731-738.

Park, M., Wu, G., Gonzalez-Sulser, A., Vaucheret, H. and Poethig, R.S. (2005). Nuclear processing and export of microRNAs in *Arabidopsis*, *Proceedings of the National Academy of Sciences*, 102(10): 3691-3696.

Phillips, J.R., Dalmay, T. and Bartels, D. (2007). The role of small RNAs in abiotic stress, *FEBS Letters*, 581(19): 3592-3597.

Ramírez, M., Flores-Pacheco, G., Reyes, J.L., Álvarez, A.L., Drevon, J.J., Girard, L. and Hernández, G. (2013). Two common bean genotypes with contrasting response to phosphorus deficiency show variations in the microRNA 399-mediated PvPHO2 regulation within the PvPHR1 signalling pathway, *International Journal of Molecular Sciences*, 14(4): 8328-8344.

Reichel, M., Li, Y., Li, J. and Millar, A.A. (2015). Inhibiting plant micro RNA activity: Molecular sponges, target MIMICs and STTMs all display variable efficacies against target micro RNAs, *Plant Biotechnology Journal*, 13(7): 915-926.

Reinhart, B.J., Weinstein, E.G., Rhoades, M.W., Bartel, B. and Bartel, D.P. (2002). MicroRNAs in plants, *Genes & Development*, 16(13): 1616-1626.

Reinhart, B.J., Weinstein, E.G., Rhoades, M.W., Bartel, B. and Bartel, D.P. (2002). MicroRNAs in plants, *Genes Dev.*, 16: 1616-1626. Doi:10.1101/gad.1004402

Reyes, J.L. and Chua, N.H. (2007). ABA induction of miR159 controls transcript levels of two MYB factors during *Arabidopsis* seed germination, *The Plant Journal*, 49(4): 592-606.

Rubio, V., Linhares, F., Solano, R., Martín, A.C., Iglesias, J., Leyva, A. and Paz-Ares, J. (2001). A conserved MYB transcription factor involved in phosphate starvation signalling both in vascular plants and in unicellular algae, *Genes & Development*, 15(16): 2122-2133.

Samad, A., Shafique, A. and Shin, Y.-H. (2017). Adsorption and diffusion of mono, di, and trivalent ions on two-dimensional TiS2, *Nanotechnology*, 28(17): 175401.

Schramke, V. and Allshire, R. (2004). Those interfering little RNAs! Silencing and eliminating chromatin, *Current Opinion in Genetics & Development*, 14(2): 174-180.

Sha, A., Zhao, J., Yin, K., Tang, Y., Wang, Y., Wei, X., Hong, Y. and Liu, Y. (2014). Virus-based microRNA silencing in plants, *Plant Physiology*, 164(1): 36-47.

Shahid, S., Kim, G., Johnson, N.R., Wafula, E., Wang, F., Coruh, C., Bernal-Galeano, V., Phifer, T., Depamphilis, C.W. and Westwood, J.H. (2018). MicroRNAs from the parasitic plant *Cuscuta campestris* target host messenger RNAs, *Nature*, 553(7686): 82-85.

Shukla, L.I., Chinnusamy, V. and Sunkar, R. (2008). The role of microRNAs and other endogenous small RNAs in plant stress responses, *Biochimica et Biophysica Acta (BBA)- Gene Regulatory Mechanisms*, 1779(11): 743-748.

Sunkar, R., Kapoor, A. and Zhu, J.-K. (2006). Posttranscriptional induction of two Cu/Zn superoxide dismutase genes in *Arabidopsis* is mediated by downregulation of miR398 and important for oxidative stress tolerance, *The Plant Cell*, 18(8): 2051-2065.

Sunkar, R. and Zhu, J.-K. (2004). Novel and stress-regulated microRNAs and other small RNAs from *Arabidopsis*, *The Plant Cell*, 16(8): 2001-2019.

Teotia, S., Singh, D., Tang, X. and Tang, G. (2016). Essential RNA-based technologies and their applications in plant functional genomics, *Trends in Biotechnology*, 34(2): 106-123.

Teotia, S. and Tang, G. (2017). Silencing of stress-regulated miRNAs in plants by short tandem target mimic (STTM) approach. *In: Plant Stress Tolerance*, pp. 337-348, Springer.

Trindade, I., Capitão, C., Dalmay, T., Fevereiro, M.P. and Santos, D.M.D. (2010). MiR398 and miR408 are up-regulated in response to water deficit in *Medicago truncatula*, *Planta*, 231(3): 705-716.

Wang, B., Sun, Y., Song, N., Wang, X., Feng, H., Huang, L. and Kang, Z. (2013). Identification of UV-B-induced microRNAs in wheat, *Genet Mol Res.*, 12(4): 4213-4221.

Wang, M., Thomas, N. and Jin, H. (2017). Cross-kingdom RNA trafficking and environmental RNAi for powerful innovative pre- and post-harvest plant protection, *Current Opinion in Plant Biology*, 38: 133-141.

Wang, Y.-G., An, M., Zhou, S.-F., She, Y.-H., Li, W.-C. and Fu, F.-L. (2014). Expression profile of maize microRNAs corresponding to their target genes under drought stress, *Biochemical Genetics*, 52(11): 474-493.

Weiberg, A., Wang, M., Lin, F.-M., Zhao, H., Zhang, Z., Kaloshian, I., Huang, H.-D. and Jin, H. (2013). Fungal small RNAs suppress plant immunity by hijacking host RNA interference pathways, *Science*, 342(6154): 118-123.

Whangbo, J.S. and Hunter, C.P. (2008). Environmental RNA interference, *Trends in Genetics*, 24(6): 297-305.

Wu, G., Park, M.Y., Conway, S.R., Wang, J.-W., Weigel, D. and Poethig, R.S. (2009). The sequential action of miR156 and miR172 regulates developmental timing in *Arabidopsis*, *Cell*, 138(4): 750-759.

Wu, L., Fan, J. and Belasco, J.G. (2006). MicroRNAs direct rapid deadenylation of mRNA, *Proceedings of the National Academy of Sciences*, 103(11): 4034-4039.

Xin, M., Wang, Y., Yao, Y., Xie, C., Peng, H., Ni, Z. and Sun, Q. (2010). Diverse set of microRNAs are responsive to powdery mildew infection and heat stress in wheat (*Triticum aestivum* L.), *BMC Plant Biology*, 10(1): 1-11.

Yan, J., Gu, Y., Jia, X., Kang, W., Pan, S., Tang, X., Chen, X. and Tang, G. (2012). Effective small RNA destruction by the expression of a short tandem target mimic in *Arabidopsis*, *The Plant Cell*, 24(2): 415-427.

Yang, J., Zhang, N., Mi, X., Wu, L., Ma, R., Zhu, X., Yao, L., Jin, X., Si, H. and Wang, D. (2014). Identification of miR159s and their target genes and expression analysis under drought stress in potato, *Computational Biology and Chemistry*, 53: 204-213.

Zhang, B. and Unver, T. (2018). A critical and speculative review on microRNA technology in crop improvement: Current challenges and future directions, *Plant Science*, 274: 193-200.

Zhang, J., Xu, Y., Huan, Q. and Chong, K. (2009). Deep sequencing of *Brachypodium* small RNAs at the global genome level identifies microRNAs involved in cold stress response, *BMC Genomics*, 10(1): 1-16.

Zhang, T., Zhao, Y.-L., Zhao, J.-H., Wang, S., Jin, Y., Chen, Z.-Q., Fang, Y.-Y., Hua, C.-L., Ding, S.-W. and Guo, H.-S. (2016). Cotton plants export microRNAs to inhibit virulence gene expression in a fungal pathogen, *Nature Plants*, 2(10): 1-6.

Zhang, Z., Yu, J., Li, D., Zhang, Z., Liu, F., Zhou, X., Wang, T., Ling, Y. and Su, Z. (2010). PMRD: Plant microRNA database, *Nucleic Acids Research*, 38(suppl_1): D806-D813.

Zhao, B., Ge, L., Liang, R., Li, W., Ruan, K., Lin, H. and Jin, Y. (2009). Members of miR-169 family are induced by high salinity and transiently inhibit the NF-YA transcription factor, *BMC Molecular Biology*, 10(1): 1-10.

Zhao, B., Liang, R., Ge, L., Li, W., Xiao, H., Lin, H., Ruan, K. and Jin, Y. (2007). Identification of drought-induced microRNAs in rice, *Biochemical and Biophysical Research Communications*, 354(2): 585-590.

Zhao, J., Liu, Q., Hu, P., Jia, Q., Liu, N., Yin, K., Cheng, Y., Yan, F., Chen, J. and Liu, Y. (2016). An efficient potato virus X-based microRNA silencing in *Nicotiana benthamiana*, *Scientific Reports*, 6(1): 1-7.

Zhao, Y., Ma, W., Wei, X., Long, Y., Zhao, Y., Su, M. and Luo, Q. (2019). Identification of exogenous nitric oxide-responsive miRNAs from alfalfa (*Medicago sativa* L.) under drought stress by high-throughput sequencing, *Genes*, 11(1): 30.

Zhou, L., Liu, Y., Liu, Z., Kong, D., Duan, M. and Luo, L. (2010). Genome-wide identification and analysis of drought-responsive microRNAs in *Oryza sativa*, *Journal of Experimental Botany*, 61(15): 4157-4168.

Zhou, M. and Luo, H. (2013). MicroRNA-mediated gene regulation: Potential applications for plant genetic engineering, *Plant Molecular Biology*, 83(1): 59-75.

CHAPTER

6

MicroRNA-mediated Regulation of Heat Stress Response

Vincent Ezin[1] and Rachael C. Symonds[2]*

[1] Faculty of Agricultural Sciences, University of Aboemy-Calavi, Benin
[2] School of Biological and Environmental Sciences, Liverpool John Moores University, UK

1. Introduction

Climate change and its impacts are now recognised as one of the greatest challenges facing the world, its peoples, its environment, its sustainable agriculture and its economies (IPCC, 2007; 2014). The development of industrial and agricultural activities, the proliferation of means of transport and the demographic explosion over the past century have led to increased concentrations of greenhouse gas (GHG) emissions – the cause of current climate change. Thus, from 1906 to 2012, the temperature of the Earth's surface increased by 0.8°C (IPCC, 2013). Globe's surface temperature for each of the past thirty years has been continually heating than in all that preceding decades since 1850. The years 1983 to 2012 are definitely the warmest periods in the history of the Northern Hemisphere (IPCC, 2013).

Global warming has become a serious threat to crop production and food security. In the tropical climates, high temperature is a most important restraining factor for plant growth and development. Heat reduces the germinating power of seeds and therefore leads to poor germination and poor plant-stand establishment (Fahad *et al.*, 2017). It also causes reduced water availability, reduction in leaf water potential, modification of root hydraulic conductance, water loss in the plant, chlorophyll degradation, decrease in photosynthesis, the activity of the PSII is partially or even completely stopped, the activity of metabolic enzymes is disturbed, leading to inactivation or denaturation of enzymes. Unlike animals, plants are sessile and under extreme environmental conditions, their growth and development are severely affected. The increase in leaf temperature accentuates the fluidity of the membranes, resulting in loss of physiological functions associated with the membranes (photosystem function and electron transport in chloroplasts and mitochondria). Heat stress alters the cellular

---

*Corresponding author: r.c.symonds@ljmu.ac.uk

metabolism of sensitive plants through damage to cell structures, impairment of membrane function, reduction of growth and death of the plants. Many episodes of high temperatures are expected to occur more often due to climate changes, which require plant physiological/biochemical adaptation for survival and productivity (Stavang *et al.*, 2009; Wigge, 2013). Among the physiological mechanisms of heat tolerance, we have (1) transpiration, (2) metabolism, (3) enzymatic activities, (4) heat shock proteins and (5) proportions of saturated/unsaturated membrane fatty acids. Transpiration is a simple and effective mechanism that uses the energy of water evaporation to cool the leaves. The closing of the stomata causes temperature increase of several degrees. Phenolic-like flavonoids, anthocyanin, lignin are important classes of secondary metabolites in plants and play a vital role in abiotic stress tolerance. Heat stress causes the accumulation of secondary metabolites of various kinds in plants, contributing to increase in heat tolerance. The increase in phenylalanine-ammonium lyase (PAL) activity is thought to be the acclimatisation response of cells to heat stress. Heat stress induces the biosynthesis of phenolic compounds and suppresses their oxidation, triggering plant adaptation to heat stress. Metabolites like carotenoids protect the cellular structure against the harmful effects of heat stress. For some plants, acclimatisation to high temperatures is associated with prominence of fatty acid saturations of membrane lipids, leading to an increase in membrane fluidity (Kumar *et al.*, 2012). For sensitive plants to high temperatures, transmembrane proteins bind more strongly with the lipid phase of the membrane. The composition and structure of the membrane is then modified, which generates ion leaks accompanied by the inactivation of photosynthesis and photorespiration.

MicroRNA involvement in plant heat stress responses have been defined in numerous plants comprising important economic crops. Their ability to respond and regulate gene expression precisely to various levels and durations of heat stress, as well as their ability to target multiple genes simultaneously, makes them novel and exciting targets for plant breeding for heat stress tolerance. Among rice miRNAs, OsmiR397 was identified as heat stress response miRNA which regulates L-ascorbate oxidase expression with regards to high temperature stress and rice adaptation (Yu *et al.*, 2012). It was also observed that the up-regulation of OsmiR397 and the down-regulation L-ascorbate oxidase lead to high yield in rice (Zhang *et al.*, 2013). OsamiR820 is involved in many abiotic stress tolerances such as salinity, heat, and drought stress (Sharma *et al.*, 2015a). MiR169 family is responsible for heat and drought tolerance sorghum (Paterson *et al.*, 2009; El Sanousi *et al.*, 2016).

This chapter begins with a discussion on the effects of high temperature stress in plants and the physiological and cellular processes that occur along with the transcriptional regulation of these processes. An overview of types of microRNAs involved in heat stress and their regulation is plants are then discussed, followed by their involvement in heat shock and heat stress memory, flowering, reproduction and yield, interaction with phytohormones and cross talk with other abiotic stresses. Finally, the use of microRNAs as a technology for plant breeding and biotechnology is discussed.

2. Heat Stress in Plants

Temperatures above the optimal become an essential environmental factor to growth, development and functions of plants. Heat stress is referred to as a condition of the environment when there is a rise in ambient temperature from 10-15°C (Wahid *et al.*, 2007) and which is a complex abiotic stress due to the following factors – intensity, duration and the period of its occurrences. Heat stress negatively affects plants at all levels including cells, tissue and whole plant. Under heat stress, marked modifications are observed in plant growth and development, not only according to the duration and intensity of the stress, but also at plant cell levels and developmental stages. Schöffl *et al.* (1999) reported that under extreme temperatures, cells are severely injured within minutes, ultimately leading to their death and the breakdown of the entire plant. There is also change in plant morphology, phenology, anatomy, chemistry and biochemistry, leading to plant damage in extreme cases and consequently huge losses of economic yields (Wahid *et al.*, 2007). Plant phenology with regards to high temperature stress is genotype- and species-dependent as their tolerance/susceptibility differ from emergence to reproductive stages (Wollenweber *et al.*, 2003; Howarth, 2005). Grain pollen and spikelet viability are significantly reduced under high temperature stress (Jagadish *et al.*, 2013). Heat causes rolling of leaves even when soil moisture is adequate, reduction in stomatal conductance and photosynthesis and rise in transpiration. There is low uptake of CO_2 for efficient and optimum photosynthesis and the enzymes responsible for photosynthesis are inhibited. High temperatures disrupt plant days to germination, to flowering, even flower failure and days to maturity. Ezin *et al.* (2022) observed that heat stress significantly affected height of the plants, length of the leaves, tillers number, internodes number, dates of flowering and maturity, 1,000-seed weight and yield of rice plant. They further stated that heat stress expressively delayed the flowering of all the rice varieties when compared to the controls. The proteins and lipids are affected through denaturation and high fluidity, respectively. The combination of extreme air temperature and soil heat significantly modifies the absorption of water from the soil into plant through the roots and it becomes difficult for the plant to meet the high atmospheric evaporative demands (Beudez and Doussan, 2013). In such conditions, plant survival becomes difficult due to damage of plant at all levels.

A tolerant plant to heat stress is the aptitude of the plant to withstand high temperatures, develop and produce biological and economic yields. It has been reported that any increase in both day temperature and night temperature affects crop production. For instance, temperatures between 25-30°C during the day and 20°C at night are ideal for tomato production (Camejo *et al.*, 2005). However, a slight rise of about 2-4°C above the optimal disrupts gamete formation, flower failure in tomato and many other crops, ultimately leading to reduce yields (Wang *et al.*, 2003; Sato et *al.*, 2001; Thomas and Prasad, 2003; Wahid *et al.*, 2007; Ribeiro *et al.*, 2009). Heat stress causes oxidative damage through huge accumulation of H_2O_2, $^{\cdot}OH$, and $^{\cdot}O_2^{-}$ which rescind many molecules, destabilise membrane structures and degrade proteins (Asada, 2006; Miller *et al.*, 2010; Niu and Xiang, 2018). Tolerant plant to heat stress is marked by undamaged chlorophyll, high rate of photosynthetic products, high viability of pollen grains and spikelet, high numbers of fruit and seed set and yield. A tolerant plant must develop morphological, phenological, physiological and biochemical mechanisms to withstand heat stress. Other strategies for tolerance include acclimation, cooling effects, avoidance and alteration in transcriptional and translational patterns for end products. For terminal heat or heat stress, crops may mature earlier to escape huge yield losses (Adams *et al.*, 2001). A high proportion of saturated fatty acids in membrane lipids provide stability under heat stress. At

MicroRNA-mediated Regulation of Heat Stress Response

molecular level, changes in gene expression are required for acclimatisation of plant to high-temperature stress. For the plant under heat stress to modify its gene expression, the plant must (1) perceive the heat stress signals (2) transduce the signals and (3) induce changes of gene expression. The response of plants to high temperatures or heat shock results in the production of specific proteins, called 'heat shock proteins' or HSPs, thus helping the cell to withstand stress (Richter *et al.*, 2010). HSPs, namely small HSPs, HSP90, HSP70, HSP100 and HSP60, are responsible for re-naturing proteins which are altered by high temperature stresses. When the plant goes from 25°C to more than 32°C, we observe the synthesis of mRNA coding for HSPs while the production of common proteins is stopped. HSPs are produced as chaperone proteins to support non-native and nascent proteins under extreme temperatures (Baniwal *et al.*, 2007; Kotak *et al.*, 2007; Richter *et al.*, 2010; Qu *et al.*, 2013). These proteins are expressed as a response to heat stress. The structure of ordinary proteins is indeed sensitive to heat: they denature and lose their biological action. Therefore, the role of chaperone proteins is to prevent potential damage caused by heat on the structure of proteins. Heat stress brings about reactive oxygen species (ROS)-scavenging enzymes and HSPs, which are target genes of heat stress responsive transcription factors (TFs) (Ohama *et al.*, 2017). Apart from HSPs produced as a response to heat stress, many other proteins are synthesised in plants and include MYB proteins, MYB TFs, Mn-POD, ubiquitin cytosolic Cu/Zn-SOD (Brown *et al.*, 1993; Herouart and Inze, 1994; Sun and Callis, 1997; Cao *et al.*, 2013; Liu *et al.*, 2013). Under heat stress, soybean produces ubiquitin and ubiquitin-conjugated proteins to counteract the negative effects of high temperatures (Ortiz and Cardemil, 2001).

3. Heat Responsive Transcriptional Regulation

The adaptations of plants to global warming necessitate transcriptional reprogramming so as to trigger response pathways for plant survival, acclimation and tolerance (Werghi *et al.*, 2021). During high temperature stresses, transcriptional reprogramming is triggered by the high temperature stress transcription factor (HSF) genes (Schramm *et al.*, 2006; Huang *et al.*, 2016; Zang *et al.*, 2019). HSF play a vital role in the heat stress response (HSR) as major regulators that are key in the stimulation of transcriptional networks (Mishra *et al.*, 2002; Yoshida, T. *et al.*, 2011). The following complex network, i.e. heat stress transcription factors, antioxidants and heat shock proteins are put in place as a heat stress response for the survival and the resistance mechanism of plants to heat stress (Sharkey and Schrader, 2006; Cottee *et al.*, 2014). HSF TFs are members of transcriptional regulatory network involved in the stimulation of heat stress candidate genes of tomato, cabbage, etc. (Song *et al.*, 2014; Guo *et al.*, 2016). Three classes of HSFs (A, B and C) are identified in plant species (Nover *et al.*, 2001). During abiotic stress, class A HSFs are transcriptionally considered as response regulators of the stress (Shim *et al.*, 2009; Anfoka *et al.*, 2016). HsfA1s modulate abiotic stresses, which are directly engaged in controlling the expressions of HsfA2 and HsfB1 (Yoshida *et al.*, 2011), while class B HSFs regulate the activity of HsfA1 as repressors in *Arabidopsis* or co-activators in tomato (Shim *et al.*, 2009; Ikeda *et al.*, 2011; Zhu *et al.*, 2012). The study of Ikeda and Ohme-Takagi (2009) showed that class B HSFs encompass an 'R/KLFGV' motif which performs as a transcriptional repressor. HsfA2 are regulators of environmental stress response by increasing the gene expression of antioxidant which encodes ascorbate peroxidase enzymes

(Panchuk, 2002; Charng *et al.*, 2006). Wang *et al.* (2016) observed that CtHsfA2b gene is responsible for enhancing heat tolerance in plant species via transcriptional regulation of HSPS, AtHsp70b, AtHsp101-3, AtHsp22.0-ER, AtHsp25.3-P and AtHsp26.5-P(r). AtHsp18.1-CI and ascorbate peroxidase which have earlier been revealed as heat tolerance (Sharkey and Schrader, 2006; Schramm *et al.*, 2006). The expression levels of HsfA2, HsfA7a, HsfBs, dehydration-responsive element-binding protein 2A (DREB2A) and multi-protein bridging factor 1C (MBF1C) are being modulated by HsfA1s under heat stress as heat stress responsive TFs (Yoshida *et al.*, 2011). The expression of HsfA3 as a target gene with a transcriptional co-regulatory complex of nuclear factor Y, subunit A2 (NF-YA2), NF-YB3 and DNA polymerase II subunit B3-1 (DPB3-1)/NF-YC10 are induced by DREB2A (Sato *et al.*, 2014). HsfA2 is vital for heat stress response in plants and directly targeted by HsfA1s gene. NAC019 is a new TF-modulating HSR, which binds to the promoters of high temperature responsive TFs and positively influences the expression levels of genes, such as HsfA1b and HsfB1, but nac019 mutant has been sensitive to heat stress (Guan *et al.*, 2014). NAC019 is also active during heat stress due to its interaction with phosphatase and C-repeat binding factor gene expression 2 (RCF2). Jungbrunnen1 (JUB1) is another NAC TF implicated in DREB2A heat stress responsive expression (Ohama *et al.*, 2017).

Many HSF family proteins respond to high temperature independent of HsfA1pathways as reported by von Koskull-Döring *et al.* (2007). Heat stress also induces HsfA4a, which regulates ascorbate peroxidase 1 expression levels while a monomer from HsfA5 with a monomer fromHsfA4 forms a dimer to repress the activity of HsfA4. HsfA9 causes the activation of many HSP genes during high temperature stresses. Another TF that modulates the HSR is bZIP28 via a route independent on HsfA1 is another. Thus under heat stress, it induces target gene expression through Site 1 Protease and Site 2 Protease and to the nucleus for the stimulation of gene, but under normal conditions, bZIP28 is bound to the main ER chaperone-binding immunoglobulin protein to become inhibited and inactivated (Srivastava *et al.*, 2014). Several HSFs, such as HSFA1b, HSFA2, HSFA3, HSFA7, HSFB1 and HSFB2b are considered as crucial regulators that demonstrate higher transcript levels at high temperatures of 39°C and 45°C in both tolerant and sensitive tomato species (Hu *et al.*, 2020a) but Rao *et al.* (2021) reported that HSFA7 was highly induced in tolerant genotype CLN compared to sensitive CA4, thus concluding that HSFA7 is a regulator of heat stress response in tomato. The transcriptomic analysis, under heat stress, revealed down-regulation of 97.4% i.e. 75 genes from the identified 77 major photosynthetic genes. Among the TFs/miRNAs identified, 6 miRNAs and 4 TFs were main transcriptional regulators which play crucial roles in mediating photosynthesis under high temperatures stress (Zhao *et al.*, 2019).

3.1 Clarification of MicroRNAs Concept and Types of MicroRNAs

The microRNA is an 18-25 nucleotides long single-stranded RNA. The miRNAs are minor no coding RNAs that post-transcriptional control gene expression (Carrington and Ambros, 2003; Bartel, 2004). It is after transcription that microRNA controls the expression of target genes. They also cause degradation and inhibition of translation of target mRNAs (Saikumar *et al.*, 2014), though it is reported that it is a mechanism through which miRNAs regulate cells for growth and development and respond to abiotic and biotic stress. It is also reported that based on miRNAs and target gene complementarity, miRNAs modulate gene expression via mRNAs splitting or inhibition of protein translation (Iwakawa and Tomari, 2013). Under heat stress, miRNAs could be either down-regulated or up-regulated as a response to the stress and according to plant species under heat stress (Zhou *et al.*, 2016). Several research reports have shown that miRNAs are responsible for growth and development, signal perception and transduction, conservation of whole genome stability and integrity, cell signalling, plant growth hormone signalling pathways, ion homeostasis, inherent immunity and mechanism response for many abiotic and biotic stresses (Sun, 2012; Xie *et al.*, 2014; Zhang and Wang, 2015; Djami-Tchatchou *et al.*, 2017; Pandita, 2019, Pandita and Wani, 2019; Pandita, 2021; 2022a; 2022b; 2022c). Through the use of high throughput sequencing technology, miRNAs expression has been depicted and understood under severe environmental conditions.

It is demonstrated that most plants have conserved their miRNAs across the evolution (Cuperus *et al.*, 2011; Sun, 2012). Two classes of miRNAs, including the ancient miRNAs and the young miRNAs, have been identified according to their diversity and conservation. It is only the ancient miRNAs that are greatly expressed and evolutionarily present in plants while the young ones are expressed under certain conditions and are not evolutionarily conserved and are not present in all plant species (Qin *et al.*, 2014). MicroRNAs involved in abiotic stress responses are shown in Table 1.

3.2 Overview of Plant miRNAs Responses to Heat Stress

Early work by Sunkar and Zhou (2004) identified many miRNAs in *Arabidopsis* with differential expression when responding to abiotic stresses including salinity or drought, or cold stress, or ABA, including members of the miR171 and miR319 families. Amongst these miRNAs differences in tissue and stress regulation were found, with different miRNAs being up- or down-regulated in response to stress, highlighting the complexity of miRNA stress responses. When plants are subjected to abiotic stress, miRNAs act as negative regulators of stress responses by down-regulating their target genes, or positive regulators, when down-regulated miRNAs result in up-regulation of their target genes (Chinnusamy *et al.*, 2007). In general, miRNAs respond to abiotic stresses by negatively controlling the expression of their target genes, post-transcriptionally (Khraiwesh *et al.*, 2012). Evidence from *Arabidopsis* confirmed that in some instances, alternative splicing is a regulatory mechanism for controlling heat stress expression of miRNAs (Yan *et al.*, 2012a). Since the study by Sunkar and Zhou (2004), miRNAs have been identified as mediating plant responses to heat stress in an ever-increasing number of crop species, including rice (Mangrauthia *et al.*, 2017), *Arabidopsis* (Steif *et al.*, 2014) and tomato (Zhou *et al.*, 2016). These miRNAs have species as well as genotype (Liu *et al.*, 2017), tissue (He *et al.*, 2019) and growth stage (Liu *et al.*, 2017) differences in expression patterns (Fig. 1). When barley (*Hordeum vulgare*) is subjected to temperatures above optimum, it loses yield and dry matter production (Hossain *et al.*, 2012). Mature miRNAs (miR5175a ,160a, 166a and 167) were up-regulated in barley in response to heat stress and this heat stress induced the splicing of intron-containing precursors pri-miR5175a and pri-miR160a, which then correlated with mature miRNA accumulation (Kruszka *et al.*, 2014). This suggests that there was post-transcriptional regulation of miRNA precursor processing as well as transcriptional regulation, demonstrating that the miRNAs and their precursors were affected by plant heat stress (Kruszka *et al.*, 2014). This contrast with

Table 1: Different types of miRNA in some plants

Plant Species	MicroRNAs	Environmental Stress Responses	References
Arabidopsis (Arabidopsis thaliana)	miR398	Heat tolerance	Guan *et al.*, 2013
Arabidopsis (Arabidopsis thaliana)	multiple miRNAs including miR529, miR896, miR1030, miR156, miR170, miR171, miR172, miR319, miR159, miR168, miR396, miR397, miR408, miR1035, miR1050, miR1088, and miR1126) were down-regulated by drought stress while it induced the overexpression of 14 miRNAs including miR159, miR474, miR845, miR851, miR854, miR896, miR169, miR171, miR319, miR395, miR901, miR903, miR1026, and miR1125), notably, miR319, miR171 and miR896	Drought stress	Zhou *et al.*, 2010
Arabidopsis (Arabidopsis thaliana)	miR158a, miR159a, miR172a,b, miR391miR156g, miR157d and miR775	Flooding stress	Moldovan *et al.*, 2010
Arabidopsis (Arabidopsis thaliana)	miR160, miR780, miR826, miR842, and miR846, miR169, miR171, miR395, miR397, miR398, miR399, miR408, miR827, and miR857	Nitrogen starvation	Liang *et al.*, 2012
Rice (*Oryza sativa*)	miR1846d-5p, osa-miR5077, osa-miR156b-3p, osa-miR531b, osa-miR5149, osa-miR168a-5p, osa-miR167e-3p, multiple miRNAs including miR827	Heat stress	Kushawaha *et al.*, 2021; Goel *et al.*, 2019
	osa-miR162, osa-miR164	Drought stress	
	multiple miRNAs	Cold stress	Jiang *et al.*, 2019
	multiple miRNAs	Abiotic stress	Sepúlveda-García *et al.*, 2020
	miR393, miR390	Multiple stress	Lu *et al.*, 2018
	Osa-miR820, osa-miR414, osa-miR164e, osa-miR408, and multiple miRNAs	Salt stress	Sharma *et al.*, 2015; Macovei and Tuteja, 2012; Goel *et al.*, 2019

MicroRNA-mediated Regulation of Heat Stress Response

Plant	miRNAs	Stress	Reference
	miR169, osa-miR444a.4-3p, miR399, miR530, miR156, miR167, miR164, miR167, miR168, miR528, miR820, miR821, miR1318	Nitrogen deficiency	Shin et al., 2018; Cai et al., 2012; Nischal et al., 2012
	multiple miRNAs	Phosphate deficiency	Secco et al., 2013
	miR529a	Oxidative stress	Yue et al., 2017
	miR408, miR528	Cadmium stress	Liu et al., 2020
	multiple miRNAs	Arsenic stress	Tang et al., 2019
Maize (Zea mays)	multiple miRNAs	Heat stress	Zhang et al., 2019
	multiple miRNAs	Drought stress	Li et al., 2019
	multiple miRNAs	Flooding	Zhai et al., 2012
	multiple miRNAs	Chilling stress	Aydinoglu, 2020
	multiple miRNAs, miR167, miR168, miR169, miR172, miR160, miR164, miR169, miR398, miR399, miR395, miR397, miR408, miR528, miR827	Nitrogen deficiency	Yang et al., 2019; Zhao et al., 2012; Xu et al., 2011
	multiple miRNAs	Phosphate deficiency	Nie et al., 2016
Cotton (Gossypium sp L.)	multiple miRNAs	Heat stress	Ding et al., 2017
	multiple miRNAs	Extreme (low and high) temperature stress	Wang et al., 2016
	gh-miRNVL5, miR414, miR397a/b, miR399a,ghr-miR399 and ghr-156e, miR156a/d/e, miR169, miR167a, miR535a/b, miR827b, multiple miRNAs	Salt stress	Wang et al., 2019; Deng et al., 2015; Yin et al., 2011; 2017

(Contd.)

Plant Species	MicroRNAs	Environmental Stress Responses	References
	miR319	Abiotic stress signalling	Yin *et al.*, 2018
	ghr-miR5272a	Immune response	Wang *et al.*, 2017
Wheat (*Triticum aestivum*)	many miRNAs	Heat and drought stress	Liu *et al.*, 2019
	miR169, miR172, miR395, miR396, miR408, miR159, miR160, miR166, miR472, miR477, miR482, miR1858, miR2118, miR5049	Drought stress	Akdogan *et al.*, 2016
	miR159, miR393, miR398	Salt, cold and wound, stress	Wang *et al.*, 2014
	TaemiR408	Salinity and phosphate deficiency Stress	Bai *et al.*, 2018
	TamiR1139	Phosphate deficiency	Liu *et al.*, 2018
	many miRNAs	Reactive oxygen species (ROS) response	Cao *et al.*, 2019
Tomato (*Solanum lycopersicum*)	many miRNAs including sly-miR169	Heat and drought stress	Zhou *et al.*, 2020
	miR482/2118	Immune response	Zhai *et al.*, 2011 ; Shivaprasad *et al.*, 2012

Sugarcane (*Saccharum* L.)	multiple miRNAs	Drought stress	Zheng *et al.*, 2016
	multiple miRNAs	Water saturation stress	Khan *et al.*, 2014
	multiple miRNAs	Low temperature stress	Yang *et al.*, 2017
Soybean (*Glycine max*)	multiple miRNAs	Water deficit	Zheng *et al.*, 2016
Potato (*Solanum tuberosum*)	many miRNAs	Nitrogen deficiency stress	Tiwari *et al.*, 2020
	miR396c, miR4233, miR2673, miR172, miR396a, miR6461	Water deficit stress	Yang *et al.*, 2013
	Stu-mi164	Osmotic stress	Zhang *et al.*, 2018
Radish (*Raphanus sativus*)	miR398a-3p, ath-miR398b-3p, ath-miR159b-3p, ath-miR159c, ath-miR165a-5p, novel_86, novel_107, novel_21, ath-miR169g-3p, ath-miR171b-3p	High temperature stress	Yang *et al.*, 2019
	multiple miRNAs	Salinity stress	Sun *et al.*, 2015
	multiple miRNAs	Cadmium stress	Xu *et al.*, 2015
Date Palm (*Phoenix dactylifera*)	multiple miRNAs	Salt stress	Yaish *et al.*, 2015
Cowpea (*Vigna unguiculata*)	multiple miRNAs	Water deficit stress	Barrera-Figueroa *et al.*, 2015
Flax (*Linum usitatissimum*)	141 miRNA including miR482/2118 and miR2275 f	Heat stress	Pokhrel and Meyers, 2022

Poplus tomentosa in which miR160a is down-regulated by heat stress highlights that common miRNAs can respond differently to heat stress in different species (Chen *et al.*, 2012).

A genome wide study using deep sequencing of *B. rapa* seedlings subjected to high temperature stress for an hour identified five heat-responsive conserved miRNA families that were shared with *Arabidopsis* (Yu *et al.*, 2012). The conserved miRNA, bra-miR398, had an inverse relationship with its target genes, BracCSD1 and BracCSD2 (Yu *et al.*, 2012). MiR398 has previously been shown to target CSD1 and CSD2 in *Arabidopsis* which regulate tolerance to oxidative stress and therefore may play a role in *B. rapa* thermotolerance via protection from oxidative damage (Sunkar *et al.*, 2006). Genome wide high-throughput sequencing of wheat (Xin *et al.*, 2010) and *P. tomentosa* (Chen *et al.*, 2012) identified 52 and 12 heat stress responsive miRNAs respectively, the majority of which were down-regulated. He *et al.* (2019) compared miRNA expression in response to heat stress in maize vegetative tissues with that of vegetative seedlings in tomato (Zhou *et al.*, 2016), rice (Mangrauthia *et al.*, 2017) and *Arabidopsis* (Barciszewska-Pacak *et al.*, 2015). Of these four crop species only of miR166 were commonly up-regulated in heat-stressed tissues. In maize and tomato, four miRNA species were up-regulated (miR166, miR166*, miR168* and miR396*) and five miRNAs were up-regulated in both maize and rice (He *et al.*, 2019). In addition to species differences, genotype differences in miRNA responses to heat stress must be considered when establishing the role of specific miRNAs and their target genes to heat stress. Two Chinese cabbage varieties with different tolerance to heat stress were exposed to elevated temperatures for a longer duration of six or 12 hours (Ahmed *et al.*, 2020). A total of 49 novel and 43 known miRNAs were differently expressed between the two varieties under heat stress. Twenty-three of these miRNAs were differentially expressed solely under heat stress, indicating that their target genes may play a key role in heat stress tolerance in Chinese cabbage (Ahmad *et al.*, 2020). Root and shoot seedlings were subjected to either heat shock or prolonged heat stress in tolerant and susceptible varieties of rice, revealing genotype specific miRNA expression patterns (Mangrauthia *et al.*, 2017). Some miRNAs were found to be expressed in both varieties, implying a general role in the heat stress response while others were only expressed in the heat tolerant variety and were considered specific.

To unravel the complex networks of miRNA regulation of plant responses to heat stress and to understand the biological processes, it is important to identify the target genes. Degradome sequencing has been used successfully in species such as wheat (Ravichandran *et al.*, 2019), *Brassica chinensis* (Jiang *et al.*, 2014), *Solanum pimpinellifolium* (Zhou *et al.*, 2016) and cultivated tomato (Zhou *et al.*, 2020). Ravichandran *et al.* (2019) surveyed both the degradome and small RNAs of wheat leaves subjected to heat stress, either immediately after the heat stress or one or four days later. From this, they identified 36 miRNAs differentially expressed after heat stress with heat responsive miRNA targets including protein kinases, HD-ZIP proteins, targeted SODs and F-box proteins (Ravichandran *et al.*, 2019). A combination of high throughput sequencing and degradome libraries were used to identify miRNAs and their target sequences in both cultivated tomato, which is sensitive to high temperatures, and its wild heat tolerant *Solanum pimpinellifolium* subjected to moderate or acute stress (Zhou *et al.*, 2016). A total of 138 conserved miRNAs with 349 target sequences and 13 targets from eight novel miRNAs were identified, providing high quality global information on miRNA targets in tomato. Many of the conserved target genes were involved in processes, such as apoptosis, protein phosphorylation, ATP binding and oxidation reduction. A greater number of miRNAs were shown to respond to acute compared to moderate temperatures in both tomato species and while more miRNAs induced by moderate temperatures were down-regulated, miRNAs responding to acute temperatures tended to be up-regulated, indicating differences in responses to different stress levels (Zhou *et al.*, 2016).

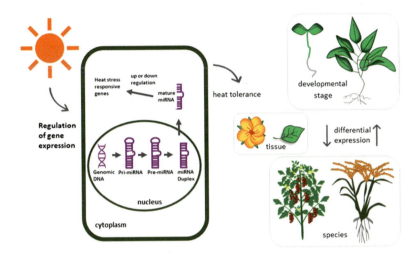

Fig. 1: miRNA-mediated regulation of gene expression under heat stress showing differential expression for developmental stage, tissue and species

3.3 Heat Shock and Heat Stress Memory

A major means of defence against heat stress and sudden heat shock in plants is the production of heat shock proteins via heat shock factors (Haslbeck *et al.*, 2015). There are five classes of heat shock proteins – sHSP, HSP60, HSP70, HSP90 and HSP100, which are all strongly induced in response to heat stress playing a key role in plant responses to high temperatures (Kotak *et al.*, 2007). Heat shock factors bind to heat shock elements in the upstream regions of heat responsive genes, inducing heat shock proteins, which are involved in maintaining and restoring protein homeostasis after plants are exposed to elevated temperatures (Gomez-Pastor *et al.*, 2018). Protein denaturation resulting from heat shock is a major source of heat stress injury in plants and heat shock proteins, together with ROS scavenging enzymes are targeted by heat shock responsive transcription factors, playing a critical role in refolding these proteins (Janni *et al.*, 2020). Several miRNAs are emerging as implicated in heat stress transcription regulation and heat stress memory (Ohama *et al.*, 2017). A diverse and complex pattern of conserved and novel miRNAs induced in response to heat shock have been identified through genome-wide studies in varied species, such as *Brassica rapa* (Yu *et al.*, 2012), wheat (Xin *et al.*, 2010), *Brassica rapa* (Yu *et al.*, 2012) and *Arabidopsis* (Pegler *et al.*, 2019). The largest miRNA family in *Arabidopsis*, miR169, is known to be responsive to both ambient temperatures and sudden heat shock playing important roles in environmental stress responses and developmental cues, such as flowering time (Lee *et al.*, 2008; Li *et al.*, 2010). Heat shock triggers a rapid increase in expression in a subset of *Arabidopsis* miR169 family members, revealing complex overlapping roles (Li *et al.*, 2010). Many miRNAs are species-specific; however, miR396 is an ancient, conserved miRNA and found in all land plants (Axtell *et al.*, 2007), involved in the development of leaves and targeting growth-regulating factor (GRF), transcription factors as part of the miR396-GRF module (Das and Nath, 2015). WRKY transcription factors are associated with responses to abiotic stresses (Rushton *et al.*, 2010) and in sunflower, miR396 mediates the cleavage of *HaWRKY6*, silencing it in response to heat shock confirmed (Giacomelli *et al.*, 2012), by transgenic plants expressing an miR396 resistant form of *HaWRKY6* suffering lethal heat shock This miR396 WRKY target site appears to be a recent evolutionary event, with a transcription factor diverging and acquiring a target site for a conserved miRNA.

Several different strategies are employed by plants to regulate responses to heat shock. *Arabidopsis* plants subjected to rapid heat shock induced the expression of miR398, which down-regulates its target superoxide dismutase genes CSD1 and CSD2 and their copper chaperone CCS (Guan *et al.*, 2013). Heat shock factors, HSFA1b and HSFA7, members of the HSFA1 which are master regulators of heat high temperature stress responses in plants (Mishra *et al.*, 2002), were responsible for the heat stress induction of miR398, revealing a previously unknown strategy for regulating thermotolerance in *Arabidopsis*. Plants with resistant forms of the target genes had reduced expression of heat shock factors and heat shock proteins (Guan *et al.*, 2013).

Many studies look at the short-term response to heat shock, but fewer compare it to longer term stress or recovery. Responses to either heat shock, a longer duration of five days' heat stress and heat shock recovery were compared in both heat tolerant and sensitive rice genotypes (Sailaja *et al.*, 2014). Genotype specific responses to the different heat stresses were seen with the majority of miRNAs being up-regulated, and then down-regulated upon recovery. This contrasted with the heat tolerant genotype in which most miRNAs were down-regulated with a longer duration of stress and then up-regulated upon recovery, indicating fine-tuned heat response mechanisms (Sailaja *et al.*, 2014).

All plants have a level of tolerance to stress dependent on genotype and species and developmental stage (Hilker *et al.*, 2016). Plants can also acquire increased levels of heat stress tolerance by exposure to non-lethal stress which protects them against future lethal levels of stress (Ling *et al.*, 2018). Plants primed with a short duration of non-lethal stress have differing stress-tolerance levels to non-primed plants, gaining a heat stress memory which allows them to adapt better to future stress (Mauch-Mani *et al.*, 2017). This heat stress memory can be maintained over several days, even if the stress is not continuous (Yeh *et al.*, 2012). Knowledge of the molecular mechanisms controlling heat stress memory and acclimation are still uncertain; however it is known that it induces heat shock proteins (Wu *et al.*, 2013). Kushawaha *et al.* (2021) thermo-primed both rice seedlings and mature rice plants with a short heat exposure followed by more prolonged heat stress, comparing them to plants only subjected to prolonged heat stress. Plants that had been thermo-primed before or after anthesis had significantly higher yields than non-primed plants. Several miRNAs, including osa-miR5149, osa-miR531a, osa-miR1846d-5p, osa-miR168a-5p, osa-miR5077, osamiR156b-3p, osa-miR167e-3p along with their respective target heat shock activators and repressors, showed differential expression at the seedling stage. An

MicroRNA-mediated Regulation of Heat Stress Response

inverse level of expression seen between the miRNAs and their target genes indicated that these miRNAs acted on their respective heat shock factors or heat shock proteins to adjust gene expression differentially in response to thermo-priming (Kushawaha *et al.*, 2021).

3.4 miRNAs and Phytohormone Responses to Heat Stress

Plant growth regulators play essential roles in the regulation of plant responses to heat stress. Ethylene, sbscisic acid (ABA) and salicylic acid are all known to protect plants from oxidative damage resulting from high temperature stress (Larkindale and Knight, 2002; Li *et al.*, 2021). The complex cross talk and interaction between plant growth regulators is still not fully understood; however, exogenous applications of plant growth regulators are known to significantly improve plant thermotolerance (Li *et al.*, 2021). miRNAs are known to control several plant growth regulator pathways, such as miR159, which targets mRNAs which encode gibberellin-MYB transcription factors (Achard *et al.*, 2004) and miR160 which targets the *A. thaliana* auxin response factors (ARFs) ARF10, ARF16 and ARF17 (Li *et al.*, 2007). Auxin plays a role in plant adaptations to high temperature stress by altering plant architecture through regulation of leaf hyponasty and hypocotyl elongation (Küpers *et al.*, 2020) and exogenous application of auxin restores high temperature induced male sterility (Oshino *et al.*, 2011). Additionally, changes to flower patterning and formation and formation of adventitious roots are a common auxin-mediated effect of heat stress and therefore, an understanding of the role of miRNAs in regulating plant growth regulators in response to high temperatures is essential. The auxin signalling pathway is regulated by auxin response transcription factors (ARF) and auxin-related miRNAs in response to heat stress (Kruszka *et al.*, 2014). ARF8, a target of miR167h is down-regulated in barley in response to heat stress (Kruszka *et al.*, 2014). In *A. thaliana*, ARF8 regulates the development of floral organs and petal vascular patterning (Tabata *et al.*, 2010) and adventitious root formation (Gutierrez *et al.*, 2012). Cotton plants over-expressing both miR157 and miR160 show increased sensitivity to heat stress and anther indehiscence (Ding *et al.*, 2017). The overexpression of miR160 was associated with ARF10 and ARF17 suppression while miR157 overexpression caused a suppression of the auxin signal. miRNA auxin signalling therefore appears to be an important component of anther fertility in heat-stressed cotton (Ding *et al.*, 2017). In *Arabidopsis*, miR160 was highly induced by heat stress, while its target auxin response genes – ARF10, ARF16 and ARF17 were repressed (Lin *et al.*, 2018). Overexpression of miR160 repressed ARF10, ARF16 and ARF17 and improved seed germination and heat stress survival. In contrast, miR160 target mimicry plants had higher levels of ARF10, ARF16 and ARF17 and worse performance under heat stress, indicating the capability of miR160 to regulate these genes (Lin *et al.*, 2018).

There is a transient and very rapid increase in ABA levels in response to heat shock with a subsequent increase in ROS levels, which enhance the plants' antioxidant capacity, increasing heat tolerance (Larkindale *et al.*, 2005). Members of the miR169 family, which regulate both developmental (Lee *et al.*, 2008) and environmental cues, are down-regulated by ABA and heat stress (Li *et al.*, 2010). Several *Arabidopsis* miR169 members have an over-representation of ABRE and HSE elements, which respond to ABA and heat stress in their upstream regions (Li *et al.*, 2010). Mi159 has been identified as being highly induced by ABA in *Arabidopsis* and regulates the MYB101 and MYB33 transcription factors in ABA-dependent germination of seeds

(Reyes and Chua et al., 2007). Similarly, miR394 is involved in regulating plants' responses to ABA, salinity and drought stress in *A. thaliana* (Song et al., 2013). High throughput sequencing identified 107miRNAs which were differentially expressed in WT and ABA-deficient rice mutants (Tian et al., 2015). High throughput sequencing of small RNAs of the rice *Osaba1* ABA-deficient mutant found 107 miRNAs that were differentially expressed (Tian et al., 2015). Thirteen of these miRNAs were confirmed by stem-loop RT PCR and ABA was found to regulate the expression of these miRNAs and their target genes. The miRNA target genes are involved in diverse processes including adaptations to abiotic stress (Tian et al., 2015).

The miRNA miR159 is a negative regulator of heat stress which targets gibberellic acid regulated MYB (GAMYB) transcription factors (Wang et al., 2012). Heat stress caused down-regulation of miR159 in wheat (Xin et al., 2010), while overexpression resulted in increased susceptibility to heat stress (Wang et al., 2012), indicating a role for miR159 in heat stress tolerance in wheat. Cucumber plants grafted on to a more heat tolerant luffa rootstock showed improved heat tolerance resulting from increased ABA biosynthesis from the rootstock, through repression of csa-miR159b and up-regulation of its target genes – mRNAs CsGAMYB1 and CsMYB29 (Li et al., 2016). *Arabidopsis* plants over-expressing csa-miR159b showed increased heat sensitivity and a down-regulation of *AtMYB*, further confirming the role of miR159 as a negative regulator of heat stress (Li et al., 2016). He et al. (2020) constructed a ceRNA regulatory network of lncRNA/circRNA-miRNA in heat stressed cucumber, which revealed that circRNAs interacted with miR9748 via phytohormone pathways in response to heat stress. The authors concluded that miR9748 is involved in the interaction between phytohormone and heat stress responsive pathways rather than acting directly on the heat stress pathways.

3.5 The Role of miRNAs in Flowering time and Reproduction: A Critical Stage for Heat Stress

Flowering is critical for plant reproductive success and therefore plant fitness and yield, and knowledge of the complex molecular processes controlling this is essential for continued crop breeding strategies (Huijser and Schmid, 2011). The flowering stage is particularly heat sensitive and the appropriate temperature is critical for flowering time regulation (Lee et al., 2008). Several ambient temperature responsive miRNAs, including miR156, miR163, miR169, miR172, miR398 and miR399, have been identified which are associated with flowering time (Lee et al., 2010). However, the predicted rise in global temperatures will implications for flowering time and fertility, impacting the yields. Unravelling the molecular mechanisms involved in both heat stress as well ambient temperature control of flowering is essential for understanding the impact of rising temperatures on anthesis and flowering time.

A key miRNA associated with ambient temperature flowering responses, miR156 (Lee et al., 2010) is highly conserved across angiosperms (Axtell et al., 2007). It plays an important role in the timing of developmental transitions, including flowering, by targeting squamosa-promoter-binding-protein-like transcription factor (SPL) levels (Wu and Poethig, 2006), which in turn activate the flowering gene fruitful (FUL) (Wang et al., 2009). The miR156-SPL module that controls flowering time by directly regulating flowering locus T (FT) expression is highly temperature sensitive (Lee et al., 2010; Kim et al., 2012). Overexpression of mi156 in *A. thaliana* resulted in delayed flowering at ambient temperature while overexpression of miR156-resistant

SPL3 resulted in early flowering, irrespective of temperature (Kim *et al.*, 2012). In addition to ambient temperature responses, this miRNA is highly induced by heat shock and is involved in promoting the sustained expression of heat shock proteins, playing an important role in heat tolerance memory in *Arabidopsis* by repressing SPL2 and SPL11 (Steif *et al.*, 2014). MiR156 is also up-regulated in response to heat stress in several other crops, including *Brassica rapa* (Yu *et al.*, 2012) and wheat (Kumar *et al.*, 2015) and down-regulated in *Saccharina japonica* (Liu *et al.*, 2015) and rice (Liu *et al.*, 2017), indicating species differences in heat responsive mechanisms.

Small RNA profiling, comparing heat tolerant and sensitive rice at the flowering stage, identified a greater number of differentially expressed miRNAs in the heat tolerant genotype, with conserved miRNAs, including miR167, miR168, miR160 and miR166, having higher expression levels in the heat tolerant genotype (Liu *et al.*, 2017). Most of these miRNAs were down-regulated by heat stress, consistent with other species, such as *Populus tomentosa* (Chen *et al.*, 2012) and *Brassica rapa* (Yu *et al.*, 2012), which saw most miRNAs down-regulated in response to heat stress. Further QTL mapping experiments identified candidate miRNAs and their associated target genes correlated with the heat tolerant phenotype, including miR169r-5p, which showed significant up-regulated in both genotypes. However, while LOC_Os12g42400, its target gene, was significantly down-regulated in the heat-tolerant genotype, it was unchanged in the heat-sensitive genotype (Liu *et al.*, 2017). Its role was confirmed by overexpressing miR169r-5p which produced plants with higher levels of spikelet fertility compared to the wild type. Studies such as this, comparing genotypes of different tolerance levels, demonstrate the value as well of the complexity of considering the genotype specific patterns of miRNA expression in response to heat stress. Heat stress can cause early flowering and overexpression of miR169 results in stress-induced early flowering in *Arabidopsis* by targeting FLC via the AtNF-YA2 transcription factor (Xu *et al.*, 2014). This suggests a major role for the miR169/NF-YA module in stress-induced flowering time.

Maize plants are susceptible to reduced pollen viability and grain yield at temperatures above 37°C and are particularly sensitive at the reproductive stage (Waqas *et al.*, 2021). RNAseq data at different vegetative and reproductive stages over both short- and longer-term heat stress (38°C) revealed miRNA and miRNAs* that had not previously been reported as responding to heat stress (He *et al.*, 2019). Longer term stress induced more miRNAs compared to shorter term stress and interestingly, a significant biological role was found for miRNA168* and miR528*. The expression patterns of miR168* and miR528* showed tissue-specific patterns of expression, and degradome analyses revealed potential gene targets indicating an important biological role for these two miRNA* in heat stress in maize by regulating their target genes. Similarly, miR169 was unaffected by heat stress in switchgrass while miR169* was down-regulated under high temperature stress (Hivrale *et al.*, 2016).

3.6 Heat Stress Cross Talk: miRNAs Responsive to Combined Stresses

Plants are simultaneously exposed to multiple stresses and to generate realistic data, the impact of different stresses must be considered. MiRNAs are known to respond differently to different stresses, for example, miR156 is strongly repressed by heat stress (Lee *et al.*, 2010) and up-regulated by drought (Reyes and Chua, 2007). The miR156, miR529 and miR535 super families in plants have very high sequence similarity and all target SPL genes (Zheng *et al.*, 2013). Overexpression of miR156 in alfalfa showed increased drought tolerance partly by silencing SPL13 (Arshad *et al.*, 2017), and increased heat tolerance using SPL13 RNAi knockdown, demonstrating that the miR156/SPL13 network increases stress tolerance in alfalfa (Matthews *et al.*, 2019). High-throughput sequencing of switchgrass leaves, subjected to either heat or drought stress, identified 13 conserved miRNA families differentially expressed between these stresses (Hivrale *et al.*, 2016). Of the miRNAs that responded in a similar way to both stresses, heat stress produced a stronger response. *Arabidopsis* seedlings subjected to drought, salt or heat stress revealed numerous miRNAs which responded to each stress (Pegler *et al.*, 2019). Different responses to the three stresses by the miRNAs indicated the presence of multiple cis elements in the miRNA promoter regions. The differential expression of several conserved miRNAs in safflower seedings, including miR156, miR162, miR164, miR166, miR172, miR398 and miR408 regulates their target gene expression, which is associated with heat, cold, salinity and drought stress (Kouhi *et al.*, 2020). Ectopic expression of *Solanum habrochaites* sha-miR319d in cultivated tomato conferred both chilling and heat stress tolerance (Shi *et al.*, 2019). The expression of genes involved in heat, chilling and reactive oxygen species signalling suggested that sha-miR319d may regulate temperature responses in tomato via heat, chilling and ROS signalling. Multiple stresses are more detrimental to the plant (Handayani and Watanabe, 2020); however, most studies focus on single stresses. A recent study profiling miRNA expression in potato genotypes with contrasting tolerance levels subjected plants to combined heat and drought stress (Öztürk Gökçe *et al.*, 2021). Common and stress specific miRNAs were identified, including several novel miRNAs which were induced by the combined stress while higher numbers of miRNAs and were expressed in a tolerant variety, indicating genotype specific miRNA profiles. Similar results were found using RNA deep sequencing which identified 335 known and 430 novel miRNAs in tomato plants subjected to simultaneous heat and drought stress (Zhou *et al.*, 2020). A unique miRNA profile was seen in plants subjected to combined stress compared to single stresses and target genes which were primarily associated with transcription factor activity, transcriptional regulation, or sequence specific DNA-binding. This unique miRNA profile in response to combined heat and drought has been reported in several other crops, including sorghum, which showed miRNA regulation was stronger for the combined stress or heat alone as compared to drought (Puli *et al.*, 2021). Studies like these highlight the advantages of combined stress experiments, which can reveal specific stress miRNA patterns and species-specific miRNAs. miRNAs, such as stu-miR160a-5p, stu-miR156a, stu-miR162a-3p, which are known to be members of heat inducible families in several plant species (Lin *et al.*, 2018; Kouhi *et al.*, 2020; Zheng *et al.*, 2013) have shown different expression patterns for single and combined stresses (Öztürk Gökçe *et al.*, 2021).

3.7 Implications for Crop Improvement

Since miRNAs were first discovered in plants (Reinhart *et al.*, 2002), there has been considerable interest in their functions in regulating plant responses to abiotic stress and their potential use as targets for breeding and crop improvement. The availability of transcriptome and degradome information for many miRNAs and their target genes in a range of plant species, offers new tools for plant biotechnology and breeding. miRNAs are promising targets for crop heat tolerance improvement as they can target multiple genes, including important heat stress transcription factors, such as HSFA1, simultaneously (Guan *et al.*, 2013; Puli *et al.*, 2021). miRNA overexpression has been successfully used in several crops to improve heat stress tolerance, including creeping bentgrass (Zhao *et al.*, 2019), alfalfa (Arshad *et al.*, 2017) and *Arabidopsis* (Lin *et al.*, 2018); however, pleiotropic effects can be a problem (Zhao *et al.*, 2019). Global up-regulation using constituent promoters, such as CAMV, can have unwanted pleiotropic effects; so tissue specific promoters can sometimes be more effective. When miR99, involved in the P starvation response (Pant *et al.*, 2008) was constituently over-expressed, P toxicity and stunting was seen in several species, including tomato (Gao *et al.*, 2010) and *Arabidopsis* (Fujii *et al.*, 2005). However, when the rd29A inducible promoter was used to overexpress ath-miR399d, it successfully improved tolerance to cold and salinity and P deficiency in tomato (Gao *et al.*, 2015). Several different strategies have been used for miRNA genetically-modified crop development, including overexpression and RNAi (Matthews *et al.*, 2019), target mimicry (Yan *et al.*, 2012) and artificial miRNAs (amiRNAs) (Schmollinger *et al.*, 2010).

Lack of knowledge of the complex miRNA gene regulation networks can hinder the effective use of miRNAs to target heat tolerance traits. Traditionally genetic mutants are used to examine gene function, but miRNA families have multiple members with overlapping roles and can regulate multiple genes, making this approach challenging. miRNAs, originally designed for silencing protein coding genes in plants (Schwab *et al.*, 2006), can circumvent these issues by directly targeting heat stress responsive miRNAs. The role of HSF1 in Chlamydomonas thermotolerance was confirmed using amiRNA (Schmollinger *et al.*, 2010). Transgenic *Arabidopsis* plants containing an artificial miR160 (MIM160), mimicking an miR160 inhibitor, showed reduced levels of seed germination and lower overall survival in response to heat stress (Lin *et al.*, 2018). Short tandem target mimics have been used successfully to degrade specific miRNA families, such as the miR165/166 family, in *Arabidopsis*, allowing an entire miRNA family to be blocked simultaneously (Yan *et al.*, 2012). CRISPR has been used effectively in rice to knock out both single miRNA genes, OsMIR408 and OsMIR528, and miR815 and miR820 gene families (Zhou *et al.*, 2017).

Several miRNAs are widely reported as responsive to heat stress and are important targets for genetic modification of crop plants for improved heat stress tolerance. The miR156-SPL module is highly expressed during heat stress and activation persists after stress recovery playing a role in heat stress memory (Stief *et al.*, 2014). Overexpression improves heat tolerance in wheat (Matthews *et al.*, 2019) and alfalfa (Arshad *et al.*, 2017). The rice OsmiR156 regulates OsSPL14 and a point mutation in this gene disrupts OsmiR156 regulation, resulting in plants with reduced tiller numbers and higher grain yield, indicating that miR156 is negatively controlling yield in rice (Jiao *et al.*, 2010). Another important heat responsive miRNA, miR398, targeting CSDs involved in ROS reduction under heat stress (Sunkar *et al.*, 2006), are an important part of the heat stress tolerance response in several crops, including *P. tomentosa* and *B. rapa* (Yu *et al.*, 2012). MiR393 has been shown to be a regulator of responses to multiple stresses in both rice and *Arabidopsis* (Gao *et al.*, 2011; Xia *et al.*, 2012) and of Osa-miR393a in creeping bentgrass enhances tolerance to heat, drought and overexpression salinity stress (Zhao *et al.*, 2019). Increased heat tolerance was associated with elevated levels of AsHSP17.0 and AsHSP26.7a and is a promising candidate for improvement in multiple abiotic stresses. Other important heat responsive miRNAs which alter plant heat tolerance when over-expressed, include miR169 (Liu *et al.*, 2017) and miR160 (Lin *et al.*, 2018).

4. Conclusion and Perspectives

Climate change and increasing global temperatures are a threat to plant productivity and yield. High temperature stress is one of the most damaging of the abiotic stresses, resulting in cellular, biochemical and physiological changes, alternations to flowering time and fertility and sensitivity to biotic stresses (Saini *et al.*, 2022). Rapid advances in several sequencing technologies have allowed enormous improvements in our understanding of the role of miRNAs in plant responses to abiotic stress. Furthermore, the use of degradome sequencing has allowed target genes to be characterised (Ravichandran *et al.*, 2019). miRNAs are central to complex gene networks responding to heat stress which makes them targets for plant improvement; however, the bottleneck now comes in understanding these targets and the wider gene networks. Rather than functioning directly in plant responses to abiotic stress, miRNAs regulate important components of plant gene networks by controlling gene expression, either by inhibiting protein translation or by targeting mRNAs for cleavage. These tend to target transcription factors, allowing for rapid and specific responses to stress (Janni *et al.*, 2020). There now needs to be a shift in focus away from surveying and identifying miRNAs under heat stress to functional analysis and unravelling the complex associated gene networks. With increasing global temperatures, more studies comparing heat stress over longer time periods would also give important insights into future crop performances. There are a few studies looking specifically at crop improvement rather than miRNA functions and designing studies with this focus would add to the understanding of the role of miRNAs in crop heat stress tolerance. The detailed molecular advances that will come from these studies will help to generate novel heat tolerant germplasm and reduce the impacts of heat-induced crop losses.

References

Achard, P., Herr, A., Baulcombe, D.C. and Harberd, N.P. (2004). Modulation of floral development by a gibberellin-regulated microRNA, *Development*, 131: 3357-3365.

Adams, S.R., Cockshull, K.E. and Cave, C.R.J. (2001). Effect of temperature on the growth and development of tomato fruits, *Ann. Bot.*, 88: 869-877.

Ahmed, W., Li, R., Xia, Y., Bai, G., Siddique, H.M., Zhang, K., Zheng, H., Yang, Y., X. and Guo, P. (2020). Comparative analysis of miRNA expression profiles between heat-tolerant and heat-sensitive genotypes of flowering Chinese cabbage under heat stress using high-throughput sequencing, *Genes*, 11: 264.

Akdogan, G., Tufekci, E.D., Uranbey, S. and Unver, T. (2016). miRNA-based drought regulation in wheat, *Funct. Integr. Genom.*, 16: 221-233.

Arshad, M., Feyissa, B.A., Amyot, L., Aung, B. and Hannoufa, A. (2017). MicroRNA156 improves drought stress tolerance in alfalfa (*Medicago sativa*) by silencing SPL13, *Plant Sci.*, 258: 122-136.

Asada, K. (2006). Production and scavenging of reactive oxygen species in chloroplasts and their functions, *Plant Physiol.*, 141: 391-396.

Axtell, M.J., Snyder, J.A. and Bartel, D.P. (2007). Common functions for diverse small RNAs of land plants, *Plant Cell*, 19: 1750-1769.

Aydinoglu, F. (2020). Elucidating the regulatory roles of microRNAs in maize (*Zea mays* L.) leaf growth response to chilling stress, *Planta*, 251: 38.

Bai, Q., Wang, X., Chen, X., Shi, G., Liu, Z., Guo, C. and Xiao, K. (2018). Wheat miRNA TaemiR408 acts as an essential mediator in plant tolerance to Pi deprivation and salt stress via modulating stress-associated physiological processes, *Front. Plant Sci.*, 9: 499.

Baniwal, S.K., Chan, K.Y., Scharf, K.D. and Nover, L. (2007). Role of heat stress transcription factor HsfA5 as specific repressor of HsfA4, *J. Biol. Chem.*, 282: 3605-3613.

Barciszewska-Pacak, M., Milanowska, K., Knop, K., Bielewicz, D., Nuc, P., Plewka, P., Pacak, A.M., Vazquez, F., Karlowski, W., Jarmolowski, A. and Szweykowska-Kulinska, Z. (2015). *Arabidopsis* microRNA expression regulation in a wide range of abiotic stress responses, *Front. Plant Sci.*, 6: 410.

Barrera-Figueroa, B.E., Gao, L., Wu, Z., Zhou, X., Zhu, J., Jin, H., Liu, R. and Zhu, J.K. (2012). High throughput sequencing reveals novel and abiotic stress-regulated microRNAs in the inflorescences of rice, *BMC Plant Biol.*, 12: 132.

Bartel, D.P. (2004). MicroRNAs: Genomics, biogenesis, mechanism, and function, *Cell*, 116: 281-297.

Beudez, N. and Doussan, C. (2013). Influence of three root spatial arrangement on soil water flow and uptake. Results from an explicit and an equivalent, upscaled, model, *Procedia Environ. Sci.*, 19: 37-46.

Brown, J.A., Li, D. and Ic, M. (1993). Heat shock induction of manganese peroxidase gene transcription in *Phanerochaete chrysosporium*, *Appl. Environ. Microbiol.*, 59: 4295-4299.

Cai, H., Lu, Y., Xie, W., Zhu, T. and Lian, X. (2012). Transcriptome response to nitrogen starvation in rice, *J. Biosci.*, 37: 731-747.

Camejo, D., Rodríguez, P., Morales, M.A., Dellamico, J.M., Torrecillas, A. and Alarcon, J.J. (2005). High temperature effects on photosynthetic activity of two tomato cultivars with different heat susceptibility, *J. Plant Physiol.*, 162: 281-289.

Cao, Z.H., Zhang, S.Z., Wang, R.K., Zhang, R.F. and Hao, Y.J. (2013). Genome wide analysis of the apple MYB transcription factor family allows the identification of MdoMYB121 gene conferring abiotic stress tolerance in plants, *PLoS ONE*, 8: e69955.

Cao, J., Gulyás, Z., Kalapos, B., Boldizsár, Á., Liu, X., Pál, M., Yao, Y., Galiba, G. and Kocsy, G. (2019). Identification of a redox-dependent regulatory network of miRNAs and their targets in wheat, *J. Exp. Bot.*, 70: 85-99.

Carrington, J.C. and Ambros, V. (2003). Role of microRNAs in plant and animal development, *Science*, 301: 336-338.

Charng, Y.-Y., Liu, H.-C., Liu, N.-Y., Chi, W.-T., Wang, C.-N., Chang, S.-H. and Wang, T.-T. (2006). A heat-inducible transcription factor, HsfA2, is required for extension of acquired thermotolerance in *Arabidopsis*, *Plant Physiology*, 143(1): 251–262. doi:10.1104/pp.106.091322

Chen, L., Ren, Y., Zhang, Y., Xu, J., Sun, F., Zhang, Z. and Wang, Y. (2012). Genome-wide identification and expression analysis of heat-responsive and novel microRNAs in *Populus tomentosa*, *Gene*, 504: 160-165.

Chinnusamy, V., Zhu, J., Zhou, T. and Zhu, J.K. (2007). Small RNAs: Big role in abiotic stress tolerance of plants. *In:* Jenks, M.A., Hasegawa, P.M., Jain, S.M. (Eds.). *Advances in Molecular Breeding toward Drought and Salt-tolerant Crops*. Springer, Dordrecht, 223-260.

Cottee, N.S., Wilson, I.W., Tan, D.K.Y. and Bange, M.P. (2014). Understanding the molecular events underpinning cultivar differences in the physiological performance and heat tolerance of cotton (*Gossypium hirsutum*), *Funct. Plant Biol.*, 41: 56-67.

Cuperus, J.T., Fahlgren, N. and Carrington, J.C. (2011). Evolution and functional diversification of miRNA genes, *Plant Cell*, 23: 431-442.

Das Gupta, M. and Nath, U. (2015). Divergence in patterns of leaf growth polarity is associated with the expression divergence of miR396, *The Plant Cell*, 27: 2785-2799.

Deng, P., Wang, L., Cui, L., Feng, K., Liu, F., Du, X., Tong, W., Nie, X., Ji, W. and Weining, S. (2015). Global identification of MicroRNAs and their targets in barley under salinity stress, *PLoS ONE*, 10: e0137990.

Ding, Y., Ma, Y., Liu, N., Xu, J., Hu, Q., Li, Y., Wu, Y., Xie, S., Zhu, L., Min, L. and Zhang, X. (2017). MicroRNAs involved in auxin signalling modulate male sterility under high-temperature stress in cotton (*Gossypium hirsutum*), *The Plant J.*, 91: 977-994.

Djami-Tchatchou, A.T., Sanan-Mishra, N., Ntushelo, K. and Dubery, I.A. (2017). Functional roles of microRNAs in agronomically important plants, potential as targets for crop improvement and protection, *Front. Plant Sci.*, 8: 378. Doi: 10.3389/fpls.2017.00378

El Sanousi, R.S., Sanan-Mishra, N., Abdelmula, A.A., Mohammed, I.A., Seif, M.G. and Hamza, N.B. (2016). Differential expression of miRNAs in sorghum bicolor under drought and salt stress, *Am. J. Plant Sci.*, 7: 870-878. Doi: 10.4236/ajps.2016.76082

Ezin, V., Ahanchede, W.W., Ayenan, M.A.T. and Ahanchede, A. (2022). Physiological and agronomical evaluation of elite rice varieties for adaptation to heat stress, *BMC Plant Biol.*, 22: 236. https://doi.org/10.1186/s12870-022-03604-x

Fahad, S., Bajwa, A.A., Nazir, U., Anjum, S.A., Farooq, A., Zohaib, A., Sadia, S., Nasim, W., Adkins, S. and Saud, S. (2017). Crop production under drought and heat stress: Plant responses and management options, *Frontiers in Plant Science*, 8: 1147.

Fujii, H., Chiou, T.J., Lin, S.I., Aung, K. and Zhu, J.K. (2005). A miRNA involved in phosphate starvation response in *Arabidopsis*, *Curr. Biol.*, 15: 2038-2043.

Gao, P., Bai, X., Yang, L., Lv, D., Pan, X., Li, Y., Cai, H., Ji, W., Chen, Q. and Zhu, Y. (2011). Osa-MIR393: A salinity- and alkaline stress-related microRNA gene, *Mol. Biol. Rep.*, 38: 237-242.

Gao, N., Qiang, X.M., Zhai, B.N., Min, J. and Shi, W.M. (2015). Transgenic tomato overexpressing ath-miR399d improves growth under abiotic stress conditions, *Russ. J. Plant Physiol.*, 62: 360-366.

Gao, N., Su, Y.H., Min, J., Shen, W.S. and Shi, W.M. (2010). Transgenic tomato overexpressing athmiR399d has enhanced phosphorus accumulation through increased acid phosphatase and proton secretion as well as phosphate transporters, *Plant Soil*, 334: 123-136.

Giacomelli, J.I., Weigel, D., Chan, R.L. and Manavella, P.A. (2012). Role of recently evolved miRNA regulation of sunflower HaWRKY6 in response to temperature damage, *New Phytol.*, 195: 766-773.

Goel, S., Goswami, K., Pandey, V.K., Pandey, M. and Sanan-Mishra, N. (2019). Identification of microRNA-target modules from rice variety Pusa Basmati-1 under high temperature and salt stress, *Funct. Integr. Genom.*, 19: 867-888.

Gomez-Pastor, R., Burchfiel, E.T. and Thiele, D.J. (2018). Regulation of heat shock transcription factors and their roles in physiology and disease, *Nat. Rev. Mol. Cell Biol.*, 19: 4.

Guan, Q., Lu, X., Zeng, H., Zhang, Y. and Zhu J. (2013). Heat stress induction of miR398 triggers a regulatory loop that is critical for thermotolerance in *Arabidopsis*, *Plant J.*, 74: 840-851.

Guan, Q., Yue, X., Zeng, H. and Zhu, J. (2014). The protein phosphatase RCF2 and its interacting partner NAC019 are critical for heat stress-responsive gene regulation and thermotolerance in *Arabidopsis*, *Plant Cell*, 26: 438-453.

Guo, M., Liu, J.H., Ma, X., Luo, D.X., Gong, Z.H. and Lu, M.H. (2016). The plant heat stress transcription factors (HSFs): Structure, regulation, and function in response to abiotic stresses, *Frontiers in Plant Science*, 7: 114.

Gutierrez, L., Mongelard, G., Floková, K., Pacurar, D.I., Novák, O., Staswick, P., Kowalczyk, M., Pacurar, M., Demailly, H., Geiss, G. and Bellini, C. (2012). Auxin controls Arabidopsis adventitious root initiation by regulating jasmonic acid homeostasis, *Plant Cell*, 24(6): 2515-2527.

Handayani, T. and Watanabe, K. (2020). The combination of drought and heat stress has a greater effect on potato plants than single stresses, *Plant Soil Environ.*, 66: 175-182.

Haslbeck, M. and Vierling, E. (2015). A first line of stress defence: Small heat shock proteins and their function in protein homeostasis, *J. Mol. Biol.*, 427: 1537-1548.

He, X., Guo, S., Wang, Y., Wang, L., Shu, S. and Sun, J. (2020). Systematic identification and analysis of heat-stress-responsive lncRNAs, circRNAs and miRNAs with associated co-expression and ceRNA networks in cucumber (*Cucumis sativus* L.), *Physiol. Plant.*, 168: 736-754.

He, J., Jiang, Z., Gao, L., You, C., Ma, X., Wang, X., Xu, X., Mo, B., Chen, X. and Liu, L. (2019). Genome-wide transcript and small RNA profiling reveals transcriptomic responses to heat stress, *Plant Physiol.*, 181: 609-629.

Herouart, D.V.M.M. and Inze, D. (1994). Developmental and environmental regulation of the *Nicotiana plumbaginifolia* cytosolic Cu/Zn-superoxide dismutase promoter in transgenic tobacco, *Plant Physiol.*, 104: 873-880.

Hilker, M., Schwachtje, J., Baier, M., Balazadeh, S., Bäurle, I., Geiselhardt, S., Hincha, D.K., Kunze, R., Mueller-Roeber, B., Rillig, M.C., Rolff, J., Romeis, T., Schmülling, T., Steppuhn, A., van Dongen, J., Whitcomb, S.J., Wurst, S., Zuther, E. and Kopka, J. (2016). Priming and memory of stress responses in organisms lacking a nervous system, *Biological Reviews of the Cambridge Philosophical Society*, 91: 1118-1133.

Hivrale, V., Zheng, Y., Puli, C.O.R., Jagadeeswaran, G., Gowdu, K., Kakani, V.G., Barakat, A. and Sunkar, R. (2016). Characterisation of drought- and heat-responsive microRNAs in switchgrass, *Plant Sci.*, 242: 214-223.

Hossain, A., Teixeira da Silva, J.A., Lozovskaya, M.V. and Zvolinsky, V.P. (2012). High temperature combined with drought affect rainfed spring wheat and barley in South-Eastern Russia: I. Phenology and growth, *Saudi J. Biol. Sci.*, 19: 473-487.

Howarth, C.J. (2005). Genetic improvements of tolerance to high temperature. *In:* Ashraf, M., Harris, P.J.C. (Eds.). *Abiotic Stresses: Plant Resistance Through Breeding and Molecular Approaches*, Howarth Press Inc., New York.

Hu, Y., Fragkostefanakis, S., Schleiff, E. and Simm, S. (2020). Transcriptional basis for differential thermosensitivity of seedlings of various tomato genotypes, *Genes*, 11: 655.

Huang, Y.C., Niu, C.Y., Yang, C.R. and Jinn, T.L. (2016). The heat stress factor HSFA6b connects ABA signalling and ABA-mediated heat responses, *Plant Physiology*, 172: 1182-1199.

Huijser, P. and Schmid, P. (2011). The control of developmental phase transitions in plants, *Development*, 138: 4117-4129.

Ikeda, M., Mitsuda, N. and Ohme-Takagi, M. (2011). *Arabidopsis* HsfB1 and HsfB2b act as repressors of the expression of heat-inducible Hsfs but positively regulate the acquired thermotolerance, *Plant Physiology*, 157: 1243-1254.

IPCC. Climate Change (2007). *The Physical Science Basis*, Contribution of Working Group I to the Fourth Assessment Report of the Intergovernmental Panel on Climate Change, Solomon, S., Qin, D., Manning, M., Chen, Z., Marquis, M., Averyt, K.B., Tignor, M. and Miller, H.L. (Eds.). Cambridge University Press, Cambridge, UK, 2007.

IPCC. Climate Change (2014). *Synthesis Report*, Contribution of Working Groups I, II and III to the Fifth Assessment Report of the Intergovernmental Panel on Climate Change Core Writing Team, Pachauri, R.K. and Meyer, L.A. (Eds.). *IPCC*, Geneva, Switzerland, 2014, p. 151.

IPCC. (2013). Working Group I Contribution to the IPCC Fifth Assessment Report on Climate Change (2013). *The Physical Science Basis, Summary for Policymakers.* www. climatechange2013. org/images/report/WG1AR5_SPM_FINAL.pdf

Iwakawa, H. and Tomari, Y. (2013). Molecular insights into microRNA-mediated translational repression in plants, *Molecular Cell*, 52: 591-601.

Jagadish K.S.V., Craufurd P.Q., Shi W. and Oane, R. (2013). A phenotypic marker for quantifying heat stress impact during microsporogenesis in rice (*Oryza sativa* L.), *Functional Plant Biology*, 41: 48-55.

Janni, M., Gulli, M., Maestri, E., Marmiroli, M., Valliyodan, B., Nguyen, H.T. and Marmiroli, N. (2020). Molecular and genetic bases of heat stress responses in crop plants and breeding for increased resilience and productivity, *J. Exp. Bot.*, 71: 3780-3802.

Jiang, J., Lv, M., Liang, Y., Ma, Z. and Cao, J. (2014). Identification of novel and conserved miRNAs involved in pollen development in *Brassica campestris* ssp. *chinensis* by high-throughput sequencing and degradome analysis, *BMC Genomics*, 15: 146.

Jiang, W., Shi, W., Ma, X., Zhao, J., Wang, S., Tan, L., Sun, C. and Liu, F. (2019). Identification of microRNAs responding to cold stress in Dongxiang common wild rice, *Genome*, 62: 635-642.

Jiao, Y., Wang, Y., Xue, D., Wang, J., Yan, M., Liu, G., Dong, G., Zeng, D., Lu, Z., Zhu, X., Qian, Q. and Li, J. (2010). Regulation of OsSPL14 by OsmiR156 defines ideal plant architecture in rice, *Nature Genetics*, 2: 541-544.

Khan, M.S., Khraiwesh, B., Pugalenthi, G., Gupta, R.S., Singh, J., Duttamajumder, S.K. and Kapur, R. (2014). Subtractive hybridisation-mediated analysis of genes and in silico prediction of associated microRNAs under waterlogged conditions in sugarcane (*Saccharum* spp.), *FEBS Open Bio.*, 4: 533-541.

Khraiwesh, B., Zhu, J.-K. and Zhu, J. (2012). Role of miRNAs and siRNAs in biotic and abiotic stress responses of plants, *Biochimica et Biophysica Acta (BBA) – Gene Regulatory Mechanisms*, 1819(2): 137-148. doi:10.1016/j.bbagrm.2011.05.001

Kim, J.J., Lee, J.H., Kim, W., Jung, H.S., Huijser, P. and Ahn, J.H. (2012). The microRNA156-squamosa promoter binding protein-like3 module regulates ambient temperature-responsive flowering via flowering locus T. in *Arabidopsis*, *Plant Physiol.*, 159: 461-478.

Kotak, S., Larkindale, J., Lee, U., von Koskull-Döring, P., Vierling, E. and Scharf, K.D. (2007). Complexity of the heat stress response in plants, *Curr. Opin. Plant Biol.*, 10: 310-316.

Kouhi, F., Sorkheh, K., Ercisli, S. and Kumar, K. (2020). MicroRNA expression patterns unveil differential expression of conserved miRNAs and target genes against abiotic stress in safflower, *PLoS ONE*, 15: e0228850.

Kruszka, K., Pacak, A., Swida-Barteczka, A., Nuc, P., Alaba, S., Wroblewska, Z., Karlowski, W., Jarmolowski, A. and Szweykowska-Kulinska, Z. (2014). Transcriptionally and post-transcriptionally regulated microRNAs in heat stress response in barley, *J. Exp. Bot.*, 65: 6123-6135.

Kumar, R.R., Goswami, S., Sharma, S.K., Singh, K., Gadpayle, K.A., Kumar, N., Rai, G.K., Singh, M. and Rai, R.D. (2012). Protection against heat stress in wheat involves change in cell membrane stability, antioxidant enzymes, osmolyte, H_2O_2 and transcript of heat shock protein, *Int. J. Plant Physiol. Biochem.*, 4: 83-91. Doi:10.5897/IJPPB12.008

Kumar, R.R., Pathak, H., Sharma, S.K., Kale, Y.K., Nirjal, M.K. and Singh, G.P. (2015). Novel and conserved heat-responsive micro RNAs in wheat (*Triticum aestivum* L.), *Funct. Integr. Genomics*, 15: 1-26.

Küpers, J.J., Oskam, L. and Pierik, R. (2020). Photoreceptors regulate plant developmental plasticity through auxin, *Plants*, 9: 940.

Kushawaha, A.K., Khan, A., Sopory, S.K. and Sanan-Mishra, N. (2021). Priming by high temperature stress induces MicroRNA regulated heat shock modules indicating their involvement in thermopriming response in rice, *Life* (Basel), 11: 291.

Larkindale, J., Hall, J.D., Knight, M.R. and Vierling, E. (2005). Heat stress phenotypes of *Arabidopsis* mutants implicate multiple signalling pathways in the acquisition of thermotolerance, *Plant Physiol.*, 138: 882-897.

Larkindale, J. and Knight, M.R. (2002). Protection against heat stress-induced oxidative damage in *Arabidopsis* involves calcium, abscisic acid, ethylene and salicylic acid, *Plant Physiol.*, 128: 682-695.

Lee, H., Yoo, S.J., Lee, J.H., Kim, W., Yoo, S.K., Fitzgerald, H., Carrington, J.C. and Ahn, J.H. (2010). Genetic framework for flowering-time regulation by ambient temperature-responsive miRNAs in *Arabidopsis*, *Nucleic Acids Res.*, 38: 3081-3093.

Lee, J.H., Lee, J.S. and Ahn, J.H. (2008). Ambient temperature signalling in plants: An emerging field in the regulation of flowering time, *J. Plant Biol.*, 51: 321-326.

Li, Y., Fu, Y., Ji, L., Wu, C.A. and Zheng, C. (2010). Characterisation and expression analysis of the *Arabidopsis* mir169 family, *Plant Sci.*, 178: 271-280.

Li, N., Euring, D., Cha, J.Y., Lin, Z., Lu, M., Huang, L.J. and Kim, W.Y. (2021). Plant Hormone-Mediated Regulation of Heat Tolerance in Response to Global Climate Change, *Front. Plant Sci.*, 11: 627969.

Li, H., Wang, Y., Wang, Z., Guo, X., Wang, F., Xia, X.J., Zhou, J., Shi, K., Yu, J.Q. and Zhou, Y.H. (2016). Microarray and genetic analysis reveals that csa-miR159b plays a critical role in abscisic acid-mediated heat tolerance in grafted cucumber plants, *Plant Cell Environ.*, 39: 1790-804.

Li, W., Jia, Y., Liu, F., Wang, F., Fan, F., Wang, J., Zhu, J., Xu, Y., Zhong, W. and Yang, J. (2019). Integration Analysis of small RNA and degradome sequencing reveals microRNAs responsive to *Dickeya zeae* in resistant rice, *Int. J. Mol. Sci.*, 20: 222.

Liang, G., He, H. and Yu, D. (2012). Identification of nitrogen starvation responsive miRNAs in *Arabidopsis thaliana*, *PLoS ONE*, 7: e48951.

Lin, J.S., Kuo, C.C., Yang, I.C., Tsai, W.A., Shen, Y.H., Lin, C.C., Liang, Y.C., Li, Y.C., Kuo, Y.W., King, Ling, Y., Serrano, N., Gao, G., Atia, M., Mokhtar, M., Woo, Y.H., Bazin, J., Veluchamy, A., Benhamed, M., Crespi, M., Gehring, C., Reddy, A.S.N. and Mahfouz, M.M. (2018). Thermopriming triggers splicing memory in *Arabidopsis*, *J. Exp. Bot.*, 69: 2659-2675.

Liu, A., Zhou, Z., Yi, Y. and Chen, G. (2020). Transcriptome analysis reveals the roles of stem nodes in cadmium transport to rice grain, *BMC Genom.*, 21: 116.

Liu, F., Wang, W., Sun, X., Liang, Z. and Wang, F. (2015). Conserved and novel heat stress responsive microRNAs were identified by deep sequencing in *Saccharina japonica* (*Laminariales phaeophyta*), *Plant Cell Environ.*, 38: 1357-1367.

Liu, H., Able, A.J. and Able, J.A. (2019). Genotypic performance of Australian durum under single and combined water-deficit and heat stress during reproduction, *Sci. Rep.*, 9: 1-17.

Liu, Q., Yang, T., Yu, T., Zhang, S., Mao, X., Zhao, J., Wang, X., Dong, J. and Liu, B. (2017). Integrating small RNA sequencing with QTL mapping for identification of miRNAs and their target genes associated with heat tolerance at the flowering stage in rice, *Front. Plant Sci.*, 8: 43.

Liu, X., Yu, W., Zhang, X., Wang, G., Cao, F. and Cheng, H. (2013). Identification and expression analysis under abiotic stress of the R2R3-MYB genes in *Ginkgo biloba* L., *Physiology and Molecular Biology in Plants*, 23: 503-516.

Lu, Y., Feng, Z., Liu, X., Bian, L., Xie, H., Zhang, C., Mysore, K.S. and Liang, J. (2018). MiR393 and miR390 synergistically regulate lateral root growth in rice under different conditions, *BMC Plant Biol.*, 18: 261.

Liu, Z., Wang, X., Chen, X., Shi, G., Bai, Q. and Xiao, K. (2018). TaMIR1139: A wheat miRNA responsive to Pi-starvation, acts a critical mediator in modulating plant tolerance to Pi deprivation, *Plant Cell Rep.*, 37: 12931309.

Macovei, A. and Tuteja, N. (2012). MicroRNAs targeting DEAD-box helicases are involved in salinity stress response in rice (*Oryza sativa* L.), *BMC Plant Biol.*, 12: 183.

Mangrauthia, S.K., Bhogireddy, S., Agarwal, S., Prasanth, V.V., Voleti, S.R., Neelamraju, S. and Subrahmanyam D. (2017). Genome-wide changes in microRNA expression during short and prolonged heat stress and recovery in contrasting rice cultivars, *J. Exp. Bot.*, 68: 2399-2412.

Matthews, C., Arshad, M. and Hannoufa, A. (2019). Alfalfa response to heat stress is modulated by microRNA156, *Physiol. Plant*, 165: 830-842.

Mauch-Mani, B., Baccelli, I., Luna, E. and Flors, V. (2017). Defence priming: An adaptive part of induced resistance, *Annual Review of Plant Biology*, 68: 485-512.

Miller, G., Suzuki, N., Ciftci-Yilmaz, S. and Mittler, R. (2010). Reactive oxygen species homeostasis and signalling during drought and salinity stresses, *Plant Cell Environ.*, 33: 453-467.

Mishra, S.K., Tripp, J., Winkelhaus, S., Tschiersch, B., Theres, K., Nover, L. and Scharf, K.D. (2002). In the complex family of heat stress transcription factors, HsfA1 has a unique role as master regulator of thermotolerance in tomato, *Genes Dev.*, 16: 1555-1567.

Moldovan, D., Spriggs, A., Yang, J., Pogson, B.J., Dennis, E.S. and Wilson, I.W. (2010). Hypoxia-responsive microRNAs and trans-acting small interfering RNAs in *Arabidopsis*, *J. Exp. Bot.*, 61: 165-177.

Nie, Z., Ren, Z., Wang, L., Su, S., Wei, X., Zhang, X., Wu, L., Liu, D., Tang, H., Liu, H., Zhang,

S. and Gao, S. (2016). Genome-wide identification of microRNAs responding to early stages of phosphate deficiency in maize, *Physiol. Plant*, 157: 161-174.

Niu, Y. and Xiang, Y. (2018). An overview of biomembrane functions in plant responses to high-temperature stress, *Front Plant Sci.*, 9: 915. Doi:10.3389/fpls.2018.00915

Nover, L., Bharti, K., Döring, P., Mishra, S.K., Ganguli, A. and Scharf, K.-D. (2001). Arabidopsis and the heat stress transcription factor world: How many heat stress transcription factors do we need? *Cell Stress & Chaperones*, 6(3): 177. doi:10.1379/1466-1268(2001)006<0177:aa thst>2.0.co

Ohama, N., Sato, H., Shinozaki, K. and Yamaguchi-Shinozaki, K. (2017). Transcriptional regulatory network of plant heats stress response, *Trends. Plant. Sci.*, 22: 53-65.

Ortiz, C. and Cardemil, L. (2001). Heat-shock responses in two leguminous plants: A comparative study, *Journal of Experimental Botany*, 52(361): 1711-1719.

Oshino, T., Miura, S., Kikuchi, S., Hamada, K., Yano, K., Watanabe, M. and Higashitani, A. (2011). Auxin depletion in barley plants under high-temperature conditions represses DNA proliferation in organelles and nuclei via transcriptional alterations, *Plant Cell Environ.*, 34: 284-290.

Öztürk Gökçe, Z.N., Aksoy, E., Bakhsh, A., Demirel, D., Çalışkan, S and Çalışkan. M.E. (2021). Combined drought and heat stresses trigger different sets of miRNAs in contrasting potato cultivars, *Funct. Integr. Genomics*, 21: 489-502.

Panchuk, I.I., Volkov, R.A. and Schöffl, F. (2002). Heat stress- and heat shock transcription factor-dependent expression and activity of ascorbate peroxidase in *Arabidopsis*, *Plant Physiol.*, 129: 838-853.

Pandita, D. (2019). Plant MIRnome: miRNA biogenesis and abiotic stress response. *In:* Hasanuzzaman, M., Hakeem, K., Nahar, K. and Alharby, H. (Eds.). *Plant Abiotic Stress Tolerance*. Springer, Cham., 449-474. Doi https://Doi.org/10.1007/978-3-030-06118-0_18

Pandita, D. and Wani S.H. (2019). MicroRNA as a tool for mitigating abiotic stress in rice (*Oryza sativa* L.). *In:* Wani, S. (Ed.). *Recent Approaches in Omics for Plant Resilience to Climate Change*. Springer, Chamz., 109-133. Doi https://Doi.org/10.1007/978-3-030-21687-0_6

Pandita, D. (2021). Role of miRNAi technology and miRNAs in abiotic and biotic stress resilience. *In:* Aftab, T. and Roychoudhury, A. (Eds.). *Plant Perspectives to Global Climate Changes*. Academic Press, Elsevier, 303-330. https://Doi.org/10.1016/B978-0-323-85665-2.00015-7

Pandita, D. (2022a). How microRNAs regulate abiotic stress tolerance in wheat? A snapshot. *In:* Roychoudhury, A., Aftab, T. and Acharya, K. (Eds.). *Omics Approach to Manage Abiotic Stress in Cereals*. Springer, Singapore, 447-464. https://Doi.org/10.1007/978-981-19-0140-9_17

Pandita, D. (2022b). MicroRNAs shape the tolerance mechanisms against abiotic stress in maize. *In:* Roychoudhury, A., Aftab, T. and Acharya, K. (Eds.). *Omics Approach to Manage Abiotic Stress in Cereals*. Springer, Singapore, 479-493. https://Doi.org/10.1007/978-981-19-0140-9_19

Pandita, D. (2022c). miRNA- and RNAi-mediated metabolic engineering in plants. *In:* Aftab, T. and Hakeem, K.R. (Eds.). *Metabolic Engineering in Plants*. Springer, Singapore, 171-186. https://Doi.org/10.1007/978-981-16-7262-0_7

Pant, B.D., Buhtz, A., Kehr, J. and Scheible, W. (2008). MicroRNA399 is a long distance signal for the regulation of plant phosphate homeostasis, *Plant J.*, 53: 731-738.

Paterson, A.H., Bowers, J.E., Bruggmann, R., Dubchak, I., Grimwood, J., Gundlach, H. *et al.* (2009). The Sorghum bicolor genome and the diversification of grasses, *Nature*, 457: 551-556. Doi: 10.1038/nature07723

Pegler, J.L., Oultram, J.M.J., Grof, C.P.L. and Eamens, A.L. (2019). Profiling the abiotic stress responsive microRNA landscape of *Arabidopsis thaliana*, *Plants*, 8: 58.

Pokhrel, S. and Meyers, B.C. (2022). Heat-responsive microRNAs and phased small interfering RNAs in reproductive development of flax, *Plant Direct.*, 6(2): e385. https://doi.org/10.1002/pld3.385

Puli, C.O.R., Zheng, Y., Li, Y.F., Jagadeeswaran, G., Suo, A., Jiang, B., Sharma, P., Mann, R., Ganesan, G., Gogoi, N., Srinivasan, A., Kakani, A., Kakani, V.G., Barakat, A. and Sunkar, R. (2021). MicroRNA profiles in Sorghum exposed to individual drought or heat or their combination, *J. Plant Biochem. Biotechnol.*, 30: 848-861.

Qin, Z., Li, C., Mao, L. and Wu, L. (2014). Novel insights from non-conserved microRNAs in plants, *Front. Plant Sci.*, 5: 586. Doi: 10.3389/fpls.2014.00586.

Qu, A.-L., Ding, Y.F., Jiang, Q. and Zhu, C. (2013). Molecular mechanisms of the plant heat stress response, *Biochem. Biophys. Res. Commun.*, 432: 203-207.

Rao, S., Balyan, S., Das, J.R., Verma, R. and Mathur, S. (2021). *Transcriptional Regulation of HSFA7 and Post-transcriptional Modulation of HSFB4a by miRNA4200 Govern General and Varietal Thermotolerance in Tomato.* Doi: https://Doi.org/10.1101/2021.02.26.433069

Ravichandran, S., Ragupathy, R., Edwards, T., Domaratzki, M. and Cloutier, S. (2019). MicroRNA-guided regulation of heat stress response in wheat, *BMC Genomics*, 20: 488.

Reinhart, B.J., Weinstein, E.G., Rhoades, M.W., Bartel, B. and Bartel, D.P. (2002). MicroRNAs in plants, *Genes Dev.*, 16: 1616-26.

Reyes, J.L. and Chua, N.H. (2007). ABA induction of miR159 controls transcript levels of two MYB factors during *Arabidopsis* seed germination, *Plant J.*, 49: 592-606.

Ribeiro, T., Viegas, W. and Morais-Cecílio, L. (2009). Epigenetic marks in the mature pollen of *Quercus suber* L. (Fagaceae), *Sex. Plant Reprod.*, 22: 1-7.

Richter, K., Haslbeck, M. and Buchner, J. (2010). The heat shock response: Life on the verge of death, *Mol. Cell.* Doi:10.1016/j.molcel.2010.10.006

Rushton, P.J., Somssich, I.E., Ringler, P. and Shen, Q.J. (2010). WRKY transcription factors, *Trends in Plant Science*, 15: 247-258.

Saikumar, K. and V.D. Kumar (2014). Plant microRNAs: An overview. *In:* Kishor, P.B.K., Bandopadhyay, R. and Suravajhala, P. (Eds.). *Agricultural Bioinformatics.* Springer, New Delhi, India, 139-159.

Sailaja, B., Anjum, N., Vishnu Prasanth, V., Sarla, N., Subrahmanyam, D., Voleti, S.R., Viraktamath, B.C. and Mangrauthia, S.K. (2014). Comparative study of susceptible and tolerant genotype reveals efficient recovery and root system contributes to heat stress tolerance in rice, *Plant Molecular Biology Reporter*, 32: 10.1007/s11105-014-0728-y

Saini, N., Nikalje, G.C., Zargar, S.M. and Suprasanna, P. (2022). Molecular insights into sensing, regulation and improving of heat tolerance in plants, *Plant Cell Reports*, 41: 799-813.

Sato, S., Peet, M.M. and Gardner, R.G. (2001). Formation of parthenocarpic fruit, undeveloped flowers and aborted flowers in tomato under moderately elevated temperatures, *Scientia Hort.*, 90: 243-254.

Sato, H., Mizoi, J., Tanaka, H., Maruyama, K., Qin, F., Osakabe, Y., Morimoto, K., Ohori, T., Kusakabe, K., Nagata, M., Shinozaki, K. and Yamaguchi-Shinozaki, K. (2014). *Arabidopsis* DPB3-1, a DREB2A interactor, specifically enhances heat stress-induced gene expression by forming a heat stress-specific transcriptional complex with NF-Y subunits, *Plant Cell*, 26: 49544973

Schöffl, F., Prandl, R. and Reindl, A. (1999). Molecular responses to heat stress. *In:* Shinozaki, K. and Yamaguchi-Shinozaki, K. (Eds.). *Molecular Responses to Cold, Drought, Heat and Salt Stress in Higher Plants.* R.G. Landes Co., Austin, Texas, 81-98.

Schramm, F., Ganguli, A., Kiehlmann, E., Englich, G., Walch, D. and von Koskull-Döring, P. (2006). The heat stress transcription factor HsfA2 serves as a regulatory amplifier of a subset of genes in the heat stress response in *Arabidopsis, Plant Molecular Biology*, 60: 759-772.

Schmollinger, S., Strenkert, D. and Schroda, M. (2010). An inducible artificial microRNA system for *Chlamydomonas reinhardtii* confirms a key role for heat shock factor 1 in regulating thermotolerance, *Curr. Genet.*, 56: 383-389.

Schwab, R., Ossowski, S., Riester, M., Warthmann, N. and Weigel, D. (2006). Highly specific gene silencing by artificial MicroRNAs in *Arabidopsis, Plant Cell*, 18: 1121-1133.

Secco, D., Jabnoune, M., Walker, H., Shou, H., Wu, P., Poirier, Y. and Whelan, J. (2013). Spatio-temporal transcript profiling of rice roots and shoots in response to phosphate starvation and recovery, *Plant Cell*, 25: 4285-4304.

Sepúlveda-García, E.B., Pulido-Barajas, J.F., Huerta-Heredia. A.A., Peña-Castro, J.M., Liu, R. and Barrera-Figueroa, B.E. (2020). Differential expression of maize and teosinte microRNAs under submergence, drought, and alternated stress, *Plants*, 9(10): 1367. https://doi.org/10.3390/plants9101367

Sharkey, T. and Schrader, S. (2006). High temperature stress. *In: Physiology and Molecular Biology of Stress Tolerance in Plants* (Eds.). Madhava Rao, K.V., Raghavendra, A.S. and Janardhan Reddy, K. Ch. 4, 101-129, Springer Netherlands. 10.1007/1-4020-4225-6_4

Sharma, N., Tripathi, A. and Sanan-Mishra, N. (2015). Profiling the expression domains of a rice-specific microRNA under stress, *Front. Plant Sci.*, 6.

Sharma, N., Tripathi, A. and Sanan-Mishra, N. (2015a). Profiling the expression domains of a rice-specific microRNA under stress, *Front.*, 6.

Shi, X., Jiang, F., Wen, J. and Wu, Z. (2019). Overexpression of *Solanum habrochaites* microRNA319d (sha-miR319d) confers chilling and heat stress tolerance in tomato (*S. lycopersicum*), *BMC Plant Biol.*, 19: 214.

Shim, D., Hwang, J.U., Lee, J., Lee, S., Choi, Y., An, G. et al. (2009). Orthologs of the class A4 heat shock transcription factor HsfA4a confer cadmium tolerance in wheat and rice, *Plant Cell*, 21: 4031-4043. 10.1105/tpc.109.066902

Shin, S.-Y., Jeong, J.S., Lim, J.Y., Kim, T., Park, J.H., Kim, J.-K. and Shin, C. (2018). Transcriptomic analyses of rice (*Oryza sativa*) genes and non-coding RNAs under nitrogen starvation using multiple omics technologies, *BMC Genom.*, 19: 1-20.

Shivaprasad, P.V., Chen, H.-M., Patel, K., Bond, D.M., Santos, B.A. and Baulcombe, D.C.A. (2012). microRNA superfamily regulates nucleotide binding site-leucine-rich repeats and other mRNAs, *Plant Cell*, 24: 859-874.

Song, J.B., Gao, S., Sun, D., Li, H., Shu, X.X. and Yang, Z.M. (2013). MiR394 and LCR are involved in *Arabidopsis* salt and drought stress responses in an abscisic acid-dependent manner, *BMC Plant Biol.*, 13: 210.

Song, X., Liu, G., Duan W., Liu, T., Huang, Z., Ren, J., Li, Y. and Hou, X. (2014). Genome-wide identification, classification and expression analysis of the heat shock transcription factor family in Chinese cabbage, *Molecular Genetics & Genomics*, 289: 541-551.

Srivastava, R., Deng, Y. and Howell, S.H. (2014). Stress sensing in plants by an ER stress sensor/transducer, bZIP28, *Front. Plant Sci.*, 5: 59.

Stavang, J.A., Gallego-Bartolome, J., Gomez, M.D., Yoshida, S., Asami, T., Olsen, J.E., Garcia-Martinez, J.L., Alabadi, D. and Blazquez, M.A. (2009). Hormonal regulation of temperature-induced growth in *Arabidopsis*, *Plant J.*, 60: 589-601.

Steif, A., Altmann, S., Hoffmann, K., Pant, B.D., Scheible, W.R. and Bäurle, I. (2014). *Arabidopsis* miR156 regulates tolerance to recurring environmental stress through SPL transcription factors, *Plant Cell*, 26: 1792-1807.

Sun, G. (2012). MicroRNAs and their diverse functions in plants, *Plant Mol. Biol.*, 80: 17-36.

Sun, X., Xu, L., Wang, Y., Keyun, Z., Zhu, X., Luo, X., Gong, Y., Wang, R., Limera, C., Zhang, K. (2015). Identification of novel and salt-responsive miRNAs to explore miRNA-mediated regulatory network of salt stress response in radish (*Raphanus sativus* L.), *BMC Genom.*, 16: 1-16.

Sun, C.W. and Callis, J. (1997). Independent modulation of *Arabidopsis thaliana* polyubiquitin mRNAs in different organs of and in response to environmental changes. *Plant J.*, 11: 1017-1027.

Sunkar, R., Kapoor, A and Zhu, J.K. (2006). Posttranscriptional induction of two Cu/Zn superoxide dismutase genes in *Arabidopsis* is mediated by down-regulation of miR398 and important for oxidative stress tolerance, *The Plant Cell*, 18: 2051-2065.

Sunkar, R. and Zhu, J.K. (2004). Novel and stress-regulated microRNAs and other small RNAs from *Arabidopsis*, *Plant Cell*, 16: 2001-2019.

Tabata, R., Ikezaki, M., Fujibe, T., Aida, M., Tian, C., Ueno, Y., Yamamoto, K.T., Machida, Y., Nakamura, K. and Ishiguro, S. (2010). Arabidopsis Auxin Response Factor6 and 8 regulate jasmonic acid biosynthesis and floral organ development via repression of class 1 KNOX genes, *Plant and Cell Physiology*, 51(1): 164-175. doi:10.1093/pcp/pcp176

Tang, Z., Xu, M., Ito, H., Cai, J., Ma, X., Qin, J., Yu, D. and Meng, Y. (2019). Deciphering the non-coding RNA-level response to arsenic stress in rice (*Oryza sativa*), *Plant Signal. Behav.*, 14: 1629268.

Thomas, J.M.G. and Prasad, P.V.V. (2003). *Plants and the Environment/Gobal Warming Effects*, University of Florida, Gainesville, FL, USA, pp. 786-794.

Tian, C., Zuo, Z. and Qiu, J.L. (2015). Identification and characterisation of ABA-responsive microRNAs in Rice, *J. Genet. Genomics*, 42: 393-402.

Tiwari, J.K., Buckseth, T., Zinta, R., Saraswati, A., Singh, R.K., Rawat, S. and Chakrabarti, S.K. (2020). Genome-wide identification and characterisation of microRNAs by small RNA sequencing for low nitrogen stress in potato, *PLoS ONE*, 15: e0233076.

von Koskull-Döring, P., Scharf, K.D. and Nover, L. (2007). The diversity of plant heat stress transcription factors, *Trends Plant Sci.*, 12: 452-457.

Wahid, A., Gelani, S., Ashraf, M. and Foolad, M. (2007). Heat tolerance in plants: An overview, *Environmental and Experimental Botany*, 61: 199 -223.

Wang, X.Y., Vinocur, P. and Altman, A. (2003). Plant responses to drought, salinity and extreme temperatures: towards genetic engineering for stress tolerance, *Planta*, 218: 1-14.

Wang, J.W., Czech, B. and Weigel, D. (2009). MiR156-regulated SPL transcription factors define an endogenous flowering pathway in *Arabidopsis thaliana*, *Cell*, 138: 738-749.

Wang, Y., Sun, F., Cao, H., Peng, H., Ni, Z., Sun, Q. and Yao, Y. (2012). TamiR159 directed wheat TaGAMYB cleavage and its involvement in another development and heat response, *PLoS ONE*, 7: e48445.

Wang, B., Sun, Y.-F., Song, N., Wei, J.-P., Wang, X.-J., Feng, H., Yin, Z.-Y. and Kang, Z.-S. (2014). MicroRNAs involving in cold, wounding and salt stresses in *Triticum aestivum* L., *Plant Physiol. Biochem.*, 80: 90-96.

Wang, Q., Liu, N., Yang, X., Tu, L. and Zhang, X. (2016). Small RNA-mediated responses to low- and high-temperature stresses in cotton, *Sci. Rep.*, 6: 35558.

Wang, C., Katherine, D., Wang, X., Zhang, S. and Guo, X. (2017). Ghr-miR5272a-mediated regulation of GhMKK6 gene transcription contributes to the immune response in cotton, *J. Exp. Bot.*, 68: 5895-5906.

Wang, W., Liu, D., Chen, D., Cheng, Y., Zhang, X., Song, L., Hu, M., Dong, J. and Shen, F. (2019). MicroRNA414c affects salt tolerance of cotton by regulating reactive oxygen species metabolism under salinity stress, *RNA Biol.*, 16: 362-375.

Waqas, M.A., Wang, X., Zafar, S.A., Noor, M.A., Hussain, H.A., Azher Nawaz, M. and Farooq, M. (2021). Thermal stresses in maize: Effects and management strategies, *Plants*, 10: 1-23.

Werghi, S., Gharsallah, C., Bhardwaj, N.K., Fakhfakh, H. and Gorsane, F. (2021). Insights into the heat-responsive transcriptional network of tomato contrasting genotypes, *Plant Genetic Resources: Characterisation and Utilisation*, 19: 44-57.

Wigge, P.A. (2013). Ambient temperature signalling in plants, *Curr. Opin. Plant Biol.*, 16: 661-666.

Wollenweber, B., Porter, J.R. and Schellberg, J. (2003). Lack of interaction between extreme high temperature events at vegetative and reproductive growth stages in wheat, *J. Agron. Crop Sci.*, 189: 142-150.

Wu, G. and Poethig, R.S. (2006). Temporal regulation of shoot development in *Arabidopsis thaliana* by miR156 and its target SPL3, *Development*, 133: 3539-3547.

Wu, T.Y., Juan, Y.T., Hsu, Y.H., Wu, S.H., Liao, H.T., Fung, R.W. and Charng, Y.Y. (2013). Interplay between heat shock proteins HSP101 and HSA32 prolongs heat acclimation memory post transcriptionally in *Arabidopsis*, *Plant Physiol.*, 161: 2075-2084.

Xia, K., Wang, R., Ou, X., Fang, Z., Tian, C., Duan, J., Wang, Y. and Zhang, M. (2012). OsTIR1 and OsAFB2 down-regulation via OsmiR393 overexpression leads to more tillers, early flowering and less tolerance to salt and drought in rice, *PLoS ONE*, 7: e30039.

Xie, F., Stewart, C.N. Jr., Taki, F.A., He, Q., Liu, H. and Zhang, B. (2014). High throughput deep sequencing shows that microRNAs play important roles in switchgrass responses to drought and salinity stress, *Plant Biotechnol. J.*, 12: 354366.

Xin, M., Wang, Y., Yao, Y., Xie, C., Peng, H., Ni, Z. and Sun, Q. (2010). Diverse set of microRNAs are responsive to powdery mildew infection and heat stress in wheat (*Triticum aestivum* L.), *BMC Plant Biol.*, 10: 123.

Xu, Z., Zhong, S., Li, X., Li, W., Rothstein, S.J., Zhang, S., Bi, Y. and Xie, C. (2011). Genome-wide identification of microRNAs in response to low nitrate availability in maize leaves and roots, *PLoS ONE*, 6: e28009.

Xu, M.Y., Zhang, L., Li, W.W., Hu, X.L., Wang, M.B., Fan, Y.L., Zhang, C.Y. and Wang, L. (2014). Stress-induced early flowering is mediated by miR169 in *Arabidopsis thaliana*, *J. Exp. Bot.*, 65: 89-101.

Xu, L., Wang, Y., Liu, W., Wang, J., Zhu, X., Zhang, K., Yu, R., Wang, R., Xie, Y., Zhang, W., Gong, Y. and Liu, L. (2015). De novo sequencing of root transcriptome reveals complex cadmium-responsive regulatory networks in radish (*Raphanus sativus* L.), *Plant Sci.*, 236: 313-323.

Yaish, M.W., Sunkar, R., Zheng, Y., Ji, B., Al-Yahyai, R. and Farooq, S.A. (2015). A genome-wide identification of the miRNAome in response to salinity stress in date palm (*Phoenix dactylifera* L.), *Front. Plant Sci.*, 6: 946.

Yan, K., Liu, P., Wu, C.A., Yang, G.D., Xu, R., Guo, Q.H., Huang, J.G. and Zheng. C.C. (2012a). Stress-induced alternative splicing provides a mechanism for the regulation of microRNA processing in *Arabidopsis thaliana*, *Mol. Cell*, 48: 521-531.

Yan, J., Gu, Y., Jia, X., Kang, W., Pan, S., Tang, X., Chen, X. and Tang, G. (2012b). Effective small RNA destruction by the expression of a short tandem target mimic in *Arabidopsis*, *Plant Cell*, 24: 415-427.

Yang, J., Zhang, N., Ma, C., Qu, Y., Si, H. and Wang, D. (2013). Prediction and verification of microRNAs related to proline accumulation under drought stress in potato, *Comput. Biol. Chem.*, 46: 48-54.

Yang, Y., Zhang, X., Su, Y., Zou, J., Wang, Z., Xu, L. and Que, Y. (2017). miRNA alteration is an important mechanism in sugarcane response to low-temperature environment, *BMC Genom.*, 18: 833.

Yang, Z., Li, W., Su, X., Ge, P., Zhou, Y., Hao, Y., Shu, H., Gao, C., Cheng, S., Zhu, G. and Wang, Z. (2019). Early response of radish to heat stress by strand-specific transcriptome and miRNA analysis, *Int. J. Mol. Sci.*, 20: 3321.

Yang, Z., Wang, Z., Yang, C., Yang, Z., Li, H. and Wu, Y. (2019). Physiological responses and small RNAs changes in maize under nitrogen deficiency and resupply, *Genes Genom.*, 41: 1183-1194.

Yeh, C.H., Kaplinsky, N.J., Hu, C. and Charng, Y.Y. (2012). Some like it hot, some like it warm: Phenotyping to explore thermotolerance diversity, *Plant Sci.*, 195: 1-23.

Yin, Z., Li, Y., Yu, J., Liu, Y., Li, C., Han, X. and Shen, F. (2011). Difference in miRNA expression profiles between two cotton cultivars with distinct salt sensitivity, *Mol. Biol. Rep.*, 39: 49614970.

Yin, Z., Han, X., Li, Y., Wang, J., Wang, D., Wang, S., Fu, X. and Ye, W. (2017). Comparative analysis of cotton small RNAs and their target genes in response to salt stress, *Genes*, 8: 369.

Yin, Z., Li, Y., Zhu, W., Fu, X., Han, X., Wang, J., Lin, H. and Ye, W. (2018). Identification, characterisation, and expression patterns of TCP genes and microRNA319 in cotton, *Int. J. Mol. Sci.*, 19: 3655.

Yoshida, T., Ohama, N., Nakajima, J., Kidokoro, S., Mizoi, J., Nakashima, K., Maruyama, K., Kim, J.-M., Seki, M., Todaka, D., Osakabe, Y., Sakuma, Y., Schöffl, F., Shinozaki, K. and Yamaguchi-Shinozaki, K. (2011). *Arabidopsis* HsfA1 transcription factors function as the main positive regulators in heat shock-responsive gene expression, *Mol. Genet. Genomics*, 286: 321-332.

Yu, X., Wang, H., Lu, Y., de Ruiter, M., Cariaso, M., Prins, M., van Tunen, A. and He, Y. (2012). Identification of conserved and novel microRNAs that are responsive to heat stress in *Brassica rapa*, *J. Exp. Bot.*, 63: 1025-1038.

Yue, E., Liu, Z., Li, C., Li, Y., Liu, Q. and Xu, J.H. (2017). Overexpression of miR529a confers enhanced resistance to oxidative stress in rice (*Oryza sativa* L.), *Plant Cell Rep.*, 36: 1171-1182.

Zang, D., Wang, J., Zhang, X., Liu, Z. and Wang, Y. (2019). *Arabidopsis* heat shock transcription factor HSFA7b positively mediates salt stress tolerance by binding to an E-box-like motif to regulate gene expression, *Journal of Experimental Botany*, 70(19): 5355-5374.

Zhai, J., Jeong, D.H., De Paoli, E., Park, S., Rosen, B.D., Li, Y., González, A.J., Yan, Z., Kitto, S.L., Grusak, M.A. *et al.* (2011). MicroRNAs as master regulators of the plant NB-LRR defence gene family via the production of phased, trans-acting siRNAs, *Genes Dev.*, 25: 2540-2553.

Zhai, L., Liu, Z., Zou, X., Jiang, Y., Qiu, F., Zheng, Y. and Zhang, Z. (2012). Genome-wide identification and analysis of microRNA responding to long-term waterlogging in crown roots of maize seedlings, *Physiol. Plant*, 147: 181-193.

Zhang, Y.C., Yu, Y., Wang, C.Y., Li, Z.Y., Liu, Q., Xu, J., Liao, J.Y., Wang, X.J., Qu, L.H., Chen, F., Xin, P., Yan, C., Chu, J., Li, H.Q. and Chen, Y.Q. (2013). Overexpression of microRNA OsmiR397 improves rice yield by increasing grain size and promoting panicle branching, *Nature Biotechnology*, 31: 848-852.

Zhang, B. and Wang, Q. (2015). MicroRNA-based biotechnology for plant improvement, *J. Cell Physiol.*, 230: 1-15.

Zhang, L., Yao, L., Zhang, N., Yang, J., Zhu, X., Tang, X., Calderón-Urrea, A. and Si, H. (2018). Lateral root development in potato is mediated by Stu-mi164 regulation of NAC transcription factor, *Front. Plant Sci.*, 9: 383.

Zhang, M., An, P., Li, H., Wang, X., Zhou, J., Dong, P., Zhao, Y., Wang, Q. and Li, C. (2019). The miRNA-mediated post-transcriptional regulation of maize in response to high temperature, *Int. J. Mol. Sci.*, 20: 1754.

Zhao, J., Yuan, S., Zhou, M., Yuan, N., Li, Z., Hu, Q., Bethea, F.G. Jr., Liu, H., Li, S. and Luo, H. (2019). Transgenic creeping bentgrass overexpressing Osa-miR393a exhibits altered plant development and improved multiple stress tolerance, *Plant Biotechnol. J.*, 17: 233-251.

Zhao, M., Tai, H., Sun, S., Zhang, F., Xu, Y. and Li, W.X. (2012). Cloning and characterisation of maize miRNAs involved in responses to nitrogen deficiency, *PLoS ONE*, 7: e29669.

Zheng, Y., Jagadeeswaran, G., Gowdu, K., Wang, N., Li, S., Ming, R. and Sunkar, R. (2013). Genome-wide analysis of microRNAs in sacred lotus, *Nelumbo nucifera* (Gaertn), *Trop. Plant Biol.*, 6: 117-130.

Zheng, Y., Hivrale, V., Zhang, X., Valliyodan, B., Lelandais-Brière, C., Farmer, A.D., May, G.D., Crespi, M., Nguyen, H.T. and Sunkar, R. (2016). Small RNA profiles in soybean primary root tips under water deficit, *BMC Syst. Biol.*, 10: 1-10.

Zhou, L., Liu, Y., Liu, Z., Kong, D., Duan, M. and Luo, L. (2010). Genome-wide identification and analysis of drought-responsive microRNAs in *Oryza sativa*, *J. Exp. Bot.*, 61: 4157-4168.

Zhou, R., Wang, Q., Jiang, F., Cao, X., Sun, M., Liu, M. and Wu, Z. (2016). Identification of miRNAs and their targets in wild tomato at moderately and acutely elevated temperatures by high-throughput sequencing and degradome analysis, *Sci. Rep.*, 6: 33777.

Zhou, J., Deng, K., Cheng, Y., Zhong, Z., Tian, L., Tang, X., Tang, A., Zheng, X., Zhang, T., Qi, Y. and Zhang, Y. (2017). CRISPR-Cas9 based genome editing reveals new insights function and regulation in rice, *Front. Plant Sci.*, 8: 1598.

Zhu, X., Thalor, S.K., Takahashi, Y., Berberich, T. and Kusano, T. (2012). An inhibitory effect of the sequence-conserved upstream open-reading frame on the translation of the main open-reading frame of HsfB1 transcripts in *Arabidopsis*, *Plant, Cell & Environmental*, 35: 2014-2030.

Zhou, R., Yu, X., Ottosen, C.O., Zhang, T., Wu, Z. and Zhao, T. (2020). Unique miRNAs and their targets in tomato leaf responding to combined drought and heat stress, *BMC Plant Biol.*, 20: 107.

CHAPTER

7

MicroRNA-mediated Regulation of Drought Stress Response

Seyed Alireza Salami[1]* and Neda Arad[2]

[1] Department of Horticultural Sciences, Faculty of Agricultural Science and Engineering, University of Tehran, Iran

[2] School of Plant Sciences, University of Arizona, Tucson, AZ 85721, USA

1. Introduction

Scientific questions and explanations have been fundamentally revised historically. While the tools were simple, growth and development studies and descriptions were all that were possible. Improvements in scientific disciplines, tools, technologies and approaches, have greatly expanded the spirit of the studies and the biological questions. For example, in the absence of non-coding RNAs, our descriptions and interpretations were possible mainly at transcriptional level and related to mRNA molecules. The discovery of new molecular elements, such as ncRNAs (snRNAs, snoRNAs, lncRNAs, LINK RNA, Xist, microRNAs, etc.) over the past decades has changed our insights into the molecular processes that take place inside the cells and tissues. Among these genetic elements, microRNAs play crucial roles. miRNAs have been proved to trigger the expression of many genes under different conditions, including abiotic and biotic stresses (Pandita, 2019; Pandita and Wani, 2019; Pandita, 2021; 2022a; 2022b; 2022c). The genome itself has a level of biological organisation and architecture.

Besides macromolecule interactions, small molecular molecules, such as microRNAs, are considered as major players of expression scenario in parallel, resulting in a unique phenotype. Understanding how the genome functions in different environmental conditions and how the gene expression is regulated at both transcriptional and post-transcriptional levels is essential. MicroRNAs are very short nucleotide molecules that regulate the expression of other genes in humans, animals and plants. These tiny genetic molecules have macro impacts on life or death in different organisms. Today, nobody can ignore the critical roles of miRNAs in different biological processes, including biotic and abiotic stresses.

---

*Corresponding author: asalami@ut.ac.ir

Besides cultivation, harvest and post-harvest management, genetics and environmental conditions determine the yield and quality of crops. However, several biotic and abiotic stresses cause severe damage and economic losses to crops yearly and can strongly reduce crop performance. The adverse effects of such stresses in all sectors of agriculture and horticulture combined with current and future adverse impacts of global warming are one of the major concerns of researchers, producers and consumers (Singroha *et al.*, 2021). In this regard, abiotic stresses, such as extreme temperatures, drought, salinity, nutrient deficiency and ultraviolet radiation (UV) are considered integral to this story. These abiotic stresses, alongside deficiency and toxicity of nutrients, cause up to 50-80% annual loss of crop yield worldwide (Francini and Sebastiani, 2019; Oshunsanya *et al.*, 2019).

Due to the unavoidable effects of global warming on the one hand and human population growth and lack of good lands to support crop productivity and food on the other hand, scientists are always looking for reliable strategies to combat hunger in future. Plants on their own, cope with these stresses by different adaptive or innate mechanisms, such as those related to photosynthesis, lowering of leaf osmotic potential, adjusting water uptake, changes in root and shoot architecture, activating the antioxidant and non-antioxidant enzymes and compounds, glycine betaine, proline, malondialdehyde (MDA) and abscisic acid (ABA) accumulation, mRNAs and regulatory trans and cis-elements and non-coding RNAs, etc. (Yadav *et al.*, 2020). On the other hand, scientists use different strategies today to overcome the adverse effects of abiotic stresses on crops. A trustable approach is adapting to new climate conditions, using a tolerant genotype which is the output of conventional and/or modern breeding approaches (Wingfield *et al.*, 2015; Naidoo *et al.*, 2019; Babaei *et al.*, 2020).

A cascade of physiological, biochemical, cellular and molecular events are triggered by prolonged water deficit and drought stress (Zandkarimi *et al.*, 2015; Jafarnia *et al.*, 2018). Adverse water potential under water-deficit stress disrupts the plant's natural activities (Babaei *et al.*, 2020). Drought-responsive genes, transcription factors, and miRNAs play significant roles under water stress conditions. As the roles of miRNAs are better understood, more and more efforts have been made to gain better insights into molecular mechanisms and networks involved in drought stress. We need more details of plant response mechanisms against drought stress to breed tolerant plants and mitigate the effects of drought. This chapter focuses on the roles, functions and implications of micro-RNAs in post-transcriptional gene regulation under drought stress and their potential as new targets for developing drought stress resistant cultivars.

2. Global Warming Concerns and Drought Stress

Many factors, such as increasing population growth, reducing suitable lands for cultivation, diseases, and pests, drought, cold, salinity and toxicity stresses have significant adverse effects on crop production. Most of these traits have been aggravated by global warming directly or indirectly, and the primary concern is providing enough high-quality foods and medicine for humans in the future. According to NOAA's National Centres for Environmental Information, the April 2022 global surface temperature was 0.85°C above the 20th-century average of 13.7°C, tying with 2010 as the fifth-warmest April in the 143-year record. Asia had its warmest April, dating back to 1910, with a temperature departure of 2.62°C above average (Fig. 1).

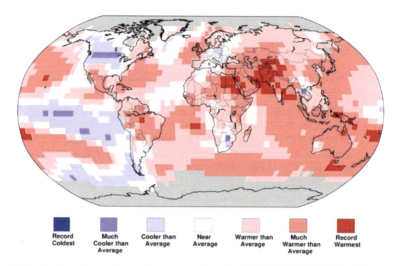

Fig. 1: Land and ocean temperature percentiles, April 2022 (NOAA's National Centres for Environmental Information)

Due to global warming and consequently limited rainfall in some parts of the world, including Middle Eastern countries, water availability is one of the significant challenges in crop production right now. The issue of water deficiency and water stress has been limiting plants' distribution and survival, particularly in arid and semi-arid regions (Morshedloo *et al.*, 2017a, b). For food and feed security in the future, crop yields need to double by 2050 (Ray *et al.*, 2013). Towards this goal, increasing crop yield through different strategies, such as using high-yield tolerant cultivars, changing the cultivation pattern using new tolerant species, proper agricultural practices, decreasing postharvest waste and controlling pests and diseases are the most sustainable methods. Although different cultural practices and agricultural improvements have been tried against abiotic stresses, the use of genetic tools, such as QTL mapping, transgenic and cisgenesis approaches and regulatory engineering using miRNAs are relatively recent scopes in this regard (Wahid *et al.*, 2007; Fahad *et al.*, 2017; Babaei *et al.*, 2020). Because of miRNAs' regulatory functions, their manipulation has significant potential for tolerance to drought stress.

Crop loss under abiotic stresses is more crucial than biotic stresses somehow. On the other hand, abiotic stresses, such as drought, indirectly affect the response and tolerance of plants to pests and diseases. In other words, due to the consequences of climate change, pests and diseases are becoming more destructive. Drought has a huge adverse impact on the growth and productivity of crops, even more than salinity and cold stresses. To better manage drought stress and its negative consequences, understanding the wide range of plant responses at molecular, physiological and biochemical levels and the critical regulatory molecules involved in this phenomenon are very important (Fahad *et al.*, 2017).

3. Key Players in Plant Drought Tolerance Scenario

The mechanisms underlying plant responses to drought are very complicated and challenging due to many different components which play key roles in plant drought tolerance scenario. In this regard, the contribution of genetics, various genetic factors and their related networks is inevitable. Drought stress triggers a cascade of physiological, biochemical, cellular and molecular events related to water relations (water use efficiency, cells and tissues water content, osmotic adjustment and turgor, tissue water potential, stomata closure and hydraulic conductivity), root-and-shoot architecture, photosynthesis, nutrient acquisition and assimilation, enzyme functioning, reactive oxygen species (ROS) scavenging and antioxidant systems, hormone signalling pathways, ABA accumulation, assimilates partitioning and transcriptional and post-transcriptional regulatory events attributed to drought related genes, transcription factors and miRNAs (Carrington and Ambros, 2003; Shinozaki and Yamaguchi-Shinozaki, 2007; Sunkar, 2010; Golldack *et al.*, 2011; Zandkarimi *et al.*, 2015; Jafarnia *et al.*, 2018; Yadav *et al.*, 2020; Xiong *et al.*, 2020; Bano *et al.*, 2021).

All these components are regulated under the genetic background of plants and modulate the regulating strategies related to transcriptional, post-transcriptional, translational and post-translational modifications to cope with drought stress. Any components involved in such gene networks can be considered a potential candidate for conventional and molecular breeding approaches to combat drought stress (Kumar *et al.*, 2021). Meanwhile, miRNAs have been considered one of the promising genetic candidates to overcome plant drought stress in the future. Although we are only at the beginning, significant achievements and advances in miRNA-mediated adaptation to drought stress have been made.

4. MicroRNAs, Tiny Molecules with Macro Impacts

MicroRNAs are small noncoding RNAs, 20-24 nucleotides in length with critical regulatory roles in gene expression at transcriptional and post-transcriptional levels (Carrington and Ambros, 2003; Gu and Kay, 2010). These tiny effective molecules target different mRNAs through imperfect sequence complementarity and consequently cleave and inhibit the related mRNA by DICER and RISC (Zhu, 2008; Gu and Kay, 2010).

MicroRNAs and drought stress are tightly related through several mechanisms (Fig. 2). Several drought-responsive genes and mainly transcription factors (TFs) are related to the wide regulatory networks mentioned above, including APETALA 2 (AP2), Growth-Regulating Factors (GRFs), NAC, MYB, NFY, WRKY, TEOSINTE BRANCHED/CYCLOIDEA/PCF (TCP) and SPL (SQUAMOSA-promoter binding transcription factor) are the major targets of specific miRNAs (Samad *et al.*, 2017).

Specific miRNAs either down-regulate negative-function mRNAs related to drought stress, or down-regulate itself, leading to the accumulation of positive-function mRNAs under stress. As each miRNA may control many genes and each gene can be controlled by many miRNAs, it may be very complex to select a proper candidate most of the time (Yang and Qu, 2013).

Several drought-responsive miRNAs and their related targets have been identified in *Arabidopsis thaliana* (Abe *et al.*, 2003; Sunkar and Zhu, 2004; Chen, 2005; Reyes and Chua, 2007; Liu *et al.*, 2008; Li *et al.*, 2008), *Oryza sativa* (Zhao *et al.*, 2007, Liu *et al.*, 2009; Zhou *et al.*, 2010; Cheah *et al.*, 2015), *Medicago truncatula* (Boualem *et al.*, 2008; Trindade *et al.*, 2010; Wang *et al.*, 2011), *Zea mays* (Griffiths-Jones *et al.*, 2006; Wei *et al.*, 2009; Xu *et al.*, 2014), *Vigna unguiculata* (Barrera-Figueroa *et al.*,

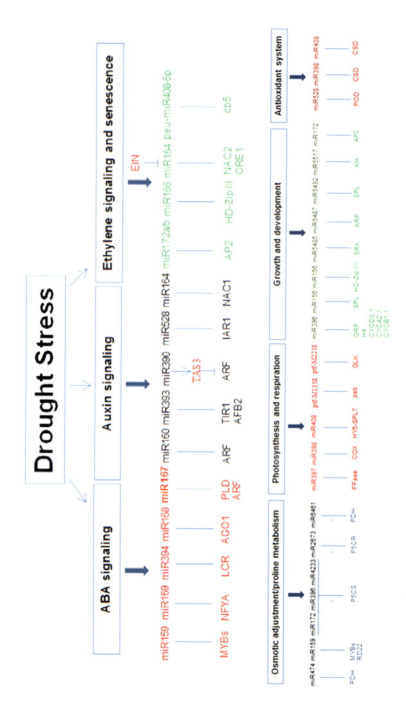

Fig. 2: miRNAs and their targets involved in microRNA-mediated regulation of drought Stress in plants

MicroRNA-mediated Regulation of Drought Stress Response   **125**

2011), *Solanum tuberosum* L. (Zhang *et al.*, 2014), *Vicia sativa* (Zhu *et al.*, 2021), *Triticum dicoccoides* (Kantar *et al.*, 2011), *Glycine max* (Li *et al.*, 2011; Kulcheski *et al.*, 2011), *Prunus persica* (Eldem *et al.*, 2012), *Populus tomentosa* (Ren *et al.*, 2012), *Populus trichocarpa* (Shuai *et al.*, 2013), *Populus euphratica* (Li *et al.*, 2011), *Solanum lycopersicum* (Zhang *et al.*, 2011), *Phaseolus vulgaris* (Arenas-Huertero *et al.*, 2009), *Gossypium hirsutum* L. (Xie *et al.*, 2015), *Triticum aestivum* L. (Ma *et al.*, 2015), *Agropyron mongolicum* (Zhang *et al.*, 2019) and *Malus* (Niu *et al.*, 2019), etc. Some specific miRNAs, such as miR169, is highly conserved across the plant kingdom, have been shown to be part of the environmental stress responses (Li *et al.*, 2008). miRNAs regulate vital processes in plants under drought, including photosynthesis and respiration, ABA, auxin and ethylene signalling, osmolyte production, antioxidant scavenging and its related defence system, etc. (Fig. 2).

Although all the above crops are economically important, most of the data about miRNAs and their functions were obtained from a study on the model plant *Arabidopsis* and strategic crops, such as wheat, maize, rice, etc. Differential expression of miRNAs in two drought-tolerant and drought-susceptible wheat genotypes revealed 233 up-regulated and 10 down-regulated miRNAs after dehydration stress (Ma *et al.*, 2015). miR160a, miR164b, miR166h, miR169d and miR444d.3 were down- and up-regulated in the drought-tolerant and drought-susceptible genotypes, respectively, whereas miR156k, miR444c.1 and wheat-miR-202 were up-regulated in both wheat genotypes after dehydration stress, and miR398 and wheat-miR-628 were expressed predominantly in only one genotype. Scientific evidence showed that these miRNAs might be related to dehydration stress tolerance in wheat by targeting several key genes and transcription factors, such as SBP, MIKC-type MADS-box, MYB, NAC, ARF, HD-ZIP4 and IF3. miR156k targets a transcription factor (SQUAMOSA promoter binding-like protein) critical for leaf growth and development (Wu and Puethig, 2006). miR444c.1 targeted MIKC-type MADS-box transcription factor involved in developmental processes and stress responses (Arora *et al.*, 2007). miR398, which targets superoxide dismutase (SOD) and wheat-miR-628, which targets alpha/beta fold hydrolase (AFH), were up- and down-regulated, respectively, in the drought-susceptible cultivar after dehydration (Ma *et al.*, 2015). In wheat, miR160a and miR164b enhance auxin-responsive genes function by targeting the auxin response factors (ARFs) and NAC transcription factors, respectively, leading to more biomass accumulation to counteract consumption and physiological and morphological adaptation to stress (Guilfoyle and Hagen, 2007; Tran *et al.*, 2010; Ma *et al.*, 2015). miR166h targets the Class III HD-ZIP protein 4 (HD-ZIP4 III), hence regulating vascular cells differentiation, enhancing the developmental process and maintaining energy supply toward developmental, metabolic and stress adaptation (Ma *et al.*, 2015). miR166h is another essential regulatory factor that targets the CCAAT-box transcription factor (CCAAT-box TF), regulating expression of ABA-responsive genes and drought stress tolerance. miR444d.3 was down-regulated in the drought-tolerant wheat genotype targets initiation factor 3 (*IF3*) gene, which regulates the translation initiation process toward metabolic and developmental adaptation to moderate water stress conditions (Ma *et al.*, 2015).

Water-deficit stress also induces genes and miRNAs expression positively and/or negatively in maize. MAPK (mitogen-activated protein kinase) and POD (peroxidase) were accumulated in maize seedlings under drought stress due to the down-regulation of miR528 and miR168 (Wei *et al.*, 2009), suggesting the association of miRNAs and antioxidant defense system under water deficit condition. Among 68 differentially expressed microRNAs in drought-tolerant maize inbred line R09, miR156, miR159 and miR319 play more important roles (Li *et al.*, 2013). miRNAs-transcription factors' interaction, such as ARF (zma-miR160, zma-miR390 and zma-miR393), HD-Zip III (zma-miR166) and NAC (zma-miR164) seem to be the major regulatory mechanisms underlying the drought stress in maize, affecting shoot and roots development (Aravind *et al.*, 2017).

Seeve *et al.* (2019) studied water-deficit responsive microRNAs in maize's primary root growth zone under mild and severe water-deficit stress. Results revealed that miR399e,i,j-3p may play an essential role under water-deficit stress in maize by regulating root growth. Under drought stress, zma-MIR168, zma-MIR156 and zma-MIR166 were highly expressed, whereas the expression of zma-MIR399, zma-MIR2218 and zma-MIR2275 was very low. In this regard, miR159, miR394 and miR319 probably modulate root growth and development and may play critical roles under drought stress in maize (Tang *et al.*, 2022).

miR156, miR167, miR168, miR169g and miR397 predominantly were reported to be up-regulated or down-regulated under drought stress in rice. miR156 negatively regulated SPL9 in rice and increased drought tolerance (Cui *et al.*, 2014). In contrast, SPL14 is up-regulated in rice due to overexpression of miR156 (Jiao *et al.*, 2010). miR167 was down-regulated, followed by ABA treatment and drought stress in rice and maize, respectively (Liu *et al.*, 2009; Wei *et al.*, 2009). miR168 targets Argonaute1 (*AGO1*) in rice, which is involved in miRNA processing under drought stress (Zhou *et al.*, 2010). miR169g was up-regulated by drought in rice. Its promoter contains two dehydration-responsive elements (DREs) (Zhao *et al.*, 2007). miR396 showed down- and up-regulation patterns under drought stress in rice, cowpea, *Arabidopsis* and tobacco, respectively (Zhou *et al.*, 2010; Barrera-Figueroa *et al.*, 2011; Liu *et al.*, 2008; Yang and Yu, 2009). miR397 was down-regulated in rice and seemed to be an important regulator of photosynthesis and CO_2 fixation, starch and sucrose metabolism (Zhou *et al.*, 2010).

Like wheat, maize and rice, soybean is also affected adversely under water-deficit conditions. In soybean plants, miRNAs are important post-transcriptional regulators of gene expression. In this regard, overexpressed *gma-miR172c Arabidopsis* transgenic lines showed an increased survival rate under stress. *Glyma01g39520* reduced the tolerance to drought stress in the *snz* mutants by reducing the expression of ABI3 and ABI5 (Li *et al.*, 2016). Among several miRNA groups, miR168 family was most abundantly miRNA-enriched in soybean primary root tips under drought stress, followed by miR396 and miR1511 (Zheng *et al.*, 2016). miR168 and miR396 regulate AGO1 and GRF transcription factors, respectively. miR1507, miR1508, miR1509 and miR1511 appear to trigger 21 nucleotide phasiRNA biogenesis in legumes (Fei *et al.*, 2013; Zheng *et al.*, 2016).

Transcriptome and degradome analysis revealed that among several miRNAs, *gma-miR398c* can negatively regulate peroxisome-related genes, such as *GmCSD1a/b*, *GmCSD2a/b/c* and *GmCCS* and affect the drought tolerance of soybean (Zhou *et al.*, 2020). *Gma-miR398c* overexpressed *Arabidopsis* plants failed to tolerate the drought stress. Expression of *GmCSD1a/b*, *GmCSD2a/b/c* and *GmCCS* and drought tolerance decreased due to overexpression of *gma-miR398c* compared with knockout miR398c and wild-type soybean plants under drought stress. It seems that *gma-miR398c* may affect the redox status to reduce soybean drought resistance (Zhou *et al.*, 2020).

MicroRNA-mediated Regulation of Drought Stress Response **127**

Chang *et al.* (2022) deeply investigated the miRNA regulatory network of abiotic stress tolerance in soybean. Although generating miRNA sequence data is straightforward to discover the functional analysis of miRNAs, they proposed that computational meta-analysis is a great approach to predict miRNA-target interactions in soybean.

Preparing a comprehensive atlas of miRNAs, their relative targets and interconnected networks of miRNAs-mRNA interactions in different plant species provides a platform for a more detailed and in-depth study of what happens during gene expression regulation at the post-transcriptional levels.

5. Resources and Approaches Towards miRNAS Discovery

Several resources and approaches are used to discover miRNAs in a wide range of plant species under drought stress, including *in silico* computational prediction (Zhang *et al.*, 2009; Yang *et al.*, 2010), comparative genome strategy (Xie *et al.*, 2010; Niu *et al.*, 2019), EST database survey (Kim *et al.*, 2011), microarray (Zhao *et al.*, 2007; Liu *et al.*, 2008; Zhou *et al.*, 2010; Li *et al.*, 2011) and genome wide identification using next generation sequencing (NGS) (Li *et al.*, 2011; Barrera-Figueroa *et al.*, 2011; Wang *et al.*, 2011; Kulcheski *et al.*, 2011; Eldem *et al.*, 2012; Zhang *et al.*, 2013; Bertolini *et al.*, 2013; Shuai *et al.*, 2013; Zhang *et al.*, 2014; Zhang *et al.*, 2019; Zhu *et al.*, 2021).

Considering that the massive omics data have been generated by NGS technology and the availability of databases, such as miRBase (Kozomara *et al.*, 2019), *in silico* analysis helps us predict novel miRNAs in different organisms under normal conditions or a specific treatment or stress. ESTs can also be considered as one of the major resources for miRNAs prediction, particularly when there is a lack of genomics and whole transcriptome data (Taheri-Dehkordi *et al.*, 2021). However, among the approaches mentioned above, high-throughput technologies, such as microarray and NGS, led to a better understanding of miRNAs, their target genes and related biological processes in plants under drought stress. Combining deep sequencing and microarray allows for the more successful discovery of novel stress-responsive miRNAs.

Today researchers have a consensus on using the benefits of high-throughput deep sequencing to discover and identify miRNAs. However, it should be noted that data processing, data analysis and interpretation are still the major bottlenecks while working with substantial sequencing reads. Furthermore, sequencing errors and validation of miRNA structure and targets are likely to be complex and should also be considered as other main issues (Taheri-Dehkordi *et al.*, 2021). To consider RNA sequences as putative miRNA candidates, they should fit the criteria related to their sequence, A+U content, match and mismatch properties and secondary structures (Ambros *et al.*, 2003; Zhang *et al.*, 2014).

Running towards a more accurate prediction of real miRNAs and reducing complexity, progressive algorithms, software and databases were developed and expanded. Bioinformatics tools are used in all steps of miRNA analysis. For example, Zhu *et al.* (2021) used Bowtie (http://bowtie.cbcb.umd.edu) to map sRNA tags to the reference sequences. To determine the potential miRNAs, draw secondary structures and predict the novel miRNAs, miRBase, modified mirdeep2 software (Friedländer *et al.*, 2012), srna-tools-cli and MiREvo (Wen *et al.*, 2012) were used. DESeq R package was used to screen differentially expressed miRNAs. Candidate target genes in the KEGG pathways were identified using KOBAS (Mao *et al*., 2005). Zhang *et al*. (2014) used SOAP (Li *et al*., 2008) and the miRBase database to analyse deep sequencing data and identify 119 and 151 up-regulated and down-regulated novel miRNAs in potatoes under drought stress. Secondary structure and Dicer cleavage site were predicted by Mireap (http://sourceforge.net/projects/mireap/).

Different tools, packages and databases were used to identify miRNAs and their possible targets. *De novo* and reference-based assembly can be performed using Trans-ABySS, Trinity, Bridger, SPAdes and Evidentialgene (Taheri-Dehkordi *et al*., 2021). Evidentialgene was the best *de novo* transcriptome assembler, followed by Trinity. The Bowtie2 and miRBase were used for mapping and surveying the miRNAs. Known miRNAs were detected, using Perl scripts, SUmirFind and SUmirFold (Lucas and Budak, 2012), and a Python script, SUmirPredictor v2 (Alptekin *et al*., 2017). psRNATarget (Dai *et al*., 2019), TAPIR (Bonnet *et al*., 2010) and psRobot (Wu *et al*., 2012) were used to identify miRNA targets (Taheri-Dehkordi *et al*., 2021).

Discovered miRNAs trigger different biological pathways and networks. Several molecules, hormones, osmolites and metabolites belonging to different biological pathways are affected and triggered by miRNAs, which eventually regulate flowering, photosynthesis, respiration, ABA regulation, auxin signalling, osmolyte production and antioxidant scavenging under stress. Considering only one major stress, such as drought, the interpretation of miRNAs-mRNAs interactions and consequent phenotypes would be less complicated. However, we should consider that most of the time, plants face different biotic and abiotic stresses, which cause more ambiguity to correctly explain pathways involved and the plant response.

6. Regulation of Drought Stress

6.1 miRNAs Involved in ABA Response

Abscisic acid accumulates automatically under different stress conditions. It plays crucial key roles not only under water deficit and drought stress, but also in other stresses as the primary plant stress-signalling hormone with an adaptive feature through ABA-mediated and ABA-independent drought stress pathways. Many drought-responsive genes, transcription factors and regulatory elements, such as miRNAs, mediate drought stress response as two sides of one coin in the context of interconnected networks (Fig. 2). Under osmotic stress, transcription factors, ABRE binding protein/ABRE binding factors (AREB/ABFs) function as the component of ABA-dependent pathway and cis-element, dehydration-response element/C-repeat (DRE/CRT) and DRE-/CRT-binding protein 2 (DREB2) transcription factors play a major role in the ABA-independent pathway (Yadav *et al*., 2020). ABA initiates and regulates stomatal conductance, which is an integral part of plant behaviour under drought stress (Wilkinson and Davies, 2002). ABA and its associated genes and signalling pathways are one of the main targets of the miRNAs.

Evidence of participating miRNAs in ABA-related responses was provided using mutants (Lu and Fedoroff, 2000; Zhang *et al*., 2008). Furthermore, the presence of ABRE (ABA-responsive element) sequences in the promoter of some miRNAs, such as miR159a, miR167c, miR168, miR393b, miR396b and miR408 suggests that miRNAs and ABA are related (Liu *et al*., 2008).

Several miRNAs, such as miR159, miR169 and miR394 target MYB, NFYA and LCR transcription factors, respectively and participate in the ABA-mediated pathway

MicroRNA-mediated Regulation of Drought Stress Response

(Li *et al.*, 2008; Song *et al.*, 2013; Singroha *et al.*, 2021). Even more miRNAs that were shown to be drought-responsive (Li *et al.*, 2008; Niu *et al.*, 2019; Kouhi *et al.*, 2020) miRNAs and ABA dependently regulate the drought stress response in plants. Up-regulation of miR159 cleaved MYB33 and MYB101, which are positive regulators of ABA signalling and reduced plant sensitivity to ABA (Reyes and Chua, 2007). Once miR169 was down-regulated, it led to higher expression of NFYA5 in *Arabidopsis* and drought resistance in transgenic lines, whereas overexpressed miR169a and nfya5 knockout plants were droughts sensitive. NFYA5 seems to play an essential role in response to environmental stresses. NFYA5 is highly expressed in vascular tissues and guard cells control stomatal aperture, or are expressed in other cells, which regulate the expression of several drought stress-responsive genes, such as glutathione transferase or peroxidase (Li *et al.*, 2008). Same as the *Arabidopsis*, miR169 was down-regulated in *M. truncatula* (Wang *et al.*, 2011), whereas miR169g, which harbours two dehydration responsive elements (DREs) in its promoter, was up-regulated by drought in rice (Zhao *et al.*, 2007). In tomatoes, tolerant overexpressed miR169c lines showed lower stomatal conductance and water loss (Zhang *et al.*, 2011). ABA treatment and drought stress caused down-regulation of miR167 in rice and maize, respectively (Liu *et al.*, 2009; Wei *et al.*, 2009), whereas it was up-regulated under drought stress in *Arabidopsis* (Liu *et al.*, 2008). miR167d targets phospholipase D (*PLD*), a positive regulator of drought stress resistance, possibly due to directing ABA response and consequent stomatal movement in guard cells (Zhang *et al.*, 2005). Such a contradictory behaviour of a conserved miRNA family in different plants could be due to different variables, such as stress features, plant condition and developmental stages, or even other unknown players which have yet to be discovered.

In *Arabidopsis* and rice, miR168 was found to target Argonaute1 (*AGO1*) involved in miRNA processing (Liu *et al.*, 2008; Zhou *et al.*, 2010). Sunkar and Zhu (2004) reported that miR393, miR397b and miR402 are involved in the ABA-mediated pathway, while miR168 impairs ABA-H_2O_2 signalling and reduces drought tolerance (Li *et al.*, 2013).

Overall, ABA and its related genes are one of the main targets of miRNAs under drought stress. Besides initiation stomatal conductance, ABA adapts roots by suppressing the lateral root growth and eventually redirects the formation of deeper roots to extract the water more efficiently. ABA is also responsible for maintaining dormancy, another adaptive ensuring avoidance of drought (Rodriguez-Gacio *et al.*, 2009).

Accumulation of ABA inhibits lateral root growth under drought stress (Xiong *et al.*, 2006). Several ABA-responsive elements are found in miRNA genes. Promoters of ABA-responsive genes harbour a cis-element and ABA-response elements (ABREs), which have been regulated under stress (Mundy *et al.*, 1990; Xu *et al.*, 1996). For example, miR167 harbour ABREs in the promoter of the corresponding gene targets two auxin response factors (ARFs) involved in root architecture (Liu *et al.*, 2008; Wu *et al.*, 2006). miR168 and miR396 also contain the ABRE elements, which could regulate drought stress (Liu *et al.*, 2008).

Different members of the miRNAs control water deficiency and drought responses in different plants through various mechanisms. Scientific evidence supports the idea that these miRNAs could be considered promising tools for drought-adaptive mechanisms.

6.2 miRNAs Modulate Auxin Signalling

Plants expend much energy on adapting to drought stress and programmed growth reduction by modulating root, leaf and stem architecture and it is one of the adaptive mechanisms to cope with drought stress and resist. Auxin, a phytohormone with an essential role in growth and development, affects these features. However, auxin is not the only phytohormone to participate in different stress conditions and miRNA is involved in the crosstalk among auxin, ABA and ethylene, etc. (Liu *et al.*, 2007; Duan *et al.*, 2016; Ilyas *et al.*, 2021). Auxin elicits drought-adaptive signalling and responses. Exogenous auxin treatments enhanced drought tolerance and increased other hormones, such as ABA and jasmonic acid (Zhang *et al.*, 2020). On the other hand, several miRNAs are involved in auxin signalling pathways under drought stress (Fig. 2). For example, *IAR1*, which regulates auxin homeostasis, is down-regulated while targeted by miR528 under drought, consequently leading to lower auxin and adventitious roots formation under drought stress (Li *et al.*, 2011). miR393 targets TIR1, a positive regulator of auxin signalling and inhibits plant growth under drought stress (Dharmasiri and Estelle, 2002; Xia *et al.*, 2012). Up-regulation of miR393 by ABA inhibits lateral root growth (Sunkar and Zhu, 2004; Chen *et al.*, 2012). In this regard, TIR1 and AFB2 are the two primary auxin receptors targeted by miR393 (Chen *et al.*, 2012).

The roots:shoots ratio and their architecture are adaptive mechanisms to overcome drought stress. The root:shoot ratio increased in *M. truncatula* and miRNAs were thought to have crucial roles in root development (Wang *et al.*, 2011). Expression of miR164 in *Arabidopsis* reduced lateral root formation and lower root:shoot ratio due to down-regulation of NAC1 and auxin signals (Guo *et al.*, 2005).

Auxin response factors (ARFs) are another major component involved in root architecture and regulation of mechanisms mediated by auxins during stress adaptation. ARFs can be either activators or repressors based on their non-conserved middle region amino acid composition (Tiwari *et al.*, 2003). Different members of the ARF family (ARF2, ARF3, ARF4, ARF6, ARF8, ARF10, ARF16 and ARF17, etc.) bind to specific cis-elements in auxin-responsive promoter elements (*AuxREs*) found in auxin-inducible genes, including *Auxin/Indole-3-Acetic Acid* (*Aux/IAA*), *SAUR* and *GH3*, are considered to be targeted for different miRNAs (miR160, miR167 and miR390, etc.) directly or indirectly (Hagen and Guilfoyle, 2002).

miR160 and miR167 directly target ARF10, ARF16, ARF17 and ARF6, ARF8, respectively, and reduce free IAA (Mallory *et al.*, 2005; Teotia *et al.*, 2008). miR390 indirectly regulates auxin signalling by tasiRNA (TAS3-derived trans-acting small interfering RNA), which regulates lateral root emergence and organ polarity establishment by targeting transcription factors, such as ARF2, ARF3 and ARF4 (Meng *et al.*, 2010, Marin *et al.*, 2010). ARF17 might be considered a possible transcriptional regulator of GH3, like early auxin-response genes. ARF17 mRNA level increased while miR160 disruption led to failure in embryonic, root, vegetative and floral development (Mallory *et al.*, 2005).

These results and evidence suggested that miRNA-mediated auxin signalling modulates plant adaptation to drought stress through different regulatory pathways and mechanisms.

6.3 miRNAs and Ethylene Signalling and Senescence during Drought

Ethylene, mostly known as senescence hormone, is another molecule that plays important roles under stress. Under extreme and stable water stress, the leaves sacrifice themselves for the survival of the whole plant, especially the reproductive organs. Such self-sacrifice is regarded as a drought-avoidance mechanism by leaf senescence and falling to reduce canopy size and transpiration. Drought stress triggers ethylene production, which consequently results in leaf senescence. Ethylene signalling regulates drought stress at both transcriptional and post-transcriptional levels. Several genes, ABA-dependent and ABA-independent transcription factors (AP2/ERF, DREB1/CBF, etc.) and small RNAs, such as miRNAs are involved in stresses (Zandkarimi *et al.*, 2015) (Fig. 2). APETALA2/Ethylene Response Element binding Factors (AP2/ERF) bind with cis-element sequence GCC box, which has a role in ethylene-responsive transcription regulation (Song *et al.*, 2013). miRNAs could also influence leaf senescence and regulate drought stress. miR172a/d was shown to regulate AP2 transcription factors with an essential role in ethylene signalling. Stress-responsive genes were overexpressed due to the down-regulation of miR172a/d under drought stress (Ecker, 1995).

In wild wheat, miR166 was down-regulated under drought stress. Two ABA-sensitive transcription factors, SHORT-ROOT (SHR) and SCARECROW (SCR), down-regulate the HD-Zip transcription factors, activating miR166 expression (Kantar *et al.*, 2011; Cui *et al.*, 2012; Carlsbecker *et al.*, 2010; Miyashima *et al.*, 2011; Williams *et al.*, 2005). Hahb-4, which codes HD-Zip TFs, harbouring an ethylene-responsive element in its promoter, was up-regulated followed by ABA treatment, ethylene-mediated leaf senescence and drought stress (Dezar *et al.*, 2005; Manavella *et al.*, 2006). Moreover, once Ath-miR164 was down-regulated by an ethylene signalling protein EIN2 (Ethylene Insensitive2), its targets mRNA, NAC2, ORE1 and At5g61430 were accumulated. EIN3 binds to miR164 promoters and triggers leaf senescence (Li *et al.*, 2013). Therefore, miR164 might be another regulator of leaf senescence (Kim *et al.*, 2009).

Considering that ABA negatively regulated ethylene production (Dong *et al.*, 2015), ABA and ethylene interact to mediate stomata movement (Wilkinson and Davies, 2010). The interaction of Cb5- peu-miR408-5p suggested another way of ethylene signalling inhibition by ABA. Scientific evidence reveals that there is crosstalk among auxin, ABA and ethylene in an interconnected network and miRNAs seem to be involved in the crosstalk among them. ABA and ethylene signalling seem to be programmed in different ways under water relations, drought stress and miRNAs' crucial roles in regulating the related pathways and networks are inevitable.

6.4 miRNAs Regulate Water Relations and Osmotic Adjustment

Regulation of water relations under normal conditions or drought stress is not detached from the innate or inductive properties of the plant for survival. In addition to specific and complex developmental, morphophysiological and biochemical mechanisms in the form of an interconnected network (plant structural apparatus, timing the crucial developmental stages, adjustment of roots/shoots architecture, decreasing cell size, maintenance of tissue water potential and cell turgor through osmotic adjustment, stomata closure and hydraulic conductivity, modulate transpiration, stabilise protein structure, osmoprotectants synthesis and accumulation, such as proline, glycine betaine, methanol, etc.), numerous genes (mRNAs), transcription factors (TFs) and negative or positive molecular regulators are involved (Fig. 2). mRNAs are one of the major players in water adjustment scenarios in plants (Ferdous *et al.*, 2015; Singh *et al.*, 2022).

One of the multifunctional molecules with crucial roles, such as osmoprotectant, free radical scavenger, and stress-related signalling agent under both biotic and abiotic stresses is proline (Szabados and Savoure, 2010; Sherwood *et al.*, 2015; Gupta *et al.*, 2016). Total proline content in cells results from its synthesis and degradation (Reddy *et al.*, 2004). Pyrroline-5-carboxylate synthetase (*P5CS*), pyrroline-5-carboxylate reductase (*P5CR*) and proline dehydrogenase (*PDH*) regulate proline metabolism. The proline metabolism pathway is targeted by several miRNAs, such as miR172, miR396a, miR396c and miR4233, which may regulate the *P5CS* gene and miR2673 and miR6461, which may regulate *P5CR* and *PDH* genes, respectively (Yang *et al.*, 2013).

Up-regulation of miR474 caused lower expression of *PDH,* higher proline content and consequently tolerance under drought conditions (Wei *et al.*, 2009). Ath-miR164c regulates genes involved in proline biosynthesis and is a key player in combined drought and pathogen stresses (Gupta *et al.*, 2020). miR159 can target and degrade the MYB transcription factors. Once MYB2 and MYC2 were overexpressed, drought tolerance improved due to the binding of MYBs to cis-elements in the promoter region of the *RD22* gene (response to dehydration 22), (Abe *et al.*, 2003). Therefore, miRNAs regulate water relations and osmotic adjustment under drought stress by targeting the critical elements involved in different related pathways.

miRNAs and their can be considered potential tools for breeding tolerant cultivars using modern biotechnological approaches.

6.5 miRNAs Mediate Photosynthesis and Respiration under Water Stress

Drought stress affects plant growth and yield by adversely affecting photosynthesis and respiration efficiency. Plants evolve mechanisms to escape or tolerate drought stress to balance and maintain photosynthetic activity under drought stress. Photosynthesis and respiration-responsive genes, transcription factors and miRNAs are considered parts of this story (Fig. 2). Photosystem II (PSII) reaction system and its significant components, D1 (*psb*A) and D2 (*psb*D) proteins, seem to be the heart of attack under drought stress (Jin *et al.*, 2012). Any damages to D1 due to stresses results in PSII collapse. *AtGLK2*, a member of the MYB family, is shown to protect chloroplasts and support D1 synthesis towards maintaining the stability of the PSII under drought stress (Waters *et al.*, 2008; 2009). Photosynthesis-related proteins, such as PSII reaction centre genes and the *GLK* are targeted negatively by ptf-M1358 and ptf-M2218 (Deng *et al.*, 2017).

miR397 is another regulator of drought-stress related to photosynthesis and CO_2 fixation, starch and sucrose metabolism (Zhou *et al.*, 2010). miR397 was down- and up-regulated in rice, tolerant soybean and *Arabidopsis under water stress, respectively (*Zhou *et al.*, 2010; Kulcheski *et al.*, 2011*). miR397 seems to target β-fructofuranosidase (FFase). It also targets a laccase gene, reducing root growth under dehydration in a knockout mutant (Cai *et al.*, 2006). Regulation of photosynthesis under water deficit is also associated with NFYs and miR167 and miR169 (Li *et al.*, 2017).

MicroRNA-mediated Regulation of Drought Stress Response **133**

SPL7 and HY5 in *Arabidopsis* implicate copper apportioning to chloroplasts and light signalling and, therefore, higher levels of plantacyanin related to photosynthesis. SPL7 and HY5 regulate miR408, which is considered a photosynthetic regulator (Zhang *et al.*, 2014).

Cytochrome C oxidase subunit V (*COX5b*) is the target of miR398 and *COX5b*-miR398 interaction negatively regulates respiration by affecting electron transport under water deficit (Sunkar and Zhu, 2004; Trindade *et al.*, 2010).

Photosynthesis and transpiration are two major crucial processes affected by drought stress and several miRNAs-target interactions modulate, maintaining a proper ratio of C-H synthesis to combat the drought stress and protect the plant against harsh water deficiency conditions.

6.6 miRNAs Regulate Growth and Development under Drought Stress

Proper growth and development ultimately ensure the production of high quantity and quality products and any factor, including biotic and abiotic stresses that affect growth and development, causes severe damage. Plant growth and development are largely regulated by phytohormones and gene networks underlying their synthesis. Similar to what happens in almost all biological processes in plants, the synthesis and function of hormones, including growth hormones and consequently cell division and plant growth, are influenced by unique gene networks, transcription factors and regulatory elements, such as miRNAs (Fig. 2).

There is a tight overlap between miRNAs and their targets. The growth and development of leaves, roots and reproductive tissues are affected by drought stress. Control of apical dominance, leaf development, apical dominance, vegetative growth and biomass production is under miRNAs' regulatory machine and specific miRNAs are considered to be a potent strategy for improving plant growth, biomass and crop yield (Zhang and Wang, 2015) (Fig. 2). It has been demonstrated that miR396 represses GRF activity (Debernardi *et al.*, 2012). miR396 targets six GRFs involved in leaf cell division and differentiation in *Arabidopsis* (Jones-Rhoades and Bartel, 2004; Wang *et al.*, 2011). In addition to GRFs, miR396 targets *CYCD3,1*, *histone H4*, *CYCA2,1* and *CYCB1,1*, which are considered cell cycle-related genes (Wang *et al.*, 2011). Evidence reveals the important role of miR396 under drought stress, although it showed a down- and up-regulation pattern under drought stress in rice, cowpea, *Arabidopsis* and tobacco, respectively (Zhou *et al.*, 2010; Barrera-Figueroa *et al.*, 2011; Liu *et al.*, 2008; Yang and Yu, 2009). SPLs regulate developmental processes in plant organs and structures. miR156 negatively regulates SPL9 in rice and *Arabidopsis* and increases drought tolerance (Cui *et al.*, 2014). In contrast, SPL14 is up-regulated in rice due to overexpression of miR156 (Jiao *et al.*, 2010). Overall, the miR156-SPL interaction regulates plant growth and development and drought stress response. Down-regulation of miR159a/b, an increase in MYB expression and growth inhibition happen in parallel (Alonso-Peral *et al.*, 2010). miR166 targets genes, such as *OsHB4* and *HD-ZIP*, which may involve leaf morphology, axillary meristem initiation, leaf polarity, vascular development and vascular shrinkage (Hawker and Bowman, 2004; Rubio-Somoza and Weigel, 2011; Ong and Wickneswari, 2012; Zhang *et al.*, 2018).

Root's architecture and its development under drought stress change the balance for survival. Roots are greatly influenced by drought stress, the same as leaves. miRNAs regulate root architecture via targeting Auxin Response Transcription Factor

(ARFs) (Khan *et al.*, 2011). It seems that miR166-mediated regulation is an important regulatory pathway involved in root architecture and drought response. Once miR166a was up-regulated in *M. truncatula*, lateral root formation was reduced under drought stress due to the down-regulation of *HD-ZIP III* (Boualem *et al.*, 2008). The miR166, however, shows controversial nature in *H. vulgaris* and *T. dicoccoides* under drought stress (Kantar *et al.*, 2010; 2011).

Like roots and leaves, reproductive tissues and organs, such as inflorescences and flowers, are also affected by drought stress. miR5485, miR5487, miR5492 and miR5517 are inflorescence-specific miRNAs that target SRK (S-domain receptor kinase), ARF, SPL11 and KH (K homology) domain, respectively. miR169d and miR169f.2 are two other non-inflorescence-specific miRNAs that target BAK1 (brassinosteroid insensitive 1-associated kinase I) and NFYA2 (Cheah *et al.*, 2017). miR172 seems to target AP2 that are involved in floral organs under biotic and abiotic stresses (Rubio-Somoza and Weigel, 2011). Other genes involved in drought-stress management in shoots, roots and flowers targeted by several miRNAs are included CK1 (Cyclin-Dependent Kinase Inhibitor), ELF (Early Flowering Protein), Glutaredoxin 2, GST (Glutathione S-Transferase), GPI-anchored Protein, OsFBX213 (F-box domain-containing protein 213) and SKP1-like protein 1B (Mutum *et al.*, 2016).

Scientific evidence confirms the important roles of miRNAs in vegetative and reproductive growth and development in many plants. These miRNAs and related target genes can be utilized as a new promising approach toward improving plant growth, biomass and yield under normal and stressful conditions.

6.7 miRNAs and Antioxidant Defence System

Once we are talking about stress, one of the significant destructive molecules produced under harsh stress conditions is ROS. Due to drought stress and prolonged water deficit, cell membrane lipid peroxidation will be upset and ROS imbalances will occur. Excessive ROS is toxic to the plant. However, plants have evolved defense systems that remove ROS through antioxidants (superoxide dismutase (SOD), catalase (CAT), peroxidase (POD), ascorbate peroxidase (APX), glutathione reductase (GR)) and genes and regulatory RNAs (Fei *et al.*, 2020). miRNAs are involved in the antioxidant defense system. miRNAs can regulate cellular redox status, target ROS and modulate antioxidant signalling capacity (Fig. 2).

Antioxidant enzymes are considered essential targets of miRNAs. For example, miR528, which targets *POD*, *was* down-regulated in maize under drought stress (Wei *et al.*, 2009). Consequently, elevated *POD* levels enhanced the removal of excessive H_2O_2 and diminished the ROS damages. They showed that down-regulated miR168 and miR528 under drought stress caused up-regulation of MAPK, which further induced the expression of antioxidant genes and antioxidant enzymes. In *M. truncatula* (Wang *et al.*, 2011) and maize (Wei *et al.*, 2009), miR398 was down-regulated under drought stress and in parallel, CSDs activity was increased. miR398 targets SODs (*CSD1* and *CSD2*), which are tightly involved in oxidative stress detoxification (Mittler, 2002). However, Trindade *et al.* (2010) reported that miR398 and miR408 were up-regulated under water deficit in *M. truncatula*, resulting lower *CSD1* and plantacyanin. Such a controversy in miRNAs behaviour may result from different experimental conditions, genetic variability, different developmental ages of plants, side effects of other unwanted situations of plants and different drought stress parameters.

7. Conclusion

Several novel OMICS-based approaches have been developed during the last two decades, led to a better understanding of genes and drought-responsive miRNAs in many plants. Enormous amounts of data have been generated using next generation sequencing and microarray platforms, enabling us to access comprehensive lists of miRNAs and their target genes. Today, nobody can ignore the critical roles of miRNAs in different biological processes related to drought stress, including phytohormones signalling (ABI, auxin, ethylene, etc.), photosynthesis and transpiration, growth and development, osmotic adjustment, and antioxidant defense system. Recent findings have changed our insights intensely toward a better understanding of miRNAs regulatory functions and mechanisms in different biological pathways, including those involved in water deficit. Comprehensive miRNAs data archives from diverse plant species help us answer many questions and ambiguities regarding miRNA-target interactions. These data can be integrated with other omics data to unravel the mysteries underlying drought tolerance toward the developing and genetic engineering resistant varieties with improved agronomic traits.

References

Abe, H., Urao, T., Ito, T., Seki, M., Shinozaki, K. and Yamaguchi-Shinozaki, K. (2003). *Arabidopsis* AtMYC2 (bHLH) and AtMYB2 (MYB) function as transcriptional activators in abscisic acid signalling, *Plant Cell*, 15: 63-78.

Alonso-Peral, M.M., Li, J., Li, Y., Allen, R.S., Schnippenkoetter, W., Ohms, S., White, R.G. and Millar, A.A. (2010). The microRNA159-regulated GAMYB-like genes inhibit growth and promote programmed cell death in *Arabidopsis*, *Plant Physiol.*, 154: 757-771.

Alptekin, B., Akpinar, B.A. and Budak, H. (2017). A comprehensive prescription for plant miRNA identification, *Front. Plant Sci.*, 7: 1-28.

Ambros, V., Bartel, B., Bartel, D.P., Burge, C.B., Carrington, J.C., Chen, X., Dreyfuss, G., Eddy, S.R., Griffiths-Jones, S., Marshall, M., Matzke, M., Ruvkun, G. and Tuschl, T. (2003). A uniform system for microRNA annotation, *RNA*, 9: 277-279.

Aravind, J., Rinku, S., Pooja, B., Shikha, M., Kaliyugam, S., Mallikarjuna, M.G., Kumar, A., Rao, A.R. and Nepolean, T. (2017). Identification, characterization, and functional validation of drought-responsive microRNAs in subtropical maize inbreds, *Front. Plant Sci.*, 8: 941.

Arenas-Huertero, C., Perez, B., Rabanal, F., Blanco-Melo, D., De la Rosa, C., Estrada-Navarrete, G., Sanchez, F., Alicia Covarrubias, A. and Luis Reyes, J. (2009). Conserved and novel miRNAs in the legume *Phaseolus vulgaris* in response to stress, *Plant Mol. Biol.*, 70: 385-401.

Arora, R., Agarwal, P., Ray, S., Singh, A.K., Singh, V.P., Tyagi, A.K. and Kapoor, S. (2007). MADS-box gene family in rice: Genome-wide identification, organisation and expression profiling during reproductive development and stress, *BMC Genom.*, 8: 242.

Babaei, M., Shokrpour, M., Sharifi, M. and Salami, S.A. (2020). Screening of different Iranian ecotypes of cannabis under water-deficit stress, *Sci. Hortic.*, 260: 108904.

Bano, H., Athar, H.R., Zafar, Z., Kalaji, H.M. and Ashraf, M. (2021). Linking changes in chlorophyll a fluorescence with drought stress susceptibility in mung bean [*Vigna radiata* (L.) Wilczek]. *Physiol. Plant*, 172: 1240-1250.

Barrera-Figueroa, B.E., Gao, L., Diop, N.N., Wu, Z.G., Ehlers, J.D., Roberts, P.A., Close, T.J., Zhu, J.K. and Liu, R. (2011). Identification and comparative analysis of drought-associated microRNAs in two cowpea genotypes, *BMC Plant Biol.*, 11: 127.

Bertolini, E., Verelst, W., Horner, D.S., Gianfranceschi, L., Piccolo, V., Inzé D. Pè, M.E. and Mica, E. (2013). Addressing the role of microRNAs in reprogramming leaf growth during drought stress in *Brachypodium distachyon*, *Mol. Plant.*, 6: 423-443.

Bonnet, E., He, Y., Billiau, K. and Van de Peer, Y. (2010). TAPIR, a web server for the prediction of plant microRNA targets, including target mimics, *Bioinformatics*, 26: 1566-1568.

Boualem, A., Laporte, P., Jovanovic, M., Laffont, C., Plet, J., Combier, J.P., Niebel, A., Crespi, M. and Frugier, F. (2008). MicroRNA166 controls root and nodule development in *Medicago truncatula*, *Plant J.*, 54: 876-887.

Cai, X., Davis, E.J., Ballif, J., Liang, M., Bushman, E., Haroldsen, V., Torabinejad, J. and Wu, Y. (2006). Mutant identification and characterisation of the laccase gene family in *Arabidopsis*, *J. Exp. Bot.*, 57: 2563-2569.

Carlsbecker, A., Lee, J.Y., Roberts, C.J., Dettmer, J., Lehesranta, S., Zhou, J., Lindgren, O., Moreno-Risueno, M.A., Vaten, A., Thitamadee, S., Campilho, A., Sebastian, J., Bowman, J.L., Helariutta, Y. and Benfey, P.N. (2010). Cell signalling by microRNA165/6 directs gene dose-dependent root cell fate, *Nature*, 465: 316-321.

Carrington, J.C. and Ambros, V. (2003). Role of microRNAs in plant and animal development, *Science*, 301: 336-338.

Chang, H., Zhang, H., Zhang, T., Su, L., Qin, Q-M., Li, G., Li, X., Wang, L., Zhao, T., Zhao, E., Zhao, H., Liu, Y., Stacey, G. and Xu, D. (2022). A multi-level iterative bi-clustering method for discovering miRNA co-regulation network of abiotic stress tolerance in soybeans, *Front. Plant Sci.*, 13: 860791.

Cheah, B.H., Nadarajah, K., Divate, M.D. and Wickneswari, R. (2015). Identification of four functionally important microRNA families with contrasting differential expression profiles between drought-tolerant and susceptible rice leaf at vegetative stage, *BMC Genom.*, 16: 692.

Cheah, B.H., Jadhao, S., Vasudevan, M., Wickneswari, R. and Nadarajah, K. (2017). Identification of functionally important microRNAs from rice inflorescence at heading stage of a qDTY4.1-QTL bearing Near Isogenic Line under drought conditions, *PLoS ONE*, 12(10): e0186382.

Chen, H., Li, Z. and Xiong, L. (2012). A plant microRNA regulates the adaptation of roots to drought stress, *FEBS Letters*, 586(12): 1742-1747.

Chen, X.M. (2005). microRNA biogenesis and function in plants, *FEBS Lett.*, 579: 923-931.

Cui, H., Hao, Y. and Kong, D. (2012). SCARECROW has a SHORT-ROOT-independent role in modulating the sugar response, *Plant Physiol.*, 158: 1769-1778.

Cui, L.G., Shan, J.X., Shi, M., Gao, J.P. and Lin, H.X. (2014). The miR156-SPL 9-DFR pathway coordinates the relationship between development and abiotic stress tolerance in plants, *Plant J.*, 80: 1108-1117.

Dai, X., Zhuang, Z. and Zhao, P.X. (2019). psRNATarget V2: A high-performance plant small RNA target analysis server. *In*: *Plant and Animal Genome XXVII Conference* (January 12-16, 2019), PAG.

Debernardi, J.M., Rodriguez, R.E., Mecchia, M.A. and Palatnik, J.F. (2012). Functional specialisation of the plant miR396 regulatory network through distinct microRNA–target interactions, *PLoS Genet.*, 8: e1002419.

Deng, M., Cao, Y., Zhao, Z., Yang, L., Zhang, Y., Dong, Y. and Fan, G. (2017). Discovery of microRNAs and their target genes related to drought in *Paulownia* 'Yuza 1' by high-throughput sequencing, *Int. J. Genomics*, 2017: 1-11.

Dezar, C.A., Fedrigo, G.V. and Chan, R.L. (2005). The promoter of the sunflower HD-Zip protein gene Hahb4 directs tissue-specific expression and is inducible by water stress, high salt concentrations and ABA, *Plant Sci.*, 169: 447-456.

Dharmasiri, S. and Estelle, M. (2002). The role of regulated protein degradation in auxin response, *Plant Mol. Biol.*, 49: 401-409.

Dong, Z., Yu, Y., Li, S., Wang, J., Tang, S. and Huang, R. (2015). Abscisic acid antagonises ethylene production through the ABI4-mediated transcriptional repression of ACS4 and ACS8 in *Arabidopsis*, *Mol. Plant*, 9: 126-135.

Duan, H., Lu, X., Lian, C., An, Y., Xia, X. and Yin, W. (2016). Genome-wide analysis of microRNA responses to the phytohormone abscisic acid in *Populus euphratica, Front. Plant Sci.*, 7: 1184.

Ecker, J.R. (1995). The ethylene signal transduction pathway in plants, *Science*, 268: 667-675.

Eldem, V., Akcay, U.C., Ozhuner, E., Bakir, Y., Uranbey, S. and Unver, T. (2012). Genome-wide identification of miRNAs responsive to drought in peach (*Prunus persica*) by high-throughput deep sequencing, *PLoS ONE*, 7: e50298.

Fahad, S., Bajwa, A.A., Nazir, U., Anjum, S.A., Farooq, A., Zohaib, A., Sadia, S., Nasim, W., Adkins, S., Saud, S., Ihsan, M.Z., Alharby, H., Wu, C., Wang, D. and Huang, J. (2017). Crop production under drought and heat stress: Plant responses and management options, *Front. Plant Sci.*, 8: 1147.

Fei, Q., Xia, R. and Meyers, B.C. (2013). Phased, secondary, small interfering RNAs in posttranscriptional regulatory networks, *Plant Cell*, 25: 2400-2415.

Fei, X., Li, J., Kong, L., Hu, H., Tian, J., Liu, Y. and Wei, A. (2020). miRNAs and their target genes regulate the antioxidant system of *Zanthoxylum bungeanum* under drought stress, *Plant PhysioL. Biochem.*, 150: 196-203.

Ferdous, J., Hussain, S.S. and Shi, B.J. (2015). Role of microRNAs in plant drought tolerance, *Plant Biotechnol. J.*, 13: 293-305.

Francini, A. and Sebastiani, L. (2019). Abiotic stress effects on performance of horticultural crops, *Horticulturae*, 5(4): 67.

Friedländer, M.R., Mackowiak, S.D., Li, N., Chen, W. and Rajewsky, N. (2012). miRDeep2 accurately identifies known and hundreds of novel microRNA genes in seven animal clades, *Nucleic Acids Res.*, 40: 37-52.

Golldack, D., Lüking, I. and Yang, O. (2011). Plant tolerance to drought and salinity: Stress regulating transcription factors and their functional significance in the cellular transcriptional network, *Plant Cell Rep.*, 30: 1383-1391.

Griffiths-Jones, S., Grocock, R.J., vanDongen, S., Bateman, A. and Nright, A.J. (2006). miRBase: MicroRNA sequences, targets and gene nomenclature, *Nucleic Acids Res.*, 34: D140-D144.

Gu, S. and Kay, M.A. (2010). How do miRNAs mediate translational repression? *Silence*, 1: 1-5.

Guilfoyle, T.J. and Hagen, G. (2007). Auxin response factors, *Curr. Opin. Struc. Biol.*, 10: 453-60.

Guo, H.S, Xie, Q., Fei, J.F. and Chua, N.H. (2005). MicroRNA directs mRNA cleavage of the transcription factor NAC1 to down-regulate auxin signals for *Arabidopsis* lateral root development, *Plant Cell*, 17(5): 1376-1386.

Gupta, A., Dixit, S.K. and Senthil-Kumar, M. (2016). Drought stress predominantly endures *Arabidopsis thaliana* to *Pseudomonas syringae* infection, *Front. Plant Sci.*, 7: 808.

Gupta, A., Patil, M., Qamar, A. and Senthil-Kumar, M. (2020). *ath-mi*R164c influences plant responses to the combined stress of drought and bacterial infection by regulating proline metabolism, *Environ. Exp. Bot.*, 172: 103998.

Hagen, G. and Guilfoyle, T. (2002). Auxin-responsive gene expression: Genes, promoters and regulatory factors, *Plant Mol. Biol.*, 49: 373-385.

Hawker, N.P. and Bowman, J.L. (2004). Roles for class III HD-Zip and KANADI genes in *Arabidopsis* root development, *Plant Physiol.*, 135: 2261-2270.

Jafarnia, Sh., Akbarinia, M., Hosseinpour, B., Sanavi, S. and Salami, S.A. (2018). Effect of drought stress on some growth, morphological, physiological, and biochemical parameters of two different populations of *Quercus brantii, Forest – Biogeosciences and Forestry*, 11(2): 212-220.

Jiao, Y., Wang, Y., Xue, D., Wang, J., Yan, M., Liu, G., Dong, G., Zeng, D., Lu, Z., Zhu, X., Qian, Q. and Li, J. (2010). Regulation of *OsSPL14* by OsmiR156 defines ideal plant architecture in rice, *Nat. Genet.*, 42: 541-544.

Jin, W.W., Wang, Y., Zhang, H.H. and Sun, G. (2012). Effects of different nitrogen rate on the functions of flue-cured tobacco seedlings photosystem PSII under drought stress, *J. Nanjing Agric. Univ.*, 35: 21-26.

Jones-Rhoades, M.W. and Bartel, D.P. (2004). Computational identification of plant microRNAs and their targets, including a stress induced miRNA, *Mol. Cell*, 14: 787-799.

Kantar, M., Unver, T. and Budak, H. (2010). Regulation of barley miRNAs upon dehydration stress correlated with target gene expression, *Func. Integr. Genomic*, 10: 493-507.

Kantar, M., Lucas, S. and Budak, H. (2011). miRNA expression patterns of *Triticum dicoccoides* in response to shock drought stress, *Planta*, 233: 471-484.

Khan, G.A., Declerck, M., Sorin, C., Hartmann, C., Crespi, M. and Lelandais-Brière, C. (2011). MicroRNAs as regulators of root development and architecture, *Plant Mol. Biol.*, 77: 47-58.

Kim, H.J., Baek, K.H., Lee, B.W., Choi, D. and Hur, C.G. (2011). *In silico* identification and characterisation of microRNAs and their putative target genes in Solanaceae plants, *Genome*, 54: 91-98.

Kim, J.H., Woo, H.R., Kim, J., Lim, P.O., Lee, I.C., Choi, S.H., Hwang, D. and Nam, H.G. (2009). Trifurcate feed-forward regulation of age-dependent cell death involving miR164 in *Arabidopsis*, *Science*, 323: 1053-1057.

Kouhi, F., Sorkheh, K. and Ercisli, S. (2020). MicroRNA expression patterns unveil differential expression of conserved miRNAs and target genes against abiotic stress in safflower, *PLoS ONE*, 15(2): e0228850.

Kozomara, A., Birgaoanu, M. and Grifths-Jones, S. (2019). miRBase: From microRNA sequences to function, *Nucleic Acids Res.*, 47: D155-D162.

Kulcheski, F.R., de Oliveira, L.F., Molina, L.G., Almerao, M.P., Rodrigues, F.A., Marcolino, J., Barbosa, J.F., Stolf-Moreira, R., Nepomuceno, A.L., Marcelino-Guimaraes, F.C., Abdelnoor, R.V., Nascimento, L.C., Carazzolle, M.F., Pereira, G.A. and Margis, R. (2011). Identification of novel soybean microRNAs involved in abiotic and biotic stresses, *BMC Genom.*, 12: 307.

Kumar, S., Tripathi, S., Singh, S.P., Prasad, A., Akter, F., Syed, M.A., Badri, J., Das, S.P., Bhattarai, R., Natividad, M.A., Quintana, M., Venkateshwarlu, C., Raman, A., Yadav, S., Singh, S.K., Swain, P., Anandan, A., Yadaw, R.B., Mandal, N.P., Verulkar, S.B., Kumar, A., Henry, A. (2021). Quintana Rice breeding for yield under drought has selected for longer flag leaves and lower stomatal density, *J. Exp. Bot.*, 72(13): 4981-4992.

Li, R., Li, Y., Kristiansen, K. and Wang, J. (2008). SOAP: Short oligonucleotide alignment programme, *Bioinformatics*, 24: 713-714.

Li, W.X., Oono, Y., Zhu, J.H., He, X.J., Wu, J.M., Iida, K., Lu, X.Y., Cui, X.P., Jin, H.L. and Zhu, J.K. (2008). The *Arabidopsis* NFYA5 transcription factor is regulated transcriptionally and post-transcriptionally to promote drought resistance, *Plant Cell*, 20: 2238-2251.

Li, T., Li, H., Zhang, Y.X. and Liu, J.Y. (2011). Identification and analysis of seven H_2O_2-responsive miRNAs and 32 new miRNAs in the seedlings of rice (*Oryza sativa* L. ssp. *indica*), *Nucleic Acids Res.*, 39: 2821-2833.

Li, H., Dong, Y., Yin, H., Wang, N., Yang, J., Liu, X., Wang, Y., Wu, J. and Li, X. (2011). Characterisation of the stress associated microRNAs in *Glycine max* by deep sequencing, *BMC Plant Biol.*, 11: 170.

Li, B., Qin, Y., Duan, H., Yin, W. and Xia, X. (2011). Genome-wide characterisation of new and drought stress responsive microRNAs in *Populus euphratica*, *J. Exp. Bot.*, 62: 3765-3779.

Li, C., Yan, J.M., Li, Y.Z., Zhang, Z.C., Wang, Q.L. and Liang, Y. (2013). Silencing the *SpMPK1*, *SpMPK2*, and *SpMPK3* genes in tomato reduces abscisic acid-mediated drought tolerance, *Int. J. Mol. Sci.*, 14: 21983-21996.

Li, J., Fu, F., An, M., Zhou, S., She, Y. and Li, W. (2013). Differential Expression of MicroRNAs in response to drought stress in maize, *J. Integr. Agric.*, 12(8): 1414-1422.

Li, W., Wang, T., Zhang, Y. and Li, Y. (2016). Overexpression of soybean *miR172c* confers tolerance to water deficit and salt stress, but increases ABA sensitivity in transgenic *Arabidopsis thaliana*, *J. Exp. Bot.*, 67(1): 175-194.

Li, Z., Peng, J., Wen, X. and Guo, H. (2013). Ethylene-insensitive3 is a senescence-associated gene that accelerates age-dependent leaf senescence by directly repressing miR164 transcription in *Arabidopsis*, *Plant Cell*, 25(9): 3311-3328.

Li, Y., Zhao, S.L., Li, J.L., Hu, X.H., Wang, H., Cao, X.L., Xu, Y.J., Zhao, Z.X., Xiao, Z.Y. and Yang, N. (2017). Osa-miR169 negatively regulates rice immunity against the blast fungus *Magnaporthe oryzae*, *Front. Plant Sci.*, 8: 2.

Liu, P.P., Montgomery, T.A., Fahlgren, N., Kasschau, K.D., Nonogaki, H. and Carrington, J.C. (2007). Repression of Auxin Responsive Factor10 by microRNA160 is critical for seed germination and post-germination stages, *Plant J.*, 52: 133-146.

Liu, H.H., Tian, X., Li, Y.J., Wu, C.A. and Zheng, C.C. (2008). Microarray-based analysis of stress-regulated microRNAs in *Arabidopsis thaliana*, *Rna-a Publ. Rna Soc.*, 14: 836-843.

Liu, Q., Zhang, Y.C., Wang, C.Y., Luo, Y.C., Huang, Q.J., Chen, S.Y., Zhou, H., Qu, L.H. and Chen, Y.Q. (2009). Expression analysis of phytohormone-regulated microRNAs in rice, implying their regulation roles in plant hormone signalling, *FEBS Lett.*, 583: 723-728.

Lu, C. and Fedoroff, N. (2000). A mutation in the *Arabidopsis* HYL1 gene encoding a dsRNA binding protein affects responses to abscisic acid, auxin, and cytokinin, *Plant Cell*, 12: 2351-2366.

Lucas, S.J. and Budak, H. (2012). Sorting the wheat from the chaff: Identifying miRNAs in genomic survey sequences of *Triticum aestivum* chromosome 1AL, *PLoS ONE*, 7: e40859.

Ilyas, M., Nisar, M., Khan, N., Hazrat, A., Khan, A.H., Hayat, K., Fahad, S., Khan, A. and Ullah, A. (2021). Drought tolerance strategies in plants: A mechanistic approach, *J. Plant Growth Regul.*, 40(3): 926-944.

Ma, X., Xin, Z., Wang, Z., Yang, Q., Guo, S., Guo, X., Cao, L. and Lin, T. (2015). Identification and comparative analysis of differentially expressed miRNAs in leaves of two wheat (*Triticum aestivum* L.) genotypes during dehydration stress, *BMC Plant Biol.*, 15: 21.

Mallory, A.C., Bartel, D.P. and Bartel, B. (2005). microRNA directed regulation of *Arabidopsis* auxin response factor17 is essential for proper development and modulates expression of early auxin response genes, *Plant Cell*, 17: 1360-1375.

Manavella, P.A., Arce, A.L., Dezar, C.A., Bitton, F., Renou, J.P., Crespi, M. and Chan, R.L. (2006). Cross-talk between ethylene and drought signalling pathways is mediated by the sunflower Hahb-4 transcription factor, *Plant J.*, 48: 125-137.

Mao, X., Cai, T., Olyarchuk, J.G. and Wei, L. (2005). Automated genome annotation and pathway identification using the KEGG Orthology (KO) as a controlled vocabulary, *Bioinforma* (Oxford, England), 21: 3787-3793.

Marin, E., Jouannet, V., Herz, A., Lokerse, A.S., Weijers, D., Vaucheret, H., Nussaume, L., Crespi, M.D. and Maizel, A. (2010). miR390, *Arabidopsis* TAS3 tasiRNAs, and their auxin response factor targets define an autoregulatory network quantitatively regulating lateral root growth, *Plant Cell*, 22: 1104-1117.

Meng, Y., Ma, X., Chen, D., Wu, P. and Chen, M. (2010). MicroRNA-mediated signalling involved in plant root development, *Biochem. Biophys. Res. Commun.*, 393: 345-349.

Mittler, R. (2002). Oxidative stress, antioxidants and stress tolerance, *Trends Plant. Sci.*, 7: 405-410.

Miyashima, S., Koi, S., Hashimoto, T. and Nakajima, K. (2011). Non-cell-autonomous microRNA165 acts in a dose-dependent manner to regulate multiple differentiation status in the *Arabidopsis* root, *Development*, 138: 2303-2313.

Morshedloo, M.R., Salami, S.A., Nazeri, V. and Craker, L.E. (2017a). Prolonged water stress on growth and constituency of Iranian of Oregano (*Origanum vulgare* L.), *J. Med. Act. Plants*, 5: 7-19.

Morshedloo, M.R., Craker, L.E., Salami, A., Nazeri, V., Sang, H. and Maggi, F. (2017b). Effect of prolonged water stress on essential oil content, compositions and gene expression patterns of mono- and sesquiterpene synthesis in two oregano (*Origanum vulgare* L.) subspecies, *Plant Physiol. Biochem.*, 111: 119-128.

Mundy, J., Yamaguchishinozaki, K. and Chua, N.H. (1990). Nuclear proteins bind conserved elements in the abscisic acid-responsive promoter of a rice rab gene, *Proc. Natl Acad. Sci.*, 87: 1406-1410.

Mutum, R.D., Kumar, S., Balyan, S., Kansal, S., Mathur, S. and Raghuvanshi, S. (2016). Identification of novel miRNAs from drought tolerant rice variety Nagina 22, *Sci. Rep.*, 6: 30786.

Naidoo, S., Slippers, B., Plett, J.M., Coles, D. and Oates, C.N. (2019). The road to resistance in forest trees, *Front. Plant Sci.*, 10: 273.

Niu, C., Li, H., Jiang, L., Yan, M., Li, C., Geng, D., Xie, Y., Yan, Y., Shen, X., Chen, P., Dong, J., Ma, F. and Quam, Q. (2019). Genome-wide identification of drought-responsive microRNAs in two sets of *Malus* from interspecific hybrid progenies, *Hortic. Res.*, 6: 75.

Ong, S.S. and Wisckneswari, R. (2012). Characterisation of microRNAs expressed during secondary wall biosynthesis in *Acacia mangium*, *PLoS ONE*. 7: e49662.

Oshunsanya, S.O., Nwosu, N.J. and Li, Y. (2019). Abiotic stress in agricultural crops under climatic conditions. *In:* Jhariya, M.K., Banerjee, A. and Meena, R.S. (Eds.). *Sustainable Agriculture, Forest and Environmental Management.* Singapore: Springer, 71-100.

Pandita, D. (2019). Plant MIRnome: miRNA biogenesis and abiotic stress response. *In:* Hasanuzzaman, M., Hakeem, K., Nahar, K. and Alharby, H. (Eds.). *Plant Abiotic Stress Tolerance.* Springer, Cham., 449-474. Doi https://Doi.org/10.1007/978-3-030-06118-0_18

Pandita, D. and Wani, S.H. (2019). microRNA as a tool for mitigating abiotic stress in rice (*Oryza sativa* L.). *In:* Wani, S. (Ed.). *Recent Approaches in Omics for Plant Resilience to Climate Change.* Springer, Chamz, 109-133. Doi https://Doi.org/10.1007/978-3-030-21687-0_6

Pandita, D. (2021). Role of miRNAi technology and miRNAs in abiotic and biotic stress resilience. *In:* Aftab, T. and Roychoudhury, A. (Eds.). *Plant Perspectives to Global Climate Changes.* Academic Press, Elsevier, 303-330. https://Doi.org/10.1016/B978-0-323-85665-2.00015-7

Pandita, D. (2022a). How microRNAs regulate abiotic stress tolerance in wheat? A snapshot. *In:* Roychoudhury, A., Aftab, T. and Acharya, K. (Eds.). *Omics Approach to Manage Abiotic Stress in Cereals.* Springer, Singapore, 447-464. https://Doi.org/10.1007/978-981-19-0140-9_17

Pandita, D. (2022b). MicroRNAs shape the tolerance mechanisms against abiotic stress in maize. *In:* Roychoudhury, A., Aftab, T. and Acharya, K. (Eds.). *Omics Approach to Manage Abiotic Stress in Cereals.* Springer, Singapore, 479-493. https://Doi.org/10.1007/978-981-19-0140-9_19

Pandita, D. (2022c). miRNA- and RNAi-mediated metabolic engineering in plants. *In:* Aftab, T. and Hakeem, K.R. (Eds.). *Metabolic Engineering in Plants.* Springer, Singapore, 171-186. https://Doi.org/10.1007/978-981-16-7262-0_7

Ray, D.K., Mueller, N.D., West, P.C. and Foley, J.A. (2013). Yield trends are insufficient to double global crop production by 2050, *PLoS ONE*, 8(6): e66428.

Reddy, A.R., Chaitanya, K.V. and Vivekanandan, M. (2004). Drought-induced responses of photosynthesis and antioxidant metabolism in higher plants, *J. Plant Physiol.*, 161: 1189-1202.

Ren, Y., Chen, L., Zhang, Y., Kang, X., Zhang, Z. and Wang, Y. (2012). Identification of novel and conserved *Populus tomentosa* microRNA as components of a response to water stress, *Funct. Integr. Genomics*, 12: 327-339.

Reyes, J.L. and Chua, N.H. (2007). ABA induction of miR159 controls transcript levels of two MYB factors during *Arabidopsis* seed germination, *Plant J.*, 49: 592-606.

Rodriguez-Gacio, M.D.C., Matilla-Vazquez, M.A. and Matilla, A.J. (2009). Seed dormancy and ABA signalling: The breakthrough goes on, *Plant Signal. Behav.*, 4: 1035-1049.

Rubio-Somoza, I. and Weigel, D. (2011). MicroRNA networks and developmental plasticity in plants, *Trends Plant Sci.*, 16: 258-264.

Samad, A.F., Sajad, M., Nazaruddin, N., Fauzi, I.A., Murad, A., Zainal, Z. and Ismail, I. (2017). MicroRNA and transcription factor: Key players in plant regulatory network, *Front. Plant Sci.*, 8: 565.

Seeve, C.M., Sunkar, R., Zheng, Y., Liu, L., Liu, Z., McMullen, M., Nelson, S., Sharp, R.E. and Oliver, M.J. (2019). Water-deficit responsive microRNAs in the primary root growth zone of maize, *BMC Plant Biol.*, 19: 447.

Sherwood, P., Villari, C., Capretti, P. and Bonello, P. (2015). Mechanisms of induced susceptibility to *Diplodia* tip blight in drought-stressed Austrian pine, *Tree Physiol.*, 35: 549-562.

Shinozaki, K. and Yamaguchi-Shinozaki, K. (2007). Gene networks involved in drought stress response and tolerance, *J. Exp. Bot.*, 58: 221-227.

Shuai, P., Liang, D., Zhang, Z., Yin, W. and Xia, X. (2013). Identification of drought-responsive and novel *Populus trichocarpa* microRNAs by high-throughput sequencing and their targets using degradome analysis, *BMC Genom.*, 14: 233.

Singh, P.K., Indoliya, Y., Agrawal, L., Awasthi, S., Deeba, F., Dwivedi, S., Chakrabarty, D., Shirke, P.A., Pandey, V., Singh, N., Dhankher, O.P., Barik, S.K. and Tripathi, R.D. (2022). Genomic and proteomic responses to drought stress and biotechnological interventions for enhanced drought tolerance in plants, *Curr. Plant Bio.*, 29: 100239.

Singroha, G., Sharma, P. and Sunkur, R. (2021). Current status of microRNA-mediated regulation of drought stress responses in cereals, *Physiol Plant*, 172(3): 1808-1821.

Song, J.B., Gao, S., Sun, D., Li, H., Shu, X.X. and Yang, Z.M. (2013). miR394 and LCR are involved in *Arabidopsis* salt and drought stress responses in an abscisic acid-dependent manner, *BMC Plant Biol.*, 13: 210.

Song, X., Li, Y. and Hou, X. (2013). Genome-wide analysis of the AP2/ERF transcription factor superfamily in Chinese cabbage (*Brassica rapa* ssp. pekinensis), *BMC Genom.*, 14(1): 1-15.

Sunkar, R. (2010). MicroRNAs with macro-effects on plant stress responses, *Semin. Cell Dev. Biol.*, 21: 805-811.

Sunkar, R. and Zhu, J.K. (2004). Novel and stress-regulated microRNAs and other small RNAs from *Arabidopsis*, *Plant Cell*, 16: 2001-2019.

Szabados, L. and Savoure, A. (2010). Proline: A multifunctional amino acid, *Trend Plant Sci.*, 15: 89-97.

Taheri-Dehkordi, A., Naderi, R., Martinelli, F. and Salami, S.A. (2021). Computational screening of miRNAs and their targets in saffron (*Crocus sativus* L.) by transcriptome mining, *Planta.*, 254: 117.

Tang, Q., Lv, H., Li, Q., Zhang, X., Li, L., Xu, J., Wu, F., Wang, Q., Feng, X. and Lu, Y. (2022). Characteristics of microRNAs and target genes in maize root under drought stress, *Int. J. Mol. Sci.*, 23: 4968.

Teotia, P.S., Mukherjee, S.K. and Mishra, N.S. (2008). Fine tuning of auxin signalling by miRNAs, *Physiol. Mol. Biol. Plants*, 14: 81-90.

Tiwari, S.B., Hagen, G. and Guilfoyle, T. (2003). The roles of auxin response factor domains in auxin-responsive transcription, *Plant Cell*, 15: 533-543.

Tran, L.S.P., Nishiyama, R., Shinozaki, K.Y. and Shinozaki, K. (2010). Potential utilisation of NAC transcription factors to enhance abiotic stress tolerance in plants by biotechnological approach, *GM Crops*, 1: 32-39.

Trindade, I., Capitao, C., Dalmay, T., Fevereiro, M.P. and dos Santos, D.M. (2010). miR398 and miR408 are up-regulated in response to water deficit in *Medicago truncatula. Planta*, 231: 705-716.

Wahid, A., Gelani, S., Ashraf, M. and Foolad, M.R. (2007). Heat tolerance in plants: An overview, *Environ. Exp. Bot.*, 61: 199-223.

Wang, L., Gu, X., Xu, D., Wang, W., Wang, H., Zeng, M., Chang, Z., Huang, H. and Cui, X. (2011). miR396-targeted AtGRF transcription factors are required for coordination of cell division and differentiation during leaf development in *Arabidopsis*, *J. Exp. Bot.*, 62: 761-773.

Wang, T., Chen, L., Zhao, M., Tian, Q. and Zhang, W. (2011). Identification of drought-responsive microRNAs in *Medicago truncatula* by genome-wide high throughput sequencing, *BMC Genom.*, 12: 367.

Waters, M.T., Moylan, E.C. and Langdale, J.A. (2008). GLK transcription factors regulate chloroplast development in a cell-autonomous manner, *Plant J.*, 56(3): 432-444.

Waters, M.T., Wang, P., Korkaric, M., Capper, R.G., Saunders, N.J. and Langdale, J.A. (2009). GLK transcription factors coordinate expression of the photosynthetic apparatus in *Arabidopsis*, *Plant Cell*, 21(4): 1109-1128.

Wei, L., Zhang, D.F., Xiang, F. and Zhang, Z.X. (2009). Differentially expressed miRNAs potentially involved in the regulation of defence mechanism to drought stress in maize seedlings, *Int. J. Plant Sci.*, 170: 979-989.

Wen, M., Shen, Y., Shi, S. and Tang, T. (2012). miREvo: An integrative microRNA evolutionary analysis platform for next-generation sequencing experiments, *BMC Bioinforma.*, 13: 140.

Wilkinson, S. and Davies, W.J. (2002). ABA-based chemical signalling: The coordination of responses to stress in plants, *Plant Cell Environ.*, 25: 195-210.

Wilkinson, S. and Davies, W.J. (2010). Drought, ozone, ABA and ethylene: New insights from cell to plant to community, *Plant Cell Environ.*, 33: 510-525.

Williams, L., Grigg, S.P., Xie, M.T., Christensen, S. and Fletcher, J.C. (2005). Regulation of *Arabidopsis* shoot apical meristem and lateral organ formation by microRNA miR166g and its AtHD-ZIP target genes, *Development*, 132: 3657-3668.

Wingfield, M., Brockerhoff, E., Wingfield, B.D. and Slippers, B. (2015). Planted forest health: The need for a global strategy, *Science*, 349: 832-836.

Wu, G. and Poethig, R.S. (2006). Temporal regulation of shoot development in *Arabidopsis thaliana* by miR156 and its target SPL3, *Development*, 133: 3539-3547.

Wu, M.F., Tian, Q. and Reed, J.W. (2006). *Arabidopsis* microRNA167 controls patterns of ARF6 and ARF8 expression, and regulates both female and male reproduction, *Development*, 133: 4211-4218.

Wu, H.J., Ma, Y.K., Chen, T., Wang, M. and Wang, X.J. (2012). PsRobot: A web-based plant small RNA meta-analysis toolbox, *Nucleic Acids Res.*, 40: W22-W28.

Xia, K., Wang, R., Ou, X., Fang, Z., Tian, C., Duan, J., Wang, Y. and Zhang, M. (2012). *OsTIR1* and *OsAFB2* down-regulation via *osmiR393* overexpression leads to more tillers, early flowering and less tolerance to salt and drought in rice, *PLoS ONE*, 7: e30039.

Xie, F., Frazier, T.P. and Zhang, B. (2010). Identification, characterisation and expression analysis of microRNAs and their targets in the potato (*Solanum tuberosum*), *Gene*, 473: 8-22.

Xie, F., Wang, Q., Sun, R. and Zhang, B. (2015). Deep sequencing reveals important roles of microRNAs in response to drought and salinity stress in cotton, *J. Exp. Bot.*, 66(3): 789-804.

Xiong, L., Wang, R.G., Mao, G. and Koczan, J.M. (2006). Identification of drought tolerance determinants by genetic analysis of root response to drought stress and abscisic acid, *Plant Physiol.*, 142: 1065-1074.

Xiong, R., Liu, S., Considine, M.J., Siddique, K.H.M., Lam, H.M. and Chen, Y. (2020). Root system architecture, physiological and transcriptional traits of soybean (*Glycine max* L.) in response to water deficit: A review, *Physiol. Plan*, 172: 405-418.

Xu, D.P., Duan, X.L., Wang, B.Y., Hong, B.M., Ho, T.H.D. and Wu, R. (1996). Expression of a late embryogenesis abundant protein gene, HVA1, from barley confers tolerance to water deficit and salt stress in transgenic rice, *Plant Physiol.*, 110: 249-257.

Xu, J., Yuan, Y., Xu, Y., Zhang, G., Guo, X., Wu, F., Wang, Q., Rong, T., Pan, G., cAo, M., Tang, Q., Gao, S., Liu, Y., Wang, J., Lan, H. and Lu, Y. (2014). Identification of candidate genes for drought tolerance by whole-genome resequencing in maize, *BMC Plant Biol.*, 14: 83.

Yadav, S., Payal, M., Dave, A., Vijapura, A., Patel, D. and Patel, M. (2020). Effect of abiotic stress on crops. *In:* Hasanuzzaman, M. *et al.* (Eds.). *Sustainable Crop Production*. IntechOpen, London.

Yang, F. and Yu, D. (2009). Overexpression of *Arabidopsis* miR396 enhances drought tolerance in transgenic tobacco plants, *Acta Bot. Yunn.*, 31: 421-426.

Yang, J., Zhang, N., Ma, C., Qu, Y., Si, H. and Wang, D. (2013). Prediction and verification of microRNAs related to proline accumulation under drought stress in potato, *Comput Biol Chem.*, 46: 48-54.

Yang, W., Liu, X., Zhang, J., Feng, J., Li, C. and Chen, J. (2010). Prediction and validation of conservative microRNAs of *Solanum tuberosum* L., *Mol. Biol. Rep.*, 37: 3081-3087.

Yang, J.H. and Qu, L.H. (2013). Discovery of microRNA regulatory networks by integrating multidimensional high-throughput data, *Adv. Exp. Med. Biol.*, 774: 251-266.

Zandkarimi, H., Ebadi, A., Salami, S.A., Alizadeh, H. and Baisakh, N. (2015). Analysing the expression profile of AREB/ABF and DREB/CBF genes under drought and salinity stresses in grape (*Vitis vinifera* L.), *PLoS ONE*, 10(7): e0134288.

Zhang, W., Yu, L., Zhang, Y. and Wang, X. (2005). Phospholipase D in the signalling networks of plant response to abscisic acid and reactive oxygen species, *Biochim. Biophys. Acta*, 1736: 1-9.

Zhang, J.F., Yuan, L.J., Shao, Y., Du, W., Yan, D.W. and Lu, Y.T. (2008). The disturbance of small RNA pathways enhanced abscisic acid response and multiple stress responses in *Arabidopsis*, *Plant Cell Environ.*, 31: 562-574.

Zhang, X., Zou, Z., Gong, P., Zhang, J., Ziaf, K., Li, H., Xiao, F. and Ye, Z. (2011). Overexpression of microRNA169 confers enhanced drought tolerance to tomato, *Biotechnol. Lett.*, 33: 403-409.

Zhang, H., Zhao, X., Li, J., Cai, H., Deng, X.W. and Li, L. (2014). MicroRNA408 is critical for the HY5-SPL7 gene network that mediates the coordinated response to light and copper, *Plant Cell*, 26: 4933-4953.

Zhang, J., Zhang, H., Srivastava, A.K., Pan, Y., Bai, J., Fang, J., Shi, H. and Zhu, J.K. (2018). Knockdown of rice microRNA166 confers drought resistance by causing leaf rolling and altering stem xylem development, *Plant Physiol.*, 176: 2082-2094.

Zhang, R., Marshall, D., Bryan, G.J. and Hornyik, C. (2013). Identification and characterisation of miRNA transcriptome in potato by high-throughput sequencing, *PLoS ONE*, 8: e57233.

Zhang, W., Luo, Y., Gong, X., Zeng, W. and Li, S. (2009). Computational identification of 48 potato microRNAs and their targets, *Comput. Biol. Chem.*, 33: 84-93.

Zhang, B. and Wang, Q. (2015). MicroRNA-based biotechnology for plant improvement, *J. Cell Physiol.*, 230: 1-15.

Zhang, X., Fan, B., Yu, Z., Nie, L., Zhao, Y., Yu, X., Sun, F., Lei, X. and Ma, Y. (2019). Functional analysis of three miRNAs in *Agropyron mongolicum* Keng under drought stress, *Agronomy, 9:* 661.

Zhang, Y., Li, Y., Hassan, M.J., Li, Z. and Peng, Y. (2020). Indole-3-acetic acid improves drought tolerance of white clover via activating auxin, abscisic acid and jasmonic acid related genes and inhibiting senescence genes, *BMC Plant Biol.*, 20(1): 1-12.

Zhao, B., Liang, R., Ge, L., Li, W., Xiao, H., Lin, L., Ruan, K. and Youxin, J. (2007). Identification of drought-induced microRNAs in rice, *Biochem. Biophys. Res. Commun.*, 354: 585-590.

Zheng, Y., Hivrale, V., Zhang, X., Valliyodan, B., Lelandais-Briere, C., Farmer, A.D., May, G.D., Crespi, M., Nguyen, H.T. and Sunkar, R. (2016). Small RNA profiles in soybean primary root tips under water deficit, *BMC Syst Biol.*, 10: 126.

Zhou, L., Liu, Y.Z., Kong, D., Duan, M. and Luo, L. (2010). Genome-wide identification and analysis of drought-responsive microRNAs in *Oryza sativa*, *J. Exp. Bot.*, 61: 4157-4168.

Zhou, Y., Liu, W., Li, X., Sun, D., Xu, K., Feng, C., Carther Kue Foka, I., Ketehouli, T., Gao, H., Wang, N., Dong, Y., Wang, F. and Li, H. (2020). Integration of sRNA, degradome, transcriptome analysis and functional investigation reveals gma-miR398c negatively regulates drought tolerance via GmCSDs and GmCCS in transgenic *Arabidopsis* and soybean, *BMC Plant Biol.*, 20: 190.

Zhu, J.K. (2008). Reconstituting plant miRNA biogenesis, *Proc. Natl Acad. Sci.*, 105: 9851-9852.

Zhu, Y., Liu, Q., Xu, W., Yao, L., Wang, X., Wang, H., Xu, Y., Li, L., Duan, C., Yi, Z. and Lin, C. (2021). Identification of novel drought-responsive miRNA regulatory network of drought stress response in common vetch (*Vicia sativa*), *Open Life Sci.*, 16(1): 1111-1121.

CHAPTER

8

MicroRNA-mediated Regulation of UV Radiation Stress Response

Sonam Dwivedi, Elhan Khan and Iffat Zareen Ahmad*

Natural Products Laboratory, Department of Bioengineering and Biosciences, Integral University, Dasauli, Kursi Road, Lucknow-226026, Uttar Pradesh, India

1. Introduction

Sessile plants are vulnerable to a wide range of environmental stresses to which they must adapt in order to thrive. Environmental stress cab be due to UV irradiation, heat, cold, salt and drought that inhibit plant productivity, development and growth. UV-B responses have been linked to autoimmune-like reactions in genome-wide association studies due to activation of defence-related genes expression (Piofczyk *et al.*, 2015). The ozone layer absorbs the majority of UV-B radiation, but about 2% escapes into the atmosphere, harming all living things. Plants have evolved a UV-B sensor and signalling mechanism that is highly conserved (Tilbrook *et al.*, 2016). Low-fluence UV-B irradiation works as an environmental trigger, allowing a variety of genes involved in development and hormone communication, such as ABA, auxin, ethylene and cell-wall-modifying genes, to be expressed (Jenkins, 2017). A conserved and highly specialised signalling cascade, including photomorphogenic receptors, is also activated, in addition to the up-regulation of free radical scavenging enzymes and proteins involved in DNA repair and cell cycle regulation (Jenkins, 2017). A few studies have also shown that photomorphogenic responses to UV exposure are regulated by a variety of miRNA activities, including the blue light receptors Cryptochrome1 and 2, which mediate the expression of miR172 (Zhou *et al.*, 2016; Facella *et al.*, 2017; Zhang and Chen, 2017), and miR408 is regulated in response to light by transcription factors (TFs) coordinated by squamosal-promoter binding protein. Plants use phenylpropanoids accumulated in the vacuoles of epidermal cells to defend themselves against UV-B damage (Dotto and Casati, 2017). These flavonoids are produced by biosynthetic enzymes that include Chalcone synthase (CHS). In a variety of mammals, UV-B activation of CHS (Jenkins, 2017) and downstream flavonoid biosynthesis genes have been widely studied.

---

*Corresponding author: iffat@iul.ac.in

MicroRNAs (miRNAs) are noncoding regulatory RNAs that suppress gene expression via post-transcriptional cleavage of mRNA, RNA-dependent RNA polymerase (RdRP)-mediated second-strand synthesis, translation inhibition and trans-acting small interfering RNAs (ta-siRNAs) triggered by miRNA and miRNA-dependent DNA methylation as shown in Fig. 1. They were first discovered in *C. elegans* and have since been found in a wide range of plants and animals (Gu *et al.*, 2018; Neumeier and Meister, 2021) with certain miRNAs or miRNA-like RNAs being found in viruses and fungi (Yang *et al.*, 2021). Signal transduction, metabolism and plant responsiveness to environmental signals, such as UV and blue light, have all been connected to miRNAs in plants (Paradiso and Proietti, 2022). The two primary types of photoreceptors present in plants are UV-B receptors and cryptochromes (Tissot and Ulm, 2020). Phototropism, anthocyanin assembly, stomatal opening and de-etiolations are all phtomorphogenic responses in the presence of UV and blue light. miRNA regulate genes in maize and *Arabidopsis* in the presence of blue and UV-B light (Casati, 2013). In *Arabidopsis*, cryptochromes 1 (CRY1) and 2 (CRY2) regulate photoperiodic flowering time through modulating miR172 expression in a Constans (CO)-independent way in blue light (Yu *et al.*, 2010). UV-B radiation activates four miRNAs (miR160, miR165/166, miR167 and miR393), which may be involved in auxin signalling pathways. Under UV-B, miRNA396 suppresses cell and leaf growth by reducing transcripts of growth-regulating factor genes (GRFs) (Zhou *et al.*, 2007; Casati 2013). In *Arabidopsis*, the miR408 and its target genes are regulated by squamosal-promoter binding protein-like7 (SBP/SPL7) and Elongated Hypocotyl 5 (HY5) and their levels alter in response to light and copper (Zhang *et al.*, 2014). When maize is exposed to UV-B radiation, miR164, miR165, miR166 and miR398 are up-regulated, whereas miR156, miR171, miR172, miR396 and miR529 are down-regulated. Furthermore, after eight hours of UV-B exposure, both miR156 and miR529 are reduced in maize leaves, but their various targets are enhanced

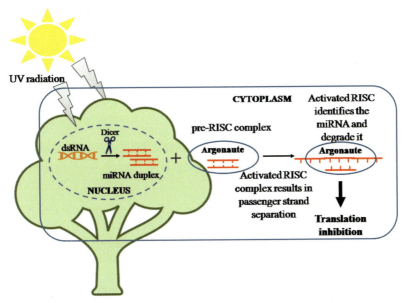

Fig. 1: Plant microRNA-mediated response to UV radiation stress

(Casati, 2013). In *B. rapa subsp. pekinensis cv. Chiifu*, high-throughput sequencing in various organs at various stages of development revealed a novel microRNA (Chinese cabbage). Furthermore, while various signalling factors have been connected to anthocyanin formation, blue and UV-A light-responsive miRNAs have yet to be found and explained at the genome level (He *et al.*, 2020). This chapter looks at how UV-B stress affects microRNA regulation in plants.

2. Conserved miRNAs in Plants

Eukaryotes are green plants that include green algae, a single-cell organism and living multicellular animals like Cyanophora and Embryophyta, mosses, ferns, gymnosperms and angiosperms which are among the members of the Embryophyta subkingdom of green plants, which includes mosses, ferns, gymnosperms and angiosperms (Taylor *et al.*, 2014). Although no strictly conserved miRNAs have been identified between the single-celled algae *Chlamydomonas* and multicellular plants (Molnar *et al.*, 2007; Zhao *et al.*, 2007), several ancient miRNAs have been found in land plants and non-flowering mosses, including miR156, miR160, miR165/166, miR167, miR319, miR390, miR395 and miR408 (Taylor *et al.*, 2014). MiR156, through miR408, are conserved miRNAs found in gymnosperms and angiosperms, including eudicotyledons (dicots) and monocotyledons (monocots) in gymnosperms and angiosperms, respectively (Cuperus *et al.*, 2011; Taylor *et al.*, 2014). Conserved miRNAs that regulate ancestral transcription factors or physiological enzymes involved in fundamental plant growth or stress tolerance are known to have biological significance in plants.

3. Non-conserved miRNAs in Plants

Many plant miRNAs and their associated targets appear to be species-specific or present in only a few closely related species. Non-conserved miRNAs are used to identify these lineage-specific miRNAs from conserved miRNAs. In contrast to conserved miRNAs with high abundance and few sequence modifications, non-conserved miRNAs are often processed imprecisely, created weakly and lack functional targets, leading to their designation as transitory products and energy wasters in the plant genome. Regardless, certain non-conserved miRNAs are abundantly generated in specific tissues or are significantly activated in specific situations, suggesting that non-conserved miRNAs in plants may play a physiological function in atypical environmental responses. Researchers will be able to figure out how lineage-specific miRNAs impact various growth stages by studying them in different plants (Qin *et al.*, 2014).

4. UV-B Responsive miRNAS

The notion that protein-coding genes targeted by the same miRNA have comparable expression patterns led to the identification of stress-responsive miRNA genes. If the expressions of its target genes are coherently suppressed and the coherence is statistically significant beyond a threshold, we consider a miRNA to be putatively stressed inducible (O'Brien *et al.*, 2018). Only true target genes reported in the literature were investigated. Pairwise cosine similarity of the expressions of each miRNA's target genes was calculated for each miRNA. We used average pairwise similarity to examine the coherence of its target genes' expressions. A P-value from a Monte Carlo simulation was used to establish the statistical significance of the coherence. For each miRNA with n target genes, the average pairwise cosine similarity of the target genes' expressions was calculated. The average pairwise cosine similarity of n genes randomly picked from the whole list of monitored genes was then computed. We ran the experiment a million times and obtained an empirical P-value based on the likelihood of seeing a similarity value greater than the target genes. The average P- Forty miRNA genes fulfil the first requirement, according to the results of a 100-time simulation for each miRNA. Anti-correlations shown by average cosine similarities of inferred expressions and target gene expressions, as well as average cosine similarities of inferred expressions and target gene expressions, are statistically significant. miR168a, MiR395c and MiR395e are three miRNA genes that may be UV-B sensitive. UV-B stress is statistically significant in the arrays of motifs in their proximal promoter regions. There are more GO (gene ontology) elements related to stress response in protein-coding genes with similar patterns. The coherence of their target gene expressions, as well as the anti-correlations between their expressions and those of their targets, could not be studied since these miRNAs had fewer than two experimentally validated target genes (Wang *et al.*, 2013). Lin-4 and let-7 are regulatory RNAs that are regarded founding members of a large class of 22-nt noncoding RNAs, known as micro RNAs (miRNAs), some of which are conserved from worms to humans. miRNAs are RNAs that have properties in common with lin-4 and let-7 RNAs: (1) the mature RNA is frequently seen on Northern blots as a 20-nt to 24-nt species. (2) The mature miRNA is trapped within an incomplete RNA duplex that is assumed to be necessary for processing from a longer precursor transcript (O'Brien *et al.*, 2018). miRNAs are also produced outside of the anticipated protein-coding areas of the genome. So far, roughly 150 short RNAs have been discovered in animals that satisfy these criteria. The number of miRNA genes, their fascinating expression patterns in different tissues or at different phases of development and their evolutionary conservation suggest that miRNAs as a class have extensive regulatory functions (Hammond, 2015). Six of the newly discovered Drosophila miRNAs are complementary to 3'-UTR regions known to confer post-transcriptional control in this species, which supports this theory. Dicer processes a range of small RNAs in addition to microRNAs. Dicer was first discovered as a nuclease involved in the animal RNA interference (RNAi) pathway. Long double-stranded RNA (dsRNA), which is commonly delivered via injection or expressed via a transgene, causes this sort of RNA silencing. Dicer cleaves the dsRNA trigger into 22-nt RNAs. SiRNAs are 22-nt RNAs that operate as guide RNAs, deleting homologous mRNA regions. Plants have been linked to post-transcriptional gene silencing (PTGS) by RNAs of fewer than 25 nucleotides with Dicer-like activity thought to be responsible (Xu *et al.*, 2019). RNAi, PTGS and Neurospora quelling use the same collection of proteins. Agronaute, the RNA-directed RNA polymerase SDE1/SGS2, which may amplify dsRNA utilised as a silencing trigger and the RNA helicase SDE3 are all required for PTGS to work (Li *et al.*, 2010). Some elements of RNA silencing, such as the RNA-directed DNA methylation necessary to sustain transgene silence in plants, may be species-specific. RNA silencing is thought to have originated as a viral defence mechanism, but it might also be employed to control endogenous genes by organisms. Endogenous siRNAs may be produced by other animals or plants for gene regulation throughout development. To learn more about the role of short RNAs in plant gene regulation, researchers cloned endogenous RNAs from *Arabidopsis*. The discovery of miRNAs in plants dramatically expands the evolutionary distribution of this class of short noncoding RNAs, implying that miRNAs appeared early in eukaryotic evolution, before the last common ancestor of plants and animals (Qin *et al.*, 2014). According to the occurrence of miRNAs in plants, miRNA processing faults might cause developmental abnormalities in the carpel factory (caf), a mutation in a Dicer homolog and mutations in Argonaute family members. The caf mutant has much decreased levels of plant miRNA. Based on their ancient origins and the probable relationship between miRNAs and development, miRNAs may have played a role in the genesis and evolution of both plant and animal multicellular life (Dexheimer and Cochella, 2020).

5. Plant MicroRNA Regulation under UV-B Stress Response

5.1 UV-B Responsive miRNAs in in vitro-grown Plantlets

Differences in miRNA expression were examined in two biological replicates from low-fluence UV-B-irradiated plants and two samples from the control treatment. UV-B sensitive miRNAs, vvi-MIR5225 and vvi-MIR160e, which target the Ca2+-ATPase10 and ARF10/17 TFs, were shown to be considerably up-regulated. In terms of evolution, MiR5225 is a 22-nt species linked to miR3627 and miR4376 (Sunitha *et al.*, 2019). According to Cleaveland analysis of miR4376 and iso-miR3627 in public grape miR5225, miR3627 and miR4376 target the 5′UTR intron2 region of the calcium-ATPase10, 14-40 nucleotides upstream of the translation start point Pantaleo *et al.* (2010). Because of the limited sample size and high read number, UV treatment had no effect on vvi-MIR482 expression in *in vitro* plantlets. A non-coding phasi-RNA putative TAS locus was discovered using Phase Tank. Zhang *et al.* designated this unique putative TAS locus as vvi-TAS11 because it was significantly up-regulated by UV treatment (2012). The uncharacterised non-coding RNA in ESTs corresponds to the 5′ end of an unannotated 87aa peptide-encoding locus VIT 13s0047g00100. tasiRNA3′-D3(-) also cleaves a phasi-RNA-producing leucine-rich-repeat (LRR) transcript, which is likely to target another phasi-RNA-producing LRR 12 NB-ARC-LRRs and the putative cation/hydrogen exchanger Vvi-CHX15 (Sunitha *et al.*, 2019). UV-B radiation has previously been shown to promote MYBA6 and MYBA7 expression in vegetative tissues (Matus *et al.*, 2017). TAS"4-3′D4 (-)-triggered phasi-RNAs are formed in leaves as a consequence of miR828/TAS4-mediated post-transcriptional regulation of MYBA6 and MYBA7 (Rock, 2013). Deep sequencing was difficult to analyse because of the extremely low amounts of mature miR828 in the sRNA libraries, including leaves (one per 20 million reads, where it is predicted to be greatest). The expression profile of MIR828 cluster siRNAs was confirmed using a more sensitive RNA blot assay using locked nucleic acid-anti-vvi-miR828 as probe (miR828* was the predominant species detected at 2.6 reads per million; quantified by short stack). Despite the fact that mature miR828 levels were greatest in leaves, there was no discernible variation in miR828 levels in plantlets in response to UV-B.

5.2 UV-B Response in Greenhouse and Fields

In both the greenhouse and the field, high-fluence UV-B treatments were observed to considerably delay the down-regulation of MYBA6 and MYBA7 at the end of the berry development phase (Matus *et al.*, 2017). The production of miR828 TAS4abc main transcripts and diced tasi-RNAs, as well as the derived MYBA6 and MYBA7 phasi-RNAs generated by TAS4ab-3'D4 (-) activity, provide strong support for a functional orthologous connection. The *Arabidopsis* MYBA6/7 auto-regulatory loop (Hsieh *et al.*, 2009; Luo *et al.*, 2012) either directly or indirectly boosts MIR828 and TAS4abc expression. We looked for a pattern of up-regulation by high UV fluence and down-regulation by low UV fluence in other highly differentially expressed miRNAs (all analysed by DESeq2). Using co-expression investigations of validated mRNA targets, meta-analysis may also be utilised to infer supportive evidence for miRNA function in connection to reported UV-B fluctuations in miRNA abundances. Three genome-wide transcriptomics datasets of grapevine berry skin tissues in response to one hour of UV-C exposure, five weeks of UV-B after veraison (Carbonell-Bejerano *et al.*, 2014) and five red-skinned varieties sampled across berry developmental stages were the subject datasets (Suzuki *et al.*, 2015; Massonnet *et al.*, 2017). It's important to note that UV-C lamps only emit roughly 5% of their energy in the UV-B band. The high- and low-fluence irradiated greenhouse fruit samples were compared against each other at a high/low ratio across developing time, including across (-) UV controls, based on the apparent coordination for miRNA expressions by high- vs low-fluence UV-B treatments. There were 18 miRNAs with substantial differential expression as a result of high-fluence UV-B treatment, 11 of which were up-regulated and seven of which were down-regulated. Three novel miRNAs were discovered to have diverse expression patterns in response to high-fluence UV-B. A number of vvi-MIR477 relatives have been discovered. Only two of these have been annotated by MiRBase: miR477a and miR477b. vvi-MIR399i, which is projected to target phosphate transporters (Mica *et al.*, 2010) and vvi-MIR167c, which targets ARF TFs, were among the miRNAs up-regulated by high-fluence UV-B as compared to low-fluence reference samples. The coordinated down-regulation of Argonaute 2 (AGO2) in dicots by high-fluence UV-B was one of the two notable discoveries made by all six miR403 family members. AGO2, which has been linked to stress reactions, such as DNA damage (Fátyol *et al.*, 2016; Wei *et al.*, 2012), is regulated by UV radiation.

6. Plant MicroRNAs: Biogenesis, Homeostasis and Degradation

Plants are constantly exposed to a range of environmental challenges that result in considerable DNA damage and genotoxic stress, which can impede growth, genome stability and crop output. Abiotic challenges that plants face on a regular basis include drought, severe temperature stress, salt stress, oxidative stress and UV irradiation damage (Tuteja *et al.*, 2011). When bacteria, viruses, fungi and insects infect plants, they are subjected to a variety of biotic stressors (Huang *et al.*, 2016). These genotoxic stressors damage and undermine the integrity of plant DNA (Tuteja *et al.*, 2009). Plants have evolved strategies for dealing with ongoing DNA damage and maintaining genomic stability. Photoreactivation is an important DNA repair mechanism that returns UV-induced DNA damage to its native form (Friedberg, 2015). Base excision repair (BER) can fix single-strand breaks as well as deaminated, oxidised or alkylated bases, whereas nucleotide excision repair (NER) can only fix CPDs. Double-strand break repair is aided by HR and NHEJ-mediated processes (Spampinato, 2017). Plants have developed systems to scavenge oxidative stress-generated free radicals through a range of enzymatic and non-enzymatic processes including catalases, peroxidases and superoxide dismutases (SODs), as well as non-enzymatic processes using ascorbic acid and secondary metabolites (Das and Roychoudhury, 2014). MicroRNAs (miRNAs) are non-coding ribonucleic acids that interfere with post-transcriptional processes to influence gene expression (Yu *et al.*, 2017). Plant miRNAs are essential for growth, development and tolerance to biotic and abiotic stressors, such as high temperatures, nutritional shortages and salt (Li *et al.*, 2016; Pandita, 2019; Pandita and Wani, 2019; Pandita 2021; 2022a; 2022b; 2022c). Plant microRNAs control gene expression by cleaving or inhibiting the translation of target mRNA (Xie *et al.*, 2015). miRNAs choose targets based on base complementarity, then cleave, destabilise or inhibit translation of those targets (Moro *et al.*, 2018). When miRNA base pairs correctly with target mRNA, target cleavage occurs, whereas target mRNA translational repression occurs when base pairing is poor (Djami-Tchatchou *et al.*, 2017). Strict control is required to keep the system working smoothly. Plant cells have evolved a DNA damage response (DDR) network and ROS scavenging mechanisms to assist them withstand genotoxic stressors. Plant miRNAs are yet to be investigated for their role in either directly or indirectly influencing the expression of genes implicated in genotoxic stress. Only a few researches have looked into the role of miRNAs in DDR and ROS scavenging; therefore, we chose to look at plant miRNAs that are specialised for these functions. Due to the lack of availability of miRNAs and direct engagement of their respective targets inside the DDR network, we have encountered a number of roadblocks in our study. Several genotoxic stressors in plants were discussed, as well as the function of plant miRNAs in combating genotoxic stresses, obstacles in employing miRNAs as a tool to combat genotoxic stresses in plants and solutions based on available data. The insights collected will be important for future practical use of miRNAs as possible instruments to ensure and stabilise crop yield, especially in the light of continuous climate change. RNA silence, which is mediated by short non-coding RNAs of 20-35 nucleotides, is an essential and required form of gene regulation in most eukaryotes. The four primary categories of short RNAs are microRNA (miRNA), small interfering RNA (siRNA), PIWI-interacting RNA (piRNA, animals only) and transfer RNA-derived small RNAs (tsRNAs) (Czech *et al.*, 2018; Zhu *et al.*, 2018; Treiber *et al.*, 2019). Heterochromatic siRNAs (hc-siRNAs), phased secondary siRNAs (phasiRNAs) and epigenetically-activated siRNAs (easiRNAs) are examples of siRNA subtypes (Borges and Martienssen, 2015; Zhang *et al.*, 2016). miRNAs repress target genes after they have been translated, unlike siRNAs and piRNAs, which may silence genes at both the transcriptional (TGS) and post-transcriptional (PTGS) stages (Borges and Martienssen, 2015; Bartel, 2018). Plant miRNAs are produced from specific stem parts of single-stranded hairpin precursors, which have features that set them apart from other short RNAs. Plant miRNA annotation guidelines were recently published in detail (Axtell and Meyers, 2018). Unless otherwise noted, plant miRNAs will be referred to as miRNAs. Plants employ miRNAs in almost every aspect of their development and growth, as well as in response to environmental changes, such as light, nutrition and a variety of abiotic and biotic stresses (Budak *et al.*, 2015; Shriram *et al.*, 2016; Li S.J. *et al.*, 2017; Brant and Budak, 2018). As a result, intracellular miRNA expression is controlled on numerous levels to guarantee that target genes are tightly regulated.

7. Specific Effects of UV-B Radiation on Plants

Plants are especially sensitive to UV-B radiation due to their sessile nature. Plant cell membranes and organelles, including chloroplasts, mitochondria and deoxyribonucleic acid (DNA) within the nucleus, are all damaged by UV-B radiation. Damage to these cell organelles has an impact on photosynthesis, respiration, growth and reproduction, either directly or indirectly. As a result of UV-B damage, crop productivity and quality suffer. UV-B radiation has a different effect based on the intensity and length of exposure, as well as the stage of plant development. Furthermore, the susceptibility to UV-B radiation varies significantly amongst plant species and cultivars within the same species (Dotto and Casati, 2017). There is a need for more study into the physical or physiological underpinnings of species variations in UV-B resistance. Many photochemical reactions occur in living organisms because most organic compounds absorb UV radiation readily. The nucleus of each cell is mostly composed of genetic material in the form of DNA. Nuclear DNA is intrinsically unstable and can be damaged by spontaneous or metabolically induced environmental changes (Vanhaelewyn, 2020). UV-B radiation is particularly susceptible to DNA damage which, if not repaired, can lead to heritable mutations and so can have a significant influence on a range of physiological processes. DNA is the primary absorbing molecule in the UV-B region of the spectrum. UV-B light can induce chromosomal breakage, abnormalities and the exchange and synthesis of dangerous and carcinogenic photoproducts (such as cyclobutane pyrimidine dimmers (CPDs), 6, 4 pyrimidine-pyrimidone or 6, 4-photoproduct, thymine glycols and pyrimidine hydrates) (Gill *et al.*, 2015). Changes in DNA alter gene transcription, replication and recombination in plants lead to significant metabolic and genetic changes. UV-B radiation can cause protein degradation and lipid peroxidation in cells by affecting proteins, membrane lipids and other essential components. These damages have an impact on protein synthesis, enzyme function and gene expression. All cellular life forms include DNA repair enzymes, which detect chemically altered bases, including those induced by UV light. Cells have also developed a variety of biochemical techniques for restoring and maintaining the integrity of genetic information after it has been damaged. These procedures are known as 'DNA repair mechanisms'. The two basic ways for DNA repair are CPDs and 6, 4-photoproduct, photo repair and dark repair. Photo repair is enabled by the enzyme photolyase, which forms a lesion-specific complex with CPDs and 6, 4-photoproducts that are stable in the dark. Dark repair removes the 6, 4-photoproducts from the DNA through nucleotide excision of the damaged oligonucleotide, gap-filling DNA synthesis and correct pairing order ligation. UV radiation may affect the ultrastructure of various cellular components in many plant species (Yamamoto *et al.*, 2013). Damage to the nuclear membrane, swelling of chloroplasts, rupture of the chloroplast wall, dilation of thylakoid membranes, disruption of the thylakoid structure and disintegration of the double membrane that envelops surrounding chloroplasts are the most common causes of ultra-structural changes, resulting in the accumulation of large starch granules. It has also been linked to enlarged cisternae in the endoplasmic reticulum, mitochondrial and plastid damage and plasmalemma and tonoplast vesiculation. UV-B radiation also changes the shape and structure of cells. Physiological activities including cell division, photosynthesis, respiration and reproduction may be affected by these changes in ultrastructure. Plants, on the other hand, have defence and repair systems that assist to cope with expected damages at the cellular level by removing or fixing damaged parts, and at the whole-plant level by producing UV-B absorption compounds and pigments in leaves. The most prevalent of these compounds are flavonoids and anthocyanins, which accumulate in the vacuoles of epidermal and subepidermal cell layers (Takshak and Agrawal, 2019).

7.1 Different Types of UV Induces DNA Damage

UV-B is extremely powerful in generating chemical alterations and DNA damage in natural situations (Rastogi *et al.*, 2010). Photoproducts are a group of DNA modifications caused by UV-B radiation. The most likely photoproducts that trigger mutations are two different lesions that join neighbouring pyrimidine residues on the same strand. The most prevalent lesions are CPDs and 6-4PPs, which account for around 75% of UV-B-induced DNA damage (Takahashi *et al.*, 2011). Minor DNA damage includes hydrated bases, oxidised bases and single-strand breaks (Ballare *et al.*, 2001; Takahashi *et al.*, 2011). CPDs, particularly 6-4PPs, induce DNA bending and unwinding, leading in growth retardation, death, or mutagenesis (Tuteja *et al.*, 2009). Such types of DNA damage are seldom seen at low UV-B rates (1 mol m-2s-1), which are adequate to induce photomorphogenic and defensive responses.

7.2 DNA Repair Possibilities in Plants

The principal mechanism for UV-B-induced repair is photorepair or photoreactivation. By absorbing blue/UV-A light, CPD photolyases and 6-4PP photolyases can monomerise UV-B produced dimers. These photolyases attach selectively to DNA lesions and remove CPDs and 6-4PPs (Sancar, 2003; Bray and West, 2005). UV-B sensitive rice cultivars were shown to be less capable of mending CPDs than resistant cultivars, although overexpression of CPD photolyase increased UV-B damage tolerance significantly (Teranishi *et al.*, 2012). UV-B regulates the expression of the CPD photolyase gene (PHR) in a UVR8-dependent manner, whilst UV-A and blue light do so in a CRY-dependent manner. In addition to the photoreactivation system, there is a light-independent or dark repair mechanism. The most frequent types of DNA repair are nucleotide excision repair (NER), mismatch repair (MMR) and base excision repair (BER) (BER). Boubriak *et al.* (2008) found that NER can repair CPDs and 6-4PPs by recognising structural alterations in DNA rather than particular types of DNA damage (Tuteja and Tuteja, 2001). Double strand break repair and homologous recombination, which have previously been studied (Gill *et al.*, 2015), are seldom detected after UV exposure.

7.3 UV Radiation-induced DNA Damage and Plant Cell Response

UV light damages plant genome stability by causing pyrimidine hydrates to form as a result of oxidative damage and cross-links between DNA and protein, restricting plant growth and development (Gill *et al.*, 2015). UV-B radiation is most harmful because it generates DNA lesions, such CPD and pyrimidine (6-4) pyrimidinone adducts (6-4 PPs) (Ries *et al.*, 2000). CPDs have been identified to obstruct transcribing complexes, causing gene expression patterns to shift. In addition to CPD-mediated damage, UV-B induces a delay in the G1-to-S phase transition within the plant cell cycle (Jiang *et al.*, 2011). UV-C produces single-stranded and double-stranded breaks in *Arabidopsis* (Abas *et al.*, 2007). Plant genome instability and UV-associated mutagenicity have also been related to oxidative DNA damage (Manova and Gruszka, 2015). Plants activate a complex DDR network when they are subjected to DNA damage, which involves DNA repair, cell cycle arrest and death (Yoshiyama *et al.*, 2013). Plants activate different DNA repair pathways in response to DNA damage, including photoreactivation, BER, NER, mismatch repair (MMR) and double-strand break (DSB) repair (Kimura and Sakaguchi, 2006). A wide range of protein components are implicated in these repair processes, with a few of them serving as potential miRNA targets. In the photoreactivation-mediated DNA repair process, CPD lyase or (6-4) photolyase cleaves thymine dimer complexes (Waterworth *et al.*, 2002). In the absence of light, photolyases connect only to the damaged regions of double-stranded DNA, but UV-A activates them for defect repair. The covalent bonds of the dimers are then split in an error-free manner (Manova and Gruszka, 2015). Lesion-specific DNA glycosylases detect aurinic/apyrimidinic sites, which break the N-glycosidic link after removing the damaged base, leading in the formation of abasic sites in plants (Manova and Gruszka, 2015). 3-methyladenine-DNA-glycosylase (MAG), formamido-pyrimidine-DNA-glycosylase (FPG), 8-oxoG DNA-glycosylases (OGG), uracil-DNA-glycosylase (UNG) and DNA glycosylase/lyase DNG701 have all been identified in *Arabidopsis*, carrot and rice (La *et al.*, 2011). The system has been studied extensively in *Arabidopsis thaliana*, where NER identifies and repairs a wide range of DNA damage caused by UV radiation and other mutagens. Xeroderma pigmentosum complementation of group C (XP-C)/AtRAD4 detect DNA damage. The transcription factor IIH (TFIIH), which includes AtXP-D, then unwinds the damaged area of DNA. Damaged oligonucleotides are removed by AtXP-F or other *Arabidopsis* homologs (e.g. AtERCC1). DNA synthesis, driven by proliferating cell nuclear antigen (PCNA) and replication factor C, fills the excised gap (RFC). DNA ligase then joins the two DNA strands together (Liu *et al.*, 2012). MMR homologs of MutS (MSH) detect mismatches produced by erroneous nucleotide insertion by DNA polymerase. The formation of nicks by MutL (MLH) homologue activity is then followed by the repair system's subsequent processes (Lario *et al.*, 2015). Plants use HR and NHEJ-mediated processes to repair DSBs. AtRad51, AtRadA, AtRad50, OsRadA, AtMre11, AtKu70, AtKu80, *Arabidopsis* DNA ligase IV, AtXRCC4, AtXP-F and AtERCC1 are only a few of the DSB repair components found in *Arabidopsis* and rice (Manova and Gruszka, 2015). ATM and ATM-Rad3-related (ATR) proteins have also been shown to play important roles in DNA repair. Checkpoint kinases (CHK), including as Chk1 and Chk2, phosphorylate CHK1 to initiate G-2 phase arrest downstream of ATM and ATR proteins. This results in DDR-induced transcriptional suppression (Culligan *et al.*, 2004).

7.4 The Role of miRNA in the DNA Repair Process

When studying the direct involvement of miRNAs in plant DNA repair pathways, it was discovered that certain helicases, essential enzymes involved in both NER and DSB repairs, are targeted by miR164, miR408 and miR414 in rice (Macovei and

Tuteja, 2013). Several studies have employed *in silico* analysis of probable miRNA-targeting mRNAs of genes implicated in DDR. Using the psRNATarget service, a computational approach for identifying miRNA targets, Brassica rapa miR838b was identified to putatively target the photolyase mRNA (Hajieghrari *et al.*, 2017). This prediction, however, must be put to test in the laboratory. Numerous studies have discovered that miR319 affects TCP (teosinte-branched1/Cincinnata/proliferating cell factor) transcription factors, which play direct roles in leaf formation (Danisman, 2016; Koyama *et al.*, 2017; Bresso *et al.*, 2018). PCF1 and PCF2, transcription factors with the non-canonical TCP domain, have been shown to control PCNA (proliferating cell nuclear antigen), a crucial component of the DDR network (Danisman, 2016; Nicolas and Cubas, 2016). MiR319, one of the 24 members of the *Arabidopsis* TCP family, targets TCP2, TCP3, TCP4, TCP10 and TCP24 genes (Koyama *et al.*, 2017). As of now, there is no concrete evidence that miR319 targeting TCPs can bind to the PCNA promoter.

7.5 Light-relevant cis-elements in the Promoters of miRNA genes

Many key cis-elements that are indicative of miRNA UV-B responsiveness have been discovered utilising the WordSpy genome-wide motif-finding tool. Some of them have been well characterised in plant motif databases, such as PLACE, while others have been investigated in the literature. The G-box (CACGTG), GT-1 site (GGTTAA), I-boxes (GATAAGA), TGA-box (TGACGT), GATA-box (GATATTT), H-box (CCTACC) and CCAAT (CCAAT) are cis-elements in light-regulated genes that have been experimentally characterised (Zhou *et al.*, 2007). The most prevalent cis-elements in miRNAs are the GT-1 site, I-box core and CCAAT-box. The promoters of these miRNA genes feature GT-1 sites, I-box score in all but one and CCAAT-box in 17 of them. Abiotic stress management has been linked to the GT-1 site, I-box and CCAAT-box. They're detected in almost all of these UV-B-induced miRNA genes, which is not surprising.

7.6 Different UV-responsive Expression Pathways Involved miRNAs in Plants

The transcription of a group of miRNA genes, which control the degradation of their target protein-coding mRNAs, can be triggered by light. Light-inducible miRNAs (excluding miR393, miR398 and miR401) have transcription factor targets, which is surprising. These specific transcription factors may have an impact on downstream gene expression. As a result, miRNAs initiate down-regulation pathways in response to moderate stress, which cascade to miRNA targets, miRNA targets and so on, as shown in Table 1. Auxin signalling pathways have been demonstrated to be influenced by many light-induced miRNA genes. Auxin (mainly indole-3-acetic acid) is a necessary hormone in plants. It has an effect on auxin response factors (ARF), a plant-specific family of DNA binding proteins that regulate a variety of plant growth and development processes. ARFs control the expression of auxin-inducible genes, such as GH3 and auxin/indole-3-acetic acid (Aux/IAA) via binding auxin response elements (AREs).

Table 1: Regulatory role of miRNA and their target genes in different plants

Sr. No.	Plants	miRNAs	Response	Target	References
1.	Triticum aestivum (Poaceae)	miR159 miR165/166 miR167 miR393	Up-regulated	HD-ZIPIII ARF TIR1 MYB	Wang et al., 2013
2.	Brassica rapa	BrmiR159 BrmiRC0418 BrmiRC191 BrmiRC0460, BrmiRC0323	Up-regulated Down-regulated	Putative target genes were identified using a psRNA target program such as SPL, ARF, MYB, bZIP, AGL, WRKY, APETALA2, NAM, ATAF (1 and 2) CUC2 (NAC) proteins and many more.	Zhou et al., 2016
3.	Prunus persica	miR171d-3p, miR3627-5p, and miR397Pp03-22, 312-3p miR399b, miR395e, miR399a, miR395d and miR7122b-5p	Up-regulated Down-regulated	Genes involved in biosynthesis of anthocyanin	Li et al., 2017
4.	Populous tremula	miR408, miR398, miR156, miR167, miR164, miR166, miR168 and miR160 miR472, miR169, miR390, miR395, miR393, miR399 and miR159	Up-regulated Down-regulated	Phytohormone signal-related proteins.	Jia et al., 2009
5.	Zea mays	miR398, miR166, miR165 and miR164 miR529, miR396, miR156, miR172 and miR171	Up-regulated Down-regulated	IDS1, GLOSSY 15, PHAP2B, an alpha-mannosidase and nonhypocotylhypocotyl 1	Casati, 2013
6.	Grapevine (Vitis vinifera L.)	miR156, miR168, miR828, miR530 and miR482 miR403	Up-regulated Down-regulated	AGO1	Sunitha et al., 2019 Yang et al., 2020
7.	Chrysanthemum (Chrysanthemum morifolium Ramat)	miR397a, PC-5p-294053_21 andmiR396f-5p	Up-regulated	Glycolysis genes	

7.7 Regulatory Role of MicroRNAs in Response to Ultraviolet Radiation Stress Tolerance of Plants

Due to ROS-induced oxidative damages, increased levels of ultraviolet-B (UV-B) radiation (280-320 nm) caused by stratospheric ozone depletion have major negative impacts on plant morphology, physiology, growth, development and production (Begum, 2022). Light influences the growth, development and pigment production of *Brassica rapa subsp. rapa cv. Tsuda*, a common vegetable crop. Despite the fact that microRNAs (miRNAs) play an important role in plant metabolism and abiotic stress responses, it is not clear whether miRNAs are involved in anthocyanin formation and development in light-exposed Tsuda seedlings. In reaction to blue and UV-A light, 17 conserved and 226 new miRNAs varied at least twofold from dark-treated values. Blue light stimulated BrmiR159, BrmiRC0191, BrmiRC0460, BrmiRC0323, BrmiRC0418 and BrmiRC0005, according to real-time PCR, but the transcription level of BrmiR167 did not differ substantially between seedlings treated with dark, blue, or UV light, according to northern blot. Only SPL9 (Bra004674) and SPL15 (Bra003305), two of BrmiR156 and BrmiR157's eight putative targets in the SPL gene family, increased in expression in response to blue or UV-A. In addition, miR157 cleaved target SPL9 mRNAs (Bra004674, Bra016891) and SPL15 mRNAs (Bra003305, Bra014599) at 10 or 11 bases from the miR157 binding region's 5′ ends. The discovery of a set of miRNAs and their targets that regulate the light-induced photomorphogenic phenotype in *Brassica rapa* seedlings revealed new information about blue and UV-A light-responsive miRNAs in Tsuda seedlings, as well as evidence of multiple miRNA targets and their various roles in plant development (Zhou *et al.*, 2016). Because shifting light quality can generate a variety of developmental reactions, some of which can be harmful, a number of screening methods have been developed to better understand how light, notably bright light, UV-A and UV-B radiation, affects miRNA accumulation and their targets. Zhou *et al.* (2016) detected 15 conserved and 226 unique differentially expressed miRNAs in *Brassica rapa* seedlings exposed to blue light and UV-A, which may target genes encoding regulators of plant growth, development and photomorphogenesis (Zhou *et al.*, 2016). In *Arabidopsis,* MiR156 and SPLs are involved in complex feedback loops (Wu *et al.*, 2009). MiR156 levels fall as plants develop, whereas SPLs (such as SPL9) increase, limiting anthocyanin production (Gou *et al.*, 2011). Blue and UV-A light activate genes involved in light signalling, some of which boost anthocyanin synthesis, according to Zhou *et al.* (2016). As anthocyanin levels rise, regulatory feedback loops, like miR156/157, keep the metabolism under control. UV-B light affects plant development, physiology and metabolism in the same way as UV-A light does (Dotto and Casati, 2017). Zhou *et al.* (2007) used a computational technique to predict miRNA genes activated by UV-B light in *Arabidopsis*, which was one of the first findings on the effect of light on miRNAs. According to researchers, UV-B radiation has been demonstrated to up-regulate miRNAs from 11 different families (Zhou *et al.*, 2007). UV-B has been shown to modify miRNA levels in aspen (*Populus tremula*), maize (*Zea mays*), and wheat (*Zea mays*); albeit these findings have not been confirmed in *Arabidopsis* (*Triticum aestivum).* According to Jia *et al.* (2009), UV-B light exposure increased the levels of eight miRNA families in aspen, whereas the levels of seven families decreased. Six of the 16 UV-B-regulated miRNA families were elevated in maize, while 10 were down-regulated (Casati, 2013). In UV-B-treated wheat plants, three known miRNAs were up-regulated and three were down-regulated, as well as a novel wheat miRNA was modestly up-regulated early after UV-B exposure and thereafter down-regulated (Wang *et al.*, 2013). At least two plant species share 13 UV-B-responsive miRNA families: miR156/157, miR159, miR160, miR164, miR165/166, miR167, miR169, miR170/171, miR172, miR393, miR395, miR398 and miR399. UV-B radiation consistently promotes (miR165/166, miR167 and miR398) or represses miR395, particularly miRNAs in three species. The relevance of these miRNAs in plant UV-B radiation responses accounts for this conservation. Many light-related motifs are discovered in the upstream regulatory regions of UV-B-regulated MiRNA genes (Zhou *et al.*, 2007; Jia *et al.*, 2009; Wang *et al.*, 2013), which is consistent with the action of light on these genes. Several promoters had stress-responsive cis-elements (Jia *et al.*, 2009; Wang *et al.*, 2013). UV-B-regulated miRNAs have a variety of targets, including transcription factors, auxin signalling factors and stress response factors (Zhou *et al.*, 2007; Jia *et al.*, 2009; Casati, 2013). In most cases, UV-B has an inverse connection with miRNAs and their corresponding target transcripts, as one would expect (Zhou *et al.*, 2007; Jia *et al.*, 2009; Casati, 2013; Wang *et al.*, 2013). An increase in miR165/166 in maize, for example, was linked to the inhibition of rolled leaf1, an HD-ZIP III transcription factor (RLD1) (Casati, 2013). Leaf arching has been related to RLD1 regulation by UV-B, implying that these miRNAs are involved in delimiting HD-ZIP III transcripts to the adaxial side of leaf primordia (Juarez *et al.*, 2004; Casati *et al.*, 2006). In aspen and wheat, overexpression of ATP sulphurylases (APS), enzymes involved in inorganic sulphate absorption, has been linked to miR395 down-regulation (Jia *et al.*, 2009; Wang *et al.*, 2013). Overexpression of the UV-B-regulated miRNA MiR395 affects *Arabidopsis'* sensitivity to abiotic stress conditions and changes APS transcript levels (Jones-Rhoades and Bartel, 2004; Kim *et al.*, 2010), linking a UV-B-regulated miRNA to other stress responses. Down-regulation of miR164 in UV-B-treated maize leaves is connected to increased levels of two stress-responsive NAC-domain protein target transcripts, as well as extostosin protein-like and an aspartyl protease target transcript (Hegedus *et al.*, 2003; Casati, 2013). UV-B radiation also inhibited miRNAs (miR171, miR172, miR156 and miR529) that target genes involved in critical developmental transitions (Lauter *et al.*, 2005; Xie *et al.*, 2006; Chuck *et al.*, 2008; Cuperus *et al.*, 2011). This demonstrates that UV-B exposure causes miRNA-mediated responses to postpone developmental transitions, allowing plants to heal UV-B-induced damage or adapt to harsh environments, as shown in Table 1. Overall, researchers discovered that a core group of UV-B-responsive miRNAs is conserved across different plant species. The findings do, however, show that different species may recruit alternative miRNAs to cope with UV-B damage. More research is needed to understand the involvement of these miRNAs in UV-B-regulated processes (Sánchez-Retuerta *et al.*, 2018).

In plants, UV-B radiation causes physiological and developmental alterations. The global miRNA expression profiles in juvenile maize leaves were studied using small RNA microarrays under control conditions in the absence of UV-B and after an eight-hour UV-B treatment, similar to ambient UV-B levels, to determine the mechanisms of UV-B gene regulation. Seventeen miRNAs were found to express differentially in response to UV-B. The expression of possible mRNA targets was studied in mRNA microarrays using the same RNA samples used for small RNA analysis. In general, miRNA expression has a strong negative correlation. In maize, UV-B radiation influences post-transcriptional regulation, a critical stage in gene expression (Casati, 2013).

8. Conclusion and Future Prospects

miRNAs have recently been identified as novel and important regulators of biological processes, such as development and disease. Knowing more about miRNAs and their targets in relation to DNA repair processes in plants is critical, given their importance. We might employ genome-editing tools, like CRISPR-Cas to modify miRNAs, to battle a wide spectrum of genotoxic stimuli once we understand the precise role of miRNAs and their targets in DNA repair and ROS scavenging. The discovery of a spate of miRNA transcription and processing actors has thrown light on pri-miRNA co-transcriptional splicing, modification and processing. The formation, composition and importance of the dicing body in co-transcriptional pri-miRNA processing are of great interest at the subcellular level. To solve these issues, novel approaches, like single-cell biology and *in vitro* re-constitution of the dicing machinery would be necessary. Although a range of parameters, such as AGOs, targets and 3′ alterations influence miRNA stability, little is known about the enzymes and modulators involved. It will also be crucial to understand the biological significance of global or sequence-specific miRNA degradation throughout developmental transitions and in response to environmental stressors. Understanding how miRNAs affect plant development is essential for crop breeding. When dominant genes are taken out during development, plants perish but miRNAs can safely control gene expression and help plants develop. The number of branches in rice determines grain yield. The genes that governed rice tillering and panicle branches were revealed to be miR156/miR529/SPL and miR172/AP2 modules. Tillering is inhibited by the SPL gene, although inflorescence meristem and spikelet transformation are enhanced. Changes in SPL expression will reduce panicle branching (Wang and Wang, 2015). In the regulation of seed size and grain yield, OsmiR397 can boost grain size, induce panicle branching and improve grain production by down-regulating its target gene OsLAC (Zhang *et al.*, 2013). The MiR1432-OsACOT modules are important for rice because they are involved in fatty acid metabolism and plant hormone production (Zhao *et al.*, 2019). MiR319s lower the number of tillers and grain production by targeting OsTCP21 and OsGAmyb (Wang *et al.*, 2021). Several 'miRNA-target gene' loops that control rice immunity and growth are affected by changes in the 'miR168-AGO1' regulatory pathway. The 'miR535-SPL14' loop controls rice yield and immunity, whereas the 'miR164-NAC11' loop controls rice growth period and immunity and rice immunity is controlled by miR1320 (Wang *et al.*, 2021). Tasselseed4 encodes miR172 in maize, which affects meristem cell fate and sex determination by targeting IDS1 (Indeterminate Spikelet1). Furthermore, during the shift from the juvenile to adult ear, miR156a-l regulates multiple SPL genes and indirectly activates miR172 via SPLs (Lauter *et al.*, 2005; Chuck *et al.*, 2007; Salvi *et al.*, 2007). Epigenetic changes in agriculture account for a significant amount of agricultural production variation. More research is needed to better understand how intermediate processes affect miRNA mobility and non-cellular autonomous function.

Also more research on the relationship between hormone concentration and miRNA mobility is needed to better understand and prepare plants for abiotic stressors. Plants can also create new sRNA in response to biotic and abiotic stressors. In response to stress, like nitrogen shortage, *A. thaliana* can create a significant number of 22 nt siRNAs that are dependent on DCL2 and RDR6. It's still a mystery why only a few gene loci in *A. thaliana* can create 22 nt siRNAs. Meanwhile, there is a significant gap between what we know about the manufacture of 22-nt siRNA and what we know about their biological activity. Due to sRNA's cellular non-autonomy, more evidence is needed to show whether 22 nt siRNAs can effect target genes in distal organs. As a result, breakthroughs in sequencing technology and miRNA research approaches are aggressively encouraged. More miRNAs, their action mechanisms and regulatory pathways will be discovered in model plants using single-cell omics and nanopore sequencing, providing an important theoretical foundation for understanding how miRNA regulates plant growth and development, which can then be applied to agriculturally important plants. miRNAs are now recognised to be among the most important gene regulators, as evidenced by the case studies presented. Plant miRNA research and characterisation have advanced significantly in recent years, with a rising number of study publications focusing on miRNAs' critical significance in agricultural plants. In practice, many transgenic techniques focus on important miRNAs and their known target genes can be applied. Only a few examples include constitutive miRNA overexpression, stress-induced or tissue-specific miRNA or target expression, evolution of miRNA-resistant target genes, artificial target mimics and artificial miRNAs (Gupta, 2015; Dong *et al.*, 2022). When the native target gene has a negative effect, constitutive overexpression of the regulating miRNA suppresses the matching mRNA – a strategy that could be used for crop improvement if the miRNAs of interest are positive stress regulators (Yang *et al.*, 2013). When the target gene has a favourable effect on the trait of interest (and the miRNA acts as a negative regulator), overexpression or miRNA-r selection can be applied (Gupta, 2015). One of the most useful approaches for crop improvement is to utilise artificial miRNAs to reduce target gene expression of a protein-coding mRNA of interest. Because synthetic nucleic acids create a single miRNA capable of silencing a single target gene, all potential unwanted target genes may be foreseen and avoided during the experimental design stages (Schwab *et al.*, 2010). Another way to achieve a desired result is to use engineered decoys to change miRNA regulatory networks by changing miRNA activity (Ivashuta *et al.*, 2011). miRNA decoys, which are endogenous RNAs that can suppress miRNA activity, are a versatile and robust tool for understanding the function of miRNA families and targeted gene expression engineering in plants (Dong *et al.*, 2022). By modifying the sequence of miRNA decoy sites, which minimises miRNA inactivation, this technique enables fine regulation of native miRNA targets and the production of a suitable range of plant phenotypes. The chemical complex that may form during the interaction between miRNA and the miRNA decoy is unknown, and it has the potential to cause miRNA instability in plants. The expression of a miRNA or its target gene is altered, resulting in unwanted pleiotropic alterations in plant development and morphology. As a result, it is vital to understand how miRNA regulates plant growth and development, as well as plant responses to various abiotic and biotic stresses (Dong *et al.*, 2022).

References

Abas, Y., Touil, N., Kirsch-Volders, M., Angenon, G., Jacobs, M. and Famelaer, I.D.H. (2007). Evaluation of UV damage at DNA level in *Nicotiana plumbaginifolia* protoplasts using single cell gel electrophoresis, *Plant. Cell, Tissue, Organ, Cult.*, 91: 145-154.

Axtell, M.J. and Meyers, B.C. (2018). Revisiting criteria for plant microRNA annotation in the era of big data, *Plant Cell*, 30: 272-284.

Ballare, C.L., Rousseaux, M.C., Searles, P.S., Zaller, J.G., Giordano, C.V., Robson, T.M., Caldwell, M.M., Sala, O.E. and Scopel, A.L. (2001). Impacts of solar ultraviolet-B radiation on terrestrial ecosystems of Tierra del Fuego (southern Argentina): An overview of recent progress, *J. Photochem. Photobiol. B: Biol.*, 62: 67-77.

Bartel, D.P. (2018). Metazoan micrornas, *Cell*, 173: 20-51.

Begum, Y. (2022). Regulatory role of microRNAs (miRNAs) in the recent development of abiotic stress tolerance of plants, *Gene*, 146283.

Borges, F. and Martienssen, R.A. (2015). The expanding world of small RNAs in plants, *Nat. Rev. Mol. Cell. Biol.*, 16(12): 727-741.

Boubriak, I.I., Grodzinsky, D.M., Polischuk, V.P., Naumenko, V.D., Gushcha, N.P., Micheev, A.N., McCready, S.J. and Osborne, D.J. (2008). Adaptation and impairment of DNA repair function in pollen of *Betula verrucosa* and seeds of *Oenothera biennis* from differently radionuclide-contaminated sites of Chernobyl, *Ann. Bot.*, 101: 267-276.

Brant, E.J. and Budak, H. (2018). Plant small non-coding RNAs and their roles in biotic stresses, *Front. Plant. Sci.*, 9: 1038.

Bray, C.M. and West, C.E. (2005). DNA repair mechanisms in plants: Crucial sensors and effectors for the maintenance of genome integrity, *New. Phytol.*, 168: 511-528.

Bresso, E.G., Chorostecki, U., Rodriguez, R.E., Palatnik, J.F. and Schommer, C. (2018). Spatial control of gene expression by miR319-regulated TCP transcription factors in leaf development, *Plant Physiol.*, 176: 1694-1708.

Budak, H., Kantar, M., Bulut, R. and Akpinar, B.A. (2015). Stress responsive miRNAs and isomiRs in cereals, *Plant Sci.*, 235: 1-13.

Carbonell-Bejerano, P., Diago, M.P., Martínez-Abaigar, J., Martínez-Zapater, J.M., Tardáguila, J. and Núñez-Olivera, E. (2014). Solar ultraviolet radiation is necessary to enhance grapevine fruit ripening transcriptional and phenolic responses, *BMC Plant Biol.*, 14: 1-16.

Casati, P. (2013). Analysis of UV-B regulated miRNAs and their targets in maize leaves, *Plant Signal. Behav.*, 8: e26758.

Casati, P., Stapleton, A.E., Blum, J.E. and Walbot, V. (2006). Genome-wide analysis of high-altitude maize and gene knockdown stocks implicates chromatin remodeling proteins in response to UV-B, *Plant. J.*, 46: 613-627.

Chuck, G., Meeley, R. and Hake, S. (2008). Floral meristem initiation and meristem cell fate are regulated by the maize AP2 genes *ids1 and sid1*, *Development*, 135: 3013-3019.

Chuck, G., Meeley, R., Irish, E., Sakai, H. and Hake, S. (2007). The maize tasselseed4 microRNA controls sex determination and meristem cell fate by targeting tasselseed6/indeterminate spikelet1, *Nat. Genet.*, 39: 1517-1521.

Culligan, K., Tissier, A. and Britt, A. (2004). ATR regulates a G2-phase cell-cycle checkpoint in *Arabidopsis thaliana*, *Plant Cell*, 16: 1091-1104.

Cuperus, J.T., Fahlgren, N. and Carrington, J.C. (2011). Evolution and functional diversification of miRNA genes, *Plant Cell*, 23: 431-442.

Czech, B., Munafò, M., Ciabrelli, F., Eastwood, E.L., Fabry, M.H., Kneuss, E. and Hannon, G.J. (2018). piRNA-guided genome defence: From biogenesis to silencing, *Annu. Rev. Genet.*, 52: 131-157.

Danisman, S. (2016). TCP transcription factors at the interface between environmental challenges and the plant's growth responses, *Front. Plant Sci.*, 7: 1930.

Das, K. and Roychoudhury, A. (2014). Reactive oxygen species (ROS) and response of antioxidants as ROS-scavengers during environmental stress in plants, *Front. Environ. Sci.*, 2: 53.

Dexheimer, P.J. and Cochella, L. (2020). MicroRNAs: From mechanism to organism, *Front. Cell Dev. Biol.*, 8: 409.

Djami-Tchatchou, A.T., Sanan-Mishra, N., Ntushelo, K. and Dubery, I.A. (2017). Functional roles of microRNAs in agronomically important plants-potential as targets for crop improvement and protection, *Front. Plant Sci.*, 8: 378.

Dong, Q., Hu, B. and Zhang, C. (2022). MicroRNAs and their roles in plant development, *Front. Plant. Sci.*, 13: 824240.

Dotto, M. and Casati, P. (2017). Developmental reprogramming by UV-B radiation in plants, *Plant Sci.*, 264: 96-101.

Facella, P., Carbone, F., Placido, A. and Perrotta, G. (2017). Cryptochrome 2 extensively regulates transcription of the chloroplast genome in tomato, *FEBS Open. Bio.*, 7: 456-471.

Fátyol, K., Ludman, M. and Burgyán, J. (2016). Functional dissection of a plant Argonaute, *Nucleic Acids Res.*, 44: 1384-1397.

Friedberg, E.C. (2015). A history of the DNA repair and mutagenesis field: I. The discovery of enzymatic photoreactivation, *DNA Repair* (*Amst*), 33: 35-42.

Gill, S.S., Anjum, N.A., Gill, R., Jha, M. and Tuteja, N. (2015). DNA damage and repair in plants under ultraviolet and ionising radiations, *Sci. World J.*, 2015.

Gou, J.Y., Felippes, F.F., Liu, C.J., Weigel, D. and Wang, J.W. (2011). Negative regulation of anthocyanin biosynthesis in *Arabidopsis* by a miR156-targeted SPL transcription factor, *Plant Cell*, 23: 1512-1522.

Gu, K., Mok, L. and Chong, M.M. (2018). Regulating gene expression in animals through RNA endonucleolytic cleavage, *Heliyon*, 4: e00908.

Gupta, P.K. (2015). MicroRNAs and target mimics for crop improvement, *Current Science*, 1624-1633.

Hajieghrari, B., Farrokhi, N., Goliaei, B. and Kavousi, K. (2017). Computational identification of microRNAs and their transcript target (s) in field mustard (*Brassica rapa* L.), *Iran. J. Biotechnol.*, 15: 22.

Hammond, S.M. (2015). An overview of microRNAs, *Adv. Drug Deliv. Rev.*, 87: 3-14.

He, Q., Lu, Q., He, Y., Wang, Y., Zhang, N., Zhao, W. and Zhang, L. (2020). Dynamic changes of the anthocyanin biosynthesis mechanism during the development of heading Chinese cabbage (*Brassica rapa* L.) and *Arabidopsis* under the control of BrMYB2, *Front. Plant Sci.*, 11: 593766.

Hegedus, D., Yu, M., Baldwin, D., Gruber, M., Sharpe, A., Parkin, I., Whitwill, S. and Lydiate, D. (2003). Molecular characterisation of Brassica napus NAC domain transcriptional activators induced in response to biotic and abiotic stress, *Plant Mol. Biol.*, 53: 383-397.

Hsieh, L.C., Lin, S.I., Shih, A.C.C., Chen, J.W., Lin, W.Y., Tseng, C.Y., Li, W.H. and Chiou, T.J. (2009). Uncovering small RNA-mediated responses to phosphate deficiency in *Arabidopsis* by deep sequencing, *Plant. Physiol.*, 151: 2120-2132.

Huang, J., Yang, M. and Zhang, X. (2016). The function of small RNAs in plant biotic stress response, *J. Integr. Plant Biol.*, 58: 312-327.

Ivashuta, S., Banks, I.R., Wiggins, B.E., Zhang, Y., Ziegler, T.E., Roberts, J.K. and Heck, G.R. (2011). Regulation of gene expression in plants through miRNA inactivation, *PLoS ONE*, 6: 21330.

Jenkins, G.I. (2017). Photomorphogenic responses to ultraviolet-B light, *Plant Cell Environ.*, 40: 2544-2557.

Jia, X., Ren, L., Chen, Q.J., Li, R. and Tang, G. (2009). UV-B-responsive microRNAs in *Populus tremula*, *J. Plant. Physiol.*, 166: 2046-2057.

Jiang, L., Wang, Y., Björn, L.O. and Li, S. (2011). UV-B-induced DNA damage mediates expression changes of cell cycle regulatory genes in *Arabidopsis* root tips, *Planta*, 233: 831-841.

Jones-Rhoades, M.W. and Bartel D.P. (2004). Computational identification of plant microRNAs and their targets, including a stress-induced miRNA, *Molecular Cell*, 14: 787-799.

Juarez, M.T., Kui, J.S., Thomas, J., Heller, B.A. and Timmermans, M.C. (2004). MicroRNA-mediated repression of rolled leaf1 specifies maize leaf polarity, *Nature*, 428: 84-88.

Kim, J.Y., Lee, H.J., Jung, H.J., Maruyama, K., Suzuki, N. and Kang, H. (2010). Overexpression of microRNA395c or 395e affects differently the seed germination of *Arabidopsis thaliana* under stress conditions, *Planta*, 232: 1447-1454.

Kimura, S. and Sakaguchi, K. (2006). DNA repair in plants, *Chem. Rev.*, 106: 753-766.

Koyama, T., Sato, F. and Ohme-Takagi, M. (2017). Roles of miR319 and TCP transcription factors in leaf development, *Plant Physiology*, 175(2): 874-885. https://doi.org/10.1104/pp.17.00732

La, H., Ding, B., Mishra, G.P., Zhou, B., Yang, H., del Rosario Bellizzi, M., Chen, S., Meyers, B.C., Peng, Z., Zhu, J.K. and Wang, G.L. (2011). A 5-methylcytosine DNA glycosylase/lyase demethylates the retrotransposon Tos17 and promotes its transposition in rice, *Proc. Natl. Acad. Sci.*, 108: 15498-15503.

Lario, L.D., Botta, P., Casati, P. and Spampinato, C.P. (2015). Role of AtMSH7 in UV-B-induced DNA damage recognition and recombination, *J. Exp. Bot.*, 66: 3019-3026.

Lauter, N., Kampani, A., Carlson, S., Goebel, M. and Moose, S.P. (2005). MicroRNA172 down-regulates glossy15 to promote vegetative phase change in maize, *Natl. Acad. Sci.*, 102: 9412-9417.

Li, H., Wang, Y., Wang, Z., Guo, X., Wang, F., Xia, X.J., Zhou, J., Shi, K., Yu, J.Q. and Zhou, Y.H. (2016). Microarray and genetic analysis reveals that csa-miR159b plays a critical role in abscisic acid-mediated heat tolerance in grafted cucumber plants, *Plant Cell Environ.*, 39: 1790-1804.

Li, J., Ren, L., Gao, Z., Jiang, M., Liu, Y., Zhou, L., He, Y. and Chen, H. (2017). Combined transcriptomic and proteomic analysis constructs a new model for light-induced anthocyanin biosynthesis in eggplant (*Solanum melongena* L.), *Plant Cell, Environ.*, 40: 3069-3087.

Li, L., Chang, S.S. and Liu, Y. (2010). RNA interference pathways in filamentous fungi, *Cell. Mol. Life Sci.*, 67: 3849-3863.

Li, S., Castillo-González, C., Yu, B. and Zhang, X. (2017). The functions of plant small RNA s in development and in stress responses, *Plant J.*, 90: 654-670.

Liu, J., Tang, X., Gao, L., Gao, Y., Li, Y., Huang, S., Sun, X., Miao, M., Zeng, H., Tian, X. and Niu, X.(2012).A role of tomato UV-damaged DNA binding protein 1 (DDB1) in organ size control via an epigenetic manner, *PLoS.One*, 7: e42621. Incom

Luo, Q.J., Mittal, A., Jia, F. and Rock, C.D. (2012). An auto-regulatory feedback loop involving PAP1 and TAS4 in response to sugars in *Arabidopsis, Plant Mol. Biol.*, 80: 117-129.

Macovei, A. and Tuteja, N. (2013). Different expression of miRNAs targeting helicases in rice in response to low and high dose rate γ-ray treatments, *Plant. Signal. Behav.*, 8: 25128.

Manova, V. and Gruszka, D. (2015). DNA damage and repair in plants – From models to crops, *Front. Plant Sci.*, 6: 885.

Massonnet, M., Fasoli, M., Tornielli, G.B., Altieri, M., Sandri, M., Zuccolotto, P., Paci, P., Gardiman, M., Zenoni, S. and Pezzotti, M. (2017). Ripening transcriptomic program in red and white grapevine varieties correlates with berry skin anthocyanin accumulation, *Plant Physiol.*, 174: 2376-2396.

Matus, J.T., Cavallini, E., Loyola, R., Höll, J., Finezzo, L., Dal Santo, S. and Arce-Johnson, P. (2017). A group of grapevine MYBA transcription factors located in chromosome 14 control anthocyanin synthesis in vegetative organs with different specificities compared with the berry colour locus, *Plant J.*, 91: 220-236.

Mica, E., Piccolo, V., Delledonne, M., Ferrarini, A., Pezzotti, M., Casati, C., Del Fabbro, C., Valle, G., Policriti, A., Morgante, M. and Pesole, G. (2010). Erratum to: High throughput approaches reveal splicing of primary microRNA transcripts and tissue specific expression of mature microRNAs in *Vitis vinifera*, *BMC Genom.*, 11: 1-15.

Molnar, A., Schwach, F., Studholme, D.J., Thuenemann, E.C. and Baulcombe, D.C. (2007). miRNAs control gene expression in the single-cell alga *Chlamydomonas reinhardtii*, *Natr.*, 447: 1126-1129.

Moro, B., Chorostecki, U., Arikit, S., Suarez, I.P., Höbartner, C., Rasia, R.M., Meyers, B.C. and Palatnik, J.F. (2018). Efficiency and precision of microRNA biogenesis modes in plants, *Nucleic Acids Res.*, 46: 10709-10723.

Neumeier, J. and Meister, G. (2021). SiRNA specificity: RNAi mechanisms and strategies to reduce off-target effects, *Front. Plant. Sci.*, 11: 526455.

Nicolas, M. and Cubas, P. (2016). TCP factors: New kids on the signalling block, *Curr. Opin. Plant Biol.*, 33: 33-41.

O'Brien, J., Hayder, H., Zayed, Y. and Peng, C. (2018). Overview of microRNA biogenesis, mechanisms of actions, and circulation, *Front. Endocrinol.*, 9: 402.

Pandita, D. (2019). Plant MIRnome: miRNA biogenesis and abiotic stress response. *In:* Hasanuzzaman, M., Hakeem, K., Nahar, K. and Alharby, H. (Eds.). *Plant Abiotic Stress Tolerance*. Springer, Cham., 449-474. Doi https://Doi.org/10.1007/978-3-030-06118-0_18

Pandita, D. and Wani, S.H. (2019). MicroRNA as a tool for mitigating abiotic stress in rice (*Oryza sativa* L.). *In:* Wani, S. (Ed.). *Recent Approaches in Omics for Plant Resilience to Climate Change*. Springer, Chamz., 109-133. Doi https://Doi.org/10.1007/978-3-030-21687-0_6

Pandita, D. (2021). Role of miRNAi technology and miRNAs in abiotic and biotic stress resilience. *In:* Aftab, T. and Roychoudhury, A. (Eds.). *Plant Perspectives to Global Climate Changes*. Academic Press, Elsevier, 303-330. https://Doi.org/10.1016/B978-0-323-85665-2.00015-7

Pandita, D. (2022a). How microRNAs regulate abiotic stress tolerance in wheat? A snapshot. *In:* Roychoudhury, A., Aftab, T. and Acharya, K. (Eds.). *Omics Approach to Manage Abiotic Stress in Cereals*. Springer, Singapore, 447-464. https://Doi.org/10.1007/978-981-19-0140-9_17

Pandita, D. (2022b). MicroRNAs shape the tolerance mechanisms against abiotic stress in maize. *In:* Roychoudhury, A., Aftab, T. and Acharya, K. (Eds.). *Omics Approach to Manage Abiotic Stress in Cereals*. Springer, Singapore, 479-493. https://Doi.org/10.1007/978-981-19-0140-9_19

Pandita, D. (2022c). miRNA- and RNAi-mediated metabolic engineering in plants. *In:* Aftab, T. and Hakeem, K.R. (Eds.). *Metabolic Engineering in Plants*. Springer, Singapore, 171-186. https://Doi.org/10.1007/978-981-16-7262-0_7

Pantaleo, V., Szittya, G., Moxon, S., Miozzi, L., Moulton, V., Dalmay, T. and Burgyan, J. (2010). Identification of grapevine microRNAs and their targets using high-throughput sequencing and degradome analysis, *Plant J.*, 62: 960-976.

Paradiso, R. and Proietti, S. (2022). Light-quality manipulation to control plant growth and photomorphogenesis in greenhouse horticulture: The state of the art and the opportunities of modern LED systems, *J. Plant. Growth. Regul.*, 41: 742-780.

Piofczyk, T., Jeena, G. and Pecinka, A. (2015). *Arabidopsis thaliana* natural variation reveals connections between UV radiation stress and plant pathogen-like defense responses, *Plant Physiol. Biochem.*, 93: 34-43.

Qin, Z., Li, C., Mao, L. and Wu, L. (2014). Novel insights from non-conserved microRNAs in plants, *Front. Plant Sci.*, 5: 586.

Rastogi, R.P., Kumar, A., Tyagi, M.B. and Sinha, R.P. (2010). Molecular mechanisms of ultraviolet radiation-induced DNA damage and repair, *J. Nucleic Acids*, 2010: 1-32.

Ries, G., Heller, W., Puchta, H., Sandermann, H., Seidlitz, H.K. and Hohn, B. (2000). Elevated UV-B radiation reduces genome stability in plants, *Nature*, 406: 98-101.

Rock, C.D. (2013). Transacting small interfering RNA4: Key to nutraceutical synthesis in grape development? *Trends Plant. Sci.*, 18: 601-610.

Salvi, S., Sponza, G., Morgante, M., Tomes, D., Niu, X., Fengler, K.A., Meeley, R., Ananiev, E.V., Svitashev, S., Bruggemann, E. and Li, B. (2007). Conserved noncoding genomic sequences associated with a flowering-time quantitative trait locus in maize, *Proc. Natl. Acad. Sci.*, 104: 11376-11381.

Sancar, A. (2003). Structure and function of DNA photolyase and cryptochrome blue-light photoreceptors, *Chem. Rev.*, 103: 2203-2238.

Sánchez-Retuerta, C., Suaréz-López, P. and Henriques, R. (2018). Under a new light: Regulation of light-dependent pathways by non-coding RNAs, *Front. Plant. Sci.*, 9: 962.

Schwab, R., Palatnik, J.F., Riester, M., Schommer, C., Schmid, M. and Weigel, D. (2005). Specific effects of microRNAs on the plant transcriptome, *Dev. Cell*, 8: 517-527.

Shriram, V., Kumar, V., Devarumath, R.M., Khare, T.S. and Wani, S.H. (2016). MicroRNAs as potential targets for abiotic stress tolerance in plants, *Front. Plant Sci.*, 7: 817.

Spampinato, C.P. (2017). Protecting DNA from errors and damage: An overview of DNA repair mechanisms in plants compared to mammals, *Cell. Mol. Life Sci.*, 74: 1693-1709.

Sunitha, S., Loyola, R., Alcalde, J.A., Arce-Johnson, P., Matus, J.T. and Rock, C.D. (2019). The role of UV-B light on small RNA activity during grapevine berry development, *G3 (Bethesda)*, 9: 769-787.

Suzuki, M., Nakabayashi, R., Ogata, Y., Sakurai, N., Tokimatsu, T., Goto, S., Suzuki, M., Jasinski, M., Martinoia, E., Otagaki, S. and Matsumoto, S. (2015). Multiomics in grape berry skin revealed specific induction of the stilbene synthetic pathway by ultraviolet-C irradiation, *Plant Physiol.*, 168: 47-59.

Takahashi, M., Teranishi, M., Ishida, H., Kawasaki, J., Takeuchi, A., Yamaya, T., Watanabe, M., Makino, A. and Hidema, J. (2011). Cyclobutane pyrimidine dimer (CPD) photolyase repairs ultraviolet-B-induced CPDs in rice chloroplast and mitochondrial DNA, *Plant. J.*, 66: 433-442.

Takshak, S. and Agrawal, S.B. (2019). Defence potential of secondary metabolites in medicinal plants under UV-B stress, *J. Photochem. Photobiol. B: Biol.*, 193: 51-88.

Taylor, R.S., Tarver, J.E., Hiscock, S.J. and Donoghue, P.C. (2014). Evolutionary history of plant microRNAs, *Trends. Plant. Sci.*, 19: 175-182.

Teranishi, M., Taguchi, T., Ono, T. and Hidema, J. (2012). Augmentation of CPD photolyase activity in japonica and Indica rice increases their UVB resistance but still leaves the difference in their sensitivities, *Photochem. Photobiol. Sci.*, 11: 812-820.

Tilbrook, K., Dubois, M., Crocco, C.D., Yin, R., Chappuis, R., Allorent, G., Schmid-Siegert, E., Goldschmidt-Clermont, M. and Ulm, R. (2016). UV-B perception and acclimation in *Chlamydomonas reinhardtii*, *Plant Cell*, 28: 966-983.

Tissot, N. and Ulm, R. (2020). Cryptochrome-mediated blue-light signalling modulates UVR8 photoreceptor activity and contributes to UV-B tolerance in *Arabidopsis*, *Nat. Commun.*, 11: 1-10.

Treiber, T., Treiber, N. and Meister, G. (2019). Regulation of microRNA biogenesis and its crosstalk with other cellular pathways, *Nat. Rev. Mol. Cell Biol.*, 20: 5-20.

Tuteja, N. and Tuteja, R. (2001). Unravelling DNA repair in human: Molecular mechanisms and consequences of repair defect, *Crit. Rev. Biochem. Mol. Biol.*, 36: 261-290.

Tuteja, N. (2009). Signaling through G protein coupled receptors, *Plant Signal. Behav.*, 4: 942-947.

Tuteja, N., Tiburcio, A.F., Fortes, A.M. and Bartels, D. (2011). Introduction to PSB Special Issue: Plant abiotic stress, *Plant Signal. Behav.*, 6: 173-174.

Vanhaelewyn, L., Van Der Straeten, D., De Coninck, B. and Vandenbussche, F. (2020). Ultraviolet radiation from a plant perspective: The plant-microorganism context, *Front. Plant Sci.*, 1984.

Wang, B., Sun, Y.F., Song, N., Wang, X.J., Feng, H., Huang, L.L. and Kang, Z.S. (2013). Identification of UV-B-induced microRNAs in wheat, *Genet. Mol. Res.*, 12: 4213-4221.

Wang, H. and Wang, H. (2015). The miR156/SPL module, a regulatory hub and versatile toolbox, gears up crops for enhanced agronomic traits, *Mol. Plants*, 8: 677-688.

Wang, H., Li, Y., Chern, M., Zhu, Y., Zhang, L.L., Lu, J.H., Li, X.P., Dang, W.Q., Ma, X.C., Yang, Z.R. and Yao, S.Z. (2021). Suppression of rice miR168 improves yield, flowering time and immunity, *Nat. Plants*, 7: 129-136.

Waterworth, W.M., Jiang, Q., West, C.E., Nikaido, M. and Bray, C.M. (2002). Characterisation of *Arabidopsis* photolyase enzymes and analysis of their role in protection from ultraviolet-B radiation, *J. Exp. Bot.*, 53: 1005-1015.

Wei, W., Ba, Z., Gao, M., Wu, Y., Ma, Y., Amiard, S., White, C.I., Danielsen, J.M.R., Yang, Y.G. and Qi, Y. (2012). A role for small RNAs in DNA double-strand break repair, *Cell*, 149: 101-112.

Wu, G., Park, M.Y., Conway, S.R., Wang, J.W., Weigel, D. and Poethig, R.S. (2009). The sequential action of miR156 and miR172 regulates developmental timing in *Arabidopsis*, *Cell*, 138: 750-759.

Xie, K., Wu, C. and Xiong, L. (2006). Genomic organisation, differential expression, and interaction of squamosal-promoter-binding-like transcription factors and microRNA156 in rice, *Plant Physiol.*, 142: 280-293.

Xie, M., Zhang, S. and Yu, B. (2015). microRNA biogenesis, degradation and activity in plants, *Cell. Mol. Life Sci.*, 72: 87-99.

Xu, W., Jiang, X. and Huang, L. (2019). RNA interference technology, *Comprehensive Biotechnology*, 560.

Yamamoto, J., Martin, R., Iwai, S., Plaza, P. and Brettel, K. (2013). Repair of the (6–4) photoproduct by DNA photolyase requires two photons, *Angew. Chem. Int. Ed.*, 52: 7432-7436.

Yang, C., Li, D., Mao, D., Liu, X.U.E., Ji, C., Li, X., Zhao, X., Cheng, Z., Chen, C. and Zhu, L. (2013). Overexpression of microRNA 319 impacts leaf morphogenesis and leads to enhanced cold tolerance in rice (*Oryza sativa* L.), *Plant Cell Environ.*, 36: 2207-2218.

Yang, X., Zhang, L., Yang, Y., Schmid, M. and Wang, Y. (2021). miRNA mediated regulation and interaction between plants and pathogens, *Int. J. Mol. Sci.*, 22: 2913.

Yang, Y., Guo, J., Cheng, J., Jiang, Z., Xu, N., An, X., Chen, Z., Hao, J., Yang, S., Xu, Z. and Shen, C. (2020). Identification of UV-B radiation responsive microRNAs and their target genes in chrysanthemum (*Chrysanthemum morifolium* Ramat) using high-throughput sequencing, *Ind. Crops Prod.*, 151: 112484.

Yoshiyama, K.O., Sakaguchi, K. and Kimura, S. (2013). DNA damage response in plants: Conserved and variable response compared to animals, *Biology*, 2: 1338-1356.

Yu, X., Liu, H., Klejnot, J. and Lin, C. (2010). The cryptochrome blue light receptors, *The Arabidopsis Book/American Society of Plant Biologists*, 8.

Yu, Y., Jia, T. and Chen, X. (2017). The 'how' and 'where' of plant microRNAs, *New Phytol.*, 216: 1002-1017.

Zhang, C., Li, G., Wang, J. and Fang, J. (2012). Identification of trans-acting siRNAs and their regulatory cascades in grapevine, *Bioinformatics*, 28: 2561-2568. (not in text)

Zhang, Y.C., Yu, Y., Wang, C.Y., Li, Z.Y., Liu, Q., Xu, J., Liao, J.Y., Wang, X.J., Qu, L.H., Chen, F. and Xin, P. (2013). Overexpression of microRNA OsmiR397 improves rice yield by increasing grain size and promoting panicle branching, *Nat. Biotechnol.*, 31: 848.

Zhang, H., Zhao, X., Li, J., Cai, H., Deng, X.W. and Li, L. (2014). MicroRNA408 is critical for the HY5-SPL7 gene network that mediates the coordinated response to light and copper, *Plant. Cell*, 26: 4933-4953.

Zhang, T., Zhao, Y.L., Zhao, J.H., Wang, S., Jin, Y., Chen, Z.Q., Fang, Y.Y., Hua, C.L., Ding, S.W. and Guo, H.S. (2016). Cotton plants export microRNAs to inhibit virulence gene expression in a fungal pathogen, *Nat. Plants*, 2: 1-6.

Zhang, J., Khan, S.A., Heckel, D.G. and Bock, R. (2017). Next-generation insect-resistant plants: RNAi-mediated crop protection, *Trends. Biotechnol.*, 35: 871-882.

Zhao, T., Li, G., Mi, S., Li, S., Hannon, G.J., Wang, X.J. and Qi, Y. (2007). A complex system of small RNAs in the unicellular green alga *Chlamydomonas reinhardtii*, *Genes Dev.*, 21: 1190-1203.

Zhao, Y.F., Peng, T., Sun, H.Z., Teotia, S., Wen, H.L., Du, Y.X., Zhang, J., Li, J.Z., Tang, G.L., Xue, H.W. and Zhao, Q.Z. (2019). MiR1432-Os ACOT (Acyl-CoA thioesterase) module determines grain yield via enhancing grain filling rate in rice, *Plant Biotechnol. J.*, 17: 712-723.

Zhou, B., Fan, P., Li, Y., Yan, H. and Xu, Q. (2016). Exploring miRNAs involved in blue/UV-A light response in Brassica rapa reveals special regulatory mode during seedling development, *BMC Plant Biol.*, 16: 1-13.

Zhou, X., Wang, G. and Zhang, W. (2007). UV-B responsive microRNA genes in *Arabidopsis thaliana*, *Mol. Syst. Biol.*, 3: 103.

Zhu, L., Ow, D.W. and Dong, Z. (2018). Transfer RNA-derived small RNAs in plants, *Sci. China Life Sci.*, 61: 155-161.

CHAPTER

9

MicroRNA-mediated Regulation of Salinity Stress

Seyed Alireza Salami* and Shirin Moradi

Department of Horticultural Sciences, Faculty of Agricultural Science and Engineering, University of Tehran, Iran

1. Introduction

Climate change, global warming, and human activities cause soil drought and consequently soil salinity which pose major challenges in the agricultural sector. Extensive saline water and soil resources are major threats to sustainable agricultural production, especially in arid and semi-arid regions of the world. The Food and Agriculture Organisation of the United Nations (FAO) celebrates 'World Soil Day' every year on fifth of December. In 2021, it warned of the challenge posed by soil salinity in world agriculture and came out with the motto – 'stop soil salinity, increase soil productivity'. Salinity is not a new phenomenon and the relationship between humans and the issue of salinity has a historical background that has led to the decline of great civilisations. It is said that the destruction of the Sumerian civilisation in Mesopotamia around 1700 BC was due to the phenomenon of salinity (Jacobsen and Adams, 1958). Soil salinity is a dynamic and expanding phenomenon and no continent is immune to it. So far more than 100 countries in arid and semi-arid regions of the world are facing this challenge. Saline soil is defined as the soil which has electrical conductivity of more than 4 decisiemens per metre (dS/m) of sodium chloride salt and about 2 megapascals (MPa) of osmotic pressure (Gupta *et al.*, 2019).

Salinity stress affects plant growth and development; so understanding how plants cope with salinity stress can be helpful in improving plants quality and yield and finding solutions to overcome this challenge. Accordingly, many studies were conducted to find various pathways of plant response to salinity stress. During salinity stress, an increase in Na^+ concentration is observed and based on which the ion imbalance occurs due to salinity. An increase in osmolyte accumulation happens. Induction of some hormones is carried out due to water deficit after the increase in osmolytes. Abscisic acid (ABA) is one of the hormones which increases after water

---

*Corresponding author: asalami@ut.ac.ir

deficit and leads to closure of the stomata. Decrease in transpiration and photosynthesis after closure of stomata leads to photoinhibition and oxidative stress. At the molecular level, salinity leads to enhanced stress-response genes and activation of osmotic faction gene (Park *et al.*, 2016). In addition to genes and transcription factors involved in salinity tolerance or susceptibility, small non-coding RNAs, such as miRNAs, are known as regulators of important processes under harsh stress conditions at post-transcriptional level on the specific tissue at a certain time (de Lima *et al.*, 2012; Pandita, 2019, Pandita and Wani, 2019; Pandita, 2021; 2022a; 2022b; 2022c).

Not only transcription factors but also different miRNAs modulate the salinity stress signalling and several studies have highlighted the importance of miRNA-based regulation of gene expression under salt stress and which will be reviewed in this chapter. A deeper insight into multiple biological processes affected by miRNAs under salinity stress and how the positive and negative regulatory elements work at the transcriptional and post-transcriptional level allow us to breed superior plants towards salinity stress tolerance. In this chapter, various miRNAs, which have been specified to different salinity stress pathways, are discussed. This chapter also focuses on the roles, functions and implications of micro-RNAs in post-transcriptional gene regulation under salinity stress. miRNAs regulate vital processes in plants under salinity stress, including ion balance (K^+/N^+ balance and calcium signalling), plant growth, hormone signalling, osmotic homeostasis and antioxidant responses and its related defence systems (Fig. 1). One possible solution to tolerate salinity stress could be identification and introduction of salt-tolerant cultivars and rootstocks (Sohrabi *et al.*, 2017). Such tolerant plants, equipped with specific salinity-responsive genes and miRNAs, regulate networks and signalling pathways underlying salinity stress tolerance.

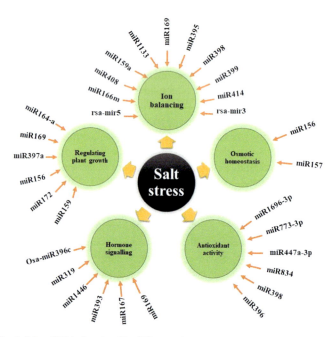

Fig. 1: MicroRNAs involved in different plant responses under salinity stress

2. Ion Imbalance

The synchronisation of salt stress and dominance of calcareous soils in warm and dry climate has led to alkaline salinity stress in the fields (Tahanian *et al.*, 2019). Salinity stress is the crucial factor that decreases plant yield and quality in arid areas (Zandkarimi *et al.*, 2015). High concentration of salt causes various changes at morphological, physiological, biochemical and molecular levels. The first change caused by salt stress is ion imbalance. Increasing salinity destroys the balance between ions, such as sodium (Na^+), calcium (Ca^+) and potassium (K^+).

Potassium (K^+) is the cation with high frequency in plant cell. It plays an important role in plant growth and development as it is essential for proteins and enzymes production and activation. Also, it plays a crucial role in photosynthesis, sugar transport and cell growth. Potassium is high movable cation in plants; based on that, it has an important role in balancing cell osmotic pressure and cation/anions in cytoplasm (Xu *et al.*, 2020). The balance between K^+ and Na^+ is essential for protecting some enzymatic activity and cell wall (Zhang *et al.*, 2018). Salt overly sensitive (SOS) is considered to be an important pathway which keeps K+/Na+ normal balance in plant cell with the help of Na^+/H^+ antiporter (Zhang *et al.*, 2018). It was observed that under salinity stress, plants increase Na^+ and decrease K^+ uptake, respectively. This destruction in K^+/N^+ balance affects enzymatic activity, cell viability and plant growth process (Zhang *et al.*, 2018).

There are also some miRNAs specified for balancing ions in plant cells (Fig. 1). It was reported in *Arabidopsis* that miR169, miR395 and miR398 have a role in causing potassium deficit. An abundance of miR399 was observed in low K^+ level, in both *Arabidopsis* and rice plants (Zhao *et al.*, 2020). The miR408 family up-regulation in potassium deficit condition in plants. Based on that, it can be predicted that miR399 has an important role in response to the ion imbalance under salt stress (Zhao *et al.*, 2020).

The *Arabidopsis K Transporter1* (*AKT1*) is an important gene that controls the selective potassium channel and up-regulates under K^+ deficit (Wang *et al.*, 2012). It was observed that the miR408 regulates the expression of *TaAKT1* gene in wheat under salt stress (Zhao *et al.*, 2020). It was also suggested that miR159a, miR398, miR319 and miR1133 have a crucial role in wheat plants under salt stress by regulating K^+ uptake pathway (Zhao *et al.*, 2020). In radish plant, miR414 regulated *KUP3p* as the target gene, which is related to potassium transport under salinity stress (Sun *et al.*, 2015).

Calcium (Ca^{2+}) has an important role in plant growth and quality as well as activation of some enzymes. It has the main role in cell wall maintenance; accordingly, it can help the plant to be more disease resistance due to the stronger cell wall. It also has a role in conversion of nitrogen to a useable form in plants. Another role of calcium is improvement of nutrition absorption from the root (Thor, 2019).

It was proven that Ca^{2+} signalling has been the main plant response to environmental changes (Manishankar *et al.*, 2018). Immediately after increasing NaCl concentration, the Ca^{2+} concentration increased in plant cytosol. It was reported that Ca^{2+} works as a signal for activation of *SOS* gene. For instant, SOS3 is the member of B-like proteins (CBLs) that act as Ca^{2+} sensor proteins (Manishankar *et al.*, 2018).

Calcium-dependent protein kinase (CDPK) is the protein member of calcium signalling pathway; it was found that the expression of *CDPK* genes increased under salt stress. The interaction between *CDPK* and two-pore K^+-channels (*TPK*) was observed under salt stress (Manishankar *et al.*, 2018). It can be explained that this is one way of calcium signalling to balanced ions level in plant cells under salinity stress.

The *CDPK1* gene is regulated by miR166m as the target genes in soybean plant under salinity stress (Li *et al.*, 2022). The miR414, rsa-mir3 and rsa-mir5 in radish plants regulate the expression of *CDPK9* as the target gene under salt stress (Sun *et al.*, 2015).

The CBL-interacting serine/threonine protein (CIPK21) is another member of calcium signalling pathway. It was reported that in radish plants, miR414 moderate the expression of *CIPK21* gene under salinity stress (Sun *et al.*, 2015).

The Ca^{2+} sensor calmodulin (CAM) has a role in plant tolerance under salt stress. In the radish plant, *CAM7* gene is a member of this family that was regulated by miR414 under salt stress (Sun *et al.*, 2015).

3. Osmotic Homeostasis

It was proven that osmolytes protect plant cells against abiotic stress. Osmolytes include sugars (sucrose, fructose, fructans, trehalose and raffinose), metabolites (proline, glycine, betaine, putrescine, glutamate, oxalate, aspartate and malate) and some amino acids (Jogawat, 2019). All of them have the same role of balancing the cytosolic potential. The increase in osmolyte accumulation is another response of plant to salinity stress. Plants try to balance cell osmotic potential and inhibit oxidative damage which is caused by water deficit after salt stress (Jogawat, 2019). Understanding osmolyte role at the molecular level and the miRNAs which target their related genes may help us to make the plant more tolerant to salt stress. It has been reported that several miRNAs are specified in salt stress (Fig. 1). miR156/miR157 regulates *GS2* as the target gene in radish plant under salt stress. This gene-encoding glutamine synthase enzyme has a role in biosynthesis of glutamine as one of the osmolytes in plant cell (Sun *et al.*, 2015). In *Arabidopsis* plants, the miR156 acts in salt stress by up-regulation of SQUAMOSA-promoter binding protein-like (*SPLs*) as the target gene (Covarrubias and Reyes, 2010).

4. Hormone Signalling

Plant phytohormones are essential for plant growth and development. Different types of hormones, including auxins, cytokinins (CKs), gibberellin (GA), strigolactones (SLs), brassinosteroids (BRs), abscisic acid (ABA), ethylene, salicylic acid (SA) and jasmonic acid (JA) regulate plant growth and development under normal and stress conditions (Yu *et al.*, 2020).

Among different phytohormones, ABA is more popular for its role under different stress conditions. Under salt stress and osmotic changes, the level of ABA increases immediately (Yu *et al.*, 2020). ABA increases osmolytes by activating some genes related to breaking down of the starch to small sugars. On the other hand, under salt stress, ABA regulates stomata closure (Yu *et al.*, 2020). Auxin regulates root plasticity under salt stress. The polar transport of auxin will be changed due to salt stress while the plants try to decrease the accumulation of auxin in the root. During salt stress, auxin promotes lateral root initiation which consequently induces plant adaptation to salt stress (Yu *et al.*, 2020). It was reported that high concentrations of NaCl increased ABA accumulation in roots with decreasing auxin polar transport

MicroRNA-mediated Regulation of Salinity Stress **171**

and further inhibition of lateral root induction (Yu *et al.*, 2020). In maize plants, it was reported that under salt stress, the miR167 down-regulated auxin receptor factor genes (Ding *et al.*, 2009). Another way to cope with salt stress is induction of genes related to GA biosynthesis, like *AtGA2ox7*, *OsGA2ox5* and *OsMYB91* that repress the GA production and inhibit plant growth (Yu *et al.*, 2020). Cytokinins decrease plant tolerance to salt and drought stresses. It was observed in tomato plants under salt stress that ABA increased, whereas CKs levels decreased. ABA-dependent genes were reported to be targets of some miRNAs under drought and salt stresses (Fig. 1). miR169 and miR319 seem to be the main miRNAs related to modulating ABA level in response to salt stress (Sun *et al.*, 2015).

The auxin response factors (ARF) family (ARF16 and ARF17) negatively regulate and inhibit plant growth in response to abiotic stress. In the pistachio plants, miR160 regulates *ARF* gene family under salt stress. In rice, miR167 is responsible for regulating expression of *ARF* under salinity stress (Sun *et al.*, 2015). The concentration of miR393 increased in rice under salt stress. This miRNA regulates auxin transporter gene (*OsAUX1*) and tiller inhibitor gene (*OsTIR1*) as the target genes. It was suggested that *Osa*-miR396c are related to regulating of ABA level under salt stress. The mutant that lacks this miRNA was sensitive to salt stress (Mittal *et al.*, 2016). In *Populus* tree, the miR1446 down-regulates the expression of genes related to biosynthesis of gibberellin under high concentrations of NaCl (Mittal *et al.*, 2016). There is a large family of transcription factors with more than 100 members, named NAC family gene, which was suggested to regulate biosynthesis of phytohormones in plants. It was reported that in the maize plant, miR164 down-regulates the NAC family gene under salts stress (Ding *et al.*, 2009).

5. Antioxidant Response

Oxidative stress is the physiological phenomenon occurring after biotic and abiotic stress. Reactive oxygen species (ROS) were over-produced and accumulated in the cells. ROS causes oxidative stress by disturbing the cell function. In the reaction to salinity stress, some changes occurred in plants – like ion imbalance, misbalancing in osmotic homeostasis, physiological and biochemical changes, water deficit in cells and finally oxidative stress. Plants make these changes to maintain their health and quality at tolerance level under salinity stress. Plants react to oxidative stress by increasing antioxidant production and accumulation (Demidchik, 2015).

Antioxidants are chemicals which stop or decrease the damage made by ROS in the cells. There are different types of antioxidants, some of which act individually and some others act in combination with other substances. Antioxidants were classified into two groups as enzymatic and non-enzymatic. Enzymatic antioxidants include superoxide dismutase (SOD), catalase (CAT), ascorbic peroxidase (APX), etc.; non-enzymatic antioxidants include vitamins, carotenoids, thiol antioxidants and proline, etc. (Sindhu *et al.*, 2022).

There are some miRNAs which regulate production or function of these antioxidants (Fig. 1). In the radish plant, miR396s regulate the *APX1* which encodes ascorbic peroxidase enzyme as the target gene. The results showed that miR396a up-regulated the *APX1* under salinity stress. In contrast, miR396b down-regulated the expression of this gene (Sun *et al.*, 2015). It was reported that *CSD1* encoded superoxide dismutase which was down-regulated by miR398 in salt stress (Sun *et al.*, 2015).

The pathway of antioxidant activity under miRNA regulation was studied in *Zanthoxylum bungeanum*. The results showed that the pathway of miRNA and antioxidant activity was similar to the model plants. They suggested that miR1696-3p regulates the *CAT* as the target gene. Also, the *GPX1* gene that encoded glutathione peroxidase was targeted by miR773b-3p under oxidative stress. The miR834 and miR447a-3p regulated the expression of *SOD* and *APX* respectively (Fei *et al.*, 2020).

6. Regulation of Plant Growth

The other response of plants under salt stress is changing the growth which includes all stages ranging from germination to flowering. Plants change their strategies of growth and development at different molecular, biochemical and morphological levels and miRNAs play an important role in these changes (Fig. 1).

At the cellular level, miR397a seems to target the laccase (*LAC*) genes with crucial roles in lignification under salinity stress. miR397a up-regulates *LACs* to increase the thickness of cell wall in radish plant (Sun *et al.*, 2015). In maize plants under salt stress, miR395 up-regulates ATP sulfurylase gene as target gene (Ding *et al.*, 2009). In wheat plants, miR169 regulates the *NF-YA* that is a nuclear gene responsible for carbohydrate metabolism and cell elongation. It was reported that this gene was highly active under abiotic stress, such as salinity (Sun *et al.*, 2015).

NAC is the large family of transcriptional factors that have a role in resistance of plant to abiotic stresses, such as salinity; the members of this family regulate different stages of plant growth and development (Sohrabi *et al.*, 2017). In the model plants, miR164a up-regulated *NAC80* and *NAC100* under salinity stress (Mittal *et al.*, 2016).

The gene named *AP2* that encoded the floral homeotic protein APETALA 2 was regulated by miR156 under salinity stress and which have a role in plant growth at flowering stage. On the other hands, miR172 targeted *SPL* which encoded squamosal-promoter-binding-like protein that is involved in plant growth. It was observed that under salinity stress, differential activation of miR156 and miR172 makes transition between vegetative phase and flowering phase (Sun *et al.*, 2015). In maize plants, miR156 regulates SPL-like transcription factor (TF) genes that have a role in plant growth and development by encoding different amino acids; in salt stress, miR156 down-regulates these genes and controls plant growth (Ding *et al.*, 2009). It has been also reported that miR159 represses the expression of *MYBs* as the target genes to delay flowering under salt stress conditions (Sun *et al.*, 2015). There is a large family of genes, named F-box genes that encode proteins which have various roles in regulating the different biological processes in plants. It was reported that in soybean, miR394 down-regulates F-box genes as target genes under salt stress (Li *et al.*, 2022). The SQUAMOSA-promoter binding protein (SBP)-box gene family is also the big family that has a critical role in plant growth and development. It was reported that in maize and soybean plants, miR166 up-regulates SBP genes as target genes (Mittal *et al.*, 2016).

7. Conclusion

Salt stress is one of the major concerns and challenges towards sustainable crop production in the whole world. Salt stress alongside drought stress causes a significant reduction in plant growth and productivity. Tolerant plants response to salinity stress

MicroRNA-mediated Regulation of Salinity Stress

through different strategies, including ion unbalancing, balance osmotic homeostasis, increasing antioxidant activity and regulating the plant growth at different morphological, biochemical and molecular levels. Several genes, transcriptions factors and miRNAs are responsible for regulating progressive interactions towards salinity tolerance in plants.

References

Covarrubias, A.A. and Reyes, J.L. (2010). Post-transcriptional gene regulation of salinity and drought responses by plant microRNAs, *Plant Cell Environ.*, 33(4): 481-489.

de Lima, J.C., Loss-Morais, G. and Margis, R. (2012). MicroRNAs play critical roles during plant development and in response to abiotic stresses, *Genet. Mol. Biol.*, 35(4): 1069-1077.

Demidchik, V. (2015). Mechanisms of oxidative stress in plants: From classical chemistry to cell biology, *Environ. Exp. Bot.*, 109: 212-228.

Ding, D ., Zhang, L., Wang, H., Liu, Z., Zhang, Z. and Zheng, Y. (2009). Differential expression of miRNAs in response to salt stress in maize roots, *Ann. Bot.*, 103: 29-38.

Fei, X., Li, J., Kong, L., Hu, H., Tian, J., Liu, Y. and Wei, A. (2020). miRNAs and their target genes regulate the antioxidant system of *Zanthoxylum bungeanum* under drought stress, *Plant J. Physiol. Biochem.*, 150: 196-203.

Gupta, A., Singh, S.K., Singh, M.K., Singh, V.K., Modi, A., Singh, P.K. and Kumar, A. (2019). Plant growth-promoting Rhizobacteria and their functional role in salinity stress management. *In:* Sing, P., Kumar, A. and Borthakur, A. (Eds.). *Abatement of Environmental Pollutants: Trends and Strategies*. Elsevier, New York, USA, 151-160.

Jacobsen, T. and Adams, R.M. (1958). Salt and silt in ancient Mesopotamian agriculture, *Science*, 128(3334): 1251-1258.

Jogawat, A. (2019). Osmolytes and their role in abiotic stress tolerance in plants. *In:* Tripathi, D.K. and Roychoudhury, A. (Eds.). *Molecular Plant Abiotic Stress: Biology and Biotechnology*. Wiley, New York, USA, pp. 91-104.

Li, C., Nong, W., Zhao, S., Lin, X., Xie, Y., Cheung, M.Y., Xiao, Z., Wong, A.Y.P., Chan, T.F., Hui, J.H.L. and Lam, H.M. (2022). Differential microRNA expression, microRNA arm switching, and microRNA: Long noncoding RNA interaction in response to salinity stress in soybean, *BMC Genom.*, 23(1): 65-65.

Manishankar, P., Wang, N., Köster, P., Alatar, A.A. and Kudla, J. (2018). Calcium signalling during salt stress and in the regulation of ion homeostasis, *J. Exp. Bot.*, 69(17): 4215-4226.

Mittal, D., Sharma, N., Sharma, V., Sopory, S.K. and Sanan-Mishra, N. (2016). Role of microRNAs in rice plant under salt stress, *Ann. Appl. Biol.*, 168(1): 2-18.

Pandita, D. (2019). Plant MiRnome: miRNA biogenesis and abiotic stress response. *In:* Hasanuzzaman, M., Hakeem, K., Nahar, K. and Alharby, H. (Eds). *Plant Abiotic Stress Tolerance*. Springer, Cham, 449-474. Doi https://Doi.org/10.1007/978-3-030-06118-0_18

Pandita, D. (2021). Role of miRNAi technology and miRNAs in abiotic and biotic stress resilience. *In:* Aftab, T. and Roychoudhury, A. (Eds.). *Plant Perspectives to Global Climate Changes*. Academic Press, Elsevier, 303-330. https://Doi.org/10.1016/B978-0-323-85665-2.00015 -7

Pandita, D. and Wani, S.H. (2019). MicroRNA as a tool for mitigating abiotic stress in rice (*Oryza sativa* L.). *In:* Wani, S. (Ed.). *Recent Approaches in Omics for Plant Resilience to Climate Change*. Springer, Chamz., 109-133. Doi https://Doi.org/10.1007/978-3-030-21687-0_6

Pandita, D. (2022a). How microRNAs regulate abiotic stress tolerance in wheat? A snapshot. *In:* Roychoudhury, A., Aftab, T. and Acharya, K. (Eds.). *Omics Approach to Manage Abiotic Stress in Cereals*. Springer, Singapore, 447-464. https://Doi.org/10.1007/978-981-19-0140-9_17

Pandita, D. (2022b). MicroRNAs shape the tolerance mechanisms against abiotic stress in maize. *In:* Roychoudhury, A., Aftab, T. and Acharya, K. (Eds.). *Omics Approach to Manage Abiotic Stress in Cereals.* Springer, Singapore, 479-493. https://Doi.org/10.1007/978-981-19-0140-9_19

Pandita, D. (2022c). miRNA- and RNAi-mediated metabolic engineering in plants. *In:* Aftab, T. and Hakeem, K.R. (Eds.). *Metabolic Engineering in Plants.* Springer, Singapore, 171-186. https://Doi.org/10.1007/978-981-16-7262-0_7

Park, H.J., Kim, W.Y. and Yun, D.J. (2016). A new insight of salt stress signalling in plant, *Mol. Cells,* 39(6): 447-459.

Sindhu, R.K., Kaur, P., Singh, H., Batiha, G.E.S. and Verma, I. (2022). Exploring multifunctional antioxidants as potential agents for management of neurological disorders, *ESPR,* 29(17): 24458-24477.

Sohrabi, S., Ebadi, A., Jalali, S. and Salami, S.A. (2017). Enhanced values of various physiological traits and *VvNAC1* gene expression showing better salinity stress tolerance in some grapevine cultivars as well as rootstocks, *Sci. Hortic.,* 225: 317-326.

Sun, X., Xu, L., Wang, Y., Yu, R., Zhu, X., Luo, X., Gong, Y., Wang, R., Limera, C., Zhang, K. and Liu, L. (2015). Identification of novel and salt-responsive miRNAs to explore miRNA-mediated regulatory network of salt stress response in radish (*Raphanus sativus* L.), *BMC Genom.,* 16(1): 1-16.

Tahanian, H.R., Ebadi, A. and Salami, A.R. (2019). Effect of rootstocks on physiological and biochemical responses of *Vitis vinifera* 'Shahroudi' to salinity and bicarbonate stress conditions, *IJHST,* 20(1): 1-10.

Thor, K. (2019). Calcium-nutrient and messenger, *Front. Plant Sci.,* 10: 440.

Wang, H., Zhang, M., Guo, R., Shi, D., Liu, B., Lin, X. and Yang, C. (2012). Effects of salt stress on ion balance and nitrogen metabolism of old and young leaves in rice (*Oryza sativa* L.), *BMC Plant Biol.,* 12(1): 1-11.

Xu, X., Du, X., Wang, F., Sha, J., Chen, Q., Tian, G., Zhu, Z., Ge, S. and Jiang, Y. (2020). Effects of potassium levels on plant growth, accumulation and distribution of carbon and nitrate metabolism in apple dwarf rootstock seedlings, *Front. Plant Sci.,* 11: 904.

Yu, Z., Duan, X., Luo, L., Dai, S., Ding, Z. and Xia, G. (2020). How plant hormones mediate salt stress responses, *Trends Plant Sci.,* 25(11): 1117-1130.

Zandkarimi, H., Ebadi, A., Salami, S.A., Alizade, H. and Baisakh, N. (2015). Analysing the Expression profile of AREB/ABF and DREB/CBF genes under drought and salinity stresses in grape (*Vitis vinifera* L.), *PLoS ONE,* 10(7): e0134288.

Zhang, Y., Fang, J., Wu, X. and Dong, L. (2018). Na^+/K^+ balance and transport regulatory mechanisms in weedy and cultivated rice (*Oryza sativa* L.) under salt stress, *BMC Plant Biol.,* 18(1): 1-14.

Zhao, Y., Xu, K., Liu, G., Li, S., Zhao, S., Liu, X., Yang, X. and Xiao, K. (2020). Global identification and characterisation of miRNA family members responsive to potassium deprivation in wheat (*Triticum aestivum* L.), *Sci. Rep.,* 10(1): 1-13.

CHAPTER

10

MicroRNA-mediated Regulation of Cold and Chilling Stress Response

Seyed Alireza Salami* and Shirin Moradi

Department of Horticultural Sciences, Faculty of Agricultural Science and Engineering, University of Tehran, Iran

1. Introduction

Cold stress causes significant damage to agricultural products around the world every year. Similar to many other abiotic stresses, cold causes a significant loss of quality and yield in different crops due to chlorosis and necrosis of seedlings, leaves, flowers and small fruits (Yadav, 2010). Low temperature negatively affects growth and development and limits the cultivation area, while the human population and need for food are increasing drastically in parallel. Avoiding cold stress by using tolerant cultivars is considered the most efficient strategy to ensure high-quality products every year. It seems necessary to improve cold-tolerant cultivars to expand the cultivation area where cold stress has a history or new forms of cold stress due to global climate change. There is a high demand for tolerant cultivars which have been introduced by conventional breeding or new biotechnology approaches.

Chilling and cold stresses trigger several cellular, physical, physiological, biochemical and molecular changes, including oxidative stress, membrane lipid peroxidation, DNA, RNA and protein degradation, and cell death. However, plants have evolved different defensive mechanisms, such as protective and regulatory substances, including soluble sugars, proline and cold-resistance proteins (e.g. LEA, AFP, CSP), mRNAs, protein kinases and phosphatases, transcription factors, miRNA-mediated gene expression reprograming which enable them to sense and tolerate cold stress (Ding *et al.*, 2019). Plant adaptation to cold occurs following physiological and biochemical changes in various cells, tissues and organs that alter the expression of specific genes, transcription factors and non-coding RNAs, such as miRNAs, lncRNAs, etc.

The more we know about the molecular mechanisms of cold and chilling tolerance and susceptibility, the more successful we will be in dealing with low-

---

*Corresponding author: asalami@ut.ac.ir

temperature damages. This chapter focuses on the roles and functions of micro-RNAs under cold stress as one of the primary critical players in susceptibility or tolerance under temperature drop. This review also compiles the current understanding of cold-responsive miRNAs and their possible targets in plants for better management of crop production, protection and distribution.

2. Key Players Involved in Cold and Chilling Tolerance

Understanding possible mechanisms and key players of cold stress is essential to achieve cold tolerance. The significant adverse effects of cold stress include growth inhibition, ice crystal formation, plasma membrane disintegration, dehydration, solute leakage, metabolite imbalance and metabolic dysfunction, which trigger signal transduction pathways of transcription factors (TFs) and cold-responsive genes to control the damage by regulating cold stress tolerance. Fundamental defence responses of plants during cold stress, such as detoxification of reactive oxygen species (ROS), restructuring plasma membrane, repair progress and accelerating osmolyte production are crucial to overcoming the cold situation (Yadav, 2010; Ding *et al.*, 2019).

Once we are talking about low temperatures, chilling tolerance and cold acclimation must be distinguished in the definition. Chilling tolerance is the ability to tolerate low temperatures not necessarily below zero (0-10°C or 15°C) without injury, while cold acclimation is an enhanced tolerance to the physical and physiochemical consequences of freezing (<0°C) stress (Sanghera *et al.*, 2011; Josine *et al.*, 2011). Once normal cellular functions are disrupted, a quick reprogramming at the molecular level is required. Consequently, activation of metabolic pathways protects plant cells from cold and freezing damage. Protection strategies involve sugars and proline accumulation, production of proteins that stabilise membranes and activation of the ROS scavenging system. At the molecular level, several genes, transcription factors and non-coding RNAs play important roles in protecting the plant against cold stress. Among these molecular players, miRNAs modulate many post-transcriptional regulatory processes (Sunkar *et al.*, 2012; Taheri-Dehkordi *et al.*, 2021; Pandita, 2019; Pandita and Wani, 2019; Pandita, 2021; 2022a; 2022b, 2022c). For example, the ICE1-CBF-COR pathway is considered to be an integral component of cold stress response in many plants. C-repeat Binding Factor/Dehydration-Responsive Element-Binding Protein1 (CBF/DREB1) genes play essential roles in cold acclimation (Stockinger *et al.*, 1997; Liu *et al.*, 1998). Several miRNAs are involved in the CBF-dependent pathway (Megha *et al.*, 2018).

Plants recognise cold stress followed by a rapid decrease in temperature (during day and night) or rapid and more gradual temperature decrease (day only) (Kidokoro *et al.*, 2017). In this regard, transcription factors regulate gene expression negatively or positively in response to cold stress.

Cold-responsive genes and transcription factors are essential because they are often the targets of different miRNAs under cold/chilling stress regulatory networks. Significant groups of transcription factors involved in cold stress include DREB1-type and CAMTAs, which interact together under rapid temperature drop (during both day and night) and another transcription factor that regulates *DREB1* expression according to the circadian rhythms (rapid and slow temperature drop during the day) (Kidokoro *et al.*, 2017).

Transcription factors and different miRNAs modulate cold stress signalling and several studies highlight their important roles under low-temperature stress, which will be reviewed in this chapter. Also provided is a better understanding of multiple biological processes affected by miRNAs under temperature drop and how the positive and negative regulatory elements work at the transcriptional and post-transcriptional levels allow us to breed superior plants toward cold stress tolerance.

2.1 MicroRNAs-mediated Regulation of Cold Stress

The mechanisms underlying plant response to temperature drop are complicated and challenging due to several morphological, anatomical, physiological, biochemicals, cellular and molecular events and components triggered by low temperature (Huang *et al.*, 2021). mRNAs, protein kinases, phosphatases, transcription factors (BREB/CBF, WRKY, MYB, etc.), and miRNAs have been proved to be involved in cold/chilling stress regulatory and signalling pathways (Ito *et al.*, 2006; Zou *et al.*, 2010; Agarwal *et al.*, 2006). No doubt that miRNAs (conserved and species-specific) are one of the most critical players in cold stress scenarios (Fig. 1).

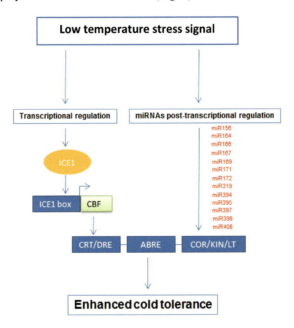

Fig. 1: The regulatory network involved in low-temperature responses including different miRNAs as positive or negative regulators

MicroRNAs are tiny 20-24 nucleotides RNAs with different regulatory roles under different conditions and treatments (Sunkar *et al.*, 2012). Cold-responsive miRNAs discovery can be accomplished using sequencing approaches, such as RNA-seq and miRNA-Seq (Zhang *et al.*, 2009; Zhou *et al.*, 2008), microarray (Liu *et al.*, 2008; Lu *et al.*, 2008; Lv *et al.*, 2010) and integration analysis or by *in silico* meta data analysis (Huang *et al.*, 2021; Vergata *et al.*, 2022).

The first related report of cold stress revealed up- and down-regulation of some miRNAs, such as miR319c, miR398a and miR393 (Sunkar and Zhu, 2004). After that, many other miRNAs discovered under chilling/cold stress were up- and down-regulated. Such a differential expression under cold stress perhaps is due to the species-specific nature of miRNAs in cold response. About 19 microRNAs were up-regulated in *Arabidopsis thaliana* by cold stress; some were differentially induced (miR165/166, miR169, miR172 and miR396) and some (miR156/157, miR159/319, miR164, miR394 and miR398) were constantly expressed under low temperatures. Among the identified miRNAs in *Arabidopsis*, some had ABRE or LTRE/DRE/C-repeat only and the rest had both (Zhou *et al.*, 2008). They showed that miR169 induction might inhibit cell wall loosening, repressing the growth under low temperatures. Other miRNAs are also involved in auxin signalling and plant growth and development (Zhou *et al.*, 2008).

Small RNAs sequencing in *Brachypoodium* under cold stress revealed some cold-induced and cold-suppressed miRNAs (Zhang *et al.*, 2009). In rice, most cold-responsive miRNAs, such as miR167 and miR319 were down-regulated under 4°C (Lv *et al.*, 2010). In another study, miR812q was overexpressed under cold stress, leading to down-regulations of its target CIPK10, which is involved in calcium-dependent CBL-CIPK signalling pathway (Jeong *et al.*, 2011; Jeong and Green, 2013). In tea plants, 31 miRNAs were up-regulated under cold stress treatments and 43 miRNAs were down-regulated (Zhang *et al.*, 2014a). In tomatoes, 192 and 205 chilling-responsive miRNAs were up- and down-regulated, respectively (Cao *et al.*, 2014).

Mutants and transgenic plants are considered robust tools to confirm the functions of miRNAs. Mutation of LOS4, which is responsible for osmotic relationship, enhanced cold stress induction of C-repeat binding factor 2 (CBF2) and its downstream target genes (Gong *et al.*, 2005). miR397a overexpressed in *Arabidopsis* plants showed increased tolerance to chilling and freezing stresses (Dong and Pei, 2014). miR319b overexpressed rice plants showed enhanced cold tolerance (Wang *et al.*, 2014). In this regard, miR319 positively regulates cold tolerance by targeting *OsPCF6* and *OsTCP21*. miR390 encodes *OsSRK*, which is responsible for response to abiotic stresses and plant development.

While miR156 was overexpressed in rice plants, cold tolerance was reduced (Cui *et al.*, 2015). miR164 transgenic *Arabidopsis* plants cope with low temperature by increasing root length and germination rate (Sobkowiak *et al.*, 2016). miR394 targets LCR (Leaf Curling Responsiveness) and miR394 overexpression in *Arabidopsis* enhanced cold stress tolerance (Song *et al.*, 2016).

Chilling stress inhibits early growth of maize plant. miRNAs seems to regulate the growth and development by regulating transcriptional factors and sustaining redox homeostasis of cells and therefore may be considered as a tool to prevent shortening of leaves under chilling stress (Aydinoglu, 2020). Chilling stress (6°C) caused up-regulation of miR408b and miRn138 and down-regulation of miR168a, miR529, miRn120, miRn44 and miRn22 in maize. Differentially expressed miRNAs seem to change nucleic acid metabolisms, hydrolase and phosphatase activities and cellular components to tolerate the chilling stress (Li *et al.*, 2016). miRNAs were induced by cold stress in four soybean varieties. A number of miRNAs in soybean have been reported to be associated with cold (Kuczyński *et al.*, 2020; Kuczyński *et al.*, 2021). The level of expression of miRNAs was dependent on variety, tissue and developmental stage. In general, the most abundant miRNAs included miR159, miR165, miR166, miR167, miR319, miR396, miR398, miR408, miR482, miR4414, miR1510 and miR3522 (Kuczyński *et al.*, 2021). miR167 and tasiRNA-ARF are

MicroRNA-mediated Regulation of Cold and Chilling Stress Response **179**

involved in auxin-signalling pathway and possibly in response to cold stress in wheat (Tang *et al.*, 2012). Differentially expressed miRNAs in the young spikes of common wheat under cold stress targeted floral-associated transcription factors which may cause abnormal reproductive organ development (Song *et al.*, 2017).

2.2 Regulation Mechanisms of MicroRNAs in Cold Stress

Different pathways and their related genes, transcription factors and signalling molecules and components are described well in many scientific reviews. However, there are a few reports related to miRNAs and their functions in cold tolerance. Genes and transcription factors associated with physiological responses to cold stress, sensing of cold signals, cell membrane fluidity, calcium channels and signalling, hormone signalling and transcriptional and post-transcriptional signal transduction may be the putative targets of several miRNAs. Some of these cold/chilling stress-related genes in plants that have been reviewed so far include *bHLH* and *ICE1* (*Fagopyrum tataricum, Dimocarpus longan, Zoysia japonica*), *FDA* (*Gossypium hirsutum, A. thaliana*), *GST* (*Sorghum bicolor*), *Trx* (*Solanum lycopersicum*), *AP2/ERF* (*A. thaliana, Vitis amurensis*), *MYB* (*Oryza sativa, Malus domestica*), *CBF-DREB* (*G. hirsutum, Vitis vinifera*), *CPK* (*Camellia sinensis*), *COR* (*Saussurea involucrata*), *LEA* (*Camellia sinensis*) and *RDM4* (*A. thaliana*) (Ritonga and Chen, 2020).

Among cold regulatory pathways, CBF/DREB transcription factor and its cold-inducible target genes, known as *COR* (cold regulated), *KIN* (cold-induced gene), *RD* (responsive gene to dehydration), or *LTI* (low-temperature-induced gene) are the best characterised ones (Lv *et al.*, 2010). The CBF-COR signalling pathway is better known under cold stress. In *Arabidopsis*, *CBFs* genes expression and CBFs proteins stability are affected by several positive and negative regulators of the CBF-dependent pathway, including ICE1/2, CAMTAs, CESTA, SIZ1, CRLK1/2, EIN3, PRRs, CRPK1, MYB, etc. Also, cold regulated (*COR*) expression is regulated by various CBF-independent regulators, including WRKY33, CZF1 related to ABI3/VP1 (RAV1), Ethylene Responsive Element-binding Factors (ERF5), CZF2, MYB, ZAT10 and Heat Shock Transcription Factor C1 (HSFC1). Several miRNAs, CBF-dependent pathway and CBF-independent regulators work together in response to low temperatures (Fig. 1) (de Lima *et al.*, 2012; Park *et al.*, 2015; Megha *et al.*, 2018; Ding *et al.*, 2019).

Other possible mechanisms involve plant reprograming for higher survival rate, better growth and development performance, regulatory expression of critical genes, such as DREB1, scavenging ROS, etc. In rice, tolerance to cold stress was achieved by overexpression of *Osa-miR319b* and consequently higher survival rate and proline content. Also, *Osa-miR319b* transgenic lines showed an increased expression of *DREB1A/B/C*, *DREB2A*, and *TPP1/2 genes* and decreased levels of *OsPCF6* and *OsTCP21* as putative targets of *Osa-miR319b which promotes* modifying active oxygen scavenging (Wang *et al.*, 2014).

3. Conclusion

Low temperatures cause moderate to severe damage annually. To guarantee food support, scientists and growers use different strategies today to overcome or diminish the adverse effects of cold and chilling stresses on crops. Among different strategies, adapting to new climate conditions using tolerant genotypes or species obtained by genetic improvement through conventional or modern breeding approaches is considered a reliable approach. Cold tolerance is an inherited genetic trait due to genes, transcription factors and regulatory RNAs, such as miRNAs, in plants. miRNAs modulate the expression of mRNA targets (up-regulation or down-regulation), leading to regulation of the processes that eventually cause the plant to tolerate low-temperature stress. Deeper insights into miRNA functions, molecular mechanisms, gene networks and signalling pathways underlying low-temperature stress improve our ability to breed tolerant plants more efficiently. This chapter summarises miRNAs involved in low-temperature stress and miRNA-mediated regulation of their target genes.

References

Agarwal, M., Hao, Y.J., Kapoor, A., Dong, C.H., Fujii, H., Zheng, X.W. and Zhu, J.K. (2006). A R2R3 type MYB transcription factor is involved in the cold regulation of CBF genes and in acquired freezing tolerance, *J. Biol. Chem.*, 28(49): 37636-37645.

Aydinoglu, F. (2020). Elucidating the regulatory roles of microRNAs in maize (*Zea mays* L.) leaf growth response to chilling stress, *Planta*, 251: 38.

Cao, X., Wu, Z., Jiang, F., Zhou, R. and Yang, Z. (2014). Identification of chilling stress-responsive tomato microRNAs and their target genes by high-throughput sequencing and degradome analysis, *BMC Genom.*, 15: 1130.

Cui, N., Sun, X., Sun, M., Jia, B., Duanmu, H., Lv, D., Duan, X. and Zhu, Y. (2015). Overexpression of *OsmiR156k* leads to reduced tolerance to cold stress in rice (*Oryza sativa*), *Mol. Breed.*, 35: 214.

de Lima, J.C., Loss-Morais, G. and Margis, R. (2012). MicroRNAs play critical roles during plant development and in response to abiotic stresses, *Genet. Mol. Biol.*, 35: 1069-1077.

Ding, Y., Shi, Y. and Yang, S. (2019). Advances and challenges in uncovering cold tolerance regulatory mechanisms in plants, *New Phytol.*, 222(4): 1690-1704.

Dong, C.H. and Pei, H.X. (2014). Overexpression of miR397 improves plant tolerance to cold stress in *Arabidopsis thaliana*, *J. Plant Biol.*, 57: 209-217.

Gong, Z., Dong, C.H., Lee, H., Zhu, J., Xiong, L., Gong, D., Stevenson, B. and Zhu, J.K. (2005). A DEAD box RNA helicase is essential for mRNA export and important for development and stress responses in *Arabidopsis*, *Plant Cell*, 17: 256-267.

Huang, X., Liang, Y., Zhang, B., Song, X., Li, Y., Qin, Z., Li, D., Chen, R., Zhou, Z., Deng, Y., Wei, J. and Wu, J. (2021). Integration of transcriptional and post-transcriptional analysis revealed the early response mechanism of sugarcane to cold stress, *Front. Genet.*, 11: 581993.

Ito, Y., Katsura, K., Maruyama, K., Taji, T., Kobayashi, M., Seki, M., Shinozaki, K. and Yamaguchi-Shinozaki, K. (2006). Functional analysis of rice DREB1/CBF-type transcription factors involved in cold-responsive gene expression in transgenic rice, *Plant Cell Physiol.*, 47(1): 141-153.

Jeong, D.H. and Green, P.J. (2013). The role of rice microRNAs in abiotic stress responses, *J. Plant Biol.*, 56: 187-197.

Jeong, D.H., Park, S., Zhai, J., Gurazada, S.G., De Paoli, E., Meyers, B.C. and Green, P.J. (2011). Massive analysis of rice small RNAs: Mechanistic implications of regulated microRNAs and variants for differential target RNA cleavage, *Plant Cell*, 23: 4185-4207.

Josine, T.L., Ji, J., Wang, G. and Guan, C.F. (2011). Advances in genetic engineering for plants abiotic stress control, *Afr. J. Biotechnol.*, 10: 5402-5413.

Kidokoro, S., Yoneda, K., Takasaki, H., Takahashi, F., Shinozaki, K. and Yamaguchi-Shinozaki, K. (2017). Different cold-signalling pathways function in the responses to rapid and gradual decreases in temperature, *Plant Cell*, 29: 760-774.

Kuczyński, J., Twardowski, T., Nawracała, J., Gracz-Bernaciak, J. and Tyczewska, A. (2020). Chilling stress tolerance of two soya bean cultivars: Phenotypic and molecular responses, *J. Agron. Crop Sci.*, 206: 759-772.

Kuczyński, J., Gracz-Bernaciak, J., Twardowski, T., Karłowski, W.M. and Tyczewska, A. (2021). Cold stress-induced miRNA and degradome changes in four soybean varieties differing in chilling resistance, *J. Agron. Crop Sci.* https://Doi.org/10.1111/jac.12557.

Li, S.P., Dong, H.X., Yang, G., Wu, Y., Su, S.Z., Shan, X.H., Liu, H.K., Han, J.Y., Liu, J.B. and Yuan, Y.P. (2016). Identification of microRNAs involved in chilling response of maize by high-throughput sequencing, *Biol. Plant*, 60: 251-260.

Liu, Q., Kasuga, M., Sakuma, Y., Abe, H., Miura, S., Yamaguchi-Shinozaki, K. and Shinozaki, K. (1998). Two transcription factors, DREB1 and DREB2, with an EREBP/AP2 DNA binding domain separate two cellular signal transduction pathways in drought- and low-temperature-responsive gene expression, respectively, in *Arabidopsis*, *Plant* Cell, 10: 1391-1406.

Liu, H.H., Tian, X., Li, Y.J., Wu, C.A. and Zheng, C.C. (2008). Microarray-based analysis of stress-regulated microRNAs in *Arabidopsis thaliana*, *RNA*, 14: 836-843.

Lu, S.F., Sun, Y.H. and Chiang, V.L. (2008). Stress-responsive microRNAs in *Populus*, *Plant J.*, 55: 131-151.

Lv, D.K., Bai, X., Li, Y., Ding, X.D., Ge, Y., Cai, H., Ji, W., Wu, N. and Zhu, Y.M. (2010). Profiling of cold-stress-responsive miRNAs in rice by microarrays, *Gene*, 459: 39-47.

Megha, S., Basu, U. and Kav, N.N.V. (2018). Regulation of low temperature stress in plants by microRNAs, *Plant Cell Environ.*, 41: 1-15.

Pandita, D. (2019). Plant miRnome: miRNA biogenesis and abiotic stress response. *In:* Hasanuzzaman, M., Hakeem, K., Nahar, K. and Alharby, H. (Eds.). *Plant Abiotic Stress Tolerance.* Springer, Cham, 449-474. Doi https://Doi.org/10.1007/978-3-030-06118-0_18

Pandita, D. (2021). Role of miRNAi technology and miRNAs in abiotic and biotic stress resilience. *In:* Aftab, T. and Roychoudhury, A. (Eds.). *Plant Perspectives to Global Climate Changes.* Academic Press, Elsevier, 303-330. https://Doi.org/10.1016/B978-0-323-85665-2.00015-7

Pandita, D. and Wani, S.H. (2019). MicroRNA as a tool for mitigating abiotic stress in rice (*Oryza sativa* L.). *In:* Wani, S. (Eds.). *Recent Approaches in Omics for Plant Resilience to Climate Change.* Springer, Chamz, 109-133. Doi https://Doi.org/10.1007/978-3-030-21687-0_6

Pandita, D. (2022a). How microRNAs regulate abiotic stress tolerance in wheat? A snapshot. *In:* Roychoudhury, A., Aftab, T. and Acharya, K. (Eds.). *Omics Approach to Manage Abiotic Stress in Cereals.* Springer, Singapore, 447-464. https://Doi.org/10.1007/978-981-19-0140-9_17

Pandita, D. (2022b). MicroRNAs shape the tolerance mechanisms against abiotic stress in maize. *In:* Roychoudhury, A., Aftab, T. and Acharya, K. (Eds.). *Omics Approach to Manage Abiotic Stress in Cereals.* Springer, Singapore, 479-493. https://Doi.org/10.1007/978-981-19-0140-9_19

Pandita, D. (2022c). miRNA- and RNAi-mediated metabolic engineering in plants. *In:* Aftab, T. and Hakeem, K.R. (Eds.). *Metabolic Engineering in Plants*, Springer, Singapore, 171-186. https://Doi.org/10.1007/978-981-16-7262-0_7

Park, S., Lee, C.M., Doherty, C.J., Gilmour, S.J., Kim, Y. and Thomashow, M.F. (2015). Regulation of the *Arabidopsis* CBF regulon by a complex low temperature regulatory network, *Plant J.*, 82: 193-207.

Ritonga, F.N. and Chen, S. (2020). Physiological and molecular mechanism involved in cold stress tolerance in plants, *Plants*, 9(5): 560.

Sanghera, G.S., Wani, S.H., Hussain, W. and Singh, N.B. (2011). Engineering cold stress tolerance in crop plants, *Curr Genomics*, 12(1): 30-43.

Sobkowiak, A., Jónczyk, M., Adamczyk, J., Szczepanik, J., Solecka, D., Kuciara, I., Hetmanczyk, K., Trzcinska-Danielewicz, J., Grzybowski, M., Skoneczny, M., Fronk, J. and Sowins, P. (2016). Molecular foundations of chilling-tolerance of modern maize, *BMC Genom.*, 17: 125-147.

Song, J.B., Gao, S., Wang, Y., Li, B.W., Zhang, Y.L. and Yang, Z.M. (2016). miR394 and its target gene LCR are involved in cold stress response in *Arabidopsis*, *Plant Gene*, 5: 56-64.

Song, G., Zhang, R., Zhang, S., Li, Y., Gao, J., Han, X., Chen, M., Wang, J. and Li, W. (2017). Response of microRNAs to cold treatment in the young spikes of common wheat. *BMC Genom.*, 18: 212.

Stockinger, E.J., Gilmour, S.J. and Thomashow, M.F. (1997). *Arabidopsis thaliana* CBF1 encodes an AP2 domain-containing transcriptional activator that binds to the C-repeat/DRE, a cis acting DNA regulatory element that stimulates transcription in response to low temperature and water deficit, *Proc. Natl. Acad. Sci. U.S.A.*, 94: 1035-1040.

Sunkar, R. and Zhu, J.K. (2004). Novel and stress-regulated microRNAs and other small RNAs from *Arabidopsis*, *Plant Cell*, 16: 2001-2019.

Sunkar, R., Li, Y.F. and Jagadeeswaran, G. (2012). Functions of microRNAs in plant stress responses, *Trends Plant Sci.*, 17: 196-203.

Taheri-Dehkordi, A., Naderi, R., Martinelli, F. and Salami, S.A. (2021). Computational screening of miRNAs and their targets in saffron (*Crocus sativus* L.) by transcriptome mining, *Planta*, 254: 117.

Tang, Z., Zhang, L., Xu, C., Yuan, S., Zhang, F., Zheng, Y. and Zhao, C. (2012). Uncovering small RNA-mediated responses to cold stress in a wheat thermosensitive genic male-sterile line by deep sequencing, *Plant Physiology*, 159(2): 721-738.

Vergata, C., Yousefi, S., Buti, M., Vestrucci, F., Gholami, M., Sarikhani, H., Salami, S.A. and Martinelli, F. (2022). Meta-analysis of transcriptomic responses to cold stress in plants, *Funct. Plant Biol.*, PMID: 35379384.

Wang, S.T., Sun, X.L., Hoshino, Y., Yu, Y., Jia, B., Sun, Z.W., Sun, M.Z., Duan, X.B. and Zhu, Y.M. (2014). MicroRNA319 positively regulates cold tolerance by targeting *OsPCF6* and *OsTCP21* in rice (*Oryza sativa* L.), *PLoS ONE*, 9(3): e91357.

Yadav, S.K. (2010). Cold stress tolerance mechanisms in plants: A review, *Agron. Sustain. Dev.*, 30: 515-527.

Zhang, J., Xu, Y., Huan, Q. and Chong, K. (2009). Deep sequencing of *Brachypodium* small RNAs at the global genome level identifies microRNAs involved in cold stress response, *BMC Genom.*, 10: 449.

Zhang, Y., Zhu, X., Chen, X., Song, C., Zou, Z., Wang, Y., Wang, M., Fang, W. and Li, X. (2014). Identification and characterisation of cold-responsive microRNAs in tea plant (*Camellia sinensis*) and their targets using high-throughput sequencing and degradome analysis, *BMC Plant Biol.*, 14: 271.

Zhou, X., Wang, G., Sutoh, K., Zhu, J.K. and Zhang, W. (2008). Identification of cold-inducible microRNAs in plants by transcriptome analysis, *Biochim Biophys Acta.*, 1779(11): 780-788.

Zou, C.S., Jiang, W.B. and Yu, D.Q. (2010). Male gametophyte-specific WRKY34 transcription factor mediates cold sensitivity of mature pollen in *Arabidopsis*, *J. Exp Bot.*, 61(14): 3901-3914.

CHAPTER

11

MicroRNA-mediated Regulation of Heavy Metal Stress in Plants

Swarnavo Chakraborty and Aryadeep Roychoudhury*

Post Graduate Department of Biotechnology, St. Xavier's College (Autonomous),
30, Mother Teresa Sarani, Kolkata - 700016, West Bengal, India

1. Introduction

Abiotic stress appears to be one of the most dreaded environmental factors, adversely affecting crop productivity globally. Heavy metal toxicity is one of such abiotic stressors involved in the alterations associated with several crucial biochemical and physiological plant processes, leading to decline in crop yield in a significant manner (Hossain *et al.*, 2010; Villiers *et al.*, 2011; Rascio and Navari-Izzo, 2011). A vast range of heavy metals, including iron (Fe), copper (Cu), zinc (Zinc), manganese (Mn), cadmium (Cd), cobalt (Co), nickel (Ni), arsenic (As), aluminium (Al) and many others have been found to cause toxicity to the plant system. Interestingly, a series of the above-mentioned heavy metals serve as cofactors of important enzymes, mediating biochemical reactions governing optimal growth and developmental phases of many organisms, including plants. However, at excessive concentrations, these crucial metals turn out to be detrimental for the living system, thereby indicating the necessity of maintenance of an optimal concentration of each of these metals within the system (Rascio and Navari-Izzo, 2011). Amongst all these metals, cadmium and mercury have been reported to cause maximum harm to the plant system (Chen *et al.*, 2009a). Currently, phytoremediation has been considered as an effective tool for the removal of these toxic heavy metals from polluted soils, via hyper accumulation of these metals present in soil, followed by their subsequent translocation within the plant system, thereby reducing the heavy metal concentrations to near optimum (Chen *et al.*, 2009b). In addition, sophisticated tools, like genetic and breeding techniques, have also been found to be effective in restricting heavy metal contamination in food (Grant *et al.*, 2008). A series of biochemical, molecular and physiological analyses have been carried out to decipher the mechanisms governing the homeostasis and tolerance against various heavy metals in plants. In addition, plants have come up

---

*Corresponding author: aryadeep.rc@gmail.com

with a diverse range of mechanisms, including chelation strategies, signalling circuits, sequestration strategies and transportation systems, in order to safeguard against the toxicity induced by these detrimental metals. Maksymiec (2007) explained the potential of plants to trigger crucial signal transduction networks upon inception of cues associated with heavy metals, thereby triggering the transcriptional activation of a range of genes responsive to these toxic metals and establishing effective stress mitigation. In response to heavy metal stress, a considerable number of meticulously characterised pathways have been found to be responsive, including the calcium-calmodulin pathway; the MAPK (mitogen activated protein kinase) pathway; the ROS (reactive oxygen species) signalling and the phytohormone cascades (Yang and Poovaiah, 2003; Yeh *et al.*, 2004; Jonak *et al.*, 2004; Maksymiec and Krupa, 2006; Romero-Puertas *et al.*, 2007; Peleg and Blumwald, 2011). However, different heavy metals trigger an entirely different set of pathways (Dal-Corso *et al.*, 2010). In addition, an array of plasma membrane-associated transporter families have been found to mediate heavy metal homeostasis, uptake and transport, which constitute the CDF (cation diffusion facilitator), ZIP family of trsnscription factors, OPTs (oligopeptide transporters), CTRs (copper transporters), HMAs (heavy metal ATPases), NRAMP (natural resistance-associated macrophage protein) and ABC (ATP binding cassette) transporters (Williams *et al.*, 2000; Guerinot, 2000; Curie *et al.*, 2001; Puig and Thiele, 2002; Baxter *et al.*, 2003; Nevo and Nelson, 2006; Kim *et al.*, 2007; Kramer *et al.*, 2007). However, the exact metal specificity, cellular localisation and precise biological role of very few transporters have been lucidly elucidated. Whenever the levels of these toxic metals shoot over their threshold in the soil, plants tend to induce the production of certain metal chelators, i.e. phytochelatins and metallothioneins, to ensure proper chelation and sequestration of these metals in respective cellular pockets (Cobbett, 2000; Cobbett and Goldsbrough, 2002; Sekhar *et al.*, 2011). In addition, a number of organic acids, phosphate derivatives and amino acids also mediate chelation of heavy metals within different cellular sub-compartments, thereby leading to detoxification of these toxic agents (Rauser, 1999).

Mostly all eukaryotic organisms constitute a group of small riboregulator, non-protein coding RNA molecules, referred to as miRNAs, which comprise 20 to 24 nucleotides. These short non-coding RNAs particularly bind to the 3'-untranslated regions of certain target mRNAs, leading to mRNA destabilisation and subsequent repression of translation. The miRNA molecules display a pivotal role in the sequence-specific modulation of gene expression, mostly at the post-transcriptional level via cleavage at specific targets or inhibition of protein synthesis (Jones-Rhoades *et al.*, 2006). Different abiotic stressors, like heavy metal toxicity, drought, temperature extremes, salinity, radiation, oxidative stress, etc. result in significant differential modulation of many crucial plant genes and miRNAs (Pandita, 2019; Pandita and Wani, 2019; Pandita 2021; 2022a; 2022b; 2022c) in order to cope up with the adversities imposed by these stressors. Hence, several attempts are undertaken to efficiently mine the conserved yet novel miRNAs, followed by the development of proper links of these miRNAs with different metal stresses in plants. The prime focus of this chapter is to emphasise on the links between various new metal-responsive miRNAs with various target molecules, including chelators, transporters, enzymes, etc. in association with different forms of abiotic stresses in plants. However, the role of various miRNAs in relation to plant signalling during heavy metal toxicity is rather a virgin field. Unravelling of this puzzle will provide a better and newer insight into the signalling mechanisms occurring in plants during heavy metal stress.

2. MicroRNA-mediated Regulation of Mercury Toxicity in Plants

Mercury ranks fourth in terms of its toxicity, amongst the top seven heavy metals, posing serious threats to life. Although mercury exists in a number of forms in the soil, its divalent cationic form seems to be the most biologically available and prevalent form, which can be readily taken up by the aerial parts and roots of the plant (Ericksen *et al.*, 2003; Fay and Gustin, 2007). Mercury toxicity in plants evokes a series of detrimental effects in plants, like stunted growth, abnormality in cell shape, reduction in intercellular spaces, vascular abnormalities, production of toxic ROS molecules and reduction in chlorophyll and other photosynthetic pigments (Shiyab *et al.*, 2009; Meng *et al.*, 2011). Therefore, it is necessary to unravel the post-translational regulatory mechanisms involved in the minimisation of mercury toxicity in plants. Using bioinformatics approaches followed by validation via qPCR, Zhou *et al.* (2008) have identified a set of mercury toxicity-responsive miRNAs in *Medicago truncatula*. Such expression profiling displayed up-regulation of particularly three miRNAs, i.e. miR529, miR393 and miR171 under 20 µM mercury exposures, while miR398 and miR166 were negatively regulated. All of the positively regulated miRNAs have potential roles in plant development, particularly plant branching patterns and also in essential signalling cascades. Currently, a greater number of such novel mercury-responsive miRNAs have been identified with the advent of newer high throughput sequencing platforms. One of such efficient technologies is the deep sequencing approach, which has aided in the investigation of the complexity and global expression patterns of many miRNAs during mercury exposure (Zhou *et al.*, 2012), ultimately leading to the discovery of 54 new and 201 known miRNAs, of which 15 miRNA candidates displayed up-regulation and four were down-regualted as a result of mercury toxicity response. Moreover, this analysis also revealed 47 species-specific sets of miRNAs, which displayed significant regulation upon mercury treatment, thereby indicating widespread modulation of miRNA biogenesis cascades via positive or negative regulation of various miRNAs, generating an efficient mercury toxicity response in plants. In addition, from degradome library approach, identification of 130 targets for *Medicago trunculata*-associated non-conserved and conserved miRNAs and 37 targets for 18 new *Medicago trunculata*-specific miRNAs have been carried out, of which miR2681 has been found to target TIR-NBS-LRR (toll-interleukin-like receptor-nucleotide-binding site-leucine-rich repeat) disease resistance protein transcripts, which in turn have been found to be induced by mercury toxicity (Zhou *et al.*, 2012). In addition, XTH (xyloglucan endotransglucosylase/hydrolase) participating in the modification of cell wall and plant abiotic stress tolerance responses, has been predicted to serve as the target of miR2681. Hence, this differential modulation of a number of crucial genes and miRNAs will provide a better insight into the regulatory mechanisms operational during mercury toxicity in plants.

3. MicroRNA-mediated Regulation of Arsenic Toxicity in Plants

Arsenic has been categorised as Class I carcinogen by virtue of its severe ill effects on global health. This metalloid is widely distributed across the earth's crust and particularly enters the human system via contaminated food crops (Srivastava *et al.*,

2011). For the proper dissection of the molecular mechanisms governing arsenite hyper accumulation in plants, it is necessary to understand the patterns of expression and regulation of arsenite-responsive genes. Accumulation of arsenite in plants promotes a series of toxic effects within plants, i.e. increased rate of generation of harmful ROS, enhanced peroxidation of membrane lipids, disruption in carbohydrate metabolism and reduction in photosynthetic efficiency of plants (Dat *et al.*, 2000; Stoeva *et al.*, 2003; Jha and Dubey, 2004). Although several reports indicate the mode of arsenite uptake, detoxification and translocation in plants, very little information exists regarding the miRNA-mediated remodelling of plant regulatory and metabolic cascades during arsenic toxicity. Therefore, proper understanding of the regulatory mechanisms associated with arsenic stress response in plants is required.

Yu *et al.* (2012) have taken resort to the high throughput sequencing technology and have identified 36 new arsenic-responsive miRNAs, upon administration of 100 μM arsenic in rice plants. Amongst these 36 candidates, 14 were found to be associated with transcripts regulating stress-related signalling, transportation and metabolism in plants. In addition, these studies also revealed the involvement of jasmonate signalling and lipid metabolism as a part of miRNA-mediated regulation of arsenic stress response in plants. Moreover, via solexa sequencing pipeline, another 67 arsenic-responsive miRNAs have been discovered in rice variety, Minghui 86, upon application of 25 μM arsenic (Liu and Zhang, 2012). During arsenic toxicity, particularly three miRNAs (miR397b, miR408 and miR528) regulating cell cycle and cell wall biogenesis, plant reproduction and development, signalling transduction and cellular homeostasis displayed an increase in expression, while miR390 and 318 exhibited suppression in their expression patterns. In addition, down-regulation of two other miRNAs (miR1318 and 1432) targeting calcium-binding proteins (mainly Ca^{+2}-ATPase), have led to increased tolerance against arsenic-mediated damages in plants, particularly due to triggering of calcium-mediated signalling response during stress imposition. It has been found that miR528 can target certain putative IAR1 (indole acetic acid-alanine resistance protein 1), which controls the cellular free auxin concentrations and also mediates the efflux of inhibitory metals (like copper, zinc, etc.) out of the endoplasmic reticulum. Hence, overexpression of this miRNA results in lowering of IAR1 expression, leading to maintenance of optimal auxin conjugated forms, rendering protection against arsenic-mediated damages in plants. Moreover, deep sequencing technology also revealed another set of 18 novel miRNAs from Minghui 86 rice plants (Liu, 2012). Two out of these 18 miRNAs involved in regulation of flowering periods and photosynthetic cascades (miR169i-3p and 6250) displayed constitutive expression in almost all plant tissues, while miR6254 demonstrated a root-specific expression.

Upon 300 μM arsenic treatment in *Brassica juncea*, Srivastava *et al.* (2012) discovered 69 miRNAs, surpassing almost 18 different plant miRNA families. Most prominent candidates identified were found to be governing the key plant developmental processes via miR172, 169 and 156; sulphur uptake, assimilation and translocation via miR854, 838 and 395; biosynthesis and functionality of hormones via miR159, 164, 167 and 319. Moreover, exogenous application of auxin (indole acetic acid) and jasmonate led to altered expression of a series of miRNAs, i.e. miR854, 319 and 167, thereby leading to the improvement of plant growth parameters during arsenic toxicity. Such data probably points at the elaborate regulatory cross-talk operational amongst miRNAs and phytohormones during arsenic exposure.

During arsenic toxicity, miR159 has been found to target the oligopeptide transporter protein 1 (OPT1), which can efficiently translocate metals like iron, cadmium, zinc, copper, nickel, manganese, etc. as a part of arsenic toxicity response (Srivastava *et al.*, 2012). Many plant hormones, like jasmonate, ethylene and salicylic acid, also can positively/negatively modulate several transcription factors, governing metabolism involving different heavy metals (Maksymiec, 2007). For instance, miR319 have been proposed to target transcription factors associated with jasmonate biosynthesis in plants under conditions of metal toxicity, including arsenic, mercury, aluminium and cadmium (Zhou *et al.*, 2008; Schommer *et al.*, 2008; Srivastava *et al.*, 2012). In addition, miR838 tends to target transcripts encoding for certain lipase, associated with oxylipin biosynthesis and jasmonate biosynthesis (Srivastava *et al.*, 2012). Therefore, jasmonate seems to play an active role in metal toxicity responses in plants.

4. MicroRNA-mediated Regulation of Cadmium Toxicity in Plants

Cadmium is one of the most dreaded heavy metals in the plant system, due to its widespread toxicity among plants and fast accumulation within the plant system, generating a series of negative effects, like cell death, wilting, chlorosis and growth reduction (Rodriguez-Serrano *et al.*, 2009). In addition, toxicity due to cadmium can also induce oxidative damages due to generation of toxic ROS and widespread protein damages via interaction with thiol and carboxyl groups in proteins (Sanitadi and Gabrielli, 1999; Schutzendubel and Poole, 2002). Apart from its severe effects on crop productivity, cadmium also poses huge risks to global food security. Therefore, scientists have made considerable progress during the last few years, in identifying various signalling molecules and novel transporters, mediating toxicity response in plants against cadmium stress. Upon exposure of *Brassica napus* (seven-day old seedlings) to 80 µM cadmium for a period of eight hours, four miRNAs, i.e. miR396a, miR393, miR171 and miR156 displayed constitutive expression, with almost no effect in case of miR399 (Xie *et al.*, 2007). All the miRNAs mentioned above were associated with target genes, regulating signalling and developmental pathways during plant cadmium stress, like heavy metal detoxification; defence responses; phosphate starvation response; translational regulation; flower, leaf and root development, etc. Huang *et al.* (2010) successfully cloned a set of well-known miRNAs which have cadmium responsiveness in *Brassica napus,* followed by validation via reverse transcriptase PCR. Upon exposure to 80 µM cadmium, miR393 exhibited quite strong up-regulation while miR160 and 164b controlling flower and root development, and signalling cascades functional during cadmium-mediated oxidative stress response, were rather down-regulated. In addition, another important miRNA, called miR395, was predicted to target a low affinity sulphate transporter (*BnSultr2.1*) and three other ATP sulphurylases (*BnAPS1, 3* and *4*), which have been further justified in *Brassica* transgenics overexpressing this miRNA. Moreover, under conditions of sulphate sufficiency, the expression of this miRNA in all the tissues was null, but upon imposition of cadmium stress, even under sufficient sulphate levels, an induction in miR395 expression was observed, indicating a complementary effect of cadmium toxicity comparable with deficiency of sulphate. In addition, a large dose of sulphur accumulation, along with up-regulation in the production of compounds harbouring sulphur was observed under cadmium toxicity in plants (Hall, 2002).

In an experiment, four small RNA libraries were constructed from *Brassica napus* seedlings undergoing 80 µM cadmium stress for an extended period of six to 48 hours; 84 non-conserved and conserved miRNAs were identified. 19 out of these 84 miRNAs were completely novel and were not identified previously (Zhou *et al.*, 2012). Interestingly, almost all of these miRNAs displayed a differential expression pattern in roots and shoots upon administration of cadmium. Particularly, three miRNAs, viz. miR172f, miR398 and miR857 were up-regulated, while miR159, miR319d, miR394, miR398b and miR2111 were down-regulated in roots. In shoots, six out of 13 miRNAs were positively regulated, while the others were negatively regulated upon cadmium exposure. Most of the miRNAs up-regulated have potential roles in plant developmental and signalling cascades, mediating plant abiotic and biotic stress tolerance. Many members of this group tend to target the transcripts associated with heavy metal signalling, transportation and sequestration. The transcripts of GGT (glutathione-Y-glutamylcysteinyl transferase) serve as the major target of miR156. GGT in association with another important player, phytochelatin synthase, forming a crucial part of the heavy metal detoxification system in plants. This system particularly chelates toxic heavy metals, like mercury and cadmium via tripeptide glutathione (composed of three amino acids; viz., glumatic acid, glycine and cysteine) or phytochelatin molecules, followed by their transportation into vacuoles (Cobbett, 2000). Cheng *et al.* (2011) also demonstrated the regulation of plant oxidative stress during cadmium toxicity, via miR164-mediated targeting of monothiol glutaredoxin. Two natural resistance-associated macrophage protein transporters and an ATP-binding cassette transporter have been discovered to form the key targets of miR167 and miR159, respectively, having important roles in heavy metal translocation and uptake within the plant system (Bovet *et al.*, 2003; Kramer *et al.*, 2007). Moreover, Ding *et al.* (2011) employed microarray assay to identify 19 rice miRNAs from seven-day old seedlings, undergoing cadmium stress (60 µM $CdCl_2$) for one day, out of which very few have been experimentally validated. Similarly, Huang *et al.* (2009) also identified 19 novel and nine other known rice miRNAs, which have been found to be responsive to cadmium toxicity. An important member belonging to this group is miR604, which was found to target LTP (lipid transfer protein), regulating important signalling events mediating heavy metal responses in plants, including salicylic acid, abscisic acid, methyl jasmonate and ethylene pathways (Verica *et al.*, 2003). Moreover, the WAK (wall associated kinase) also forms a target of lipid transfer protein, which was found to be associated with plant defence mechanisms and heavy metal toxicity responses in plants. In addition, the rice miR602 was reported to target the XETs (xyloglucan endotransglucosylase/hydrolases), mediating the loosening of the plant cell walls due to endolytic cleavage of xyloglucan chains, followed by the re-joining of newly-formed free ends with water or other existing free xyloglucan ends (Rose *et al.*, 2002). Therefore, these data indicate the potentiality of miRNAs to regulate cadmium stress responses in various plants. Meticulous research involving heavy metal-associated transcription factors, proteins and enzymes is necessary to explore newer and efficient strategies to manage the adversities of cadmium toxicity in plants.

5. MicroRNA-mediated Regulation of Aluminium Toxicity in Plants

Aluminium bioavailability and its subsequent toxic effects in plants become more

MicroRNA-mediated Regulation of Heavy Metal Stress in Plants **189**

pronounced in soil having lower pH, i.e. below pH 5.5. Aluminium toxicity results in widespread reduction in crop yield and production, via binding to phosphate and carboxyl moieties of the cell wall of root cells, generating havoc structural alterations, like disrupted expansion of cellular walls, reduction of elongation of roots, fall in mineral, nutrient and water uptake via plant roots (Ma *et al.*, 2001; Ma *et al.*, 2004). In addition, aluminium stress in plants also imposes negative impact on crucial plant physiological processes, including damages incurred due to aluminium-induced oxidative stress, alteration in callose deposition and disruption in the levels of cytoplasmic calcium ions. However, the detailed and exact regulatory mechanism underlying aluminium toxicity response in plants still remains poorly deciphered.

Chen *et al.* (2012) used high throughput sequencing pipeline to demonstrate about 23 miRNAs showing responsiveness to aluminium, in seven-day old *Medicago tranculata* seedlings, upon administration of 10 μM aluminium chloride. Post-identification, these miRNAs have been categorised into three groups, viz, rapid responsive (18 miRNAs), sustained responsive (four miRNAs) and late responsive (only mtr-miR390). Upon increment of the aluminium stress load to 50 μM, four miRNAs, including miR529, miR393, miR319 and miR171 (involved in branching patterns of shoots; cell division and differentiation; development of embryonic structures; leaf, root and flower development; defence responses and other crucial plant signalling and developmental pathways) displayed an up-regulated expression, while miR398 and miR166 depicted a down-regulated expression pattern (Zhou *et al.*, 2008). By the application of the same technique, Zeng *et al.* (2012) successfully discovered 30 aluminium-responsive miRNAs from wild varieties of soybean plants. These findings hint at the fact that miRNAs specific to wild soybean might play a better and crucial role in the tolerance responses against aluminium toxicity. Another important miRNA, miR160, targets auxin receptor factors 10 and 16, which control the root cap formation and development (Wang *et al.*, 2005). Apart from miR160, miR390 also displays the potential of regulating various auxin receptor factors, thereby controlling optimal root growth and development (Wang *et al.*, 2005; Yoon *et al.*, 2010). Interestingly, during aluminium stress, mtr-miR160 down-regulation was observed, which in turn led to hampered root elongation due to up-regulated expression of auxin response factors. The biosynthesis of TAS3-derived trans-acting short interfering RNA (tasiRNA) was found to be associated with miR390. This miRNA negatively modulates the emergence of lateral roots via targeting an array of transcription factors; particularly the auxin response factors 2, 3 and 4 (Yoon *et al.*, 2010). Also, two newly discovered miRNAs (pmiR003 and 008) during aluminium stress response, probably target certain TIR-NBS-LRR resistance proteins, thereby mediating disease resistance mechanisms in plants (Chen *et al.*, 2012).

It has been found that miR393b can target and mediate the subsequent degradation of transcripts of an auxin-binding protein (TIR1), leading to modulation of the auxin-signalling cascades associated with the process of formation of lateral roots (Xie *et al.*, 2000). Moreover, in rice plants, upon down-regulation of this miRNA during aluminium toxicity, an increase in proteasome-associated processing of proteins occurs. However, miR160 up-regulation during aluminium stress establishes a rather inverse relation with miR393b via targeting of auxin response factors involved in auxin-signalling pathway. In *Arabidopsis*, miR528 tends to control branching pattern in shoots, while in the early phases of growth in rice seeds, this miRNA regulates cellular division (Xue *et al.*, 2009). In addition, the rise in the accumulation of miR528 in rice seedlings experiencing aluminium toxicity, indicates the alterations in protein turnover as a part of metal toxicity response. Another potential candidate revealed via *in silico* analysis, i.e. miR166k can modulate the formation of lateral roots via targeting of gene transcripts belonging to the ZIP (zinc-regulated, iron-regulated transporter like protein) family in *Arabidopsis* (Nagasaki *et al.*, 2007). Thus, positive regulation of miRNA160, miRNA166k and miRNA528, and negative regulation of miR393b probably establish the delicate control of aluminium toxicity-mediated responses within the plant system. Overexpression of an enzyme named fucosyl transferase (target of miR808), via down-regulation of miR808 probably results in altered biosynthesis of certain cell wall-associated sugar, thereby generating structural modifications of the root cell wall in rice plants during aluminium stress (Vanzin *et al.*, 2002). In *Arabidopsis*, the protein PHO2 (regulator of phosphate homeostasis) forms the target of miR399; accumulation of this miRNA in roots of rice plants leads to reduction in root-mediated uptake of minerals and subsequently leads to impairment of root growth and development during aluminium stress. Like almost all forms of abiotic stress, aluminium-mediated damages induce higher production of ROS, which in turn triggers superoxide dismutase (SOD) activity as a part of counter defence in plants (Sharma and Dubey, 2007). Lima *et al.* (2011) demonstrated rise in SOD activity and subsequent lowering of toxic ROS levels associated with negative regulation of miR398a, b in roots of rice seedlings during aluminium toxicity. Interestingly, application of aluminium oxide nanoparticles progressively increased the expression of set of four miRNAs, viz, miR399, 398, 397 and 395 in *Nicotiana tabacum*, probably highlighting the role of these miNAs in mediating aluminium stress tolerance (Burklew *et al.*, 2012). To summarise, miRNAs play a key role in various transportation events, developmental stages and signalling cascades operational in plants as a part of the protective response against the detrimental effects of aluminium in plants.

6. MicroRNA-mediated Regulation of Manganese Toxicity in Plants

Manganese generally serves as an inorganic catalyst of many important plant enzymatic reactions. However, incidences of manganese toxicity in plants have sometimes been reported. Thus, information regarding the miRNA-mediation regulation of manganese toxicity in plants is rather limited, particularly due very less research conducted in this regard. Valdes-Lopez *et al.* (2010) demonstrated the expressional profile of 68 miRNAs regulated by abiotic stressors, like manganese, iron, phosphorus, nitrogen, etc. via microarray and qRT-PCR, in nodules, roots and leaves of common bean plants. Amongst the above-mentioned miRNAs, 11 were found to be very strongly up-regulated in the tissues of nodules, while another set of 11 miRNAs have been found to be strongly down-regulated, upon inception of 200 µM magnesium chloride. It has been speculated that those miRNAs associated with the regulatory mechanisms during mercury, cadmium, aluminium toxicity in plants (Zhou *et al.*, 2008) might be involved in plant manganese toxicity, particularly due to the lack of data on the expression profiles and functional patterns of manganese responsive miRNAs in plants. In addition, miR2118 and 1510 tend to target the NBS-LRR (nucleotide binding site-leucine-rich repeat) resistance-like proteins, mediating tolerance against metal toxicity (including manganese), along with disease resistance pathways in plants.

7. MicroRNA-mediated Regulation of Other Metal Toxicity in Plants

Sometimes, essential plant micronutrients regulating crucial plant physiological processes pose a threat to the plant system under excessive concentrations. Iron and copper belong to this category of micronutrients, which, when present in optimal concentrations, render protection against oxidative damages, regulate metabolism of cell wall, respiratory electron transport chain and photosynthetic machinery, aid in perception of the phytohormone ethylene, but evoke serious toxic symptoms during the above optimal application. Excessive administration of these micronutrients in plants results in inhibited elongation of plant roots, disruption in the production of chlorophyll and functioning of photosystem II, and over-accumulation of harmful ROS in cellular organelles, like chloroplast (Patsikka *et al.*, 2002). Some abundant copper harbouring proteins, including cytochrome c oxidase, copper/zinc superoxide dismutase and plastoquinone have been reported to mediate alleviation of oxidative stress-induced damages and can also regulate the respiratory and photosynthetic machineries in plants (Marschner, 1995). Moreover, SODs can be categorised into three groups on the basis of metal cofactors present, viz. MnSOD (manganese superoxide dismutase); FeSOD (iron superoxide dismutase) and Cu/Zn-SOD (copper/ zinc superoxide dismutase). Each of these is compartmentalised into different cellular organelles (Mittler, 2002). Hence, it is necessary to emphasise on the cross-talk operational amongst various miRNAs and homeostatic responses triggered due to iron and copper toxicity in plants.

Ding and Zhu (2009) demonstrated the role of miR398 in modulating the functionality of Cu/Zn-SOD during oxidative stress. Waters *et al.* (2012) identified the link existing between iron and copper homeostasis, via miRNA profiling in *Arabidopsis*. For instance, plants exhibiting iron deficiency, tend to accumulate high doses of copper in their leaves (Valdes-Lopez *et al.*, 2010). In addition, the genes encoding for CuSOD (*CSD1* and *2*) and FeSOD (*FSD1*) have been positively regulated by the copper concentration within the system. The transcripts associated with the genes encoding for CuSOD and that of copper chaperone for SODs form the target of miR398a, b and c (Beauclair *et al.*, 2010). This can explain the fact that during copper sufficiency, accumulation of CuSOD protein and its transcripts takes place due to lower miR398s expression. On the other hand, during deficiency of copper, these miRNAs are abundantly expressed, leading to degradation of these target transcripts, encoding for CuSOD (Beauclair *et al.*, 2010). Interestingly, during copper insufficiency in plants, the levels of *FSD1* transcript and protein display a sharp rise, in turn allowing the functional replacement of iron-containing SODs with that of copper harbouring ones. As expected, during iron limitation, CuSODs formed the functional replacement of FeSODs. In iron-deficient environments, miR397a possibly led to an alteration in the production of laccase enzyme (containing copper) in *Arabidopsis* roots. Similar down-regulation in this miRNA expression was observed in iron-deficient *Brassica* leaves, root and phloem tissues, while an up-regulation in this miRNA expression occurred during copper deficiency in plant root and phloem tissues (Buhtz *et al.*, 2010). Therefore, association of miRNAs with copper and iron homeostasis setting up an inverse regulatory network serves as a way to cope up with the toxicity of these metals in plants.

8. Conclusion and Future Perspectives

With the progressive rise in human population, it has become a challenge to sustain crop yield and productivity under conditions of various abiotic stresses, including heavy metal stress. Although plants display some specialised strategies, like restricted uptake of metals, detoxification, chelation, sequestration, etc. to cope up with such stressful conditions, an increased understanding of the mode of regulation of miRNAs in context to heavy metal toxicity will surely pave a better path, ensuring an improvement in the tolerance limits of various crops exhibiting heavy metal stress. For the sake of such improvements, it is mandatory to identify new miRNAs, responsive to toxic heavy metals, followed by the linking of the expression profiles of these miRNAs with their associated targets. Therefore, researchers have already initiated the process of mining the conserved yet novel miRNA candidates involved in different heavy metal stresses (some of which have been summarised in Table 1) and have also established the links with their potential targets. However, the regulatory mechanisms operational at the background of these potential links has not been clearly deciphered and demands extensive experimental validation as well. In addition, identification of intermediate pathways mediating heavy metal tolerance in plants has not been entirely carried out. Some of the researchers have reported very interesting targets (mostly transcription factors and transporter proteins) involved in the regulation of pathways rendering plant tolerance against heavy metals. In order to establish the links between miRNAs with its targets, the reverse genetic approach (i.e. knockout/down) will

Table 1: miRNA-associated with different forms of heavy metal stress in plants

Heavy Metal Stress	miRNAs Involved	References
Aluminium	miR808; miR529; miR399; miR398; miR397; miR395; miR393; miR390; miR319; miR171; miR160	Vanzin *et al.*, 2002; Wang *et al.*, 2005; Zhou *et al.*, 2008; Burklew *et al.*, 2012
Arsenic	miR6254; miR6250; miR1432; miR1318; miR854; miR838; miR528; miR408; miR397; miR395; miR319; miR172; miR169; miR167; miR164; miR159; miR156	Zhou *et al.*, 2008; Srivastava *et al.*, 2012; Liu and Zhang, 2012; Liu, 2012
Cadmium	miR857; miR604; miR602; miR398; miR397; miR395; miR393; miR172; miR171; miR167; miR164; miR160; miR159; miR156	Bovet *et al.*, 2003; Xie *et al.*, 2007; Huang *et al.*, 2010; Chen *et al.*, 2011; Zhou *et al.*, 2012
Copper	miR398; miR397	Beauclair *et al.*, 2010; Buhtz *et al.*, 2010
Manganese	miR2118; miR1510	Valdes-Lopez *et al.*, 2010
Mercury	miR2681; miR529; miR393; miR171	Zhou *et al.*, 2008; Zhou *et al.*, 2012
Iron	miR398; miR397	Beauclair *et al.*, 2010; Buhtz *et al.*, 2010

provide better outcomes in determining the involvement of various miRNAs in heavy metal toxicity responses. Moreover, the development of miRNA-resistant targets via alteration in the nucleotide sequences of the targets, that make them unable to bind to the associated miRNAs, will indicate the functioning of these miRNAs during heavy metal toxicity response and will also aid in the establishment of the possible regulatory linkages. Knowledge about alterations in DNA methylation profiles under heavy metal stress and their association with various miRNAs claims to become an interesting research field in near future.

Unfortunately, the mechanisms depicting the possible ways by which miRNAs themselves are modulated upon inception of heavy metal stress still remains unknown. Hence, a combinatorial approach summing up all the existing and new information in this regard, tagged with meticulous research, might accelerate the process of management of heavy metal stress within the plant system in a much better way.

Acknowledgements

Financial assistance from Science and Engineering Research Board, Government of India through the grant [EMR/2016/004799] and Department of Higher Education, Science and Technology and Biotechnology, Government of West Bengal, through the grant [264(Sanc.)/ST/P/S&T/1G-80/2017] to Dr. Aryadeep Roychoudhury is gratefully acknowledged.

References

Baxter, I., Tchieu, J., Sussman, M.R., Boutry, M., Palmgren, M.G., Gribskov, M., Harper, J.F. and Axelsen, K.B. (2003). Genomic comparison of P-Type ATPase ion pumps in *Arabidopsis* and rice, *Plant Physiol.*, 132: 618-628.

Beauclair, L., Yu, A. and Bouche, N. (2010). MicroRNA-directed cleavage and translational repression of the copper chaperone for superoxide dismutase mRNA in *Arabidopsis*, *The Plant J.*, 62: 454-462.

Bovet, L., Eggmann, T., Meyland-Bettex, M., Polier, J., Kammer, P., Marin, E., Feller, U. and Martinoia. E. (2003). Transcript levels of AtMRPs after cadmium treatment, induction of AtMRP3, *Plant Cell Environ.*, 26: 371-381.

Buhtz, A., Pieritz, J., Springer, F. and Kehr, J. (2010). Phloem small RNAs, nutrient stress responses, and systemic mobility, *BMC Plant Biol.*, 10: 64.

Burklew, C.E., Ashlock, J., Winfrey, W.B. and Zhang, B. (2012). Effects of aluminium oxide nanoparticles on the growth, development, and microRNA expression of tobacco (*Nicotiana tabacum*), *PLoS ONE*, 7: e34783.

Chen, J., Shiyab, S., Han, F.X., Monts, D.L., Waggoner, C.A., Yang, Z. and Su, Y. (2009a). Bioaccumulation and physiological effects of mercury in *Pteris vittata* and *Nephrolepis exaltata*, *Ecotoxicology*, 18: 110-121.

Chen, J., Yang, Z.M., Su, Y., Han, F.X. and Monts, D.L. (2009b). Phytoremediation of heavy metal/metalloid-contaminated soils. *In:* Steinberg, R.V. and Steinberg, R.V. (Eds.). *Contaminated Soils, Environmental Impact, Disposal and Treatment*, Nova Science Publishers, New York.

Chen, L., Wang, T., Zhao, M., Tian, Q. and Zhang, W.H. (2012). Identification of aluminum-responsive microRNAs in *Medicago truncatula* bygenome-wide high-throughput sequencing, *Planta*, 35: 375-386.

Cheng, N.H., Liu, J.Z., Liu, X., Wu, Q., Thompson, S.M., Lin, J., Chang, J., Whitham, S.A., Park, S. and Cohen, J.D. (2011). *Arabidopsis* monothiolglutaredoxin, AtGRXS17, is critical for temperature-dependent post-embryonic growth and development via modulating auxin response, *J. Biol. Chem.*, 286: 20398-20406.

Cobbett, C. and Goldsbrough, P. (2002). Phytochelatins and metallothioneins: Roles in heavy metal detoxification and homeostasis, *Ann. Rev. Plant Biol.*, 53: 159-182.

Cobbett, C.S. (2000). Phytochelatin biosynthesis and function in heavy metal detoxification, *Curr. Opinion Plant Biol.*, 3: 211-216.

Curie, C., Panaviene, Z., Loulergue, C., Dellaporta, S.L., Briat, J.F. and Walker, E.L. (2001). Maize yellow stripe1 encodes a membrane proteindirectly involved in Fe(III) uptake, *Nature*, 409: 346-349.

Dal-Corso, G., Farinati, S. and Furini, A. (2010). Regulatory networks of cadmium stress in plants, *Plant Signal. Behav.*, 5: 663-667.

Dat, J., Vandenabeele, S., Vranova, E., Van Montagu, M., Inze, D. and Van Breusegem, F. (2000). Dual action of the active oxygen species during plant stress responses, *Cell Mol. Life Sci.*, 57: 779-795.

Ding, Y., Chen, Z. and Zhu, C. (2011). Microarray-based analysis of cadmium-responsive microRNAs in rice (*Oryza sativa*), *J. Exp. Bot.*, 62: 3563-3573.

Ding, Y.F. and Zhu, C. (2009). The role of microRNAs in copper and cadmium homeostasis, *Biochem. Biophys. Res. Comm.*, 386: 6-10.

Ericksen, J.A., Gustin, M.S., Schorran, D.E., Johnson, D.W., Lindberg, S.E. and Coleman, J.S. (2003). Accumulation of atmospheric mercury in forest foliage, *Atmosph Environ.*, 37: 1613-1622.

Fay, L. and Gustin, M.S. (2007). Assessing the influence of different atmospheric and soil mercury concentrations on foliar mercury concentrations in a controlled environment, *Water Air Soil Poll.*, 181: 373-384.

Grant, C.A, Clarke, J.M., Duguid, S. and Chaney, R.L. (2008). Selection and breeding of plant cultivars to minimise cadmium accumulation, *Sci. Total Environ.*, 390: 301-310.

Guerinot, M.L. (2000). The ZIP family of metal transporters, *Biochim. Biophy. Acta.*, 1465: 190-198.

Hall, J.L. (2002). Cellular mechanisms for heavy metal detoxification and tolerance, *J. Exp. Bot.*, 53: 1-11.

Hossain, M.A., Hasanuzzaman, M. and Fujita, M. (2010). Up-regulation of antioxidant and glyoxalase systems by exogenous glycine betaine and proline in *mung* bean confer tolerance to cadmium stress, *Physiol. Mol. Biol. Plants*, 16: 259-272.

Huang, S.Q., Peng, J., Qiu, C.X. and Yang, Z.M. (2009). Heavy metal-regulated new microRNAs from rice, *J. Inorg. Biochem.*, 3: 282-287.

Huang, S.Q., Xiang, A.L., Che, L.L., Chen, S., Li, H., Song, J.B. and Yang, Z.M. (2010). A set of miRNAs from *Brassica napus* in response to sulphate deficiency and cadmium stress, *Plant Biotechnol. J.*, 8: 887-899.

Jha, A.B. and Dubey, R.S. (2004). Carbohydrate metabolism in growing rice seedlings under arsenic toxicity, *J. Plant Physiol.*, 161: 867-872.

Jonak, C., Nakagami, H. and Hirt, H. (2004). Heavy metal stress. Activation of distinct mitogen-activated protein kinase pathways by copper and cadmium, *Plant Physiol.*, 136: 3276-3283.

Jones-Rhoades, M.W., Bartel, D.P. and Bartel, B. (2006). MicroRNAs and their regulatory roles in plants, *Ann. Rev. Plant Biol.*, 57: 9-53.

Kim, D.Y., Bovet, L., Maeshima, M., Martinoia, E. and Lee, Y. (2007). The ABC transporter AtPDR8 is a cadmium extrusion pump conferring heavy metal resistance, *Plant, J.*, 50: 207-218.

Kramer, U., Talke, I.N. and Hanikenne, M. (2007). Transition metal transport, *FEBS Lett.*, 581: 2263-2272.

Lima, J,C., Arenhart, R.A., Margis-Pinheiro1, M. and Margis, R. (2011). Aluminum triggers broad changes in microRNA expression in rice roots, *Genet. Mol. Res.*, 10(4): 2817-2832, doi.org/10.4238/2011

Liu, Q. (2012). Novel miRNAs in the control of arsenite levels in rice, *Funct. Integr. Genomics*, 12: 649-658.

Liu, Q. and Zhang, H. (2012). Molecular identification and analysis of arsenite stress-responsive miRNAs in rice, *J. Agric. Food Chem.*, 60: 6524-6536.

Ma, J.F., Ryan, P.R. and Delhaize, E. (2001). Aluminium tolerance in plants and the complexing role of organic acids, *Trends Plant Sci.*, 6: 273-278.

Ma, J.F., Shen, R., Nagao, S. and Tanimoto, E. (2004). Aluminium targets elongating cells by reducing cell wall extensibility in wheat roots, *Plant Cell Physiol.*, 45: 583-589.

Maksymiec, W. (2007). Signalling responses in plants to heavy metal stress, *Acta. Physiol. Plant*, 29: 177-187.

Maksymiec, W. and Krupa, Z. (2006). The effects of short-term exposition to Cd, excess Cu ions and jasmonate on oxidative stress appearing in *Arabidopsis thaliana*, *Environ. Exp. Bot.*, 57: 187-194.

Marschner, H. (1995). *Mineral Nutrition of Higher Plants*, Academic Press Inc., London.

Meng, D.K., Chen, J. and Yang, Z.M. (2011). Enhancement of tolerance of Indian mustard (*Brassica juncea*) to mercury by carbon monoxide, *J. Hazard Mat.*, 186: 1823-1829.

Mittler, R. (2002). Oxidative stress, antioxidants and stress tolerance, *Trends Plant Sci.*, 7: 405-410.

Nagasaki, H., Itoh, J., Hayashi, K., Hibara, K., Satoh-Nagasawa, N., Nosaka, M., Mukouhata, M., Ashikari, M., Kitano, H. and Matsuoka, M. (2007). The small interfering RNA production pathway is required for shoot meristem initiation in rice, *Proc. Natl. Acad. Sci.*, 104: 14867-14871.

Nevo, Y. and Nelson, N. (2006). The NRAMP family of metal-ion transporters, *Biochem. Biophys. Acta.*, 1763: 609-620.

Pandita, D. (2019). Plant MIRnome: miRNA biogenesis and abiotic stress response. *In:* Hasanuzzaman, M., Hakeem, K., Nahar, K. and Alharby, H. (Eds.). *Plant Abiotic Stress Tolerance*. Springer, Cham. 449-474. Doi https://Doi.org/10.1007/978-3-030-06118-0_18

Pandita, D. and Wani, S.H. (2019). MicroRNA as a tool for mitigating abiotic stress in rice (*Oryza sativa* L.). *In:* Wani, S. (Ed.). *Recent Approaches in Omics for Plant Resilience to Climate Change*. Springer, Chamz, 109-133. Doi https://Doi.org/10.1007/978-3-030-21687-0_6

Pandita, D. (2021). Role of miRNAi technology and miRNAs in abiotic and biotic stress resilience. *In:* Aftab T. and Roychoudhury A. (Eds.). *Plant Perspectives to Global Climate Changes*. Academic Press, Elsevier, 303-330. https://Doi.org/10.1016/B978-0-323-85665-2.00015-7

Pandita, D. (2022a). How microRNAs regulate abiotic stress tolerance in wheat? A snapshot. *In:* Roychoudhury, A., Aftab, T. and Acharya, K. (Eds.). *Omics Approach to Manage Abiotic Stress in Cereals*. Springer, Singapore, 447-464. https://Doi.org/10.1007/978-981-19-0140-9_17

Pandita, D. (2022b). MicroRNAs shape the tolerance mechanisms against abiotic stress in maize. *In:* Roychoudhury, A., Aftab, T. and Acharya, K. (Eds.). *Omics Approach to Manage Abiotic Stress in Cereals*. Springer, Singapore, 479-493. https://Doi.org/10.1007/978-981-19-0140-9_19

Pandita, D. (2022c). miRNA- and RNAi-mediated metabolic engineering in plants. *In:* Aftab, T. and Hakeem, K.R. (Eds.). *Metabolic Engineering in Plants*. Springer, Singapore, 171-186. https://Doi.org/10.1007/978-981-16-7262-0_7

Patsikka, E., Kairavuo, M., Sersen, F., Aro, E.M. and Tyystjarvi, E. (2002). Excess copper enhances photoinhibition *in vivo* by outcompeting iron and causing chlorophyll deficiency, *Plant Physiol.*, 129: 1359-1367.

Peleg, Z. and Blumwald, E. (2011). Hormone balance and abiotic stress tolerance in crop plants, *Curr. Opinion Plant Biol.*, 14: 290-295.

Puig, S. and Thiele, D.J. (2002). Molecular mechanisms of copper uptake and distribution, *Curr. Opi. Chem. Biol.*, 6: 171-180.

Rascio, N. and Navari-Izzo, F. (2011). Heavy metal hyperaccumulating plants: How and why do they do it? And what makes them so interesting? *Plant Sci.*, 180: 169-181.

Rauser, W.E. (1999). Structure and function of metal chelators produced by plants: The case for organic acids, amino acids, phytin and metallothioneins, *Cell Biochem. Biophys.*, 31: 19-48.

Rodriguez-Serrano, M., Romero-Puertas, M.C., Pazmino, D.M., Testillano, P.S., Risueno, M.C., del Rio, L.A. and Sandalio, L.M. (2009). Cellular response of pea plants to cadmium toxicity: Cross talk between reactive oxygen species, nitric oxide and calcium, *Plant Physiol.*, 150: 229-243.

Romero-Puertas, M.C., Corpas, F.J., Rodriguez-Serrano, M., Gomez, M., del Rıo, L.A. and Sandalio, L.M. (2007). Differential expression and regulation of antioxidative enzymes by cadmium in pea plants, *J. Plant Physiol.*, 164: 1346-1357.

Rose, J.K., Braam, J., Fry, S.C. and Nishitani, K. (2002). The XTH family of enzymes involved in xyloglucan endotrans glucosylation and endohydrolysis: Current perspectives and a new unifying nomenclature, *Plant Cell Physiol.*, 43: 1421-1435.

Sanitadi, T.L. and Gabbrielli, R. (1999). Response to cadmium in higher plants, *Environ. Exp. Bot.*, 41: 105-130.

Schommer, C., Palatnik, J.F., Aggarwal, P., Chetelat, A., Cubas, P., Farmer, E.E., Nath, U. and Weigel, D. (2008). Control of jasmonate biosynthesis and senescence by miR319 targets, *PLoS Biol.*, 6: e230.

Schutzendubel, A. and Polle, A. (2002). Plant responses to abiotic stress: Heavy metal induced oxidative stress and protection by mycorrhisation, *J. Exp. Bot.*, 53: 1351-1365.

Sekhar, K., Priyanka, B., Reddy, V.D. and Rao, K.V. (2011). Metallothionein 1 (CcMT1) of pigeonpea (*Cajanus cajan*, L.) confers enhanced tolerance to copper and cadmium in *Escherichia coli* and *Arabidopsis thaliana*, *Environ. Exp. Bot.*, 72: 131-139.

Sharma, P. and Dubey, R.S. (2007). Involvement of oxidative stress and role of antioxidative defence system in growing rice seedlings exposed to toxic concentrations of aluminium, *Plant Cell Rep.*, 26: 2027-2038.

Shiyab, S., Chen, J., Han, F.X., Monts, D.L., Matta, F.B., Gu, M. and Su, Y. (2009). Phytotoxicity of mercury in Indian mustard (*Brassica juncea* L.), *Ecotoxicol. Environ. Saf.*, 72: 619-625.

Srivastava, S., Srivastava, A.K., Suprasanna, P. and D'Souza, S.F. (2012). Identification and profiling of arsenic stress-induced microRNAs in *Brassica juncea*, *J. Exp. Bot.*, 64: 303-315.

Srivastava, S., Suprasanna, P. and D'Souza, S.F. (2011). Mechanisms of arsenic tolerance and detoxification in plants and their application in transgenic technology: A critical appraisal, *Int. J. Phytoremed.*, 14: 506-517.

Stoeva, N., Berova, M. and Zlatev, Z. (2003). Physiological response of maize to arsenic contamination, *Biol. Plantarum.*, 47: 449-452.

Valdes-Lopez, O., Yang, S.S., Aparicio-Fabre, R., Graham, P.H., Reyes, J.L., Vance, C.P. and Hernandez, G. (2010). MicroRNA expression profile in common bean (*Phaseolus vulgaris*) under nutrient deficiency stresses and manganese toxicity, *New Phytol.*, 187: 805-818.

Vanzin, G.F., Madson, M., Carpita, N.C., Raikhel, N.V., Keegstra, K. and Reiter, W.D. (2002). The mur2 mutant of *Arabidopsis thaliana* lacks fucosylated xyloglucan because of a lesion in fucosyltransferase AtFUT1, *Proc. Natl. Acad. Sci.*, 99: 3340-3345.

Verica, J.A., Chae, L., Tong, H., Ingmire, P. and He, Z.H. (2003). Tissue-specific and developmentally regulated expression of a cluster of tandemly arrayed cell wall-associated kinase-like kinase genes in *Arabidopsis*, *Plant Physiol.*, 133: 1732-1746.

Villiers, F., Ducruix, C., Hugouvieux, V., Jarno, N., Ezan, E., Garin, J., Junot, C. and Bourguignon, J. (2011). Investigating the plant response to cadmium exposure by proteomic and metabolomic approaches, *Proteomics*, 11: 1650-1663.

Wang, J.W., Wang, L.J., Mao, Y.B., Cai, W.J., Xue, H.W. and Chen, X.Y. (2005). Control of root cap formation by microRNA-targeted auxin response factors in *Arabidopsis*, *Plant Cell*, 17: 220-221.

Waters, B.M., McInturf, S.A. and Stein, R.J. (2012). Rosette iron deficiency transcript and microRNA profiling reveals links between copper and iron homeostasis in *Arabidopsis thaliana*, *J. Exp. Bot.*, 63: 5903-5918.

Williams, L.E., Pittman, J.K. and Hall, J.L. (2000). Emerging mechanisms for heavy metal transport in plants, *Biochem Biophys. Acta.*, 1465: 104-126.

Xie, F.L., Huang, S.Q., Guo, K., Xiang, A.L., Zhu, Y.Y., Nie, L. and Yang, Z.M. (2007). Computational identification of novel microRNAs and targets in *Brassica napus*, *FEBS Lett.*, 581: 1464-1474.

Xie, Q., Frugis, G., Colgan, D. and Chua, N.H. (2000). *Arabidopsis* NAC1 transduces auxin signal downstream of TIR1 to promote lateral root development, *Genes Dev.*, 14: 3024-3036.

Xue, L.J., Zhang, J.J. and Xue, H.W. (2009). Characterisation and expression profiles of miRNAs in rice seeds, *Nucl. Acids Res.*, 37: 916-930.

Yang, T. and Poovaiah, B.W. (2003). Calcium/calmodulin-mediated signal network in plants, *Trends Plant Sci.*, 8: 505-512.

Yeh, C.M., Hsiao, L.J. and Huang, H.J. (2004). Cadmium activates a mitogen activated protein kinase gene and MBP kinases in rice cell, *Plant Cell Physiol.*, 45: 1306-1312.

Yoon, E.K., Yang, J.H., Lim, J., Kim, S.H., Kim, S.K. and Lee, W.S. (2010). Auxin regulation of the microRNA390-dependent transacting small interfering RNA pathway in *Arabidopsis* lateral root development, *Nucl Acids Res.*, 38: 1382-1391.

Yu, T.J., Luo, Y.F., Liao, B., Xie, L.J., Chen, T., Xiao, S., Li, J.T., Hu, S.N. and Shu, W.H. (2012). Comparative transcriptome analysis of transporters, phytohormone and lipid metabolism pathways in response to arsenic stress in rice (*Oryza sativa*), *New Phytol.*, 195: 97-112.

Zeng, Q.Y., Yang, C.Y., Ma, Q.B., Li, X.P., Dong, W.W. and Nian, H. (2012). Identification of wild soybean miRNAs and their target genes responsive to aluminium stress, *BMC Plant Biol.*, 12: 182.

Zhou, Z.S., Huang, S.Q. and Yang, Z.M. (2008). Bioinformatic identification and expression analysis of new microRNAs from *Medicago truncatula*, *Biochem. Biophys. Res. Comm.*, 374: 538-542.

Zhou, Z.S., Zeng, H.Q., Liu, Z.P. and Yang, Z.M. (2012). Genome-wide identification of *Medicago truncatula* microRNAs and their targets reveals their differential regulation by heavy metal, *Plant Cell Environ.*, 35: 86-99.

CHAPTER

12

MicroRNA-mediated Regulation of Osmotic and Oxidative Stress

Ramachandra Reddy Pamuru[1]*, T. Chandrasekhar[2] and Arifullah Mohammed[3]

[1] Department of Biochemistry, Yogi Vemana University, Kadapa - 516005, A.P., India
[2] Department of Environmental Science, Yogi Vemana University, Kadapa - 516005, A.P., India
[3] Department of Agriculture Sciences, Faculty of Agro Based Industry (FIAT), Universiti Malaysia Kelantan, 17600 Jeli, Kelantan, Malaysia

1. Introduction

Stress is a condition (phenomenon) of altering the natural physiological functions of the all organisms. Biotic and abiotic stresses show the negative impact on physiological aspects of living organisms, including plants. Many factors of the environment naturally create stress in plants. But the organisms try to overcome the stress and continue to exist without any altered physiology. Certain mechanisms have been evolved to exert the stress and vary from organism to organism. The mechanistic role of improved stress tolerance varies from organism to organism.

Abiotic and biotic stresses are the factors that lower the yield of crop and ornamental plants on the globe. Quality food production plays a crucial role in feeding the ever growing population in the world. Food scarcity can be overcome through agricultural yield improvement in the available area. A number of factors causing biotic and abiotic stresses in crop plants have been identified. Knowing stress tolerance factors and their mechanism of action in crop plants and other ornamental plants may help to improve the crop yield in future. Many studies are in this line and working to find the best possible factor(s)/mechanism(s) to improve the crop yield even during unfavourable conditions.

Plants have their own mechanisms to show stress tolerance for various biotic and abiotic stress causatives. It is possible through gene activation only. Plants carry a number of stress tolerance genes in their genome and are evolved through acclimation and adaptation of plants to stress conditions naturally (Yamaguchi-Shinozaki and Shozaki, 2006). A major part of this is under the control of micro-RNAs (mi-RNAs)

---

*Corresponding author: reddyprbiotech@gmail.com

which are stress responsive elements. The mi-RNAs are endogenous, small RNAs with 21 to 25 nucleotides, existing in a double stranded form (Voinnet, 2009). Moreover, in plants these are non-coding and regulate the expression of genes through repression of target m-RNA, especially during and after transcription (post-transcription) stages (Kawaguchi *et al.*, 2004). Stages like DNA methylation, chromatin remodelling and cleavage of mRNA are some other sites where mi-RNAs exhibit their functions. Barciszewska-Pacak *et al.* (2015) and Pandita (2019) listed the crop plants (wheat, rice, barley, tomato, potato, legumes, sugarcane, etc.) that produce mi-RNAs in response to stress.

In recent days, a lot of significance has been given to mi-RNA-mediated gene regulation. Especially in ornamental and crop plant improvement, this has become a prime area to work and look for breakthroughs in the scientific world. The significance of the present chapter is to gather information associated with mi-RNAs and their response in controlling various stress conditions with special focus on osmotic and oxidative stress.

2. MicroRNAs in Plants

During early 1990s, the first mi-RNA named small temporal RNA (stRNA) was isolated and characterised as *Caenorhabditis elegans*, a nematode (Lee *et al.*, 1993). Later, many mi-RNAs have been identified and cloned from a variety of organisms, including plants. In the year 2000, the first plant mi-RNA was isolated from *Arabidopsis thaliana*. They named them as mi-R156, mi-R159, mi-R164, mi-R171 etc. (Reinhart *et al.*, 2002). In the same species, several mi-RNA molecules, holding 21 to 24 nucleotides, have been cloned and reported (Llave *et al.*, 2002b). The conservation of mi-RNA sequence is another aspect to find the similarity between genes among the plant species. Conservation of 8 mi-RNA sequences isolated from *Arabidopsis thaliana* is reported with *Oryza sativa* (Mette *et al.*, 2002). Several mi-RNAs are reported from other plant species. The isolated mi-RNAs are available in the database mi-RBase (www.mirbase.org) and plant mi-RNAs in PMRD_plant mi-RNA (http://bioinformatics.cau.edu.cn/PMRD).

3. Genesis and Mechanistic Action of Bio-miRNAs

The mi-RNA biogenesis is a process where transcription, precursor processing for the modification or maturation, transportation and execution of its functions through RNA-induced silencing complex (RISC) formation are included. The genes that code for mi-RNAs are mostly non-coding type and are located predominantly in the intra-genic and inter-genic regions of genes. Some of them are located in the exons of m-RNAs and 3' untranslated regions (3'UTRs) or 5' untranslated regions (5'UTRs). According to Lee *et al.* (2014) mi-RNA regions on genomic DNA have their own regulatory and other elements like regular genes. The components such as promoter sequence, the nucleotides that are complimentary to specific m-RNA sequences and terminator sequence are common in mi-RNA genes.

The known polymerases that transcribe the mi-RNA genes are RNA Polymerase II in plants and induced by a coactivator (Kim *et al.*, 2011). The mi-RNA gene product immediately after transcription can be named as pri-mi-RNA (primary mi-RNA transcript) which requires maturation through its processing. The processing has been done by dicing complex, which holds the components like DCL-1 (Dicer-like-1), HYL-1 (Hyponastic leaves-1) and SE (serrate) (Fukudome and Fukuhara, 2017) and where the pri-mi-RNA get converted into pre-mi-RNA (pre-mi-RNA transcript), the presence of predominant endonuclease enzyme DCL-RNase III. The pre-mi-RNA is a precursor for matured mi-RNA, which holds 70 nucleotides with hairpin structure (Rajagopalan et al., 2006). Finally, hairpin duplex mi-RNA by the action of RNA-specific HEN1 methyltransferase get methylated. This whole process of mi-RNA synthesis and its processing happen in the nucleus. Methylation of mi-RNA exports it into the cytosol through HST export in five nuclear transporters (Park et al., 2002) (Fig. 1).

There are two ways that RNA interference silent the gene expression in the cytosol. In one way mature mi-RNA activates RNA-induced silencing complex (RISC) through Argonaute protein 1 (AGO1), thereby cleavage occurs of homologues m-RNA molecules present in the cytosol. The second pathway is mediated by various types of exogenous or endogenous small RNA molecules, such as hc-RNA, ra-si-RNA, nat-si-RNA, ta-si-RNA and ds-RNA (Zheng et al., 2018) (Fig. 1).

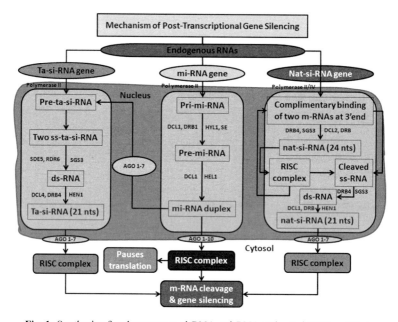

Fig. 1: Synthesis of endogenous ta-si-RNA, mi-RNA and nat-si-RNA and their mechanism of action in gene silencing

4. Role of miRNAs in Plant Stress Response

The miRNAs are tiny in size but play a crucial role in controlling physiological aspects, especially in adverse conditions. They release upon adverse conditions in the plants and facilitate the system to overcome these conditions. Most predominantly known adverse conditions cause physiological stress in any organism. The known adverse conditions can be basically divided into internal and external. The external factors are also further classified as biotic, abiotic and mechanical conditions. A number of identifying adverse conditions are well reported in many studies with the elevation of mi-RNA in the plants. The classification of various factors that cause stress and the release of mi-RNAs is presented in Fig. 2.

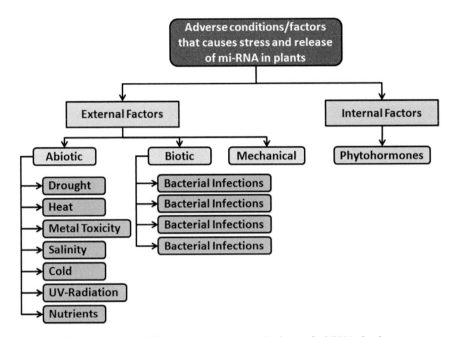

Fig. 2: Factors/conditions that cause stress and release of mi-RNAs in plants

In plants, stress causatives illustrate a pessimistic impact on various activities, like morphology, growth, reproduction, physiology, productivity, biochemistry, development and various molecular mechanisms. Moreover, the known fact of elevated mi-RNAs to drop down the stress generated by various factors suggests the development of transgenic crops and ornamental plants using different tools of genetic engineering. Development of transgenic plants with mi-RNA silencing gene(s) mediated mechanism to withstand at various adverse conditions may help to improve the crop yield and longevity of life in ornamental plants. Furthermore, these transgenic plants may adapt to the climatic changes taking place from time to time in the world.

4.1 Abiotic Stress and mi-RNAs

Predominant stress that occurs in plants and varies with seasons and areas on the Earth is abiotic stress. Though the plants evolve to survive in unfavourable conditions, it is so in all the plants. Abiotic stress is the main causative and does not allow crops to grow on every place on the globe. An adaptation to various abiotic stresses is a known natural phenomenon in plants. Many groups of scientists are working to fish out the mechanism of abiotic stress tolerance in order introduce this in regular crop plants to yield more and grow plants in every corner of the world. Stress response in plants is mostly resisted by expressing large numbers of genes-regulating protein metabolism and these genes vary with species (Turan and Tripathy, 2013; Khare *et al.*, 2015).

The best example that elevates the stress-induced protein synthesis is the response of genes that are involved in proline biosynthesis (Kumar *et al.*, 2010) and MSRB7 gene (Lee *et al.*, 2014). On the other hand, these proteins are under the regulation of mi-RNAs. A number of mi-RNAs have been recognised during abiotic stress in crop plants (Pandita, 2019; Pandita and Wani, 2019; Pandita, 2021; 2022a; 2022b; 2022c). The abiotic stress is caused by a number of factors, such as temperature, drought, hypoxia, oxidative stress, metal toxicity, salinity, UV radiation, cold, nutrients, etc. (Liang *et al.*, 2015; Gao *et al.*, 2011; Casadevall *et al.*, 2013; Goswami *et al.*, 2014; Gupta *et al.*, 2014).

4.1.1 Heat and Drought and mi-RNAs

It is a well-known fact that heat-shock proteins are the responsive molecules for heat stress in all organisms, including plants. The variety of heat-shock proteins have been identified when plants are exposed to high temperatures (heat) (Iba *et al.*, 2002). These are molecular chaperones crucial for protein folding, congregation, translocation and degradation (Vierling *et al.*, 1991; Wang *et al.*, 2004). Basically, they are grouped into five classes, namely HSP60, HSP70, HSP90, HSP100 and small HSPs (Iba *et al.*, 2002). Recent studies evidence that the synthesis of heat-shock proteins and heat-shock transcription factors during stress conditions in plants is mediated by mi-RNA molecules. Studies in this direction are vital and the details are tabulated in Table 1. Heat-shock transcription factors are classified into A, B and C groups and are predominantly studied gene families that belong to HSF group (HSFA1-9).

Table 1: Mediation of mi-RNAs in heat-shock protein and heat-shock transcription factor synthesis during heat stress in plants

S.No.	Name of the plant	Type of HSP/HSF	Species of mi-RNAs	References
1.	Tomato	HSP70	Moderately high temperature: spi-mi-R6300	Zhou *et al.*, 2016
		HSP60-3A	Acutely high temperature: spi-mi-R166g-3p and spi-mi-R166c-3p	
2.	*Arabidopsis*	HSP22, HSP 17.6A and HSFA2	mi-R156	Yu *et al.*, 2014
		HSP70B, HSP21, HSP17.6II and HSP17.6A	Elevated expression of mi-R160	Lin *et al.*, 2018
3.	Rice	HSP70 and HSP90	mi-R166e-3p	Liu *et al.*, 2017
		4 HSPs	mi-R169r-5P	
		HSP20, HSFB-2c, HSP70 and HSP-90-6	osa-mi-R5077, osa-mi-R397b and osa-mi-R167e	Kushawaha *et al.*, 2021
4.	*Raphanus sativus*	HSP81-83 and HSP91	mi-R5293 and mi-R5292	Wang *et al.*, 2014
5.	Wheat	HSP17 and HSP70	Elevated expression of mi-R164 and mi-R160	Kumar *et al.*, 2014

HSP: Heat-Shock Protein; HSF: Heat-Shock Protein Transcription Factor

MicroRNA-mediated Regulation of Osmotic and Oxidative Stress **203**

Excess water evaporation, water deficiency in soil and precipitation shortage are the reasons for drought stress in plants (Shukla *et al.*, 2008). Though there are mechanisms to overcome drought stress, recent studies have revealed that mi-RNAs are responsive against drought stress. The crop plants that have elevated mi-RNAs under drought stress have been reported in tobacco, cowpea and soybean (Frazier *et al.*, 2011; Barrera-Figueroa *et al.*, 2011; Kulcheski *et al.*, 2011). Liu *et al.* (2008) found that 12 mi-RNAs in *Arabidopsis* (mi-R165, mi-R396, mi-R156, mi-R159, mi-R169, mi-R171, mi-R319, mi-R158, mi-R168, mi-R393, mi-R167 and mi-R167) responded to drought stress. Out of 12, only three, i.e. mi-R397, mi-R319 and mi-R393 are elevated and the others are down-regulated (Sunkar and Zhu, 2004). Up-regulation of these genes lowers the growth and development of the plant, whereas in rice, 30 mi-RNAs were identified with drought stress (Zhou *et al.*, 2010). Table 2 shows the drought-stress induced mi-RNAs in various plant species.

Table 2: Expression of different mi-RNAs during drought stress in various plant species

S.No.	Name of the Plant	No. of mi-RNAs	Species of mi-RNAs	References
1.	Rice	30	16 Nos. down-regulated: mi-R397, -R156, -R159, -R396, -R168, -R408, -R1035, -R319, -R172, -R171, -R529, -R170, -R1126, -R896, -R1050 and -R1030 14 Nos. up-regulated: mi-R854, -R159, -R896, -R1125, -R851, -R169, -R319, -R474, -R901, -R171, -R845, -R903, -R395 and -R1026	Zhou *et al.*, 2010
2.	*Arabidopsis*	12	mi-R165, -R396, -R156, -R159, -R169, -R171, -R319, -R158, -R168,-R393, -R167 and -R167	Liu *et al.*, 2008
3.	*Populus*	16	16 Nos. moderately up-regulated: mi-R1447, -R1450, -R1444a, -R171-1_n, -R1446-a_e, -R1445, -R482.2, -R827, -R530-a and -R1448	Lu *et al.*, 2008
4.	*Phaseolus vulgaris*	6	3 Nos. moderately up-regulated: mi-R2119, -RS1 and -R1514a. 3 Nos. highly up-regulated: mi-R2118, -R159.2 and -R393	Arenas-Huertero *et al.*, 2009
5.	*Medicago truncatula*	4	Down-regulated: mi-R169. 3 Nos. up-regulated: mi-R408, -R398a and b	Trindade *et al.*, 2010
6.	*Triticum turgidum*	13	Differentially regulated: mi-R166, -R1432, -R1867, -R1450, -R1881, -R398, -R396, -R156, -R474, -R896, -R528, -R171 and -R894	Kantar *et al.*, 2011

4.1.2 Metal Toxicity and mi-RNAs

Metals are naturally occurring elements in the Earth's crust. A few heavy metals are essential and majority are toxic, non-essential for life (Rascio and Navari-Izzo, 2011;

Gielen *et al.*, 2012) (Fig. 3). Very low amounts of these metals are naturally distributed as a source for the survival of all organisms. Major amounts of metals are unavailable and are not directly in contact with most of the organisms living on the Earth. In very few places, high amounts of metals accumulate and only a few bio-accumulators are rich in these areas. Because of anthropogenic activities and usage of large areas of land for industrial extraction of metals and inorganic compounds releasing heavy metals into the environment and creating pollution, toxic effects on the health of humans and survival of other organisms is experienced. Heavy metals enter into the food chain and accumulate in various parts of the human body, thereby threatening human health (Gupta *et al.*, 2014). Heavy metal toxicity is one of the major abiotic stress causative leading to dis-regulation of physiological and metabolic functions in humans and other organisms (Gupta *et al.*, 2014; Rascio and Navari-Izzo, 2011). Yang and Chen (2013) reported that the heavy metal toxicity stress elevates mi-RNA binding to its genes. These mi-RNAs target the corresponding m-RNA molecules at transcriptional and post-transcriptional levels. The following table shows the effects of heavy metal toxicity and response of corresponding mi-RNAs in plants (Table 3).

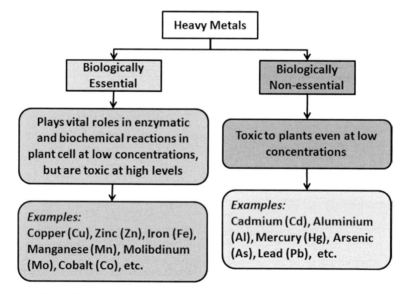

Fig. 3: Essential and non-essential heavy metals and their role in plant cell

4.1.3 Salinity and mi-RNAs

Salty soils are strong enough to restrict the growth and development of organisms, including plants through the reverse osmosis process. Saline soils of the Earth occupy a large area (approximately 20% of agricultural lands) and have many disadvantages including disallowing the growth of crops (Qadir *et al.*, 2014). Plants that grow in saline soils are of two types: i.e. halophytes and glycophytes. Salinity-sensitive plants are glycophytes, which may not survive in a saline environment (Horie *et al.*, 2012). The majority of the crops fall under this category. The saline-adapted plants that grow easily in salty soils are halophytes and these yield moderately also (Su *et al.*, 2020).

MicroRNA-mediated Regulation of Osmotic and Oxidative Stress **205**

Table 3: Heavy metal toxicity and RNAs response to its stress in plants

Heavy Metal	Toxicity in Plants	Response of mi-RNA	References
Cadmium (Cd)	Wilting, chlorosis, cell death, arrest of growth, protein damage, oxidative stress and crop productivity.	Elevated levels in *Brassica napus* leaves – *bna-mi-R393*; roots and leaves - *bna-mi-R167a/c* and *bna-mi-R156a;* all tissues - *bna-mi-R394a/b/c* and *bna-mi-R164b;* down regulation of *bna-mi-R160*	Zhao *et al.*, 2012; Rizwan *et al.*, 2016; Huang *et al.*, 2010
		Conserved and non-conserved 84 mi-RNAs in roots and shoots of *Brassica napus* expressed differentially.	Zhou *et al.*, 2012a
		Constructed mi-RNA libraries in *Raphnus sativus.*	Xu *et al.*, 2013
Mercury (Hg)	Reduced content of chlorophyll, stunted growth, vascular abnormalities, elevated reactive oxygen species and loss of cell shape.	mi-RNAs differential expression in variety of plant species.	Chen and Yang, 2012
		Elevated levels of *mi-R398* in *Arabidopsis.*	Sunkar *et al.*, 2006
Manganese (Mn)	Beneficial at low concentration but toxic at high concentration	About 37 mi-RNAs of *Phaseolus vulgaris* expressed differentially. The mi-R1532, mi-R1515, mi-R1510 or 2110, mi-R1508 were characterised. Targets for these mi-RNAs are heat-shock proteins, Ca-dependent protein kinase, receptor kinase and proteins.	Valdés-Lopez *et al.*, 2010
Arsenic (As)	Lowered photosynthetic rate, lipid peroxidation, elevation of ROS and alteration in carbohydrate metabolism.	About 36 mi-RNAs regulate metabolism, transportation and signalling of lipid metabolism and jasmonic acid.	Requejo and Tena, 2005; Yu *et al.*, 2012
		Roots of indica rice showed 67 mi-RNAs with elevated levels of mi-R397b, mi-R528 and mi-R408; lowered levels of mi-R390 and mi-R1316.	Liu and Zhang, 2012
		Identified mi-R319.	Srivastava *et al.*, 2012
Aluminum (Al)	Binds with phosphate and carboxylic groups of root cell wall and inhibits expansion of cell wall and structural changes; callose deposition, calcium imbalance in cytoplasm, oxidative stress.	Seedlings of *M. truncatula* differentially expressed 23 mi-RNAs; elevated mi-R390 and mi-R396.	Ma *et al.*, 2004; Chen *et al.*, 2012
		Elevated mi-R390 and mi-R396 out of 30 mi-RNAs identified in soybean.	Zeng *et al.*, 2012

Saline soils are rich in sodium, which limits the uptake of nutrients and water from the soil and creates salt stress. Under this condition, plants rapidly perceive excess sodium concentrations and begin the lowering mechanism of salt stress (Gong, 2021). Besides activating signalling of reactive oxygen species, cytosolic calcium accumulation and modification in the composition of membrane phospholipids prevents the salt loss due to salinity stress in plants (Stephan *et al.*, 2016; Jiang *et al.*, 2019). Meanwhile, different metabolic pathways are initiated against salt stress in plants, including mi-RNA-mediated stress relief. The signalling of abscisic acid plays a crucial role in salt stress response which is mediated by the expression of mi-RNAs. A number of salt responsive mi-RNAs have been identified in a variety of plant species (Table 4).

Table 4: Showing the details of mi-RNAs that respond to salt stress in various plant species

S.No.	Name of the Plant	Species of mi-RNA	References
1.	*Arabidopsis thaliana*	Up-regulation: mi-R156, mi-R159, mi-R167, mi-R168, mi-R171, mi-R319, mi-R393, mi-R397 and mi-R396. Down-regulation: mi-R398	Liu *et al.*, 2008; Sunkar and Zhu, 2004; Jia *et al.*, 2009; Jagadeeswaran *et al.*, 2009
2.	*Zea mays*	Down-regulation: mi-R156, mi-R167 and mi-R396. Up-regulation: mi-R168 and mi-R395	Ding *et al.*, 2009
3.	*Oryza sativa*	Up-regulation: mi-R169	Zhao *et al.*, 2009
4.	*Populus tremula*	Up-regulation: mi-R168, mi-R398, mi-R399 and mi-R395 Down-regulation: mi-R169	Jia *et al.*, 2009
5.	*Phaseolus vulgaris*	Up-regulation: mi-R393	Arenas-Huertero *et al.*, 2009

4.1.4 Cold and mi-RNAs

Cold is an extreme condition that freezes cells and causes death of organisms. Due to low temperatures (below 0°C), water gets frozen and expands inside the cell which in turn breaks the cell membrane, thereby cell death. Plants have their own mechanism to protect cells from cold stress. Sugar accumulation and production of cold-active proteins are the known methods of plant protection against cold stress. Differential expression of various proteins at different time points are regulated by various transcription factors which respond to changes in the environmental conditions. In recent years, stress responsive mi-RNAs have come into the picture and are not exempted during cold stress.

The elevated expression of mi-RNAs during cold stress is well reported by various groups. At first, in response to cold stress expression of mi-R319c, mi-R397b, mi-R393 and mi-R402 from *Arabidopsis* was reported (Sunkar and Zhu, 2004). Later, Palatnik *et al.* (2007) reported mi-R319c in *Arabidopsis*. Cold stress response leads to expression of 18 mi-RNAs (mi-R156k, mi-R444a.1, mi-R1320, mi-R1876, mi-R1435, mi-R1868, mi-R1850, mi-R1884b, mi-R319a/b, mi-R535, mi-R171a, mi-R169e, mi-R169h, mi-R169f, mi-R168b, mi-R167a, mi-R167a or b or c) in rice (Lv *et al.*, 2010).

MicroRNA-mediated Regulation of Osmotic and Oxidative Stress

The differential expression of mi-RNA genes has been reported in many plant species. In *Arabidopsis* and *Brachypodium*, mi-R172 is elevated, while mi-R169 and mi-R397 lower its expression during cold stress (Liu *et al.*, 2008; Zhang *et al.*, 2009). Zhang *et al.* (2009) reported that mi-R159 or 319, mi-R394, mi-R157 or 156, mi-R398 and mi-R164 are differentially regulated while mi-R397, mi-R169, mi-R166 or 165,mi-R396, mi-R408, mi-R393 and mi-R172 are regulated at high levels during cold stress. Similarly, Lu and Huang (2008) reported the elevated levels of mi-R477a-b and mi-R168a-b and down-regulation of mi-R476a, mi-R475a and mi-R156g to j against cold stress in *Populus*.

4.1.5 UV Radiation and mi-RNAs

Excessive ultra-violet radiation on the Earth creates aberrations in the genome of organisms. Depletion of the ozone layer allows high amounts of UV-B radiation, which in turn, grounds ROS-mediated oxidative stress, thereby causing suffering to the growth and development of plants (Kruszka *el.*, 2012). Since UV-B creates stress in the plants, the response of mi-RNAs to this stress has been demonstrated and has found a novel Tae-mi-R6000 (Wang *et al.*, 2013). In addition to Tae-mi-R6000, also found are mi-R365, mi-R159, mi-R171, mi-R156, mi-167a and mi-R164 in the same study. UV-B treatment caused differential expression of these genes at different time points of UV exposure. UV-B exposure elevates the expression of mi-R171, mi-R167a and mi-R159 genes, whereas mi-R164, mi-R156 and mi-R365 are down-regulated (Wang *et al.*, 2013).

4.1.6 Nutrients and mi-RNAs

Nutrients are the basic constituents for the survival of any organism. Requirement of nutrients varies from one to another. These are required in traces (micro), smaller (moderate) and larger (macro) amounts as per the utilisation for growth, reproduction, germination against pathogens and pests and finally lead to health issues. Excess or non-availability of nutrients, called nutrient stress, gravely influences growth, development, quality and yield of crop plants. Similar to other stress adoptions, plants show their own mechanisms to mitigate the nutrient stress (Pandey *et al.*, 2021). One of such mechanisms recently popularised is adaptation through mi-RNA-mediated stress response. Homeostasis of plant nutrients is mediated by induction of mi-RNA synthesis. This was reported by Khraiwesh *et al.* (2010) with mi-R398 (copper), mi-R395 (sulphate) and mi-R399 (phosphate) as markers of nutrient stress. During phosphate deficiency stress, elevated levels of mi-R399 was observed with lowered PHO_2 transcripts that regulate phosphate uptake channels (regulators) in roots of the plants. The mi-R399 is synthesised in shoots and finally reaches the roots via phloem where it degrades PHO_2 and opens the phosphate uptake channels via its regulatory gene expression (Bari *et al.*, 2006; Chiou, 2007; Pant *et al.*, 2008) (Fig. 4).

4.2 Biotic Stress and mi-RNAs

In plants, stress caused by infection of pathogenic organisms can be called biotic stress. A number of microbes include bacterial, viral, fungal, and nematode pathogens that create biotic stress in plants. Table 5 shows the mi-RNAs response against infection of various pathogens.

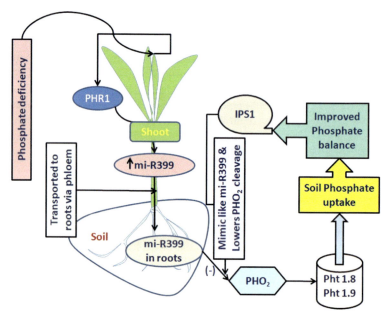

Fig. 4: The PHO$_2$ and mi-R399-mediated homeostasis in plants where PHR1: phosphate starvation response 1; PHO$_2$: phosphate gene; Pht 1.8 and Pht 1.9: phosphate transporters; IPS1: induced phosphate starvation

4.2.1 Bacterial Infections and mi-RNAs

Plant anti-bacterial activity has been reported in *Arabidopsis* with a special pattern of bacteria-associated molecular activity due to elevated mi-R393. This is possible due to down-regulation of signalling of auxin biosynthesis by mi-R393 and is a first report showing the mi-RNA role in bacterial infections (Navarro *et al.*, 2006). In the same species, down-regulation of mi-R398 was reported against bacterial disease (Jagadeeswaran *et al.*, 2009). Later, in tomato, Fahlgren *et al.* (2007) demonstrated the *Pseudomonas syringae* infection induced up-regulation (mi-R393, mi-R167 and mi-R160) and down-regulation (mi-R825) of mi-RNAs.

4.2.2 Viral Infections and mi-RNAs

Arabidopsis, a model plant species for experiments, was used to test the mi-R171-induced RNA silencing of turnip mosaic virus repressor HC-Pro/P1 infection (Zhou and Luo, 2013). Similarly, He *et al.* (2008) showed infection of turnip mosaic virus-induced mi-R1885 in *Brassica*. In tomato, down-regulation of mi-R171 and mi-R164 and up-regulation of mi-R159 were reported with infection of tomato leaf curl virus (Naqvi *et al.*, 2008). Naqvi *et al.* (2010) also reported the infection of tomato leaf curl in New Delhi virus-induced expression of mi-R172 and mi-R319 or mi-R159. This research area is at the primary stage and is open for scientists to take up.

4.2.3 Fungal Infections and mi-RNAs

The differential expression of wheat mi-RNA (up-regulated: mi-R827, miR-393 and mi-R444; down-regulated: mi-R396, mi-R159, mi-R156 and me-R164) has been

MicroRNA-mediated Regulation of Osmotic and Oxidative Stress   **209**

Table 5: Biotic stress response of mi-RNAs against different pathogens

S. No.	Species of mi-RNAs	Type of Biotic Stress	References
1.	Down-regulated: mi-R825	Bacteria	Fahlgren *et al.*, 2007
	Up-regulated: mi-R167 and mi-R160		Jagadeeswaran *et al.*, 2009
	Down-regulated: mi-R398		
	Up-regulated: mi-R393		Navarro *et al.*, 2006
2.	Up-regulated: mi-R7696	Fungi	Campo *et al.*, 2013
	Up-regulated: mi-R827, mi-R393 and mi-R444		Xin *et al.*, 2010
	Down-regulated: mi-R396, mi-R156, mi-R164, mi-R159 and mi-R171		
3.	Up-regulated: mi-R1885	Virus	He *et al.*, 2008
	Up-regulated: mi-R164 and mi-R156		Kasschau *et al.*, 2006
	Down-regulated: mi-R171		Kasschau *et al.*, 2006; Zhou and Luo, 2013
	Down-regulated: mi-R171 and mi-R164		Naqvi *et al.*, 2008
	Up-regulated: mi-R159		
4.	Down-regulated: mi-R398a, mi-R161, mi-R172c, mi-R167a, mi-R164 and mi-R396a	Nematode	Hewezi *et al.*, 2008

reported with powdery mildew infection of *Blumeria graminis* (Xin *et al.*, 2010). Blast infection induced elevated mi-RNA (osa-mi-R7696) expression in rice as reported by Campo *et al.* (2013). Lu *et al.* (2007) reported the differential expression of 10 families of mi-RNAs due to fusiform rust disease caused by rust fungus (*Cronarium quercuum*) in loblolly pine trees.

4.2.4 Nematode Infections and mi-RNAs

Hewezi *et al.* (2008) and Khraiwesh *et al.* (2012) demonstrated that nematode (*Heterodera schachtii*) infection lowered gene expression of mi-RNA species, such as mi-R172c, mi-R161, mi-R398a, mi-R164, mi-R396a and b and mi-R167a in *Arabidopsis*. *Heterodera flycines*, a cyst nematode, infects soybean- induced expression of 40 families of mi-RNAs (101 numbers) (Li *et al.*, 2012).

4.3 Phytohormone Stress and mi-RNAs

Phytohormones are plant growth regulators and control various physiological functions. Not all the phytohormones respond against stress and the only one that improves tolerance in plants during stress conditions are abscisic acid and auxins, but the release of these hormones is linked to the expression of mi-RNAs. Abscisic acid signalling pathway regulated by mi-R159 was reported in *Arabidopsis*. Hormone signalling desensitisation restores the growth of the plant (Reyes and Chua, 2007). In response to abscisic acid, it moderately elevates the mi-R2119, mi-R1514 and mi-RS1 whereas its treatment induces expression of mi-R2118, mi-R393 and mi-R159.2 in *Phaseolus vulgaris* (Arenas-Huertero *et al.*, 2009). Liu *et al.*, (2007), during germination of seed under abscisic acid stress, found a transcription regulatory factor named as auxin response factor 10 which is regulated by me-R160. Table 6 shows the regulation of mi-RNAs and hormone responses under stress conditions.

Table 6: Showing the details of mi-RNAs that respond to salt stress in various plant species

S.No.	Name of the Plant	Regulation of Abscisic Acid	Species of mi-RNA	References
1.	*Arabidopsis thaliana*	Up-regulated	mi-R159, mi-R393, mi-R397	Reyes and Chua, 2007; Sunkar and Zhu, 2004;
		Down-regulated	mi-R169, mi-R398	Liu *et al.*, 2008; Jia *et al.*, 2009
2.	*Oryza sativa*	Down-regulated	mi-R167, mi-R169	Liu *et al.*, 2009
		Up-regulated	mi-R319	
3.	*Populus tremula*	Up-regulated	mi-R168, mi-R395, mi-R398, mi-R399	Jia *et al.*, 2009
		Down-regulated	mi-R169	
4.	*Phaseolus vulgaris*	Up-regulated	mi-R393	Arenas-Huertero *et al.*, 2009

4.4 Mechanical Stress and mi-RNAs

Plants are exposed to many natural calamities like rain and cyclones and their aerial parts are frequently exposed to wind and animals. This leads to bending and rubbing of stem or brushing/shaking the total shoot, which causes mechanical stress in plants (Biddington, 1986). The adoptive procedures against mechanical stress are not much studied in plants. However, mi-RNA response has been identified during mechanical stress. There is up-regulation of mi-R408 and down-regulation of mi-R481, mi-R156, mi-R475, mi-R162, mi-R480 and mi-R164 in *P. trichocarpa* of stressed tissues (Khraiwesh *et al.*, 2012). Lu *et al.* (2005) also reported the up (mi-R168) and down (mi-R172 and mi-R160) regulation of mi-RNAs in compressed tissue due to mechanical stress in *Arabidopsis*.

5. Osmotic and Oxidative Stress and its Regulation by miRNAs

A number of stress conditions that affect the growth and yield of plants have been discussed in the above sections. Osmotic stress in plants invariably by salinity leads to increase in cellular toxicity. Unlike other stress adaptations, plants also exhibit adaptation to salt stress by altering the physiological, cellular and biochemical pathways through differential gene expression of corresponding molecules that regulate pathways/processes (Vinocur and Altman, 2005). These gene products are responsible for the maintenance of osmotic protection by accumulation of compatible solutes and polyamines, homeostasis of ion, protection of oxidative damage and finally regulation through transcription (Gupta and Huang, 2014). Failure in osmoregulation damages the living cells, where cellular contents come out from the cell wall damage. Cell wall damage usually happens when there is either high amounts of water uptake or complete loss of water from cell. Regulation of osmotic stress in plants is mediated through mi-RNA expression. It is found that mi-R-167 targets Indole-3-acetic acid-Ala and auxin response factors under osmotic stress in *Arabidopsis* and *Manihot esculenta* (Kinoshita *et al.*, 2012; Phookaew *et al.*, 2014). The mi-RNA genes that are responsive to salt stress are also responsible for osmotic stress in plants (Table 4).

Adverse conditions like high temperature, cold, salinity, metal toxicity, UV-radiation, ozone, nutrient availability, diseased conditions, additional hormone release, etc. are responsible for creating oxidative stress in plants. Oxidative stress originates not only due to adverse conditions, but also due to hypoxia. In plants, studies are focused to elucidate the mi-RNA response in both the conditions. Hypoxia is a condition of non-availability of oxygen for mitochondrial oxidative phosphorylation leading to low oxygen stress in plants (Agarwal and Grover, 2006). Moldovan *et al.* (2010) suggested that prolonged hypoxia causes a metabolic shift where anaerobic respiration establishes with a differential gene expression. They also found the high level expression of mi-R158a, mi-R775, mi-R157d, mi-R159a, mi-R156g, mi-R172a and b, mi-R391 and mi-R159a in the roots of *Arabidopsis* at condition of hypoxia. On the other hand, oxidative stress in *Arabidopsis* lowers the mi-R398 expression and elevates SOD (super oxide dismutase) levels (Sunkar *et al.*, 2006). Li *et al.* (2010) studied the H_2O_2 mediated oxidative stress response of mi-RNAs in rice seedlings. They found differential expression of mi-RNAs in H_2O_2 treated seedlings and also reported high levels of mi-R1425, mi-R167, mi-R827 and mi-R397 and low levels

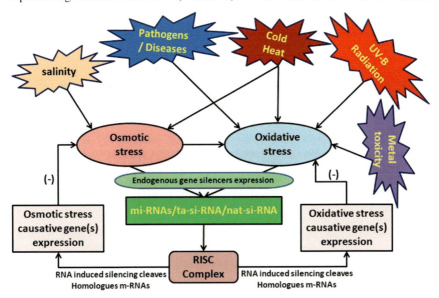

Fig. 5: The response of mi-RNA against osmotic and oxidative stress in plants where (-) indicates negative regulation

of mi-R528. Ozone fumigation in *Arabidopsis thaliana* expresses differentially in 22 families of mi-RNAs in an hour. However, these expressed mi-RNAs are also found with UV-B exposure (Iyer *et al.*, 2012). Mase and Tsukagoshi (2021) found the mediation of oxidative stress through reactive oxygen species and their link to regulatory networks of genes in the roots of *Arabidopsis*. They explained that the ROS homeostasis is maintained in the stressed plants through mi-RNA mediation. The regulation of mi-RNA mediated osmotic and oxidative stress in plants is depicted in Fig. 5.

6. Conclusion

Stress tolerance is a method of adaptation that allows the plant to survive, develop and reproduce for next generation. Among all stress-adoptive methods, gene silencing by mi-RNAs has become one of the prominent methods. A number of examples elucidate the functional targets of mi-RNAs as stress responsive and nullify the adverse condition that has been created by various biotic and abiotic stressors in plants. Though oxidative and osmotic stress is most effective in damaging cellular functions, mi-RNA responsiveness has been reported to suppress the stress conditions in plants. Using available data on mi-RNA responsiveness against various stressors, biotic or abiotic stress tolerance has been achieved through mi-RNA gene manipulations in the crop and ornamental plants. Besides using as stress biomarkers, mi-RNAs can also be used to identify the stressed conditions, molecular mechanism(s) of adaptation and species-specific environmental and seasonal risk factors of crops. Developing suitable technology of mi-RNA mediated anti-stress responses will play a key role in future agriculture.

Acknowledgements

Authors are grateful to the institutions where they are working for providing facilities to accomplish the writing of this book chapter successfully.

References

Agarwal, S. and Grover, A. (2006). Molecular biology, biotechnology and genomics of flooding-associated low O_2 stress response in plants, *Crit. Rev. Plant Sci.*, 25: 1-21. Doi: 10.1080/07352680500365232

Arenas-Huertero, C., Pérez, B., Rabanal, F., Blanco-Melo, D., De la Rosa, C., Estrada-Navarrete, G., Sanchez, F., Covarrubias, A. and Reyes, J. (2009). Conserved and novel miRNAs in the legume *Phaseolus vulgaris* in response to stress, *Plant Molecular Biology*, 70: 385-401. http://dx.Doi.org/10.1007/s11103-009-9480-3

Barciszewska-Pacak, M., Milanowska, K., Knop, K., Bielewicz, D., Nuc, P., Plewka, P., Pacak, A.M., Vazquez, F., Karlowski, W., Jarmolowski, A. and Szweykowska-Kulinska, Z. (2015). *Arabidopsis* microRNA expression regulation in a wide range of abiotic stress responses, *Front. Plant Sci.*, 6: 410.

Bari, R., Datt Pant, B., Stitt, M. and Scheible, W.R. (2006). PHO_2, microRNA399 and PHR1 define a phosphate-signalling pathway in plants, *Plant Physiology*, 141: 988-999. http://dx.Doi.org/10.1104/pp.106.079707

MicroRNA-mediated Regulation of Osmotic and Oxidative Stress   **213**

Barrera-Figueroa, B.E., Gao, L., Diop, N.N., Wu, Z., Ehlers, J.D., Roberts, P.A., Close, T.J., Zhu, J.K. and Liu, R.Y. (2011). Identification and comparative analysis of drought-associated microRNAs in two cowpea genotypes, *BMC Plant Biology*, 11: 127. http://dx.Doi.org/10.1186/1471-2229-11-127

Biddington, N.L. (2086). The effects of mechanically-induced stress in plants – A review, *Plant Growth Regul.*, 4: 103-123. https://Doi.org/10.1007/BF00025193

Campo, S., Peris-Peris, C., Siré, C., Moreno, A.B., Donaire, L., Zytnicki, M., Notredame, C., Llave, C. and San Segundo, B. (2013). Identification of a novel microRNA (miRNA) from rice that targets an alternatively spliced transcript of the Nramp6 (Natural Resistance Associated Macrophage Protein 6) gene involved in pathogen resistance, *New Phytologist.*, 199: 212-227. http://dx.Doi.org/10.1111/nph.12292

Casadevall, R., Rodriguez, R.E., Debernardi, J.M., Palatnik, J.F. and Casati, P. (2013). Repression of growth regulating factors by the microRNA396 inhibits cell proliferation by UV-B radiation in *Arabidopsis* leaves, *Plant Cell*, 25: 3570-3583. Doi: 10.1105/tpc.113.117473

Chen, J. and Yang, Z.M. (2012). Mercury toxicity, molecular response and tolerance in higher plants, *Biometals*, 25: 847-857. Doi: https://doi.org/10.1007/s10534-012-9560-8

Chen, L., Wang, T., Zhao, M., Tian, Q. and Zhang, W.H. (2012). Identification of aluminum-responsive microRNAs in *Medicago truncatula* by genome-wide high-throughput sequencing, *Planta.*, 235: 375-386. Doi: 10.1007/s00425-0111514-9

Chiou, T.J. (2007). The Role of microRNAs in sensing nutrient stress, *Plant, Cell and Environment*, 30: 323-332. http://dx.Doi.org/10.1111/j.1365-3040.2007.01643.x

Ding, D., Zhang, L., Wang, H., Liu, Z., Zhang, Z. and Zheng, Y. (2009). Differential expression of miRNAs in response to salt stress in maize roots, *Annals of Botany*, 103: 29-38.

Fahlgren, N., Howell, M.D., Kasschau, K.D., Chapman, E.J., Sullivan, C.M., Cumbie, J.S., Givan, S.A., Law, T.F., Grant, S.R., Dangl, J.L. and Carrington, J.C. (2007). High-throughput sequencing of *Arabidopsis* microRNAs: Evidence for frequent birth and death of miRNA genes, *PLoS ONE*, 2Article ID: e219. http://dx.Doi.org/10.1371/journal.pone.0000219

Frazier, T.P., Sun, G., Burklew, C.E. and Zhang, B. (2011). Salt and drought stresses induce the aberrant expression of microRNA genes in tobacco, *Molecular Biotechnology*, 49: 159-165. http://dx.Doi.org/10.1007/s12033-011-9387-5

Fukudome, A. and Fukuhara, T. (2017). Plant dicer-like proteins: Double-stranded RNA cleaving enzymes for small RNA biogenesis, *J. Plant. Res.*, 130: 33–44. doi:10.1007/s10265-016-0877-1

Gao, P., Bai, X., Yang, L., Lv, D., Pan, X., Li, Y., Cai, H., Ji, W., Chen, Q. and Zhu, Y. (2011). osa-MIR393: A salinity and alkaline stress-related microRNA gene, *Mol. Biol. Rep.*, 38: 237-242. Doi: 10.1007/s11033-010-0100-8

Gielen, H., Remans, T., Vangronsveld, J. and Cuypers, A. (2012). MicroRNAs in metal stress: Specific roles or secondary responses? *Int. J. Mol. Sci.*, 13: 15826-15847. Doi:10.3390/ijms131215826

Gong, Z. (2021). Plant abiotic stress: New insights into the factors that activate and modulate plant responses, *J. Integr. Plant Biol.*, 63: 429-430. Doi: 10.1111/jipb.13079

Goswami, S., Kumar, R.R. and Rai, R.D. (2014). Heat-responsive microRNAs regulate the transcription factors and heat shock proteins in modulating thermo-stability of starch biosynthesis enzymes in wheat (*Triticum aestivum* L.) under the heat stress, *Aust. J. Crop Sci.*, 8: 697-705.

Gupta, O.P., Sharma, P., Gupta, R.K. and Sharma, I. (2014). MicroRNA mediated regulation of metaltoxicity in plants: Present status and future perspectives, *Plant Mol. Biol.*, 84: 1-18. Doi: 10.1007/s11103-013-0120-6

He, X.F., Fang, Y.Y., Feng, L. and Guo, H.S. (2008). Characterisation of conserved and novel microRNAs and their targets, including a TuMV-induced TIR-NBS-LRR Class R gene-derived novel miRNA in *Brassica*, *FEBS Letters*, 582: 2445-2452. http://dx.Doi.org/10.1016/j.febslet.2008.06.011

Hewezi, T., Howe, P., Maier, T.R. and Baum, T.J. (2008). *Arabidopsis*: Small RNAs and their targets during cyst nematode parasitism, *Mol. Plant Microbe Interact.*, 21: 1622-1634. http://dx.Doi.org/10.1094/MPMI-21-12-1622

Horie, T., Karahara, I. and Katsuhara, M. (2012). Salinity tolerance mechanisms in glycophytes: An overview with the central focus on rice plants, *Rice*, 5: 1-18. Doi: 10.1186/1939-8433-5-11

Huang, S.Q., Xiang, A.L., Che, L.L., Chen, S., Li, H., Song, J.B. and Yang, Z.M. (2010). A set of miRNAs from *Brassica napus* in response to sulfate deficiency and cadmium stress, *Plant Biotechnol. J.*, 8: 887-899. Doi: 10.1111/j.1467-7652.2010.00517.x

Iba, K. (2002). Acclimative response to temperature stress in higher plants: Approaches of gene engineering for temperature tolerance, *Ann. Rev. Plant Biol.*, 53: 225-245.

Iyer, N.J., Jia, X., Sunkar, R., Tang, G. and Mahalingam, R. (2012). microRNAs responsive to ozone-induced oxidative stress in *Arabidopsis thaliana*, *Plant Signal. Behav.*, 7: 484-491. 10.4161/psb.19337

Jagadeeswaran, G., Saini, A. and Sunkar, R. (2009). Biotic and abiotic stress down-regulate miR398 expression in *Arabidopsis*, *Planta.*, 229: 1009-1014. http://dx.Doi.org/10.1007/s00425-009-0889-3

Jia, X., Wang, W.X., Ren, L., Chen, Q.J., Mendu, V., Willcut, B., Dinkins, R., Tang, X. and Tang, G. (2009). Differential and dynamic regulation of miR398 in response to ABA and salt stress in *Populus tremula* and *Arabidopsis thaliana*, *Plant Molecular Biology*, 71: 51-59.

Jiang, Z., Zhou, X., Tao, M., Yuan, F., Liu, L., Wu, F., Wu, X., Xiang, Y., Niu, Y., Liu, F., et al. (2019). Plant cell-surface GIPC sphingolipids sense salt to trigger Ca^{2+} influx, *Nature*, 572: 341-346. doi: 10.1038/s41586-019-1449-z

Kantar, M., Lucas, S. and Budak, H. (2011). miRNA expression patterns of *Triticum dicoccoides* in response to shock drought stress, *Planta*, 233: 471-484.

Kasschau, K.D., Xie, Z., Allen, E., Llave, C., Chapman, E.J., Krizan, K.A. and Carrington, J.C. (2006). P1/HC-Pro: A viral suppressor of RNA silencing, interferes with *Arabidopsis* development and miRNA function, *Developmental Cell*, 4: 205-217. http://dx.doi.org/10.1016/S1534-5807(03)00025-X

Kawaguchi, R., Girke, T., Bray, E.A. and Bailey-Serres, J. (2004). Differential mRNA translation contributes to gene regulation under non-stress and dehydration stress conditions in *Arabidopsis thaliana*, *Plant J.*, 38: 823-839.

Khare, T., Kumar, V. and Kavi Kishor, P.B. (2015). Na+ and Cl- ions show additive effects under NaCl stress on induction of oxidative stress and the responsive antioxidative defense in rice, *Protoplasma*, 252: 1149-1165. Doi: 10.1007/s00709-014-0749-2

Khraiwesh, B., Arif, M.A., Seumel, G.I., Ossowski, S., Weigel, D., Reski, R. and Frank, W. (2010). Transcriptional control of gene expression by microRNAs, *Cell*, 140: 111-122. http://dx.Doi.org/10.1016/j.cell.2009.12.023

Khraiwesh, B., Zhu, J.K. and Zhu, J. (2012). Role of miRNAs and siRNAs in biotic and abiotic stress responses of plants, *Biochimica et Biophysica Acta*, 1819: 137-148. http://dx.Doi.org/10.1016/j.bbagrm.2011.05.001

Kim, Y.J., Zheng, B., Yu, Y., Won, S.Y., Mo, B. and Chen, X. (2011). The role of mediator in small and long noncoding RNA production in *Arabidopsis thaliana*, *EMBO J.*, 30: 814-822.

Kinoshita, N., Wang, H., Kasahara, H., Liu, J., MacPherson, C., Machida, Y., Kamiya, Y., Hannah, M.A. and Chua, N.H. (2012). IAA-Ala Resistant3, an evolutionarily conserved target of miR167, mediates *Arabidopsis* root architecture changes during high osmotic stress, *Plant Cell*, 24: 3590-3602.

Kruszka, K., Pieczynski, M., Windels, D., Jarmolowski, A., Kulinsk, S.Z. and Vazquez, F. (2012). Role of microRNAs and other sRNAs of plants in their changing environments, *J. Plant Physiol.*, 16: 1664-1672. Doi: 10.1016/j.jplph.2012.03.009

Kulcheski, F.R., de Oliveira, L.F., Molina, L.G., Almerao, M.P., Rodrigues, F.A., Marcolino, J., Barbosa, J.F., Stolf-Moreira, R., Nepomuceno, A.L., Marcelino-Guimarães, F.C., Abdelnoor, R.V., Nascimento, L.C., Carazzolle, M.F., Pereira, G.A. and Margis, R. (2011).

Identification of novel soybean microRNAs involved in abiotic and biotic stresses. *BMC Genomics*, 12: 307. http://dx.Doi.org/10.1186/1471-2164-12-307

Kumar, R.R., Pathak, H., Sharma, S.K., Kale, Y.K., Nirjal, M.K. and Singh, G.P. (2014). Novel and conserved heat-responsive microRNAs in wheat (*Triticum aestivum* L.), *Funct. Integr. Genomics,* 15: 1-26. Doi: 10.1007/s10142-014-0421-0

Kumar, V., Shriram, V., Kavi Kishor, P.B., Jawali, N. and Shitole, M.G. (2010). Enhanced proline accumulation and salt stress tolerance of transgenic Indica rice by overexpressing P5CSF129A gene, *Plant Biotechnol. Rep.*, 4: 37-48. Doi: 10.1007/S11816-009-0118-3

Kushawaha, A.K., Khan, A., Sopory, S.K. and Sanan-Mishra, N. (2021). Priming by high temperature stress induces microRNA regulated heat shock modules indicating their involvement in thermopriming response in rice, *Life*, 11: 291. https://Doi.org/10.3390/life11040291

Lee, R.C., Feinbaum, R.L. and Ambros, V. (1993). The *C. elegans* heterochronic gene lin-4 encodes small RNAs with antisense complementarity to lin-14, *Cell*, 75: 843-854.

Lee, S.H., Li, C.W., Koh, K.W., Chuang, H.Y., Chen, Y.R., Lin, C.S. and Chan, M.T. (2014). MSRB7 reverses oxidation of GSTF2/3 to confer tolerance of *Arabidopsis thaliana* to oxidative stress, *J. Exp. Bot.*, 65: 5049-5062, Doi: 10.1093/jxb/eru270

Li, H., Deng, Y., Wu, T., Subramanian, S. and Yu, O. (2010). Mis-expression of miR482, miR1512 and miR1515 increases soybean nodulation, *Plant Physiol.*, 153: 1759-1770. Doi: 10.1104/pp.110.156950

Li, X., Wang, X., Zhang, S., Liu, D., Duan, Y. and Dong, W. (2012). Identification of soybean microRNAs involved in soybean cyst nematode infection by deep sequencing, *PLoS ONE*, 7, Article ID: e39650. http://dx.Doi.org/10.1371/journal.pone.0039650

Liang, G., Ai, Q. and Yu, D. (2015). Uncovering miRNAs involved in crosstalk between nutrient deficiencies in *Arabidopsis*, *Sci. Rep.*, 5: 11813. Doi: 10.1038/srep11813

Lin, J.S., Kuo, C.C., Yang, I.C., Tsai, W.A., Shen, Y.H., Lin, C.C., Liang, Y.C., Li, Y.C., Kuo, Y.W., King, Y.C., Lai, H.M. and Jeng, S.T. (2018). MicroRNA160 modulates plant development and heat shock protein gene expression to mediate heat tolerance in *Arabidopsis*, *Front. Plant Sci.*, 9: 68.

Liu, P.P., Montgomery, T.A., Fahlgren, N., Kasschau, K.D. and Nonogaki, H. (2007). Repression of auxin response factor 10 by microRNA160 is critical for seed germination and post-germination stages, *Plant Journal*, 52: 133-146. http://dx.Doi.org/10.1111/j.1365-313X.2007.03218.x

Liu, H.H., Tian, X., Li, Y.J., Wu, C.A. and Zheng, C.C. (2008). Microarray-based analysis of stress-regulated microRNAs in *Arabidopsis thaliana*, *RNA*, 14: 836-843. http://dx.Doi.org/10.1261/rna.895308

Liu, Q., Zhang, Y.C., Wang, C.Y., Luo, Y.C., Huang, Q.J., Chen, S.Y., Zhou, H., Qu, L.H. and Chen, Y.Q. (2009). Expression analysis of phytohormone-regulated microRNAs in rice, implying their regulation roles in plant hormone signalling, *FEBS Letters*, 583: 723-728.

Liu, Q. and Zhang, H. (2012). Molecular identification and analysis of arsenite stress-responsive miRNAs in rice, *J. Agric. Food Chem.*, 60: 6524-6536. Doi: 10.1021/jf300724t

Liu, Q., Yan, S., Yang, T., Zhang, S., Chen, Y. and Liu, B. (2017). Small RNAs in regulating temperature stress response in plants, *J. Integ. Plant Biol.*, 59: 774-791.

Llave, C., Xie, Z., Kasschau, K.D. and Carrington, J.C. (2002b). Cleavage of scarecrow-like mRNA targets directed by a class of *Arabidopsis* miRNA, *Science*, 297: 2053-2056.

Lu, X.Y. and Huang, X.L. (2008). Plant miRNAs and abiotic stress responses, *Biochemical and Biophysical Research Communications*, 368: 458-462. http://dx.Doi.org/10.1016/j.bbrc.2008.02.007

Lu, S.F., Sun, Y.H., Shi, R., Clark, C., Li, L.G. and Chiang, V.L. (2005). Novel and mechanical stress responsive microRNAs in *Populus trichocarpa* that are absent from *Arabidopsis*, *Plant Cell*, 17: 2186-2203. http://dx.Doi.org/10.1105/tpc.105.033456

Lu, S., Sun, Y.H., Amerson, H. and Chiang, V.L. (2007). MicroRNAs in Loblolly pine (*Pinustaeda* L.) and their association with fusiform rust gall development, *The Plant Journal*, 51: 1077-1098. http://dx.Doi.org/10.1111/j.1365-313X.2007.03208.x

Lu, S.F., Sun, Y.H. and Chiang, V.L. (2008). Stress-responsive microRNAs in *Populus*, *Plant Journal*, 55: 131-151. http://dx.Doi.org/10.1111/j.1365-313X.2008.03497.x

Lv, D.K., Bai, X., Li, Y., Ding, X.D., Ge, Y., Cai, H., Ji, W., Wu, N. and Zhu, Y.M. (2010). Profiling of cold-stress-responsive miRNAs in rice by microarrays, *Gene*, 459: 39-47. http://dx.Doi.org/10.1016/j.gene.2010.03.011

Ma, J.F., Shen, R., Nagao, S. and Tanimoto, E. (2004). Aluminum targets elongating cells by reducing cell wall extensibility in wheat roots, *Plant Cell Physiol.*, 45: 583-589. Doi: 10.1093/pcp/pch060

Mase, K. and Tsukagoshi, H. (2021). Reactive oxygen species link gene regulatory networks during *Arabidopsis* root development, *Front Plant Sci.*, 12: 660274. Doi: 10.3389/fpls.2021.660274

Mette, M.F., van der Winden, J., Matzke, M. and Matzke, A.J. (2002). Short RNAs can identify new candidate transposable element families in *Arabidopsis*, *Plant Physiol.*, 130: 6-9.

Moldovan, D., Spriggs, A., Yang, J., Pogson, B.J., Dennis, E.S. and Wilson, I.W. (2010). Hypoxia-responsive microRNAs and trans-acting small interfering RNAs in *Arabidopsis*, *J. Exp. Bot.*, 61: 165-177. Doi: 10.1093/jxb/ erp296

Naqvi, A.R., Choudhury, N.R., Haq, Q.M.R. and Mukherjee, S.K. (2008). MicroRNAs as biomarkers in tomato leaf curl virus (ToCLV), *Nucleic Acids Symposium*, Series No. 52, 507-508.

Naqvi, A.R., Haq, Q.M.R. and Mukherjee, S.K. (2010). MicroRNA profiling of tomato leaf curl New Delhi Virus (ToLCNDV) infected tomato leaves indicates that deregulation of mir159/319 and mir172 might be linked with leaf curl disease, *Virology Journal*, 7: 281. http://dx.Doi.org/10.1186/1743-422X-7-281

Navarro, L., Dunoyer, P., Jay, F., Arnold, B., Dharmasiri, N., Estelle, M., Voinnet, O. and Jones, J.D. (2006). A plant miRNA contributes to antibacterial resistance by repressing auxin signalling, *Science*, 312: 436-439. http://dx.Doi.org/10.1073/pnas.0510928103

Palatnik, J.F., Wollmann, H., Schommer, C., Schwab, R., Boisbouvier, J., Rodriguez, R., Warthmann, N., Allen, E., Dezulian, T., Huson, D., Carrington, J.C. and Weigel, D. (2007). Sequence and expression differences underlie functional specialisation of *Arabidopsis* microRNAs miR159 and miR319, *Developmental Cell*, 13: 115-125. http://dx.Doi.org/10.1016/j.devcel.2007.04.012

Pandey, R., Vengavasi, K. and Hawkesford, M.J. (2021). Plant adaptation to nutrient stress, *Plant Physiol. Rep.*, 26: 583-586. https://Doi.org/10.1007/s40502-021-00636-7

Pandita, D. (2019). Plant MIRnome: miRNA biogenesis and abiotic stress response. *In:* Hasanuzzaman, M., Hakeem, K., Nahar, K. and Alharby, H. (Eds.). *Plant Abiotic Stress Tolerance*. Springer, Cham., 449-474. Doi https://Doi.org/10.1007/978-3-030-06118-0_18

Pandita, D. and Wani, S.H. (2019). MicroRNA as a tool for mitigating abiotic stress in rice (*Oryza sativa* L.). *In:* Wani, S. (Ed.) *Recent Approaches in Omics for Plant Resilience to Climate Change*. Springer, Chamz, 109-133. Doi https://Doi.org/10.1007/978-3-030-21687-0_6

Pandita, D. (2021). Role of miRNAi technology and miRNAs in abiotic and biotic stress resilience. *In:* Aftab, T. and Roychoudhury, A. (Eds.). *Plant Perspectives to Global Climate Changes*. Academic Press, Elsevier, 303-330. https://Doi.org/10.1016/B978-0-323-85665-2.00015-7

Pandita, D. (2022a). How microRNAs regulate abiotic stress tolerance in wheat? A snapshot. *In:* Roychoudhury, A., Aftab, T. and Acharya, K. (Eds.). *Omics Approach to Manage Abiotic Stress in Cereals*. Springer, Singapore, 447-464. https://Doi.org/10.1007/978-981-19-0140-9_17

Pandita, D. (2022b). MicroRNAs shape the tolerance mechanisms against abiotic stress in maize. *In:* Roychoudhury, A., Aftab, T. and Acharya, K. (Eds.). *Omics Approach to Manage Abiotic Stress in Cereals*. Springer, Singapore, 479-493. https://Doi.org/10.1007/978-981-19-0140-9_19

Pandita, D. (2022c). miRNA- and RNAi-mediated metabolic engineering in plants. *In:* Aftab, T. and Hakeem, K.R. (Eds.). *Metabolic Engineering in Plants*, Springer, Singapore, 171-186. https://Doi.org/10.1007/978-981-16-7262-0_7

Pant, B.D., Buhtz, A., Kehr, J. and Scheible, W.R. (2008). MicroRNA399 is a long-distance signal for the regulation of plant phosphate homeostasis, *Plant Journal*, 53: 31-38. http://dx.Doi.org/10.1111/j.1365-313X.2007.03363

Park, W., Li, J., Song, R., Messing, J. and Chen, X. (2002). Carperl factory, a dicer homolog, and HEN1, a novel protein, act in microRNA metabolism in *Arabidopsis thaliana*, *Curr Biol.*, 12: 1484-1495.

Phookaew, P., Netrphan, S., Sojikul, P. and Narangajavana, J. (2014). Involvement of miR164- and miR167-mediated target gene expressions in responses to water deficit in cassava, *Biologia Plantarum*, 58(3): 469-478. doi: 10.1007/s10535-014-0410-0

Qadir, M.., Quillérou, E., Nangia, V., Murtaza, G., Singh, M., Thomas, R.J., Drechsel, P. and Noble, A.D. (2014). Economics of salt-induced land degradation and restoration, *Nat. Resour. Forum*, 38: 282-295. Doi: 10.1111/1477-8947.12054

Rajagopalan, R., Vaucheret, H., Trejo, J. and Bartel, D.P. (2006). A diverse and evolutionarily fluid set of microRNAs in *Arabidopsis thaliana*, *Genes Dev.*, 20: 3407-3425.

Rascio, N. and Navari-Izzo, F. (2011). Heavy metal hyperaccumulating plants: How and why do they do it? And what makes them so interesting? *Plant Sci.*, 180: 169-181. Doi:10.1016/j.plantsci.2010.08.016

Reinhart, B.J., Weinstein, E.G., Rhoades, M.W., Bartel, B. and Bartel, D.P. (2002). MicroRNAs in plants, *Genes Dev.*, 16: 1616-1626.

Requejo, R. and Tena, M. (2005). Proteome analysis of maize roots reveals that oxidative stress is a main contributing factor to plant arsenic toxicity. *Phytochemistry*, 66: 1519-1528. Doi: 10.1016/j.phytochem.2005.05.003

Reyes, J.L. and Chua, N.H. (2007). ABA Induction of miR159 controls transcript levels of two MYB factors during *Arabidopsis* seed germination, *Plant Journal*, 49: 592-606. http://dx.Doi.org/10.1111/j.1365-313X.2006.02980.x

Rizwan, M., Ali, S., Adrees, M., Rizvi, H., Zia-ur-Rehman, M., Hannan, F., Qayyum, M.F., Hafeez, F. and Ok, Y.S. (2016). Cadmium stress in rice: Toxic effects, tolerance mechanisms, and management: A critical review, *Environ. Sci. Pollut. Res. Int.* Doi: 10.1007/s11356-016-6436-4

Shukla, L.I., Chinnusamy, V. and Sunkar, R. (2008). The role of microRNAs and other endogenous small RNAs in plant stress responses, *Biochimica et Biophysica Acta*, 1779: 743-748.

Srivastava, S., Srivastava, A.K., Suprasanna, P. and D'Souza, S.F. (2012). Identification and profiling of arsenic stress-induced microRNAs in *Brassica juncea*, *J. Exp. Bot.*, 64: 303-315. Doi: 10.1093/jxb/ers333

Stephan, A.B., Kunz, H.H., Yang, E. and Schroeder, J.I. (2016). Rapid hyperosmotic-induced Ca^{2+} responses in *Arabidopsis thaliana* exhibit sensory potentiation and involvement of plastidial KEA transporters, *Proc. Natl. Acad. Sci.*, USA, 113: E5242–E5249. Doi: 10.1073/pnas.1519555113

Su, T., Li, X., Yang, M., Shao, Q., Zhao, Y., Ma, C. and Wang, P. (2020). Autophagy: An intracellular degradation pathway regulating plant survival and stress response, *Front. Plant Sci.*, 11: 164. Doi: 10.3389/fpls.2020.00164

Sunkar, R. and Zhu, J.K. (2004). Novel stress-regulated microRNAs and other small RNAs from *Arabidopsis*, *Plant Cell*, 16: 2001-2019. http://dx.doi.org/10.1105/tpc.104.022830

Sunkar, R., Kapoor, A. and Zhu, J.K. (2006). Post-transcriptional induction of two Cu/Zn superoxide dismutase genes in *Arabidopsis* is mediated by down regulation of miR398 and important for oxidative stress tolerance, *Plant Cell*, 18: 2051-2065. Doi: 10.1105/tpc.106.041673

Trindade, I., Capitão, C., Dalmay, T., Fevereiro, M.P. and Santos, D.M. (2010). miR398 and miR408 are up-regulated in response to water deficit in *Medicago truncatula*, *Planta*, 231: 705-716. http://dx.Doi.org/10.1007/s00425-009-1078-0

Turan, S. and Tripathy, B.C. (2013). Salt and genotype impact on antioxidative enzymes and lipid peroxidation in two rice cultivars during de-etiolation, *Protoplasma*, 250: 209-222. Doi:10.1007/s00709-012-0395-5

Valdés-López, O., Yang, S.S., Aparicio-Fabre, R., Graham, P.H., Reyes, J.L., Vance, C.P. and Hernández, G. (2010). MicroRNA expression profile in common bean (*Phaseolus vulgaris*) under nutrient deficiency stresses and manganese toxicity, *New Phytol.*, 187: 805-818. Doi: 10.1111/j.1469-8137.2010.03320.x

Vierling, E. (1991). The roles of heat shock proteins in plants, *Ann. Rev. Plant Biol.*, 42: 579-620.

Vinocur, B. and Altman, A. (2005). Recent advances in engineering plant tolerance to abiotic stress: Achievements and limitations, *Curr. Opin. Biotechnol.*, 16: 123-132.

Voinnet, O. (2009). Origin, biogenesis and activity of plant microRNAs, *Cell*, 136: 669-687.

Wang, B., Sun, Y.F., Song, N., Wang, X.J., Feng, H., Huang, L.L. and Kang, Z.S. (2013). Identification of UV-B-induced microRNAs in wheat, *Genet. Mol. Res.*, 12: 4213-4221. Doi: 10.4238/2013.October.7.7

Wang, R., Xu, L., Zhu, X., Zhai, L., Wang, Y., Yu, R., Gong, Y., Limera, C. and Liu, L. (2014). Transcriptome-wide characterisation of novel and heat stress-responsive microRNAs in radish (*Raphanus sativus* L.) Using next-generation sequencing, *Plant Mol. Biol. Rep.*, 33: 867-880. https://Doi.org/10.1007/s11105-014-0786-1.

Wang, W., Vinocur, B., Shoseyov, O. and Altman, A. (2004). Role of plant heat-shock proteins and molecular chaperones in the abiotic stress response, *Trends Plant Sci.*, 9: 244-252.

Xin, M., Wang, Y., Yao, Y., Xie, C., Peng, H., Ni, Z. and Sun, Q. (2010). Diverse set of microRNAs are responsive to powdery mildew infection and heat stress in wheat (*Triticum aestivum* L.), *BMC Plant Biology*, 10: 123. http://dx.Doi.org/10.1186/1471-2229-10-123

Xu, L., Wang, Y., Zhai, L., Xu, Y., Wang, L., Zhu, X., Gong, Y., Yu, R., Limera, C. and Liu, L. (2013). Genome wide identification and characterisation of cadmium-responsive microRNAs and their target genes in radish (*Raphanus sativus* L.) roots, *J. Exp. Bot.*, 64: 4271-4287. Doi: 10.1093/jxb/ert240

Yamaguchi-Shinozaki, K. and Shinozaki, K. (2006). Transcriptional regulatory networks in cellular responses and tolerance to dehydration and cold stresses, *Annu. Rev. Plant Biol.*, 57: 781-803.

Yang, Z.M. and Chen, J. (2013). A potential role of microRNAs in regulating plant response to metaltoxicity, *Metallomics*, 5: 1184-1190. Doi:10.1039/c3mt0 0022b

Yu, A., Altmann, S., Hoffmann, K., Pant, B.D., Scheible, W.R. and Bäurle, I. (2014). *Arabidopsis* miR156 regulates tolerance to recurring environmental stress through SPL transcription factors, *Plant Cell*, 26: 1792-1807.

Yu, L.J., Luo, Y.F., Liao, B., Xie, L.J., Chen, L., Xiao, S., Li, J.T., Hu, S.N, and Shu, W.S. (2012). Comparative transcriptome analysis of transporters, phytohormone and lipid metabolism pathways in response to arsenic stress in rice (*Oryza sativa*), *New Phytol.*, 195: 97-112. Doi: 10.1111/j.1469-8137.2012.04154.x

Zeng, Q.Y., Yang, C.Y., Ma, Q.B., Li, X.P., Dong, W.W. and Nian, H. (2012). Identification of wild soybean miRNAs and their target genes responsive to aluminum stress, *BMC Plant Biol.*, 12: 182. Doi: 10.1186/1471-2229-12-182

Zhang, J., Xu, Y., Huan, Q. and Chong, K. (2009). Deep sequencing of Brachypodium small RNAs at the global genome level identifies microRNAs involved in cold stress response, *BMC Genomics*, 10: 449. http://dx.Doi.org/10.1186/1471-2164-10-449

Zhao, B., Ge, L., Liang, R., Li, W., Ruan, K., Lin, H. and Jin, Y. (2009). Members of miR-169 family are induced by high salinity and transiently inhibit the NF-YA transcription factor, *BMC Molecular Biology*, 10: 29.

Zhao, X., Ding, C., Chen, L., Wang, S., Wang, Q. and Ding, Y. (2012). Comparative proteomic analysis of the effects of nitric oxide on alleviating Cd-induced toxicity in rice (*Oryza sativa* L.), *Plant Omics*, 5: 604-614.

Zheng, X., Yang, L., Li, Q., Ji, L., Tang, A., Zang, L., Deng, K., Zhou, J. and Zhang, Y. (2018). MIGS as a simple and efficient method for gene silencing in rice, *Front. Plant Sci.*, 9: 662. Doi: 10.3389/fpls.2018.00662

Zhou, L., Liu, Y., Liu, Z., Kong, D., Duan, M. and Luo, L. (2010). Genome-wide identification and analysis of drought-responsive microRNAs in *Oryza sativa*, *J. Exp. Bot.*, 61: 4157-4168. Doi: 10.1093/jxb/erq237

Zhou, Z.S., Song, J.B. and Yang, Z.M. (2012a). Genome-wide identification of *Brassica napus* microRNAs and their targets in response to cadmium, *J. Exp. Bot.*, 63(12): 4597-4613. doi: 10.1093/jxb/ers136

Zhou, M. and Luo, H. (2013). MicroRNA-mediated gene regulation: Potential applications for plant genetic engineering, *Plant Molecular Biology*, 83: 59-75. http://dx.Doi.org/10.1007/s11103-013-0089-1

Zhou, R., Wang, Q., Jiang, F.L., Cao, X., Sun, M.T., Liu, M. and Wu, Z. (2016). Identification of miRNAs and their targets in wild tomato at moderately and acutely elevated temperatures by high-throughput sequencing and degradome analysis, *Sci Rep.*, 6: 33777.

CHAPTER

13

MicroRNA-mediated Regulation of Herbicide Resistance

Rafiq Lone[1]*, Parvaiz Yousuf[2], Shahid Razzak[2] and Semran Parvaiz[2]

[1] Department of Botany, Central University of Kashmir, Ganderbal, Jammu and Kashmir, India
[2] Department of Zoology, Central University of Kashmir, Ganderbal, Jammu and Kashmir, India

1. Introduction

As per the estimates, the world population will rise to 9 billion by 2050, resulting in increased demand for agricultural output due to the increasing consumption of crops, meat, dairy and biofuels. As per the researcher, to fulfil this demand, it is necessary to increase agricultural production around the world by 100-110% (Tilman *et al.*, 2011). At the same time, there isn't any chance of increasing the overall agricultural land in the following decades. Instead of extending land area, scientists say that increasing agricultural yield is the best sustainable solution to overcome the problem (Foley *et al.*, 2011; Phalan *et al.*, 2011; Tscharntke *et al.*, 2012; Godfray *et al.*, 2010). Since the first commercialisation of genetically-modified crops 20 years ago, they have been aggressively embraced to increase agricultural productivity. Within 17 years, the global land area devoted to genetically modified crops, especially having herbicide or insect tolerance, has significantly increased from a lower value of 1.7 million to 175 million hectares (Clive, 2013; Privalle *et al.*, 2012). At this point in time, experts are focusing on certain agricultural features, such as higher resistance to abiotic and biotic challenges, greater crop yield and a better nutritional value (Wang *et al.*, 2003; Knauf, 1987; Lucca *et al.*, 2006). Due to the various pressures affecting the growth of plants in the field, genetically-modified crops with the agronomic features listed above are attractive. Scientists have started using new technologies, such as microRNA (miRNA)-based genetic modifications that act as negative regulators of messenger RNA (mRNA), to enhance crop yield. miRNAs have a significant role in some essential plant biological processes, including the development of plants, defending

---

*Corresponding author: rafiqlone@gmail.com

MicroRNA-mediated Regulation of Herbicide Resistance   **221**

against pathogens and responding to environmental stimuli (Seo *et al.*, 2013; Pattanaik *et al.*, 2014; Zhou and Luo, 2014; Sun, 2012). The methods of controlling weeds have been revolutionised by using glyphosate-resistant (GR) crops (herbicide-resistant (HR) crops) in recent times. Even though billions of dollars and decades of time have been invested, the number of transgenic herbicide-resistant crops that are available in the market is quite low (CaJacob *et al.*, 2004). This chapter deals with how miRNAs are responsible for regulating herbicide resistance and how their regulation can help farmers to get varieties that are naturally resistant to herbicides.

2. Plant MicroRNAs Biogenesis: An Overview

Drosha is in charge of creating pre-miRNAs from pri-miRNAs during the miRNA biogenesis process, whereas Dicer is in charge of converting pre-miRNAs to mature miRNAs (Ha and Kim, 2014). Dicer-like (DCL) proteins in plants are responsible for both Drosha and Dicer activities (Voinnet, 2009: Ha and Kim, 2014). An in-depth phylogenetic 53 analysis of DCLs revealed that antiviral immunity was the driving mechanism underlying plant DCL evolution (Mukherjee *et al.*, 2013). This discovery was based on the fact that DCLs began in the same location but evolved separately in plants and mammals. Some plants, such as *Chlamydomonas reinhardtii,* which is a green alga that differs from the higher plant forms, possess CrDCL3 as a gene that is significant for miRNA synthesis (Valli *et al.*, 2016). Surprisingly, like Drosha, CrDCL3 has a proline-rich domain but no PAZ domain. This is a fascinating pairing. These discrepancies between algal and higher plant DCLs imply that the miRNA machinery evolved in parallel throughout plant lineages. Higher plants, on the other hand, do not contain any Drosha-like proteins. The discovery of a Drosha-like protein within Chlamydomonas, on the other hand, suggests that the DCL genes present in humans and plants evolved from the same parent ((Valli *et al.*, 2016). According to recent research, the target identification properties of Chlamydomonas miRNAs are equivalent to those of animal miRNAs (Yamasaki *et al.*, 2013). This conclusion is consistent with the fact that Chlamydomonas and mammals, both manufacture similar quantities of miRNA. Complementarity between the destination and nucleotides 2 to 8 of the miRNA, often known as the seed region, allows miRNA to locate target transcripts in animals. The seed area is another name for the seed region. To inhibit a target transcript, only the seed region of a Chlamydomonas microRNA is required (Yamasaki *et al.*, 2013). This is in contrast to the requirement for practically perfect base matching imposed on higher plant species miRNAs in order to identify their targets (Tang *et al.*, 2003). Argonaute (AGO) family proteins bind to short RNAs and act as mediators of their actions. The AGO75 plant family is made up of four basic higher plant lineages (AGO1/5/10, AGO2/3/7 and AGO4/6/8/9) and one algal lineage (Singh *et al.*, 2015). It is unclear where the grass-specific AGO subfamily, known as AGO18, originated (Zhang *et al.*, 2015), nor how closely related it is to the AGO1/5/10 lineage. According to a study (Wu *et al.*, 2015), rice's AGO18 can bind to miR168 and operate as a decoy for the miR168 target, AGO1. ZmAGO18b is in charge of binding 24-nt phased primary siRNAs in maize during the male reproductive development process (Zhai *et al.*, 2014). AGO10 in *Arabidopsis* works as a decoy for AGO1, preventing the latter from binding to miR165/166 (Zhu *et al.*, 2011; Ji *et al.*, 2011). Despite the fact that AGO18 is exclusively found in grasses, the presence of its miRNA partner (miR168) and the miRNA target gene (AGO1) in land plants (Aguilar *et al.*, 2021) suggests that AGO18 evolved later. AGO18 in grains and AGO10 in *Arabidopsis* may have evolved in parallel due to their similarity in inhibiting AGO1-miRNA interactions. Despite the fact that AGO18 and AGO10 interact with distinct miRNA partners and perform diverse biological activities, this is the case.

The miRNAs play a lot of vital functions within the cells. Studies have found that miRNAs have been playing important roles since their progression in plant cells. In evolutionary history, it is the miRNAs that emerged first, while the proteins (AGO proteins) that bound them emerged later. The addition of these proteins has solidified the functions of these miRNAs in plant cells (Aguilar *et al.*, 2021; Fang and Qi *et al.*, 2016). The studies have also found that these AGO proteins became more specialised with time as the miRNAs diversified. It's plausible that early DCLs and AGOs collaborated with siRNAs to improve immunity and that the DCLs and AGOs that function with miRNAs developed from them. It is unknown if the formation of the machinery that processes miRNA occurred during the split of living forms. There are several plausible origins for MIR genes, as indicated by the lack of conservation of miRNAs across plants and animals, as well as between algae and higher plants. In plants, RNA silencing, also known as co-suppression or post-transcriptional gene silencing, was identified in the early 1990s (Napoli *et al.*, 1990; van der Krol *et al.*, 1990). Small RNAs were connected to gene silencing in an increasing number of eukaryotic models by the end of the decade, indicating that suppressing RNAs is a common adaptive process (Zamore *et al.*, 2000; Hamilton and Baulcombe, 1999; Cogoni and Macino, 2000; Fire *et al.*, 1998). A decade following the findings of silencing of RNAs, multiple groups of endogenously formed short RNA molecules grounded on varied sources have been identified, the majority of which are miRNA and siRNA molecules (Llave *et al.*, 2002; Chapmann and Carrington, 2007). After scientists focused on other parameters, such as the complex nature and variety of regulatory work of such shorter RNA molecules, it has led to a steady exploration of the small RNA universe. The next section discusses how the siRNAs and miRNAs are formed and how they work.

2.1 Formation and Mechanism of Working of miRNAs

miRNAs are 21 nucleotide-long RNA molecules that have regulatory functions inside the cells. Polymerase II enzyme transcribes these miRNA molecules from endogenously located miRNA loci to form primary miRNAs which are essential for the functioning of the cells (Kurihara and Watanabe, 2004; Aukerman and Sakai, 2003; Kurihara and Watanabe, 2004). After that, the formation of precursor miRNAs (pre-miRNAs) is initiated as soon as the hairpin-like structures are created by the folding back of pri-miRNAs on themselves. To help with the formation of pri-miRNAs, several proteins are required. For instance, two important ones include C2H2-type zinc finger domain-containing protein SE and a nuclear cap-binding complex (CBC) (Yang *et al.*, 2006; Laubingerr *et al.*, 2008; Gregory *et al.*, 2008). Similarly, other proteins, such as a forked-associated domain-containing protein DDL and a double-strand are needed too (Kurihara *et al.*, 2006). In addition, there are other molecules, such as DCL1, that perform the functions of converting pre-miRNAs to double-strand duplex (21nt miRNA/miRNA*, to be more specific). After that, the miRNAs must be saved from an alternative end modification and this is where the role of adding a methyl group comes

MicroRNA-mediated Regulation of Herbicide Resistance

in. This is why HEN1, which is a small RNA methyltransferase, adds a methyl group to these double-stranded duplexes at their 3' end ribose molecule (Yang *et al.*, 2006; Park *et al.*, 2002). After that, the plant's ortholog of exportin 5 (Hasty) is given the charge of transferring the double-strand duplexes to the cytoplasmic space (Bollman *et al.*, 2003). In the end, the translational inhibitory and/or post-transcriptional cleavage activity is exerted by the RNA-induced silencing complex (RISC) after one of the working miRNA strands of the double helix that can be either miRNA*, miRNA, or two of them integrated into RISC (Li *et al.*, 2013; Brodersen *et al.*, 2008).

2.2 Biogenesis and Mechanisms of siRNA

The formation of siRNAs and miRNAs differ considerably from each other. For instance, the single transcript, which has the hairpin-shaped precursor, is responsible for creating the microRNAs, while for the formation of siRNAs, the double-helix RNAs (dsRNAs) duplexes are mainly responsible (Voinnet, 2009). Similarly, other differences exist between the two types of RNA molecules. SiRNA works at the transcription or post-transcription level via mRNA degradation, DNA methylation, or histone modification, while it is the miRNA that is responsible for regulating the expression of genes on a post-transcriptional basis (Baulcombe, 2004; Allshiree, 2002; Almeida and Allshiree, 2005; Sunkar *et al.*, 2006; Vaucheret, 2006). This chapter is mainly concerned with negative regulators of mRNAs. At the same time, it is responsible for exploring the siRNA classes that have the role of silencing the post-transcriptional genes. This includes long siRNA (lsiRNA), naturally formed anti-sense siRNAs (nat-siRNAs) and transacting siRNAs (ta-siRNAs).

The endogeneous TAS loci(s) are responsible for the formation of ta-siRNA precursor and miRNA-guided Argonaute (AGO) cleavage processes are the same later. From the one helical ta-siRNAs (which act as precursors), the phased 21-nt ta-siRNA(s) duplex strands are then generated with the aid of a DCL family member(s). These target RNAs are cleaved due to the interaction of AGO with one of the strands of the ta-siRNA duplexes, which happens because of the absolute complementarity in sequences (Gasciollii *et al.*, 2005; Allen *et al.*, 2005). So, let's talk about the source of plant nat-siRNA. As per experts, environmental stress-induced RNA and a duo of convergently transcribed RNAs that consist of RNA generated constitutively act as a plant nat-siRNA source (Borsanii *et al.*, 2005; Katiyar-Agarwall *et al.*, 2006). By processing SGS3, DCL2 and DCL2, RdRP 6 and Pol IV, the overlap areas among the sense transcripts and naturally formed anti-sense transcripts generate 21 to 24-nt nat-siRNAs (Katiyar-Agarwal *et al.*, 2006; Katiyar-Agarwal *et al.*, 2007; Zhang and Trudeau, 2008; Borsani *et al.*, 2005). The constitutively produced sense transcript is cleaved by the nat-siRNA, which is a means of responding to environmental stresses. Similarly, if the plants are stressed, then they will produce lsiRNA from an RNA pair that develops after the convergent transcription. These lsiRNAs have wider roles in plant defence mechanisms. They protect a plant against invading pathogens by creating several defence mechanisms. At the same time, the lsiRNAs considerably differ from other types of RNA molecules. For instance, they are longer (30-40 nucleotides) than the siRNA molecules (Katiyar-Agarwal *et al.*, 2007). Their formation requires a lot of proteins that include Pol IV, AGO7, RDRP6, HYL1 and DCL1 (Katiyar-Agarwal *et al.*, 2007). Moreover, researchers believe that lsiRNA works by destabilising the target by de-capping and degrading the mRNAs (Katiyar-Agarwal *et al.*, 2007).

3. The Roles of miRNA in Plant Abiotic or Biotic Responses

3.1 Stress Responses

Plants cannot prevent environmental stressors that include stresses due to sunlight, higher salinity in the soil, higher stresses of water, harsh temperatures, nutritional restriction, oxidative stress and stresses caused due to higher metal concentration since they are sessile organisms. Water shortage, rise in temperature as well as pollution of air, water and soil are all made worse by the ongoing climate change and the expanding global population. Under these miserable situations, plant biota has adapted a combination of abiotic stress mechanisms, which include molecular and physiological processes to adapt, live and reproduce. For creating crop yield that is resistant to different stress forms by using engineering techniques, it's crucial that we must have a good understanding of different molecular-level pathway(s) that underlie plant response to abiotic stress. Although presently, numerous upstream regulatory proteins and downstream functional proteins have a role in responding to abiotic stresses are clearly identified, but how these proteins regulate the response of plant biota to various forms of stress is not clear. Many researchers have shown the miRNA to have significant roles in regulating the post-transcriptional processes in plants, which are a kind of responses to a variety of environmental stressors (Pandita, 2019; Pandita and Wani, 2019; Pandita, 2021; 2022a; 2022b; 2022c) which further complicate the regulatory networks.

3.2 Drought Stress

An evolutionary maintained, controlled, mechanical and water-loss stressor is miR408 (Kantar *et al.*, 2010; Lu *et al.*, 2005). A new experiment on chickpeas supports its function in the plant's reaction to dehydration (Hajyzadeh *et al.*, 2015). Through the direct management of a Cu^{2+} binding protein, i.e. plantacyanin, overexpressing of miR408 enhances drought stress resistance (Hajyzadeh *et al.*, 2015). This suggests a better ability to maintain optimum Cu levels when the plants are exposed to stresses caused by droughts. Additionally, transgenic chickpea overexpressing miR408 exhibit drought stress-induced activation of certain genes that work against drought stress. These include RD22, RD29A, RD17, DREB1A and DREB1A in contrast to vector controlled plants. The finding suggests that miR408s and components responding to droughts interact and they may work together to improve tolerance to drought stress. Overexpression of miR480 by transgenic chickpea also has a pleiotropic effect of a dwarf phenotype in addition to improved drought tolerance. The outcome suggests that miR408, which controls the growth rate, must have additional targets.

By identifying and modifying the precise target genes with suitable and specific agronomic properties, it is crucial to evade the pleiotropic consequences of miRNAs in the phenomenon of genetic engineering of crops. The five NAC transcription factor genes – At5g61430, At5g07680, NAC1, CUC1 and CUC2 that miR614 regulates – have many other functions too. For instance, it has been found to regulate the growth and developmental processes of *Arabidopsis* plants (Mallory *et al.*, 2004; Guo *et al.*, 2005; Laufs *et al.*, 2004; Kim *et al.*, 2009b; Rhoades *et al.*, 2002). Numerous researchers have demonstrated that miR164s participating in plants' responses to stress are caused by abiotic factors. At the same time, researchers were not exactly clear as

MicroRNA-mediated Regulation of Herbicide Resistance

to how it works at the molecular level, but it became clear with researchers recently characterising the genes (NAC-OTMN) of the rice plant, wherein targeting was done by miR164 (Fang *et al.*, 2014). Moreover, the reproductive stages are increasingly sensitive to droughts, as seen in genetically engineered rice showing overexpression of OMTN6, OMTN3, OMTN2 and OMTN4, which is linked to relatively low spikelet fertility (Fang *et al.*, 2014). Additional genome expressions outline analysis indicates that the reduction of several regulatory and functionally important genes in reaction stresses caused by droughts is because of miR164 overexpression targeting OMTNS (Fang *et al.*, 2014). The research proves a link between miR164 and the response to abiotic stress. Additionally, it opens up the prospect of modifying OMTNs to help rice and the great crops become more drought tolerant.

3.3 Salt Stress

MiR394 suppresses the target gene, leaf curling responsiveness (LCR), which contributes to the plant's response to abiotic stress. LCR overexpression improves resistance to salts in comparison to the wild plants in transgenic *Arabidopsis* plants that are miR394 and LCR loss of function mutant (lcr) are susceptible to salinity (Palatnik *et al.*, 2003). Interestingly, transgenic plants overexpressing LCR are extremely sensitive to stresses caused by droughts, in contrast to those plants which show overexpression of miR394 and are lcr mutant, which is resistant to stresses caused by droughts (Palatnik *et al.*, 2003).

Additionally, miR394 controls the expression of the genes, such as RD22, K1N1, ABF3, ABF4, RD29A, AB15, AB13 and AB14, which are stress- and ABA-responsive genes. It is generally known that plants respond to salt and drought conditions with comparable osmotic adjustment techniques. It is suggested that a functional equilibrium of miR394-LCR pairings controls the different pathways in response to salt stress and drought. Additionally, it is conceivable that miR394s have a significant role in controlling targets (including both direct and indirect) that have a role in salt-specific adaptations to salt excretion and compartmentalisation as opposed to ionic imbalance. When using the tactic of modifying miR394 or its target for a given trait, there is a need for additional studies in the future that could help us understand the miR394's mediated molecular mechanism.

3.4 Cold Stress

The teosinte branched/cycloidea/PCF (TCP) transcription factor genes are the target of the conserved miRNA, miR319 (Schommer *et al.*, 2008; Koyama *et al.*, 2007; Ori *et al.*, 2007; Koyamaa *et al.*, 2010; Schwab *et al.*, 2005; Palatnik *et al.*, 2007; Koyama *et al.*, 2010). It is demonstrated to have vital functions in *Arabidopsis*, rice and sugarcane plants' responses to cold stress (Thiebaut *et al.*, 2012; Liu *et al.*, 2008; Zhou *et al.*, 2010). Recently performed research shows that overexpressing levels of miR319bs to its target OsTCP21, OsPCF8, OsPCF5 and OsPCF6 results in increased cold-stress resistance in genetically-engineered rice plants (Yang *et al.*, 2013; Wang *et al.*, 2014b). Additionally, OsPCF5, OsPCF6, OsPCF8 or OsTCP21 down-regulation in RNAi transgenic plants results in increased cold tolerance, whereas OsTCP21 or OsPCF6 overexpression considerably reduces cold plant resistance, possibly as a result of their opposite roles in scavenging and formation of ROS (Yang *et al.*, 2013; Wang *et al.*, 2014b). Along with its direct target(s), miR319s also control gene expression that responds to cold, like DREB2A, DREB1A/B/C & TPP1/2, resulting in increased cold resistance (Wang *et al*., 2014b). It is likely that coordinated regulation of various cold responsiveness pathways is necessary for miR319-mediated cold tolerance.

3.5 Heat Stress

miR159 is one of the conserved miRNAs in plants that has a role in controlling the GAMYB genes' cleavage. These GAMYB genes act as a target for miR159 in wheat plants. For instance, two such genes include TaGAMYB1 and TaGAMYB2 (Wang *et al*., 2012b). According to reports, TamiR159 in wheat plants is down-regulated when under stresses caused by heat (Wang *et al*., 2012b). Additionally, there is more sensitiveness to heat in TamiR159 overexpression in transgenic rice plants (Wang *et al*., 2012b). Its targets in *Arabidopsis* are the functionally redundant GAMYB-like genes AtMYB65 and AtMYB33. *Arabidopsis* myb33myb65 double mutants exhibit heat stress sensitivity (Wang *et al*., 2012b). The findings suggest that the miR159/GAMYB module may be involved in a pathway connected to heat. It is interesting to note that TaGAMYB1 overexpression does not result in increased heat stress tolerance (Wang *et al*., 2012b). It is conceivable that TAGAMYB1's constitutive expression leads to the induction of additional components that cancel out the effects of heat tolerance. To further improve our knowledge of how miR159 mediates resistance to heat, there is a need for more research. For instance, we should focus on identifying and characterising the TamiR159 targets while responding to the heat stresses. Furthermore, their targets, such as CSD1, CSD2 and CCS lower while responding to stress caused by heat in *Arabidopsis* plants and MiR398 is quickly activated during heat stress (Guan *et al*., 2013). Comparing mutant plants – ccs, cd2 and cd1 to wild-type controls, they have improved heat tolerance (Guan *et al*., 2013). Plants that overexpress the usual forms of CSD2 have a role in lowering the levels of heat-shock proteins and heat-resistant transcription factors. In comparison, there is a greater heat sensitivity which is caused by CSD2 variant-miR398 (Lu *et al*., 2013). MiR398-targeted CSD2 manipulation may be an impractical method for developing crops that are tolerant to high temperatures. This is because of the fact that many plant species possess miR398s and their targets, which are conserved to a great extent.

3.6 UV-B Radiation

Exposure of a plant to UV-B radiations is responsible for the inhibition of cell expansion and cell proliferation (Kakanii *et al*., 2003; Hectorss *et al*., 2007; Wargent *et al*., 2009; Robson and Aphalo, 2012). It is shown by research that through the regulation of GRFs in leaf primordia, leaf cell expansion and proliferation is regulated by miR396. The function of GRFs or miR396 module in UV-B-mediated inhibition of leaf growth in the case of *Arabidopsis* has been demonstrated in a recent study (Casadevall *et al*., 2013). The research depicts that whereas UV-B radiation decreases the transcripts of GRF3, GRF2 and GRF1 in proliferating tissues, it increases the expression of miR396 (Casadevall *et al*., 2013). According to additional research, UV-B radiation activates a protein kinase called mitogen-activated protein kinase three, which is responsible for miR396's induction. Transgenic plants that express miR396 resistant GRFs or less endogenous miR396 through engineered targets exhibited decreased vulnerability to UV-B-mediated growth suppression (Casadevall *et al*., 2013). The findings imply that crop species may benefit from increasing their resistance to UV-B radiation by altering the amounts of miR396 or GRFs, which would increase crop yields.

3.7 Heavy Metal Stress

Like animals, plants are in need of various macronutrients, such as Zn, Cu, Mn, and Fe as co-factors of some important proteins and enzymes. Toxic metals, including mercury (Hg), cadmium (Cd) and (Pb) lead, can harm plants if they are consumed in large quantities or in excess. The first step in reducing heavy metal stress is to identify the genes and regulatory pathways that are particularly vulnerable to the toxicity of toxic metals. miRNAs and their targets may have a role in the body's response to heavy metals. Researchers analysed how different types of microRNAs are identified along with their specific targets. Their definitive roles in offering tolerance to heavy metal ions have also been found. Transgenic rapeseed (*Brassica napus*) with a mutant miR395 gene exhibits improved Cd tolerance, according to a recent study (Zhang *et al.*, 2013a). Toxic Cd contamination has become a major environmental issue because of its huge emission due to human activities. Agricultural productivity and human health are affected by trophic chains because of Cd accumulation in plants, as it disrupts enzyme activity, limits cell division and stunts root development (Chaney *et al.*, 1999; Clemens, 2006; Chen *et al.*, 2009). MiR395 is a conserved microRNA that governs plant sulphate absorption and distribution via modulating the ATP sulfurylases genes: APS1, APS3 and APS4 and the sulphate transporter gene SULTR2:1 (Kawashima *et al.*, 2009). MiR395 activates by Cd stress in rapeseed (Zhang *et al.*, 2013a). Researchers have compared transgenic and wild rapeseed for several factors and found that lower stress damage due to oxidation and higher concentration of glutathione and chlorophyll occurs in the transgenic varieties of rapeseed (Zhang *et al.*, 2013a). Sulphur, cadmium and biomass accumulate in transgenics under Cd stress, despite the fact that Cd translocation in plants is restricted. MiR395 strongly influences the metal-tolerance genes: BnHO1, BnPCS1 and Sultr1;1 through additional molecular investigation (Zhang *et al.*, 2013a). Cd and sulphur-containing compounds had previously been proposed as a mechanism for plant tolerance to Cd, according to previous research (Khan *et al.*, 2008). Additionally, heavy metals influence the activity of numerous sulphate transporters (Xue Mei *et al.*, 2007). Because of this, it is possible that miR395-mediated sulphate assimilation plays a role in the sulphur-containing compound synthesis and Cd detoxification and that this signalling pathway may be involved.

3.8 Nutrition Deprivation

We know that phosphorus and nitrogen are the most vital macronutrients that have a significant role in the reproduction, growth and development of plants. The monocot-specific short RNA, miR444, has been implicated in rice nitrate build up and phosphate deficiency responses (Yan *et al.*, 2014). Researchers have focused on studying certain genes in rice too that have a role in nutrition deprivation. For instance, the four validated miR444 targets in rice plants include MIKC-type MADS-box genes OsMADS23, OsMADS27b, OsMADS27a and OsMADS57 (Sunkar *et al.*, 2005). These genes have a great homology to *Arabidopsis* ANR1, which is an important part of nitrate signalling controlling the development of lateral roots.

Under conditions of N- and P-deficiency, miR444 is positively regulated in rice (Yan *et al.*, 2014). Under normal growth conditions, overexpression of miR444 increases nitrate accumulation in rice. However, this does not happen when nitrogen is present in lower concentrations (Yan *et al.*, 2014). As determined by analysing the gene expressions, four nitrate transporter transcripts are elevated in rice plants that overexpress miR444 (Yan *et al.*, 2014). At the same time, when there is a scarcity of nitrogen, then the movement of nitrates from dying leaves to the young ones is stopped in these transgenic rice plants, increasing the susceptibility of miR444 to the limitation of nitrogen (Yan *et al.*, 2014). Under varying phosphate supply quantities, overexpression of the miR444 gene in these plants leads to changes in root architecture and activation of the phosphate transporter genes, which is intriguing (Yan *et al.*, 2014). MiR444 may have a role in rice's nitrate signal system and its response to phosphate shortage, according to the findings. It is more practical to evaluate and alter miR444 targets in order to improve N- and P-starvation tolerance properties. Furthermore, in response to a deficiency in nutrients, additional miRNAs play an important role. Sulphur distribution and homeostasis are affected by miR395 (Kawashima *et al.*, 2011), the phosphate-starvation signalling pathway is activated by miR399 (Chiou *et al.*, 2006), miR393 and miR167 operate in the nitrate-signalling pathway (Gifford *et al.*, 2008) and miR169s have a great role in the scarcity of nitrogen (Zhao *et al.*, 2011). Using miRNA-mediated plant responses to biotic stresses for genetic engineering purposes, these stress-responsive genes are able to rapidly change expression levels at the molecular level when infected with pathogens like bacteria or viruses. Moreover, the functions of miRNA in these defence systems are something that became clear to the researchers just 10 years back (Llave, 2004). Thus, numerous scientists out there are trying their best to find and characterise microRNA targets that respond to environmental stress.

3.9 Bacteria

miR393 is responsible for certain plant responses against bacteria as it inhibits auxin signalling. According to research, miRNAs have great functions in many immunity-induced mechanisms upon infection by pathogenic organisms (Navarro *et al.*, 2006; Navarro *et al.*, 2008). For instance, miR472, which is responsible for targeting the disease-resistant gene groups, belongs to the coiled-coil nucleotide-binding leucine-rich repeats (CNL) family. Similarly, a new study found that RDRP6 (RNA silencing factor) and *Arabidopsis* miR472 are responsible for regulating certain immunological responses against plants (Boccara *et al.*, 2014).

3.10 Fungi

Fungal diseases are responsible for a lot of damage to crops worldwide. For instance, *Puccinia striiformis* f. sp. tritici (Pst) is responsible for causing one of the worst fungal diseases, the wheat stripe disease. In recent times, newer genes, such as miRNA PN-2013-targeted TaMDHAR gene are important for seedling Pst response in wheat (Feng *et al.*, 2014).

The regulation of ROS metabolism is done mainly by the TaMDHAR through the ascorbate-glutathione cycle. Studies have found that there is increased resistance in TaMDHAR knockdown mutants (Feng *et al.*, 2014). There is an up-regulation of genes related to pathogenicity in TaMDHAR mutants that affect the greater resistance to Pst. We can use this mechanism of regulation of PN-2013 against Pst for resistance to pathogens. This may be a great way to cope with the pathogenic stresses in plants.

3.11 Virus

A lot of plant viruses affect plant productivity considerably. For instance, tomato leaf curls New Delhi virus (ToLCNDV) is one of the viruses that infects tomato plants and leads to significant crop losses. This virus causes upward leaf curling in such plants. As ToLCNDV agroinfection progresses in tomato cv Pusa Riby, chili plants and tomato cv JK Asha, miR319 miR172 and miR159 accumulate (Naqvi *et al.*, 2010). Many studies have shown that these miRNAs can be used as biomarkers to detect the infection of ToLCNDV in plants. Recently, several studies have shown many miRNAs, such as Tom 17; Tom 21; Tom 4; Tom 14; Tom 43 and Tom 29, as well as the targets after the ToLCNDV infection occurs (Pradhan *et al.*, 2015). This may help us understand the relationship between miRNAs and ToLCNDV in a better way. More recent studies have focused on discovering more and more plant miRNAs that work against viral infections (Amin *et al.*, 2011). However, the techniques that employ endogenous plant miRNAs for promoting resistance to viruses are very low. However, researchers have been focusing more on the production of artificial miRNAs that could target the genome of viruses. This is becoming an efficient weapon for silencing the viral genome (Kung *et al.*, 2012; Fahim *et al.*, 2012).

3.12 Insect

Recently, a study related the miRNA-mediated secondary mechanisms of metabolic defence with resistance to aphids (Kettles *et al.*, 2013). It has been found that the new generation of peach aphids is significantly reduced because of certain miRNAs in Ago1, dcl1 and hen1 mutants. The mechanism behind this is that it leads to the formation of camalexin, which significantly affects the reproduction of aphids (Kettles *et al.*, 2013). Whenever aphids affect the green peach plants, the camalexin synthesis increases because of dcl1 mutants. They do so by activating certain genes in the biosynthetic pathway of camalexin (Kettles *et al.*, 2013).

Apart from this, there are other RNAs, such as small artificial RNAs and siRNAs, which have been found to be present in plants upon attack by herbivores (Pandey *et al.*, 2008; Pandey and Baldwin, 2007). These mRNAs are a great way to counteract the insect attack on plants in the near future.

3.13 Nematode

Researchers have found miRNAs while investigating the soybean miRNAs by deep sequencing technology. To be more specific, the mRNA was soybean cyst nematode (SCN)-responsive miRNA (Li *et al.*, 2012). Plants use several types of miRNAs to fight against nematode infections. For instance, there are 101 miRNAs exhibiting variable expression and include many conserved stress-responsive miRNAs, such as miR390, miR319 and miR169 (Li *et al.*, 2012). We do not have great knowledge about nematodes affecting plants right now. Thus, there is a need for many more studies to find more nematode-responsive miRNAs in plant biota. Identifying SCN-responsive miRNA/target modules is a great way to help fight against nematode infections.

4. Herbicide Tolerance in Plants

Farmers manage the weeds in different ways, but the use of herbicide-resistant crops, especially the use of glyphosate-resistant crops, has modified their way of weed management. A lot of time and money had been invested in research to develop herbicide-resistant plants, but unfortunately, only some herbicide-resistant traits are found which are commercially available. Researchers have focused on finding various transgenes that are responsible for insensitiveness to glyphosate. For instance, a transgene that has been extracted from corn is the mutated gene zm-2mepsps and the other one extracted from *Agrobacterium tumefaciens* is the cp4 epsps gene 5-enolpyruvylshikimate-3-phosphate synthase (EPSPS; EC 2.5.1.19). Other than this, glufosinate is inactivated by N acetyl transferases which are in turn encoded by two genes (bar and pat from *Streptomyces hygroscopicus* and *Streptomyces viridochromogenes*) that are homologous to each other. Apart from these two genes, another gene (glyphosate oxidoreductase-GOX) is extracted from the *Ochrobactrum anthropi* strain LBAA. All the three mentioned genes are responsible for inactivating metabolic processes. Later, the farmers were not hesitant to adopt these GR crops because of several reasons. This is because, in fact, they helped farmers with weed control on a glyphosate basis, which was quite easier to use and much cheaper. These unique technologies were vital to making new inventions that protect the large investments required for new technology development, and farmers and growers lauded the clarity and amenity of glyphosate-based crop systems. When farmers and growers notice the simplicity, convenience and effectiveness of glyphosate-based crop systems, they start their use in an unplanned manner and rely only on this system for weed control management. When the researchers noticed these things, they became concerned about things and expected the evolution of resistance. However, no case of GR weeds has grown after long use in non-crop situations. This gives a moment of relief to growers and weed scientists and they think that the use of GR crops would never create any problem. But all this made a U-turn when, in Australia, the discovery of GR rigid ryegrass (*Lolium rigidum* Gaudin) was confirmed.

Today, everyone agrees that the evolution of GR weeds poses a danger to the success of GR crops and the long-term viability of glyphosate. Resistance to glyphosate has arisen in 19 weeds, with roughly half of them developing in GR crops. Changes in the EPSPS target site, sequestration in the vacuole, reduced translocation or cellular transport to the plastid and gene amplification are the main causes of resistance. GR weeds raise weed control costs and decrease the effectiveness of glyphosate-based weed management methods. In retrospect, the evolution of GR weeds was unavoidable. The success of glyphosate made it a victim of its own success. It does not matter how much the herbicide is effective; weed management cannot solely depend on single tactics. Otherwise, weed will find its way to adopt, thrive and survive in a vast number.

In essence, an ideal storm is created by GR crops for weeds to evolve resistance. To control genetically variable and prolific weeds, cultivators make use only of glyphosate over large cropping areas, year after year. Numerous weeds had already developed resistance to another herbicide mode of action, so when these weeds displayed resistance against glyphosate, there was no other alternative herbicide present. This problem became more serious at the time when a case of highly competitive, fast-growing and prolific Palmer amaranth (*Amaranthus palmeri* S. Wats) came into existence. It caused such an explosion and damage in south-east United States that it was named as 'pigweed disaster'. Due to this crisis, the GR crops are urging the growers to change their way of weed management methods and increase the efforts to control the weed; even hand weeding is also prescribed. Due to problems and less effectiveness of glyphosate, its use gets reduced day by day and it is recommended that it should be used in addition to other herbicides. Now is a need for cultivators to diversify their herbicides to mitigate the dispersion of GR weeds. But unfortunately, the mode of action of herbicides is most primitive and the herbicide industry has not evolved any new strategies. This can be due to the reason that a number of chemicals has increased from less than 1,000 in 1950 to more than 500,000 today to be tested to evolve new herbicides. Also, the chemical industries invest less amounts of money in discovering novel herbicides because their market value has reduced due to the widespread use of GR crops. Now industries are developing new herbicide-resistance traits to address the GR weed problem so that they will expand the use of currently available herbicides. However, it should be kept in mind that these solutions are of short-term use and only for a current weed problem. Industries are unable to replace the long-term efforts to discover novel herbicides with new modes of action and give new weed management techniques.

5. MicroRNA-mediated Herbicide Resistance

5.1 RNA Interference (RNAi) Technology

Advancement in technology is essential to modify and improve plant physiology for better adaptations to biotic as well as biotic stressors, including herbicide application. There are various internal mechanisms in plant cells that help them to turn off harmful gene expression, to maintain the genome integrity, modify responses to biotic and biotic stressors and regulate and modify the developmental processes via non-coding, small RNA molecules, which differ in their biosynthesis (Khraiwesh *et al.*, 2012).

5.2 Advances in Weed Management

Researchers are trying to find a way to use RNA interference to manage weeds. For attaining RNA interference, two types of RNAs (siRNA and miRNA) are essential (Sanan-Mishra *et al.*, 2013). The process of RNA interference is a vital process occurring in numerous eukaryotes. Now, researchers are trying to modify this process to change the genetic composition of many plant species (Rutz and Scheffold, 2004). Moreover, researchers have also found the role of miRNA in regulating genes. They do so by binding to complementary sequences (reverse), resulting in several changes in the targets. For instance, they lead to translational inhibition and cleavage of the RNA molecules that they target (Khraiwesh *et al.*, 2012). Both miRNAs and siRNAs, performing the same set of functions, are formed through the same processes and have the same structural anomalies. However, siRNA is a longer molecule derived from dsRNA and has a significant role in methylating DNA at specific sequences (Khraiwesh *et al.*, 2012).

Some miRNAs act as regulatory molecules which regulate the expression of some genes in response to needs of plant development or stress factors. Among them, some act as positive trait regulators, which increase the expression of a particular product needed by the plant and some act as negative trait regulators, which decrease or turn off the expression of some particular gene products which are not needed by plants (Zhou and Luo, 2013). Thus, the expression of miRNA can be modulated artificially to alter plant traits. In weed management, this technology can be used in crops to alter the level of tolerance to oxidative stress. For example, tolerance of oxidative stresses gets increased when overexpression of a negative regulator miR398 form of *CSD2* (Cu/Zn SOD gene) by RNAi occurs (Sunkar *et al.*, 2006). The resultant trait will show an increased level of crop tolerance to some herbicides. In the same way, the expression of these miRNAs can be modulated to control nitrogen metabolism, to enhance the competitive capability of crops (Fischer *et al.*, 2013), salt tolerance (Zheng and Qu, 2015), drought tolerance (Ferdous *et al.*, 2017) and other tolerances and stresses also. For manipulating the genetic makeup of plants, the initial step is to identify the target site in miRNA, which helps in understanding the effects of modulating the expression of the target gene on the phenotype of the plant across key species. Over the last decade, the strategy in herbicide-resistance weed management had changed significantly by focusing mainly on reducing the weed seed bank and low seed-bank maintenance by any possible means. To check the impact of weed density and to determine the time of emergence regarding crop-on-crop revenue loss, a lot of research, from the 1960s to the 1990s, has been devoted to it. Moreover, the results obtained from it have been used to describe the models and decision-making systems for herbicide use.

The early weed economic models with a focus on single crop season have been changed to more sophisticated models, which include weed seed-bank implications and longer time periods, i.e. injury levels or optimum economic thresholds. Today, these threshold models are not used by weed scientists or practitioners, but still, they are used in pathology and entomology. Today, most growers have weed seed-banks in their area which are resistant to one or more than one herbicide site of action. Another policy was implicated, known as the zero-tolerance policy (take no prisoners), but economically and agronomically, this policy may not be feasible for all weed species to achieve the required goal. Various new developments, strategies and trends regarding HR weed management have been described. This includes the recent discovery of various new herbicides by the agrochemical industries, following a long period in which no new herbicide site of action was commercialised. In this prolonged period, the strategy of industries was to use existing chemistry for newly discovered ones, introducing various single or combined (stacked) HR traits into main agronomic crops, such as cotton (*Gossypium hirsutum* L.), maize (*Zea mays* L.) and soybean [*Glycine max* (L.) Merr.]. Although much flexibility has been offered to cultivators by HR trait stacking for HR weed management, the opinion of weed academics is that it can be used for a long period of time and can lead to a high risk of multiple HR populations.

Another major trend in herbicide use is the rise in popularity of pre-emergence (PRE) herbicides with soil residual activity to fill the gap left by post-emergence (POST) products, like acetyl-CoA carboxylase (ACCase) or acetolactate synthase (ALS) inhibitors, which have lost efficacy due to widespread evolved resistance. PRE herbicide treatment, that requires soil integration, fell out of favour in the 1970s and 1980s in preference to post-herbicide application that was better suited to low soil disturbance, no-tillage cropping systems.

There is also some non-herbicidal approach towards weed management that cannot be ignored at all and has shown a great impact on weed management. The approach 'harvest weed seed control, HWSC' has already shown a significant effect and 'weed competitive crop cultivars' is supposed to do so in the future. For almost 20 years, research and development have been ongoing on site-specific weed management (SSWM), but this technology has not yet been adopted completely in the agronomic fields. Nevertheless, the significant environmental and economic benefits of SSWM can't be ignored.

5.3 Improvised Gene Functions for Resistance against Herbicides

Researchers are consistently identifying putative driver genes for connecting them to certain resistant phenotypes. It has resulted from increased sequencing efforts to understand the causes of resistance. Unfortunately, many researchers have stopped functionally validating these potential genes, leaving the true and main reason for resistance unknown. Scientists used *Agrobacterium tumefaciens*-mediated transformation of candidate genes in tobacco (*Nicotiana benthamiana*), budding yeast (*Saccharomyces cerevisiae*), transgenic rice and rice (*Oryza sativa*) and *Arabidopsis thaliana* in successful validation studies. The reason why these technologies for weed management do not reach farmers is lower investments are being made in the research. Method development in the field of plant transformation is critical. Transient expression systems can be used to examine plant gene function by knocking out or overexpressing a candidate gene variation. In non-model species, there are currently several techniques for investigating gene function by RNA interference (RNAi), like virus-induced gene silencing (VIGS). This approach has recently been used to silence CYP749A16 in trifloxysulfuron-tolerant cotton and a GST gene cluster in Verticillium wilt-resistance cotton, both of which are relevant to herbicide resistance. Inoculation by the modified virus can be done in plants to transcribe the antisense RNA and subsequent target mRNA breakage, as seen in the cereal, barely mosaic virus system. Other various techniques are there to suppress target mRNA, like double-stranded RNA (dsRNA) spray, long dsRNA application and small interfering RNAs (siRNAs) complexed with a protein carrier. In comparison to reverse genetics ways that knock out the gene function by means of antisense transcript silencing, transient infection with Agrobacterium and RNA-directed DNA methylation is something researchers are trying to use for gaining the gene functions. Mutant plants are being generated by other techniques, such as transposon insertional mutagenesis and alternative transfer DNA (t-DNA). They are doing so to study phenotypes that can be used as herbicide-tolerant ones. With the help of these techniques, there is a great possibility of creating and maintaining germplasm for later use. Moreover, this will happen only if a large number of researchers focus on one species. Furthermore, it may be better for researchers to use other techniques to perform such studies. Two such techniques that can be used are transcription activator-like effector nucleases (TALEN) and zinc finger nucleases (ZFNs). Besides, one of the quicker ways for achieving the targeted gene editing is when small RNAs are used for guiding the editing of genes with clustered, regularly interspaced, short, palindromic repeats/CRISP-associated protein 9 (Cas9) rather than the protein molecules. CRISP systems have been proven to work with great efficiency and specificity, both transiently and stably. Using such a technique for validating the gene functions would be extremely beneficial to the weed science community, but, as with other ways of studying gene functions, investments in plant transformation technologies are required to properly enable gene editing in weeds.

6. Conclusion

According to the worldwide background of HR weed management, manufacturers respond to problems after they have occurred rather than using preventative steps (Beckie, 2006). Despite the fact that efforts may have been taken to prevent the situation in the first place, this is the case. Furthermore, the complexity (cross- and multiple-resistance patterns) and magnitude (distribution and abundance) of the herbicide-resistant weed population identified in their fields will dictate the number of modifications that need to be made to their agricultural system. Because there are just a few herbicides that are effective against HR weeds, a growing number of farmers are being compelled to explore crop rotation or shifting their crops to keep their HR weeds under control. For example, in the United Kingdom, the number of farmers using spring cropping as a weed management approach for HR grasses, like blackgrass, increased from 32% in 2000 to 81% in 2016 (Moss, 2019). Between the years 2000 and 2016, there was significant growth in the number of people who used the internet (Moss, 2019). Moreover, the leguminous crops may be a weak choice for crop rotation because it does not have a stronger weed performance, a limited number of herbicides approved for use on them, or reliance on a restricted number of herbicide systems of action (SOAs), such as ALS inhibitors. Furthermore, the number of herbicides approved for use on them is restricted. It is probable that both the planting area and the production of these crops may decline in the not-too-distant future. Grain crop performance and low weed seed bank levels in Australia have finally been achieved by combining effective PRE herbicides with agronomic practices that encourage crop competition and limit weed seed set, according to Australian experiences during the past decade. Regardless of the fact that efficient PRE herbicides have only been on the marketplace for a short time, this has been the case. This was true, regardless of the agronomic practices implemented to obtain the intended results.

It is critical that any long-term HR weed control policy applied anywhere on the planet prioritises reducing the use of herbicides, particularly glyphosate. Non-herbicidal options are more commonly used to compensate for decrease in herbicide efficacy caused by a rise in resistance prevalence (Moss, 2019). This is because it is simpler to adjust for reduced herbicide efficacy with non-herbicidal alternatives than it is to employ herbicides themselves. Low weed seed banks must be a prerequisite for success in integrated weed management as well as decreased pesticide use. Only then can we call integrated weed management a success. People are already using organic and conventional means of managing weeds around the world. The main reason for this is the growing popularity of HR weed crops, as well as the pressure on farmers to use synthetic pesticides in their fields. Other important factors why people are avoiding the use of pesticides is because of its bad impact on health, the greater costs, as well as the restrictions that are imposed by governments on the usage of pesticides. Furthermore, this is occurring as a result of growing social pressures to reduce the utilisation of pesticides in feed and food preparation blocks. This is happening as a result of a number of factors, including the following: (1) increase in the prevalence rate of HR weed population; (2) increase in the influence of HR weed populaces; and (3) increase in temperature. Due to the ubiquitous prevalence of multiple-HR populations in many large agricultural systems throughout the world, herbicides may become a 'once in a century' strategy for weed management (Davis and Frisvold, 2017). One of the most important questions to investigate in a future study is 'how much weed control is necessary to consistently achieve the goal of reduced weed seed banks?' (To what extent the weed management is vital to help reach the lower weed seed back goals). This question highlights the prospect of a reduction in herbicide application, which may be calculated by multiplying the treated area by the total number of treatments and the herbicide loading (kg ha^{-1}). The adoption of a standardised methodology allows for a more exact assessment of the economic allowance, effect, as well as consequences of lower utilisation of pesticides (Frisvold, 2019). It is critical to conduct multi-locational, medium-term (four to eight years), large-plot, or landscape-level agricultural system research studies in order to provide an acceptable response to this issue. Comparing the most important economic, agronomic and environmental variables of diverse agricultural systems, each of which has its own individual crop sequence and unique combinations of weed control tactics, can aid in the discovery of more environmentally-sustainable weed management practices. Each of these systems has its own particular crop sequence and weed control strategy combinations. Furthermore, weed surveys and the accompanying farmer management questionnaires can aid in establishing the best management measures for reducing the quantity of weed seed banks present in herbicide-free farming systems. This may be done by establishing the best management options for reducing the number of weed seed banks on the ground.

References

Aguilar, A.T., Grimanelli, D., Garcia, G.A., Calzada, J.P.V., Badillo-Corona, J.A. and Duran-Figueroa, N. (2021). miR822 modulates monosporic female gametogenesis through an ARGONAUTE9-dependent pathway in *Arabidopsis thaliana*, *BioRxiv.*, 46: 10-18.

Allen, E., Xie, Z., Gustafson, A.M. and Carrington, J.C. (2005). microRNA-directed phasing during trans-acting siRNAs biogenesis in plants, *Cell*, 121: 207-221.

Allshire, R. (2002). RNAi and heterochromatin: A hushed-up affair, *Science*, 297(5588): 1818-1819.

Almeida, R. and Allshire, R.C. (2005). RNA silencing and genome regulation, *Trends in Cell Biology*, 15(5): 251-258.

Amin, I., Patil, B.L., Briddon, R.W., Mansoor, S. and Fauquet, C.M. (2011). A common set of developmental miRNAs are up-regulated in *Nicotiana benthamiana* by diverse begomoviruses, *Virol. J.*, 8: 300.

Aukerman, M.J. and Sakai, H. (2003). Regulation of flowering time and floral organ identity by a microRNA and its APETALA2-like target genes, *Plant Cell*, 15: 2730-2741.

Baulcombe, D. (2004). RNA silencing in plants, *Nature*, 431(7006): 356-363.

Beckie, H.J. (2006). Herbicide-resistant weeds: Management tactics and practices, *Weed Technology*, 20(3): 793-814.

Boccara, M., Sarazin, A., Thiebeauld, O., Jay, F., Voinnet, O., Navarro, L. and Colot, V. (2014). The *Arabidopsis* miR472-RDR6 silencing pathway modulates PAMP- and effector-triggered immunity through the post-transcriptional control of disease resistance genes, *PLoS Pathogens*, 10: e1003883.

Bollman, K.M., Aukerman, M.J., Park, M.Y., Hunter, C., Berardini, T.Z. and Poethig, R.S. (2003). Hasty, the *Arabidopsis* ortholog of exportin 5/MSN5, regulates phase change and morphogenesis, *Development*, 130: 1493-1504.

Borsani, O., Zhu, J., Verslues, P.E., Sunkar, R. and Zhu, J. (2005). Endogenous siRNAs derived from a pair of natural cis-antisense transcripts regulate salt tolerance in *Arabidopsis*, *Cell*, 123: 1279-1291.

Brodersen, P., Sakvarelidze-Achard, L., Bruun-Rasmussen, M., Dunoyer, P., Yamamoto, Y.Y., Sieburth, L. and Voinnet, O. (2008). Widespread translational inhibition by plant miRNAs and siRNAs, *Science*, 320: 1185-1190.

CaJacob, C.A., Feng, P.C.C., Heck, G.R., Alibhai, M.F., Sammons, D.R. and Padgette, S.R. (2004). Engineering resistance to herbicides. *In:* Christou, P. and Klee, H. (Eds.). *Handbook of Plant Biotechnology*, 34: 353-373.

Casadevall, R., Rodriguez, R.E., Debernardi, J.M., Palatnik, J.F. and Casati, P. (2013). Repression of growth regulating factors by the microRNA396 inhibits cell proliferation by UV-B radiation in *Arabidopsis* leaves, *Plant Cell*, 25: 3570-3583.

Chaney, R., Ryan, J., Li, Y. and Brown, S. (1999). Soil cadmium as a threat to human health, *Cadmium in Soils and Plants*, Springer, pp. 219-256.

Chapman, E.J. and Carrington, J.C. (2007). Specialisation and evolution of endogenous small RNA pathways, *Nature Reviews Genetics*, 8: 884-896.

Chen, J., Yang, Z., Su, Y., Han, F. and Monts, D. (2009). Phytoremediation of heavy metal/metalloid-contaminated soils. *In:* Steinberg, R.V. (Ed.). *Contaminated Soils: Environmental Impact, Disposal, and Treatment*, Nova Science Publishers, Inc. NY, USA.

Chiou, T.J., Aung, K., Lin, S.I., Wu, C.C., Chiang, S.F. and Su, C.L. (2006). Regulation of phosphate homeostasis by MicroRNA in *Arabidopsis*, *Plant Cell*, 18: 412-421.

Clemens, S. (2006). Toxic metal accumulation, responses to exposure and mechanisms of tolerance in plants, *Biochemis.*, 88: 1707-1719.

Clive, J. (2013). Global status of commercialised biotech/GM crops, *ISAAA Brief*, 46:

Cogoni, C. and Macino, G. (2000). Posttranscriptional gene silencing across kingdoms, *Curr. Opin. Genet. Dev.*, 10: 638-643.

Davis, A.S. and Frisvold, G.B. (2017). Are herbicides a once in a century method of weed control? *Pest Management Science*, 73(11): 2209-2220.

Fahim, M., Millar, A.A., Wood, C.C. and Larkin, P.J. (2012). Resistance to wheat streak mosaic virus generated by expression of an artificial polycistronic microRNA in wheat, *Plant Biotechnol. J.*, 10: 150-163.

Fang, X. and Qi, Y. (2016). RNAi in plants: An Argonaute-centred view, *The Plant Cell*, 28(2): 272-285.

Fang, Y., Xie, K. and Xiong, L. (2014). Conserved miR164-targeted NAC genes negatively regulate drought resistance in rice, *J. Exp. Bot.*, 65: 2119-2135.

Feng, H., Wang, X., Zhang, Q., Fu, Y., Feng, C., Wang, B. … and Kang, Z. (2014). Monodehydroascorbate reductase gene, regulated by the wheat PN-2013 miRNA, contributes to adult wheat plant resistance to stripe rust through ROS metabolism, *Biochimica et Biophysica Acta (BBA)-Gene Regulatory Mechanisms*, 1839(1): 1-12.

Ferdous, J., Sanchez-Ferrero, J.C., Langridge, P., Milne, L., Chowdhury, J., Brien, C. and Tricker, P.J. (2017). Differential expression of microRNAs and potential targets under drought stress in barley, *Plant, Cell & Environment*, 40(1): 11-24.

Fire, A., Xu, S., Montgomery, M.K., Kostas, S.A., Driver, S.E. and Mello, C.C. (1998). Potent and specific genetic interference by double-stranded RNA in *Caenorhabditis elegans*, *Nature*, 391: 806-811.

Fischer, J.J., Beatty, P.H., Good, A.G. and Muench, D.G. (2013). Manipulation of microRNA expression to improve nitrogen use efficiency, *Plant Science*, 210: 70-81.

Foley, J.A., Ramankutty, N., Brauman, K.A., Cassidy, E.S., Gerber, J.S., Johnston, M., Mueller, N.D., O'Connell, C., Ray, D.K., West, P.C., Balzer, C., Bennett, E.M., Carpenter, S.R., Hill, J., Monfreda, C., Polasky, S, Rockstrom, J., Sheehan, J., Siebert, S., Tilman, D. and Zaks, D.P. (2011). Solutions for a cultivated planet. *Nature*, 478: 337-342.

Frisvold, G.B. (2019). How low can you go? Estimating impacts of reduced pesticide use, *Pest Management Science*, 75(5): 1223-1233.

Gasciolli, V., Mallory, A.C., Bartel, D.P. and Vaucheret, H. (2005). Partially redundant functions of *Arabidopsis* Dicer-like enzymes and a role for DCL4 in producing trans-acting siRNAs, *Current Biology*, 15: 1494-1500.

Gifford, M.L., Dean, A., Gutierrez, R.A., Coruzzi, G.M. and Birnbaum, K.D. (2008). Cell-specific nitrogen responses mediate developmental plasticity, *Proc. Natl. Acad. Sci.*, USA, 105: 803-808.

Godfray, H.C., Beddington, J.R., Crute, I.R., Haddad, L., Lawrence, D., Muir, J.F., Pretty, J., Robinson, S., Thomas, S.M. and Toulmin, C. (2010). Food security: The challenge of feeding 9 billion people, *Science*, 327: 812-818.

Gregory, B.D., O'Malley, R.C., Lister, R., Urich, M.A., Tonti-Filippini, J., Chen, H., Millar, A.H. and Ecker, J.R. (2008). A link between RNA metabolism and silencing affecting *Arabidopsis* development, *Developmental Cell*, 14: 854-866.

Guan, Q., Lu, X., Zeng, H., Zhang, Y. and Zhu, J. (2013). Heat stress induction of miR398 triggers a regulatory loop that is critical for thermotolerance in *Arabidopsis*, *The Plant Journal*, 74: 840-851.

Guo, H.S., Xie, Q., Fei, J.F. and Chua, N.H. (2005). MicroRNA directs mRNA cleavage of the transcription factor NAC1 to down-regulate auxin signals for *Arabidopsis* lateral root development, *Plant Cell*, 17: 1376-1386.

Ha, M. and Kim, V.N. (2014). Regulation of microRNA biogenesis, *Nature Reviews Molecular Cell Biology*, 15(8): 509-524.

Hajyzadeh, M., Turktas, M., Khawar, K.M. and Unver, T. (2015). miR408 overexpression causes increased drought tolerance in chickpea, *Gene*, 555(2): 186-193.

Hamilton, A.J. and Baulcombe, D.C. (1999). A species of small anti-sense RNA in post-transcriptional gene silencing in plants, *Science*, 286: 950-952.

Hectors, K., Prinsen, E., De Coen, W., Jansen, M.A. and Guisez, Y. (2007). *Arabidopsis thaliana* plants acclimated to low dose rates of ultraviolet B radiation show specific changes in morphology and gene expression in the absence of stress symptoms, *New Phytol.*, 175: 255-270.

Ji, L., Liu, X., Yan, J., Wang, W., Yumul, R.E., Kim, Y.J., … and Chen, X. (2011). Argonaute10 and Argonaute1 regulate the termination of floral stem cells through two microRNAs in *Arabidopsis*, *PLoS Genetics*, 7(3): e1001358.

Kakani, V.G., Reddy, K.R., Zhao, D. and Mohammedm, A.R. (2003). Effects of ultraviolet-B radiation on cotton (*Gossypium hirsutum* L.) morphology and anatomy, *Ann. Bot.*, 91: 817-826.

Kantar, M., Unver, T. and Budak, H. (2010). Regulation of barley miRNAs upon dehydration stress correlated with target gene expression, *Functional & Integrative Genomics*, 10: 493-507.

Katiyar-Agarwal, S., Gao, S., Vivian-Smith, A. and Jin, H. (2007). A novel class of bacteria-induced small RNAs in *Arabidopsis*, *Genes Dev.*, 21: 3123-3134.

Katiyar-Agarwal, S., Morgan, R., Dahlbeck, D., Borsani, O., Villegas, A., Jr, Zhu, J.K., Staskawicz, B.J. and Jin, H. (2006). A pathogen-inducible endogenous siRNA in plant immunity, *Proc. Natl. Acad. Sci.*, USA, 103: 18002-18007.

Kawashima, C.G., Yoshimoto, N., Maruyama Nakashita, A., Tsuchiya, Y.N., Saito, K., Takahashi, H. and Dalmay, T. (2009). Sulphur starvation induces the expression of microRNA395 and one of its target genes but in different cell types, *The Plant Journal*, 57: 313-321.

Kawashima, C.G., Matthewman, C.A., Huang, S., Lee, B., Yoshimoto, N., Koprivova, A., Rubio Somoza, I., Todesco, M., Rathjen, T. and Saito, K. (2011). Interplay of SLIM1 and miR395 in the regulation of sulphate assimilation in *Arabidopsis*, *The Plant Journal*, 66: 863-876.

Kettles, G.J., Drurey, C., Schoonbeek, H., Maule, A.J. and Hogenhout, S.A. (2013). Resistance of *Arabidopsis thaliana* to the green peach aphid, *Myzus persicae*, involves camalexin and is regulated by microRNAs, *New Phytol.*, 198: 1178-1190.

Khan, N.A., Singh, S. and Umar, S. (2008). *Sulphur Assimilation and Abiotic Stress in Plants*, Springer-Verlag, Berlin Heidelberg, Germany, 372 p.

Khraiwesh, B., Zhu, J.K. and Zhu, J. (2012). Role of miRNAs and siRNAs in biotic and abiotic stress responses of plants, *Biochimica. Biophysica. Acta*, 1819: 137-148.

Kim, J.H., Woo, H.R., Kim, J., Lim, P.O., Lee, I.C., Choi, S.H., Hwang, D. and Nam, H.G. (2009b). Trifurcate feed-forward regulation of age-dependent cell death involving miR164 in *Arabidopsis*, *Science*, 323: 1053-1057.

Knauf, V.C. (1987). The application of genetic engineering to oilseed crops, *Trends Biotechnol.*, 5: 40-47.

Koyama, T., Furutani, M., Tasaka, M. and Ohme-Takagi, M. (2007). TCP transcription factors control the morphology of shoot lateral organs via negative regulation of the expression of boundary-specific genes in *Arabidopsis*, *Plant Cell*, 19: 473-484.

Koyama, T., Mitsuda, N., Seki, M., Shinozaki, K. and Ohme-Takagi, M. (2010). TCP transcription factors regulate the activities of asymmetric Leaves1 and miR164, as well as the auxin response, during differentiation of leaves in *Arabidopsis*, *Plant Cell*, 22: 3574-3588.

Kung, Y.J., Lin, S.S., Huang, Y.L., Chen, T.C., Harish, S.S., Chua, N.H. and Yeh, S.D. (2012). Multiple artificial microRNAs targeting conserved motifs of the replicase gene confer robust transgenic resistance to negative-sense single-stranded RNA plant virus, *Molecular Plant Pathology*, 13(3): 303-317.

Kurihara, Y. and Watanabe, Y. (2004). *Arabidopsis* micro-RNA biogenesis through Dicer-like 1 protein functions, *Proc. Natl. Acad. Sci.*, USA, 101: 12753-12758.

Laubinger, S., Sachsenberg, T., Zeller, G., Busch, W., Lohmann, J.U., Ratsch, G. and Weigel, D. (2008). Dual roles of the nuclear cap-binding complex and SERRATE in pre-mRNA splicing and microRNA processing in *Arabidopsis thaliana, Proc. Natl. Acad. Sci.*, USA, 105: 8795-8800.

Laufs, P., Peaucelle, A., Morin, H. and Traas, J. (2004). MicroRNA regulation of the CUC genes is required for boundary size control in *Arabidopsis meristems, Development*, 131: 4311-4322.

Li, S., Liu, L., Zhuang, X., Yu, Y., Liu, X., Cui, X., Ji, L., Pan, Z., Cao, X. and Mo, B. (2013). MicroRNAs inhibit the translation of target mRNAs on the endoplasmic reticulum in *Arabidopsis, Cell*, 153: 562-574.

Li, X., Wang, X., Zhang, S., Liu, D., Duan, Y. and Dong, W. (2012a). Identification of soybean microRNAs involved in soybean cyst nematode infection by deep sequencing, *PLoS ONE*, 7: e39650.

Liu, H.H., Tian, X., Li, Y.J., Wu, C.A. and Zheng, C.C. (2008). Microarray-based analysis of stress-regulated microRNAs in *Arabidopsis thaliana, RNA*, 14: 836-843.

Llave, C., Kasschau, K.D., Rector, M.A. and Carrington, J.C. (2002). Endogenous and silencing-associated small RNAs in plants, *Plant Cell*, 14: 1605-1619.

Llave, C. (2004). MicroRNAs: More than a role in plant development? *Mol. Plant Pathol.*, 5: 361-366.

Llave, Lucca, P., Poletti, S. and Sautter, C. (2006). Genetic engineering approaches to enrich rice with iron and vitamin A, *Physiol. Plantarum.*, 126: 291-303.

Lu, S., Sun, Y.H., Shi, R., Clark, C., Li, L. and Chiang, V.L. (2005). Novel and mechanical stress-responsive MicroRNAs in *Populus trichocarpa* that are absent from *Arabidopsis*, *Plant Cell*, 17: 186-220.

Lu, X., Guan, Q. and Zhu, J. (2013). Down-regulation of CSD2 by a heat-inducible miR398 is required for thermotolerance in *Arabidopsis, Plant Signal Behav.*, 8: e24952.

Lucca, P., Poletti, S. and Sautter, C. (2006). Genetic engineering approaches to enrich rice with iron and vitamin A, *Physiologia Plantarum*, 126(3): 291-303.

Mallory, A.C., Dugas, D.V., Bartel, D.P. and Bartel, B. (2004). MicroRNA regulation of NAC-domain targets is required for proper formation and separation of adjacent embryonic, vegetative, and floral organs, *Current Biology*, 14: 1035-1046.

Moss, S. (2019). Integrated weed management (IWM): Why are farmers reluctant to adopt non-chemical alternatives to herbicides? *Pest Management Science*, 75(5): 1205-1211.

Mukherjee, K., Campos, H. and Kolaczkowski, B. (2013). Evolution of animal and plant dicers: Early parallel duplications and recurrent adaptation of antiviral RNA binding in plants, *Molecular Biology and Evolution*, 30(3): 627-641.

Naqvi, A.R., Haq, Q. and Mukherjee, S.K. (2010). MicroRNA profiling of tomato leaf curl New Delhi virus (tolcndv) infected tomato leaves indicates that deregulation of mir159/319 and mir172 might be linked with leaf curl disease, *Virol. J.*, 7: 281.

Napoli, C., Lemieux, C. and Jorgensen, R. (1990). Introduction of a Chimeric Chalcone synthase gene into petunia results in reversible co-suppression of homologous genes in trans, *Plant Cell*, 2: 279-289.

Navarro, L., Dunoyer, P., Jay, F., Arnold, B., Dharmasiri, N., Estelle, M., Voinnet, O. and Jones, J.D. (2006). A plant miRNA contributes to antibacterial resistance by repressing auxin signalling, *Science*, 12: 436-439.

MicroRNA-mediated Regulation of Herbicide Resistance 239

Navarro, L., Jay, F., Nomura, K., He, S.Y. and Voinnet, O. (2008). Suppression of the microRNA pathway by bacterial effector proteins, *Science*, 321: 964-967.

Ori, N., Cohen, A.R., Etzioni, A., Brand, A., Yanai, O., Shleizer, S., Menda, N., Amsellem, Z., Efroni I. and Pekker, I. (2007). Regulation of Lanceolate by miR319 is required for compound-leaf development in tomato, *Nat. Genet.*, 39: 787-791.

Palatnik, J.F., Allen, E., Wu, X., Schommer, C., Schwab, R., Carrington, J.C. and Weigel, D. (2003). Control of leaf morphogenesis by microRNAs, *Nature*, 425: 257-263.

Palatnik, J.F., Wollmann, H., Schommer, C., Schwab, R., Boisbouvier, J., Rodriguez, R., Warthmann, N., Allen, E., Dezulian, T. and Huson, D. (2007). Sequence and expression differences underlie functional specialisation of *Arabidopsis* microRNAs miR159 and miR319, *Dev. Cell*, 13: 115-125.

Pandey, S.P. and Baldwin, I.T. (2007). RNA-directed RNA polymerase 1 (RdR1) mediates the resistance of *Nicotiana attenuata* to herbivore attack in Nature, *The Plant Journal*, 50: 40-53.

Pandey, S.P., Shahi, P., Gase, K. and Baldwin, I.T. (2008). Herbivory-induced changes in the small-RNA transcriptome and phytohormone signalling in *Nicotiana attenuate*, *Proc. Natl. Acad. Sci.*, USA, 105: 4559-4564.

Pandita, D. (2019). Plant miRnome: miRNA biogenesis and abiotic stress response. *In:* Hasanuzzaman, M., Hakeem, K., Nahar, K. and Alharby, H. (Eds.). *Plant Abiotic Stress Tolerance*. Springer, Cham, 449-474. Doi https://Doi.org/10.1007/978-3-030-06118-0_18

Pandita, D. and Wani, S.H. (2019). MicroRNA as a tool for mitigating abiotic stress in rice (*Oryza sativa* L.). *In:* Wani, S. (Ed.). *Recent Approaches in Omics for Plant Resilience to Climate Change*. Springer, Chamz., 109-133. Doi https://Doi.org/10.1007/978-3-030-21687-0_6

Pandita, D. (2021). Role of miRNAi technology and miRNAs in abiotic and biotic stress resilience. *In:* Aftab, T. and Roychoudhury, A. (Eds.). *Plant Perspectives to Global Climate Changes*. Academic Press, Elsevier, 303-330. https://Doi.org/10.1016/B978-0-323-85665-2.00015-7

Pandita, D. (2022a). How microRNAs regulate abiotic stress tolerance in wheat? A snapshot. *In:* Roychoudhury, A., Aftab, T. and Acharya, K. (Eds.). *Omics Approach to Manage Abiotic Stress in Cereals*. Springer, Singapore, 447-464. https://Doi.org/10.1007/978-981-19-0140-9_17

Pandita, D. (2022b). MicroRNAs shape the tolerance mechanisms against abiotic stress in maize. *In:* Roychoudhury, A., Aftab, T. and Acharya, K. (Eds.). *Omics Approach to Manage Abiotic Stress in Cereals*, Springer, Singapore, 479-493. https://Doi.org/10.1007/978-981-19-0140-9_19

Pandita, D. (2022c). miRNA- and RNAi-mediated metabolic engineering in plants. *In:* Aftab, T. and Hakeem, K.R. (Eds.). *Metabolic Engineering in Plants*, Springer, Singapore, 171-186. https://Doi.org/10.1007/978-981-16-7262-0_7

Park, W., Li, J., Song, R., Messing, J. and Chen, X. (2002). Carpel factory, a Dicer homolog, and HEN1, a novel protein, act in microRNA metabolism in *Arabidopsis thaliana*, *Current Biology*, 12: 1484-1495.

Pattanaik, S., Patra, B., Singh, S.K. and Yuan, L. (2014). An overview of the gene regulatory network controlling trichome development in the model plant, *Arabidopsis*, *Frontiers in Plant Science*, 5.

Phalan, B., Balmford, A., Green, R.E. and Scharlemann, J.P.W. (2011). Minimising the harm to biodiversity of producing more food globally, *Food Policy* 36 (Supplement 1): S62-S71.

Pradhan, B., Naqvi, A.R., Saraf, S., Mukherjee, S.K. and Dey, N. (2015). Prediction and characterisation of tomato leaf curl New Delhi virus (ToLCNDV) responsive novel microRNAs in *Solanum lycopersicum*, *Virus Res.*, 195: 183-195.

Privalle, L.S., Chen, J., Clapper, G., Hunst, P., Spiegelhalter, F. and Zhong, C.X. (2012). Development of an agricultural biotechnology crop product: Testing from discovery to commercialisation, *J. Agric. Food Chem.*, 60: 10179-10187.

Rhoades, M.W., Reinhart, B.J., Lim, L.P., Burge, C.B., Bartel, B. and Bartel, D.P. (2002). Prediction of plant microRNA targets, *Cell*, 110: 513-520.

Robson, T.M. and Aphalo, P.J. (2012). Species-specific effect of UV-B radiation on the temporal pattern of leaf growth, *Physiol. Plantarum*, 144: 146-160.

Rutz, S. and Scheffold, A. (2004). Towards in vivo application of RNA interference – New toys, old problems, *Arthritis Research Therapy*, 6: 78-85.

Sanan-Mishra, N., Varanasi, S.P. and Mukherjee, S.K. (2013). Micro-regulators of auxin action, *Plant Cell Reports*, 32(6): 733-740.

Schommer, C., Palatnik, J.F., Aggarwal, P., Chételat, A., Cubas, P., Farmer, E.E., Nath, U. and Weigel, D. (2008). Control of jasmonate biosynthesis and senescence by miR319 targets, *PLoS Biology*, 6: e230.

Schwab, R., Palatnik, J.F., Riester, M., Schommer, C., Schmid, M. and Weigel, D. (2005). Specific effects of microRNAs on the plant transcriptome, *Dev. Cell.*, 8: 517-527.

Seo, J., Wu, J., Lii, Y., Li, Y. and Jin, H. (2013). Contribution of small RNA pathway components in plant immunity, *Mol. Plant-Microbe Interact.*, 26: 617-625.

Shi, Y., Guo, J., Zhang, W., Jin, L., Liu, P., Chen, X. and Wang, R. (2015). Cloning of the lycopene β-cyclase gene in *Nicotiana tabacum* and its overexpression confers salt and drought tolerance, *International Journal of Molecular Sciences*, 16(12): 30438-30457.

Singh, R.K., Gase, K., Baldwin, I.T. and Pandey, S.P. (2015). Molecular evolution and diversification of the Argonaute family of proteins in plants, *BMC Plant Biology*, 15(1): 1-16.

Sun, G. (2012). MicroRNAs and their diverse functions in plants, *Plant Mol Biol.*, 80: 17-36.

Sunkar, R., Girke, T., Jain, P.K. and Zhu, J.K. (2005). Cloning and characterisation of microRNAs from rice, *Plant Cell*, 17: 1397-1411.

Sunkar, R., Kapoor, A. and Zhu, J.K. (2006). Post-transcriptional induction of two Cu/Zn superoxide dismutase genes in *Arabidopsis* is mediated by down regulation of miR398 and important for oxidative stress tolerance, *Plant Cell*, 18: 2051-2065.

Tang, G., Reinhart, B.J., Bartel, D.P. and Zamore, P.D. (2003). A biochemical framework for RNA silencing in plants, *Genes and Development*, 17(1): 49-63.

Thiebaut, F., Rojas, C.A., Almeida, K.L., Grativol, C., Domiciano, G.C., Lamb, C.R.C., de Almeida Engler, J., Hemerly, A.S. and Ferreira, P.C. (2012). Regulation of miR319 during cold stress in sugarcane, Plant, *Cell Environ.*, 35: 502-512.

Tilman, D., Balzer, C., Hill, J. and Befort, B.L. (2011). Global food demand and the sustainable intensification of agriculture, *Proc. Natl. Acad. Sci.*, USA, 108: 20260-20264.

Tscharntke, T., Clough, Y., Wanger, T.C., Jackson, L., Motzke, I., Perfecto, I., Vandermeer, J. and Whitbread, A. (2012). Global food security, biodiversity conservation and the future of agricultural intensification, *Biol Conserv.*, 151: 53-59.

Valli, A.A., Santos, B.A., Hnatova, S., Bassett, A.R., Molnar, A., Chung, B.Y. and Baulcombe, D.C. (2016). Most microRNAs in the single-cell alga *Chlamydomonas reinhardtii* are produced by Dicer-like 3-mediated cleavage of introns and untranslated regions of coding RNAs, *Genome Research*, 26(4): 519-529.

van der Krol, A.R., Mur, L.A., Beld, M., Mol, J.N. and Stuitje, A.R. (1990). Flavonoid genes in petunia: Addition of a limited number of gene copies may lead to a suppression of gene expression, *Plant Cell*, 2: 291-299.

Vaucheret, H. (2006). Post-transcriptional small RNA pathways in plants: Mechanisms and regulations, *Genes & Development*, 20(7): 759-771. https://doi.org/10.1101/gad.1410506

Voinnet, O. (2009). Origin, biogenesis, and activity of plant microRNAs, *Cell*, 136(4): 669-687.

Wang, W., Vinocur, B. and Altman, A. (2003). Plant responses to drought, salinity and extreme temperatures towards genetic engineering for stress tolerance, *Planta*, 218: 1-14.

Wang, Y., Sun, F., Cao, H., Peng, H., Ni, Z., Sun, Q. and Yao, Y. (2012b). TamiR159 directed wheat TaGAMYB cleavage and its involvement in another development and heat response, *PLoS ONE*, 7: e48445.

Wang, S., Sun, X., Hoshino, Y., Yu, Y., Jia, B., Sun, Z., Sun, M., Duan, X. and Zhu, Y. (2014b). MicroRNA319 positively regulates cold tolerance by targeting OsPCF6 and OsTCP21 in rice (*Oryza sativa* L.), *PLoS ONE*, 9: e91357.

Wargent, J.J., Moore, J.P., Roland Ennos, A. and Paul, N.D. (2009). Ultraviolet radiation as a limiting factor in leaf expansion and development, *Photochem Photobiol.*, 85: 279-286.

Wu, J., Yang, Z., Wang, Y., Zheng, L., Ye, R., Ji, Y., … and Li, Y. (2015). Viral-inducible Argonaute18 confers broad-spectrum virus resistance in rice by sequestering a host microRNA, *Elife*, 4: e05733.

Xue, Mei, S., Bo, L., Huang, Si Qi, Mehta, Surya Kant and Lai Lang, X. (2007). Coordinated expression of sulphate transporters and its relation with sulphur metabolites in *Brassica napus* exposed to cadmium, *Botanical Studies*, 48: 43-54.

Yamasaki, T., Voshall, A., Kim, E.J., Moriyama, E., Cerutti, H. and Ohama, T. (2013). Complementarity to an miRNA seed region is sufficient to induce moderate repression of a target transcript in the unicellular green alga *Chlamydomonas reinhardtii*, *The Plant Journal*, 76(6): 1045-1056.

Yan, Y., Wang, H., Hamera, S., Chen, X. and Fang, R. (2014). miR444a has multiple functions in the rice nitrate signalling pathway, *The Plant Journal*, 78: 44-55.

Yang, L., Liu, Z., Lu, F., Dong, A. and Huang, H. (2006). Serrate is a novel nuclear regulator in primary microRNA processing in *Arabidopsis*, *The Plant Journal*, 47: 841-850.

Yang, Z. (2013). A potential role of microRNAs in plant response to metal toxicity, *Metallomics*, 5: 1184-1190.

Zamore, P.D., Tuschl, T., Sharp, P.A. and Bartel, D.P. (2000). RNAi: Double-stranded RNA directs the ATP-dependent cleavage of mRNA at 21 to 23 nucleotide intervals, *Cell*, 101: 25-33.

Zhai, L., Sun, W., Zhang, K., Jia, H., Liu, L., Liu, Z., … and Zhang, Z. (2014). Identification and characterisation of Argonaute gene family and meiosis-enriched Argonaute during sporogenesis in maize, *Journal of Integrative Plant Biology*, 56(11): 1042-1052.

Zhang, D. and Trudeau, V.L. (2008). The XS domain of a plant specific SGS3 protein adopts a unique RNA recognition motif (RRM) fold, *Cell Cycle*, 7: 2268-2270.

Zhang, L.W., Song, J.B., Shu, X.X., Zhang, Y. and Yang, Z.M. (2013a). miR395 is involved in detoxification of cadmium in *Brassica napus*, *J. Hazard Mater.*, 250: 204-211.

Zhang, H., Xia, R., Meyers, B.C. and Walbot, V. (2015). Evolution, functions, and mysteries of plant Argonaute proteins, *Current Opinion in Plant Biology*, 27: 84-90.

Zhao, M., Ding, H., Zhu, J., Zhang, F. and Li, W. (2011). Involvement of miR169 in the nitrogen starvation responses in *Arabidopsis*, *New Phytol.*, 190: 906-915.

Zhou, L, Liu, Y., Liu, Z., Kong, D., Duan, M. and Luo, L. (2010). Genome-wide identification and analysis of drought-responsive microRNAs in *Oryza sativa*, *J. Exp. Bot.*, 61: 4157-4168.

Zhou, M. and Luo, H. (2013). MicroRNA-mediated gene regulation: Potential applications for plant genetic engineering, *Plant Molecular Biology*, 83: 59-75.

Zhou, M. and Luo, H. (2014). Role of microRNA319 in creeping bentgrass salinity and drought stress response, *Plant Signalling and Behaviour*, 9: 1375-1391.

Zhu, H., Hu, F., Wang, R., Zhou, X., Sze, S.H., Liou, L.W., … and Zhang, X. (2011). *Arabidopsis* Argonaute10 specifically sequesters miR166/165 to regulate shoot apical meristem development, *Cell*, 145(2): 242-256.

CHAPTER

14

MicroRNA-mediated Regulation of Plant Viral Disease Development

Saurabh Pandey[1]*, Suresh H. Antre[2], Saumya Kumari[2] and Ashutosh Singh[3]

[1] Department of Agriculture, Guru Nanak Dev University, Amritsar,
Punjab - 143005, India
[2] University of Agricultural Sciences, GKVK Campus, Bangalore - 560065, India
[3] Centre for Advanced Studies on Climate Change, RPCAU, Pusa,
Samastipur - 848125, India

1. Introduction

Crop diseases and pests are significant constraints to agricultural production and quality (Savary *et al.*, 2019) at the household and national and global levels. These constraints results in major economic losses and threaten the food security of global population. Yield loss projected at a worldwide level per hotspot for different crops have been reported and includes rice [30.0% (24.6-40.9%)], wheat [21.5% (10.1-28.1%)], maize [22.5% (19.5-41.1%)], soybean [21.4% (11.0-32.4%)] and potato [17.2% (8.1-21.0%)]. This suggests highest losses associated with the developing world, with rapid population growth and food-deficit regions, with a higher frequency of emergence or re-emergence for diseases and pests (Savary *et al.*, 2019). Therefore, plant protection and crop disease management play an indispensable role in accomplishing the food quality and quantity demand for human population (Strange and Scott, 2005).

Among plant pathogens, viruses are obligate parasites with smaller genomes dependent upon the host cell for replication and multiplication (Pallas and Garcia, 2011). The genome of any virus consists of DNA or RNA packaged inside the protein coat and this unit is known as the 'virion' (Ding and Voinnet, 2007). Single-stranded RNA viruses with smaller protein repertoire are more common among plant viruses. These virus particles disassemble in the process of infection inside the plant cell. The nucleic acid will be transcribed/translated to produce more virus particles, encapsulating and moving between cell to cell through long-distance or cell-to-cell transport (Wang, A., 2015). Generally, viruses encode replication-related movement and coat protein for their multiplication (Mandahar, 2006).

---

*Corresponding author: pandey.saurabh784@gmail.com

The different plant disease control methods were first classified by Whetzel (1929) as exclusion, eradication, protection and immunisation. Further advances in plant pathology led to two newer methods; avoidance and therapy (Spadaro and Gullino, 2005). Briefly exclusion deals with preventing the introduction of inoculum, eradication deals with elimination, destruction, or inactivation of the inoculum. Protection/immunisation deals with preventing infection by employing a toxic molecule/other barrier against invading pathogen. In resistance, we exploit resistant or tolerant cultivars against pathogen infection. Similarly, avoidance employs selecting a time of the year or a site with no inoculum or an unfavourable environment for infection. At last, therapy deals with curing plants that are already infected. Recently, regulation of plant virus multiplication or essential factors and susceptibility genes with the help of microRNAs (miRNA) is an attractive and successful approach utilised for plant virus protection (Pandey *et al.*, 2020a; b).

Multicellular organism gene regulatory pathways involve 20-24 nucleotide long non-coding miRNAs regulating the gene expression, both at the transcriptional and post-transcriptional levels (Liu and Chen, 2009; Liu *et al.*, 2017; Pérez-Quintero *et al.*, 2010). These types of regulatory RNAs have already been reported in plant development and different plant virus interactions (Liu *et al.*, 2017; Pérez-Quintero *et al.*, 2010). Apart from miRNAs, other small RNAs, such as small interfering RNA (siRNA), viral siRNAs (vsiRNAs), natural-antisense RNAs (nat-siRNAs), virus-activated siRNAs (vasiRNAs), trans-acting siRNAs (ta-siRNAs) and repeat-associated siRNAs (ra-siRNAs) also act to regulate gene expression. These small RNA categories are also reported to be important in plant stress resistance or tolerance, for example, plant virus infection (Singh *et al.*, 2015).

Various miRNA (MIR) genes were encoded and categorised as intergenic or intronic in the plant genome based on their location (Wang *et al.*, 2019). These repertoires of genome-encoded small RNAs control the biological processes in the plant, whereas loss of function of these regulatory miRNAs results in disease development or resistance (Wang and Cui, 2012). Therefore, these molecules are crucial for ascertaining the plant phenotypes under stress conditions (Yang *et al.*, 2011). These molecules operate by the RNA silencing mechanism, which involves double-stranded RNA (dsRNA) induction, processing into sRNA, 3'-O-methylation of these sRNAs and incorporation into RNA-induced silencing complex (RISC), which targets the complementary strand (Voinnet, 2009). Therefore, this mRNA transcript manipulation through miRNA offers a genetic engineering strategy to target plant disease resistance traits (Djami-Tchatchou *et al.*, 2017). In this direction, several attempts have been made to understand resistance mechanisms at the molecular level to improve yield by producing disease-resistance crops. This chapter discusses the miRNA-based approach regarding viral infections in different crops.

2. MicroRNAs in Plants

2.1 Origins and Biogenesis

MicroRNAs are short, endogenous, about 21-24 nucleotides-long single-stranded RNAs derived from hairpin transcripts. miRNA is classified as intronic or intergenic, based on their location in the genome. Intronic miRNA is processed from the intron of their host region. At the same time, intergenic is processed between the part of two genes by DNA-dependent RNA polymerase II enzyme (Wang *et al.*, 2019).

miRNA biogenesis is an essential multistep process in the eukaryotic gene regulation. It initiates from the RNA polymerase II/III mediated post- or co-transcriptional processing (Ha *et al.*, 2014). Around 50% of reported miRNAs are intragenic, which are mainly processed from introns. However, a few are derived from protein-coding genes, whereas the rest are intergenic and independently transcribed through their promoters (Kim *et al.*, 2007; De Rie *et al.*, 2017). miRNAs are also transcribed by clusters that are one long transcript with similar seed regions and are considered single-family members (Tanzer *et al.*, 2004). This section will concisely deliberate the non-canonical and canonical pathways of miRNA biogenesis.

2.1.1 miRNA Biogenesis: Canonical Pathway

The canonical pathway of miRNA processing is prominent in eukaryotic systems. This pathway starts with miRNA genes being transcribed into pri-miRNAs. Further, it is administered by the microprocessor complex into pre-miRNAs, involving an RNA binding protein DGCR8 (DiGeorge Syndrome Critical Region 8) and a ribonuclease III enzyme, Drosha. Afterwards, DGCR8 identified N6-methyl adenylated specific motifs, like GGAC and others inside the pri-miRNA, and led to cleaving at the pri-miRNA hairpin structure bases by the Drosha enzyme. In the end, 2 nucleotide 3′ overhang on pre-miRNA is formed. This pre-miRNA is now exported through the exportin 5 (XPO5)/RanGTP complex into the cytoplasm. Subsequently, it is cleaved by an RNase III endonuclease, known as Dicer. The Dicer enzyme involves the cleavage and elimination of the terminal loop, entailing to the mature miRNA duplex formation. This mature form of miRNA is defined through the directionality of the miRNA strand. The 5p and 3p strands arise from the pre-miRNA hairpin's 5′ and 3′ ends. These two strands are derivatives of the mature miRNA duplex that will then be processed through the AGO (Argonaute) proteins in an energy-dependent manner. The cell type and cellular environment affect the proportion of AGO-loading on either strand, ranging from almost equal amounts to one. The 5′ ends thermodynamic stability of the miRNA duplex or presence of a 5′ U at nucleotide position 1 determines the selection of either strand. Usually, the strand with lesser 5′ uracil or 5′ stability is considered the guide strand and loaded into AGO. The remaining strand is known as passenger strand. Based upon complementarity, this strand will further unwind from the guide strand by diverse mechanisms. For example, entirely complementary passenger strands of miRNA are cleaved by AGO2 and further processed by the cellular machinery. Else, miRNA duplexes with partial complementarity or central mismatches are not processed by AGO2 and these miRNAs passively unwound and degrade (O'Brien *et al.*, 2018).

2.1.2 miRNA Biogenesis: Non-canonical Pathways

These pathways use different groups of proteins with combinations in the canonical way, such as Dicer, Drosha, AGO2 and exportin 5. Based upon the processing mode, they are further grouped as Drosha-independent and Dicer-independent pathways of miRNA biogenesis. Generally, Pre-miRNAs are synthesised through Drosha independent pathway that resembles Dicer substrates. For example, mirtron production from splicing of the introns of mRNA and 7-methylguanosine (m7G)-capped pre-miRNA (Babiarz *et al.*, 2008; Ruby *et al.*, 2007). With the help of exportin 1, the above-mentioned nascent RNAs are exported directly to the cytoplasm without cleavage from the Drosha enzyme. However, strong 3p strand bias was observed in this pathway, probably due to m7G cap-mediated prevention of 5p strand loading on to AGO protein. At the same time, endogenous short hairpin RNA (shRNA) transcripts were processed by Drosha in a Dicer-independent processing of miRNAs. Because of the lack of Dicer-substrates in the non-canonical pathway AGO2 is essential for these pre-miRNAs maturation within the cytoplasm. This processing, in turn, encourages the entire pre-miRNA loading on to AGO2. Also it promotes the AGO2-dependent slicing of the 3p strand. Ultimately, trimming of the 5p strand in the $3'\text{-}5'$ direction finishes their maturation in the pathway.

3. MiRNA and Plant Disease Regulation

Since the discovery of miRNAs, there have been plenty of efforts at figuring out their role in various plant disease regulations. Disturbance in miRNA expression is often crucial for disease progression. Therefore, expression profiling of miRNA at genome level could provide useful diagnostic and prognostic insights for disease management. During infection, the host plant and virus interactions lead to changes in plant physiology and appearance, causing disease symptoms. In general, to manage viral diseases, plants are prevented from contacting disease inoculum or favourable conditions to attack by pathogens. In the arms race between plant and viruses, plants have deployed diverse mechanisms to resist viruses, comprising small RNAs-mediated defence gene regulation (Tables 1 and 2). Interestingly, most of these miRNAs target

Table 1: Different miRNA used for crop improvement against the fungal pathogens in different crops

S. No.	miRNA	Crops	Resistance against	References
1.	miR482	Potato	*Verticillium dahlia*	Yang *et al.*, 2015
2.	miR482	Cotton	*Verticillium dahlia*	Zhu *et al.*, 2013
3.	miR482	Tomato	*Fusarium oxysporum*	Ouyang *et al.*, 2014
4.	miR472a	Poplar	*Colletotrichum gloeosporioides* and *Cytospora chrysosperma*	Su *et al.*, 2018
5.	miR160	Rice	*Magnaporthe oryzae*	Li *et al.*, 2014
6.	miR396	Rice	*Plectosporium cucumerina, Fusarium oxysporum, Colletototrichum higginianum*	Soto-Suarez *et al.*, 2017
7.	miR773	*Arabidopsis*	*Fusarium oxysporum, Colletototrichum higginianum, Plectosporium cucumerina*	Salvador-Guirao *et al.*, 2018
8.	miR844	*Arabidopsis*	*Pseudomonas syringae* pv. tomato *DC3000/ Botrytis cinerea*	Lee *et al.*, 2015
9.	miR482e-3p	Tomato	*Fusarium oxysporum f.* sp. *Lycopersici*	Gao *et al.*, 2021
10.	miR858	*Arabidopsis*	*Plectosphaerella cucumerina* or *Fusarium oxysporum* and *Colletotrichum higginsianum*	Camargo-Ramírez *et al.*, 2018

Table 2: Different miRNA used for crop improvement against the viral pathogens

S. No.	Crops	miRNA Backbone	Targeted Gene	Resistance against	References
1.	Rice	miR528	Middle segment 3′ end	RSV	Sun *et al.*, 2016
2.	Maize	miR159a	Conserved region	RBSDV	Xuan *et al.*, 2015
3.	*Arabidopsis*	miR159	P69 HC-Pro (coat protein)	TYMV TuMV	Niu *et al.*, 2006
4.	Tobacco	miR159a	HC-Pro	PVY	Ai *et al.*, 2011
5.	Tomato	miR159a	2a and 2b viral genes 3′-UTR	CMV	Zhang *et al.*, 2011
6.	*Arabidopsis*	miR162	DCL1	CMV	Zhang *et al.*, 2006
7.	*Arabidopsis*	miR168	AGO1	TCV	Várallyay *et al.*, 2010
8.	*Brassica rapa*	miR1885	TIR-NBS-LRR gene	TuMV	He *et al.*, 2008
9.	*N. tabacum*	miR6020/ miR6019	Receptor N	TMV	Li *et al.*, 2012
10.	*Oryza sativa*	miR444	OsMADS23/27a/57	RSV	Wang *et al.*, 2016
11.	*Oryza sativa*	miR528	OsAO	RSV	Yao *et al.*, 2019
12.	*Gossypium hirsutum*	miR168/ miR395ad	C1, C3, C4, V1, and V2	CLCuD	Akhter and Khan, 2018

TYMV - *Turnip yellow mosaic virus*; TuMV - *Turnip mosaic virus*; RSV - *Rice stripe virus*; RBSDV - *Rice black streaked dwarf virus*; CMV - *Cucumber mosaic virus*; TCV - *Turnip crinkle virus*; TMV - *Tobacco mosaic virus*; CLCuD - *Cotton leaf curl Burewala virus*

NBS-LRR resistance genes (Park *et al.*, 2015), having nucleotide-binding site (NBS) and a leucine-rich repeat (LRR) for activation of plant defence upon interaction with viral effectors. These resistance genes are key components of the innate defence system by activation of effector-triggered immunity (ETI). It is reported that miRNAs are the key modulators of the resistance gene family in many fungal diseases. For example, during *Verticillium dahlia* infection, miR482 modulates potato resistance by suppressing NBS-LRR genes (Yang *et al.*, 2015). miR482 also confers resistance to *V. dahliae* in cotton, by increasing NBS-LRR transcripts (Zhu *et al.*, 2013). Interestingly, in *Fusarium oxysporum* infection also, similar responses were also observed in tomato plants (Ouyang *et al.*, 2014). Moreover, poplar miR472a is also involved in targeting NBS-LRR transcripts for defence against fungal pathogens (Su *et al.*, 2018). These reports suggest possible role of miRNA-mediated resistance gene targeting for viral defence in plants as discussed in this section.

3.1 miRNA as Key Player in Plant Antiviral Immunity

The ability of plants to use a variety of defence mechanisms against attacking pathogens or harsh environmental conditions is critical to their survival. Plants use

MicroRNA-mediated Regulation of Plant Viral Disease Development

siRNA-mediated gene silencing as their primary protection strategy against viral infections. Small RNA-mediated plant defence is effective because of the spread of siRNA signals and its transitive nature (Eamens *et al.*, 2008; Lu *et al.*, 2008). Nevertheless, infected plant cells cannot spread a warning signal until the virus has attacked. As a result, only siRNA-mediated gene silencing mechanism might not be enough to protect against attacking viral pathogens, necessitating the deployment of a proactive strategy. miRNAs are endogenous RNAs; therefore, they are present in cell before the viral attack. In contrast, pathogen-induced miRNAs have been also reported, demonstrating their potential to counteract the attack virus pathogen (Lu *et al.*, 2008). Interestingly, plant miRNAs can target multiple variants of pathogen because of their evolution for optimising cleavage efficiency to their targets (Jones-Rhoades, 2012; Voinnet, 2009). This partial complementarity to target allows three or more mismatches and greatly expands the range of targets and releases processed target transcripts from the RISC complex in much more accessible way. miRNAs in antiviral defensive responses in plants have been proposed in two ways: direct viral RNAs targeting and antiviral response eliciting the biogenesis of siRNA in indirect manner accountable for the plant viral defence.

Endogenous miRNAs in mammals have been demonstrated to act as a critical factor in inhibiting plant viruses (Gottwein and Cullen, 2008). Interestingly, plant miRNA target sequences against viruses could be functional in different hosts, for example plum pox virus (PPV) chimeras possess target sequences of plant miRNA that functions in *Arabidopsis* and other plants (Simón-Mateo and Garcia, 2006). Furthermore, numerous investigations have revealed miRNA-mediated post-transcriptional gene regulation imparts plant defence reactions to viruses (Pacheco *et al.*, 2012; Li *et al.*, 2012; Amin *et al.*, 2011). Prevention of fungal virulence genes expression through cotton miRNAs against *Verticillium dahlia* has been shown (Zhang *et al.*, 2016). Researchers discovered that miR166 and miR159 target isotrichodermin C-15 hydroxylase (HiC-15) and Ca^{2+}-dependent cysteine protease (Clp1), required for fungus pathogenicity. These studies only give circumstantial indication for the plant miRNA function in pathogen defence, which could be utilised for directly targeting viral RNAs. Further research is needed to deeply understand their mechanism of antiviral defence in plants.

3.2 Application of Plant miRNAs against Virus Infections

RNA-mediated silencing via a homology-dependent model has been utilised to provide antiviral resistance by expressing viral sequences in the double-stranded forms and sense or antisense orientation in plants (Smith *et al.*, 1994; Waterhouse *et al.*, 1998, Helliwell and Waterhouse, 2005). In these cases, co-suppression of both the attacking viral RNA and virus-derived transgenic RNA would follow through classical RNA silencing, which leads to plant resistance and recovery against the viruses (Lindbo *et al.*, 1993; Guo and Garcia, 1997). Many versatile techniques, like RNA interference (RNAi), transcriptional gene silencing (TGS), antisense suppression, host-induced gene silencing (HIGS), virus-induced gene silencing (VIGS) and spray-induced gene silencing (SIGS) are now being utilised for plant antiviral engineering. Moreover, microRNA (miRNA) precursors-based short-hairpin RNAs expressing the artificial miRNAs (amiRNAs) could also provide virus resistance in plants by following the silencing mechanism. Specific miRNAs have been effectively designed for targeting gene transcripts in different plants, for example *Arabidopsis* (Schwab *et al.*, 2006), tomato, tobacco (Alvarez *et al.*, 2006), rice (Warthmann *et al.*, 2008), alga (Molnar *et al.*, 2009) as well as moss (Khraiwesh *et al.*, 2008).

It is evident from studies that miRNAs have essential roles in antiviral immunity. The prominent targeting genes for this type of defence include R-genes, RNA silencing components and defence hormone signalling pathway genes (Mlotshwa *et al.*, 2008; Naqvi *et al.*, 2010; Zhang *et al.*, 2016). For example, Wang *et al.* (2016) found up-regulation of miR444 transcript level by *rice stripe virus* (RSV) infection. This up-regulation diminished the MADS-box genes suppression on RDR1 transcription, thereby inducing the RDR1-dependent antiviral immunity. Similarly, in rice, AGO18-mediated miR168 sequestration leads to antiviral RNAi. This ultimately suppresses the main antiviral effector, AGO1 expression. Also, AGO18 binding with miR528 results in de-repression of its target L-ascorbate oxidase (AO) gene, providing enhanced resistance against the viral pathogen (Wu *et al.*, 2017). Besides, miR403a also provides tolerance against *tomato mosaic virus* (ToMV) in *Nicotiana benthamiana* by modulating AGO2 expression, which is a critical effector acting in plant antiviral defence that can be targeted in other crops (Diao *et al.*, 2019).

Taking these cues from natural RNA silencing mechanisms, antiviral strategies in plants mimicking the processes were engineered. They benefit from the miRNA precursor backbone and introduce several base pairs mutations achieving differential target identification and processing (Ossowski *et al.*, 2008). Initial evidence was documented in *Arabidopsis* where two plant viruses, TYMV (*turnip yellow mosaic virus*) and TuMV (*turnip mosaic virus*), were targeted by an artificial pre-miRNAs159 (pre-amiRNAs159) comprising complementary sequences against them. The targeting sequences encode the silencing suppressors P69 of TYMV and HC-Pro of TuMV. The resistance was particular to the amiRNA/virus pair and, thus, cannot be described by the generalised host defence pathway activation. This amiRNA-mediated strategy generates resistance which was identified at the cell level and was heritable (Niu *et al.*, 2006). Simon *et al.* (2006) reported the behaviour of PPV chimeras with different miRNA target sequences (miR159, miR171 and miR167), which are functional in *Arabidopsis*, and were not identical in other host plants (*N. clevelandii* and *N. benthamiana*) tested. However, usually not all miRNAs equally work against a given mRNA target, because it is affected by different factors, such as virus infectivity, miRNA target sequence insertion site in the viral genome, etc. This indicates certain sites are further available to the miRNA-RISC-mediated cleavage than others.

Generally, to avoid the resistance breakdown due to mutation, most conserved viral genomic regions were chosen as targets. The relationship among the transcript level of the miRNA and virus resistance determines the selection of the highly-resistant promising lines for virus resistance. Similarly, Ai *et al.* (2011) designed amiRNAs targeting silencing suppressor the TGBp1/p25 (p25) of *potato virus X* (PVX) and HC-pro of *potato virus Y* (PVY) by utilising different miRNA precursors as backbones (miR159a, miR167b and miR171a). This amiRNAs proficiently conferred precise resistance to counter PVX or PVY infection in transgenic *Nicotiana tabacum*. Similarly, Kung *et al.* (2012) against *watermelon silver mottle virus* (WSMoV) targeted L (replicase) gene conserved motifs by artificial miRNAs (amiRNAs) with *Arabidopsis* pre-miRNA159a as the backbone in *Nicotiana benthamiana*. Further, Mitter *et al.* (2016) also developed a miRNA strategy against tomato spotted wilt virus (TSWV), directing the silencing suppressor (NSs) and nucleoprotein (N) genes degradation by using miR159a precursor backbone (athmiR159a).

MicroRNA-mediated Regulation of Plant Viral Disease Development   **249**

Plant viruses also advanced approaches to counteract plant resistance mechanism by encoding viral suppressors of RNA silencing (VSRs). These assist viruses in numerous processes, like interfering with host-RNA silencing machinery via multiple pathways and help in viral replication, movement or encapsidation (Wang *et al.*, 2012; Burgyan and Havelda, 2011). CMV 2b is one of the first studied VSRs among all the viral suppressors of RNA silencing (Brigneti *et al.*, 2015). It was reported to interfere in the transfer of the systemic silencing signal, host's salicylic acid-mediated antiviral immunity (Ji and Ding, 2001) as well as interrupt DNA methylation-induced silencing (Guo and Ding, 2002). Qu *et al.* (2007) tried to answer the question whether the artificial sRNAs expression for directing the suppressor 2b sequences can neutralise the suppressor function of 2b. They found that indeed transgenic expression of the 2b-specific miRNA in plants acts as an effective method of protection from infection by CMV (*cucumber mosaic virus*). Further, two amiRNAs were designed from *cucumber mosaic virus* (CMV) highly conserved 3′ untranslated region (UTR) for targeting the common coding regions of 2a and 2b (Zhang *et al.*, 2010). After introducing this construct into the susceptible tomato, authors found transgenic plants expressing amiRNA exhibited effective repression of CMV RNA accumulation than transgenic plants. This strategy was also utilised to counter *wheat streak mosaic virus* (WSMV), by combining five amiRNAs sequences within one polycistronic amiRNA precursor. Here natural miRNA was replaced by polycistronic rice miR395, generating amiRNA precursor, FanGuard (FGmiR395), which was used for wheat transformation (Fahim *et al.*, 2012). Three different response types were observed in T1 plants: completely immune, primarily susceptible, followed by plant recovery and primarily resistant with breaking down of resistance over time. Jelly *et al.* (2012) established two amiRNA precursors (pre-amiRNAs) target the coat protein (CP) gene of *grapevine fan leaf virus* (GFLV) by using pre-miR319a of *Arabidopsis thaliana* as a backbone and characterised them in grapevine somatic embryos.

In cotton, based upon miRNA169a two amiRNA constructs were developed comprising 21 nucleotides of the *cotton leaf curl Burewala virus* (CLCuBuV) V2 gene sequence. These sequences were further used for plant transformation into *Nicotiana benthamiana*. Among them, the first amiRNA construct was similar to miR169a sequence except for substituted 21 nucleotides. In the second construct, miRNA169a backbone was changed to reinstate hydrogen bonding of the mature miRNA duplex. First miRNA construct displayed an overall resistance to CLCuBuV compared to second one (Ali *et al.*, 2013). For targeting PVY, multiple amiRNA target sites were designated based on the sequence information of virus nuclear inclusion a and b protein (NIa, NIb), cylindrical inclusion protein (CI) and coat protein (CP) genes. It was found that the amiRNA which targeted the *NIb* and *CP* genes showed a greater silencing efficiency as compared to the other two gene sequence targeting (Song *et al.*, 2014). In rice, dimeric amiRNA precursor expression vectors were used, based on the rice osa-MIR528 precursor structure for targeting *CP* genes of *rice black streaked dwarf virus* (RBSDV) and RSV. Interestingly, transformed transgenic plants displayed great resistance against both RBSDV and RSV infection in rice (Sun *et al.*, 2016). Transgenic plants affected by *Ugandan cassava brown streak virus* (UCBSV) and *cassava brown streak virus* (CBSV) isolates, showed resistance levels generated amiRNA directing conserved sequences within the genomes of CBSV and UCBSV. The resistance ranged between ~20-60% against CBSV and UCBSV which was correlated with transgenically-derived miRNAs transcript levels (Wagaba *et al.*,

2016). Liang *et al.* (2019) developed three amiRNAs targeting *cucumber green mottle mosaic virus* (CGMMV) RNA targeting viral coat, movement and replication genes (amiR1-CP, amiR4-MP and amiR6-Rep) and found reduction of CGMMV replication and disease resistance in transgenic plants.

RNA silencing, due to plant miRNAs and viral genes in plant genomes, is assembled as a database named PAmiRDB (Satish *et al.*, 2019). This database consists of information about identified plant miRNAs, with their putative virus genomes targets. The latest version of database contains information about more than 2,600 different plant miRNAs and their precise interactions with predicted targets in almost 500 viral species (largely geminiviruses and potyviruses).

4. Signalling in Antiviral Defences related to miRNAS

During infection, host-virus interactions result in the physiological disorders accountable for plant disease symptoms, significantly impacting agricultural productivity worldwide. To prevent this viral reproduction and spread, plants use a variety of defence mechanisms. These include viral gene silencing, hormonal activated defence, immune receptor signalling, autophagy-mediated degradation, ubiquitin-mediated protein breakdown and metabolic regulation (Incarbone and Dunoyer, 2013). The primary tool for plant antiviral immunity in virus–plant interactions is RNA silencing, frequently reduced by counter-evolving viral suppressors, thus boosting pathogenicity in vulnerable hosts. Another strategy to counter virus infection is the production of resistance proteins containing nucleotide-binding leucine-rich repeat (NB-LRR) domains. For example, the tobacco *N* gene was demonstrated to provide resistance against the *tobacco mosaic virus (*TMV) and was the first R gene discovered for virus resistance (Whitham *et al.*, 1994). This discovery led to the identification of many *R* genes in various plants providing antiviral resistance (Mandadi and Scholtor, 2013; Gururani *et al.*, 2012). These domains help detect viral effectors and, in turn, trigger effector-triggered immunity (ETI) for plant defence (Mandadi and Scholthof, 2013). Apart from ETI, innate pathogen-associated molecular pattern (PAMP)-triggered immunity (PTI) has also been discovered in helping to combat the plant viruses (Kørner *et al.*, 2013). Recently, it has been demonstrated that a plant has deployed a transmembrane immunological receptor to fight DNA viruses. This receptor activates host translation suppression and physically works similar to the co-receptor-like kinases operated in PTI, revealing a new antiviral defence mechanism in plants (Zorzatto *et al.*, 2015). Apart from dominant resistance, plant chaperones, such as RNA helicases and transcription factors, act as virus defence molecules that can be targeted with the miRNA for durable resistance (Pandey *et al.*, 2019; Sharma *et al.*, 2017; Pandey *et al.*, 2018). These factors activate plant defence machinery to provide the plant virus resistance.

5. miRNA-based Molecular Marker in Disease Resistance

The non-coding RNAs are reported to be crucial in plant development and adaptation. This non-coding region of the plant genome caught the attention of plant scientists for the improvement of molecular markers. Further, public domain accessibility of the whole genome sequence information of model and non-model plants has accelerated the effort for the miRNA-families genome-wide studies in various plant species. The most critical markers developed from non-coding RNAs were SSRs identified in many crops, such as rice, wheat and brassica. These SSRs markers were identified from the conserved region's miRNA genes, and primers were designed. These markers were further used to characterise the crop plants for different traits. These miRNA-SSRs have the potential to be utilised in marker-aided selection (MAS) programmes to breed different crops. The highly conserved nature of miRNA-SSRs makes them transferable across other species using advanced biotechnological interventions (Tyagi *et al.*, 2021).

6. Conclusion and Future Perspective

Plant viruses are a significant and evolving constraint in global food production. However, these tiny obligate parasites modulate the host enzymes for their replication and regulatory pathways with limited coding capacity. The plant deploys a miRNA-based silencing mechanism to regulate plant development and plant–virus interaction. Most of these miRNAs are promising antiviral molecules which regulate plant virus resistance. Moreover, viruses evolve suppressors to block the RNA silencing machinery, which inhibits the assembly of the RISC complex. The plant then overcomes these pathways by employing different Argonaute proteins. miRNA-based protection is widely used for protection against various plant pathogens, specifically plant viruses because of their specificity, stability and promoter compatibility. Future research needs to focus on in-depth analysis of miRNA target, its regulation and characterisation that will serve as the template for large-scale amiRNA development for deploying resistance against ever-evolving plant viruses. With CRISPR-Cas9 technology, miRNA-mediated resistance applications will provide broader spectrum resistance.

References

Ai, T., Zhang, L., Gao, Z., Zhu, C.X. and Guo, X. (2011). Highly efficient virus resistance mediated by artificial microRNAs that target the suppressor of PVX and PVY in plants, *Plant Biol.* (*Stuttg*), 13: 304-316. Doi: 10.1111/j.1438-8677. 2010.00374.x

Akhter, Y. and Khan, J.A. (2018). Genome wide identification of cotton (*Gossypium hirsutum*)-encoded microRNA targets against cotton leaf curl Burewala virus, *Gene*, 638: 60-65.

Ali, I., Amin, I., Briddon, R.W. and Mansoor, S. (2013). Artificial microRNA-mediated resistance against the monopartite begomovirus *cotton leaf curl Burewala virus*, *Virol. J.*, 10: 231. https://Doi.org/10.1186/1743-422X-10-231

Alvarez, J.P., Pekker, I., Goldshmidt, A., Blum, E., Amsellem, Z. and Eshed, Y. (2006). Endogenous and synthetic microRNAs stimulate simultaneous, efficient and localised regulation of multiple targets in diverse species, *Plant Cell*, 18: 1134-1151.

Amin, I., Patil, B.L., Briddon, R.W., Mansoor, S. and Fauquet, C.M. (2011). A common set of developmental miRNAs are up-regulated in *Nicotiana benthamiana* by diverse begomoviruses, *Virol. J.*, 8(1): 1-9.

Babiarz, J.E., Ruby, J.G., Wang, Y., Bartel, D.P. and Blelloch, R. (2008). Mouse ES cells express endogenous shRNAs, siRNAs, and other microprocessor-independent, Dicer-dependent small RNAs, *Genes Dev.*, 22: 2773-2785. Doi: 10.1101/gad.1705308

Brigneti, G., Voinnet, O., Li, W.X., Ji, L.H., Ding, S.W. and Baulcombe, D.C. (2015). Retraction: Viral pathogenicity determinants are suppressors of transgene silencing in *Nicotiana benthamiana*, *EMBO J.*, 34(20): 2595. https://doi.org/10.15252/embj.201570030

Burgyan, J. and Havelda, Z. (2011). Viral suppressors of RNA silencing, *Trends Plant Sci.*, 16: 265-272. Doi: 10.1016/j.tplants.2011.02.010

Camargo-Ramírez, R., Val-Torregrosa, B. and San Segundo, B. (2018). MiR858-mediated regulation of flavonoid-specific MYB transcription factor genes controls resistance to pathogen infection in *Arabidopsis*, *Plant Cell Physiol.*, 59(1): 190-204.

De Rie, D., Abugessaisa, I., Alam, T., Arner, E., Arner, P., Ashoor, H., Åström, G., Babina, M., Bertin, N., Burroughs, A.M., Carlisle, A.J., Daub, C.O., Detmar, M., Deviatiiarov, R., Fort, A., Gebhard, C., Goldowitz, D., Guhl, S., Ha, T.J., Harshbarger, J., … de Hoon, M.J.L. (2017). An integrated expression atlas of miRNAs and their promoters in human and mouse, *Nat. Biotechnol.*, 35: 872-878. Doi: 10.1038/nbt.3947

Diao, P.F., Zhang, Q.M., Sun, H.Y., Ma, W.J., Cao, A.P., Yu, R.N., Wang, J.J., Niu, Y.D. and Wuriyanghan, H. (2019). miR403a and SA are involved in NbAGO2 mediated antiviral defences against TMV infection in *Nicotiana benthamiana*, *Genes*, 10.

Ding, S. and Voinnet, O. (2007). Antiviral immunity directed by small RNAs, *Cell*, 130: 413-426.

Djami-Tchatchou, A.T., Sanan-Mishra, N., Ntushelo, K. and Dubery, I.A. (2017). Functional roles of microRNAs in agronomically important plants – Potential as targets for crop improvement and protection, *Front Plant Sci.*, 8: 378.

Eamens, A., Wang, M.B., Smith, N.A. and Waterhouse, P.M. (2008). RNA silencing in plants: Yesterday, today and tomorrow, *Plant Physiol.*, 147(2): 456-468.

Fahim, M., Millar, A., Wood, C. and Larkin, P. (2012). Resistance to *wheat streak mosaic virus* generated by expression of an artificial polycistronic microRNA in wheat, *Plant Biotechnol J.*, 10(2): 150-163.

Gao, Y., Li, S.J., Zhang, S.W., Feng, T., Zhang, Z.Y., Luo, S.J., … and Ouyang, S.Q. (2021). SlymiR482e-3p mediates tomato wilt disease by modulating ethylene response pathway, *Plant Biotechnol J.*, 19(1): 17.

Gottwein, E. and Cullen, B.R. (2008). Viral and cellular microRNAs as determinants of viral pathogenesis and immunity, *Cell Host & Microbe*, 3(6): 375-387.

Guo, H.S. and Garcia, J.A. (1997). Delayed resistance to *plum pox potyvirus* mediated by a mutated RNA replicase gene: Involvement of a gene-silencing mechanism, *Mol. Plant-Microbe Interact.*, 10: 160-170.

Guo, H.S. and Ding, S.W. (2002). A viral protein inhibits the long range signalling activity of the gene silencing signal, *EMBO J.*, 21: 398-407.

Gururani, M.A., Venkatesh, J., Upadhyaya, C.P., Nookaraju, A., Pandey, S.K. and Park, S.W. (2012). Plant disease resistance genes: Current status and future directions, *Physiol. Mol. Plant Pathol.*, 78: 51-65. 10.1016/j.pmpp.2012.01.00

Ha, M. and Kim, V.N. (2014). Regulation of microRNA biogenesis, *Nat. Rev. Mol. Cell Biol.*, 15: 509-524. Doi: 10.1038/nrm3838

He, X.F., Fang, Y.Y., Feng, L. and Guo, H.S. (2008). Characterisation of conserved and novel microRNAs and their targets, including a TuMV-induced TIR–NBS–LRR class R gene-derived novel miRNA in Brassica, *FEBS Letters*, 582(16): 2445-2452.

Helliwell, C.A. and P.M. Waterhouse (2005). Constructs and methods for hairpin RNA-mediated gene silencing in plants, *Methods Enzymol.*, 392: 24-35.

Incarbone, M. and Dunoyer, P. (2013). RNA silencing and its suppression: Novel insights from in planta analyses. *Trends Plant Sci.*, 18(7): 382-392. https://doi.org/10.1016/j.tplants.2013.04.001

Jelly, N.S., Schellenbaum, P., Walter, B. and Maillot, P. (2012). Transient expression of artificial microRNAs targeting *grapevine fan leaf virus* and evidence for RNA silencing in grapevine somatic embryos, *Transgenic Res.*, 21: 1319-1327. https://Doi.org/10.1007/s11248-012-9611-5

Ji, L.H. and Ding, S.W. (2001). The suppressor of transgene RNA silencing encoded by *cucumber mosaic virus* interferes with salicylic acid-mediated virus resistance, *Mol. Plant-Microbe Interact.*, 14: 715-724.

Jones-Rhoades, M.W. (2012). Conservation and divergence in plant microRNAs, *Plant Mol. Biol.*, 80(1): 3-16.

Khraiwesh, B., Ossowski, S., Weigel, D., Reski, R. and Frank, W. (2008). Specific gene silencing by artificial MicroRNAs in *Physcomitrella patens*: An alternative to targeted gene knockouts, *Plant Physiol.*, 148: 684-693.

Kim, Y.K. and Kim, V.N. (2007). Processing of intronic microRNAs, *EMBO J.*, 26: 775-783. Doi: 10.1038/sj.emboj.7601512

Kørner, C.J., Klauser, D., Niehl, A., Domínguez-Ferreras, A., Chinchilla, D., Boller, T., ... and Hann, D.R. (2013). The immunity regulator BAK1 contributes to resistance against diverse RNA viruses, *Mol. Plant-Microbe Interact.*, 26(11): 1271-1280.

Kung, Y.J., Lin, S.S., Huang, Y.L., Chen, T.C., Harish, S.S., Chua, N.H. and Yeh, S.D. (2012). Multiple artificial microRNAs targeting conserved motifs of the replicase gene confer robust transgenic resistance to negative-sense single-stranded RNA plant virus, *Molecular Plant Pathol.*, 13(3): 303-317. Doi: 10.1111/j.1364-3703.2011.00747.x

Lee, H.J., Park, Y.J., Kwak, K.J., Kim, D., Park, J.H., Lim, J.Y., ... and Kang, H. (2015). MicroRNA844-guided downregulation of cytidinephosphate diacylglycerol synthase3 (CDS3) mRNA affects the response of *Arabidopsis thaliana* to bacteria and fungi, *Mol. Plant-Microbe Interact.*, 28(8): 892-900.

Li, F., Pignatta, D., Bendix, C., Brunkard, J.O., Cohn, M.M., Tung, J., ... and Baker, B. (2012). MicroRNA regulation of plant innate immune receptors, *PNAS*, 109(5): 1790-1795.

Li, Y., Lu, Y.G., Shi, Y., Wu, L., Xu, Y.J., Huang, F., ... and Wang, W.M. (2014). Multiple rice microRNAs are involved in immunity against the blast fungus *Magnaporthe oryzae*, *Plant Physiol.*, 164(2): 1077-1092.

Liang, C., Hao, J., Li, J., Baker, B. and Luo, L. (2019). Artificial microRNA-mediated resistance to cucumber green mottle mosaic virus in *Nicotiana benthamiana*, *Planta*, 250: 1591-1601. https://Doi.org/10.1007/s00425-019-03252-w

Lindbo, J.A., Silva-Rosales, L., Proebsting, W.M. and Dougherty, W.G. (1993). Induction of a highly specific antiviral state in transgenic plants: Implications for regulation of gene expression and virus resistance, *Plant Cell*, 5: 1749-1759.

Liu, Q. and Chen, Y. (2009). Insights into the mechanism of plant development: Interactions of miRNAs pathway with phytohormone response, *Biochem. Biophys. Res. Commun.*, 384: 1-5.

Liu, S., Zhou, J., Hu, C., Wei, C. and Zhang, J. (2017). MicroRNA-mediated gene silencing in plant defence and viral counter-defence, *Front Microbiol.*, 8: 1801.

Lu, C., Jeong, D.H., Kulkarni, K., Pillay, M., Nobuta, K., German, R., Thatcher, S.R., Maher, C., Zhang, L., Ware, D., Liu, B., Cao, X., Meyers, B.C. and Green, P.J. (2008). Genome-wide analysis for discovery of rice microRNAs reveals natural antisense microRNAs (nat-miRNAs), *PNAS USA*, 105(12): 4951-4956. https://doi.org/10.1073/pnas.0708743105

Mandadi, K.K. and Scholthof, K.B.G. (2013). Plant immune responses against viruses: How does a virus cause disease? *Plant Cell*, 25(5): 1489-1505.

Mandahar, C.L. (2006). *Multiplication of RNA Plant Viruses*, Springer, New York, 353 pp.

Mitter, N., Zhai, Y., Bai, A.X., Chua, K., Eid, S., Constantin, M., Mitchell, R. and Pappu, H.R. (2016). Evaluation and identification of candidate genes for artificial microRNA mediated resistance to *tomato spotted wilt virus*, *Virus Res.*, 211: 151-158. Doi: 10.1016/j.virusres.2015.10.003

Mlotshwa, S., Pruss, G.J. and Vance, V. (2008). Small RNAs in viral infection and host defence, *Trends Plant Sci.*, 13: 375-382.

Molnar, A., Bassett, A., Thuenemann, E., Schwach, F., Karkare, S., Ossowski, S., Weigel, D. and Baulcombe, D. (2009). Highly specific gene silencing by artificial microRNAs in the unicellular alga *Chlamydomonas reinhardtii*, *Plant J.*, 58: 165-174.

Naqvi, A.R., Haq, Q.M.R. and Mukherjee, S.K. (2010). MicroRNA profiling of *tomato leaf curl New Delhi virus* (ToLCNDV) infected tomato leaves indicates that deregulation of mir159/319 and mir172 might be linked with leaf curl disease, *Virol. J.*, 2010: 7.

Spadaro, D. and Gullino, M.L. (2005). Improving the efficacy of biocontrol agents against soilborne pathogens, *Crop Protection*, 24(7): 601-613.

Niu, Q.W., Lin, S.S., Reyes, J.L., Chen, K.C., Wu, H.W., Yeh, S.D. and Chua, N.H. (2006). Expression of artificial microRNAs in transgenic *Arabidopsis thaliana* confers virus resistance, *Nat. Biotechnol.*, 24: 1420-1428. Doi: 10.1038/ nbt1255

O'Brien, J., Hayder, H., Zayed, Y. and Peng, C. (2018). Overview of microRNA biogenesis, mechanisms of actions, and circulation, *Frontiers in Endocrinol.*, 9: 402.

Ossowski, S., Schwab, R. and Weigel, D. (2008). Gene silencing in plants using artificial microRNAs and other small RNAs, *Plant J.*, 53: 674-690.

Ouyang, S., Park, G., Atamian, H.S., Han, C.S., Stajich, J.E., Kaloshian, I. and Borkovich, K.A. (2014). MicroRNAs suppress NB domain genes in tomato that confer resistance to *Fusarium oxysporum*, *PLoS Pathogens*, 10(10): e1004464.

Pacheco, R., García-Marcos, A., Barajas, D., Martiáñez, J. and Tenllado, F. (2012). PVX–potyvirus synergistic infections differentially alter microRNA accumulation in *Nicotiana benthamiana*, *Virus Res.*, 165(2): 231-235.

Pallas, V. and García, J.A. (2011). How do plant viruses induce disease? Interactions and interference with host components, *J. Gen. Virol.*, 92: 2691-2705.

Pandey, S., Muthamilarasan, M., Sharma, N., Chaudhry, V., Dulani, P., Shweta, S., Jha, S., Mathur, S. and Prasad, M. (2019). Characterization of DEAD-box family of RNA helicases in tomato provides insights into their roles in biotic and abiotic stresses, *Environ. Exp. Bot.*, 158: 107-116.

Pandey, S., Prasad, A., Sharma, N. and Prasad, M. (2020a). Linking the plant stress responses with RNA helicases, *Plant Sci.*, 299: 110607. https://Doi.org/10.1016/j.plantsci.2020.110607

Pandey, S., Sahu, P., Kulshreshtha, R. and Prasad, M. (2018). Role of host transcription factors in modulating defense response during plant virus interaction. *In:* Patil, B. (Ed.). *Genes, Genetics and Transgenics for Virus Resistance in Plants*. Caister Academic Press, UK, 25-54.

Pandey, S., Sharma, N. and Prasad, M. (2020b). Role of RNA-interacting proteins in modulating plant-microbe interactions, *Adv. Genet.*, 105: 67-94. https://Doi.org/10.1016/bs.adgen.2019.12.001

Park, J.H. and Shin, C. (2015). The role of plant small RNAs in NB-LRR regulation, *Brief Funct. Genomics*, 14(4): 268-274.

Pérez-Quintero, Á.L., Neme, R., Zapata, A. and Lopez, C. (2010). Plant microRNAs and their role in defence against viruses: A bioinformatics approach, *BMC Plant Biol.*, 10: 138.

Qu, J., Ye, J. and Fang, R. (2007). Artificial microRNA-mediated virus resistance in plants, *J. Virol.*, 81(12): 6690-6699. https://Doi.org/10.1128/JVI.02457-06

Ruby, J.G., Jan, C.H. and Bartel, D.P. (2007). Intronic microRNA precursors that bypass Drosha processing, *Nature*, 448: 83-86. Doi: 10.1038/nature05983

Salvador-Guirao, R., Baldrich, P., Weigel, D., Rubio-Somoza, I. and San Segundo, B. (2018). The microRNA miR773 is involved in the *Arabidopsis* immune response to fungal pathogens, *Mol. Plant-Microbe Interact.*, 31(2): 249-259.

Satish, D., Mukherjee, S.K. and Gupta, D. (2019). PAmiRDB: A web resource for plant miRNAs targeting viruses, *Sci. Rep.*, 9: 4627. https://Doi.org/10.1038/s41598-019-41027-1

Savary, S., Willocquet, L., Pethybridge, S.J., Esker, P., McRoberts, N. and Nelson, A. (2019). The global burden of pathogens and pests on major food crops, *Nat. Ecol. Evol.*, 3(3): 430-439.

Schwab, R., Ossowski, S., Riester, M., Warthmann, N. and Weigel, D. (2006). Highly specific gene silencing by artificial microRNAs in *Arabidopsis*, *Plant Cell*, 18: 1121-1133. Doi: 10.1105/tpc.105.039834

Sharma, S., Pandey, S., Muthamilarasan, M., Chaudhry, V., Dulani, P. and Prasad, M. (2017). Genomics resources for abiotic stress tolerance in Solanaceae crops. *In:* Chakrabarti, S.K.,

Xie, C. and Tiwari, J.K. (Eds.). *The Potato Genome*, Springer, Doi: 10.1007/978-3-319-66135-3_12

Simón-Mateo, C. and García, J.A. (2006). MicroRNA-guided processing impairs *Plum pox virus* replication, but the virus readily evolves to escape this silencing mechanism, *J. Virol.*, 80(5): 2429-2436. Doi:10.1128/JVI.80.5.2429-2436.2006

Singh, A., Taneja, J., Dasgupta, I. and Mukherjee, S.K. (2015). Development of plants resistant to tomato geminiviruses using artificial trans-acting small interfering RNA, *Mol. Plant. Pathol.*, 16: 724-734.

Smith, H.A., Swaney, S.L., Parks, T.D., Wernsman, E.A. and Dougherty, W.G. (1994). Transgenic plant virus resistance mediated by untranslatable sense RNAs: Expression, regulation and fate of nonessential RNAs, *Plant Cell*, 6: 1441-1453.

Song, Y.Z., Han, Q.J., Jiang, F., Sun, R.Z., Fan, Z.H., Zhu, C.X. and Wen, F.J. (2014). Effect of the sequence characteristics of miRNAs on multi-viral resistance mediated by single amiRNAs in transgenic tobacco, *Plant Physiol. Biochem.*, 77: 90-98. Doi: 10.1016/j.plaphy.2014.01.008

Strange, R.N. and Scott, P.R. (2005). Plant disease: A threat to global food security, *Annu. Rev. Phytopathol.*, 43: 83-116.

Su, Y., Li, H.G., Wang, Y., Li, S., Wang, H.L., Yu, L., ... and Xia, X. (2018). Poplar miR472a targeting NBS-LRRs is involved in effective defence against the necrotrophic fungus *Cytospora chrysosperma*, *J. Exp Bot.*, 69(22): 5519-5530.

Sun, L., Lin, C., Du, J., Song, Y., Jiang, M., Liu, H., ... and Zhu, C. (2016). Dimeric artificial microRNAs mediate high resistance to RSV and RBSDV in transgenic rice plants, *Plant Cell Tissue Organ Cult.*, 126(1): 127-139.

Tanzer, A. and Stadler, P.F. (2004). Molecular evolution of a microRNA cluster, *J. Mol. Biol.*, 339(2): 327-335.

Tyagi, S., Kumar, A., Gautam, T., Pandey, R., Rustgi, S. and Mir, R.R. (2021). Development and use of miRNA-derived SSR markers for the study of genetic diversity, population structure, and characterisation of genotypes for breeding heat tolerant wheat varieties, *PLoS ONE*, 16(2): e0231063.

Várallyay, É., Válóczi, A., Ágyi, Á., Burgyán, J. and Havelda, Z. (2010). Plant virus-mediated induction of miR168 is associated with repression of Argonaute1 accumulation, *EMBO J.*, 29(20): 3507-3519.

Voinnet, O. (2009). Origin, biogenesis and activity of plant microRNAs, *Cell*, 136(4): 669-687.

Wagaba, H., Patil, B.L., Mukasa, S., Alicai, T., Fauquet, C.M. and Taylor, N.J. (2016). Artificial microRNA-derived resistance to *cassava brown streak disease*, *J. Virol. Methods*, 231: 38-43. Doi: 10.1016/j.jviromet.2016.02.004

Wang, A. (2015). Dissecting the molecular network of virus-plant interactions: The complex roles of host factors, *Annu. Rev. Phytopathol.*, 53: 45-66.

Wang, H., Jiao, X., Kong, X., Hamera, S., Wu, Y., Chen, X., Fang, R. and Yan, Y. (2016). A signalling cascade from miR444 to RDR1 in rice antiviral RNA silencing pathway, *Plant Physiol.*, 170: 2365-2377.

Wang, J. and Cui, Q. (2012). Specific roles of microRNAs in their interactions with environmental factors, *J. Nucleic Acids*. https://Doi.org/10.1155/2012/978384

Wang, J., Mei, J. and Ren, G. (2019). Plant microRNAs: Biogenesis, homeostasis, and degradation, *Front Plant Sci.*, 10: 360.

Warthmann, N., Chen, H., Ossowski, S., Weigel, D. and Herve, P. (2008). Highly specific gene silencing by artificial miRNAs in rice, *PLoS ONE*, 3: e1829.

Waterhouse, P.M., Graham, M.W. and Wang, M.B. (1998). Virus resistance and gene silencing in plants can be induced by simultaneous expression of sense and antisense RNA, *Proc. Natl. Acad. Sci., USA*, 95: 13959-13964.

Whetzel, H.H. 1929. The terminology of plant pathology, *Proc. Int. Cong. Plant Science*, Ithaca, NY, 1926: 1204-1215.

Whitham, S., Dinesh-Kumar, S.P., Choi, D., Hehl, R., Corr, C. and Baker, B. (1994). The product of the tobacco mosaic virus resistance gene N: Similarity to toll and the interleukin-1 receptor, *Cell*, 78(6): 1101-1115.

Wu, J., Yang, R., Yang, Z., Yao, S., Zhao, S., Wang, Y., Li, P., Song, X., Jin, L., Zhou, T. and Lan, Y. (2017). ROS accumulation and antiviral defence control by microRNA528in rice, *Nature Plants*, 3: 16203.

Xuan, N., Zhao, C., Peng, Z., Chen, G., Bian, F., Lian, M., ... and Bi, Y. (2015). Development of transgenic maize with anti-rough dwarf virus artificial miRNA vector and their disease resistance, *Chinese Journal of Biotechnology*, 31(9): 1375-1386.

Yang, Q., Qiu, C., Yang, J., Wu, Q. and Cui, Q. (2011). miR environment database: Providing a bridge for microRNAs, environmental factors and phenotypes, *Bioinformatics*, 27: 3329-3330.

Yang, L., Mu, X., Liu, C., Cai, J., Shi, K., Zhu, W. and Yang, Q. (2015). Overexpression of potato miR482e enhanced plant sensitivity to *Verticillium dahliae* infection, *J. Integrative Plant Biol.*, 57(12): 1078-1088.

Yao, S., Yang, Z., Yang, R., Huang, Y., Guo, G., Kong, X., ... and Li, Y. (2019). Transcriptional regulation of miR528 by OsSPL9 orchestrates antiviral response in rice, *Molecular Plant*, 12(8): 1114-1122.

Zhang, C., Ding, Z., Wu, K., Yang, L., Li, Y., Yang, Z., Shi, S., Liu, X., Zhao, S., Yang, Z. and Wang, Y. (2016). Suppression of jasmonic acid-mediated defense by viral-inducible microRNA319 facilitates virus infection in rice, *Mol. Plant*, 9: 1302-1314.

Zhang, X., Li, H., Zhang, J., Zhang, C., Gong, P., Ziaf, K., ... and Ye, Z. (2010). Expression of artificial microRNAs in tomato confers efficient and stable virus resistance in a cell-autonomous manner, *Transgenic Research*, 20(3): 569-581. Doi:10.1007/s11248-010-9440-3

Zhang, X., Li, H., Zhang, J., Zhang, C., Gong, P., Ziaf, K., Xiao, F. and Ye, Z. (2011). Expression of artificial microRNAs in tomato confers efficient and stable virus resistance in a cell-autonomous manner, *Transgenic Research*, 20(3): 569-581.

Zhang, X., Yuan, Y.R., Pei, Y., Lin, S.S., Tuschl, T., Patel, D.J. and Chua, N.H. (2006). Cucumber mosaic virus-encoded 2b suppressor inhibits *Arabidopsis* Argonaute1 cleavage activity to counter plant defence, *Genes Devel.*, 20(23): 3255-3268.

Zhang, Y., Xia, R., Kuang, H. and Meyers, B.C. (2016). The diversification of plant NBS-LRR defense genes directs the evolution of microRNAs that target them, *Mol. Biol. Evol.*, 33(10): 2692-2705.

Zhu, Q.H., Fan, L., Liu, Y., Xu, H., Llewellyn, D. and Wilson, I. (2013). miR482 regulation of NBS-LRR defence genes during fungal pathogen infection in cotton, *PLoS ONE*, 8(12): e84390.

Zorzatto, C., Machado, J.P.B., Lopes, K.V., Nascimento, K.J., Pereira, W.A., Brustolini, O.J., ... and Fontes, E.P. (2015). NIK1-mediated translation suppression functions as a plant antiviral immunity mechanism, *Nature*, 520(7549): 679-682.

CHAPTER

15

Crosstalk of MicroRNAs with Phytohormone Signalling Pathways

Shilpy Singh[1], Ruth Assumi[2] and Pooja Bhadrecha[3]*

[1] Department of Biotechnology and Microbiology, Noida International University, Gautam Budh Nagar, U.P., India

[2] Scientist, ARS, Division of System Research and Engineering, ICAR Research Complex for NEH Region, Umiam, Meghalaya, India

[3] University Institute of Biotechnology, Chandigarh University, Mohali, Punjab, India

1. Introduction

Phytohormones are signalling molecules that play a key part in practically every biological event throughout a plant's entire life (Foo *et al.*, 2019; Dubois *et al.*, 2018). Major phytohormones found to date include auxin (AUX), cytokinin (CK), gibberellic acid (GA), ethylene (ET), abscisic acid (ABA), brassinosteroids (BR), jasmonic acid (JA), strigolactones (SL) and salicylic acid (SA). They are generated in a number of processes and are recognised by receptor proteins, which then trigger intracellular signalling (Blazquez *et al.*, 2020). The bulk of signalling eventually makes its way to transcriptional factors (TFs), which influence the hormones response downstream. Phytohormones collaborate to regulate an extensive diversity in terms of life events, like the growth of cells, vascular root morphogenesis and reception of stress by several stress situations of biotic and/or abiotic factors (Yang *et al.*, 2019; Li *et al.*, 2019). The existence of functional relationship among phytohormones and microRNA (miRNA) was demonstrated in a plant with the hyl1 mutant, which displayed various developmental abnormalities, decreased miRNA accumulation and was sensitive to plant hormones, like AUX, ABA and CK (Han *et al.*, 2004). Several varieties of miRNAs were shown to stimulate the assembly of trans-acting small interfering RNAs, also known as the tasiRNAs that operate equal to the mobile signalling molecules (Marin *et al.*, 2010; Si-Ammour *et al.*, 2011). Plant growth and development are masterly regulated by microRNAs, which, after various researches aimed at their involvement in plant growth and survival, have now been well known as the master for regulating expression of genes. Different recent studies have suggested

---

*Corresponding author: pbhadrecha.pb@gmail.com

that miRNAs responses to different stresses in various plant species. Additionally, the gene expressions directed by various miRNAs required for the growth, development and survival are recorded to be exclusively altered while stress situations faced by the plants, whether it is of biotic origin or abiotic origin (Pandita, 2019; Pandita and Wani, 2019; Pandita, 2021; 2022a; 2022b; 2022c). In this chapter, we present advancement of studies on mechanisms of miRNA action in plant as a response in biotic and/or abiotic stress situations. Figure 1 represents various stages a plant undergoes and in which phytohormones are necessary.

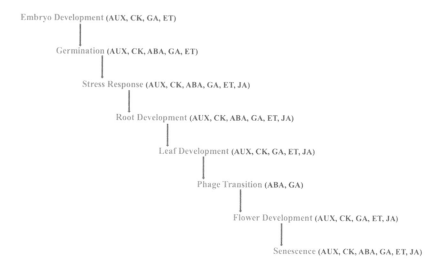

Fig. 1: Developmental processes of plant and required phytohormones (Curaba *et al.*, 2014)

miRNA is a highly conserved type of non-translated RNA molecule that works by suppressing or cleaving transcripts of specific genes. They were identified in early 1990s and are generally 20-24 nucleotides long (Lee *et al.*, 1993). The RNA pol II enzyme transcribe primary miRNA (pre-miRNA), using targeted DNA sequence. It generally suppresses mRNA expression by adhering to 3' untranslated region, but also the 5' untranslated region, coding sequences and promoters. Dicer-like1 (DCL-1) recognises single-stranded RNAs with stem loop structures and chops the pre-miRNAs up into precursor microRNAs (pre-miRNAs), which it then converts to miRNAs. After that, the miRNAs are then placed into Argonaute-linked microRNA-induced silencing complex (miRISC) for further activation. MiRNAs can be located inside the nucleus, or nucleolus, or mitochondrion or even in the endoplasmic reticulum (ER), wherein they regulate silencing of the genes, after the process of translation via migrating between various sub-cellular components (Makarova *et al.*, 2016). Abiotic stress resistance, nutritional balance and transcriptional regulation of gene expression are all regulated via miRNAs within plants. Some of the miRNAs are very précise towards the functions they perform in abiotic stress situations due to a complex network of action, though this field is rapidly increasing presently, with more miRNAs being characterised. Stress responses are triggered by a variety of physical and chemical variables that impact plants all over the world, including excessive soil salts, drought, floods, severe temperatures, heavy metal (HM) toxicity, ultraviolet (UV) radiation, nutrient shortages, etc. (Wang *et al.*, 2003; Wani *et al.*, 2016). Various signalling mechanisms involving several phytohormones are also triggered in relation to stress. Even though miRNAs and plant hormones possess separate metabolic and signalling routes, new research suggests that miRNA processes and phytohormone actions interact under a variety of stress conditions. Such connection leads to overcoming variety of stresses, either by modifying miRNAs with phytohormones or managing phytohormone levels utilising miRNAs as an intermediate. So, during stress responses, phytohormone homeostasis and miRNA control therefore go hand in hand, implying a linked network controlling the genes that control stress tolerance (Noman and Aqeel, 2017). Furthermore, miRNAs are noticed for behaving on their own, allowing for the creation of a regulatory gradient across neighbouring tissues. Reports of various miRNAs have demonstrated their affinity to promoting assembly of trans-acting small interfering RNAs (tasiRNAs), known for working as mobile signals (Si-Ammour *et al.*, 2011; Marin *et al.*, 2010). Due to its crucial activities in development and trans-regulation, miRNAs have been proved as appealing targets to synchronise numerous hormone responses. In this chapter, we discuss how miRNAs may act as possible mediators of hormone crosstalk at different stages of development, since a few current researches have been related to the molecular links between miRNAs associated with numerous hormonal responses. We have also discussed how miRNAs control the action of phytohormones in response to different challenges that plants face. Figure 2 describes the process of miRNA production followed by silencing of respective gene. As depicted in the figure, the MIR gene codes for the miRNA which binds to RISC complex and cleaves the target mRNA.

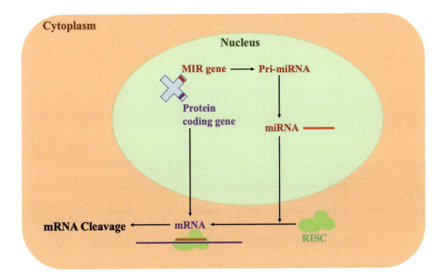

Fig. 2: Biogenesis and function of plant miRNA (RISC: RNA-induced silencing complex)

Plants exposed to various stress (biotic/abiotic) situations possess the capability of regulating and utilising various gene regulatory mechanisms, encompassing synchronisation of gene expression post-translation (Kawaguchi *et al.*, 2004).

miRNAs refer to a few very small, 20-24 nt., non-coding RNA molecules (Carrington and Ambros, 2003) and though very few miRNAs are present in plants, they cannot be de-emphasised because their main targets are the transcription factors that eventually effect the transcription of mRNAs, hence regulating gene expression (Jones-Rhoades and Bartel, 2006). Moreover, they are the major factors which possibly take part in most evolution strategies, starting from the germination of seed till its maturation. Hence, miRNAs synchronise particularly the processes involved in development. The direct involvement of miRNAs with plant stress responses has been recorded since a last few decades, e.g. miR398 was found to be responsible for targeting CSD1 and CSD2 which are Cu/Zn superoxide dimutases and miR395 and miR399 were found to aim AST68 (sulphate transporter) as well as PHO1 (phosphate transporter), respectively (Jones-Rhoades and Bartel, 2004; Sunkar and Zhu, 2004). A few studies reported that the miRNA response to plant stress is linked, e.g. hyl1 and cbp80/abh1 (a cap-binding protein80/ABA hypersensitive 1) mutants which couldn't synthesise miRNA were found hypersensitive to ABA (Kim et al., 2008; Lu and Fedoroff, 2000). Moreover, those miRNAs which directly influence the transcription factors have been recorded to have exclusively conserved expression profiles, which, as per the stress situations, may get modified. Therefore, to discuss various findings which have shown the significance miRNAs' function for survival of the plants in severe biotic as well as abiotic stress conditions, this chapter focusses on the mechanisms by which miRNAs act in response to stress, the major reason which has drawn the focus of researchers to discover which miRNAs targeted which genes as a response to various stresses conditions, to help the plant survive. Figure 3 depicts how miRNAs regulate plant hormones at several stages of plant growth.

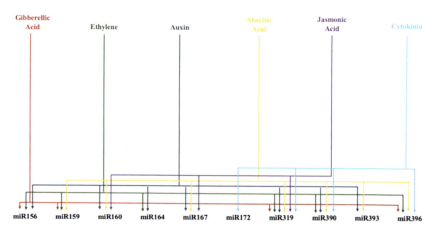

Fig. 3: Regulation of plant hormones by respective miRNAs

2. miRNAS and Plant Hormones in Response to Stress

Plants are regularly subjected to many challenges at the same time in Nnature, resulting in a range of responses, depending on kind and level of the stress. Reactions might be stress-related or convergent, as in the instances of cold and drought stresses, which cause dehydration in plants. Whenever the stress factor becomes detected, it

Crosstalk of MicroRNAs with Phytohormone Signalling Pathways

must be communicated to controllers via secondary mediators, resulting in synthesis of effectors, like late embryogenesis abundant (LEA) proteins, reactive oxygen radicals, chaperones, enzyme inhibitors and plant hormones (Patel *et al.*, 2019). Phytohormones are signalling factors especially essential for almost every biological function in plants, including physiology, morphology and environmental adaption (Peleg and Blumwald, 2011). AUX, ABA, BR, CK, ET, GA, JA, SA and SL have all been found as distinctive phytohormones that influence transcriptional factor (TF)-regulated hormone response. Abscisic acid and ethylene are the most essential phytohormones in regulating environmental stress resistance. Several miRNAs are up- or down-regulated in response to environmental stress, implying that they may play a significant function for stress regulation by addressing particular genes or transcription factors (Wani *et al.*, 2016; Sunkar and Zhu, 2004). miRNAs either up-regulate or down-regulate differently in various plant species under various situations, hence targeting the same miRNAs to increase stress resistance within different plant species would result in varied results (Djami-Tchatchou *et al.*, 2017). Activation of the effector genes is controlled by the promoter region of miRNA and it has been discovered to contain hormone and abiotic stress response factors, like MYB binding elements, heat shock-responsive element (CCA AAT), copper-responsive elements (CuREs), MYC binding elements, gibberellic acid responsive elements (GARE), early response to dehydration (ERD) and ethylene-responsive elements (ERE) (Zhou *et al.*, 2013).

2.1 microRNA Responses during Biotic Stress

2.1.1 Viral Infection

In response to viral infection, miRNAs biomarkers viz., miR171 and miR164 were reported to be down-regulated (Naqvi *et al.*, 2008) whereas, miR156, miR164, miR159 and miR1885 were found to be up-regulated (He *et al.*, 2008; Kasschau *et al.*, 2003). Earlier research reports state that biomarker miRNA159 altered from *Arabidopsis thaliana* was observed to attack the viral miRNA sequences which encode two different suppressors for silencing the gene, which were P69 of TYMV (turnip yellow mosaic virus) and HC-Pro of TuMV (turnip mosaic virus) (Niu *et al.*, 2006). Therefore, biomarker miRNA159 expressing P69 and HC-Pro were found to be resisting both TYMV and TuMV, respectively.

2.1.2 Bacterial Infection

In response to bacterial infection, biomarker miR393 modified from *Arabidopsis* instigated with PAMP (pattern associated molecular pattern) of bacteria, was the very first miRNA discovered with a significant role in plant antibacterial PTI (pattern triggered immunity) that can asymmetrically regulate signalling pathway of AUX (Navarro *et al.*, 2006). Moreover, three miRNA biomarkers, viz. miR (160, 167 and 393) have been recorded as highly regulated whereas ones, miR398 and miR825 were observed down-regulated post infection with *Pseudomonas syringae* pv. tomato (Pst) DC3000 (Fahlgren *et al.*, 2007).

2.1.3 Fungal Infection

As a response to infection caused by *Cronartium quercuum* f. sp. *fusiforme* (known as rust fungus) which leads to fusiform rust disease in pine trees, several miRNA families exhibited dissimilar function as a response to the infection (Lu *et al.*, 2007). Earlier research conducted by Campo *et al.* reported that when *Blumeria graminis* sp. *tritici* attacked wheat and caused powdery mildew disease, some miRNAs (156, R159, 164, 171 and 396) got up-regulated, whereas few miRNAs (393, 444 and 827) were recorded as down-regulated. In the case of rice, miRNAs Osa-miR7696 exhibited overexpressing during rice blast infection (Campo *et al.*, 2013).

2.1.4 Nematode Infection

The fact that miRNAs regulate expression of various genes is not restricted to biotic stress caused by microorganisms only, in fact, several miRNAs (161, miR164, miR167a, miR172c, 396a, b and 398a) were observed down-regulated in *Arabidopsis*, as a response to nematode infection caused by *Heterodera schachtii* (Khraiwesh *et al.*, 2012; Hewezi *et al.*, 2008). Moreover, more than hundred miRNAs from 40 different families were observed to be highly reactive to infection from soybean cyst nematode *Heterodera glycines* (Li *et al.*, 2012).

2.2 MicroRNAs Responses during Abiotic Stress

2.2.1 Oxidative Stress

Some of the studies have reported that miR398 gets down-regulated in oxidative stress. Other studies reported that in *A. thaliana*, the miR398 family members were represented by three different loci at miRNA (398a, 398b and 398c) (Phillips *et al.*, 2007; Jones-Rhoades and Bartel, 2004; Sunkar and Zhu, 2004; Bonnet *et al.*, 2004).

2.2.2 Drought Stress

A total of 16 miRNAs were reported as down-regulated in response to the drought conditions in rice plant and these were miR156, 159, 168, 170, 171, miR172, miR319, miR396, miR397, miR408, miR529, miR896, miR1030, miR1035, miR1050, miR1088 and 1126; whereas 14 miRNAs were up-regulated when the same drought conditions reoccurred. These were miR159, 169, 171, miR319, 395, miR474, miR845, miR851, miR854, miR896, miR901, miR903, miR1026 and miR1125 (Zhou *et al.*, 2010). In case of plant *Medicago truncatula,* when it faced drought conditions, miR169 was noted to be down-regulated only in the plant roots, whereas, miR398a, miR398b and miR408 were recorded to be highly up-regulated (Trindade *et al.*, 2010). Moreover, miR159.2, miR393 and miR2119 showed up-regulation in *P. vulgaris* (Arenas-Huertero *et al.*, 2009). Therefore, from the studies it was reported that 13 significantly regulated miRNAs, viz. miR156, 166, 171, 396, 398, 474, miR528, miR894, miR896, miR1432, miR1450, miR1867 and 1881 were found in response to drought resistant on wild emmer wheat (*Triticum turgidum* ssp. *dicoccoides*) (Kantar *et al.*, 2011).

2.2.3 Salinity Stress

In response to salinity stress, several members from miR169 family, viz. miR169g, miR169n, miR169o and 393 were reported to induce stress response in rice plant (Zhao *et al.*, 2009; Gao *et al.*, 2011). Moreover, in case of *A. thaliana*, members of miRNAs, like miR156, 158, 159, 165, 167, 168, 169, 171, 319, 393, 394, 396 and 397 were reported to be up-regulated whereas, miR398 was down-regulated (Liu *et al.*, 2008). A recent study on salt-tolerance and salt-sensitivity has been reported on *Zea mays* where 98 miRNAs belonging to 27 families were identified. Family members of miR156, 164, 167 and 396 were reported as down-regulated, whereas miR162, 168, 395 and 474 families were noticed as up-regulated when maize roots faced salt-shock (Ding *et al.*, 2009).

2.2.4 Cold Stress

Various miRNAs were recorded as significantly up-regulated in cold stress conditions and these were miR165/166, 169, 172, 393, 396, 397 and 408 (Zhou *et al.*, 2008). In a study on *Populus*, miR168 a, b and 477 a, b were recorded to be up-regulated whereas miR156g-j, 475a, b and 476a were down-regulated (Lu *et al.*, 2008). Therefore, the earlier studies reported that a total of 18 miRNAs, viz. miR156k, 166k, 166m, 167a/b/c, 168b, 169e, 169f, 169h, 171a, 535, 319a/b, 1884b, 444a.1, 1850, 1868, 1320, 1435 and 1876 were reported in response to cold stress (Lv *et al.*, 2010).

2.2.5 Phosphate Stress

As a response to phosphate stress, plants respond mainly via transcriptional regulation, though post-transcriptional regulation through miRNA-mediated is also responsible as a significant response to phosphate stress. Some researchers have reported that miRNA399 was found to get highly induced in response to phosphate stress (Fuji *et al.*, 2005; Chiou *et al.*, 2006).

2.2.6 Copper Stress

In response to copper stress, miRNA395 was found to be up-regulated in *Arabidopsis* (Huang *et al.*, 2010; Liang and Yu, 2010; Sunkar *et al.*, 2007; Allen *et al.*, 2005; Jones-Rhoades and Bartel, 2004). Moreover, from the previous studies, three miRNA families, viz. miR397, 408 and 857 were recorded to be up-regulated in response to copper stress (Abdel-Ghany and Pilon, 2008).

2.2.7 Sulphate Stress

As a response to sulphate stress, miRNA398 has been recorded to get highly induced (Yamasaki *et al.*, 2007; Beauclair *et al.*, 2010). Similar response was showed by miR857, miR397 and miR408 reported by Abdel-Ghany and Pilon (2008) whereas miR156, 162, 164, 475, 480 as well as miR481 were found down-regulated in a situation of sulphate stress (Lu *et al.*, 2005; Khraiwesh *et al.*, 2012). Therefore, from the studies it was observed that along with miR395, regulation levels of miR156, 160, 164, 167, 168 and 394 varied during sulphate stress in *Brassica napa* (Huang *et al.*, 2010).

2.2.8 Mechanical Stress

Even in the situation of mechanical stress, miRNAs get regulated, functioning for the proper fitness of plants in terms of structural and mechanical. An earlier study conducted on *Populus trichocarpa*, reported that miR156, 162, 164, 475, 480 and 481 were found to be down-regulated on exposure to UV-B radiations, whereas, miR408 was up-regulated in the same situation. miR160 and 172 were recorded as down-regulated only in compressed-stressed tissue, though miR168 was up-regulated in tension-stressed tissues (Jia *et al.*, 2009).

Respective researches on miRNAs have highlighted a few miRNAs as the main players of regulating phytohormones and the following section discusses the most studied miRNAs in plants.

3. Most Studied miRNAs Regulating Phytohormones in Plants

3.1 miR159

The relative expression of miR159 was changed in relation to drought tolerance (Liu *et al.*, 2008). MiR159 inhibition promotes the growth of lateral and adventitious roots via activating ARF, HD-ZIP and GAMYB transcriptional factors (Xue *et al.*, 2017). ABA or water scarcity promotes the levels of miR159a that break down the genes expressing MTB33 and MYB101 in *A. thaliana* seedlings. The altered MYB33 and MYB101 variants which were resistant to cleavage displayed ABA hypersensitivity in miR159a overexpressing variants, whereas those variants which were susceptible to cleavage showed hyposensitivity (Reyes and Chua, 2007). Cold stress causes transient or moderate expression of miR159 in *A. thaliana*. C-repeat binding F factors (CBF)'s transcriptional factors bound to dehydration-responsive element (DRE) within *A. thaliana* throughout cold stress, were found to be leading to up-regulation of cold-responsive genes. ABA signalling is controlled by a number of miRNAs, which allow many plants to withstand cold stress. Furthermore, according to investigations cold-tolerant cultivars produce more ABA throughout cold stress than cold-sensitive ones (Kumar, 2014). miR159 significantly down-regulated in maize during flood conditions (Liu *et al.*, 2012). By targeting ACC and GAMYB synthase, miR159 alters ET, GA and ABA production, lead to increased defensive response and root architectural diversity (Zhang *et al.*, 2008). UV-B exposure increased the up-regulation of miR159 in *A. thaliana* (Zhou *et al.*, 2007). miR159 also gets activated in wheat to give UV-B protection via modulating the GA and ABA signalling pathways. ABA induces MiR159, which inhibits MYB33 and MYB101 during seed development. Chromium (Cr), is the second-most hazardous HM discharged into the atmosphere as a result of human activity. Cr stops plants from growing and developing by blocking water and nutrient uptake and by also interfering with photosynthesis and respiration. Cr stress caused differential expression level of 16 new and 54 recognised miRNAs in *R. sativus*. Among them, miR159 was down-regulated during Cr stress (Liu *et al.*, 2015). Researchers revealed that miR159, which targets multiple SPLs-SPL3/6/9/13/15 in radish, has been shown to have a function in controlling Cr homeostasis. miR159, which targets MYB3/13/101/104/305, may also play a function in regulation of numerous genes during Cr stress. Consequently, miRNA-mediated control of Cr regulation in radish reduces Cr stress by targeting different transcriptional factors (Liu *et al.*, 2015). Under HM stress, relationship among MAPK and AUX signalling is unclear; nevertheless, MAPK signalling has been found to negatively influence IAA, ARF and PIN (Zhao *et al.*, 2014). During Cr stress, another rice miRNA, osa-miR159 targets and supresses MAPK signalling, promoting AUX signalling and serving as an AUX response regulator (Dubey *et al.*, 2020).

3.2 miR160 and miR167

miR160 and miR167 influence the expression of many AUX responsive factors (ARFs), which stand for transcriptional factors that significantly affect the expression

Crosstalk of MicroRNAs with Phytohormone Signalling Pathways **265**

of major AUX-responsive proteins, making them dynamical components of AUX response mechanisms. In *A. thaliana*, miR160 controls AtARF10/16/17 (Liu *et al.*, 2007; Mallory *et al.*, 2005) while miR167 controls AtARF6/8 (Wu *et al.*, 2006). At least a few of these regulators have been consistent across monocots, with AtARF orthologs in barley being mediated by miR167 (Curaba *et al.*, 2012). Pleitropic developmental abnormalities have been reported in plants which express miRNA-resistant version of several of such ARFs, indicating that these regulators are critical for optimal plant development. There is additional feedback regulation, in which AUX influences miR160 and 167 expression via ARFs (Gutierrez *et al.*, 2009). Such regulatory feedback seems to be retained in *Brassica* species. When plants were subjected to arsenic exposure, the frequency of miR167 rose in relation to AUX treatments (Srivastava *et al.*, 2013). Overall, the miR160/miR167 and ARFs network is a powerful regulatory circuit via which AUX may regulate both ABA and JA routes. Throughout the plant life cycle, these regulations recognise to engage in a wide variety of developmental activities. Drought stress also triggers miR160 expression and lateral root growth via variations in AUX levels mediated by miR391 (Bustos-Sanmamed *et al.*, 2013; Yoon *et al.*, 2010). Activation of AUX responsive factors ARF10/16/17 driven by miR160 and 167 appeared in increased salt tolerance in cotton under severe salinity stress (Yin *et al.*, 2017). miR160 and miR167 modulate AUX signalling by targeting different ARFs in relation to UV stress, impacting plant growth and survival. UV-B (280-320 nm), one of the several radiations that sessile plants are exposed to, has been reported to be exclusively influencing plant growth and survival, owing to membrane damage induced by reactive oxygen species formation (Mckenzie *et al.*, 2007). UV-B exposure was recorded to enhance miR160 and miR167 expressions in *A. thaliana* (Zhou *et al.*, 2007). miR160 and miR167 modulate AUX signalling by targeting different ARFs in response to UV stress, impacting plant growth and survival. Cadmium (Cd) is among the most harmful contaminants that reduces plant productivity via induction of chlorosis, wilting, stunted development and even death. Cd stress causes up-regulation of miR167a/c (Huang *et al.*, 2010) but also down-regulation of miR160 in *B. napus* (Zhou *et al.*, 2012). Aluminium (Al) toxicity is severe issue in acidic soil with the pH of 5.5 or below, which affects 30-40% of the world agricultural land (Gupta *et al.*, 2014). It interacts to carboxyl as well as phosphate groups inside root cell walls, reducing root biomass and restricting water and nutrient absorption, as well as altering callose deposition, cytoplasmic Ca^{2+} imbalances and even in conditions of oxidative stress (Ma *et al.*, 2004; Silva, 2012). AUX response factors (ARFs) govern root development via Al-responsive miRNAs, such as miR160. During Al stress, down-regulated miR160 targets ARF10 and ARF16, which affect the development of root caps (Wang *et al.*, 2005). Up-regulation of miR160e and targeting of ARFs in the AUX signalling pathway are antagonistic to miR393b. As a result, miR160e is up-regulated and miR393b is down-regulated in rice to counteract Al stress. Under Cr stress, miR160 was suppressed (Liu *et al.*, 2015). Cr stress causes plants to produce reactive oxygen species (ROS); thus they activate HM-responsive signalling molecules and phytohormones like AUX, ET and JA signalling F-box proteins, mitogen-activated protein kinase-kinase (MEKK) and Ca-dependent protein kinases (CDPK). As a result, plants with a high number of miRNAs that target different hormones and signalling biomolecules have less Cr toxicity. miR160 is down-regulated in rice, which causes an increase in AUX expression by targeted AUX responsive factors (ARF), reducing Cr toxicity. In the *A. thaliana*, osa-miR160 has been demonstrated to induce comparable effect on formation of the root cap (Wang *et al.*, 2005). Small RNA sequencing investigations have revealed several drought-responsive miRNAs in plants *Arabidopsis*, tomato, common bean, barley, maize and rice (Aravind *et al.*, 2017; Wu *et al.*, 2017). The expression levels of miR167 was changed in relation to drought stress. MiR167 is up-regulated in *A. thaliana* under drought response; however, miR167 is down-regulated in rice and maize after ABA therapy (Liu *et al.*, 2008; Zou *et al.*, 2009). Water stress on maize suppressed miR167 expression but increased phospholipase D (PLD) transcript abundance, indicating that PLD seems important for ABA-dependent stomatal movement and is a miR167 target (Wei *et al.*, 2009; Zhang *et al.*, 2005). miR167 was induced in maize during flooding conditions (Liu *et al.*, 2012). The miR167 regulates AUX signalling by modulating free AUX concentrations and adventitious roots growth by targeting AUX-responsive factors (ARFs). Exogenous administration of AUX and JA modulated the transcription of miR167 in response to arsenic exposure, favourably impacting plant development and regulating crosstalk among plant hormones and miRNAs. As a consequence, miRNAs play a critical role in JA biosynthesis additionally in relation to HM stress (Gupta *et al.*, 2014).

3.3   miR156 and miR172

miR156 and miR172 function at the molecular level by down-regulating genes that code for squamosa promoter binding-like (SPL) as well as Apetala-2 like (AP2-like) transcriptional factors family, respectively; therefore, subject to feedback regulation by respective targets. Moreover, numerous SPL genes affected by miR156 have been found as linked with miR172, including AtSPL9/10, which precisely up-regulates AtMIR172b transcription and AtSPL3/4/5, whose expression is up-regulated indirectly via miR172 (Wu *et al.*, 2009; Jung *et al.*, 2011), indicating a molecular connection between the two miRNAs. In a recent tomato research, strigolactones were described to operate as molecular link among drought and miR156, with miR156 acting as mediators of strigolactones' ABA-dependent action on stomatal pores which were affected by drought conditions (Visentin *et al.*, 2020). Cold stress persuades temporary or moderate expression of miR156/157 in *A. thaliana*. CBF (C-repeat binding F factors) transcriptional factors bound to DRE (dehydration-responsive element) in *A. thaliana* throughout cold stress conditions, regulating cold-responsive genes. ABA signalling is influenced by a numerous variety of miRNAs, which allow many plants to resist cold stress. Furthermore, according to investigations, cold-tolerant cultivars produce more ABA under cold stress than cold-sensitive ones (Kumar, 2014). Cold stress raised miR172 expression in *A. thaliana*. In *R. sativus*, suppressing miR172c increased the AP2-like ethylene-responsive factor, SNZ and SAM-dependent methyltransferase genes throughout high salt stress (Sun *et al.*, 2015). UV-B exposure enhanced the transcription of miR156 and miR172 in *A. thaliana* (Zhou *et al.*, 2007). miR156s are up-regulated in relation to Cd stress (Huang *et al.*, 2010) and miR156 down-regulation (Zhou *et al.*, 2012). Cr stress caused differential expression of miRNAs in radish. In the presence of Cr stress, miR156 was down-regulated, whereas miR172 was up-regulated (Liu *et al.*, 2015). miR156/157, which target multiple SPLs – SPL3/6/9/13/15 in radish, might play a crucial role in maintaining Cr homeostasis, according to research. As a consequence, miRNA-mediated control of Cr stress in radish reduces Cr stress by addressing different transcriptional factors (Liu *et al.*, 2015).

3.4 miR164 and miR319

Cold stress causes miR164 and miR319 to express transiently or minimally in *A. thaliana*. CBF (C-repeat binding F factors) transcriptional factors linked to DRE (dehydration-responsive element) in *A. thaliana* under cold stress, leading to up-regulation of cold-responsive factors. ABA signalling is mediated by numerous miRNAs, which allow many plants to withstand cold stress. Moreover, according to investigations, cold-tolerant cultivars produce more ABA throughout cold stress than cold-sensitive ones (Kumar, 2014). miR164b is up-regulated in relation to Cd stress (Huang *et al.*, 2010). By modulating the TCP transcription factors, osa-miR319a overexpressing *Agrostis stolonifera* demonstrated increased resistance to stress conditions, like salinity and drought (Zhou *et al.*, 2013). miR319 expression level was increased by UV-B exposure (Zhou *et al.*, 2007). Though jasmonic acid has been known to function exclusively in response to arsenic stress, only some miRNAs regulate its synthesis (Singh *et al.*, 2017; Gupta *et al.*, 2014). miR319 was revealed as up-regulated throughout arsenic stress and to impact JA biosynthesis by directing the TCP transcriptional factor under As, Hg, Cd and Al stress, according to a research (Liu *et al.*, 2012). This shows that miRNAs regulate JA synthesis and, as an output, plant response to metal stress. Extracellular administration of AUX and JA modulated the regulation miR319 in respect to arsenic exposure, promoting plant development and modulating hormone-miRNA interaction. As a result, miRNAs are important in JA production as well as the response to HM stress (Gupta *et al.*, 2014). Cr stress caused differential regulation of miRNAs in radish. miR319 was one of them and it was down-regulated in relation to Cr stress. miR319, which targets MYB3/13/101/104/305, may also have a role in various genes expression under Cr stress. As a consequence, miRNA-mediated Cr stress regulation in radish reduces Cr stress by targeting different transcriptional factors (Liu *et al.*, 2015).

3.5 miR390

Antibacterial resistance induced via bacterial flagellin involves AtTIR1 and AtAFB2/3 down-regulation by miR393-mediated suppression of AUX signals (Navarro *et al.*, 2006). Because ABA has a function in stress reactions, a rise in ABA level during bacterial infection might activate the miR393-AUX anti-bacterial resistant mechanism. These findings imply that the ABA-miR393-AUX regulation pathway is implicated in both abiotic and biotic stress mechanisms. The AUX signalling mediated miR390-TAS3-ARF2/3/4 mechanism in *A. thaliana* is governed by miR390, that targeted the generation of trans-acting small interfering RNAs (tasi-RNA) derived through TAS3, that affect adventitious root growth and polarisation through targeting these ARF2/3/4 transcriptional elements (Meng *et al.*, 2010). Drought stress significantly triggers miR160 regulation and lateral plant root growth by altering AUX levels controlled by miR390 (Bustos-Sanmamed *et al.*, 2013; Yoon *et al.*, 2010). AUX responsive factors (ARFs) govern root growth and are regulated by Al-responsive miRNAs, miR390 and miR160. The miR390 regulates lateral root growth emergence via modulating tasiRNA synthesis, which targets ARF2/3/4 (Yoon *et al.*, 2010). During Cr stress, MiR390 was up-regulated. As an outcome, miRNA-mediated control of Cr stress within radish reduces Cr stress by addressing different transcriptional factors (Liu *et al.*, 2015).

3.6 miR393

The crucial role of miR393 in unidirectional regulation of ABA to AUX signalling has been suggested. The biogenesis of miR393 is boosted via ABA treatment (Sunkar and Zhu, 2004). miR393 inhibits AUX signal reception by down-regulating the transcription of four members of TIR-1/AFB (transport inhibitor response1/AUX signalling F-box) family of TAAR (AUX receptor) proteins found in *A. thaliana* (AtTIR1, AtAFB1/2/3) (Si-Ammour *et al.*, 2011; Chen *et al.*, 2012). During drought circumstances, an increase in ABA levels has been associated to reduce adventitious root growth, resulting in miR393-dependent suppression of AUX signalling mechanism via down-regulation of AtTIR1 and AtAFB2 (Chen *et al.*, 2012). In rice, the overexpression of miR393 decreases regulation of OsTIR1 and OsAFB2, resulting in reduction in root development and drought tolerance (Bian *et al.*, 2012; Xia *et al.*, 2012). Antibacterial resistance generated by bacterial flagellin also includes the suppression of AUX signal by miR393, which results in suppression of AtTIR1 and AtAFB2/3 (Navarro *et al.*, 2006). Because ABA is involved in stress reactions, a surge in ABA levels during bacterial infection might activate the miR393-AUX antibacterial resistant mechanism. The ABA-miR393-AUX regulatory pathway is implicated in abiotic and biotic stress reactions, according to these findings. *A. thaliana* overexpressing ABA-responsive miR393 showed resistance through drought stress, indicating that miRNA plays a crucial role in ABA-mediated unfavourable environmental adaptation (Baek *et al.*, 2016). In *A. thaliana*, mutant investigations of important genes have a function in synthesis of miRNA and these are Dcl1, Hyl1, Hen1, Hasty, and Se, indicating that hyl1, dcl1, and hen1 variants are hypersensitive to ABA, but the hasty as well as Se variants did not, establish a relationship among miRNA and ABA sensitivity (Cambiagno *et al.*, 2021; Zhang *et al.*, 2008). ABA appears to stimulate the regulation of miR393 in *A. thaliana*, based on many findings (Jung and Kang, 2007; Jia *et al.*, 2009). By modulating AUX receptors, OsTIR1 and OsAFB2, up-regulation of miR393 in rice plants leads to early maturity as well as higher count of tiller, besides drought and salt tolerance, showing the function of AUX during response to the stress situation (Bian *et al.*, 2012). Furthermore, miR393-mediated degradation of TAAR mRNAs yields tasiRNAs, which may exacerbate TAAR gene down-regulation (Si-Ammour *et al.*, 2011). Eventually, miR393 may be a powerful and sensitive pathway via which ABA might control in AUX signal. Cold stress causes the regulation of miR393 in *A. thaliana*. The transcription factors for C-repeat binding F factors (CBF) transcriptional factors were reported to interact with dehydration-responsive element (DRE) in *A. thaliana* during cold stress, leading to up-regulation of cold-responsive genes. ABA signalling is controlled by numerous miRNAs, which allow many plants to withstand cold stress. Furthermore, investigations revealed that cold-tolerant types produce more ABA during cold stress than cold-sensitive kinds (Kumar, 2014). UV-B exposure enhanced the miR393 expression in *A. thaliana* (Zhou *et al.*, 2007). MiR393 regulates AUX signalling by targeting multiple ARFs in relation to UV stress, impacting growth and survival of plants. Cd stress causes up-regulation of miR393 in *B. napus* (Huang *et al.*, 2010). miR393 is expected to target proteins of F-box, like E3 ubiquitin ligases and TIR1 (transport inhibitor response 1), which are critical for the ubiquitination process. TIR1 modulates AUX signalling via exacerbating AUX/indole acetic acid molecules (Chinnusamy *et al.*, 2007). Up-regulation of miR393 at conditions of cadmium stress points to higher TIR1 mRNA levels which inhibits AUX signalling and reduces E3 ubiquitin ligase activity. As an output, there might

Crosstalk of MicroRNAs with Phytohormone Signalling Pathways **269**

be a link between Cd stress and AUX signalling. In rice, Al toxicity causes miR393b to be down-regulated, which is involved with proteasome-mediated processing of proteins. MiR393b is inhibited by up-regulation of miR160e and targeting of ARFs in the AUX signalling pathway. As a consequence, miR393b is down-regulated in rice to counteract Al stress.

3.7 miR396

miR396-GRF has been recorded to regulate an expression of diverse ethylene responsive factors (ERFs) and ABA-responsive genes, indicating that its control of ABA and ET mechanisms is conserved. ERFs are genes that code for transcriptional factors and have been associated with adaptation to stress (Mizoi *et al.*, 2012). ERF11, a key molecular connection among ABA and ET biogenesis (Li *et al.*, 2011), was depicted to considerably elevate among grfl/2/3 triple mutants. Up-regulation of miR396 by salinity stress (Liu *et al.*, 2008) or also overexpression of AtMIR396a/b has been detected to raise drought resistance via change in leaf shape (Liu *et al.*, 2009), indicating that miR396 may have a role in ABA-mediated drought adaptation in plants. In rice, salinity treatment decreases OsMIR396c expression and transgenic overexpression reduces salinity stress resistance in both *A. thaliana* and *O. sativa* (Gao *et al.*, 2010). The mature sequence of Osa-miR396c and Ath-miR396b, in contrast, were determined to be similar. This finding emphasises on the complexities of stress-mediated control of miR396 which indicates that precursor elements that overexpress the miRNAs (Ath-primiR396b and Osa-pri-miR396c) may be processed differently in relation to drought stress, potentially resulting in the generation of possible small RNA species. These findings suggest that miR396 may be a critical regulator of cell proliferation in relation to biotic as well as abiotic stress. Cold stress causes miR396 to be expressed in *A. thaliana*. The C-repeat binding F factors (CBF) transcriptional factors interact to dehydration-responsive element (DRE) in *A. thaliana* during cold stress, leading to up-regulation of cold-responsive genes. ABA signalling is controlled by a variety of miRNAs, which allow many plants to withstand cold stress. Furthermore, investigations have shown that cold-tolerant cultivars produce more ABA during cold stress than cold-sensitive cultivars (Kumar, 2014). Cadmium (Cd) was one of utmost harmful contaminants that harm plant production by producing chlorosis, wilting, stunted development and even death. Cd stress causes miR396a to be down-regulated in *B. napus* (Zhou *et al.*, 2012). The findings indicate that miR396 may be a key pathway for ABA and ET by regulating cellular proliferation during response to biotic and abiotic stresses.

4. Conclusion

miRNAs control gene regulation after transcription and are engaged in a range of growth or metabolic processes of plants. These miRNAs develop stress tolerance via modulating numerous stress-responsive transcripts, proteins, TFs and phytohormones despite harsh environmental circumstances. Hormonal control is also engaged in survival and growth of plants as and when they face any kind of biotic or abiotic stress. miRNAs, in general, influence hormone responses by controlling initial hormone-responsive proteins, and hence primarily affect small portion of growth and survival. In contrast to the major hormones, such as abscisic acid, AUX, CK, ET, GA, JA and other hormonal mechanisms like SA, BR, polyamines, plant peptide hormones, nitric oxide and SL are also involved with miRNA regulation. This chapter also looks at how MAPK and calcium signalling interact with phytohormones in relation to various abiotic stressors, including HM stress, implying that MAPK plays a role in signalling molecule regulation, both upstream and downstream. Interestingly, the majority of miRNAs engaged with hormones crosstalk link via AUX/ABA/ET pathways. Land plants evolved to necessitate adaptation with environmental changes, such as seasonal and stress resistance that necessitated interaction between AUX (primary hormone directing plant organogenesis), ABA/ET (stress-related hormone), implying that miRNAs played a role in plant stress adaptation. miRNAs govern hormonal responses primarily downstream of phytohormone transmission signal via influencing the expression of initial phytohormone-responsive transcripts, impacting just a portion of the hormonally controlled developmental processes. Exactly four miRNA groups, i.e. miR167, miR319, miR393 and miR396 were identified to function upstream via modulating genes associated with phytohormone production, transport and perception; hence they may play an important role in hormonal efficient regulation. According to this approach, miR167, miR319 and miR393 might serve as crucial ABA gateways to the AUX signal. The frequency of hormones regulated miRNAs and respective targets implicated in hormonal signalling keep growing as more research employing miRNAs and mRNA degradome resources across various plant taxa are published. miRNA/hormonal network becomes increasingly complicated as research develops, necessitating more precise investigations, such as losing miRNA function, may uncover genetic controls underlying hormonal interaction within spatiotemporal framework of the biological event. Under abiotic stress, especially HMs, the interaction among miRNAs and plant hormones is not well known. To get a better knowledge of the miRNA-phytohormone interaction, comprehensive recognition of miRNAs, interconnections among miRNAs and different metabolic or signalling mechanisms and discovery of regulatory factors upstream from miRNAs are necessary. As a result, a more in-depth investigation in future will be beneficial in developing better methods for managing diverse challenges.

References

Abdel-Ghany, S.E. and Pilon, M. (2008). MicroRNA-mediated systemic down-regulation of copper protein expression in response to low copper availability in *Arabidopsis*, *Journal of Biological Chemistry*, 283(23): 15932-15945.

Allen, E., Xie, Z., Gustafson, A.M. and Carrington J.C. (2005). MicroRNA-directed phasing during transacting siRNA biogenesis in plants, *Cell*, 121(2): 207-221.

Aravind, J., Rinku, S., Pooja, B., Shikha, M., Kaliyugam, S., Mallikarjuna, M.G., Kumar, A., Rao, A.R. and Nepolean, T. (2017). Identification, characterisation, and functional validation of drought-responsive microRNAs in subtropical maize inbreds, *Frontiers in Plant Science*, 8: 941.

Arenas-Huertero, C., Perez, B., Rabanal, F., Blanco-Melo, D., De la Rosa, C., Estrada-Navarrete, G., Sanchez, F., Covarrubias, A.A. and Reyes, J.L. (2009). Conserved and novel miRNAs in the legume *Phaseolus vulgaris* in response to stress, *Plant Molecular Biology*, 70: 385-401.

Baek, D., Chun, H.J., Kang, S., Shin, G., Park, S.J., Hong, H., Kim, C., Kim, D.H., Lee, S.Y., Kim, M.C. and Yun, D.J. (2016). A role for *Arabidopsis* miR399f in salt, drought, and ABA signalling, *Molecules and Cells*, 39(2): 111.

Crosstalk of MicroRNAs with Phytohormone Signalling Pathways

Beauclair, L., Yu, A. and Bouche, N. (2010). MicroRNA-directed cleavage and translational repression of the copper chaperone for superoxide dismutase mRNA in *Arabidopsis*, *Plant Journal*, 62(3): 454-462.

Bian, H., Xie, Y., Guo, F., Han, N., Ma, S., Zeng, Z., Wang, J., Yang, Y. and Zhu, M. (2012). Distinctive expression patterns and roles of the miRNA393/TIR1 homolog module in regulating flag leaf inclination and primary and crown root growth in rice (*Oryza sativa*), *New Phytologist.*, 196(1): 149-161.

Blazquez, M.A., Nelson, D.C. and Weijers, D. (2020). Evolution of plant hormone response pathways, *Annual Review of Plant Biology*, 71: 327-353.

Bonnet, E., Wuyts, J., Rouze, P. and Van de Peer, Y. (2004). Detection of 91 potential conserved plant microRNAs in *Arabidopsis thaliana* and *Oryza sativa* identifies important target genes, *Proceedings of the National Academy of Sciences of the United States of America*, 101(31): 11511-11516.

Bustos-Sanmamed, P., Mao, G., Deng, Y., Elouet, M., Khan, G.A., Bazin, J., Turner, M., Subramanian, S., Yu, O., Crespi, M. and Lelandais-Brière, C. (2013). Overexpression of miR160 affects root growth and nitrogen-fixing nodule number in *Medicago truncatula*, *Functional Plant Biology*, 40(12): 1208-1220.

Cambiagno, D.A., Giudicatti, A.J., Arce, A.L., Gagliardi, D., Li, L., Yuan, W., Lundberg, D.S., Weigel, D. and Manavella, P.A. (2021). HASTY modulates miRNA biogenesis by linking pri-miRNA transcription and processing, *Molecular Plant*, 14(3): 426-439.

Campo, S., Peris-Peris, C., Sire, C., Moreno, A.B., Donaire, L., Zytnicki, M., Notredame, C., Llave, C. and San Segundo, B.S. (2013). Identification of a novel microRNA (miRNA) from rice that targets an alternatively spliced transcript of the Nramp6 (Natural resistance associated macrophage protein 6) gene involved in pathogen resistance, *New Phytologist.*, 199(1): 212-227.

Carrington, J.C. and Ambros, V. (2003). Role of microRNAs in plant and animal development, *Science*, 301(5631): 336-338.

Chen, H., Li, Z. and Xiong, L. (2012). A plant microRNA regulates the adaptation of roots to drought stress, *FEBS Letters*, 586(12): 1742-1747.

Chinnusamy, V., Zhu, J. and Zhu, J.K. (2007). Cold stress regulation of gene expression in plants, *Trends in Plant Science*, 12(10): 444-451.

Chiou, T.J., Aung, K., Lin, S.I., Wu, C.C., Chiang, S.F. and Su, C.L. (2006). Regulation of phosphate homeostasis by micro-RNA in *Arabidopsis*, *Plant Cell*, 18(2): 412-421.

Curaba, J., Singh, M.B. and Bhalla, P.L. (2014). miRNAs in the crosstalk between phytohormone signalling pathways, *Journal of Experimental Botany*, 65(6): 1425-1438.

Curaba, J., Spriggs, A., Taylor, J., Li, Z. and Helliwell, C. (2012). miRNA regulation in the early development of barley seed, *BMC Plant Biology*, 12(1): 1-16.

Ding, D., Zhang, L., Wang, H., Liu, Z., Zhang, Z. and Zheng, Y. (2009). Differential expression of miRNAs in response to salt stress in maize roots, *Annals of Botany*, 103: 29-38.

Djami-Tchatchou, A.T., Sanan-Mishra, N., Ntushelo, K. and Dubery, I.A. (2017). Functional roles of microRNAs in agronomically important plants: Potential as targets for crop improvement and protection, *Frontiers in Plant Science*, 8: 378.

Dubey, S., Saxena, S., Chauhan, A.S., Mathur, P., Rani, V. and Chakrabaroty, D. (2020). Identification and expression analysis of conserved microRNAs during short and prolonged chromium stress in rice (*Oryza sativa*), *Environmental Science and Pollution Research*, 27(1): 380-390.

Dubois, M., Van den Broeck, L. and Inzé, D. (2018). The pivotal role of ethylene in plant growth, *Trends in Plant Science*, 23(4): 311-323.

Fahlgren, N., Howell, M.D., Kasschau, K.D., Chapman, E.J., Sullivan, C.M., Cumbie, J.S., Givan, S.A., Law, T.F., Grant, S.R., Dangl, J.L. and Carrington, J.C. (2007). High-throughput sequencing of *Arabidopsis* microRNAs: Evidence for frequent birth and death of miRNA genes, *PLoS ONE*, 2(2): e219.

Foo, E., Plett, J.M., Lopez-Raez, J.A. and Reid, D. (2019). The role of plant hormones in plant-microbe symbioses, *Frontiers in Plant Science*, 10(1391).

Fuji, H., Chiou, T.J., Lin, S.I., Aung, K. and Zhu, J.K. (2005). A miRNA involved in phosphate starvation response in *Arabidopsis*, *Current Biology*, 15(22): 2038-2043.

Gao, P., Bai, X., Yang, L., Lv, D., Li, Y., Cai, H., Ji, W., Guo, D. and Zhu, Y. (2010). Overexpression of osa-MIR396c decreases salt and alkali stress tolerance, *Planta*, 231(5): 991-1001.

Gao, P., Bai, X., Yang, L., Lv, D., Pan, X., Li, Y., Cai, H., Ji, W., Chen, Q. and Zhu, Y. (2011). *osa-MIR393*: A salinity and alkaline stress-related microRNA gene, *Molecular Biology Reports*, 38: 237-242.

Gupta, O.P., Sharma, P., Gupta, R.K. and Sharma, I. (2014). MicroRNA-mediated regulation of metal toxicity in plants: Present status and future perspectives, *Plant Molecular Biology*, 84(1-2): 1-18.

Gutierrez, L., Bussell, J.D., Pacurar, D.I., Schwambach, J., Pacurar, M. and Bellini, C. (2009). Phenotypic plasticity of adventitious rooting in *Arabidopsis* is controlled by complex regulation of auxin response factor transcripts and microRNA abundance, *The Plant Cell*, 21(10): 3119-3132.

Han, M.H., Goud, S., Song, L. and Fedoroff, N. (2004). The *Arabidopsis* double-stranded RNA-binding protein HYL1 plays a role in microRNA-mediated gene regulation, *Proceedings of the National Academy of Sciences*, 101(4): 1093-1098.

He, X.F., Fang, Y.Y., Feng, L. and Guo, H.S. (2008). Characterisation of conserved and novel microRNAs and their targets, including a TuMV-induced TIR-NBS-LRR Class R Gene-derived novel miRNA in *Brassica*, *FEBS Letters*, 582(16): 2445-2452.

Hewezi, T., Howe, P., Maier, T.R. and Baum, T.J. (2008). *Arabidopsis* small RNAs and their targets during cyst nematode parasitism, *Molecular Plant-Microbe Interactions*, 21(12): 1622-1634.

Huang, S.Q., Xiang, A.L., Che, L.L., Chen, S., Li, H., Song, J.B. and Yang, Z.M. (2010). A set of miRNAs from *Brassica napus* in response to sulphate deficiency and cadmium stress, *Plant Biotechnology Journal*, 8(8): 887-899.

Jia, X., Ren, L., Chen, Q.J., Li, R. and Tang, G. (2009). UV-B-responsive microRNAs in *Populus tremula*, *Journal of Plant Physiology*, 166(18): 2046-2057.

Jia, X., Wang, W.X., Ren, L., Chen, Q.J., Mendu, V., Willcut, B., Dinkins, R., Tang, X. and Tang, G. (2009). Differential and dynamic regulation of miR398 in response to ABA and salt stress in *Populus tremula* and *Arabidopsis thaliana*, *Plant Molecular Biology*, 71(1-2): 51-59.

Jones-Rhoades, M.W. and Bartel, D.P. (2004). Computational identification of plant microRNAs and their targets, including a stress-induced miRNA, *Molecular Cell*, 14(6): 787-799.

Jones-Rhoades, M.W., Bartel, D.P. and Bartel, B. (2006). MicroRNAs and their regulatory roles in plants, *Annu. Rev. Plant. Biol.*, 57: 19-53.

Jung, H.J. and Kang, H. (2007). Expression and functional analyses of microRNA417 in *Arabidopsis thaliana* under stress conditions, *Plant Physiology and Biochemistry*, 45(10-11): 805-811.

Jung, J.H., Seo, P.J., Kang, S.K. and Park, C.M. (2011). miR172 signals are incorporated into the miR156 signalling pathway at the SPL3/4/5 genes in *Arabidopsis* developmental transitions, *Plant Molecular Biology*, 76(1): 35-45.

Kantar, M., Lucas, S.J. and Budak, H. (2011). miRNA expression patterns of *Triticum dicoccoides* in response to shock drought stress, *Planta*, 233(3): 471-484.

Kasschau, K.D., Xie, X., Allen, E., Llave, C., Chapman, E.J., Krizan, K.A. and Carrington, J.C. (2003). P1/HC-Pro, a viral suppressor of RNA silencing, interferes with *Arabidopsis* development and miRNA unction, *Development Cell*, 4(2): 205-217.

Kawaguchi, R., Girke, T., Bray, E.A. and Bailey-Serres, J. (2004). Differential mRNA translation contributes to gene regulation under non-stress and dehydration stress conditions in *Arabidopsis thaliana*, *Plant Journal*, 38(5): 823-839.

Khraiwesh, B., Zhu, J.K. and Zhu, J. (2012). Role of miRNAs and siRNAs in biotic and abiotic stress responses of plants, *Biochimica et Biophysica Acta*, 1819(2): 137-148.

Kim, S., Yang, J., Xu, J., Jang, I.C., Prigge, M.J. and Chua, N.H. (2008). Two cap-binding proteins CBP20 and CBP80 are involved in processing primary microRNAs, *Plant Cell Physiology*, 49(11): 1634-1644.

Kumar, R. (2014). Role of microRNAs in biotic and abiotic stress responses in crop plants, *Applied Biochemistry and Biotechnology*, 174(1): 93-115.

Lee, R.C., Feinbaum, R.L. and Ambros, V. (1993). The *C. elegans* heterochronic gene lin-4 encodes small RNAs with antisense complementarity to lin-14, *Cell*, 75(5): 843-854.

Li, N., Han, X., Feng, D., Yuan, D. and Huang, L.J. (2019). Signalling crosstalk between salicylic acid and ethylene/jasmonate in plant defence: Do we understand what they are whispering? *International Journal of Molecular Sciences*, 20(3): 671.

Li, X., Wang, X., Zhang, S., Liu, D., Duan, Y. and Dong, W. (2012). Identification of soybean microRNAs involved in soybean cyst nematode infection by deep sequencing, *PLoS ONE*, 7(6): e39650.

Li, Z., Zhang, L., Yu, Y., Quan, R., Zhang, Z., Zhang, H. and Huang, R. (2011). The ethylene response factor AtERF11 that is transcriptionally modulated by the bZIP transcription factor HY5 is a crucial repressor for ethylene biosynthesis in *Arabidopsis*, *The Plant Journal*, 68(1): 88-99.

Liang, G. and Yu, D. (2010). Reciprocal regulation among miR395 APS and SULTR2; 1 in *Arabidopsis thaliana*, *Plant Signalling and Behaviour*, 5(10): 1257-1259.

Liu, D., Song, Y., Chen, Z. and Yu, D. (2009). Ectopic expression of miR396 suppresses GRF target gene expression and alters leaf growth in *Arabidopsis*, *Physiologia Plantarum.*, 136(2): 223-236.

Liu, H.H., Tian, X., Li, Y.J., Wu, C.A. and Zheng, C.C. (2008). Microarray-based analysis of stress-regulated microRNAs in *Arabidopsis thaliana*, *RNA*, 14(5): 836-843.

Liu, P.P., Montgomery, T.A., Fahlgren, N., Kasschau, K.D., Nonogaki, H. and Carrington, J.C. (2007). Repression of auxin response factor10 by microRNA160 is critical for seed germination and post-germination stages, *The Plant Journal*, 52(1): 133-146.

Liu, W., Xu, L., Wang, Y., Shen, H., Zhu, X., Zhang, K., Chen, Y., Yu, R., Limera, C. and Liu, L. (2015). Transcriptome-wide analysis of chromium-stress responsive microRNAs to explore miRNA-mediated regulatory networks in radish (*Raphanus sativus* L.), *Scientific Reports*, 5(1): 1-17.

Liu, Z., Kumari, S., Zhang, L., Zheng, Y. and Ware, D. (2012). Characterisation of miRNAs in response to short-term waterlogging in three inbred lines of *Zea mays*, *PLoS ONE*, 7(6): e39786.

Lu, C. and Fedoroff, N. (2000). A mutation in the *Arabidopsis* HYL1 gene encoding a dsRNA binding protein affects responses to abscisic acid, auxin, and cytokinin, *Plant Cell*, 12(12): 2351-2366.

Lu, S., Sun, Y.H., Amerson, H. and Chiang, V.L. (2007). MicroRNAs in loblolly pine (*Pinus taeda* L.) and their association with *fusiform* rust gall development, *The Plant Journal*, 51(6): 1077-1098.

Lu, S., Sun, Y.H., Shi, R., Clark, C., Li, L. and Chiang, V.L. (2005). Novel and mechanical stress responsive microRNAs in *Populous trichocarpa* that are absent from *Arabidopsis*, *Plant Cell*, 17(8): 2186-2203.

Lu, X.Y. and Huang, X.L. (2008). Plant miRNAs and abiotic stress responses, *Biochemical and Biophysical ResearchCommunications*, 368(3): 458-462.

Lv, D.K., Bai, X., Li, Y., Ding, X.D., Ge, Y., Cai, H., Ji, W., Wu, N. and Zhu, Y.M. (2010). Profiling of cold-stress responsive miRNAs in rice by microarrays, *Gene*, 459(1-2): 39-47.

Ma, J.F., Shen, R., Nagao, S. and Tanimoto, E. (2004). Aluminium targets elongating cells by reducing cell wall extensibility in wheat roots, *Plant and Cell Physiology*, 45(5): 583-589.

Makarova, J.A., Shkurnikov, M.U., Wicklein, D., Lange, T., Samatov, T.R., Turchinovich, A.A. and Tonevitsky, A.G. (2016). Intracellular and extracellular microRNA: An update on localisation and biological role, *Progress in Histochemistry and Cytochemistry*, 51(3-4): 33-49.

Mallory, A.C., Bartel, D.P. and Bartel, B. (2005). MicroRNA-directed regulation of *Arabidopsis* auxin response factor17 is essential for proper development and modulates expression of early auxin response genes, *The Plant Cell*, 17(5): 1360-1375.

Marin, E., Jouannet, V., Herz, A., Lokerse, A.S., Weijers, D., Vaucheret, H., Nussaume, L., Crespi, M. D. and Maizel, A. (2010). miR390, *Arabidopsis* TAS3 tasiRNAs, and their auxin response factor targets define an auto-regulatory network quantitatively regulating lateral root growth, *The Plant Cell*, 22(4): 1104-1117.

McKenzie, R.L., Aucamp, P.J., Bais, A.F., Bjorn, L.O. and Ilyas, M. (2007). Changes in biologically-active ultraviolet radiation reaching the Earth's surface, *Photochemical & Photobiological Sciences*, 6(3): 218-231.

Meng, Y., Ma, X., Chen, D., Wu, P. and Chen, M. (2010). MicroRNA-mediated signalling involved in plant root development, *Biochemical and Biophysical Research Communications*, 393(3): 345-349.

Mizoi, J., Shinozaki, K. and Yamaguchi-Shinozaki, K. (2012). AP2/ERF family transcription factors in plant abiotic stress responses, *Biochimica et Biophysica Acta (BBA)-Gene Regulatory Mechanisms*, 1819(2): 86-96.

Naqvi, A.R., Choudhury, N.R., Haq, Q.M.R. and Mukherjee, S.K. (2008). MicroRNAs as biomarkers in tomato leaf curl virus (ToCLV), *Nucleic Acids Symposium Series*, 52(1): 507-508.

Navarro, L., Dunoyer, P., Jay, F., Arnold, B., Dharmasiri, N., Estelle, M., Voinnet, O. and Jones, J.D. (2006). A plant miRNA contributes to antibacterial resistance by repressing auxin signalling, *Science*, 312(5772): 436-439.

Niu, Q.W., Lin, S.S., Reyes, J.L., Chen, K.C., Wu, H.W., Yeh, S.D. and Chua, N.H. (2006). Expression of artificial microRNAs in transgenic *Arabidopsis thaliana* confers virus resistance, *Nature Biotechnology*, 24(11): 1420-1428.

Noman, A. and Aqeel, M. (2017). miRNA-based heavy metal homeostasis and plant growth, *Environmental Science and Pollution Research*, 24(11): 10068-10082.

Pandita, D. (2019). Plant MIRnome: miRNA biogenesis and abiotic stress response. *In:* Hasanuzzaman, M., Hakeem, K., Nahar, K. and Alharby, H. (Eds.). *Plant Abiotic Stress Tolerance*. Springer, Cham., 449-474. Doi https://Doi.org/10.1007/978-3-030-06118-0_18

Pandita, D. and Wani, S.H. (2019). MicroRNA as a tool for mitigating abiotic stress in rice (*Oryza sativa* L.). *In:* Wani, S. (Ed.). *Recent Approaches in Omics for Plant Resilience to Climate Change*. Springer, Chamz., 109-133. Doi https://Doi.org/10.1007/978-3-030-21687-0_6

Pandita, D. (2021). Role of miRNAi technology and miRNAs in abiotic and biotic stress resilience. *In:* Aftab, T. and Roychoudhury, A. (Eds.). *Plant Perspectives to Global Climate Changes*. Academic Press, Elsevier, 303-330. https://Doi.org/10.1016/B978-0-323-85665-2.00015-7

Pandita, D. (2022a). How microRNAs regulate abiotic stress tolerance in wheat? A snapshot. *In:* Roychoudhury, A., Aftab, T. and Acharya, K. (Eds.). *Omics Approach to Manage Abiotic Stress in Cereals*. Springer, Singapore, 447-464. https://Doi.org/10.1007/978-981-19-0140-9_17

Pandita, D. (2022b). MicroRNAs shape the tolerance mechanisms against abiotic stress in maize. *In:* Roychoudhury, A., Aftab, T. and Acharya, K. (Eds.). *Omics Approach to Manage Abiotic Stress in Cereals*. Springer, Singapore, 479-493. https://Doi.org/10.1007/978-981-19-0140-9_19

Pandita, D. (2022c). miRNA- and RNAi-mediated metabolic engineering in plants. *In:* Aftab, T. and Hakeem, K.R. (Eds.). *Metabolic Engineering in Plants*. Springer, Singapore, 171-186. https://doi.org/10.1007/978-981-16-7262-0_7

Patel, P., Yadav, K., Ganapathi, T.R. and Penna, S. (2019). Plant miRNAome: Crosstalk in abiotic stressful times, 1: 25-52. *In: Genetic Enhancement of Crops for Tolerance to Abiotic Stress: Mechanisms and Approaches*. Springer, Cham.

Peleg, Z. and Blumwald, E. (2011). Hormone balance and abiotic stress tolerance in crop plants, *Current Opinion in Plant Biology*, 14(3): 290-295.

Crosstalk of MicroRNAs with Phytohormone Signalling Pathways

Phillips, J.R., Dalmay, T. and Bartels, D. (2007). The role of small RNAs in abiotic stress, *FEBS Letters*, 581(18): 3592-3597.

Reyes, J.L. and Chua, N.H. (2007). ABA induction of miR159 controls transcript levels of two MYB factors during *Arabidopsis* seed germination, *The Plant Journal*, 49(4): 592-606.

Si-Ammour, A., Windels, D., Arn-Bouldoires, E., Kutter, C., Ailhas, J., Meins Jr, F. and Vazquez, F. (2011). miR393 and secondary siRNAs regulate expression of the TIR1/AFB2 auxin receptor clade and auxin-related development of *Arabidopsis* leaves, *Plant Physiology*, 157(2): 683-691.

Silva, S. (2012). Aluminium toxicity targets in plants, *Journal of Botany*, 219462. Doi:10.1155/2012/219462

Singh, P.K., Indoliya, Y., Chauhan, A.S., Singh, S.P., Singh, A.P., Dwivedi, S., Tripathi, R.D. and Chakrabarty, D. (2017). Nitric oxide mediated transcriptional modulation enhances plant adaptive responses to arsenic stress, *Scientific Reports*, 7(1): 1-13.

Srivastava, S., Srivastava, A.K., Suprasanna, P. and D'souza, S.F. (2013). Identification and profiling of arsenic stress-induced microRNAs in *Brassica juncea*, *Journal of Experimental Botany*, 64(1): 303-315.

Sun, X., Xu, L., Wang, Y., Yu, R., Zhu, X., Lou, X., Gong, Y., Wang, R., Limera, C., Zhang, K. and Liu, L. (2015). Identification of novel and salt-responsive miRNAs to explore miRNA-mediated regulatory network of salt stress response in radish (*Raphanus sativus* L.), *BMC Genomics*, 16: 197.

Sunkar, R. and Zhu, J.K. (2004). Novel and stress-regulated microRNAs and other small RNAs from *Arabidopsis*, *The Plant Cell*, 16(8): 2001-2019.

Sunkar, R., Chinnusamy, V., Zhu, J. and Zhu, J.K. (2007). Small RNAs as big players in plant abiotic stress responses and nutrient deprivation, *Trends in Plant Science*, 12(7): 301-309.

Trindade, I., Capitão, C., Dalmay, T., Fevereiro, M.P. and Santos, D.M.D. (2010). miR398 and miR408 are up-regulated in response to water deficit in *Medicago truncatula*, *Planta*, 231: 705-716.

Visentin, I., Pagliarani, C., Deva, E., Caracci, A., Tureckova, V., Novak, O., Lovisolo, C., Schubert, A. and Cardinale, F. (2020). A novel strigolactone-miR156 module controls stomatal behaviour during drought recovery, *Plant, Cell & Environment*, 43(7): 1613-1624.

Wang, J.W., Wang, L.J., Mao, Y.B., Cai, W.J., Xue, H.W. and Chen, X.Y. (2005). Control of root cap formation by microRNA-targeted auxin response factors in *Arabidopsis*, *The Plant Cell*, 17(8): 2204-2216.

Wang, W., Vinocur, B. and Altman, A. (2003). Plant responses to drought, salinity and extreme temperatures: Towards genetic engineering for stress tolerance, *Planta*, 218(1): 1-14.

Wani, S.H., Kumar, V., Shriram, V. and Sah, S.K. (2016). Phytohormones and their metabolic engineering for abiotic stress tolerance in crop plants, *The Crop Journal*, 4(3): 162-176.

Wei, L., Zhang, D., Xiang, F. and Zhang, Z. (2009). Differentially expressed miRNAs potentially involved in the regulation of defence mechanism to drought stress in maize seedlings, *International Journal of Plant Sciences*, 170(8): 979-989.

Wu, G., Park, M.Y., Conway, S.R., Wang, J.W., Weigel, D. and Poethig, R.S. (2009). The sequential action of miR156 and miR172 regulates developmental timing in *Arabidopsis*, *Cell*, 138(4): 750-759.

Wu, J., Chen, J., Wang, L. and Wang, S. (2017). Genome-wide investigation of WRKY transcription factors involved in terminal drought stress response in common bean, *Frontiers in Plant Science*, 8: 380.

Wu, M.F., Tian, Q. and Reed, J.W. (2006). *Arabidopsis* microRNA167 controls patterns of ARF6 and ARF8 expression, and regulates both female and male reproduction, *Development*, 133: 4211-4218.

Xia, K., Wang, R., Ou, X., Fang, Z., Tian, C., Duan, J., Wang, Y. and Zhang, M. (2012). OsTIR1 and OsAFB2 down-regulation via OsmiR393 overexpression leads to more tillers, early flowering and less tolerance to salt and drought in rice, *PLoS ONE*, 7(1): e30039.

Xue, T., Liu, Z., Dai, X. and Xiang, F. (2017). Primary root growth in *Arabidopsis thaliana* is inhibited by the miR159 mediated repression of MYB33, MYB65 and MYB101, *Plant Science*, 262: 182-189.

Yamasaki, H., Abdel-Ghany, S.E., Cohu, C.M., Kobayashi, Y., Shikanai, T. and Pilon, M. (2007). Regulation of copper homeostasis by micro-RNA in *Arabidopsis*, *Journal of Biological Chemistry*, 282(22): 16369-16378.

Yang, J., Duan, G., Li, C., Liu, L., Han, G., Zhang, Y. and Wang, C. (2019). The crosstalks between jasmonic acid and other plant hormone signalling highlight the involvement of jasmonic acid as a core component in plant response to biotic and abiotic stresses, *Frontiers in Plant Science*, 10: 1349.

Yin, Z., Han, X., Li, Y., Wang, J., Wang, D., Wang, S., Fu, X. and Ye, W. (2017). Comparative analysis of cotton small RNAs and their target genes in response to salt stress, *Genes*, 8(12): 369.

Yoon, E.K., Yang, J.H., Lim, J., Kim, S.H., Kim, S.K. and Lee, W.S. (2010). Auxin regulation of the microRNA390-dependent transacting small interfering RNA pathway in *Arabidopsis* lateral root development, *Nucleic Acids Research*, 38(4): 1382-1391.

Zhang, J.F., Yuan, L.J., Shao, Y.I., Du, W.E.I., Yan, D.W. and Lu, Y.T. (2008). The disturbance of small RNA pathways enhanced abscisic acid response and multiple stress responses in *Arabidopsis*, *Plant, Cell & Environment*, 31(4): 562-574.

Zhang, X., Garreton, V. and Chua, N.H. (2005). The AIP2 E3 ligase acts as a novel negative regulator of ABA signalling by promoting ABI3 degradation, *Genes & Development*, 19(13): 1532-1543.

Zhao, B., Ge, L., Liang, R., Li, W., Ruan, K., Lin, H. and Jin, Y. (2009). Members of miR-169 family are induced by high salinity and transiently inhibit the NF-YA transcription factor, *BMC Molecular Biology*, 10: 29.

Zhao, F.Y., Wang, K., Zhang, S.Y., Ren, J., Liu, T. and Wang, X. (2014). Crosstalk between ABA, auxin, MAPK signalling, and the cell cycle in cadmium-stressed rice seedlings, *Acta Physiologiae Plantarum*, 36(7): 1879-1892.

Zhou, H., Liu, Q., Li, J., Jiang, D., Zhou, L., Wu, P., Lu, S., Li, F., Zhu, L., Liu, Z. and Zhuang, C. (2012). Photoperiod- and thermo-sensitive genic male sterility in rice are caused by a point mutation in a novel noncoding RNA that produces a small RNA, *Cell Research*, 22(4): 649-660.

Zhou, L., Liu, Y., Liu, Z., Kong, D., Duan, M. and Luo, L. (2010). Genome-wide identification and analysis of drought-responsive microRNAs in *Oryza sativa*, *Journal of Experimental Botany*, 61(15): 4157-4168.

Zhou, M., Li, D., Li, Z., Hu, Q., Yang, C., Zhu, L. and Luo, H. (2013). Constitutive expression of a miR319 gene alters plant development and enhances salt and drought tolerance in transgenic creeping bentgrass, *Plant Physiology*, 161(3): 1375-1391.

Zhou, X., Wang, G. and Zhang, W. (2007). UV-B responsive microRNA genes in *Arabidopsis thaliana*, *Molecular Systems Biology*, 3(1): 103.

Zhou, X., Wang, G., Sutoh, K., Zhu, J.K. and Zhang, W. (2008). Identification of cold-inducible microRNAs in plants by transcriptome analysis, *Biochimica et Biophysica Acta*, 1779: 780-788.

Zou, J., Liu, A., Chen, X., Zhou, X., Gao, G., Wang, W. and Zhang, X. (2009). Expression analysis of nine rice heat shock protein genes under abiotic stresses and ABA treatment, *Journal of Plant Physiology*, 166(8): 851-861.

CHAPTER

16

Role of MicroRNAs in Plant-Microbe Interactions

Bushra Hafeez Kiani

Department of Biological Sciences (Female Campus), International Islamic University, Islamabad, 44000, Pakistan

1. Introduction

MicroRNAs are internal, regulatory RNAs that are not involved in coding. miRNAs consist of 19 to 28 nucleotide sequences and are produced from double-stranded RNAs. They are involved in sequential controlling of gene expressions to stop translating target mRNA sequences. RNA polymerase II transcribes plant MIR genes to primary-miRNAs that result in a turned-back structure which then forms a hairpin pre-microRNA.

Dawdle (DDL) recruits Dicer-like protein 1 (DCL1) for a crucial step in the down-streaming process of precursor miRNA and primary miRNA. Hyponastic leaves 1 (HYL1), DCL1 and serrate RNA effector molecule (SE) form a double-stranded micro RNA that later undergoes 3′ methylation by S-adenosylmethionine-dependent RNA methyltransferase. Exportin-like protein Hasty exports this duplex from the nuclear cytoplasm into the cytosol. The main regulator of RISC protein, Argonaute binds with a single strand of the duplex. Argonaute uses post-transcriptional gene silencing to either cleave or stop the translation of target mRNA. Using high-throughput sequencing technologies and computational approaches, many miRNAs have been studied in different plant species (Kumar *et al.*, 2017; Luan *et al.*, 2015; Devi *et al.*, 2016; 2018; Chakraborty *et al.*, 2016).

There are more chances for plants to face different biological and non-living stress factors, as they are sessile and live in unstable environments. RNA silencing is crucial for regulating genomic expression and remaining adaptive under stressed conditions. Plants have been greatly evolved and are frequently subjected to different disease-causing microorganisms. To face these microorganisms, they have developed many layers of immunity against them at different levels. Plants identify such microorganisms with the help of pattern recognition receptors, which identify pathogen-associated

---

*Corresponding author: bushrahafeez.kiani@gmail.com

molecular patterns and microbes-associated molecular patterns (MAMPs) or elicitors. Pathogen-associated molecular patterns turn on pathogen-triggered immunity (PTI) that increases callose deposition, free radical generation, protein phosphorylation and pathogen-related (PR) gene expression (Nejat and Mantri, 2017). As a result, proteins that code for resistance genes develop secondary immunity in plants by recognising the pathogen effector proteins and inducing resistance towards them, which is known as effector-triggered immunity (ETI). Resistance proteins commonly initiate fast and specific responses, like hypersensitivity response (HR), which starts the death of cells at the site of the attack to stop the growth of the pathogen.

Plants develop novel resistance proteins encoded by R genes for neutralising effectors whereas pathogens have diversified effectors to start effector-triggered susceptibility (ETS) pathways. Abscisic acid (ABA), 2-hydroxybenzoic acid (SA), jasmonate (JA) and ethene (ET) are well-developed phytochromes that have key roles in pathogen resistance in crops (Baldrich et al., 2015). This interaction of plants and pathogens creates comprehensive pathogen effectors as well as an interaction between resistance genes. MicroRNAs have important roles in different plant mechanisms, like catabolism, anabolism (Boke et al., 2015; Singh et al., 2016), growth (Yang et al., 2007) and stress responses towards factors, both biotic as well as abiotic (Rajwanshi et al., 2014; Pandita, 2019; Pandita and Wani, 2019; Pandita 2021; 2022a; 2022b; 2022c).

Different studies have reported the impacts of microRNAs on countering bacterial, fungal, viruses and herbivorous attacks (Table 1) (Navarro et al., 2006; Bazzini et al., 2009; Chen et al., 2004; Barah et al., 2013; Sattar et al., 2016). miRNA involvement has been reported by different authors in plant signal transduction pathways during both beneficial and harmful infections which therefore became the cause of providing insightful information on the functions of miRNA in interactions of plants and pathogens (Fig. 1). Various microorganisms regulate miRNAs in leaves and roots during microbial infections. Present information on miRNA as plant defence regulators in reply to pathogenic infection has been briefed in this chapter.

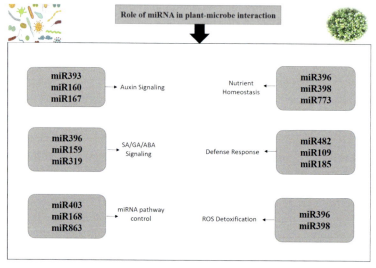

Fig. 1: Role of miRNA in plant-microbes interaction

Role of MicroRNAs in Plant-Microbe Interactions

Table 1: Role of miRNAs in plant-microbe interactions

miRNA	Plant	Tissue	Pathogen	Role
miR863-3p	*Arabidopsis*	Leaf	Pst (avrRpt2)	Immune signalling; miRNA biogenesis
miR2118	Rice	Leaf	Rice stripe virus	Defence response
miR1507	Soybean	Root	*Bradyrhizobium japonicum*	Defence response
miR403	Maize	Root	*Pseudomonas syringae*	RNA silencing
miR408	Maize	Leaf	*Herbaspirillum seropedicae, Azospirillum brasilense*	Facilitate colonisation
miR398	*Arabidopsis*	Leaf	*Pseudomonas syringae*	ROS detoxification
	Rice	Leaf	Rice black-streaked dwarf virus	ROS detoxification
	Cassava	Leaf	*Xanthomonas axonopodis*	ROS detoxification
	Maize	Root	*Herbaspirillum seropedicae*	Copper homeostasis
	Beans	Root	*Rhizobium tropici*	Facilitate colonisation
miR482	Beans	Leaf	Mungbean yellow mosaic India virus	Defence response
	Cassava	Leaf	*Xanthomonas axonopodis*	Defence response
	Potato	Leaf	*Verticillium dahliae*	Defence response
miR396	Rice	Leaf	Southern rice black-streaked dwarf virus	Leaf symptoms
	Arabidopsis	Leaf	*Puccinia cucumerina*	ROS detoxification
	Beans	Leaf	Mungbean yellow mosaic India virus	SA signalling pathway
miR171	Soybean	Leaf	Soybean mosaic virus	Signalling pathway
	Barrelclover	Root	*Rhizophagus irregularis*	Root colonisation
	Birdsfoot trefoil	Root	*Mesorhizobium loti*	Nodulation and signalling pathway
miR167	*Arabidopsis*	Leaf	*Pseudomonas syringae*	Auxin signalling
	Cassava	Leaf	*Xanthomonas axonopodis*	Auxin signalling
	Melon	Leaf	*Aphis gossypii*	Auxin signalling

(Contd.)

Table 1: (*Contd.*)

miRNA	Plant	Tissue	Pathogen	Role
miR164	Rice	Root	Southern rice black-streaked dwarf virus	Plant development
	Rice	Root	Rice black-streaked dwarf virus	Plant development
	Turmeric	Rhizome	*Phythium aphanidermatum*	Plant development
miR393	*Arabidopsis*	Leaf	*Pseudomonas syringae*	Auxin signalling
	Cassava	Leaf	*Xanthomonas axonopodis*	Auxin signalling
	Citrus	Leaf	*Candidatus Leberibcater asiaticus*	Auxin signalling
	Melon	Leaf	*Aphis gossypii*	Auxin signalling
	Chrysanthemum	Leaf	*Macrosiphoniella sanbourni* (Gillette)	Auxin signalling
	Beans	Root	Mungbean yellow mosaic India virus	Auxin signalling
	Soybean	Roots	Soybean mosaic virus	Auxin signalling
	Oncidium	Leaf	Piriformospora indica	Auxin signalling
miR156	Rice	Roots	Rice black-streaked dwarf virus	Auxin signalling
	Beans	Leaf	Mungbean yellow mosaic India virus	Auxin signalling
	Apple	Leaf	*Alternaria alternaria* f. sp. *mali* (ALT1)	Defence response

2. Role of MicroRNAs in Plant Microbe Interactions

2.1 Bacteria

The first miRNA that was to play a major role in the interaction between plants and bacteria is R393. Flg22, present in the N terminus of *Eubacterium flagellin*, recruits auxin signalling F-box proteins 2 and 3 as well as transport inhibitor response 1 (TIR1), which trigger an instantaneous down-regulating process of auxin response genes; whereas, AFB1 repression is independent of miRNA because it expresses resistance partially towards R393-induced cleavage as a result of the presence of one mismatch between R393 complementary sites. R393 increased expression leads to an increase in resistance against bacteria by reducing auxin signalling, expressing the role of down-regulation of auxin signalling in immunological plant response (Navarro *et al.*, 2006). R393 complementary strand is observed having an anti-bacterial role in immunity through negative regulation of membrin-12 (SNARE), which is one of the proteins required in fusing membranes and hence promotes pathogenesis-related protein (PR1) exocytosis. The profiling of *Arabidopsis* leaves affected with *Pseudomonas syringae pv. Solanum lycopersicum* (DC3000hrcC) consisted of small RNAs that showed

Role of MicroRNAs in Plant-Microbe Interactions

increased levels of R393, two added microRNA families R160 and R167 responsible for targeting transport inhibitor response 1 and the sequence-specific DNA-binding factor ARF family, respectively (Fahlgren *et al.*, 2007). On infection caused by *Pst* DC3000hrcC mutant (Fahlgren *et al.*, 2007) or due to harmful *Pst* DC3000 strain (Zhang *et al.*, 2011), miRNA, R400, whose targeted sequence is Penta Trico peptide repeat (PPR) mRNA, decreases in number. Plant susceptibility towards *P. syringae* PV. tomato (Pst) DC3000 decreases as a result of increased expression of R400. Pst DC3000 indication suggests down-regulation of PPR2 due to regulation of R400 and hence negatively affects defence response of *Arabidopsis*. Hence decreased number of R400 and increased number of pentatricopeptide repeat 1 and pentatricopeptide repeat 2 is the *Arabidopsis* defence mechanism against bacterial strains. Photoparoxysmal response proteins also play important functions in regulating genes post-transcription, seed germination, circadian rhythm and developing seeds. These proteins of high diversity are found in the chloroplast or mitochondria (Park *et al.*, 2014).

Xanthomonas axonopodis pv. manihotis (*Xam*) is a gram-negative bacterial infection that badly influences the output of cassava (*Manihot esculenta*) by causing blight disease. Cassava is an agriculturally and industrially important plant in torrid zones (El-Sharkawy, 2004). Ten conserved microRNA families increased in number following *Xanthomonas axonopodis pv. manihotis* infection, i.e. R167a, miRE, R165, R390, R160, R393, R167b, R171, R197b and R394. Seven microRNA families decreased in number, i.e. R482, R408, R395, R397, R1507, R535 and R399. Preliminary research suggests that R160, R167, R390 and R393 are linked with auxin signalling regulation, whereas R171, R165 and R394 target *Arabidopsis thaliana* home box (ATHB) transcription factor, SlSCL3 and F-box family protein. R408, R397 and R398, copper superoxide dismutases, target laccases and plant cyanin, therefore showing involvement in Cu regulation. On the other hand, R535, R395 and R482 target LRR disease resistance and NB-LRR genes (Pérez-Quintero *et al.*, 2012).

Pseudomonas syringae attack on plant foliage leads to the collection of reactive oxygen species which are detoxified by increased quantities of copper/zinc superoxide dismutase (SOD) (Jagadeeswaran *et al.*, 2009). Experiments suggest that if *Arabidopsis* is infected by non-virulent species of Pst DC3000 (avrRpm1) or Pst DC3000 (avrRpt2), microR398 reduction is not linked with CSD2 but rather with copper-zinc superoxide dismutase, which is its target and shows increased transcription levels (Jagadeeswaran *et al.*, 2009). According to one piece of research, callose deposition, seedling growth inhibition, defence gene expression and callose deposition induced by PAMP-induced callose deposition were up-regulated by the disease resistance induced through AGO1 assisting in pathogen-associated molecular patterns in *P. syringae*. The PAMP elicited callose deposition is up-regulated by increased expression of microR160a, whereas flg22 represses R398a and R773 expression, thereby decreasing deposition of callose and enhancing susceptibility to *P. syringae* DC3000 and DC3000 hrcC− (Li *et al.*, 2010). miRNAs 15, 27 and 30 were reported by Zhang *et al.* (2011) which were expressed differentially on an attack by *Pst* DC3000 hrcC, *Pst* DC3000 EV and *Pst* DC3000 avrRpt2. The connection between the three strains, microR158, target genes expressing pentatricopeptide repeat domain-containing proteins and glycosyltransferases, whereas R403 aims for Argonaute2 and Argonaute3 reduced abundance referring to the suspected disease resistance suspected function of Argonaute2 and Argonaute3. Different microRNAs are linked to the hormonal signal transduction pathways in plants as well as biological syntheses, like R390, R160, R393 and R167 that had been differentially expressed upon infection by bacteria. Infection by bacteria down-regulates R159 by targeting sequence-specific DNA-binding factors, MYC10, MYB33 and MYB65, that activate *Arabidopsis'* gibberellic acid signalling mechanism and positive regulators of the abscisic acid signalling mechanism. Upon bacterial infection, microRNAs linked with stress like R398 and R408 also decrease (Zhang *et al.*, 2011). One study shows that microRNAs show specific regulation in compatible, incompatible as well as non-pathogenic interactions among host and bacteria. miRNA induction levels were greater by non-virulent strain *Pst* DC3000 (avrRpt2) in comparison with avirulent *Pst* DC3000 (hrcC) as well as Pst DC3000 (EV). Bacterial wilt in ginger is induced by *Ralstonia solanacearum* (Smith) and is the prime factor that enforces constraints on production in areas of rich ginger cultivation worldwide (Kumar and Sarma, 2012). Till now, resistance sources against bacterial wilt by *Ralstoniaand Pythium* soft root and *Fusarium* yellow have been seen in the *Zingiber* genus, while *Curcuma amada Roxb* and the Indian mango ginger had been resistant towards species from *R. solanacearum* and *Pythium*; hence, offering a potential to develop wilt-resistant *Zingiber officinale* among almost 200 microRNA targets that are potentially expected to contain genes related to immunity, nucleotide-binding site-leucine-rich repeat, cytochrome P450, almost 2 and S-adenosyl methionine along with sequence-specific DNA-binding factors. Hormone signalling and its importance in immune responses have been suggested, using transcriptome data (Prasath *et al.*, 2014).

The most harmful disease of all the commercially available citrus like grapefruit, tangerines and oranges is huanglongbing or citrus-greening disease is due to *Candidatus Liberibacter* and vector *Diaphorina citri Kuwayama* (Asian citrus psyllid) (Bové, 2006). *Candidatus Liberibacter asiaticus* (CaLas) has been more common whereas *Ca. L. africanus* (CaLaf) and *Ca. L. americanus* (CaLam) have been limited on a geographic basis. Citrus small RNAs expression in reply to huanglongbing has been published by Zhao *et al.* (2013) and have recognised 76 conserved and 10 newly discovered csi-miRNAs. On Las infection, increased abundance was noted in csi-R159, 399 and 393 csi-R159, 399 and 393 whereas csi-R396 decreased. The introduction of csi-R399 and CSI-R159 was seen to be particular towards huanglongbing as no particular alteration was observed in csi-R399 and csi-R159 expression after inoculation of *Spiroplasma citri* – a bacteria that is limited to the phloem. Plants having Las infections show significant induction of R399 and reduced levels of phosphorus. Las infection can result in P deficiency that induces R399 expression and results in target gene PHO2 silencing sequentially, resulting in enhanced expression of P transporters (PTs), showing the factor of insufficiency of phosphorus as an HLB link (Zhao *et al.*, 2013). Five HLB-responsive classes of miRNA, i.e. csi-R393, csi-R166, csi-R156, csi-R167 and csi-R172 were reported (Rawat *et al.*, 2015). Cationic plant minerals, like potassium, iron, calcium, zinc, magnesium, manganese and copper have been noticed to perform a significant function in the interaction between HLB and citrus plants.

A nested network of miRNA expresses a linkage of csi-R167 with a hub probe set which codes potassium transporting proteins and can be linked further with phosphate nodes, sulphate, peptide, metal and metabolic transport-related probe sets. In callus infection, citrus proteases are controlled by csi-R396 whereas regulation of ion transport in susceptible plants is done by csi-R167, sharing roles of csi-R396 and R167 in transporting ions and immunity (Rawat *et al.*, 2015). R863-3p has been

Role of MicroRNAs in Plant-Microbe Interactions **283**

investigated to be greatly influenced by a non-virulent strain infection of bacteria *Pst* (avrRpt2), by using two distinct modes, i.e. silencing positive and negative regulators of the plant defence system. According to research, R863-3p is a true miRNA as its collection is DCL1 and SE dependent RDR6 independent. In the early stages, in order to increase defence response quickly during infection pseudokinase1 (ARLPK1) and ARLPK2 (negative regulators-typical receptor) are repressed by R863-3p by mRNA degradation. Atypical receptor-like pseudo-kinase 1 and 2 belong to the subfamily of LRR-RLK (Shiu *et al.*, 2004). The expression of proteins expressed during a microbial attack is the marker for disease responses dependent on salicylic acid which was greater in arlpk1–1 arlpk2–1 double mutant plants on *Pst* (EV) and *Pst* (avrRpt2) infection, consequently implying the negative regulation of immunological reaction by atypical receptor-like pseudo-kinase 1 and 2. In later phases of microbial attack, defensive signalling is lowered through SE negative regulation by inhibiting R863-3p during translation. Hence reduce levels of R863-3p, as it is SE dependent. SE is a C2H2 is a zinc finger-type protein that plays an important role in microRNA maturation and accumulation. Hence, lowering of immune response after effective defence in order to rapidly up-regulate the defence response of plants due to pathogenic attack is controlled by R863-3p microRNA (Niu *et al.*, 2016).

2.2 Aphids

Aphids are insects that harm crops and wild plants and secret metabolites having pathogenic microorganisms in host plants (Foyer *et al.*, 2016). Aphids stylet is a specified mouth part that has evolved for efficient plant part piercing and is used to obtain food from tissues of phloem. During feeding, they infect plants that start different processes for effective aphid colonisation and activate defence signalling pathways involving jasmonic acid, salicylic acid, abscisic acid, ethylene and gibberellic acid (Morkunas *et al.*, 2011; Louis and Shah, 2013).

The function of herbivore-induced RNA silencing defence response regulation had been examined in *Nicotiana attenuata* using RNA-directed RNA polymerase silencing that expresses increased *Manduca sexta* susceptibility. Systemic signalling can be seen in plants by siRNA which is achieved by double-stranded RNA through RNA-dependent RNA polymerases. MicroRNAs, as well as RNA-dependent RNA polymerases functions are interconnected. MicroRNA guided cleavage of primary transcripts initiates Ta-siRNA biogenesis with RdR6-dependent formation of double-stranded RNA and processed by Dicer-like protein to form ta-siRNA that is regulated negatively by different genes (Allen *et al.*, 2005). The formation of nicotine (generally produced defence compound) is lowered in *Nicotiana attenuata*. An inverted repeat RdR1 construct irRdR1 is also reduced because of ornithine decarboxylase (ODC), arginine decarboxylase and nitrate reductase (NR) down-regulation. Research proves amplification of siRNA biogenesis due to RNA-dependent RNA polymerases activity which is elicited by an attack of herbivores. The alkaloid production repressors that are targeted by small interfering RNAs are not degraded completely in RdR1-silenced plants; therefore irRdR1 plants are not able to produce nicotine causing herbivorous susceptibility in plants. The iRdR1 plants are greatly exposed to herbivorous attacks. It has been suggested that RdR1 expresses resistance against aphids by causing changes in the substrates through small RNAs synthesised in RdR1 to live through herbivorous infestations (Pandey and Baldwin, 2007).

Evaluating small RNAs as a response to the oral secretions of *M. sexta* larvae to wild plants that have silenced RdR1 (irRdR1) expresses nine microRNAs, i.e. R157/R156, R159, R396, R396, R894, R168, R894 and R397 are controlled diversely among two genotypic variants. Transcription of genes related to plant hormone signal transduction, like lipoxygenase 2, LOX3, allene oxide synthase, jasmonate-resistant 4 and ACC synthase 3a (ACS3a) was lowered, whereas the transcription of threonine deaminase (TD), JAR6 and ACC oxidase 3 (ACO3) was enhanced in irRdR1 plants that supported an association of novel small RNAs in OS-elicited irRdR1 plants, which distinctively control the genes related to plant hormone signal transduction (Pandey *et al.*, 2008). DCLs are linked with RNA silencing by the synthesis of smRNAs and accordingly linked with the interactions of plants and herbivores. DCLs that were involved in the silencing of lines of *Nicotiana attenuate*, i.e. ir-DCL2, ir-DCL3, ir-DCL4 elucidate the role of DCL in immunity during the herbivorous attack. Increased susceptibility to *M. sexta* is exhibited by ir-DCL3 and ir-DCL4 *Nicotiana attenuata* lines and ir-DCL2 does not undergo a powerful impact. DCL3 and DCL4 are important DCLs that have a major function in the immunological responses of plants (Bozorov *et al.*, 2012).

Myzus persicae, the green peach aphid comes out to be among the most dangerous pest for cultivated crops and can act as a vector for various viruses that affect plants (Blackman and Eastop, 2000). MicroRNA can be useful in providing resistance against colonisation of green peach aphid and was implied from the aphid fecundity on RNA-dependent RNA polymerases, Dicer-like protein and Argonaute mutants of *Arabidopsis*. Progeny construction of green peach aphid on *Arabidopsis* with a mutant pathway of miRNA was considerably less. *Arabidopsis* involves SA, ET, JA, camalexin and glucosinolate pathways to respond to aphid infestation. In dcl1 mutant plants (*Arabidopsis* with miRNA pathway mutants) phytol (PAD3), mitogen-activated protein kinase3, CYP81F2, PR1, LOX2, JA biosynthesis, camalexin biosynthesis pathway-related genes, indolic glucosinolate pathway genes, defense-related genes and SA signalling genes are actively induced. This indicates an association between these mechanisms in resistance against different aphid species. Pad3 mutant showed higher aphid fecundity in comparison with cyp81f2 mutant which indicates camalexin is crucial for resistance against aphids. Dihydrocamalexic acid camalexin conversion is carried out by Pad3 which is a major *Arabidopsis* phytoalexin in the phloem stream. The aphid infected mutant of the *Arabidopsis* miRNA pathway accumulates an additional quantity of camalexin as compared with the one which is raised in controlled plants. An artificial diet containing camalexin also produced less aphid progeny which shows camalexin's property of lowering the reproductive capability of the green peach aphid (Kettles *et al.*, 2013). Various microRNAs are linked with regulating both transcription factor and secondary target genes which relate to salicylic acid, ethylene, along with jasmonic acid transduction pathways in the interactions between green peach aphid and *Arabidopsis* susceptibility (Barah *et al.*, 2013).

MicroRNAs' importance in response to *Aphis gossypii* glover infestation was studied by Sattar *et al.* (2012) in resistant and susceptible melons. Aphids resistance is manifested as antixenosis (non-preferentially aphids), antibiosis (reduced aphid performance) and plant tolerance (Bohn *et al.*, 1972; Klingler *et al.*, 1998). The resistance in melon against *Aphis gossypii* is due to the existence of a single dominant virus aphid transmission (Vat) gene that is from the coiled-coil NBSLRR plant resistance gene family and its expression occurs mainly in plant tissues throughout the

Role of MicroRNAs in Plant-Microbe Interactions   **285**

developing stages of the melon (Garzo *et al.*, 2002). The most conserved microRNA, i.e. R160, R164, R165, R166, R167, R169, R2111, R2911, R390, R393, R396, R397, R398, R408 and R894 are exhibited differently in developmental stages of aphid infestation.

Increased abundance of R156, R157, R159 and R162 is noticed in resistant interaction (Vat+), while R166 and R2111 increase and R156, R162 and R159 decrease in advanced stages of infestation by an aphid. In susceptible interaction (Vat−), an increased quantity of R408 in the early stages is noticed, whereas R56, R166, R168, R169, R171, R172 and R396 increase in quantity in the latter stages. MicroRNAs from R159, R160, R172, R168, R393 and R397 families decrease considerably in quantity during the early stages, whereas miRNA from families R408 and R398 also decrease in quantity in the latter stages of attack by aphids. Increased microRNA expression is noticed in Vat+ interaction; meanwhile, expression pattern decreases in Vat− as a result of variations in the synthesis of microRNA in the susceptible and resistant melon lines. 70 target genes that were regulated differently in melon microRNAs by degradome sequencing in resistant melon lines had been identified by Sattar *et al.* (2016). Phytochromes were chiefly target genes that were related to plant defence against aphids and in auxin perception and signalling. In both resistant and susceptible interaction of aphids, infestation auxin accumulates in a model aimed to justify R167 and R393's role in gene expression regulated by auxin. In resistant interaction leaves R393 increases in abundance and silences the expression of target genes TIR-1 and AFB2 receptors, therefore hindering ubiquitin ligase complex formation which is composed of cullin, TIR-1, SKP1, F-box (SCF) and following degeneration of the Aux/IAA proteins through the complex. The consequent increased R167 expression stops ARF6 and ARF8 expression, finally resulting in the hindrance of transcription of auxin-responsive genes because ARF activators dimerise with Aux/IAA absence (Sattar *et al.*, 2016).

Chrysanthemum morifolium Ramat is a common decorative and medicinal plant that is badly damaged on *Macrosiphoniella sanbourni* invasion from a plantlet to the adult stage and influences the plant's quantity and quality. Increased expression of various MIR genes is noticed in both mock puncture treatment as well as aphid infestation. GAMYB-like 2 gene is an expected target for R159a whose manifestation was elicited on infestation by aphid and mock puncture treatment. R2R3 MYB domain transcription factors are encoded by GAMYB-like 2 genes and are involved in the GA signalling pathway. Programmed cell death caused due to aphid attack was linked with R159a. Decreased abundance of R160a and R393a cause an increase in transcript quantity of its target's auxin response factors and transport inhibitor response, correspondingly and consequently imply an affiliation of gene regulation mediated by microRNA in the interaction between *chrysanthemum* and aphids (Xia *et al.*, 2015).

2.3 Viral Infection

During viral infection, microRNAs have an essential role in influencing different gene expressions. On viral infection due to changes in the matching targeted messenger RNA sequence, the level of endogenous miRNAs is adjusted. Deep sequencing opens new horizons for plant defence induced by viruses. The interference with processes mediated by microRNA had been proposed as a standard pathogenesis aspect, in which microRNA pathways are modified by the suppressor's expression of gene silencing after transcription, resulting in developmental abnormalities in affected crops (Chapman *et al.*, 2004; Chen *et al.*, 2004). According to past data, regulation after transcription implies that suppressor silencing leads toward microRNA genetically modified viral protein expression.

The severity of symptoms is linked with changes in R156, R160, R164, R166, R169 and R171 levels. Post-transcriptional gene silencing suppressors are viral proteins that hinder steps of the post-transcriptional gene silencing pathway and therefore work as a counter-defensive strategy developed by various viruses. Post-transcriptional gene silencing suppressor was encoded by TEV and PVY whereas, TMV, ToMV and (PVX) encode weak post-transcriptional gene silencing suppressor. Five viruses study showed that the ability of miRNAs is changed in viruses that process poor or no suppression action whereas the one which can strongly stop gene silencing modified microRNAs' ability to the same levels. Hence viral proteins that cannot stop post-transcriptional gene silencing are efficient in miRNA pathway alteration of infected plant species. TMV and ToMV produce acute symptoms on tobacco with a changed accumulation of microRNA; the general trait of disease to a significant extent than TEV and PVY viruses which produced slight symptoms. But the link between how viral infection interferes with microRNA pathways is still not confirmed (Bazzini *et al.*, 2007). The detailed analysis expresses the changes in the microRNA pathway during transcription and utilises micro RNAs promoter subsequently to the virus attack. In preliminary reports, following the viral attack and GA3 treatment, P-R164a transcription is increased, hence finding a new plant microRNA biogenesis system by the interference of the virus (Bazzini *et al.*, 2009). An alteration in 9 microRNA patterns of expression and their targeted messenger RNAs on CMV and TAV attack was examined in fruit and flower growth of tomato. According to research, the level of changes in single microRNA on infection by various strains differs, proposing differences in phenotypes of symptom severity (Feng *et al.*, 2013). The homeostasis of AGO1, which is liable for inhibiting the translation of targeted messenger RNA, has been reported to be influenced through R168. Many pieces of research have shown the increased quantity of R168 and raised expression of AGO1 messenger RNA in plant species infected by the virus. The suppression of AGO1 protein and accumulation by virus-mediated induction of R168 had been proposed as a defence countering strategy and the region occupied by the virus increased R168 accumulation, showing the inclusion of products derived by the virus during the process (Varallyay *et al.*, 2010).

Southern rice black-streaked dwarf virus SRBSDV is a disastrous virus that affects plants and is transferred by *Sogatella furcifera,* affecting the rice yield. Microarray profiling of rice miRNAs was used to get an understanding of miRNA-mediated SRBSDV-rice interaction on infection by this virus. Reported miRNAs that target rice genes were miR164, R396, R530 and R1846; hence these miRNAs could be involved in respective target regulation and might be liable for different symptoms that appear on rice plants infected by SRBSDV. miR164 applied regulation of three target genes, vizNA counter-defensive induced protein in the infected plant can be because of indications like the existence of aerial rootlets, salicylic acid-mediated virus resistance, branches on stem nodes and undue tillering. On SRBSDV infection osa-R396 targets three transcription factors that are responsible for orthologous gene regulation of growth-regulating factor (AtGRF5) and AtGRF4, which play important roles in cell proliferation (Xu *et al.*, 2014a). High-throughput sequencing is used to examine the expression pattern of microRNAs in both affected and healthy leaves

Role of MicroRNAs in Plant-Microbe Interactions

and roots. Mainly miRNA was expressed more in leaves as compared to roots in which distribution of conserved microRNA patterns is affected more as compared to non-conserved ones. Upon RBSDV increased abundance of R156a-j, R166abcdf and R169hijklm were observed, while R408, R827 and R1428e decreased in roots and leaves. An expression pattern specific to tissues was also seen in which R164abf and R398 increased in quantity in leaves affected by RBSDB, excluding the roots of the plant (Sun *et al.*, 2015).

Rice stripe antivirus of the genus *Lentivirus*, induces widespread illness in rice and has hampered rice yield in a number of Asian nations. RSV is spread via insects, such as *Laodelphax striatellus*, whose virulence is closely linked to outbreaks of rice stripe illness. Prior studies on rice stripe virus-infected rice found that seven potential unique pn-microRNAs (5 pn-microRNAs rose in frequency whereas 2 pn-microRNAs reduced) were created in rice and that their protein expression changed in rice diseased by RSV (Guo *et al.*, 2012). Research found that when rice was infected with RSV, particular miRNAs are activated or suppressed at various points. The sum of 49 discovered miRNAs targeted 399 rice genomic sequences involved in signalling pathways, sequence-specific DNA-binding factors, diversification and disease tolerance. This suggested that RSV contamination may suppress genes associated with rice development and maturation, as well as disrupt the virus's defensive mechanisms, resulting in viral infection characteristics. Two recently discovered miRNAs (pm4242 and pm5124) and other existing miRNAs, were expected to target 15 disease resistance genes (osa-MIR395y, osa-MIR2118n, osa-MIR399g and osa-MIR2118q). RSV infection decreased the volume of R2118 family members that were known for targeting NBS-LRR containing genetic families (Lian *et al.*, 2016).

Mungbean yellow mosaic India virus is the main cause of the disease yellow mosaic in a variety of commercially significant agricultural plants, including *Vigna mungo*, which is spread by whiteflies and belongs to the bipartite become viral family. Once mungbean, the non-model leguminous crop, was diseased with the MYMIV, high-throughput sRNA-sequencing showed the expression profile of disease-sensitive VmiRNAs. Resistance to MYMIV in *Vigna mungo* is linked with increased gene expression implicated in the SA transmission system. MYMIV infection results in an increase in R396 expression. LOX plays a part in the generation of JA in crops and R396 targets it. A rise in R396 concentration has been shown to trigger the SA-mediated mechanism while repressing the JA mechanism, implying a function for R396 in the immunological responses of *V. mungo*. Additional stress-responsive miRNA possibilities, including R1514, R319, R159, R169 and R166 were investigated for their function in imparting tolerance in *V. mungo*. Following MYMIV inoculation, the expression of R159, which is targeting a zinc finger, was found to be considerably increased, whilst variants of R166 as well as R319, have been observed to down-regulate. Disease resistance (R) protein receptor, which comprises LRR-NBS and a putative amino-terminal signalling domain, recognises viral effector molecules in crops. The presence of a reasonable amount of complementarity among R482 and its equivalent target NB-LRR suggests that R482 is involved in the regulation of resistance to MYMIV. R482 transcript levels decreased after MYMIV infections, but NB-LRR prevalence increased.

The auxin signalling mechanism that is involved in the growth, maturation and physiologic characteristics of plants may likewise govern plant-pathogen interactions, according to the findings. Following MYMIV infection, R160 and R393 adversely affect auxin detection as well as the target auxin response factors family of transcriptional factors, resulting in a decrease in typical plant development and growth. The yellow-colored chlorotic patches on the foliage of susceptible *Vigna mungo* are caused by chlorophyll breakdown and disturbance of the photosynthetic machinery. Two microRNAs (vmu-Rn11 and vmu-Rn12) were expected to be targeting genomic sequences responsible for coding the components of the photosynthesis system, implying that increased concentration of Rn13 and Rn11 lead to destabilised reaction center of photosystem I and binding proteins of chlorophyll a/b, correspondingly (Kundu *et al.*, 2017).

Soya bean mosaic virus wreaks havoc on soya bean yield, and it belongs to the Potyvirus genus. To investigate miRNA families relevant to SMV infection, researchers looked at small RNAs from inoculated mock and soybean foliage infected by soybean mosaic virus. As a defensive reaction to SMV infections, the levels of R160 and R393 rose, resulting in miRNA-mediated inhibition of auxin signalling mechanism components (Yin *et al.*, 2013). Determining the fundamental mechanism of virus-host interaction is required for creating succeeding antiviral therapies in crops.

A genetic strategy was utilised to investigate the relationship between SMV and soybean once more. In soybeans infected with three SMV strains, miRNA profiling found 40 and 15 microRNAs either became ample or decreased in response to Soybean mosaic virus infection, accordingly. The components of the R171 (R171c/j-5p) and R156 (R156t-5p) families enhanced in quantity after infection with either of the 3 Soybean mosaic virus strains, whereas R160, R2118a/b-5p, as well as R11510, continued to increase in vast quantities after infection with strains G2-L and G2-LRB. R393 also continued to increase in quantity after infection with strains G2L and G7, respectively. The identified miRNAs have been discovered to target sequence-specific DNA-binding factors like HD-ZIP nuclear transcription factor Y, and GRAS, along with resistance genes for diseases like LRR protein 1 kinases and TIR-NBS-LRR resistance genes, according to the degradome assessment. The transcriptome assessment revealed that SMV infections affected the expression of numerous differentially expressed genes (DEGs), and also their transcription response, which was unique to distinct strains of SMV infections (Chen *et al.*, 2016).

Cotton leaf curl disease is characterised by a single-stranded begomovirus pathogen complex which is spread by *Bemisia tabaci* whitefly as well as a unique DNA beta satellite (DNA-) molecule named Cotton leaf curl Multan beta satellite. Beta satellites code for C1, a post-transcriptional gene silencing inhibitor that could have a role in viral transportation inside plants cells and increasing levels of the virus genome in plants (Qazi *et al.*, 2007). When viruses invade crops, Dicer-like protein detects and cleaves viral double-stranded RNA and single-stranded RNA, these were later converted to virus-derived small interfering RNAs, inducing antiviral silencing. Following viral infection, 2 types of vsiRNAs were generated: primary small interfering RNAs are derivatives of DCL-mediated cleavage of the original trigger RNA, as well as secondary small interfering RNAs which need RNA dependent RNA polymerases for biogenesis. Preliminary research suggested that vsiRNAs might activate antiviral responses via post-transcriptional gene silencing and targeted host genes via the plant's RNA silencing system. CLCuD infection resulted in a population of 21-nt vsiRNAs that was somewhat greater than the 22-nt equivalent, indicating a synergic effect of DCL4 and DCL2 within a generation of vsiRNAs as well as a

Role of MicroRNAs in Plant-Microbe Interactions **289**

change in the length of smRNAs following CLCuD infections. It has been found that the 5′ terminal sequence influences its unique grouping into various AGO complexes. In CLCuD-infected cotton, vsiRNA with the 5′ terminal U was packed into AGO1, indicating that this might be a key process of RNA silencing toward Cotton leaf curl disease. In the CLCuD-infected cotton crop, expected vsiRNA targets had been down-regulated, implying that viral-induced gene silencing is the primary regulation pathway (Wang *et al.*, 2016).

2.4 Fungal Infection

As plants are sessile, in order to survive, they have developed complex immune responses against different fungal agents. To start defence response plants, identify pathogens entering their body by cell surface and intracellular receptors. The plant genome contains various genes that encode for NB-LRR immune receptors, in which unchecked stimulation of receptors leads to autoimmune responses that affect the growing and development of plants.

Plant miRNAs have a crucial part in controlling gene expression by post-transcriptional gene silencing, which is liable for the growth, development, and induction of anti-pathogenic responses. As miRNAs are important molecules for regulating genes after transcription are involved in Pathogen and effector-triggered immunity, react towards infection caused due to microbes, and change the expression of a gene by suppression or repression of targeted genes (Katiyar-Agarwal and Jin, 2010). That is why connections have been established among mycosis and microRNAs like in species of *Populus szechuanica*. Expression profile of microRNAs has been established by high-throughput sequencing technique where out of ninety microRNAs that have been studied, belonging to forty-two families, and three hundred and seventy-eight novel miRNAs were detected on infection with rust fungus, *Melampsora larici-populina* incompatible isolate Sb052, M. larici-populina compatible isolate Th053, and uninfected controls. According to one of the study miRNAs expression was completely distinct after infection with various isolates and various microRNAs were repressed during rust fungal infection. An important role is being played by these miRNAs in controlling genes that are liable for proteins involved in disease resistance, transcription factors, serine/threonine kinases, along with different proteins. Of the 27 microRNAs, i.e. seven studied microRNAs and 20 newly discovered microRNAs were linked with increased disease resistance. R2118, R472, R482, and eight novel RNAs i.e. novel_mir_11, novel_mir_166, novel_mir_211, novel_mir_244, novel_mir_248, novel_mir_250, novel_mir_357, and novel_mir_403 had been identified that target proteins responsible for imparting resistance against disease. These include Disease resistance protein RPS2, Disease resistance protein RPS5, and Disease resistance protein RPM1. Different newly discovered microRNAs i.e. novel_mir_191, novel_mir_198, novel_mir_368, novel_mir_93 had been allegedly linked with the regulation of LRR receptor-like serine/threonine kinase, whereas new miRNAs, r290, r530 were engaged in regulating WRKY and MYC2 sequence-specific DNA-binding factors that have a significant part in plant defence pathway response. Research revealed that many microRNAs in *P. szechuanica* infection along with compatible Th053 and incompatible Sb052 would increase or decrease in quantity (Chen and Cao, 2015). 32 miRNAs have increased abundance whereas 40 miRNAs decrease in quantity in *Populus* szechuanica incompatible infection. Thirty-five micro RNAs increased significantly whereas fifty-three were decreased in case of compatible infection.

Among these R1444 expresses considerably negative regulation of target genes that encode for polyphenol oxidase (an enzyme involved in resistivity of living and non-living stresses during infections that are incompatible) (Lu *et al.*, 2008). The level of R1444 decreases and causes increased polyphenol oxidase enzyme expression that increases resistance in *Solanum lycopersicum L.* against the pathogen *Pseudomonas syringae* (Li and Steffens, 2002).

Pythium soft rot of turmeric is caused by *Pythium aphanidermatum*, a deadly necrotrophic (*Curcuma longa* L.). Following infection by *P. aphanidermatum*, out of 28 turmeric microRNA (Eighteen are conserved while 3 miRNAs are novel) were described, with the majority of miRNAs decreasing in abundance as well as expression levels changing in reaction to fungal stress, confirming the theory that microRNAs are involved in the turmeric-*P. aphanidermatum* interactions. Plant resistance to fungal infection was aided by R164, R159, R482 as well as R393. A decline in R164 concentration results in an increased NAC level, a protein that is involved in plant immune response, which may inhibit the oversensitive reaction, enabling fungal infection, whereas a reduction in R393, as well as R160/ R167 abundance, resulting in the stimulation of auxin-responsive genomic sequences. The abundance of a distinct miRNA, clo-novel-miRx2 rose considerably in diseased root systems in comparison to controls, however, its targeting pathogen pattern receptor (PPR), a protein that leads to reactive oxygen species generation, was suppressed, resulting in a sensitive phenotype (Chand *et al.*, 2016). Microbes need mechanisms for enabling infections as well as reproduction within the host species as plants identify and evolve ways to combat microbial invasions. *Phytophthora infestans* R1918 (pi-R1918) was discovered to possess similarities in sequences to tomatoes R1918 (sly-R1918), which increases tomatoes' susceptibility to getting infected by *P. infestans.*

The significance of R1918 in tomato-*P. infestans* interaction was investigated utilizing a synthetic pi-1918 created with *Arabidopsis* pre-R159a. When recombinant tomatoes which had overexpressed synthetic pi-R1918 were infected with *Phytophthora infestans*, they exhibited disease symptoms as opposed to the wild-type tomatoes. The decreased quantity of sly-R1918 was found in *Solanum lycopersicum* L. leaf tissues infected with *Phytophthora infestans*. Following fungus infections, there was a decrease in the amount of RxLR effector peptides, sly-R1918, pi-R1918 targets, and trafficked protein particle complex components. Sly-TG2 codes for a RING finger protein that regulates plant development and formation, hormonal signalling, and abiotic and biotic stress mechanisms. The negative association among both amiR1918 as well as its target genomic sequences revealed that amiR1918 plays a function in suppressing genomic sequences that code for the RING finger, making tomatoes more susceptible to *Phytophthora infestans* (Luan *et al.*, 2016).

Powdery mildew disease is due to *Blumeria graminis f. sp. hordei* (Bgh), common ascomycete fungi that produce a fatal foliage infection in *Hordeum vulgare* L. and therefore hinders barley yield. In Barley, R398 controls Mla (mildew resistance locus) which codes for disease-resistant protein molecules by decreasing chloroplast copper/zinc HvSOD1. It also restores Mla resistance 1 (rom1) genetic transcription. HvSOD1 has a role in increasing effector-induced HR in barley as a consequence of fungal infection. In barley interactions, plants containing the resistant protein Mla show HR and generation of reactive oxygen species (Vanacker *et al.*, 2000). The principal acceptor of the electron transport chain (PS I) connected with the plant's thylakoid membrane is the best location for the formation of superoxides (O_2-)

Role of MicroRNAs in Plant-Microbe Interactions

and chloroplasts are among the key sites for transmitting reactive oxygen species (Zurbriggen et al., 2009; Gill and Tuteja, 2010). The plastids Copper/Zinc superoxide dismutase is the most important type of superoxide dismutase (Asada, 1999), and it is found in the stroma of chloroplasts, where it tends to convert superoxide radicals into hydrogen peroxide (Ogawak et al., 1995). As the hydrogen peroxide is more stable, it may readily enter the cell and operates as an intercellular transmission molecule, amplifying a variety of metabolic pathways and eliciting the HR responses in plant and microbe interactions (Liu et al., 2007). Following the invasion of a fungus, R398, also known as hv-miR398, was shown to control hvSOD transcription, reducing hvSOD accumulation and thereby causing an increased vulnerability to infections (Xu et al., 2014b). According to current studies, R398 targets the genes coding for a copper chaperone for CCS 1 peptide, which serves an important function in Cu/Zn superoxide dismutase stabilisation (Chu et al., 2005). It has been shown that restoring Mla resistant 1 (rom 1) can boost hvSOD1 concentration by down-regulating hv-miR398 expression and initiating resistive activities through up-regulating Mildew resistance locus-mediating genes (Xu et al., 2014b). Mildew resistance locus alleles were discovered by Maekawa et al. (2011) and they cause rare and valuable immunity in barley and Mla1 in A. thaliana, suggesting that its function inside the CC-NB-LRR-triggered immunological response is maintained. TmMla1 and its ortholog Sr33 have subsequently been discovered to provide resistance in *Triticum aestivum* towards *Blumeria graminis* f. sp. *Tritici* and Ug99 stem rust infection, implying the use of R398 for controlling SOD1 by Mla.

P. cucumerina, which is the necrotrophic pathogen commonly called *Fusarium tablinum,* harms *Arabidopsis thaliana* and other dicots. R396 action of modifying targeted sequence-specific DNA-binding factors, mainly GRFs that enhance immunological response, was discovered in research addressing microarray examination of microRNA regulation in *Arabidopsis* and *Puccinia cucumerina* interactions. Flowers, seeds, root growth, as well as crop lifespan, are all the functions in which GRFs play pivotal roles. R396 is expressed by two sites in *A. thaliana*: MiR396A and MiR396B. Following fungal infection, increased expression of R396 shows a differential reduction in the frequency of GRF targets, showing serious threats to plant development and vulnerability to disease.

For preventing these outcomes, researchers created an artificial recombinant line that resembles the R396 targets (MIM396) and confers broad tolerance to harmful fungi. When MIM396 crops were infected by *P. cucumerina*, its amount of R396 was minimal, allowing for a surge in targeted expression to rise, as well as a significant buildup of reactive oxygen species, cellulose accumulation, as well as associated strong transcriptional reprogramming. This change in genomic sequences associated with cell wall dynamics which caused resistance in *Arabidopsis* following *P. cucumerina* infections is sometimes referred to as transcriptional reprogramming (Hernandez-Blanco et al., 2007; Sanchez-Rodriguez et al., 2009). Plants containing MIM396 defense mechanisms such as chromatin remodelling and deposition of dormant polypeptides (polypeptides controlled by calcium ion concentrations, cell redox status, or phosphorylation activities) function in the body's immune system also were boosted with a stronger generation of responses. Only harmful disease activates those polypeptides, allowing crops to elicit an immediate or powerful immunological response (Soto-Suarez et al., 2017).

A pathogen called *Sclerotinia sclerotiorum* produces white mold in *Brassica napus*. Following inoculation by *S. sclerotiorum* and un-inoculated foliage, high-throughput degradome sequencing detected 280 miRNA families in *B. napus*, comprising 53 new and 227 recognised miRNAs. Apart from these, miRNA microarray analysis showed that 68 microRNAs were expressed uniquely differently following infection at 3 and twelve hours after infection (hpi). The complexities of microRNA-mediated control in *Brassica napus-Sclerotinia sclerotiorum* interactions are shown by the variety of micro RNA expression profiles in response to *S. sclerotiorum* infections. Six categories were formed from the overall expression patterns of sixty-eight responsive miRNAs. Following hazardous immunisation, Group I miRNAs appeared to be strongly expressed but gradually decreased in number. Group II microRNAs displayed unique expression variations, with a drop in frequency at 3 hpi as well as a large rise at 12 hpi. At 3 hpi and 12 hpi, microRNAs in Group III exhibited little notable change; however, they expressed downregulation at 3 hpi. Different miRNAs were reported in groups IV and V than in groups I and II. Just at 12 hpi, micro RNAs in Group VI have been discovered to control transcription. The assessment of expression levels suggests that various miRNAs were engaged in the initiation and termination of Sclerotinia sclerotiorum-Brassica napus interaction. The targets of these responsive microRNAs were important to investigate to understand their involvement in interactions between pathogenic organisms and plants. These targets played a range of roles in biochemical and intracellular functions, including proliferation and maturation metabolisms, transcription, cell signalling, redox homeostasis, as well as the immune response. Nearly 50% of the miRNAs in this group targeted genomic sequences that encode transcription factors associated with plant defence and development. These defense and resistance genes have been the most significant target sequences listed. Three miRNAs with ETI response target sequences were discovered during research. The microRNAs were bra-miR1885a, which targeted TIR-NBS-LRR, which is a type of R gene; bna-miR6030, which targets proteins similar to At1g12290; and ath-miR168a L + 1R 1, which attacks SGT1a-phosphatase-like proteins. Following fungal inoculation, a reduction in the number of such miRNAs increased the resistance provided by R genes-mediated miRNAs.

Furthermore, a great contribution to antifungal resistance mechanisms is by 2 microRNAs, ath-miR397a and rco-miR397 R + 1, Both miRNA levels increase following *S. sclerotiorum-B. napus* interactions with a decline in expression of target nitrate reductase (NO-forming). Ath-miR408 highly controlled the target expression of L-ascorbate oxidase by reducing its levels after infections at latter intervals of 12 hpi; cme-miR166e mediates the target sequence expression, copper chaperone of Cu/Zn SOD; and ath-miR5021 continuously increases regulation of its target Cu/Zn.

At 12 hpi, the new microRNA PC-14-5p negatively controls the antioxidant gene producing thioredoxin reductase, demonstrating that the reduced state is regulated at the end of the fungal infections. Other 3 microRNAs targeting the two post-transcriptional gene silencing elements, AGO1 and AGO2, were discovered. Both bna-miR403 and aly-miR403a-3p L + 1 specifically target the AGO2 elements, whereas both ath-miR168a target the AGO1 elements, culminating in a reduction in target expression and disease susceptibility. The findings show that broadly implicated microRNAs in *Brassica napus-Sclerotinia sclerotiorum* interactions are tightly regulated (Cao *et al.*, 2016).

The pathogenic *Fusarium oxysporum f. sp. lycopersici* (Sacc.) induces vascular wilting in tomatoes. The deep-sequencing analysis confirmed that various miRNAs are liable for adjusting their corresponding targets in the root system of distinct *Solanum lycopersicum L.* varieties including Moneymaker and Motelle against microbial

Role of MicroRNAs in Plant-Microbe Interactions **293**

infections, demonstrating the importance of plant miRNAs in plant immunity. The P-loop area of genes producing NB domains has been found to be targeted by two micro RNAs, sl-miR482f and sl-miR5300. SolyC08g075630 (NB and CC domain) and solyC08g076000 (NB and three LRR domains) have been the targets of Sl-miR482f, whereas solyC05g008650 (NB domain and CC motifs) and solyC09g018220 have been the targets of sl-miR5300. Moneymaker exhibited vulnerability to fungus infection by lowering targeting gene expression, while Motelle, whose target expression of genes was high following *F. oxysporum* inoculation, increased the cultivar's resistance to infection. Negative regulation of the targeting genes translates the Motelle cultivar's resistance activities into vulnerability to fungus infections, as per findings (Ouyang *et al.*, 2014).

2.5 Symbiotic Interactions

The simultaneous interaction phenomena of ammonia-fixing rhizobacterial and leguminous plants are known as symbiotic nitrogen fixation (SNF). The symbiotic nitrogen fixation that occurs in the roots developed with a specialised organ known as a nodule. The modulation mechanism is complex and requires a molecular signalling link between rhizobacteria and a particular leguminous plant. In matured nodules, 32 conserved miRNAs from 11 families were discovered, with 20 of them being unique to the nodule of soybean (Wang *et al.*, 2009). Direct contact between rhizobacteria and *Bradyrhizobium japonicum*, 55 microRNA families were reportedly extracted from the soybean roots, 35 of which were discovered to be new. At 3 hpi, the infection robustly stimulates R168 and R172 regulation, but by 12 hpi, the modulation was reported to decrease. Upon rhizobacterial invasion, many miRNAs were modulated, including R160, R393, R164, and R168, which target auxin response factors (ARF10, ARF16, and ARF17), transport initiator response 1, NAC1, and AGO1 (Katiyar-Agarwal and Jin, 2010; Khraiwesh *et al.*, 2012). The amount of miRNA in soybean root hairs after rhizobial infection was studied by Yan *et al.* (2016). Out of 48 microRNAs controlling the level of expression in root hairs on *B. japonicum* infection was discovered using a high-throughput sequencing approach. There were 38 conserved miRNAs amongst them, with 16 of them being exclusive to soybean.

Scientists found the target of R166 as HD-ZIP sequence-specific DNA-binding factor Glyma05g30000, which has a specialised function in nodulation and has lowered levels by two-thirds following rhizobial infection, using degradome sequencing or parallel analysis of RNA ends study. R1507, which targets the NB-LRR-type resistance gene Glyma04g29220, was found in soybean roots, stems, leaves, and ripe greenish seeds, and exhibited a significant drop after being infected (Yan *et al.*, 2016). The increased expression of two microRNAs in soybean roots led to a change in nodules. The gma-miR2606b controls the mns1 transcription, and gma-miR4416 controls the rhizobium-induced peroxidase1 (RIP1)-like peroxidase, GmRIP1. R172 members have been also found to be variably produced in soybean roots, leaves, and nodules, indicating that they play a function in the development and maturation of a variety of soybean tissue. R172c, for example, is a crucial regulatory element in soybean nodulation and is enhanced at the commencement of infection throughout nodule growth by inhibiting the Nodule Number Control1 targeted gene, which encodes proteins that binds with the promoters of the nodulin gene, ENOD40 (Wang *et al.*, 2014).

The most significant symbiosis of plants with fungus *Rhizophagus irregularis* belonging to phylum Glomeromycota is Arbuscular mycorrhizal (AM). The exchange of nutrients is the basic benefit of this symbiotic relationship in which the host plant is provided phosphorus by fungus through penetration into roots and creates a structure known as arbuscular inside cortical cell branches. Additionally, both biotic and abiotic stress response is enhanced due to this relation. In 2016, the R171 family was investigated and expressed their regulation in symbiosis interactions. R171a, R171b, R171c, R171d, R171e, R171f and R171h are members of the R171 family. According to prior research in *M. truncatula*, the R171 family was linked to the development of root and Arbuscular Mycorrhiza symbiont relationship, and the targeted *LOST MERISTEM 1* gene is also downregulated. *LOST MERISTEM 1* significantly reduces root colonisation and arbuscular formation by *Rhizophagus irregularis* (mycorrhizal fungi) and hence stops AM symbiosis (Lauressergues *et al.*, 2012). The study illustrates the important function of R171b in the R171 family in mycorrhizal interactions regulation. The reporter GUS gene was expressed under influence of the promoter region of R171b in root tissues and cells having the arbuscular of *M. trunctula* after interaction with fungus (Formey *et al.*, 2014). Whereas R171a and R171c are shown in each tissue of roots, R171e and R171f are shown in a central cylinder and lastly, R171d is shown in a central cylinder and sometimes in parts of roots when fungus colonises regardless of their mycorrhizal consequences.

According to a study, these miRNAs can induce target *Lost meristem1* gene silencing in the majority of cells of the root. Expression of R171b was reported by Couzigou *et al.* in 2017, particularly in cells of root and contains mismatch cleavage site which is conserved in mycotrophs and are not for down-regulating the target family R171, *Lost meristem1*, resulting in the creation of Arbuscular Mycorrhiza symbiont relationship (Couzigou *et al.*, 2017).

Piriformospora indica is an endophyte that colonises the root of *Sebacinales* that conquers the dancing-lady orchid, and roots and builds different advantages for host plants. *Oncidium* grows on other plants and shows delayed growth and development for example, two to three years are required for blooming (Chugh *et al.*, 2009). The research focuses on processes like microRNA modulation involved in growth, metabolism of the cell wall, phytochrome signalling, and regulation of transcription factors during interactions of *P. indica* and orchid roots and also starts resistance responses for different living and nonliving stress factors (Varma *et al.*, 1999; Waller *et al.*, 2005). Different conserved and novel miRNAs have been studied by Ye *et al.* in 2014 that have been involved in auxin signalling factors regulation. Moreover, R390 was seen linked with the regulation of the target gene that encodes a Leucine-Rich Repeats-kinase that helps in regulating the lateral growth of the root.

miRNA89, R166a, R166h, R167, R168, R397, R535a, and R2950 are some of those miRNAs that have been seen to have a greater manifestation in root tissues in *P. indica* during colonisation. According to recent studies these miRNAs are linked with auxin mediating genes like CUC, TIR1, and ARF that are known to control target gene expression by increasing or decreasing levels of miRNA during different symbiosis stages. R159 regulates MYB factors that are associated with the GA signalling pathway and reduces expression levels in the process of symbiosis. Moreover, R397 and R408 can prominently increase anti-oxidative capacity and can enable resistance to stress by controlling the transport of copper. Moreover, when *Oncidium* and *P. indica* interact during the growth and development process they have an important role in cell wall metabolism (Ye *et al.*, 2014). *Herbaspirillum seropedicae* is a diazotroph that can do nitrogen fixation. These bacteria are connected with gramineous plants that have agricultural importance like rice, maize, and sugarcane, as they can conquer the intercellular areas and plant tissues deprived of any defect in the plants (James and Olivares, 1998). The idea of microRNA for controlling different levels of expression in maize in linkage with diazotroph bacteria was examined by deep-sequencing methods. From 25 conserved families, 15 newly discovered miRNA have been detected during association in *Zea mays-H. seropedicae*. As R397, R398, R408, and R528 stop target gene expression linked with the homeostasis of copper like tetra-copper, lactase, and Cupredoxin superfamily protein, so they are classified as Cu-miRNAs. Increased expression of these miRNAs during inoculation of *H. seropedicae* results in reduced targeted regulation of Cu-proteins that increase during the decline of the defense system, contributing to colonisation. According to recent studies, R397 is accountable for the regulation of nitrogen fixation linked with copper homeostasis in *L. japonicus* (De Luis *et al.*, 2012). Examination of the colonised corn roots disclosed that R397 decreases target gene expression that encodes for laccase protein which is part of copper homeostasis (Thiebaut *et al.*, 2014).

3. Conclusion

In order to understand the response of plants against different microorganisms and insects, miRNA-mediated gene regulation is crucial. A number of studies have proved that miRNAs are involved in plant response against pathogens and beneficial microbes and changes in miRNAs expression can cause disease resistance or disease susceptibility. The miRNAs responsible for a response towards pathogens can induce posttranscriptional gene silencing by division or by stopping target mRNA translation. Research areas like the action mechanism, miRNAs target, linked with plant-microbe interaction, and biogenesis have been opened by studying the association of miRNA in stress response.

References

Allen, E., Xie, Z., Gustafson, A.M. and Carrington, J.C. (2005). MicroRNA-directed phasing during transacting siRNA biogenesis in plants, *Cell*, 121(2): 207-221. https://Doi.org/10.1016/j. cell.2005.04.004

Baldrich, P., Campo, S., Wu, M.T., Liu, T.T., Hsing, Y.L. and San, B.S. (2015). MicroRNA-mediated regulation of gene expression in the response of rice plants to fungal elicitors, *RNA Biol.*, 12(8): 847-863. https://Doi.org/10.1080/15476286.2015.1050577

Barah, P., Winge, P., Kusnierczyk, A., Tran, D.H. and Bones, A.M. (2013). Molecular signatures in *Arabidopsis thaliana* in response to insect attack and bacterial infection, *PLoS ONE*, 8(3): e58987. https://Doi. org/10.1371/journal.pone.0058987

Bazzini, A.A., Hopp, H.E., Beachy, R.N. and Asurmendi, S. (2007). Infection and co-accumulation of tobacco mosaic virus proteins alter microRNA levels, correlating with symptom and plant development, *Proc. Natl. Acad. Sci.*, USA, 104(29): 12157-12162. https://Doi.org/10.1073/pnas.0705114104

Bazzini, A.A., Almasia, N.I., Manacorda, C.A., Mongelli, V.C., Conti, G., Maroniche, G.A., Rodriguez, M.C., Distefano, A.J., Hopp, H.E., del Vas, M. and Asurmendi, S. (2009). Virus infection elevates transcriptional activity of miR164a promoter in plants, *BMC Plant Biol.*, 9: 152. https://Doi. org/10.1186/1471-2229-9-152

Blackman, R.L. and Eastop, V.F. (2000). Aphids on the world's crops. *In: An Identification and Information Guide*, 2nd ed., John Wiley & Sons, Chichester, pp. 414.

Bohn, G.W., Kishaba, A.N. and Toba, H.H. (1972). Mechanisms of resistance to melon aphid in a muskmelon line, *Hort. Sci.*, 7: 281-282.

Boke, H., Ozhuner, E., Turktas, M., Parmaksiz, I., Ozcan, S. and Unver, T. (2015). Regulation of the alkaloid biosynthesis by miRNA in opium poppy, *Plant Biotechnol. J.*, 13(3): 409-420. https://Doi. org/10.1111/pbi.12346

Bové, J.M. (2006). Huanglongbing: A destructive, newly-emerging, century-old disease of citrus, *J. Plant Pathol.*, 88(1): 7-37.

Bozorov, T.A., Pandey, S.P., Dinh, S.T., Kim, S.G., Heinrich, M., Gase, K. and Baldwin, I.T. (2012). Dicer-like proteins and their role in plant-herbivore interactions in *Nicotiana attenuate, J. Integr. Plant Biol.*, 54(3): 189-206. https://Doi.org/10.1111/j.1744-7909.2012.01104.x

Cao, J.Y., Xu, Y.P., Zhao, L., Li, S.S. and Cai, X.Z. (2016). Tight regulation of the interaction between *Brassica napus* and *Sclerotinia sclerotiorum* at the microRNA level, *Plant Mol. Biol.*, 92(1-2): 39-55. https://Doi.org/10.1007/s11103-016-0494-3

Chakraborty, S., Devi, K.J., Deb, B. and Rajwanshi, R. (2016). Identification and characterisation of novel microRNAs and their targets in *Cucumis melo* L.: An *in silico* approach, *Focus Sci.*, 2(1).

Chand, S., Nanda, K.S., Rout, E., Mohanty, J., Mishra, R. and Joshi, R.K. (2016). Identification and characterisation of microRNAs in turmeric (*Curcuma longa* L.) responsive to infection with the pathogenic fungus *Pythium aphanidermatum*, *Physiol. Mol. Plant Pathol.*, 93: 119-128.

Chapman, E.J., Prokhnevsky, A.I., Gopinath, K., Dolja, V.V. and Carrington, J.C. (2004). Viral RNA silencing suppressors inhibit the microRNA pathway at an intermediate step, *Genes Dev.*, 18(10): 1179-1186. https://Doi.org/10.1101/gad.1201204

Chen, M. and Cao, Z. (2015). Genome-wide expression profiling of microRNAs in poplar upon infection with the foliar rust fungus *Melampsora larici-populina*, *BMC Genomics*, 16: 696. https://Doi. org/10.1186/s12864-015-1891-8

Chen, J.W., Li, X., Xie, D., Peng, J.R. and Ding, D.W. (2004). Viral virulence protein suppresses RNA silencing-mediated defense but up-regulates the role of microRNA in host gene expression, *Plant Cell*, 16(5): 1302-1313. https://Doi.org/10.1105/tpc.018986

Chen, H., Arsovski, A.A., Yu, K. and Wang, A. (2016). Genome-wide investigation using sRNA-Seq, Degradome-Seq and Transcriptome-Seq reveals regulatory networks of microRNAs and their target genes in soybean during soybean mosaic virus infection, *PLoS ONE*, 11(3): e0150582. https://Doi.org/10.1371/journal.pone.0150582

Chu, C.C., Lee, W.C., Guo, W.Y., Pan, S.M., Chen, L.J., Li, H-M. and Jinn, T.L. (2005). A copper chaperone for superoxide dismutase that confers three types of copper/zinc superoxide dismutase activity in *Arabidopsis*, *Plant Physiol.*, 139(1): 425-436.

Chugh, S., Guha, S. and Rao, I.U. (2009). Micro-propagation of orchids: A review on the potential of different explants, *Sci. Hortic.*, 122(4): 507-520.

Couzigou, J.M., Lauressergues, D., Andre, O., Gutjahr, C., Guillotin, B., Becard, G. and Combier, J.P. (2017). Positive gene regulation by a natural protective miRNA enables arbuscular mycorrhizal symbiosis, *Cell Host Microbe*, 21(1): 106-112. https://Doi.org/10.1016/j. chom.2016.12.001

De Luis, A., Markmann, K., Cognat, V., Holt, D.B., Charpentier, M., Parniske, M., Stougaard, J. and Voinnet, O. (2012). Two microRNAs linked to nodule infection and nitrogen-fixing ability in the legume *Lotus japonicus*, *Plant Physiol.*, 160(4): 2137-2154. https://Doi. org/10.1104/pp.112.204883

Devi, K., Chakraborty, J.S., Deb, B. and Rajwanshi, R. (2016). Computational identification and functional annotation of microRNAs and their targets from expressed sequence tags (ESTs) and genome survey sequence (GSSs) of *Coffea arabica* L., *Plant Gene*, 6: 30-42.

Role of MicroRNAs in Plant-Microbe Interactions   **297**

Devi, K., Saha, J.P., Chakraborty, S. and Rajwanshi, R. (2018). Computational identification and functional annotation of microRNAs and their targets in three species of kiwifruit (*Actinidia* spp.), *Indian J. Plant Physiol.*, 23(1): 179-191.

El-Sharkawy, M.A. (2004). Cassava biology and physiology, *Plant Mol Biol.*, 56(4): 481-501.

Fahlgren, N., Howell, M.D., Kasschau, K.D., Chapman, E.J., Sullivan, C.M., Cumbie, J.S., Givan, S.A., Law, L.F., Grant, S.R., Dangl, J.L. and Carrington, J.C. (2007). High-throughput sequencing of *Arabidopsis* microRNAs: Evidence for frequent birth and death of miRNA genes, *PLoS ONE*, 2: e219.

Feng, J., Lin, R. and Chen, J. (2013). Alteration of tomato microRNAs expression during fruit development upon cucumber mosaic virus and Tomato aspermy virus infection, *Mol. Biol. Rep.*, 40(5): 3713-3722. https://Doi.org/10.1007/s11033-012-2447-5

Formey, D., Sallet, E., Lelandais-Brière, C., Ben, C., Bustos-Sanmamed, P., Niebel, A., Frugier, F., Combier, J.P., Debellé, F. and Hartmann, C. (2014). The small RNA diversity from *Medicago truncatula* roots under biotic interactions evidences the environmental plasticity of the miRNAome, *Genome Biol.*, 15(9): 457.

Foyer, C., Rasool, H.B., Davey, J.W. and Hancock, R.D. (2016). Cross-tolerance to biotic and abiotic stresses in plants: A focus on resistance to aphid infestation, *J. Exp. Bot.*, 67(7): 2025-2037. https://Doi. org/10.1093/jxb/erw079

Garzo, E., Soria, C., Gómez-Guillamón, M.L. and Fereres, A. (2002). Feeding behavior of *Aphis gossypii* on resistant accessions of different melon genotypes (*Cucumis melo*), *Phytoparasitica*, 30(2): 129-140. https://Doi.org/10.1007/bf02979695

Gill, S.S. and Tuteja, N. (2010). Reactive oxygen species and antioxidant machinery in abiotic stress tolerance in crop plants, *Plant Physiol. Biochem.*, 48(12): 909-930.

Guo, W., Wu, G., Yan, F., Lu, Y., Zheng, H., Lin, L., Chen, H. and Chen, J. (2012). Identification of novel *Oryza sativa* miRNAs in deep sequencing-based small RNA libraries of rice infected with rice stripe virus, *PLoS ONE*, 7(10): e46443. https://Doi.org/10.1371/journal.pone.0046443

Hernandez-Blanco, C., Feng, D.X., Hu, J., Sanchez-Vallet, A., Deslandes, L., Llorente, F., Berrocal-Lobo, M., Keller, H., Barlet, X., Sanchez-Rodriguez, C., Anderson, L.K., Somerville, S., Marco, Y. and Molina, A. (2007). Impairment of cellulose synthases required for *Arabidopsis* secondary cell wall formation enhances disease resistance, *Plant Cell*, 19(3): 890-903. https://Doi.org/10.1105/ tpc.106.048058

Jagadeeswaran, G., Saini, A. and Sunkar, R. (2009). Biotic and abiotic stress down-regulate miR398 expression in *Arabidopsis*, *Planta*, 229(4): 1009-1014. https://Doi.org/10.1007/s00425-009-0889-3

James, E.K. and Olivares, F.L. (1998). Infection and colonisation of sugar cane and other graminaceous plants by endophytic diazotrophs, *Crit. Rev. Plant Sci.*, 17(1): 77-119.

Katiyar-Agarwal, S. and Jin, H. (2010). Role of small RNAs in host-microbe interactions, *Annu. Rev. Phytopathol.*, 48: 225-246. https://Doi.org/10.1146/annurev-phyto-073009-114457

Kettles, G., Drurey, J.C., Schoonbeek, H.J., Maule, A.J. and Hogenhout, S.A. (2013). Resistance of *Arabidopsis thaliana* to the green peach aphid, *Myzus persicae*, involves camalexin and is regulated by microRNAs, *New Phytol.*, 198(4): 1178-1190. https://Doi.org/10.1111/nph.12218

Khraiwesh, B., Zhu, J.K. and Zhu, J. (2012). Role of miRNAs and siRNAs in biotic and abiotic stress responses of plants, *Biochim Biophys Acta*, 1819(2): 137-148. https://Doi.org/10.1016/j. bbagrm.2011.05.001

Klingler, J., Powell, G., Thompson, G.A. and Isaacs, R. (1998). Phloem specific aphid resistance in *Cucumis melo* line AR 5: Effects on feeding behaviour and performance of *Aphis gossypii*, *Entomol. Exp. Appl.*, 86(1): 79-88. https://Doi.org/10.1046/j.1570-7458.1998.00267.x

Kumar, A. and Sarma, Y. (2012). Characterisation of *Ralstonia solanacearum* causing bacterial wilt in ginger, *Indian Phytopathol.*, 57: 12-17.

Kumar, D., Dutta, S., Singh, D., Prabhu, K., Kumar, M. and Mukhopadhyay, K. (2017). Uncovering leaf rust responsive miRNAs in wheat (*Triticum aestivum* L.) using high-throughput sequencing and prediction of their targets through degradome analysis, *Planta*, 245(1): 161-182. https://Doi. org/10.1007/s00425-016-2600-9

Kundu, A., Paul, S., Dey, A. and Pal, A. (2017). High throughput sequencing reveals modulation of microRNAs in *Vigna mungo* upon mungbean yellow mosaic India virus inoculation highlighting stress regulation, *Plant Sci.*, 257: 96-105. https://doi.org/10.1016/j.plantsci.2017.01.016

Lauressergues, D., Delaux, P.M., Formey, D., Lelandais-Brière, C., Fort, S., Cottaz, S., Bécard, G., Niebel, A., Roux, C. and Combier, J.P. (2012). The microRNA miR171h modulates arbuscular mycorrhizal colonisation of *Medicago truncatula* by targeting NSP2, *Plant J.*, 72(3): 512-522.

Li, L. and Steffens, J.C. (2002). Overexpression of polyphenol oxidase in transgenic tomato plants results in enhanced bacterial disease resistance, *Planta*, 215(2): 239-247. https://Doi.org/10.1007/ s00425-002-0750-4

Li, Y., Zhang, Q., Zhang, J., Wu, L., Qi, Y. and Zhou, J.M. (2010). Identification of microRNAs involved in pathogen-associated molecular pattern-triggered plant innate immunity, *Plant Physiol.*, 152(4): 2222-2231. https://Doi.org/10.1104/pp.109.151803

Lian, S., Cho, W.K., Kim, S.M., Choi, H. and Kim, K.H. (2016). Time-course small RNA profiling reveals rice miRNAs and their target genes in response to rice stripe virus infection, *PLoS ONE*, 11(9): e0162319. https://Doi.org/10.1371/journal.pone.0162319

Liu, Y., Ren, D., Pike, S., Pallardy, S., Gassmann, W. and Zhang, S. (2007). Chloroplast-generated reactive oxygen species are involved in hypersensitive response-like cell death mediated by a mitogen-activated protein kinase cascade, *Plant J.*, 51(6): 941-954.

Louis, J. and Shah, J. (2013). *Arabidopsis thaliana-Myzus persicae* interaction: Shaping the understanding of plant defence against phloem-feeding aphids, *Front Plant Sci.*, 4: 213. https://Doi. org/10.3389/fpls.2013.00213

Lu, S., Sun, Y.H. and Chiang, V.L. (2008). Stress-responsive microRNAs in *Populus*, *Plant J.*, 55(1): 131- 151. https://Doi.org/10.1111/j.1365-313X.2008.03497.x

Luan, Y., Cui, J., Zhai, J., Li, J., Han, L. and Meng, J. (2015). High-throughput sequencing reveals differential express ion of miRNAs in tomato inoculated with *Phytophthora infestans*, *Planta*, 241(6): 1405-1416. https://Doi.org/10.1007/s00425-015-2267-7

Luan, Y., Cui, J., Wang, W. and Meng, J. (2016). MiR1918 enhances tomato sensitivity to *Phytophthora infestans* infection, *Sci. Rep.*, 6: 35858. https://Doi.org/10.1038/srep35858

Maekawa, T., Cheng, W., Spiridon, L.N., Töller, A., Lukasik, E., Saijo, Y., Liu, P., Shen, Q.H., Micluta, M.A., Somssich, I.E., Takken, F.L.W., Petrescu, A.J., Chai, J. and Schulze-Lefert, P. (2011). Coiled-coil domain dependent homodimerisation of intracellular barley immune receptors defines a minimal functional module for triggering cell death, *Cell Host Microbe*, 9(3): 187-199.

Morkunas, I., Mai, V.C. and Gabryś, B. (2011). Phytohormonal signalling in plant responses to aphid feeding, *Acta Physiol. Plant*, 33(6): 2057-2073. https://Doi.org/10.1007/s11738-011-0751-7

Navarro, L., Dunoyer, P., Jay, F., Arnold, B., Dharmasiri, N., Estelle, M., Voinnet, O. and Jones, J.D. (2006). A plant miRNA contributes to antibacterial resistance by repressing auxin signalling, *Science*, 312(5772): 436-439. https://Doi.org/10.1126/science.1126088

Nejat, N. and Mantri, N. (2017). Plant immune system: Crosstalk between responses to biotic and abiotic stresses the missing link in understanding plant defence, *Curr. Issues Mol. Biol.*, 23: 1-16. https:// Doi.org/10.21775/cimb.023.001

Niu, D., Lii, Y.E., Chellappan, P., Lei, L., Peralta, K., Jiang, C., Guo, J., Coaker, G. and Jin, H. (2016). miRNA863-3p sequentially targets negative immune regulator ARLPKs and positive regulator SERRATE upon bacterial infection, *Nat. Commun.*, 7: 11324. https://Doi.org/10.1038/ncomms11324

Ogawak, K., Takabe, S.K. and Asada, S. (1995). Attachment of CuZn-superoxide dismutase to thylakoid membranes at the site of superoxide generation (PSI) in spinach chloroplasts: Detection by immune gold labeling after rapid freezing and substitution method, *Plant Cell Physiol.*, 36: 565-573.

Role of MicroRNAs in Plant-Microbe Interactions

Ouyang, S., Park, G., Atamian, H.S., Han, C.S., Stajich, J.E., Kaloshian, I. and Borkovich, K.A. (2014). MicroRNAs suppress NB domain genes in tomato that confer resistance to *Fusarium oxysporum, PLoS Patho.*, 10(10): e1004464. https://Doi.org/10.1371/journal.ppat.1004464

Pandey, S.P. and Baldwin, I.T. (2007). RNA-directed RNA polymerase 1 (RdR1) mediates the resistance of *Nicotiana attenuata* to herbivore attack in Nature, *Plant J.*, 50(1): 40-53. https://Doi. org/10.1111/j.1365-313X.2007.03030.x

Pandey S.P., Shahi, P., Gase, K. and Baldwin, I.T. (2008). Herbivory-induced changes in the small-RNA transcriptome and phytohormone signalling in *Nicotiana attenuata, Proc. Natl. Acad. Sci.*, USA, 105(12): 4559-4564. https://Doi.org/10.1073/pnas.0711363105

Pandita, D. (2019). Plant miRnome: miRNA biogenesis and abiotic stress response. *In:* Hasanuzzaman M., Hakeem K., Nahar K. and Alharby H. (Eds.). *Plant Abiotic Stress Tolerance.* Springer, Cham, 449-474. Doi https://Doi.org/10.1007/978-3-030-06118-0_18

Pandita, D. and Wani, S.H. (2019). MicroRNA as a tool for mitigating abiotic stress in rice (*Oryza sativa* L.). *In:* Wani, S. (Ed.). *Recent Approaches in Omics for Plant Resilience to Climate Change.* Springer, Chamz. 109-133. Doi https://Doi.org/10.1007/978-3-030-21687-0_6

Pandita, D. (2021). Role of miRNAi technology and miRNAs in abiotic and biotic stress resilience. *In:* Aftab, T. and Roychoudhury, A. (Eds.). *Plant Perspectives to Global Climate Changes.* Academic Press, Elsevier, 303-330. https://Doi.org/10.1016/B978-0-323-85665-2.00015-7

Pandita, D. (2022a). How microRNAs regulate abiotic stress tolerance in wheat? A snapshot. *In:* Roychoudhury, A., Aftab, T. and Acharya, K. (Eds.). *Omics Approach to Manage Abiotic Stress in Cereals.* Springer, Singapore, 447-464. https://Doi.org/10.1007/978-981-19-0140-9_17

Pandita, D. (2022b). MicroRNAs shape the tolerance mechanisms against abiotic stress in maize. *In:* Roychoudhury, A., Aftab, T. and Acharya, K. (Eds.). *Omics Approach to Manage Abiotic Stress in Cereals.* Springer, Singapore, 479-493. https://Doi.org/10.1007/978-981-19-0140-9_19

Pandita, D. (2022c). miRNA- and RNAi-mediated metabolic engineering in plants. *In:* Aftab, T. and Hakeem, K.R. (Eds.). *Metabolic Engineering in Plants*, Springer, Singapore, 171-186. https://Doi.org/10.1007/978-981-16-7262-0_7

Park, Y.J., Lee, H.J., Kwak, K.J., Lee, K., Hong, S.W. and Kang, H. (2014). MicroRNA400-guided cleavage of pentatricopeptide repeat protein mRNAs renders *Arabidopsis thaliana* more susceptible to pathogenic bacteria and fungi, *Plant Cell Physiol.*, 55(9): 1660-1668. https://Doi.org/10.1093/pcp/pcu096

Pérez-Quintero, Á., Quintero, L.A., Urrego, O., Vanegas, P. and López, C. (2012). Bioinformatic identification of cassava miRNAs differentially expressed in response to infection by *Xanthomonas axonopodis* pv. Manihotis, *BMC Plant Biol.*, 12(1): 29. https://Doi.org/10.1186/1471-2229-12-29

Prasath, D., Karthika, R., Habeeba, N.T., Suraby, E.J., Rosana, O.B., Shaji, A., Eapen, S.J., Deshpande, U. and Anandaraj, M. (2014). Comparison of the transcriptomes of ginger (*Zingiber officinale* Rosc.) and mango ginger (*Curcuma amada* Roxb.) in response to the bacterial wilt infection, *PLoS ONE*, 9(6): e99731. https://Doi.org/10.1371/journal.pone.0099731

Qazi, J., Amin, I., Mansoor, S., Iqbal, M.J. and Briddon, R.W. (2007). Contribution of the satellite encoded gene βC1 to cotton leaf curl disease symptoms, *Virus Res.*, 128(1-2): 135-139.

Rajwanshi, R., Chakraborty, S., Jayanandi, K., Deb, B. and Lightfoot, D.A. (2014). Orthologous plant microRNAs: Microregulators with great potential for improving stress tolerance in plants, *Theor. Appl. Genet.*, 127(12): 2525-2543. https://Doi.org/10.1007/s00122-014-2391-y

Rawat, N.S., Kiran, P., Du, D., Gmitter, F.G. and Deng, Z. (2015). Comprehensive meta-analysis, co-expression, and miRNA nested network analysis identifies gene candidates in citrus against huanglongbing disease, *BMC Plant Biol.*, 15: 184. https://Doi.org/10.1186/s12870-015-0568-4

Sanchez-Rodriguez, C., Estevez, J.M., Llorente, F., Hernandez-Blanco, C., Jorda, L., Pagan, I., Berrocal, M., Marco, Y., Somerville, S. and Molina, A. (2009). The ERECTA receptor-like kinase regulates cell wall-mediated resistance to pathogens in *Arabidopsis thaliana, Mol. Plant-Microbe Interact.*, 22(8): 953-963. https://Doi.org/10.1094/MPMI-22-8-0953

Sattar, S., Song, Y., Anstead, J.A., Sunkar, R. and Thompson, G.A. (2012). *Cucumis melo* microRNA expression profile during aphid herbivory in a resistant and susceptible interaction, *Mol. Plant-Microbe Interact.*, 25(6): 839-848. https://Doi.org/10.1094/MPMI-09-11-0252

Sattar, S., Addo-Quaye, C. and Thompson, G.A. (2016). miRNA-mediated auxin signalling repression during Vat-mediated aphid resistance in *Cucumis melo, Plant Cell Environ.*, 39(6): 1216-1227. https://Doi.org/10.1111/pce.12645

Shiu, S.H., Karlowski, W.M., Pan, R., Tzeng, Y.H., Mayer, K.F. and Li, W.H. (2004). Comparative analysis of the receptor-like kinase family in *Arabidopsis* and rice, *Plant Cell*, 16(5): 1220-1234.

Singh, N., Srivastava, S., Shasany, A.K. and Sharma, A. (2016). Identification of miRNAs and their targets involved in the secondary metabolic pathways of Mentha spp., *Comput. Biol. Chem.*, 64: 154-162. https://Doi.org/10.1016/j.compbiolchem.2016.06.004

Soto-Suarez, M., Baldrich, P., Weigel, D., Rubio-Somoza, I. and San Segundo, B. (2017). The *Arabidopsis* miR396 mediates pathogen-associated molecular pattern-triggered immune responses against fungal pathogens, *Sci. Rep.*, 7: 44898. https://Doi.org/10.1038/srep44898

Sun, Z., He, Y., Li, J., Wang, X. and Chen, J. (2015). Genome-wide characterisation of rice black streaked dwarf virus-responsive microRNAs in rice leaves and roots by small RNA and degradome sequencing, *Plant Cell Physiol.*, 56(4): 688-699. https://Doi.org/10.1093/pcp/pcu213

Thiebaut, F., Rojas, C.A., Grativol, C., Motta, M.R., Vieira, T., Regulski, M., Martienssen, R.A., Farinelli, L., Hemerly, A.S. and Ferreira, P.C. (2014). Genome-wide identification of microRNA and siRNA responsive to endophytic beneficial diazotrophic bacteria in maize, *BMC Genomics*, 15: 766. https://Doi.org/10.1186/1471-2164-15-766

Vanacker, H., Carver, T.L. and Foyer, C.H. (2000). Early H_2O_2 accumulation in mesophyll cells leads to induction of glutathione during the hyper-sensitive response in the barley-powdery mildew interaction, *Plant Physiol.*, 123(4): 1289-1300.

Varallyay, E., Valoczi, A., Agyi, A., Burgyan, J. and Havelda, Z. (2010). Plant virus-mediated induction of miR168 is associated with repression of Argonaute1 accumulation, *EMBO J.*, 29(20): 3507-3519. https://Doi.org/10.1038/emboj.2010.215

Varma, A., Verma, S., Sahay, N., Bütehorn, B. and Franken, P. (1999). *Piriformospora indica*, a cultivable plant-growth-promoting root endophyte, *Appl. Environ. Microbiol.*, 65(6): 2741-2744.

Waller, F., Achatz, B., Baltruschat, H., Fodor, J., Becker, K., Fischer, M., Heier, T., Hückelhoven, R., Neumann, C. and von Wettstein, D. (2005). The endophytic fungus *Piriformospora indica* reprograms barley to salt-stress tolerance, disease resistance, and higher yield, *Proc. Natl. Acad. Sci.*, USA, 102(38): 13386-13391.

Wang, Y., Li, P., Cao, X., Wang, X., Zhang, A. and Li, X. (2009). Identification and expression analysis of miRNAs from nitrogen-fixing soybean nodules, *Biochem. Biophys. Res. Commun.*, 378(4): 799-803. https://Doi.org/10.1016/j.bbrc.2008.11.140

Wang, Y., Wang, L., Zou, Y., Chen, L., Cai, Z., Zhang, S., Zhao, F., Tian, Y., Jiang, Q., Ferguson, B.J., Gresshoff, P.M. and Li, X. (2014). Soybean miR172c targets the repressive AP2 transcription factor NNC1 to activate ENOD40 expression and regulate nodule initiation, *Plant Cell*, 26(12): 4782-4801. https://Doi.org/10.1105/tpc.114.131607

Wang, J., Tang, Y., Yang, Y., Ma, N., Ling, X., Kan, J., He, Z. and Zhang, B. (2016). Cotton leaf curl Multan virus-derived viral small RNAs can target cotton genes to promote viral infection, *Front Plant Sci.*, 7: 1162. https://Doi.org/10.3389/fpls.2016.01162

Xia, X., Shao, Y., Jiang, J., Du, X., Sheng, L., Chen, F., Fang, W., Guan, Z. and Chen, S. (2015). MicroRNA expression profile during Aphid feeding in chrysanthemum (*Chrysanthemum morifolium*), *PLoS ONE*, 10(12): e0143720, https://Doi.org/10.1371/journal.pone.0143720

Xu, W., Meng, Y. and Wise, R.P. (2014). Mla- and Rom1-mediated control of microRNA398 and chloroplast copper/zinc superoxide dismutase regulates cell death in response to the barley powdery mildew fungus, *New Phytol.*, 201(4): 1396-1412. https://Doi.org/10.1111/nph.12598

Xu, D., Mou, G., Wang, K. and Zhou, G. (2014b). MicroRNAs responding to southern rice black-streaked dwarf virus infection and their target genes associated with symptom development in rice, *Virus Res.*, 190: 60-68. https://Doi.org/10.1016/j.virusres.2014.07.007

Yan, Z., Hossain, M.S., Valdes-Lopez, O., Hoang, N.T., Zhai, J., Wang, J., Libault, M., Brechenmacher, L., Findley, S., Joshi, T., Qiu, L., Sherrier, D.J., Ji, T., Meyers, B.C., Xu, D. and Stacey, G. (2016). Identification and functional characterisation of soybean root hair microRNAs expressed in response to *Bradyrhizobium japonicum* infection, *Plant Biotechnol. J.*, 14(1): 332-341. https://Doi.org/10.1111/pbi.12387

Yang, T., Xue, L. and An, L. (2007). Functional diversity of miRNA in plants, *Plant Sci.*, 172(3): 423-432. https://Doi.org/10.1016/j.plantsci.2006.10.009

Yang, L., Mu, X., Liu, C., Cai, J., Shi, K., Zhu, W. and Yang, Q. (2015). Overexpression of potato miR482e enhanced plant sensitivity to *Verticillium dahliae* infection, *J. Integr. Plant Biol.*, 57(12): 1078-1088. https://Doi.org/10.1111/jipb.12348 (not in text)

Ye, W., Shen, C.H., Lin, Y., Chen, P.J., Xu, X., Oelmuller, R., Yeh, K.W. and Lai, Z. (2014). Growth promotion-related miRNAs in Oncidium orchid roots colonised by the endophytic fungus *Piriformospora indica*, *PLoS ONE*, 9(1): e84920. https://Doi.org/10.1371/journal.pone.0084920

Yin, X., Wang, J., Cheng, H., Wang, X. and Yu, D. (2013). Detection and evolutionary analysis of soybean miRNAs responsive to soybean mosaic virus, *Planta*, 237(5): 1213-1225. https://Doi.org/10.1007/s00425-012-1835-3

Zhang, W., Gao, S., Zhou, X., Chellappan, P., Chen, Z., Zhou, X., Zhang, X., Fromuth, N., Coutino, G., Coffey, M. and Jin, H. (2011). Bacteria-responsive microRNAs regulate plant innate immunity by modulating plant hormone networks, *Plant Mol. Biol.*, 75(1-2): 93-105. https://Doi.org/10.1007/s11103-010-9710-8

Zhang, Q., Li, Y., Zhang, Y., Wu, C., Wang, S., Hao, L. and Li, T. (2017). Md-miR156ab and Md-miR395 target WRKY transcription factors to influence apple resistance to leaf spot disease, *Front Plant Sci.*, 8: 526. https://Doi.org/10.3389/fpls.2017.00526

Zhao, H., Sun, R., Albrecht, U., Padmanabhan, C., Wang, A., Coffey, M.D., Girke, T., Wang, Z., Close, T.J., Roose, M., Yokomi, R.K., Folimonova, S., Vidalakis, G., Rouse, R., Bowman, K.D. and Jin, H. (2013). Small RNA profiling reveals phosphorus deficiency as a contributing factor in symptom expression for citrus huanglongbing disease, *Mol. Plant*, 6(2): 301-310. https://Doi.org/10.1093/mp/sst002

Zurbriggen, M.D., Carrillo, N., Tognetti, V.B., Melzer, M., Peisker, M., Hause, B. and Hajirezaei, M.R. (2009). Chloroplast-generated reactive oxygen species play a major role in localised cell death during the non-host interaction between tobacco and *Xanthomonas campestris* PV vesicatoria, *Plant J.*, 60(6): 962-973.

CHAPTER
17

Micro-RNA: A Versatile Tool as Molecular Markers in Plants

Parthasarathy Seethapathy[1]*, Reena Sellamuthu[2], Dhivyapriya Dharmaraj[3], Harish Sankarasubramanian[4], Anandhi Krishnan[5], Anu Pandita[6] and Deepu Pandita[7]*

[1] Department of Plant Pathology, Amrita School of Agricultural Sciences, Amrita Vishwa Vidyapeetham, Coimbatore - 642109, India
[2] Department of Plant Biotechnology, Amrita School of Agricultural Sciences, Amrita Vishwa Vidyapeetham, Coimbatore - 642109, India
[3] Department of Plant Breeding and Genetics, Amrita School of Agricultural Sciences, Amrita Vishwa Vidyapeetham, Coimbatore - 642109, India
[4] Department of Plant Pathology, Tamil Nadu Agricultural University, Coimbatore - 641003, India
[5] Department of Plant Breeding and Genetics, Tamil Nadu Agricultural University, Coimbatore - 641003, India
[6] Vatsalya Clinic, Krishna Nagar, New Delhi - 110051, India
[7] Government Department of School Education, Jammu, Jammu and Kashmir - 180001, India

1. Introduction

Crop stress describes environmental factors that harm a plant's propensity for growth, development, or yield. Plants are constantly exposed to several biotic and abiotic stressors, such as pests, diseases, extreme temperatures, radiation, high salinity, heat, heavy metals, flooding and drought (Basso *et al.*, 2019; Tiwari *et al.*, 2022). Many plant reactions are triggered when a plant is subjected to stress, including alterations in gene expression, metabolic reactions, growth rates, crop production, etc. We have had an array of marker technologies since the era of plant breeding began. Markers are generally considered to be chromosomal landmarks. The environment highly influences classical markers, *viz.* morphological, cytological and biochemical markers and tissue-specific contributions that can identify the trait of interest. Molecular markers are able to be detected on the genome, remain unaltered by the external conditions and can be detected in any organ or stage of growth (Nadeem *et al.*, 2018).

---

*Corresponding authors: spsarathyagri@gmail.com; deepupandita@gmail.com

Micro-RNA: A Versatile Tool as Molecular Markers in Plants **303**

It will be easier to develop innovative strategies to make plants more resistant to stress if we have a better understanding of the physiological responses to the various agents of stress. There is a significant shift in the transcriptional patterns of genes connected with stress during the reaction to abiotic stress. Recent studies have demonstrated that indigenous short RNAs, such as miRNAs and nat-siRNAs, affect the growth processes of crops and the reactions of plants to environmental or biological stress. This is the case despite stress-inducible signalling pathways and transcription factors. Researchers have proven that micro-RNAs are involved in a wide variety of stressors and serve an essential role in the development of plants, even though these molecular markers are used extensively (Ma *et al.*, 2022).

According to their length, microsatellites are typically divided into two categories: class I, for satellites longer than 20 nucleotides and class II, for spacecraft shorter than 20 microsatellites. MicroRNAs, also known as miRNAs, are single-stranded fragments of non-coding RNAs that range in size from 21 to 25 nucleotides and play a significant part in regulating gene expression. The first microRNA, *lin-4*, was found in the soil-dwelling nematode (*Caenorhabditis elegans*) in 1993, which regulated mRNA (lin-14) translation through base-pairing complementarity. Since then, many microRNAs have been found in eukaryotic organisms that including plants, that benefit from the constant improvement of high-throughput sequencing tools and analytical and algorithmic approaches to prediction (Rabuma *et al.*, 2022). Increasing transcript homeostasis in cultivated species, miRNAs regulate virtually every biological and cellular function (Tyagi *et al.*, 2019). In recent years, microRNAs have gained a significant amount of interest. MicroRNAs have a significant impact on regulating various activities within the plant system. This category includes processes, such as cell growth, molecular mechanisms, cellular metabolism, sexual differentiation and the control of resistance mechanisms. Plants utilise short RNAs to modulate gene expression, thereby mitigating the impacts of environmental stress. The nucleus and cytoplasm of the cell must first digest the short RNA sequences, known as pre-miRNAs, to generate the functional microRNAs required by the cell (Hazra and Das, 2021). As a result, various experiments have defined the activities of pre-miRNA and miRNA in plants, and a miRNA database has also been developed. Therefore, this information can be used for different purposes, such as target prediction, functional analysis, metabolic pathways and even the construction of microsatellite tools (Biswas *et al.*, 2021). Representing the geographically and temporally controlled transcriptional regulation of plant miRNAs, the abundance of mature miRNAs encoding miRNA genes differs dramatically amongst miRNAs, tissue types and developmental phases (Ražná *et al.*, 2020). The increased expression of miRNA has significant repercussions for a wide variety of biotic and abiotic stressors experienced by plants (Pandita 2019; Pandita and Wani 2019; Pandita 2021; 2022a; 2022b; 2022c). Recent research in microRNA expression profiling shows that it is physiologically relevant to the diagnosis, development and therapy of specific plant stresses.

2. Markers in Crop Improvement

2.1 Conventional Markers

Understanding existing crop genetic divergence is crucial for the proper planning of crop breeding programmes. The diversity seen in a species' genome can be categorised into visible and non-visible traits. In the early days of cultivated plant varietal development projects, markers played a significant role in visible features, like flower colour, stem length, seed shape, awn type and length, pod colour, pubescence colour, hilum colour, etc. These characteristics are used to identify desired varieties for cultivation. It can be thought of as the direct form of a phenotype. However, it is subject to a great deal of sway from the environment surrounding it (Nadeem *et al.*, 2018). The markers display an array of differences in position, order, number, banding pattern, size and shape due to variations that are caused by distributions of euchromatin and heterochromatin. The detection is due to cytogenetic (Q or G banding), fluorescent, or radioactive tags (Chesnokov *et al.*, 2020). A gene that encodes proteins can be employed as a morphological variant or optional feature of enzymes with the same catalytic characteristics or activity but distinct molecular structure and electrophoretic polarisability. Since the change in electrophoretic polarisability is caused by amino acid substitution as a point mutation, resulting in many isozymes, which are the products of distinct alleles rather than genetic variants (Xu, 2010). It is strongly advised to perform morphological characterisation before conducting a biochemical or molecular genetics study. Despite the limits of morphological markers, it is possible to infer variations based on the morphological diversity of individuals using these markers. Due to these constraints, molecular markers were successfully produced. Molecular marker methods depend on naturally existing differences in nucleotide sequences present in plants. Utilising molecular markers as distinct from the more conventional morphological characterisation offers benefits.

2.2 Molecular Markers

Genomic markers are the most prevalent form of markers due to their efficiency in discriminating diversity. They result from various alterations, including point mutations and errors in the synthesis of tandemly-repeated DNA (Garrido-Cardenas *et al.*, 2018). A genomic marker is a gene fragment at a specified place on a chromosomal portion that can be used to recognise crop plant entities. DNA markers lack pleiotropic effects, enabling more markers to be utilised per population and can be performed at every stage of a crop cycle (Nadeem *et al.*, 2018). Microsatellite techniques have become a crucial component of genetic analysis. They would be essential for determining the inheritance of genetic variants, identifying the genes responsible for these characteristics and finding novel variations. Marker-assisted breeding, or MAB, is a technique that uses genomic analysis and molecular markers found in linkage maps to change and improve plant traits (Biswas *et al.*, 2021) in the fields of population genetics, genetic diversity analysis, simple trait mapping/primary gene mapping, marker-assisted selection (MAS), marker-assisted pyramiding, microarray-based mapping, linkage map construction, quantitative trait loci (QTL) mapping, marker-assisted back-crossing (MABC), marker-assisted recurrent selection (MARS), genomic selection (GS), genome-wide selection (GWS) and genome-wide association study (GWAS) (Ribaut *et al.*, 2010). These molecular markers introduced a new era in genetic discrimination and crop breeding (Ditta *et al.*, 2018). These molecular markers examine the established relationships between individuals, generate linkage maps, and evaluate the genetic variants among cultivars and germplasms.

Traditional morphological and biochemical indicators in plants have significant disadvantages compared to molecular markers. Using the markers approach in F2 and back-cross lines, near-isogenic lines, doubled haploids, recombinant inbred lines

Micro-RNA: A Versatile Tool as Molecular Markers in Plants **305**

and other types of lines, along with marker techniques, makes it possible to select a large number of desirable traits at the same time, which saves much time. It is classifiable based on the length of the DNA sequence and the number of motifs. These markers are developed, based on much earlier mapping to polymorphism-detecting techniques or methods designed with the support of southern blotting, nuclear acid hybridisation, thermal cycling, and nucleic acid sequencing, including AFLP, CAPS, DAF, DArT, EST, IRAP, RBIP, RFLP, RAPD, REMAP, SPLAT, SRAP and SCARs. Later, the existence of microsatellites as repetitive DNA sequences was identified in all living organisms. Microsatellites, also known as simple sequence repeats (SSRs), are widespread throughout the genome, including genes, intergenic regions and transposable elements. It can be further categorised as SSR, cpSSR, ISSR, iPBS, SAMPL, RAMP, SNP, STR, SSCP, SSLP and VNTR. This can be due to changes in the base due to insertions/deletions/SSR repeat length/number of motifs (Naeem, 2014). In addition, the following DNA markers' methods and practices have been spotlighted in the recent works of literature: CDDP, CoRAP, DALP, DAMD, ILPs, ISAP, iSNAP, PAAP, PARMS, PBA, SCoT, S-SAP, TRAP and TBP. These markers assist the advancement of the plant breeding process, the generation of high-density linkage maps of traits and marker and their applicability in diverse genetic backgrounds. Plant and animal metagenomic assemblages, transcriptome sequences and genome-surveyed sequences were searched to discover SSRs that might be utilised to construct genetic markers. The functional genes govern crop growth, reproduction and abiotic and biotic stress tolerance. Even during the regulation of gene transcription, only 2% of the eukaryotic genomes transcribed RNA encodes transcription factors. Together with the development of targeted genomics and gene expression analysis enabled by RNA-seq, a large number of non-coding RNAs (ncRNAs), that do not translate for functional proteins have been distinguished as being involved in numerous developmental and physiological responses (Wu *et al.*, 2013).

3. Non-coding RNAs as Markers in Plants

Eukaryotic gene expression and regulatory systems benefit from layers significantly influenced by exogenous influences. The names for these layers are transcriptional and post-transcriptional layers (Chuong *et al.*, 2013). The latter are classified as small non-coding RNAs (sRNAs; 200 to 300 nucleotides) and extended non-coding RNAs (lncRNAs; 300 to 600 nts) devoid of coding sequences. Classes of tiny ncRNAs differ significantly in their origins, structures, effector proteins and bioactivities. Intergenic areas of chromosomal DNA are the origin of plant non-coding RNAs, which influence the transcriptional, post-transcriptional and epigenetic expression of genes involved in developmental and stress responses in plants (Farrokhi and Hajieghrari, 2020; Ma *et al.*, 2022). Plant miRNAs affect numerous morphogenesis, including leaf morphogenesis and development, flowering, growth transition, root development, reproductive stage, disease resistance and hormonal signalling. During the process of a plant's response to environmental conditions, both biotic and abiotic, significant variations are seen in the accumulation of microRNAs (e.g. cold, drought, deficiency, disorder, disease, metallic toxicity and salinity) (Chaudhary *et al.*, 2021). Plant microRNAs are copied from MIR genes by the RNA polymerase II enzyme as long single-stranded RNA precursors with distinct stem-loop constructions. These are primary-miRNAs (or pri-miRNAs having two nucelotide 3′ overhangs). They are then processed in two steps by the dicer-like protein 1 (ATP-dependent RNAse III protein) to generate double-stranded miRNA duplexes. Both strands of the pair miRNA/miRNA* are 3′-methylated before being carried to the cytoplasm by the HST transport protein. In the cytoplasm, the functional miRNAs are incorporated into the RNA-induced silencing complex (RISC). The Argonaute1 (AGO1) protein is considered the core component of the RISC complex. The post-transcriptional silencing of genes is controlled by microRNAs, which target messenger RNAs for either sequence-specific fragmentation or translational repression (Ražná *et al.*, 2020). There are three different sections of plant genomes that contain pre-miRNAs: intronic, exonic, and intergenic. There is not much documentation regarding satellites' identification, distribution, or frequency in pre-microRNAs (miRNAs) in plants.

ncRNAs, a large portion of the eukaryotic transcriptome, have an inevitable role in developmental, physiological and pathological processes in plants and animals. ncRNAs, including small ncRNA, long ncRNAs (lncRNAs) and circular RNAs (circRNAs), are essential regulators of transcriptional and post-transcriptional expression on its targeted downstream pathway under various stress responses and pathological conditions. Emerging evidence reveals that ncRNAs are integral components of growth, reproductive development, immunity and adaptive response to detrimental stresses in plants (Tiwari and Rajam, 2022). In commercially important plants like *Arabidopsis*, *Brachypodium*, *Brassica*, citrus, cotton, ginseng rusty root, maize, rice, pea, potato, soybean, sugarbeet, tomato and wheat, several non-coding RNAs have been discovered (Patanun *et al.*, 2013). With recent advancements, particularly next-generation sequencing technologies, numerous non-coding RNAs have been identified and characterised and their functions have also been studied. Many ncRNAs have tissue-, time- and location-specific expression patterns. Therefore, the detection of their expression and the assessment of their levels are beneficial as markers for a particular stress situation and increase crop tolerance to abiotic and biotic stresses (Summanwar *et al.*, 2021).

Genome-wide association studies have found that numerous lncRNAs, the transcripts with >200 nucleotides and lacking protein-coding potential, are expressed in many plant species, including tea, capsicum, barley, maize, grapevine, rice, etc. and most of lcRNAs are differentially regulated under abiotic stress factors (Baruah *et al.*, 2021). They supply us with new knowledge, which bolsters our existing understanding of the plant stress response. These can serve as markers for improving a crop's resistance to environmental stress. For example, the well-characterised lncRNAs cold-assisted intronic noncoding RNA (COLDAIR) and cold-induced long antisense intragenic RNA (COOLAIR) mediate flowering response by epigenetic silencing and inactivation of flowering locus C (FLC). FLC is a MADS-box gene closely associated with vernalisation-regulated flowering in plants. Down-regulating floral genes, such as flowering locus T (FT) and suppressor of overexpression of constans 1 are one of the roles that FLC plays as a repressor of flowering (SOC1). The lncRNA COOLAIR is an antisense transcript from FLC. Interestingly, COOLAIR lncRNA is positionally conserved in FLC homologs across plants which are closely related to Brassicaceae and are distantly related monocot species (Cagirici *et al.*, 2017). In addition, since cold stress is a significant factor affecting the productivity of crops, the detection of COOLAIR helps assess the flowering time and manipulation of the expression of this lncRNA could help to improve flowering and yield in crops growing in cold and hilly regions. LRK antisense intergenic RNA, also known as LAIR, is an lncRNA

Micro-RNA: A Versatile Tool as Molecular Markers in Plants

transcript produced by the LRK gene cluster in rice. LAIR controls plant growth and contributes to an increase in grain yield. The deficiency of LAIR reduces plant growth and grain yield in rice crops. This reveals that the suppression of growth and yield can be detected by assessing the expression of LAIR (Wang *et al.*, 2018).

In commercial crops, such as tea and coffee, the aroma is an essential factor that determines the quality of the beverage predominant through the enhancement of pathways associated with the production of aroma-forming compounds. The quality and aroma of tea are mainly due to plant metabolites (catechins, caffeine, etc.). The leaf-specific expression of lncRNAs such as TCONS_0000018, TCONS_00046841, TCONS_00050120, TCONS_00052382, TCONS_00052690, TCONS_0013388, TCONS_00090099 and TCONS_0013388 target the catechin biosynthesis pathway. TCONS_00090099, TCONS_00046841 and TCONS_0000018 enhance the production of catechin and epicatechin by positively modulating the flavonoid pathways. Thus, a set of these lncRNAs could be used as markers to detect/assess the quality of tea (Varshney *et al.*, 2019). Similarly, many long non-protein-coding RNAs (npcRNA) are expressed under stress conditions in plants. Genome-wide analysis in *Arabidopsis* found that the expression of several npcRNAs is altered under abiotic stress. The npcRNAs, such as npc34, npc83, npc351, npc375, npc520, npc521 and npc523 are precursors for several endogenous siRNAs or miRNAs. Besides, npc536 can modulate the length of lateral roots under certain stress conditions. It is common knowledge that an adaptive response to soil environmental stress, particularly during periods of drought and salt stress, is the suppression of the growth of lateral roots (Armor *et al.*, 2009).

CircRNAs, a distinct type of long non-coding RNA, are more conserved than linear non-coding RNAs. CircRNAs are single-stranded and covalently closed transcripts that mainly serve as 'sponges' of miRNA. They bind with miRNA and inhibit their activity. A SEP3 exon 6 circRNA has been known to mediate the floral homeotic phenotype (Conn *et al.*, 2017). In tomato crops, the expression of circRNA45 and circRNA47 provides resistance against *Phytophthora infestans*, which is a common pathogen that damages foliage and fruits. By sponging miR477-3p in a tomato plant, these two circRNAs (*SpRLK1* and *SpRLK2*) rescue leucine-rich repeat receptor-like serine/threonine-protein kinases give resistance to infection. It is common knowledge that the RLK pathway plays an integral part in developing innate immunity and disease resistance in plants (Hong *et al.*, 2020). circRNA45 and circRNA47 are helpful in detecting the degree of protection against *P. infestans* and they can be utilised as tools to produce disease-resistant tomato crops. In food crops, such as maize, the alteration in expressions of circRNAs, circCPK (GRMZM2G032852_C1) and circCKX (GRMZM5G817173_C1) influences drought tolerance. These circRNAs are generated from genes encoding calcium-dependent protein kinase (CPK) family proteins and cytokinin oxidase/dehydrogenase (CKX), respectively, and their activity enhances plant drought tolerance. The elevated expressions of circCPK and circCKX in maize after being subjected to drought stress increase the survival rate. According to these findings, measuring the levels of expression of the genes circCPK and circCKX in maize is an effective way to evaluate the plant's ability to withstand drought (Zhang *et al.*, 2019). The deficiency of macronutrients, deficient phosphorus, drastically affects crop species' growth and yield, including soybean. Several CircRNAs (circ_000013, circ_000349, circ_000351, and circ_000277) function as sponges of many miRNAs associated with response to phosphorus deficiency in soybean. This finding suggests that these circRNAs are useful phosphorus deficiency-induced stresses in soyabean crops (Lv *et al.*, 2020).

In plants, miRNAs regulate many growths and developmental processes, including stress response. miRNA is involved in gene silencing by either cleaving the target RNA or inhibiting translation through an effector protein Argonaute guided by miRNAs (Li *et al.*, 2021; Pandita, 2019; Pandita and Wani, 2019; Pandita, 2021; 2022a; 2022b; 2022c). Expression of miR156 under stress promotes flowering and maintains the plants at the earlier stage for a more extended period (Cui *et al.*, 2014). These miRNAs can be effectively utilised for developing molecular markers in crop improvement programmes. In the foxtail millet, 176 miRNA-based markers were developed by utilising the pre-miRNAs identified in the foxtail millet and the pre-mRNAs identified from the miRbase data of rice, maize, wheat, sorghum and *Brachypodium* sp. (Yadav *et al.*, 2014). These will serve as functional markers in molecular breeding for improving crop varieties.

4. Development of miRNA-based Markers

Generally, the miRNA structures of plants and arthropods (insect pests) are similar but have distinguishing features. RNA interference reflects the immunological and defensive systems of plants. A unique family of regulatory RNAs, short non-coding RNAs (ncRNAs), regulate various biochemical functions in relevant living organisms. The ncRNAs, particularly microRNAs (miRNAs) and small interfering RNAs (siRNAs), have emerged as critical plant abiotic stress response regulators. These non-coding RNAs are the fundamental building blocks of RNA interference (RNAi). They are responsible for the up-regulation and down-regulation of the gene expression, which is engaged in critical biological processes. As a result, microRNA interference has developed into a ground-breaking method for improving crop yields (Tiwari and Rajam, 2022). The germination of seeds, growth, development, preservation of the shoot apical meristem (SAM), recognition of floral organs, development of leaf morphogenesis and finally, the maturation of the plant is all regulated by miRNAs, which are essential regulators and play a crucial role in the process (Ma *et al.*, 2022). In recent years, noncoding RNAs, also known as ncRNAs, have become increasingly recognised as important biologically active compounds promoting genomic and phenotypic variation. Almost the entire genome is transcribed into RNA in eukaryotic cells, but only around 2% of the transcribed RNAs are translated into proteins (Waititu *et al.*, 2020). Experimentation demonstrated that the remaining noncoding RNA transcripts play a wide variety of transcriptional and post-transcriptional regulatory activities (Peschansky *et al.*, 2014). Plant development and the plant's response to abiotic stress are controlled by microRNAs' ability to target functional genes for transcript breaking and translation suppression (Djami-Tchatchou *et al.*, 2017).

The variability level displayed by microRNA-based markers is suitable for use within the genotype of a given population. They describe the miRNA loci based on the genotyping data. The indicators that are based on miRNA have a nature that is tissue- and developmentally-specific. According to Raná *et al.* (2019), developing primers that depend on the sequences of mature miRNAs, referred to as stem-loop structures, is necessary to realise the full potential of miRNA-based markers. As a novel kind of marker system, microRNA was the basis for the first microRNA-based genotyping technique to be discovered (Sabzehzari and Naghavi, 2019). Since that time, it

Micro-RNA: A Versatile Tool as Molecular Markers in Plants　　　**309**

has found application in the genotyping of foxtail millet and many different types of grass (Chakraborty *et al.*, 2020). The miRNA-based markers were distinguished by their high stability as a result of a direct PCR-based marker system, their good reproducibility and sequence specificity as a result of a high annealing temperature, their relatively high polymorphism as a result of likely random combinations of primers, their putative features as a result of their polymorphic nature and the ability to predict morphology controlled by miRNAs. Their cross-genera transferability potential due to the high level of these molecular markers based on micro-RNA has several advantages, including a high level of polymorphism, stability, functionality, reproducibility and transferability between different plant species.

5.　Applications of miRNA Markers in Crop Management

The microRNAs, often known as miRNAs, are a family of small non-coding RNAs that regulate post-transcriptional gene expression by inhibiting or preventing the translation of the genes they are regulating. Many different molecular pathways are implicated in the molecular responses of plants to abiotic and biotic stressors. Regulatory systems controlled by microRNAs are one example of this (Ma *et al.*, 2022; Pandita, 2019; Pandita and Wani 2019; Pandita, 2021; 2022a; 2022b; 2022c). MicroRNAs are responsible for modifying mRNA expression by regulating a change in self-concentration. They do so by reacting to particular physiological occurrences in plants, binding to the mRNAs of their targets based on sequence homology and then inhibiting transcription of the target sequence (Chakraborty *et al.*, 2020). Therefore, microRNAs are extremely important for crop development and management and for the modification of protein expression in the body's reaction to stress (Tyagi *et al.*, 2019). Mutagenesis of gene expression for miRNA binding sites enables bioengineering and molecular breeding uses of microRNAs to evolve attributes, such as production, quality and stress resistance. Targeting fragmentation and regulation of translation of functional gene transcripts, microRNAs govern plant growth and abiotic stress reactions (Table 1). Variability in miRNA precursor regions is frequently utilised to generate genetic markers with similar functionality in plants. These markers consist of potential SNPs or SSRs, depending on the miRNA sequence and AFLP and SRAP (Fu *et al.*, 2013). The miRNA recognises the target binding sites of the mRNA through perfect or close sequence homology. This recognition causes cleavage at the tenth nucleotide of complementary miRNA sites or inhibition of translation, depending on which occurs first. The overexpression of miRNA in a trans-activation consists of miRNA-induced gene silencing, artificial miRNA, and the external application of substances that up-regulate or down-regulate the production of miRNAs or their targets (Basso *et al.*, 2019; Tiwari and Rajam, 2022). Utilising e-PCR, the *in-silico* reproducibility and variability of miRNA SSRs were estimated. Totally, 16,892 pre-miRNA transcripts were collected from 292 plants belonging to six taxonomic groups and 51% of the pre-miRNA sequences examined had SSRs (Biswas *et al.*, 2021). Following di- and tri-nucleotide repetitions, mononucleotide repeats comprised most of the sequences. It was rare to find tetra-, penta-, or hexa-repetitions. At the time of the analysis, there were 9,498 microsatellite loci, of which 57.46% were pre-miRNA microsatellite markers. 3,573 markers, or 37.62% of the total were non-redundant, and 2,341 primer pairs, or 65.51% of the total might be translated into at least one of the plant taxonomic categories (Biswas *et al.*, 2021). This finding supported the concept

Table 1: The miRNAs associated with stress tolerance and significance in crop and model plants

S. No.	miRNA Family	Crop and Model Plants	Physiological Significance	References
1.	*miR156*	Apple	Interact in numerous developmental processes, including floral development (Md-miR156h).	Sun *et al.*, 2013
2.	*miR156*	Sorghum	Targeting SBP-box proteins/squamosa promoter binding protein-like (SPL) transcription factors for shoot development and accelerating the process of biomass metabolism.	Katiyar *et al.*, 2012
3.	*miR156*	Rice	Targeting OsSPL3-an SBP-domain protein associated with cold stress tolerance and crown root development.	Zhou and Tang, 2019
4.	*miR156, miR172, miR319*	Cotton	Targeting SBP/SPL transcription factors, AP2, SPL3 and MYB protein for crop growth and stress reaction.	Wang and Zhang, 2015
5.	*miR156, miR157, miR169, miR166*	Peanut	This involved hypersensitive-induced response protein, leucine-rich repeat (LRR) receptor-like serine/threonine-protein kinase, glycoprotein receptor-associated serine/threonine-protein kinase (GRAS), aquaporin, lipid transfer protein (LTP), auxin response factors (ARF), *v-Myb* myeloblastosis viral oncogene homolog (MYB) transcription factors and major latex-like protein (MLP) for bacterial wilt disease resistance.	Zhao *et al.*, 2015
6.	*miR156, miR157, miR159, miR160*	Cassava	The cis-regulatory elements that are important in drought stress and the plant hormone response were found in the promoter regions of these miRNA genes.	Patanun *et al.*, 2013
7.	*miR156, miR159, miR164, miR390, miR398, miR408*	Wheat	Targeting CSD (Cu–Zn superoxide dismutase), squamosa promoter-binding protein (SBP), MYB3 transcription factor, NAC (NAM, ATAF and CUC) domain transcription factor, and chemocyanin-like protein (CLP) associated with heavy metal stress tolerance.	Qiu *et al.*, 2016

(Contd.)

Micro-RNA: A Versatile Tool as Molecular Markers in Plants

311

Table 1: (*Contd.*)

S. No.	miRNA Family	Crop and Model Plants	Physiological Significance	References
8.	*miR159*	Strawberry	Undertakes strawberry fruit ripening.	Vallarino *et al.*, 2015
9.	*miR159, miR319*	Grapevine	Activation *v-Myb* myeloblastosis viral oncogene homolog (MYB) is a transcription factor that is necessary for the shift from vegetative to reproductive development as well as the reaction to stress.	Han *et al.*, 2014
10.	*miR160, miR396*	Wheat	Targeting auxin response factors (ARFs) and growth-regulating factors (GRF) related to drought stress reaction.	Akdogan *et al.*, 2016
11.	*miR160, miR166, miR167, miR171, miR396*	Rice	Possibly implicated in establishing viral resistance against *rice stripe virus* entry in plants by targeting viral RNA silencing suppressors (VSRs).	Du *et al.*, 2011
12.	*miR166*	Tobacco	Targeting leucine-rich (LRR) repeat proteins for plant disease resistance.	Guo *et al.*, 2011
13.	*miR167*	Coffee	They are specifically targeting the auxin response factor to promote vegetative development and stress response.	Chaves *et al.*, 2015
14.	*miR167, miR393*	Sorghum	Targeting transcription factors ARF8, TIR1 and AFB2 associated with drought tolerance.	Hamza *et al.*, 2016
15.	*miR167, miR169, miR172, miR393, miR397*	Tomato	Targeting the transcription factors known as auxin response factors (ARFs) as well as the transcription factors known as NF-YA in order to combat the effects of drought and cold on plants.	Koc *et al.*, 2015
16.	*miR169*	*Arabido* *psis*, maize	Targeting ABA-dependent pathway inhibited the up-regulation of genes in response to drought and salt stress.	Li *et al.*, 2008; Sheng *et al.*, 2015
17.	*miR169*	Sugarcane	Increasing salt tolerance by targeting transcription factors containing the HAP12-CCAAT box.	Carnavale-Bottino *et al.*, 2013

(Contd.)

18.	*miR169a,* *miR160e,* *miR167b,g,* *miR168a,b*	Apple	It specifically targeted inhibitors of the auxin response factor (ARFs) elements of transcription that contribute to resistance to fire blight.	Kaja *et al.,* 2015
19.	*miR169,* *miR319,* *miR390,* *miR393,* *miR396,* *miR403*	Cowpea	The targeting of metabolic pathways that are related to physiological changes brought on by drought stress.	Shui *et al.,* 2013
20.	*miR171*	Coffee	Glycoprotein receptor-associated serine/threonine-protein kinase (GRAS) transcription factors for development and metabolism of plants.	Chaves *et al.,* 2015
21.	*miR172*	Barley	Regulates lodicule formation and cleistogamous flowering.	Nair *et al.,* 2010
22.	*miR0377,* *miR3358,* *miR2452,* *miR0750*	Barley	Targeting NAC transcription factor, ARF for drought stress response.	Fard *et al.,* 2017
23.	*miR390*	Coffee	TAS3 for crop development and cellular signalling pathways.	Chaves *et al.,* 2015
24.	*miR393*	*Arabidopsis*	The inhibition of auxin signalling, which imparts antibacterial resistance in plants, is accomplished by the induction of plant microRNA by the flagellin-derived peptide.	Navarro *et al.,* 2006
25.	*MiR393*	*Arabidopsis*	It is up-regulated during nitrogen availability and targets transcripts that code for a basic helix-loop-helix (bHLH) transcription factor and auxin receptors TIR1. Maintaining root system architecture (RSA) in response to nitrate supply is controlled by miR393, which targets auxin signalling F-Box proteins 3 (AFB3) transcript.	Vidal *et al.,* 2010
26.	*miR393*	Rice	Drought response by targeting OsTIR1/OsAFB2 auxin co-receptors mediate diverse responses to the plant hormone auxin.	Zhou *et al.,* 2010
27.	*miR393a*	Rice	Targeting OsTIR1/ OsAFB2 auxin co-receptors associated with hypoxia.	Guo *et al.,* 2016

(Contd.)

Micro-RNA: A Versatile Tool as Molecular Markers in Plants

Table 1: (*Contd.*)

S. No.	miRNA Family	Crop and Model Plants	Physiological Significance	References
28.	*miR395*	Tobacco, rice, maize	The improvement of drought resistance through the manipulation of ATP-sulfurylase and sulphate transporter.	Akdogan *et al.*, 2016
29.	*miR395* *miR397a/b* *miR399a*	Sorghum	Targeting APS1, laccase gene, UBC24 enzyme for salt stress tolerance.	Begum, 2022
30.	*miR396*	*Arabidopsis*, rice, cowpea, maize, peach, tobacco	Targeting GRF for drought stress response and adoption.	Akdogan *et al.*, 2016
31.	*miR396*	Citrus	Using 1-aminocyclopropane-1-carboxylic acid oxidase (ACO) as a target for the cold stress response and adaptation.	Zhang *et al.*, 2016
32.	*miR397*	Rice	Targeting L-ascorbate oxidase for heat stress response and adaptation.	Jeong *et al.*, 2011
33.	*miR397,* *miR399,* *miR408*	Tea	Targeting laccase, ubiquitin-conjugating enzyme, plastocyanin-like for cold and other stress response.	Zhang *et al.*, 2014
34.	*miR397,* *miR437,* *miR395,* *miR1435/* *miR51812*	Wheat	Targeting L-ascorbate oxidase and ATP sulfurylase genes for abiotic stress response.	Han *et al.*, 2013
35.	*miR398*	Wheat	Cu/Zn superoxide dismutases (CDS1 and CDS2), as well as copper proteins COX5b, are the targets of this miRNA.	Akdogan *et al.*, 2016
36.	*miR399b*	Barley	Targeting phosphatase transporter gene for salinity and drought stress response.	Deng *et al.*, 2015
37.	*miR408*	Chickpea	Targeting DREB transcription factor associated with drought stress tolerance.	Hajyzadeh *et al.*, 2015
38.	*miR408*	Barley, cowpea, medicago, wheat	Targeting plantacyanin associated with drought stress tolerance.	Akdogan *et al.*, 2016

that microsatellites play crucial roles in pre-miRNAs organisation, function and development. *Arabidopsis thaliana* was the first plant species to discover microRNAs, followed by an abundance of other plant species. Rice types can be classified as indica or japonica, based on the presence or absence of a GG/AA polymorphism in the ending loop domain of Osa-miR2923a. This polymorphism is also connected with grain morphological characteristics (Hazra and Das, 2021). A one-of-a-kind miR-SNP (C/T transition) in the mature region of miR2926 in mustard resulted in a twisted and destabilising stem loop and a loss of all target specificity (Hazra and Das, 2021). It has been demonstrated that nucleotide changes in the cis-regulatory promoter sequences of miRNA genes in numerous *Arabidopsis* and rice species control the expression profile of corresponding miRNA genes (Basso *et al.*, 2019). It has been postulated that variations in the sequence of the miRNA gene are responsible for the morphological diversity of plants. Notably, Tiwari and Rajam (2022) found that a natural variation in the miR164a gene regulates the activity of various other loci in the *Arabidopsis* genome. This activity, in turn, affects the shape of the leaf and the shoot architecture of the plant. Among the many types of microsatellite markers, also known as SSR, used for genetic research on plants, those based on miRNA transcripts are a more recent development. The research that is now available is insufficient in determining the precise functional influence that SSRs have on miRNA precursor sequences. SSRs in pri-miRNA could play an essential part in the splicing process that leads to stress-induced miRNA variations (Fu *et al.*, 2013). On the other hand, miRNA-SSRs are currently being utilised as a feasible marker set for reliable genotyping and trait correlation investigations in important crop species. Numerous plant species, including *Arabidopsis*, barrel clover, cleistogenes, pomegranate, rice, tea and wheat, have exhibited this trait occasionally (Tyagi *et al.*, 2021). There is a rise in the miRNA pool, as indicated by the fact that the most recent version of miRbase (v22) reports 38,589 hairpin precursors and 48,860 mature microRNA transcripts originating from 271 different species. Around 8433 miRNAs, representing 121 different plant species, are conserved in the plant miRNA database (miRBase). Moreover, the plant microRNA online encyclopaedia (PmiREN, which may be found at http://www.pmiren.com/) has been updated with 16,422 unique microRNAs from 88 different plant species. The most recent version of PmiREN, v.2.0, has an anticipated 141,327 miRNA-target linkages in 179 plants. This makes it easier to design alternative crop markers; 38,186 known miRNAs are categorised into 7,838 groups.

6. MicroRNAs as Markers for Plant Stress

MicroRNAs (miRNAs) are essential non-coding genes in eukaryotes (Shah and Ullah, 2021), which govern numerous biological and physiological processes. They participate in RNA silencing and post-transcriptional gene expression control (Bartel, 2018). MiRNA expression profiling demonstrates that homologous miRNAs targeting diverse abiotic stress-associated gene transcription and regulatory proteins are highly altered (up- or down-regulated) under agricultural stress circumstances. These homologous miRNAs are targeted by the miRNAs being profiled. Recently, in genetic engineering, microRNAs (miRNAs) have emerged as a novel target. They have been used to develop high-productivity and stress-resistant crop types (Kung *et al.*, 2012; Chaudhary *et al.*, 2021). The stress tolerance of plants can be significantly altered by manipulating a single miRNA; as a result, multiple miRNAs have been discovered

Micro-RNA: A Versatile Tool as Molecular Markers in Plants

as attractive targets for creating transgenes with increased abiotic stress tolerance (Shriram *et al.*, 2016). In recent years, researchers have successfully developed trait-specific miRNA-based molecular markers, also known as miRNA-SSRs and SNPs, for a wide variety of quantitative plant characteristics. These molecular markers have the potential to be utilised in the plant breeding programmes of the future (Tyagi *et al.*, 2019). Therefore, it is possible to build markers to further recognise certain traits of interest. Consequently, it has been proven that miRNAs influence plant innate immunity on several levels since gene regulation by miRNAs is a crucial mechanism that supports the plant's resistance to abiotic and biotic stress.

6.1 Drought Stress

Drought is the most crucial external factor restricting global agricultural productivity and seedling growth. Because of water scarcity, plants have evolved a wide variety of physiological, biochemical and molecular survival strategies. Researchers have identified several genes, proteins and chemicals that might be genetically altered to confer drought tolerance on crop plants without reducing production. This would be possible because of the work that these researchers have done. In addition, increasing the stress resistance of plants can be accomplished by overexpressing stress-responsive genes (transcription factors) activated during drought or by altering other regulatory genes connected with stress signalling pathways. It has been demonstrated that several microRNAs have a significant impact on the ability of crop plants to respond to drought. Stress caused by drought can change the expression of miRNAs that are functionally conserved in various plant species (Shah and Ullah, 2021). Genomes of susceptible and drought-resistant wheat genotypes were analysed with the help of miRNA-based biomarkers (*hvu-miR408*, and *hvu-miR827*). Using the conserved type marker *hvu-miR156*, which is involved in regulating plant developmental mechanisms, a reduction in the activity of this type of marker was observed in both sensitive and resistant genotypes. This reduction in the activity of this type of marker suggests that the adaptive capacity mechanism of plants to deal with stressful conditions occurs at the expense of growth processes (Raná *et al.*, 2020). Chaudhary *et al.* (2021) conducted recent research investigating the functional importance of miRNA as a significant candidate in modulating the response of plants to abiotic stress. In alfalfa (*Medicago sativa*), the families *miR3512, miR3630, miR5213, miR5294, miR5368* and *miR6173* are drought-responsive. In *Arabidopsis*, the tasiRNAARF (trans-acting short-interfering RNA-auxin response factor) supports the healthy morphogenesis of flowers despite the presence of drought and high salinity. For instance, increased expression of miR169 during the early development of tomato (*Solanum lycopersicum*) results in increased water deficit tolerance in the plant (Zhang *et al.*, 2011). A genome-wide search for drought-responsive microRNAs led to several such genes in two different sets of the interspecific hybrid progeny of *Malus* sp. (Niu *et al.*, 2019). Likewise, Tibetan wild barley has non-coding microRNAs that are sensitive to the effects of drought (Qiu *et al.*, 2019). To improve barley's ability to withstand drought, multiple mi-RNAs, such as *Ath-miR169b, Hv-miRx5, Hv-miR166b/c, Hv-miR827* and *Osa-miR1432*, were artificially induced to be overexpressed in the plant (Hackenberg *et al.*, 2015). Numerous unique drought-responsive microRNAs in a high-yielding, drought-tolerant Indian cowpea variety were validated, using a transgenic method. They discovered that the transcript levels of laccases, which are involved in lipid catabolic processes and are the targets of microRNA 408 (*miR408*), decreased in drought-stressed wild-types and transgenic lines, indicating that the modulation of lignin concentration is likely a crucial trait for drought tolerance in cowpea. Wheat microRNAs, like *miR159, miR160, miR166, miR169* and *miR172*, as well as *miR395, miR396, miR408, miR472, miR477, miR483, miR1858* and *miR2118*, are responsive to drought stress when either overexpressed or taken out (Akdogan *et al.*, 2016). miRNAs have a part to play in regulating drought resistance mechanisms, including those that are mediated by ABA and those that are mediated by other mechanisms. Examples of modules involved in an ABA-dependent pathway include the *miR159-MYB* module and the *miR169-NFYA* module. On the other hand, several miRNA-target modules are not dependent on ABA (*miR156-SPL; miR393-TIR1; miR160-ARF10, ARF16, ARF17; miR167-ARF6* and *ARF8; miR390/TAS3siRNA-ARF2, ARF3, ARF4*). These miRNAs regulate the expression of several drought-resistant genes. Interactions with ABA functions include regulating stomatal closure and the induction of drought adaptation. ABA- and non-ABA-mediated drought tolerance pathways are regulated by microRNAs. Consequently, the drought response mediated by miRNAs reveals genomic, molecular and functional events (Singroha *et al.*, 2021). The roles of the miRNA families, *miR156, miR159, miR319, miR393, miR394, miR395, miR396, miR402, miR417* and *miR828* in the abiotic stress response, have been verified in a wide variety of plant species by the use of the transgenic approach (Chaudhary *et al.*, 2021). It is possible to build molecular markers for drought tolerance based on miRNA to build trait-specific (miRNA-SSRs/SNP).

6.2 Submergence/ Flooding/ Waterlogging Stress

It has been proven by genome-wide investigations in response to hypoxia (Betti *et al.*, 2020), waterlogging and submergence that micro-RNAs (miRNAs) play an active role in the responses of plants to floods (Meena *et al.*, 2020). In order to silence two mRNAs in the roots of maize that code for *GAMYBs, MYB33*, and *MYB101* homologs, *MiR159* is necessary. The waterlogging led to an increase in the expression of miR159 (Fukao *et al.*, 2019). In *Arabidopsis, MiR159* is responsible for suppressing leading root development by controlling *MYB33, MYB101* and *MYB65* (Xue *et al.*, 2017). *MiR159* was down-regulated by flooding in other tissues, including submerged lotus seedlings, where degradome and transcriptome analyses identified multiple GAMYB targets of *miR159* (Jin *et al.*, 2017). These GAMYB targets may be implicated in regulating the petiole elongation that GA mediates. In contrast, *miR159* was up-regulated by flooding in other tissues, including submerged lotus seedlings. Hypoxia caused a change in the abundance of 46 microRNAs originating from 19 families and all three tasiRNA families. Chemical suppression of mitochondrial respiration resulted in similar expression alterations in most hypoxia-responsive short RNAs studied (Moldovan *et al.*, 2010). There were submergence and drought stress studies undertaken on maize and teosinte. Some miRNAs were prevalent in the network of drought and submergence tolerance patterns of expression for 67 miRNA sequences from 23 miRNA families in maize, especially *miR159a, miR166b, miR167c* and *miR169c* are found in both plants, although to a greater extent, in maize. Drought elevated *miR156k* and *miR164e* in teosinte but down-regulated them in maize (Sepúlveda-Garca *et al.*, 2020). It is possible to build trait-specific (miRNA-SSRs/SNP) miRNA-based markers for submergence or hypoxia resistance.

6.3 Salinity Stress

Crop plants experience a considerable production reduction due to salinity stress. The consequences of salt stress can vary greatly from one crop species to another, as well as depending on the salt concentration. Salinity affects morphological, biochemical and physiological processes and eventually diminishes crop production. It is known that microRNAs regulate abiotic stress at the post-transcriptional stage. A few miRNAs in rice appear to be sensitive to salt tolerance. Their expression levels and patterns varied among plant organs (Parmar *et al.*, 2020). Under salinity stress, *miR12474, miR12475* and *miR12479* are down-regulated in the rice roots of the tolerant cultivar, but *miR12476, miR12477, miR12478, miR12480, miR12481* and *miR12482* are up-regulated. In shoot tissue, *miR12481* and *miR12478* are down-regulated. *miR12482* and *miR12480* are increased in both the plant's shoot and root tissues. These microRNAs are known to affect salt tolerance in Pokkali, a salinity-tolerant rice cultivar. The overexpression of microRNA in the tolerance cultivar demonstrates that the down-regulation of the target gene enhances tolerance. In particular, MiR393, which regulates the salinity response in rice and *Arabidopsis*, targets three genes, namely *LOC Os02g06260, LOC Os05g41010* and *LOC Os05g05800* (Gao *et al.*, 2011).

The *miR398*, expressed in stem tissue and *miR156* is expressed in both the root and stem of tomato plants under salinity stress (Cakir *et al.*, 2021). In response to salt stress, *miR398* was discovered to target the genes CSD1 and CD2 (Cu/Zn superoxide dismutase) in *Arabidopsis* (Li *et al.*, 2017). Similarly, GhFSD1, a cotton iron superoxide dismutase gene, is the target of *miR414c*, which affects salt response in cotton (Wang *et al.*, 2019). *OsmiR535* adversely regulates salinity and drought resistance in rice. *OsSPL19* has been found as the *OsmiR535* target gene (Yue *et al.*, 2020). Utilising this microRNA as a molecular identifier will aid in identifying drought-resistant lines. In addition, *OsmiR535* can be utilised in genome editing experiments to generate abiotic stress-tolerant plant varieties. Under salinity stress, the salinity-tolerant wheat cultivar exhibited overexpression of seven miRNAs and down-regulation of 39 miRNAs (Zeeshan *et al.*, 2021). Ten known and 14 new miRNAs displayed distinct expression patterns in salt-tolerant and salt-sensitive plants. In the salinity-tolerant cultivar, *miR156, miR160, miR171a, miR171b, miR395b, miR7757-5p, miR9653b, miR9666a-3p, miR9666c-5p, miR9666b-5p* and *miR9672b* are reported to be down-regulated. It is known that these differentially expressed miRNAs affect salt tolerance through target genes associated with growth, ion homeostasis, stress signalling, hormone signalling, etc. If their regulatory role and effectiveness in salt stress tolerance are understood, then microRNAs can be used as molecular markers and also when their regulatory function and efficiency in salt stress tolerance are understood. By combining the expression patterns discovered in a subsequent study, we can develop a collection of molecular markers to evaluate a cultivar's tolerance level to salinity.

6.4 Heat Stress

Like any other kind of abiotic stress, heat stress can harm the growth and yield of a crop. The consequences of being subjected to high temperatures can be observed in various morphological and physiological systems. The shift in the average daily temperature can be attributed to the changing climate. The effect of heat stress on crop plants is dependent on both the timing and the duration of heat stress. Heat stress during the reproductive development stage affects the fertility of the plant, which in turn affects grain production. Heat stress at the terminal stage is a significant issue in many crop species. Studies suggest the role of *miR156* and *miR390* in heat stress tolerance (Mangrauthia *et al.*, 2017). Many additional miRNAs exhibited varied expression patterns in response to heat stress in tolerant and sensitive reactions. Some of these have been utilised as molecular markers by developing SSR markers based on miRNA (Sihag *et al.*, 2021). Seventy microRNA-based simple sequence repeat markers have been designed and validated against a selection of heat-tolerant and heat-sensitive wheat cultivars. These miRNA-based SSR markers were differentially expressed in response to heat stress. Among the polymorphic markers, two miRNAs, viz. *miR159c* and *miR165b* were highly informative as this was able to distinguish the susceptible and tolerant cultivars, and can be developed as SSR marker.

6.5 Biotic Stress

In plants, a one-third of yield loss has been attributed to protozoan, bacterial, fungal, viral pathogens, parasitic nematodes and insect pests. Insects feed on plants by chewing and piercing and are capable of transmitting a variety of diseases, including fungus, bacteria and viruses. Pattern recognition receptors, also known as PRRs, are a subclass of transmembrane receptor proteins and a family member of wall-associated kinases (WAKs). The pattern recognition receptors (PRRs) are in charge of receiving signals, such as pathogen-associated molecular patterns (PAMPs), microbe-associated molecular patterns (MAMPs) and nematode-associated molecular patterns (NAMPs) and damage-associated molecular patterns (DAMPs). PAMPs-triggered immunity (PTI) is caused due to the detection of M/PAMPs, whereas DAMPs are created internally by the plant in response to physical injury or wounds. PTI and DAMPs are both referred to as 'protective antimicrobial peptides'. However, through the utilisation of a variety of effector proteins, plants can considerably mitigate the impact of the elicitors, as mentioned above. It has been demonstrated that several disease-resistant proteins, denoted by the letter R, are capable of recognising pathogen effectors, which result in effector-triggered immunity, or ETI (Shamrai, 2022). According to a novel and provocative notion, it also involves the signal cross-talk and cross-kingdom migration of plant microRNAs to influence the biosynthesis of the targeted gene. These plant microRNAs have emerged as possible therapeutic indicators for several pests and diseases (Rabuma *et al.*, 2022). It was once believed that PTI and ETI responses to bacterial and fungal infections were protein-based defences mediated at the transcriptional level. These responses are practically independent of the RNA-based mechanisms in antiviral defence. However, new evidence gleaned from recent studies demonstrates that these presumptions are not accurate (Saiyed *et al.*, 2022). However, a growing body of research suggests that immunological responses brought on by biotic stress in plants are controlled post-transcriptionally. In addition, the critical role that endogenous short RNAs play in regulating gene expression in the processes connected to PTI and ETI responses is becoming more widely acknowledged. The *Arabidopsis thaliana* plant was used to demonstrate one of the earliest reports on miRNAs' role in disease resistance. Navarro and colleagues (2006) revealed that bacterial flagellin flg22 polypeptide sensing increased *At-miR393*, suppressed auxin signalling and increased plant resistance. The microRNA known as *MiR393* has been linked to disease resistance in plants. The overexpression of this gene in *Arabidopsis thaliana* makes the plant more sensitive to leaf blight.

In contrast, the activity of this gene on auxin receptors makes the plant more resistant to downy mildew. The conserved microRNAs all have target sequences that are identical to one another, which makes it much easier to figure out what their targets are. In *Rhizoctonia solani*-infected rice plants, a change in the expression of eight putative conserved miRNAs and related target transcripts was observed (Sasani *et al.*, 2020). The putatively conserved miRNAs include both early (Osa-miR-156-SBP domain-containing protein) and late miRNAs (*Osa-miR167*-auxin response factor, *Osa-miR171*-scarecrow gene regulation, *Osa-miR408*-plastocyanin-like protein, *Osa-miR444*-MADS-box transcription factor). Rice plants that were susceptible to the illness had a higher level of *R. solani*-responsive miRNAs in their genomes. These microRNAs are used as a marker for diagnosing various diseases. Gupta *et al.* (2012) demonstrated that *Puccinia graminis* f. sp. *tritici*-infected bread wheat had a significantly higher microRNA *miR1138* (62G29-1) level.

Conjecture from the past lends credence to the theory that microRNAs target pathogen-responsive genes in crops during an invasion. On the other hand, pathogen-associated short RNA is thought to affect the regulation of host defence-related genes to subvert the resistance (Inal *et al.*, 2014). The movement of short RNA molecules from the plant host to the microbial pathogen can lead to the synthesis of microRNAs, which then target the genes associated with the infection (Islam *et al.*, 2018). The first observation of inter-transfer of small RNA from plant to pathogen and vice versa in the *Arabidopsis-Botrytis cinerea* and tomato-*Botrytis cinerea* pathosystems has opened up a new arena for research on small RNA-based plant-pathogen interactions. This observation was made in the *Arabidopsis-Botrytis cinerea* and tomato-*Botrytis cinerea* pathosystems, respectively. This has made it possible to conduct significant research on inter-kingdom politics over the past decade. Up- or down-regulation of target gene expression and knock-in of transcribed miRNA gene sequences is how miRNAs involved in disease sensitivity affect the expression of target genes when a fungal infection occurs. For example, overexpression of *miR164* and *miR396* resulted in a significant increase in resistance against cyst nematodes (Hewezi *et al.*, 2008). Rice plants with activated polycistronic MIR166k-166h expression increased resistance to the fungal pathogens *Magnaporthe oryzae* and *Fusarium fujikuroi*. This resistance induced up-regulation of miR166 targeting the HD-ZIP III transcription factor genes in both sensitive and resistant rice genotypes, suggesting basal response regulators (Salvador-Guirao *et al.*, 2018). In addition, a rise in the concentration of miR166 and miR159 was observed in cotton plants as a response to infection with the fungal wilt disease. This was accomplished by inhibiting the expression of the target fungal virulence genes Ca2+-dependent cysteine protease (Clp-1) and an isotrichodermin C-15 hydroxylase (HiC-15) in the hyphae of *Verticillium dahlia* (Zhang *et al.*, 2016). PAMP (flg22) is responsible for inducing the miR393, which suppresses auxin signalling and promotes host resistance; this demonstrates that auxin signalling is necessary for plant-induced immunological reactivity (Naseem *et al.*, 2015). An increase in the expression of miR393, which is a negative regulator of TIR1/AFB auxin receptors, is produced in *A. thaliana* by a fragment of the flagellin-derived elicitor peptide flg22 from *Pseudomonas syringae*. The inhibition of auxin signalling by miR393 is responsible for the increased resistance of bacteria to pathogens (Navarro *et al.*, 2006). This is the first indication that microRNAs can modify the immune response to pathogens (Luo *et al.*, 2022). In addition to *miR160a, miR393, miR398b, miR400, miR472* and *miR863*, the *Arabidopsis miR393 miR398b, miR400, miR472* and *miR863* are known to act in disease resistance (Waheed *et al.*, 2022).

The direct effect that microRNAs have when it comes to regulating the expression of either the primary class of R genes or the aberrant ARLPK gene has also been described. A primary concern, microRNAs can act either negatively or positively as moderators in PTI and ETI by activating either positively or negatively regulated defence regulators (Yang *et al.*, 2021). Almost *miR159* and *miR166*, found in Mexican cotton, are responsible for transferring miRNA to fungal pathogens and conferring tolerance to *V. dahliae* (Zhang *et al.*, 2016). They have been recognised in mycelium taken from cotton stems afflicted with illness. Confirmation of the targets was achieved by transient production of miRNA-resistant forms of HiC-15 and Clp-1 in tobacco and *V. dahliae*. Following this, further research on *V. dahliae* mutants revealed that the targeted fungal genes played a significant part in the pathogenicity of the fungus. These genes were also explicitly targeted by the miRNAs that were transferred from diseased cotton to healthy cotton to achieve resistance to plant pathogens.

Probably, the presence of a methyl ester on the ribose of the final nucleotide, interactions with RNA-binding proteins and packaging into exosomes all contribute to the stability of plant miRNAs and their transmission between kingdoms. This is something that has been hypothesised (Rabuma *et al.*, 2022). Most instances of miRNA being transferred between kingdoms are due to interactions between plants and pathogens. It is conceivable that the transmission of endogenous miRNAs from a host to pathogens or parasites can reduce the pathogen's or parasite's ability to penetrate host defences. On the other hand, the transmission of miRNAs from parasitic eukaryotic organisms to plant species can reduce the plant species' resistance. The transmission of microRNA from plants to animals in the case of mutualistic connections has the potential to influence essential processes, like growth and development. It is possible to see the development of artificial microRNAs, also known as amiRNAs, as a significant addition to the technology that is currently available in the fight against anthropogenic climate catastrophe and the concomitant difficulties in farming. Multiple organisms have shown evidence of silencing based on miRNA, which indicates that this mechanism has been conserved throughout evolution. It has also been demonstrated that alterations to the region of the precursor miRNA that forms the mature miRNA do not have any effect on the processing of the precursor miRNA. Because of this, a previously unconsidered potential arose, which ultimately led to the invention of amiRNA biotechnology. This approach expresses the necessary 21-nt amiRNA against the target sequence by utilising the native precursor architecture. amiRNAs are short noncoding RNAs. According to several reports, both naturally occurring and artificially produced miRNAs are equally effective at silencing genes. The yield of virus-resistant transgenic plants demonstrated the efficacy of this approach through alteration of the endogenous plant precursor architecture (Ravelonandro *et al.*, 2019). This approach was recently utilised to overexpress a miRNA from a pest in a plant, which ultimately led the plant to develop resistance to the insect pest that was sought. The amiRNA gene silencing process has been proven to be accurate, according to the findings of an investigation into the control of genes that spans the entirety of the genome (Lu *et al.*, 2008). The ecdysone receptor (EcR) of the pest *Helicoverpa armigera* and the ecdysone receptors of many other polyphagous lepidopteran pests were inhibited with the help of amiRNA technology. This is an effective method for the management of pests. The steroid hormone known as 20-hydroxyecdysone is responsible for regulating all stages of an insect's growth and development throughout its lifetime. This steroid interacts with its receptor, which is known as the ecdysone receptor (EcR), as well as its protein partner, ultraspiracle (usp), thereby activating the initial genes BroadComplex (BR-C), E74 and E75. These genes encode proteins responsible for the transcriptional regulation of genes (Van Lommel *et al.*, 2022). Following this, a succession of late genes involved in crucial aspects of the metamorphosis process is activated by the initial genes. These genes are responsible for apoptosis, cell proliferation and differentiation, and cuticle production. Any interference with ecdysone signalling impairs development, disrupts metamorphosis and affects the moulting of larva-to-larva and larva-to-pupal. Yogindran and Rajam (2021) demonstrated the utilisation of the insect precursor framework to generate the targeted sequence-specific *amiRNA-319a-HaEcR* duplex against the EcR gene of *H. armigera*. Using dsRNA to silence EcR in plants has also been shown to be an effective way to suppress the overall growth and survival of this polyphagous pest. In addition, the research field on the prospective applications of intermigration of miRNAs in plant protection is becoming more familiar as molecular comprehension advances rapidly. Considering the daily flood of knowledge on the trans-mobility of small RNAs, this strategy will address the problems and opportunities in crop protection against biotic stressors.

7. Future Perspectives

Recent advances in synthetic biology research, primarily plant genetic transformations emphasising crop protection, have sped up the development process and recognised crop variations with higher yields and resistance to pests, diseases and other environmental cues. These advances have been made possible by recent research progress in synthetic biology. miRNA-based suppression with locked nucleic acids/ peptide nucleic acids modifications, blocked by miRNA sponges and knockdown techniques through genomic editing are developing molecule delivery systems. Also, the targeted studies on responsive confirmation of the miRNA in influencing gene transcription using reverse gene-editing technology, such as CRISPR/Cas 9 technology, will be a promising tool to expand the current scope of miRNA. Its genome-editing technology has considerable constraints, such as successful, safe and efficient target distribution in plant cells. In addition, molecular methods, such as miRNA, CRISPR-Cas, and RNAi have made site-specific genome alterations in plants more accurate. The transfer of CRISPR, miRNA and siRNA into crop plants, arbitrated by nano-carriers, was recently achieved and required additional investigation. Much research has been done on how microRNAs control plant growth and metabolic processes. The study of microRNAs is comparable to the emergence of a novel tool; it has gradually progressed from an early stage to a point where microRNAs are rapidly entering agriculture as significant markers and promising diagnostic tools. Even though the complete genome sequences of certain crops have been determined, the sequences of microRNAs linked with biotic (protozoan, fungal, bacterial and viral diseases) and abiotic (drought, cold/chilling, heavy metals, hypoxia, nutrient deficiency, salinity, temperature, etc.) stresses have still not been fully annotated. Using high-throughput sequencing approaches and computational tools, it is possible to identify novel microRNAs and their sites that promote or inhibit plant life in response to various stressors; plants that have been genetically modified to contain microRNA or the gene that they target will be resistant to biotic and abiotic stress. Increasing a plant's resistance to a wide variety of stresses can be accomplished in unique and useful ways by manipulating the control of gene expression that is led by microRNA. By interfering with viral suppressor transcripts, the synthesis of virus-specific amiRNAs in transgenic crops confers resistance to viruses. As a result, the amiRNA-mediated approach may find significant application in generating multiple virus-resistant crop plants. This chapter provides an overview of the several approaches developed for measuring, profiling and detecting microRNA targets. Additionally, this chapter provides essential evidence for future research in this area on applied features of microRNA in clinical settings. An investigation of particular miRNA expression patterns may result in significant developments in creating markers, making classification, monitoring and treatment much more superficial.

8. Conclusion

To tolerate biotic and abiotic challenges, plants have evolved several resistance mechanisms, including storage of suitable solute molecules, synthesis of stress-related proteins and stimulation of ROS scavenging systems. Because the gene sequences of many crop species have been mapped, it is now possible to conduct sequence analysis on the microRNAs associated with biotic and abiotic stress. It is anticipated that the vast majority of these recently developed multifunctional molecular markers will have significant future applications in crop breeding efforts. This is because they possess multiple functions. The advent of functional omics techniques and the cumulative knowledge of discovered miRNA markers will accurately validate the concept of genotype-phenotype connections and may benefit future plant breeding programmes. Also, a method based on microRNAs could help plants grow and be monitored in harsh environments.

References

Ahmed, S., Rashid, M.A.R., Zafar, S.A., Azhar, M.T., Waqas, M., Uzair, M., Rana, I.A., Azeem, F., Chung, G., Ali, Z. and Atif, R.M. (2021). Genome-wide investigation and expression analysis of APETALA-2 transcription factor subfamily reveals its evolution, expansion and regulatory role in abiotic stress responses in Indica Rice (*Oryza sativa* L. ssp. indica), *Genomics*, 113(1): 1029-1043.

Akdogan, G., Tufekci, E.D., Uranbey, S. and Unver, T. (2016). miRNA-based drought regulation in wheat, *Funct. Integr. Genomics*, 16: 221-233.

Armor, B.B., Wirth, S., Merchan, F., Laporte, P., d'Aubenton-Carafa, Y., Hirsch, J., Maizel, A., Mallory, A., Lucas, A., Deragon, J.M. and Vaucheret, H. (2009). Novel long non-protein coding RNAs involved in *Arabidopsis* differentiation and stress responses, *Genome Res.*, 19: 57-69.

Bartel, D.P. (2018). Metazoan microRNAs, *Cell*, 173(1): 20-51.

Baruah, P.M., Krishnatreya, D.B., Bordoloi, K.S., Gill, S.S. and Agarwala, N. (2021). Genome wide identification and characterisation of abiotic stress responsive lncRNAs in *Capsicum annuum*, *Plant Physiol. Biochem.*, 162: 221-236.

Basso, M.F., Ferreira, P.C.G., Kobayashi, A.K., Harmon, F.G., Nepomuceno, A.L., Molinari, H.B.C. and Grossi-de-Sa, M.F. (2019). MicroRNA s and new biotechnological tools for its modulation and improving stress tolerance in plants, *Plant Biotechnol. J.*, 17: 1482-1500.

Begum, Y. (2022). Regulatory role of microRNAs (miRNAs) in the recent development of abiotic stress tolerance of plants, *Gene*, 821: 146283.

Betti, F., Ladera-Carmona, M.J., Perata, P. and Loreti, E. (2020). RNAi-mediated hypoxia stress tolerance in plants, *Int. J. Mol. Sci.*, 21: 9394.

Biswas, M.K., Biswas, D., Bagchi, M., Yi, G. and Deng, G. (2021). A comprehensive plant microRNA simple sequence repeat marker database to accelerate genetic improvements in crops, *Agronomy*, 11: 2298.

Cagirici, H.B., Alptekin, B. and Budak, H. (2017). RNA sequencing and co-expressed long non-coding RNA in modern and wild wheats, *Sci. Rep.*, 7: 10670.

Çakır, Ö., Arıkan, B., Karpuz, B. and Turgut-Kara, N. (2021). Expression analysis of miRNAs and their targets related to salt stress in *Solanum lycopersicum* H-2274, *Biotechnol. Biotechnol. Equip.*, 35: 283-290.

Carnavale-Bottino, M., Rosario, S., Grativol, C., Thiebaut, F., Rojas, C.A., Farrineli, L., Hemerly, A.S. and Ferreira, P.C.G. (2013). High-throughput sequencing of small RNA transcriptome reveals salt stress regulated microRNAs in sugarcane, *PLoS ONE*, 8: e59423.

Chakraborty, A., Viswanath, A., Malipatil, R., Rathore, A. and Thirunavukkarasu, N. (2020). Structural and functional characteristics of miRNAs in five strategic millet species and their utility in drought tolerance, *Front. Genet.*, 1498.

Chaudhary, S., Grover, A. and Sharma, P.C. (2021). MicroRNAs: Potential targets for developing stress-tolerant crops, *Life*, 11: 289.

Chaves, S.S., Fernandes-Brum, C.N., Silva, G.F.F., Ferrara-Barbosa, B.C., Paiva, L.V., Nogueira, F.T.S., Cardoso, T.C.S., Amaral, L.R., de Souza Gomes, M. and Chalfun-Junior, A. (2015). New insights on Coffea miRNAs: Features and evolutionary conservation, *Appl. Biochem. Biotechnol.*, 177: 879-908.

Chesnokov, Y.V., Kosolapov, V.M. and Savchenko, I.V. (2020). Morphological genetic markers in plants, *Russ. J. Genet.*, 56: 1406-1415.

Chuong, E.B., Rumi, M.A., Soares, M.J. and Baker, J.C. (2013). Endogenous retroviruses function as species-specific enhancer elements in the placenta, *Nature Genetics*, 45(3): 325-329.

Conn, V.M., Hugouvieux, V., Nayak, A., Conos, S.A., Capovilla, G., Cildir, G., Jourdain, A., Tergaonkar, V., Schmid, M., Zubieta, C. and Conn, S.J. (2017). A circRNA from SEPALLATA3 regulates splicing of its cognate mRNA through R-loop formation, *Nat. Plants*, 3: 1-5.

Cui, L.G., Shan, J.X., Shi, M., Gao, J.P. and Lin, H.X. (2014). The miR156-SPL 9-DFR pathway coordinates the relationship between development and abiotic stress tolerance in plants, *Plant J.*, 80: 1108-1117.

Deng, P., Wang, L., Cui, L., Feng, K., Liu, F., Du, X., Tong, W., Nie, X., Ji, W. and Weining, S. (2015). Global identification of microRNAs and their targets in barley under salinity stress, *PLoS ONE*, 10: e0137990.

Ditta, A., Zhou, Z., Cai, X., Wang, X., Okubazghi, K.W., Shehzad, M., Xu, Y., Hou, Y., Sajid I.M., Khan, M.K.R. and Wang, K. (2018). Assessment of genetic diversity, population structure and evolutionary relationship of uncharacterised genes in a novel germplasm collection of diploid and allotetraploid Gossypium accessions using EST and genomic SSR markers, *Int. J. Mol. Sci.*, 19: 2401.

Djami-Tchatchou, A.T., Sanan-Mishra, N., Ntushelo, K. and Dubery, I.A. (2017). Functional roles of microRNAs in agronomically important plants-potential as targets for crop improvement and protection, *Front. Plant Sci.*, 8: 378.

Du, P., Wu, J., Zhang, J., Zhao, S., Zheng, H., Gao, G., Wei, L. and Li, Y. (2011). Viral infection induces expression of novel phased microRNAs from conserved cellular microRNA precursors, *PLoS Pathog.*, 7: e1002176.

Fard, E.M., Bakhshi, B., Keshavarznia, R., Nikpay, N., Shahbazi, M. and Salekdeh, G.H. (2017). Drought-responsive microRNAs in two barley cultivars differing in their level of sensitivity to drought stress, *Plant Physiol. Biochem.*, 118: 121-129.

Farrokhi, N. and Hajieghrari, B. (2020). Chronicles of Dolos and Apate in plant microRNAs, *Biologia.*, 75: 2441-2463.

Fu, D., Ma, B.I., Mason, A.S., Xiao, M., Wei, L. and An, Z. (2013). Micro RNA-based molecular markers: A novel PCR-based genotyping technique in Brassica species, *Plant Breed.*, 132: 375-381.

Fukao, T., Barrera-Figueroa, B.E., Juntawong, P. and Peña-Castro, J.M. (2019). Submergence and waterlogging stress in plants: A review highlighting research opportunities and understudied aspects, *Front. Plant Sci.*, 340.

Gao, P., Bai, X., Yang, L., Lv, D., Pan, X., Li, Y., Cai, H., Ji, W., Chen, Q. and Zhu, Y. (2011). osa-MIR393: A salinity- and alkaline stress-related microRNA gene, *Mol. Biol. Rep.*, 38: 237-242.

Garrido-Cardenas, J.A., Mesa-Valle, C. and Manzano-Agugliaro, F. (2018). Trends in plant research using molecular markers, *Planta*, 247: 543-557.

Guo, F., Han, N., Xie, Y., Fang, K., Yang, Y., Zhu, M., Wang, J. and Bian, H. (2016). The miR393a/target module regulates seed germination and seedling establishment under submergence in rice (*Oryza sativa* L.), *Plant Cell Environ.*, 39: 2288-2302.

Guo, H., Kan, Y. and Liu, W. (2011). Differential expression of miRNAs in response to topping in flue-cured tobacco (*Nicotiana tabacum*) roots, *PLoS ONE*, 6: e28565.

Gupta, O.P., Permar, V., Koundal, V., Singh, U.D. and Praveen, S. (2012). MicroRNA regulated defense responses in *Triticum aestivum* L. during *Puccinia graminis* f. sp. *tritici* infection, *Mol. Biol. Rep.*, 39: 817-824.

Hackenberg, M., Gustafson, P., Langridge, P. and Shi, B.J. (2015). Differential expression of micro RNAs and other small RNAs in barley between water and drought conditions, *Plant Biotechnol. J.*, 13: 2-13.

Hajyzadeh, M., Turktas, M., Khawar, K.M. and Unver, T. (2015). miR408 overexpression causes increased drought tolerance in chickpea, *Gene*, 555: 186-193.

Hamza *et al.* (2016).

Han, J., Kong, M.L., Xie, H., Sun, Q.P., Nan, Z.J., Zhang, Q.Z. and Pan, J.B. (2013). Identification of miRNAs and their targets in wheat (*Triticum aestivum* L.) by EST analysis, *Genet. Mol. Res.*, 12: 3793-3805.

Han *et al.* (2014).

Hazra, A. and Das, S. (2021). Implications of microRNA variant markers in agriculture-paradigm and perspectives, *Plant Gene*, 25: 100267.

Hewezi, T., Howe, P., Maier, T.R. and Baum, T.J. (2008). *Arabidopsis* small RNAs and their targets during cyst nematode parasitism, *Mol. Plant-Microbe Interact.*, 21: 1622-1634.

Hong, Y.H., Meng, J., Zhang, M. and Luan, Y.S. (2020). Identification of tomato circular RNAs responsive to *Phytophthora infestans*, *Gene*, 746: 144652.

Inal, B., Türktaş, M., Eren, H., Ilhan, E., Okay, S., Atak, M., Erayman, M. and Unver, T. (2014). Genome-wide fungal stress-responsive miRNA expression in wheat, *Planta*, 240: 1287-1298.

Islam, W., Noman, A., Qasim, M. and Wang, L. (2018). Plant responses to pathogen attack: Small RNAs in focus, *Int. J. Mol. Sci.*, 19: 515.

Jeong, D.H., Park, S., Zhai, J., Gurazada, S.G.R., De Paoli, E., Meyers, B.C. and Green, P.J. (2011). Massive analysis of rice small RNAs: Mechanistic implications of regulated microRNAs and variants for differential target RNA cleavage, *Plant Cell*, 23: 4185-4207.

Jin, Q., Xu, Y., Mattson, N., Li, X., Wang, B., Zhang, X., Jiang, H., Liu, X., Wang, Y. and Yao, D. (2017). Identification of submergence-responsive microRNAs and their targets reveals complex miRNA-mediated regulatory networks in lotus (*Nelumbo nucifera* Gaertn), *Front. Plant Sci.*, 8: 6.

Kaja, E., Szczésniak, M.W., Jensen, P.J., Axtell, M.J., McNellis, T. and Makałowska, I. (2015). Identification of apple miRNAs and their potential role in fire blight resistance, *Tree Genet. Genomes*, 11: 812.

Katiyar, A., Smita, S., Chinnusamy, V., Pandey, D.M. and Bansal, K.C. (2012). Identification of miRNAs in sorghum by using a bioinformatics approach, *Plant Signal. Behav.*, 7: 246-259.

Kung, Y.J., Lin, S.S., Huang, Y.L., Chen, T.C., Harish, S., Chua, N.H. and Yeh, S.D. (2012). Multiple artificial microRNAs targeting conserved motifs of the replicase gene confer robust transgenic resistance to negative-sense single-stranded RNA plant virus, *Mol. Plant Pathol.*, 13: 303-317.

Koc, I., Filiz, E. and Tombuloglu, H. (2015). Assessment of miRNA expression profile and differential expression pattern of target genes in cold-tolerant and cold-sensitive tomato cultivars, *Biotechnol. Biotechnol. Equip.*, 29: 851-860.

Li, L., Yi, H., Xue, M. and Yi, M. (2017). miR398 and miR395 are involved in response to SO_2 stress in *Arabidopsis thaliana*, *Ecotoxicology*, 26(9): 1181-1187.

Li, M. and Yu, B. (2021). Recent advances in the regulation of plant miRNA biogenesis, *RNA Biol.*, 18: 2087-2096.

Li, W.X., Oono, Y., Zhu, J., He, X.J., Wu, J.M., Iida, K., Lu, X.Y., Cui, X., Jin, H. and Zhu, J.K. (2008). The *Arabidopsis* NFYA5 transcription factor is regulated transcriptionally and post-transcriptionally to promote drought resistance, *Plant Cell*, 20: 2238-2251.

Lu, C., Jeong, D.H., Kulkarni, K., Pillay, M., Nobuta, K., German, R., Thatcher, S.R., Maher, C., Zhang, L., Ware, D. and Liu, B. (2008). Genome-wide analysis for discovery of rice microRNAs reveals natural antisense microRNAs (nat-miRNAs), *PNAS*, 105: 4951-4956.

Luo, P., Di, D.W., Wu, L., Yang, J., Lu, Y. and Shi, W. (2022). MicroRNAs are involved in regulating plant development and stress response through fine-tuning of TIR1/AFB-dependent auxin signalling, *Int. J. Mol. Sci.*, 23: 510.

Lv, L., Yu, K., Lü, H., Zhang, X., Liu, X., Sun, C., Xu, H., Zhang, J., He, X. and Zhang, D. (2020). Transcriptome-wide identification of novel circular RNAs in soybean in response to low-phosphorus stress, *PLoS ONE*, 15: e0227243.

Ma, X., Zhao, F. and Zhou, B. (2022). The characters of non-coding RNAs and their biological roles in plant development and abiotic stress response, *Int. J. Mol. Sci.*, 23: 4124.

Mangrauthia, S.K., Bhogireddy, S., Agarwal, S., Prasanth, V.V., Voleti, S.R., Neelamraju, S. and Subrahmanyam, D. (2017). Genome-wide changes in microRNA expression during short and prolonged heat stress and recovery in contrasting rice cultivars, *J. Exp. Bot.*, 68: 2399-2412.

Meena, M.R., Kumar, R., Chinnaswamy, A., Karuppaiyan, R., Kulshreshtha, N. and Ram, B. (2020). Current breeding and genomic approaches to enhance the cane and sugar productivity under abiotic stress conditions, *3 Biotech.*, 10: 1-18.

Moldovan, D., Spriggs, A., Yang, J., Pogson, B.J., Dennis, E.S. and Wilson, I.W. (2010). Hypoxia-responsive microRNAs and trans-acting small interfering RNAs in *Arabidopsis*, *J. Exp. Bot.*, 61: 165-177.

Nadeem, M.A., Nawaz, M.A., Shahid, M.Q., Doğan, Y., Comertpay, G., Yıldız, M., Hatipoğlu, R., Ahmad, F., Alsaleh, A., Labhane, N. and Özkan, H. (2018). DNA molecular markers in plant breeding: Current status and recent advancements in genomic selection and genome editing, *Biotechnol. Biotechnol. Equip.*, 32: 261-285.

Naeem, R. (2014). Molecular markers in plant genotyping, *JBMS*, 2: 78-85.

Nair, S.K., Wang, N., Turuspekov, Y., Pourkheirandish, M., Sinsuwongwat, S., Chen, G., Sameri, M., Tagiri, A., Honda, I., Watanabe, Y. and Kanamori, H. (2010). Cleistogamous flowering in barley arises from the suppression of microRNA-guided HvAP2 mRNA cleavage, *PNAS*, 107(1): 490-495.

Naseem, M., Kaltdorf, M. and Dandekar, T. (2015). The nexus between growth and defence signalling: Auxin and cytokinin modulate plant immune response pathways, *J. Exp. Bot.*, 66: 4885-4896.

Navarro, L., Dunoyer, P., Jay, F., Arnold, B., Dharmasiri, N., Estelle, M., Voinnet, O. and Jones, J.D. (2006). A plant miRNA contributes to antibacterial resistance by repressing auxin signalling, *Science*, 312: 436-439.

Niu, C., Li, H., Jiang, L., Yan, M., Li, C., Geng, D., Xie, Y., Yan, Y., Shen, X., Chen, P. and Dong, J. (2019). Genome-wide identification of drought-responsive microRNAs in two sets of Malus from interspecific hybrid progenies, *Hortic. Res.*, 6: 75.

Pandita, D. (2019). Plant miRnome: miRNA biogenesis and abiotic stress response. *In:* Hasanuzzaman, M., Hakeem, K., Nahar, K. and Alharby H. (Eds.). *Plant Abiotic Stress Tolerance*. Springer, Cham, 449-474. Doi https://Doi.org/10.1007/978-3-030-06118-0_18

Pandita, D. and Wani, S.H. (2019). MicroRNA as a tool for mitigating abiotic stress in rice (*Oryza sativa* L.). *In:* Wani, S. (Ed.). *Recent Approaches in Omics for Plant Resilience to Climate Change.* Springer, Cham, 109-133. Doi https://Doi.org/10.1007/978-3-030-21687-0_6

Pandita, D. (2021). Role of miRNAi technology and miRNAs in abiotic and biotic stress resilience. *In:* Aftab, T. and Roychoudhury, A. (Eds.). *Plant Perspectives to Global Climate Changes.* Academic Press, Elsevier, 303-330. https://Doi.org/10.1016/B978-0-323-85665-2.00015-7

Pandita, D. (2022a). How microRNAs regulate abiotic stress tolerance in wheat? A snapshot. *In:* Roychoudhury, A., Aftab, T. and Acharya, K. (Eds.). *Omics Approach to Manage Abiotic Stress in Cereals.* Springer, Singapore, 447-464. https://Doi.org/10.1007/978-981-19-0140-9_17

Pandita, D. (2022b). MicroRNAs shape the tolerance mechanisms against abiotic stress in maize. *In:* Roychoudhury, A., Aftab, T. and Acharya, K. (Eds.). *Omics Approach to Manage Abiotic Stress in Cereals.* Springer, Singapore, 479-493. https://Doi.org/10.1007/978-981-19-0140-9_19

Pandita, D. (2022c). miRNA- and RNAi-mediated metabolic engineering in plants. *In:* Aftab, T. and Hakeem, K.R. (Eds.). *Metabolic Engineering in Plants.* Springer, Singapore, 171-186. https://Doi.org/10.1007/978-981-16-7262-0_7

Parmar, S., Gharat, S.A., Tagirasa, R., Chandra, T., Behera, L., Dash, S.K. and Shaw, B.P. (2020). Identification and expression analysis of miRNAs and elucidation of their role in salt tolerance in rice varieties susceptible and tolerant to salinity, *PLoS ONE*, 15: e0230958.

Patanun, O., Lertpanyasampatha, M., Sojikul, P., Viboonjun, U. and Narangajavana, J. (2013). Computational identification of microRNAs and their targets in Cassava (*Manihot esculenta* Crantz.), *Mol. Biotechnol.*, 53: 257-269.

Peschansky, V.J. and Wahlestedt, C. (2014). Non-coding RNAs as direct and indirect modulators of epigenetic regulation, *Epigenetics*, 9: 3-12.

Qiu, C.W., Zhao, J., Chen, Q. and Wu, F. (2019). Genome-wide characterisation of drought stress-responsive long non-coding RNAs in Tibetan wild barley, *Environ. Exp. Bot.*, 164: 124-134.

Qiu, Z., Hai, B., Guo, J., Li, Y. and Zhang, L. (2016). Characterisation of wheat miRNAs and their target genes responsive to cadmium stress, *Plant Physiol. Biochem.*, 101: 60-67.

Rabuma, T., Gupta, O.P. and Chhokar, V. (2022). Recent advances and potential applications of cross-kingdom movement of miRNAs in modulating plant's disease response, *RNA Biol.*, 19: 519-532.

Ravelonandro, M., Scorza, R. and Briard, P. (2019). Innovative RNAi strategies and tactics to tackle Plum Pox Virus (PPV) genome in *Prunus domestica* – Plum, *Plants*, 8: 565.

Ražná, K., Ablakulova, N., Žiarovská, J., Kyseľ, M., Kushiev, K.K., Gafurov, M.B. and Cagáň, Ľ. (2020). Molecular characterisation of the effect of plant-based elicitor using microRNAs markers in wheat genome, *Biologia*, 75: 2403-2411.

Ribaut, J.M., De Vicente, M.C. and Delannay, X. (2010). Molecular breeding in developing countries: Challenges and perspectives, *Curr. Opin. Plant Biol.*, 13: 213-218.

Sabzehzari, M. and Naghavi, M.R. (2019). Phyto-miRNA: A molecule with beneficial abilities for plant biotechnology, *Gene*, 683: 28-34.

Saiyed, A.N., Vasavada, A.R. and Johar, S.R. (2022). Recent trends in miRNA therapeutics and the application of plant miRNA for prevention and treatment of human diseases, *Future J. Pharm. Sci.*, 8: 1-20.

Salvador-Guirao, R., Hsing, Y.I. and San Segundo, B. (2018). The polycistronic miR166k-166h positively regulates rice immunity via post-transcriptional control of EIN2, *Front. Plant Sci.*, 9: 337.

Sasani, S.T., Soltani, B.M., Rahim Mehrabi, H.S. and Padasht-Dehkaei, F. (2020). Expression alteration of candidate rice miRNAs in response to Sheath Blight Disease, *Iran. J. Biotechnol.*, 18: e2451.

Sepúlveda-García, E.B., Pulido-Barajas, J.F., Huerta-Heredia, A.A., Peña-Castro, J.M., Liu, R. and Barrera-Figueroa, B.E. (2020). Differential expression of maize and teosinte microRNAs under submergence, drought, and alternated stress, *Plants*, 9: 1367.

Shah, S.M.S. and Ullah, F. (2021). A comprehensive overview of miRNA targeting drought stress resistance in plants, *Braz. J. Biol.*, 83.

Shamrai, S.M. (2022). Recognition of pathogen attacks by plant immune sensors and induction of plant immune response, *Cytol. Genet.*, 56: 46-58.

Shriram, V., Kumar, V., Devarumath, R.M., Khare, T.S. and Wani, S.H. (2016). MicroRNAs as potential targets for abiotic stress tolerance in plants, *Front. Plant Sci.*, 7: 817.

Shui, X.R., Chen, Z.W. and Li, J.X. (2013). MicroRNA prediction and its function in regulating drought-related genes in cowpea, *Plant Sci.*, 210: 25-35.

Sihag, P., Sagwal, V., Kumar, A., Balyan, P., Mir, R.R., Dhankher, O.P. and Kumar, U. (2021). Discovery of miRNAs and development of heat-responsive miRNA-SSR markers for characterisation of wheat germplasm for terminal heat tolerance breeding, *Front. Genet.*, 12.

Singroha, G., Sharma, P. and Sunkur, R. (2021). Current status of microRNA-mediated regulation of drought stress responses in cereals, *Physiol. Plant.*, 172: 1808-1821.

Summanwar, A., Basu, U., Rahman, H. and Kav, N.N. (2021). Non-coding RNAs as emerging targets for crop improvement, *Plant Sci.*, 297: 110521.

Sun, C., Zhao, Q., Liu, D.D., You, C.X. and Hao, Y.J. (2013). Ectopic expression of the apple Md-miRNA156h gene regulates flower and fruit development in *Arabidopsis*, *Plant Cell Tissue Organ Cult.*, 112: 343-351.

Tiwari, R. and Rajam, M.V. (2022). RNA- and miRNA-interference to enhance abiotic stress tolerance in plants, *J. Plant Biochem. Biotechnol.*, 1-16.

Tyagi, S., Kumar, A., Gautam, T., Pandey, R., Rustgi, S. and Mir, R.R. (2021). Development and use of miRNA-derived SSR markers for the study of genetic diversity, population structure and characterisation of genotypes for breeding heat tolerant wheat varieties, *PLoS ONE*, 16: e0231063.

Tyagi, S., Sharma, S., Ganie, S.A., Tahir, M., Mir, R.R. and Pandey, R. (2019). Plant microRNAs: Biogenesis, gene silencing, web-based analysis tools and their use as molecular markers, *3 Biotech.*, 9: 1-12.

Vallarino, J.G., Osorio, S., Bombarely, A., Casañal, A., Cruz-Rus, E., Sánchez-Sevilla, J.F., Amaya, I., Giavalisco, P., Fernie, A.R., Botella, M.A. and Valpuesta, V. (2015). Central role of Fa GAMYB in the transition of the strawberry receptacle from development to ripening, *New Phytol.*, 208: 482-496.

Van Lommel, J., Lenaerts, C., Delgouffe, C. and Broeck, J.V. (2022). Knockdown of ecdysone receptor in male desert locusts affects relative weight of accessory glands and mating behaviour, *J. Insect Physiol.*, 138: 104368.

Varshney, D., Rawal, H.C., Dubey, H., Bandyopadhyay, T., Bera, B., Kumar, P.M., Singh, N.K. and Mondal, T.K. (2019). Tissue specific long non-coding RNAs are involved in aroma formation of black tea, *Ind. Crops Prod.*, 133: 79-89.

Vidal, E.A., Araus, V., Lu, C., Parry, G., Green, P.J., Coruzzi, G.M. and Gutiérrez, R.A. (2010). Nitrate-responsive miR393/AFB3 regulatory module controls root system architecture in *Arabidopsis thaliana*, *Proceedings of the National Academy of Sciences*, 107(9): 4477-4482.

Waheed, S., Anwar, M., Saleem, M.A., Wu, J., Tayyab, M. and Hu, Z. (2022). The critical role of small RNAs in regulating plant innate immunity, *Biomolecules*, 11: 184.

Waititu, J.K., Zhang, C., Liu, J. and Wang, H. (2020). Plant non-coding RNAs: Origin, biogenesis, mode of action and their roles in abiotic stress, *Int. J. Mol. Sci.*, 21: 8401.

Wang, Q. and Zhang, B. (2015). MicroRNAs in cotton: An open world needs more exploration, *Planta*, 241: 1303-1312.

Wang, W., Liu, D., Chen, D., Cheng, Y., Zhang, X., Song, L., Hu, M., Dong, J. and Shen, F. (2019). MicroRNA414c affects the salt tolerance of cotton by regulating reactive oxygen species metabolism under salinity stress, *RNA Biol.*, 16: 362-375.

Wang, Y., Luo, X., Sun, F., Hu, J., Zha, X., Su, W. and Yang, J. (2018). Overexpressing lncRNA LAIR increases grain yield and regulates neighbouring gene cluster expression in rice, *Nat. Commun.*, 9: 1-9.

Wu, J., Liu, Q., Wang, X., Zheng, J., Wang, T., You, M., Sheng Sun, Z. and Shi, Q. (2013). mirTools 2.0 for non-coding RNA discovery, profiling, and functional annotation based on high-throughput sequencing, *RNA Biol.*, 10: 1087-1092.

Xu, Y. (2010). *Molecular Plant Breeding*, CABI, Wallingford, United Kingdom.

Xue, T., Liu, Z., Dai, X. and Xiang, F. (2017). Primary root growth in *Arabidopsis thaliana* is inhibited by the miR159 mediated repression of MYB33, MYB65 and MYB101, *Plant Sci.*, 262: 182-189.

Yadav, C.B., Muthamilarasan, M., Pandey, G., Khan, Y. and Prasad, M. (2014). Development of novel microRNA-based genetic markers in foxtail millet for genotyping applications in related grass species, *Mol. Breed.*, 34: 2219-2224.

Yang, X., Zhang, L., Yang, Y., Schmid, M. and Wang, Y. (2021). miRNA mediated regulation and interaction between plants and pathogens, *Int. J. Mol. Sci.*, 22: 2913.

Yogindran, S. and Rajam, M.V. (2021). Host-derived artificial miRNA-mediated silencing of ecdysone receptor gene provides enhanced resistance to *Helicoverpa armigera* in tomato, *Genomics*, 113: 736-747.

Yue, E., Cao, H. and Liu, B. (2020). OsmiR535, a potential genetic editing target for drought and salinity stress tolerance in *Oryza sativa*, *Plants*, 9: 1337.

Zafar, S., Iqbal, A., Azhar, M.T., Atif, R.M., Rana, I.A., Rehman, H.M., Nawaz, M.A. and Chung, G. (2019). GM maize for abiotic stresses: Potentials and opportunities. *In: Recent Approaches in Omics for Plant Resilience to Climate Change* (pp. 229-249). Springer, Cham.

Zeeshan, M., Qiu, C.W., Naz, S., Cao, F. and Wu, F. (2021). Genome-wide discovery of miRNAs with differential expression patterns in responses to salinity in the two contrasting wheat cultivars, *Int. J. Mol. Sci.*, 22: 12556.

Zhang, P., Fan, Y., Sun, X., Chen, L., Terzaghi, W., Bucher, E., Li, L. and Dai, M. (2019). A large-scale circular RNA profiling reveals universal molecular mechanisms responsive to drought stress in maize and *Arabidopsis*, *Plant J.*, 98: 697-713.

Zhang, T., Zhao, Y.L., Zhao, J.H., Wang, S., Jin, Y., Chen, Z.Q., Fang, Y.Y., Hua, C.L., Ding, S.W. and Guo, H.S. (2016). Cotton plants export microRNAs to inhibit virulence gene expression in a fungal pathogen, *Nat. Plants*, 2: 1-6.

Zhang, X., Wang, W., Wang, M., Zhang, H.Y. and Liu, J.H. (2016). The miR396b of *Poncirus trifoliata* functions in cold tolerance by regulating ACC oxidase gene expression and modulating ethylene–polyamine homeostasis, *Plant Cell Physiol.*, 57: 1865-1878.

Zhang, Y., Zhu, X., Chen, X., Song, C., Zou, Z., Wang, Y., Wang, M., Fang, W. and Li, X. (2014). Identification and characterization of cold-responsive microRNAs in tea plant (*Camellia sinensis*) and their targets using high-throughput sequencing and degradome analysis. *BMC Plant Biology*, 14(1): 1-18.

Zhang, X., Zou, Z., Gong, P., Zhang, J., Ziaf, K., Li, H., Xiao, F. and Ye, Z. (2011). Overexpression of microRNA169 confers enhanced drought tolerance to tomato, *Biotechnol. Lett.*, 33: 403-409.

Zhao, C., Xia, H., Cao, T., Yang, Y., Zhao, S., Hou, L., Zhang, Y., Li, C., Zhang, X. and Wang, X. (2015). Small RNA and degradome deep sequencing reveals peanut microRNA roles in response to pathogen infection, *Plant Mol. Biol. Rep.*, 33: 1013-1029.

Zhou, M. and Tang, W. (2019). MicroRNA156 amplifies transcription factor-associated cold stress tolerance in plant cells, *Mol. Genet. Genom.*, 294: 379-393.

Zhou, M., Gu, L., Li, P., Song, X., Wei, L., Chen, Z. and Cao, X. (2010). Degradome sequencing reveals endogenous small RNA targets in rice (*Oryza sativa* L. ssp. *indica*), *Front. Biol.*, 5: 67-90.

CHAPTER

18

MicroRNAs and other Non-coding RNAs in Plant Epigenetics

Auqib Manzoor[1], Tabasum Ashraf[1], Humaira Shah[1], Rouf Maqbool[1], Rachna Kaul[2] and Ashraf Dar[1]*

[1] Department of Biochemistry, University of Kashmir, Srinagar - 190006, India
[2] Bombay College of Pharmacy, Kalina, Mumbai, Maharashtra - 400098, India

1. Introduction

Epigenetics is the study of molecular mechanism of variations in the gene expression patterns and cell phenotype that do not involve alterations in the primary gene coding sequences (Tollefsbol, 2011). The field of plant epigenetics has received a significant attention over the years. Plants rely extensively on gene expression variations to adapt to different environmental changes. The chromatin-based gene expression control is extremely critical for plants in responding to the changed environmental conditions (Kohler and Springer, 2017). The variations in chromatin are caused by changes in DNA methylation patterns and differential histone modifications which regulate various physiological processes in plants, like growth, senescence and adaptation to various biotic and abiotic stresses (Table 1) (Pikaard and Mittelsten Scheid, 2014; De-la-Peña *et al.*, 2015). Over the years, the plant epigenetics area has evolved considerably with the studies on non-Mendelian gene expression patterns and discovery of chromatin-modifying enzymes and small double-stranded RNAs which regulate gene expression (Pikaard and Mittelsten Scheid, 2014). Various types of double-stranded non-coding RNAs, including miRNAs, have been shown to play a significant role in regulating the chromatin dynamics in plants (Pikaard and Mittelsten Scheid, 2014; De-la-Peña *et al.*, 2015). Most of the mechanisms involving epigenetic regulations in plants have been understood from different studies in *Arabidopsis thaliana*. Which is the first plant whose whole genome was sequenced, widely used model in plant genetic research (Pikaard and Mittelsten Scheid, 2014). The understanding of epigenetic control of different physiological pathways leading to metastable variations in gene expression patterns has a great significance in revealing the developmental, ecological and evolutionary histories of plants. This chapter highlights various aspects of non-coding

*Corresponding author: ashrafdar@kashmiruniversity.ac.in

RNAs which regulate epigenetic changes in plants, leading to altered gene expression under different conditions.

Table 1: Epigenetic modifications, and the stresses regulated by these modifications

Modification	Species	Regulator Name	Stress Type
Methylation	*A. thaliana*	ATX1	Water stress
	A. thaliana	ATX4/5	Drought
	A. thaliana	CAU1/PRMT5/SKB1	Salt stress and drought
Demethylation	*A. thaliana*	JMJ15	Salt stress
	A. thaliana	JMJ17	Water stress
Acetylation	*A. thaliana*	AtHAC1	Temperature
	Poplar		Drought
	A. thaliana	GCN5	Temperature and salt stress
Deacetylation	*A. thaliana*	HDA9	Salt stress and drought
	A. thaliana	HDA19	Drought, high temperature and salt stress
	A. thaliana	HDA6	Low and high temperature, drought and salt stress
	Rice	HDA705	Salt stress
	Brachypodium	BdHD1	Drought
Phosphorylation	*A. thaliana*	MLK1/2	Salt stress and Drought
Ubiquitination	Cotton	AtHUB2	Drought
	Rice	OsHUB2	Drought

2. DNA (Cytosine) Methylation

The process of adding a methyl group to DNA sequences, e.g. fifth carbon of the cytosine pyrimidine ring in CG, CHH or CHG motifs (where H represents A, T, or C) is called as DNA methylation. This chromatin modification is most studied and trickles down from mother to daughter cells and to some extent, through generations. 5-methylcytosine (5mC) is a common epigenetic marker for gene silencing and heterochromatin formation in plants (Jia *et al.*, 2011). The cytosine methylation is non-random, particularly in repeated sequences that are abundant in centromeric repeats, transposable elements or arrays of silent 5S or 45S rRNA gene repeats. A great majority of transposons are heavily methylated, making them an obvious target for epigenetic silencing (Pikaard and Mittelsten Scheid, 2014). The cytosine methylation acts as a mechanism to suppress transposons, which are otherwise considered as 'genomic parasites' that invade their host's genome. However, there are some differently regulated promoters and the protein-coding regions of highly expressed genes that have significant levels of cytosine methylation (Zilberman *et al.*, 2007).

MicroRNAs and other Non-coding RNAs in Plant Epigenetics **331**

Different mechanisms are involved in addition, preservation and erasing of cytosine methyl groups in various gene segments in contexts of the plant genome. The *de novo* methylation at cytosine bases is mediated by RNA directed DNA methylation (RdDM) pathway while the preservation of this cytosine methylation mark in varied sequence contexts is dependent on distinct DNA methyltransferases. The decrease in cytosine methylation may be caused by a variety of reasons, e.g. methyltransferase dysfunction, a limitation of methyl group donor (S-adenosylmethionine) during passive DNA demethylation, or by the process of active DNA demethylation (Kumar and Mohapatra, 2021). The RdDM system promotes *de novo* DNA methylation and it makes use of small interfering RNAs (siRNAs) and several auxiliary proteins during the process.

Plants possess homologues of mammalian DNA methyl transferases Dnmt1, Dnmt2, and Dnmt3 (Li and Zhang, 2014). The counterpart of human Dnmt1 in plants is DNA methyltransferase1 (MET1). It is the predominant CG maintenance methyltransferase and may potentially contribute to *de novo* methylation in CG context. Furthermore, plants have DNMT2 homologs that have RNA methylase activity (on tRNA) but no catalytic activity on DNA (Goll *et al.*, 2006). The plant domains-rearranged methyltransferases (DRM) are the main *de novo* methyltransferases in plants, catalysing methylation of cytosine bases in all sequence contexts (Pikaard and Mittelsten Scheid, 2014).

3. Histone Modifications

Modification of histone tails is another typical epigenetic signature. Unlike DNA methylation, which always occurs at the cytosine bases, histones (H2A, H2B, H3 and H4) can be modified by covalently linking to distinct chemical markers, like acetylation, methylation, ADP-ribosylation, biotinylation, ubiquitination and phosphorylation (mainly at arginine and lysine residues) (Johnson *et al.*, 2004; Zhang *et al.*, 2007). The acetylation and methylation of histones are two forms of epigenetic changes that are now regarded as essential and ubiquitous in gene regulation (Xu *et al.*, 2017). The most well-studied histone methylations in plants are mono-, di-, or tri-methylation of H3K4 (H3K4me1, H3K4me2 and H3K4me3), tri-methylation of H3K27 (H3K27me3) and di-methylation of H3K9 (H3K9me2) (Feng and Jacobsen, 2011). The histone modifications are catalysed and kept in check by the enzymes known as 'writers' and 'erasers'. The writers include acetyltransferases (HATs), methyltransferases, kinases and ubiquitin ligases, which catalyse the transfer of chemical groups onto histone tails or core domains. Erasers, which include demethylases (HDMs), deacetylases (HDACs), deubiquitinases, and phosphatases are enzymes that remove these groups (Xu *et al.*, 2017).

Acetylation of histones (specifically H3 and H4) neutralise their basic charge and enhance the accessibility to DNA by decreasing histones-DNA interaction (Onufriev and Schiessel, 2019). Both HATs and HDACs have several gene families in plants (Pandey *et al.*, 2002). For example, HAG1, a GCN5 homolog, controls various developmental events in *Arabidopsis* by selectively methylating H3K14, like it does in yeast and humans (Servet *et al.*, 2010). The plant-specific HD2 or HDT family of histone deacetylases has been linked to gene repression in plants. Studies have also found that HDA1 and HDA6 act as epigenetic regulators. HDA6, involved in the maintenance of methylation status at CG and CHG sites, interacts with the MET1 (DNA methyltransferase) and mediates silencing of transposon and repression of rRNA (Kim *et al.*, 2012).

In *Arabidopsis*, histone methylation marks can act as repressors of transcription (H3K9me2/3, H4R3me2 and H3K27me3) or activators of transcription (H3K4me3, H3K36me2/3 and H4R3me2) (Wang *et al.*, 2016; Liu *et al.*, 2010). ChIP-chip analysis of *Arabidopsis* H3K4me1, H3K4me2 and H3K4me3 revealed that they appear primarily in promoters and gene bodies (~2/3 of all genes) and are mostly missing from transposons and repetitive DNA regions. This finding is in agreement with the idea that H3K4 methylation defines active chromatin. The distributions of the three H3K4 marks within genes differ: H3K4me1 is abundant in the gene bodies, whereas H3K4me2 and H3K4me3 are predominantly present in the promoters and at 5′ end of genes. Importantly, H3K4me3 are always present upstream of H3K4me2. Moreover, only H3K4me3 is linked to activation of transcription, whereas H3K4me1 and H3K4me2 have no effect on gene transcription (Zhang *et al.*, 2003).

Besides acetylation and methylation, another key histone post-translational modification is phosphorylation (Houben *et al.*, 2007). The phosphorylation of histones (H2AX) has a role in DNA damage repair as well as in chromosomal segregation and cell division control. Other epigenetic marks can also have an impact on histone phosphorylation, e.g. phosphorylation of H3S10 is controlled by the modification status of the nearby lysine, H3K9. In plants, histones can also undergo ADP-ribosylation, although this modification has received little attention (Pikaard and Mittelsten Scheid, 2014). The ubiquitination of histones H2A and/or H2B has crucial regulatory roles in plants. H2A (H2Aub) and H2B (H2Bub) monoubiquitination is also thought to be an activation as well as repression mark of transcription. It has been found that in *Arabidopsis* H2AK121 monoubiquitination occurs independent of H3K27me3 and polycomb repressive complex2 (PRC2) (Bratzel *et al.*, 2010). DNA methylation and heterochromatic histone H3 methylation, both require H2B deubiquitination (Sridhar *et al.*, 2007). Furthermore, H2B monoubiquitination is known to increase the transcription via H3K4me3 deposition (Geng *et al.*, 2012).

4. miRNAs Regulate Plant Epigenetics

Various types of small RNAs (sRNAs) have emerged as significant factors in directing transcriptional networks in organisms by altering DNA methylation or histone modification. MicroRNAs, piwi interacting RNAs and small interfering RNAs are the three types of sRNAs. Among these sRNA types, siRNAs have been linked to changes in DNA methylation, but only in the context of transposons and repetitive DNA. The function of miRNAs in DNA methylation, on the other hand, has been described in a significantly lower number of instances. Conventional miRNAs control target gene transcription by target mRNA cleavage or by translational repression mechanism. In a few situations, conventional small miRNAs have been reported to cause DNA methylation (Table 2). However, another class of longer miRNAs (lmiRNAs) (23-27 nt) have been implicated in promotion of DNA methylation in a greater majority of instances. These lmiRNAs can cause both cis and trans cytosine methylation of a locus, resulting in transcriptional suppression of the target genes (Fig. 1).

Table 2: miRNA-stimulated DNA methylation (DM) in plants

Small RNAs mediating methylation	Examples	Species	Loci Methylated	DICER Involved	Polymerase Involved	DM Related Proteins	Pattern of DM
Conventional miRNAs	miR166, miR166	*Arabidopsis*	PBH and PHV	DCL1 is involved	RDR2 not needed	ND	A few hundred nucleotides downstream from the miRNA-binding site
	miR1026 and miR166	Moss	PpC3HD7IPl, PpbHLH and PpHB10	PpDCL1a in biogenesis of miRNA PpDCL1b negatively regulates DM	ND	miRNA is loaded in RITS complex	Target gene promoters and flanking regions surrounding miRNA binding sites in target genes.
siRNAs produced from miRNA locus	miR2831, miR2328 etc	*Arabidopsis thaliana*	At5g08490 and At4g16580	DCL3	PolIV and RDR2 is required	AGO4	RdDM of target loci in trans that also lowered the deposition of target gene transcripts.
lmiRNAs	miR820.2, miR1876, miR1683 and miR1873.1	Rice	Os06g38480, Os02g05890, Os01g01790 etc	DCL3	RDR2 not required	AGO4	Can guide DNA methylation at their origin (in cis) and target gene loci (in trans). DM binds to target genes within an 80-nt area around their binding sites, suppressing their expression.
Ta-siRNAs derived from TAS gene transcripts	miR390 and miR173	*Arabidopsis*	TAS1, TAS3 and TAS2	DCL1 with DCL2/3/4	RDR6	PolV and AGO4/6	In the proximity of miRNA cleavage sites at the TAS locus, DM is found in cis.
phasiRNAs derived from PHAS transcripts	miR2275 and miR2118	Zea mays	PHAS loci	DCL5 or DCL1	RDR6	AGO180/140	At the PHAS locus of origin, DM is in cis.

(Contd.)

Table 2: *(Contd.)*

Small RNAs mediating methylation	Examples	Species	Loci Methylated	DICER Involved	Polymerase Involved	DM Related Proteins	Pattern of DM
easiRNA derived from transposons	miR159, miR169, miR156 and numerous others.	*Arabidopsis thaliana*	Many transposon loci such as vandal, athila and numerous others	DCL4/1	RDR6	AGO1	RDR2 and DCL3 assembly is halted, preventing RdDM caused by 24 nt siRNAs. This occurs exclusively in ddm1 mutant tissues or germ line cells from WT tissues.
hc-RNA derived from transposons or other repetitive sequences	Variable types of miRNAs	Various	Different types of loci	DCL3	Pol IV and RDR2 needed	Pol V and AGO4	The classic RdDM route causes de novo DM in cis extending to hundreds to thousands of nucleotides.

MicroRNAs and other Non-coding RNAs in Plant Epigenetics   335

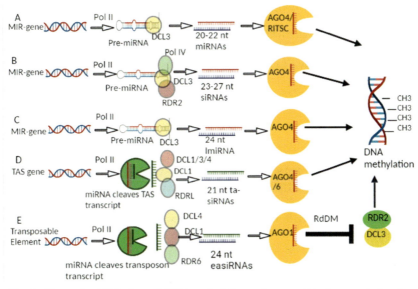

Fig. 1: Plant microRNA-directed DNA methylation pathways: (A) via canonical pathway (*Arabidopsis* and Moss), (B) via siRNAs derived from miRNA loci (*Arabidopsis*), (C) via lmiRNAs (rice), (D) via ta-siRNAs derived from micro RNA cleaved TAS gene transcripts (*Arabidopsis*), (E) via easiRNAs derived from micro RNA cleaved transposon transcripts (*Arabidopsis*)

3.1 DNA Methylation Regulated by Conventional miRNAs in Arabidopsis

In *Arabidopsis* miR165/166s target two transcriptional factors for phabulosa (PHB) and phavoluta (PHV) which regulate a number of developmental processes. In differentiated tissues, the exons of PHB and PHV genes are heavily methylated downstream to the miRNA-binding region, but not in proliferative or undifferentiated cells (Bao *et al.*, 2004). Methylation is decreased in phv-1d and phb-1d gain-of-function mutants that are tolerant to miR165/166 promoted cleavage. The pullulating transcripts of PHV and PHB might be near the DNA of such genes and miR165/166 could cause DNA methylation by binding to the mRNAs of PHB and PHV. Only the WT allele is methylated in heterozygous PHB/phb or PHV/phv plants, bearing both the WT and mutant alleles. This further implies that DNA methylation is dependent on the capacity of the miRNA to bind to the transcribed mRNAs of PHB and PHV (Bao *et al.*, 2004). An intron disrupts the miR165/166 location in the PHB and PHV loci. If miR165/166 is unable to interact with such genes and if miRNA biogenesis mutants, like AGO1 and dcl1, have no effect on DNA methylation, it is unclear how exactly miRNAs induces DNA methylation (Ronemus *et al.*, 2005). A group of 24 nt longer miRNA processed by DCL3, discovered which like in rice is responsible for methylation of target genes such as PHV and PHB (Vazquez *et al.*, 2008). This is supported by the fact that in dcl1 and AGO1 mutants, which are primarily important in the synthesis of 21 nt miRNA isoforms, the methylation driven by miR165/166 is unaffected (Bao *et al.*, 2004). The actual mechanism of miR165/166-based DNA methylation, however, is yet unknown.

3.2 DNA (cytosine) Methylation in Moss

In *Physcomitrella patens*, miRNA-directed methylation of DNA has been observed (Khraiwesh *et al.*, 2010). PpDCL1a and PpDCL1b are the Dicer-like orthologue proteins in moss. PpDCL1a promotes miRNA biosynthesis, whereas PpDCL1b contributes to target cleavage. miRNA processing and accumulation are normal in the PpDCL1b mutant, but miRNA targets remain uncleaved, demonstrating that PpDCL1b is required for miRNA-guided mRNA cleavage. Unexpectedly, in PpDCL1b, the concentration of target mRNA transcripts decreases, pointing to another regulatory route. This mechanism has proven to be epigenetic (Khraiwesh *et al.*, 2010). In the wild-type moss, the target gene promoters were found to be unmethylated while they were methylated in PpDCL1b. The regions flanking the miR166-binding motifs in the target genes, PpC3HDZIP1, and PpHB10, were reported to be methylated in the PpDCL1b mutant, but not in the wild type. The miRNA-binding motifs in PpC3HDZIP1 and PpHB10, like those in PHV and PHB, are separated by introns, rendering the formation of miRNA:DNA hybrid less likely. Rather, the miRNA-loaded RITS complex might engage with target mRNAs to generate miRNA : mRNA hybrids, which could then be directed to connect with target genic DNA to activate methylation. Another moss miRNA, miR1026, is activated by ABA (abcisic acid) and degrades the mRNA of the PpbHLH target gene. The methylation of PpbHLH is activated by ABA induction, demonstrating that miR1026 regulates PpbHLH expression via two pathways: methylation and mRNA breakage. In moss, DNA methylation coordinated by conventional miRNAs suggests that conventional miRNAs have dual function in this species. The two different types of DCL1 proteins, PpDCL1a and PpDCL1b, are thought to be responsible for miRNAs dual functioning. Methylation has been observed at cytosines at CG sites in all of the genes studied in PpDCL1b, although additional cytosine methylation sites cannot be ruled out for other genes (Chellappan *et al.*, 2010).

3.3 siRNAs-derived from miRNA Loci Methylate DNA

The methylation of DNA is found to be induced by siRNAs derived from miRNA sites. Certain siRNAs arise from normal miRNA sequences. Besides traditional (20-22 nt) miRNAs, some miRNA loci generate longer (23-27 nt) sRNA species (Chellappan *et al.*, 2010). These 23-27 nt sRNA species have a similar biogenesis mechanism as that of heterochromatic siRNAs (hc-siRNA). These sRNAs are produced by DCL3, RDR2 and Pol IV pathways. RdDM is triggered when hc-siRNAs are sorted out by AGO4 in the cytoplasm (Ye *et al.*, 2012).

miRNAs are hairpin-folded single-stranded RNAs produced from miRNA genes, whereas siRNAs are usually large double-stranded RNAs (dsRNAs) (Hamilton *et al.*, 2002). The two sRNA species: 21 nt long canonical miRNA and 23-27 nt long sRNAs, are formed by miRNAs 2883, 2831 and 2328. The 23-27 nt siRNAs are generated, using the positive strand of precursor-miRNAs and emanate from the same miRNA-producing sites as conventional miRNAs (Vaucheret, 2008). Similar to miR165/166, all these longer siRNA-like sRNAs may cause DNA methylation around their target gene-binding sites. These siRNAs were not found in nrdp1-3 or rdr2 mutants, indicating that they are not long miRNAs (lmiRNAs). At the siRNA binding sites in nrpd1-3, DNA methylation of At5g08490 (miR2831 target) and At4g16580 (miR2328

MicroRNAs and other Non-coding RNAs in Plant Epigenetics   **337**

target) is drastically reduced. This implies that siRNAs originating from miRNA sites guide RdDM of target loci in trans. The build-up of transcripts from these target genes is likewise decreased as a result of the DNA methylation. Both miR165 and miR166 create 23-26 nt siRNAs, which are largely organized into AGO4- and AGO7-guided RISC complexes, in addition to the conventional miRNAs. As previously stated, these siRNAs are the most likely participants in inducing DNA methylation at the target loci in PHB and PHV, as this methylation was unaltered in the dcl1 and ARGO1 mutants.

A large number of MIR genes in rice, moss and *Arabidopsis* have been found to generate two types of sRNA species from the same gene locus: 23-27 nt siRNA-like and 20-22 nt miRNA-like. These miRNA sites produce two kinds of sRNA species. These sRNA species regulate the target gene expression in different ways as found in dual mode of target gene expression control in moss. The smaller species suppress gene expression via mRNA cleavage or translational inhibition whereas the longer species promote DNA methylation and silence the corresponding genes. Further, two types of sRNA groups derived from a miRNA locus have been discovered in Medicago plants (Lelandais-Briere *et al.*, 2009).

3.4 DNA (Cytosine) Methylation by lmiRNAs

The methylation of DNA has been observed in some instances to be activated by lmiRNAs, in addition to conventional miRNAs. Many miRNA loci have been discovered in rice that generate both 21 nt canonical miRNAs and 24 nt lmiRNAs. The 21 nt miRNAs cleave target transcripts and the 24 nt lmiRNAs promote DNA methylation. There are certain loci that produce only lmiRNAs (Wu *et al.*, 2010). These lmiRNAs control cytosine DNA methylation at their target genes in trans as well as at their locus in cis. By coordinating the actions of DCL1 and DCL3, several miRNA precursors, such as miR168, miR396, miR1850 and miR820, may create both short and long miRNAs. Methylation of CHG and CHH occurs in miR1863, not just in sequences found in miRNA and miRNA* regions, but also in locations outside the stem-loop of miRNA precursor. lmiRNAs, like miR1863, do not break their target genes but promote direct methylation within an 80 nt band surrounding their binding sites, inhibiting the expression of the target genes. Subsequent research in rice has revealed that 65 genes display hypermethylation at CHH sequence surrounding miRNA-binding sites (Hu *et al.*, 2014) out of 325 sites targeted by 24 nt miRNAs. Hypermethylation discovered at the miRNA-binding regions of miR1862c, miR1863b, miR812, miR2121b, miR1867, miR5831 and miR5150 targets indicating that many miRNAs can target the same gene. It is well elaborated that dual-function miRNA genes are found in almost all plant species required for controlling DNA methylation as a conserved function. Additionally, the synthesis and functionality of long miRNAs is predicted to differ amongst plant species. RDR2 is not required for the synthesis of 24 nt lmiRNAs in rice and these miRNAs can drive DNA methylation at their origin (in cis) and target gene loci (in trans). Longer miRNAs/siRNAs (23-27 nt) in *Arabidopsis*, on the other hand, need RDR2 for synthesis and predominantly induce DNA methylation at target gene loci in trans (Chellappan *et al.*, 2010). In this scenario, deciphering the role of miRNAs in DNA methylation requires examining sRNA species of various size proportions (originating from a miRNA locus) and their influence on DNA methylation of distinct loci in cis and trans.

3.5 TAS/PHAS Loci-derived siRNAs that are miRNA-triggered also Methylate DNA

There are three types of endogenous siRNAs in plants: hc-siRNAs, trans-acting-siRNAs (ta-siRNAs) and natural antisense transcript-derived siRNAs (nat-siRNAs) (Ghildiyal *et al.*, 2009). TAS gene transcripts are cleaved by miRNAs to form ta-siRNAs capable of degrading mRNAs. TAS loci are frequently methylated, in addition to miRNA-directed cleavage (Lister *et al.*, 2008). The TAS1a, TAS1c and TAS3a loci each have a considerable amount of cytosine methylation at the CG, CHG and CHH positions, but the TAS2 locus has little methylation. At the TAS loci, the methylation sites are close to miRNA cleavage sites and to confirm whether this TAS loci methylation is not via RDM pathway involving hc-RNAs, DNA methylation has been examined in dcl3-1 and rdr2-2 mutants and it was found that 24nt ta-siRNAs and hc-RNAs levels were greatly reduced but the TAS loci methylation was not compromised (Wu *et al.*, 2012). It was also observed that cytosine methylation cites were reduced or lost in the Pol V mutant (nrpd1b-11), suggesting Pol V's interaction with TAS1c and TAS3a scaffold mRNAs required for methylation of DNA at their loci. Since the involvement of 24 nucleotide ta-siRNAs in TAS loci, DNA methylation has been obviated and it has been debated whether 21 nucleotide ta-siRNAs are engaged in this process. TAS transcripts have been demonstrated to be converted into 21 nucleotide ta-siRNAs by the activity of DCL1. In mutants linked to the biogenesis pathway, such as, zip-1, rdr6-11, dcl1-7, dcl4-2 and sgs3-11, TAS3a and TAS1c 21 nucleotides ta-siRNAs dramatically decrease (Wu *et al.*, 2012; Yoshikawa *et al.*, 2005). Except for dcl4-2 mutants, cytosine methylation dramatically decreases in dcl1-7, zip-1, rdr6-11 and sgs3-11 mutants in all settings, demonstrating the role of 21 nucleotide ta-siRNAs at TAS loci DNA methylation. DCL1, in combination with DCL2/3/4, is important for methylation of DNA at the TAS1c and TAS3a loci in *A. thaliana*. In the DNA methylation of distinct TAS loci, different DCL combinations are required. Furthermore, the same 21-nucleotide ta-siRNAs are found to be associated with AGO4/6 and promote TAS loci methylation.

Phased small interfering RNAs (phasiRNAS) provide are one more instance of DNA methylation by siRNAs. PhasiRNAs are made up of PHAS loci, which are found in ta-siRNA loci, protein-coding gene loci, or non-coding RNA loci (Fei *et al.*, 2013). PhasiRNAs are found in 21 or 24 nt phased segments, with 1-2 miRNA-binding locations guiding their phasing. The processing of these molecules might take place either downstream or upstream of the cleavage target location. miR2118 and miR2275 have been discovered to stimulate phasiRNA production in maize male meiocytes. PhasiRNA loci in meiocytes exhibit more methylation of cytosines compared to seedlings and anthers, with the largest increment in the CHH context (Dukowic-Schulze *et al.*, 2016). This hypermethylation in meiocytes is highly prominent in 21 nucleotide phasiRNA loci compared to 24 nucleotide phasiRNA loci. AGO104 or AGO18b are two potential AGO proteins implicated in phasiRNA-induced methylation of DNA (Dukowic-Schulze *et al.*, 2016). However further experimental proof is required to determine whether the proposed induction of this DNA methylation by phasiRNAs in cis at their loci of origin. If accurate, this methylation follows the DNA methylation caused by the previously described 21 nt ta-siRNA in cis.

4. miRNA Trigger easiRNA Biogenesis to Inhibit RDR2-Dependent RdDM

miRNAs linked by AGO7 and AGO1 particularly target and cleave a variety of transposon transcripts. They cleave to create 21-nucleotide 'epigenetically-activated small interfering RNAs' (easiRNAs) similar to ta-siRNA, dependent on RDR6, DCL4, or DCL1 for biogenesis. This effect has been observed in certain cells, such as dedifferentiated plant cell cultures and the vegetative nuclei of pollen grains, as well as in the backgrounds of mutants like ddm1 and met1. Terminal inverted repeats of many DNA transposons and transposable element promoters near long terminal repeats of retrotransposons are silenced by methylation (Slotkin *et al.*, 2005). As a result, the TEs are effectively transcribed in mutant ddm1 where methylation is low and are targeted by miRNAs. The pericentromeric and euchromatic areas in *Arabidopsis* contain the transposable elements (TEs) ATHILA6, ATGYPSY, ATCOPIA93, ATMU5, CACTA and VANDAL, among others. A large number of TEs (~ 3,662) have been predicted to be possible miRNA targets across the whole genome, and PARE-sequencing data shows that 2371 of those 3,662 TEs are found to have been cleaved by miRNAs (Creasey *et al.*, 2014). Numerous well-known miRNAs, including miR169, miR156, miR159, miR823, miR172, miR399, miR390, miR859 and miR169 target these TEs. However, not every TE transcript, that is cleaved, produces an easiRNA. In pollen, two novel epigenetically active miRNAs, ea-miR1 and ea-miR2, were discovered which originated from transposable elements and also in mutant ddm1, similar to easiRNAs. The build up of 21 nucleotide easiRNAs decreases in mutants of ddm1 dcl1 and ddm1 rdr6, whereas that of 24 nucleotide hc-siRNAs increases. 24 nt hc-siRNAs steer asymmetric RDR2-dependent methylation at CHH sites via RdDM, whereas DNA methyltransferases and histone modifications preserve CHG and CG symmetric methylation contexts (Law *et al.*, 2010). When TE transcription is activated by the reduction of DNA methylation or remodeling of the gamete germ line, miRNAs target and cleave the TE transcripts. The activities of DCL4 and RDR6 cause the transposon cleavage fragments to create 21 nucleotide easiRNAs. easiRNAs thus generated are subsequently loaded onto AGO1, which inhibits RDR2 and DCL3 from assembling, halting RdDM caused by 24 nt siRNAs. RDR6 and RDR2's activities are hence antagonistic for a particular transposon. As a result, miRNA lead easiRNAs block RDR2-initiated RdDM in germ line cells and help prevent RDR2-mediated RdDM silencing over the long run.

5. miRNAs Directly Regulate Methylation Players

Several miRNAs have direct influence over the expression of methylation-related target genes. CMT3 is targeted by miR823 and DRM2 is targeted by miR-773a in *Arabidopsis* (Jha *et al.*, 2014). Involved in *de novo*1 (IDNl1) and *de novo* 2 (IDN2) are necessary for *de novo* methylation of *A. thaliana* DNA (Ausin *et al.*, 2012). IDNl1 is targeted by miR781a and miR837, while IDN2 by miR413 and miR169 in *Arabidopsis thaliana*. These miRNAs are methylated by their own targets, indicating a negative feedback loop regulation. However, DRM2 mediates DNA methylation through these target genes. miRNAs, like miRNA153, target DNMT1 methyltransferases and inhibit methylation of DNMT1-targets in mammals (Das *et al.*, 2010). This implies that this

340   *Plant MicroRNAs and Stress Response*

trait of miRNAs is universal throughout kingdoms and miRNAs influence important components of the DNA methylation process to regulate overall DNA methylation.

6. Role of siRNAs in Plant Epigenetics

Short interfering RNAs (siRNAs) have emerged as the most valuable components of genetic and epigenetic regulation of eukaryotic genomes (Lagana *et al.*, 2015). Small interfering RNA (siRNA), also known as short interfering RNA or silencing RNA, typically 20-24 (normally 21) base pairs in length is a type of double-stranded RNA, similar to miRNA that function within the RNA interference (RNAi) pathway (Monga *et al.*, 2017). They alter the expression of specific genes having complementary nucleotide sequences by degrading mRNA after transcription, thus preventing translation. siRNAs can guide and mediate epigenetic changes in plants. This siRNA directed regulatory complex is involved in DNA methylation, histone and chromatin changes, all of which eventually influence transcription (Stacey *et al.*, 2011). The discovery and functional studies of elements of the siRNA-mediated DNA methylation pathway have revealed novel clues into heterochromatin-forming epigenetic pathways as well as paramutation, chromatin-based silencing of genes, epigenetic reprogramming and genetic imprinting. The dynamic interplay of siRNAs, methylation of DNA and histone changes, together control transcriptional gene silencing, mediating epigenetic regulation in plants. siRNAs play this role of regulating transcriptional gene silencing by inducing sequence specific DNA methylation and histone H3K9 dimethylation in a pathway known as RNA-directed DNA methylation (RdDM) (Onodera *et al.*, 2005). Despite a lot of obscurities, technological breakthroughs have shown the various steps in siRNA-directed silencing mechanisms. Ostensible mechanisms have been investigated for the link between histone methylation and DNA methylation and understanding the physiological impacts of siRNA-directed epigenetic changes has taken a lot of work. Below we discuss the process of siRNA-directed epigenetic modification, which comprises the generation of siRNA and the recruitment of histone methyltransferases and DNA methyl transferases to targets, which is aided by complementarity between 24-nucleotide siRNAs and nascent scaffold RNAs. We also discuss the impact of siRNA-directed epigenetic alteration in preserving the stability of genome and regulation of gene expression as well as the recently discovered participants in siRNA-directed silencing pathways.

6.1 siRNAs and DNA Methylation

In plants, methylation can take effect at any cytosine and in any one of the three sequence positions. 'Uniform methylation' corresponds to CHG and CG sites, while 'unsymmetrical methylation' corresponds to CHH sites in every separate case; the H represents T, A, or C (Cokus *et al.*, 2008). The mechanism of siRNA mediated DNA methylation is orchestered in three phases: siRNA synthesis, siRNA scaffold formation and silencing complex recruitment on target sequences.

6.2 De Novo DNA Methylation of Cytosines Begins with the Biosynthesis of siRNAs

RdDM begins with the production of regulatory siRNAs via the RNA interference pathway which involves a possible chromatin remodeling protein, CLASSY1 and

MicroRNAs and other Non-coding RNAs in Plant Epigenetics   **341**

employs the RNA-dependent RNA polymerase RDR2 and DNA-dependent RNA polymerases Pol IV and Pol V, specific to plants (Onodera *et al.*, 2005). Upstream of Pol V, RNA polymerase IV interacts with CLASSY1 and RDR2 to form a complex, which converts ssRNA into dsRNA. To assist in guiding chromatin modifications to homologous DNA sequences, the effector complex, which includes either one of the two AGO4 or AGO6, recruits the dsRNA molecules after DCL3 cleaves them into 24 nucleotide heterochromatic siRNAs (Xie *et al.*, 2004). It is thought that Pol V in conjunction with the aforementioned auxiliary components transcribes only those genomic sequences which are specifically chosen for interaction with siRNAs. The AGO4-attached siRNA complex facilitates in recruitment of histone altering complexes and *de novo* DNA methylation factors, to the target loci, either by engaging with the target DNA or emerging polymerase V procured RNA (Wierzbicki *et al.*, 2008). The above-mentioned Pol IV-mediated siRNA production displays primary RdDM and produces the most common type of siRNAs. These siRNAs can be augmented by a turnover mechanism in which methylated DNA template is transcribed by Pol IV, leading to the production of an abnormal or perhaps improperly processed RNA and can be duplicated by RDR2, thereby producing additional 1° siRNAs which set off methylation at the target region (Herr *et al.*, 2005). Next essential feature of RdDM uses 2° siRNAs to cause methylation to extend beyond and adjacent to the primary siRNA-targeted locations (Daxinger *et al.*, 2009). Possibly a few Pol IV primary siRNAs could function in trans at distant, connected regions, guiding secondary RdDM in a Pol V-dependent way.

In secondary RdDM, polymerase IV is responsible for transcribing a methylated target template and the subsequent sequence, resulting in an abnormal RNA being copied by RDR2 and eventually degraded by DCL3, to yield secondary siRNAs that cause downstream target site to get methylated. The production and augmentation of 18 siRNAs in 1° RdDM select and fortify methylation at the first site that generates siRNA. However, secondary siRNAs are generated to promote methylation spreading in the vicinity of the 1° RdDM. This whole pathway of primary RdDM is at no phase dependent on 2° RdDM (Fig. 2) (Daxinger *et al.*, 2009). The discovery of the machinery involved in siRNA production and siRNA-mediated DNA methylation has revealed a diverse set of RNA-directed epigenetic regulatory mechanisms in plants.

DNA methyl transferases (DMTs) are involved in establishing and maintaining of RdDM among which the notable ones are: domains rearranged methyltransferase 1 and 2, through which CHH methylation is established, chromomethylase 3 (CMT3), which maintains CHG methylation and methyltransferase 1 (MET1), which maintains CG methylation (Cao and Jacobsen, 2002a; 2002b). siRNAs are believed to regulate epigenetic alterations also through demethylation. The activity of DNA glycosylase proteins repressor of silencing 1 and 3 promotes DNA demethylation in plants, demeter (DME) and DME-like (DML) (Penterman *et al.*, 2007). Recently, RDM1 was identified to have a regulatory role in DNA methylation as is related to the aggregation of 24 nucleotide siRNAs (Gao *et al.*, 2010), silencing at target loci and DNA methylation. RdDM is a prime fit for representing a component of the AGO4-effector complex of RdDM as it encodes a protein that binds with RNA polymerase II, DRM2 and AGO4 and also binds to single-stranded methylated DNA. It is also believed to generate the DDR complex (DMS3–DRD1–RDM1) as it was co-purified with DRD1 and DMS3 (Law and Jacobsen, 2010). Also it is proposed that the binding activity of RDM1 with single-stranded

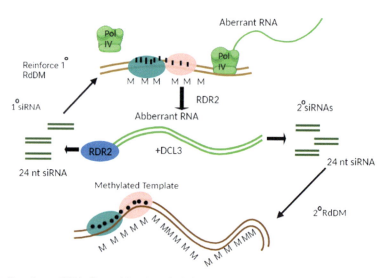

Fig. 2: By primary RNA-directed DNA methylation, RNA polymerase IV-dependent synthesis of short RNAs (1 siRNAs) strengthens existing heterochromatic areas (1 RdDM). Primary RdDM can result in the creation of 2 siRNAs, which cause methylation to spread to nearby areas, culminating in 2 RdDM

methyl-DNA facilitates the targeting of AGO4. Furthermore, it was established that RDM1 is needed for the transcription of polymerase V and that polymerase V and RDM1 co-localize in the peri-nucleolar processing centre (Gao et al., 2010). RDM1 may also facilitate polymerase V recruitment to RdDM target locations. RDM1 may therefore be crucial in tying transcription siRNAs, and DNA methylation together.

6.3 Production of Scaffold RNAs

DNA dimethylation targets the recruitment of chromatin methylation enzymes and the AGO-siRNA complex. Long, non-coding RNAs (lncRNAs), which function as structural RNAs, are important components in this recruitment, which is independent of siRNA biogenesis, as demonstrated in *Arabidopsis* mutants missing RDR2, DCL3 or NRPD1 (Wierzbicki et al., 2008). RDM1, a single-stranded methylated DNA binding protein, the chromosomal hinge domain protein DMS3 and DRD1, a chromatin-remodeling protein, make up the DDR complex required for the generation of Pol V-dependent scaffold RNAs in addition to a GHKL-type ATPase, DMS11 (Kanno et al., 2008; Kanno et al., 2005; Law and Jacobsen, 2010).

6.4 Recruitment of siRNA Silencing Machinery on Target Sites

The mechanism by which silencing machinery is recruited to required sites is still not fully clear, though several models have been presented to understand the same (Wierzbicki et al., 2008). 24-nt siRNAs should have complementary pairing with nascent scaffold RNAs for recruitment. By one of the models, scaffold RNAs engage with 24-nt siRNAs generated by DCL3 via base pairing are integrated into AGO4. IDN2, which binds dsRNA with 5' overhangs, has been recently discovered to be a member of the RdDM pathway and is essential for stabilising connections between

MicroRNAs and other Non-coding RNAs in Plant Epigenetics **343**

siRNA bound with Argonaute and a scaffold RNA generated at a set locus by the Pol V complex (Zhang *et al.*, 2012). Later, the siRNA-scaffold RNA interface guides the silencing complex, including DRM2, the cytosine methyltransferase to the target sites, resulting in their silencing.

6.5 siRNAs Induce Histone Modifications

One of the most complicated histone modifications, both in terms of its nature and biological effects, is methylation (Lachner and Jenuwein, 2002). Histones can be methylated at numerous residues (lysine and arginine), at different positions and varying levels. It is important to note that the number of methyl groups present on a single residue, in addition to the kinds and locations of the residues, all affect the molecular effects of genome expression. Lysine methylation of histone H3 in *Arabidopsis* occurs mostly at Lys9 (K9), Lys4 (K4), Lys27 (K27), and Lys36 (K36) (Cartagena *et al.*, 2008; Saleh *et al.*, 2008). These modifications are controlled by different mechanisms. For instance, H3K9 and H3K27 are related with gene repression; however, H3K27 is accumulated at genes by polycomb group (PcG) proteins, whereas H3K9 is deposited by heterochromatic siRNA (Turck *et al.*, 2007).

H3K9me2 and CHG methylation were found to be highly co-localized in a genome-wide investigation (Bernatavichute *et al.*, 2008), implying that these epigenetic marks are linked. The exclusion of H3K9me2 from genomic areas by an increase in histone demethylase bonsai methylation1 (IBM1) and the recruitment of CMT3-KYP to specific regions have been proposed as potential reasons for the connection (Inagaki *et al.*, 2010). H3K9 may be linked to DNA methylation because methylated cytosine and H3K9 residues are recognized by protein modules respectively (Johnson *et al.*, 2007). Among the three protein domains which recognize methylated cytosines is SRA domain of KYP. It specifically binds double-stranded DNA with cytosines that are methylated in a non-CG environment suggesting that the histone modifier could identify DNA methylation marks directly. Also, CMT3, with a chromodomain, recognizes and binds K9 dimethylated tails of H3 (Lindroth *et al.*, 2004; Du *et al.*, 2012). A positive feedback loop links the control of histone H3K9 dimethylation and DNA methylation by the interdependent engagement of KYP and CMT3 (Lindroth *et al.*, 2004; Johnson *et al.*, 2007). Another possible mechanism is that IBM1 eliminates H3K9me2, which encodes a histone demethylase in *Arabidopsis* that removes H3K9 monomethylation and dimethylation. Down-regulating of IBM1 leads to external H3K9 methylation at the bonsai site, resulting in CMT3- and KYP-dependent non-CG DNA hypermethylation and gene knockdown (Saze *et al.*, 2008). The genome-wide sequencing IBM1 hypermethylates histone H3K9 and non-CG cytosines in a significant number of genes (Miura *et al.*, 2009; Inagaki *et al.*, 2010), implying that the protein-coding genes are protected by IBM1 against repression via non-CG and H3K9 and DNA methylation. According to a recent study, IBN1 can also regulate genomic expression via indirect silencing through siRNA-mediated gene repression (Fan *et al.*, 2012).

6.6 Architecture of siRNAs in the Methylation and Chromatin Modifications

Heterochromatin is made up of retrotransposons, transposons and other repetitive DNA elements that are kept transcriptionally inactive state by methylation and post-translational modifications of histones (Bender, 2004). In plants, a large portion of

344   *Plant MicroRNAs and Stress Response*

the siRNA population is produced from repeats and transposons which play crucial roles in turning down transposons and other repeat elements, thus representing the epigenetic 'landscape' of plant genes. Loss of any of the RdDM pathway components, which include RDR2, DCL3, DRM2, AGO4, Pol IV and Pol V, have been shown to influence chromatin (Onodera *et al.*, 2005; Pontier *et al.*, 2005; Pontes *et al.*, 2006). Reportedly methylation directed by siRNAs and the organization of heterochromatin in chromocentres is linked (Pontes *et al.*, 2009). In a separate study, the methylation pattern of the epigenome in wheat was studied and it was discovered that a substantial number of sRNAs matched with transposable elements (TEs) (Cantu *et al.*, 2010). Because TEs account for more than 80% of the wheat genome, epigenetic silencing mediated by sRNAs plays a critical role in decreasing TE mutagenesis activity. With the availability of large, whole-genome database for histone modifications and DNA methylation, heterochromatin can now be identified by its marks and not by the presence of specific repeats.

6.7 siRNAs and Paramutation

The epigenetic process of paramutation was first explained in maize. Two alleles of a gene with the same sequence but distinct functions can interact with each other, and pass their silent status previously active allele (Chandler, 2007; Pikaard and Tucker, 2009). Through forward genetic screening, several genes required for paramutation and influencing plant coloring, have been identified (Alleman *et al.*, 2006; Sidorenko and Chandler, 2008). Paramutation defective mutants identified in later stages include those required to maintain repression 1 (RMR1) and mediator of paramutation 1 (MOP1). The relation between RdDM and paramutation was first illustrated by mutants rmr6 and mop1 and the maize orthologs of *Arabidopsis* Pol IV subunit and RDR2 (Alleman *et al.*, 2006; Sidorenko and Chandler, 2008; Nobuta, 2008). Both these mutants reported lost siRNA-directed DNA methylation, decline in levels of 24 nucleotide siRNA, and transposon activation along with several developmental phenotypes. The mutant phenotype is attributed to the rejuvenation of suppressed transposons. Allelic maize mutants, mop2 and rmr7, have developmental defects, although not to the similar severity as mop1 and rmr6. This might be because maize NRPD2/NRPE2-like genes are partially redundant (Sidorenko, 2009).

6.8 Epigenetic Modifications Modulated by siRNAs

The consequences of epigenetic modifications mediated by siRNAs are well known. These are necessary for the maintaining stability of the genome and gene expression regulation. The main role of siRNA-mediated epigenetic alteration is to keep genome integrity maintained by inactivating transposable elements. To create a complex that facilitates the recruitment of DNA methyltransferases and histone H3K9 methyltransferases to the site of action, the 24-nucleotide siRNAs produced by DCL3 are loaded first into AGO4 (Qi *et al.*, 2006). It has been found that AGO4 loss-of-function mutations repress histone HI3K9 methylation and DNA methylation, leading to heterochromatic loci reactivation (Zilberman *et al.*, 2003; Xie *et al.*, 2004). In addition to repressing transposons and repetitive DNA sites and maintaining genomic integrity, additionally it plays a significant role in the expression of a significant count of endogenous genes. In *Arabidopsis* flowering time variation is due to epialleles of flowering locus C (FLC) created by siRNA-mediated gene inactivation. siRNAs cause transposon methylation, which extends to adjoining FLC gene, resulting in

MicroRNAs and other Non-coding RNAs in Plant Epigenetics   **345**

early blooming and poor expression of FLC. A sine transposon inclusion in the FWA gene promoter is the second example of gene regulation. These transposons produce siRNAs, which cause DNA methylation of the promoter, resulting in FWA suppression (Soppe *et al.*, 2000; Chan *et al.*, 2006). The role of siRNA-mediated methylation has also been reported in the alteration of SDC gene which produces an F-box protein, the promoter of which comprises seven direct repeats. siRNAs generated by SDC repeats initiate transcriptional inactivation of genes and DNA methylation (Henderson and Jacobsen, 2008). Studies have found that inhibition of SDC suppression causes changes in phenotype of leaves from flat to downward-curled in *A. thaliana*.

7. Long Non-coding RNAs Regulate Epigenetic Programming in Plants

Long non-coding (lncRNAs) RNA transcripts, comprising of 200 or more nucleotides have a low level of evolutionary conservation and are made up of genes that are typically shorter than protein-coding genes and have lesser exons. They do, however, share several characteristics with transcripts that code proteins, such as being processed by RNA polymerase II and being polyadenylated, capped and spliced. The capacity of lncRNAs to crease together into secondary or higher thermodynamically stable structures are highly preserved as one of their primary properties. The longer the lncRNA is, the more likely it is to construct such structures. Because lncRNAs may connect to one other via bonds, they can fold into structures like double helix, hairpins, loops, pseudonodes and more. They may attach to multiple molecules simultaneously because of their intricate architectures, controlling expression of genes at multiple levels via RNA-RNA, RNA-protein and RNA-DNA complexes.

7.1 Long Non-coding RNAs are the Progenitors of RdDM Silencing Pathway

LncRNAs have the potential to regulate gene expression at genetic and epigenetic levels. They can operate near the site of RNA synthesis, thus acting as cis-elements, directly on adjacent genes on the same strand or as trans-acting elements, further from the site of synthesis. LncRNAs may prevent transcription factors from binding to the promoter areas (Csorba *et al.*, 2014). They may even be used as miRNA target mimics (Shuai *et al.*, 2014) (Fig. 1E), miRNA and trans-acting small interfering RNA (tasiRNA) precursors (Zhang *et al.*, 2014) and siRNA processing (Wunderlich *et al.*, 2014). Surprisingly, the bulk of lncRNAs are transcribed by RNA polymerase II (Pol II), though some RNA polymerases specific to plants, such as Pol IV and Pol V, can also have a role in lncRNA production, mostly through RdDM-mediated epigenetic control (Wierzbick *et al.*, 2008). Moreover, lncRNAs have been shown to affect epigenetic processes, such as methylation of DNA (Ariel *et al.*, 2014) and histone modifications (Heo and Sung, 2011). In the RdDM silencing pathway, lncRNAs of plants play a crucial role. The functionality of Pol IV generated Pol IV-dependent RNAs (P4RNAs) is the basis for this regulatory pathway (Blevins *et al.*, 2015). RNA-dependent RNA polymerase 2 (RDR2) processes these precursor RNAs to double-stranded RNAs (dsRNAs), that are cleaved largely by Dicer-like 3 (DCL3) to yield 24-nt siRNAs (Xie *et al.*, 2004). These siRNAs generate an AGO-siRNA complex when they interact with Argonaute 4 (AGO4) (Holoch and Moazed, 2015). Concurrently, Pol V-transcribed lncRNAs serve as scaffold RNAs that siRNA-AGO complex recognizes due to complementarity in sequence (Böhmdorfer *et al.*, 2016). Once the AGO4–siRNA–lncRNA complex has been generated, it is transported to the chromatin-specific site along with a DNA methylation enzyme, DNA methyltransferase domains rearranged methyltransferase 2 (DRM2) (Gao *et al.*, 2010). This enzyme initiates gene silencing by *de novo* methylation of cytosines in every kind of sequence constraint at the targeted region (Wierzbicki *et al.*, 2008). As a result, RdDM refers to a *de novo* DNA methylation process, specific to plants that use lncRNAs to designate targeted loci of genome (Wierzbicki *et al.*, 2009). RdDM has made a significant contribution to realizing the significance of lncRNAs as precursors in epigenetic silencing (Chen *et al.*, 2018). Numerous studies have found that lncRNAs of plants regulate root organogenesis (Chen *et al.*, 2019), embryogenesis, reproduction (Ding *et al.*, 2012) and gene silencing. Biologists have also looked into the possibility of stress-induced lncRNAs triggering DNA methylation in response to environmental factors. In *Arabidopsis*, the auxin regulated promoter loop (APOLO) has been discovered as an auxin-induced lncRNA (Ariel *et al.*, 2014). The double transcription by Pol II and V of APOLO has been linked to the creation of a chromatin loop that engages the promoter of its nearby gene, pinoid (PID), which is a critical polar auxin transport regulator, resulting in translational down-regulation, which may also distinguish distal non-associated site by forming R-loops, alternatively. Like heterochromatic protein 1 (LHP1) decoy mediated by APOLO may activate transcription of target loci while also co-regulating auxin-responsive genes and modifying chromatin structure (Ariel *et al.*, 2020). The RdDM system, which is the principal epigenetic mechanism regulated by siRNA in plants, represents an amazing increase in eukaryotic organisms' transcriptional ability (Matzke and Mosher, 2014). The classic pathway of RdDM includes Pol IV being engaged to target site to transcribe single-stranded RNAs (ssRNAs). The RDR2 makes dsRNAs by copying ssRNAs. DCL3 converts dsRNAs to 24 nt siRNA. Lastly, *de novo* methylation takes place, which necessitates the use of AGO4-bound 24-nt siRNAs, DRM2 and Pol V-dependent scaffold RNAs (Mosher *et al.*, 2008). Since it has been discovered that transposons and tasiRNAs are originally transcribed by Pol II, replicated by RDR6 and cleaved by DCL2 and DCL4 into 21–22-nt siRNAs, non-canonical RdDM pathways offer a connection between *de novo* methylation of transposon DNA and PTGS of transposon transcripts (Matzke *et al.*, 2015). Furthermore, it has been demonstrated in *Arabidopsis* that methylation of DNA at several of the RdDM target loci did not match with 24-nt siRNAs and did not depend on dcl1/2/3/4 mutants. Instead, the predominant sRNAs class produced from maximum RdDM loci in dcl plants was found to be 25-50 nt RNAs. RdDM's biological effects have been reported to control silencing of transposons (La *et al.*, 2011), biotic interactions, plant development and gene expression (Satgé *et al.*, 2016). The roles of lncRNAs as crucial modulators in plant responses to stressors have been suggested by functional analyses. An increasing amount of research suggests that plant lncRNAs may play an important role in abiotic stress responses via RdDM. Despite restricted research, it is expected that RdDM-associated lncRNAs have a lot of potential to be explored.

7.2 Role of lncRNA in FLC Repression Mediated by Polycomb in Vernalisation

Plants just answer natural boosts to start formative changes, i.e. blooming during a particular season. Vernalization, or the lengthy cold of winter is one illustration of such natural signs (Sung and Amasino, 2004). FLC is epigenetically quietened by vernalization, and the concealment of FLC is steady even after the colder time of the year. The enactment and hindrance of chromatin-renovating edifices are both embroiled in the control of FLC expression, as per molecular studies (Kim *et al.*, 2009). Blossoming is postponed when FLC articulation is high, while blooming is empowered when FLC is smothered by vernalization (Fig. 3). With hereditary strategies it has been found that numerous players are involved in for the steady concealment of FLC by vernalization (Kim and Sung, 2012). Vernalization insensitive 3 (VIN3) has been identified as a critical gene for vernalization-induced FLC suppression (Sung and Amasino, 2004). VIN3 is a plant homeodomain (PHD) finger protein that is only activated when it becomes chilly. VIN3's PHD finger motif is frequently seen in chromatin-remodeling complex components (Kim and Sung, 2013). VIN3 and PRC2 have been isolated together biochemically (De Lucia *et al.*, 2008). This finding implies that PRC2's function is dependent on the PHD-PRC2 relationship. PHD-PRC2 connection with FLC chromatin increases with vernalization, as does H3K27me3 deposition at FLC chromatin (Kim and Sung, 2013). The steady suppression of FLC by vernalization is characterized by enrichments of PRC2 and H3K27me3. In higher eukaryotes, PRC2 is ubiquitous repressive chromatin remodeling complex. Suppressor' of Zeste-12 [Su (Z)12], enhancer of zeste [E(Z)], ESC and P55 are the four main subunits of PRC2 in mammals. VRN2 in *Arabidopsis* communicates a homolog of Su (Z)12, a critical part of PRC2 (Wood *et al.*, 2006; De Lucia *et al.*, 2008). There are about three E(Z) homologs (CLF, SWN, and MEDEA) and five p55 homologs (MSI1-5). CLF, SWN, MSI1 and FIE have been recognized as parts of the vernalization PRC2 complex. In *Arabidopsis*, the sole PRC2 part addressed by a solitary part is FIE, a homolog of ESC (De Lucia *et al.*, 2008). Vernalization-induced FLC suppression is mediated by two groups of lncRNAs. COOLAIR is a family of (1100bp long) lncRNAs that are transcribed from the 3' end of FLC in the opposite direction as FLC mRNA. Based on two distinct polyadenylation sites, six different COOLAIR transcripts may be divided into two groups (Swiezewski *et al.*, 2009). COOLAIR expression levels rise fleetingly

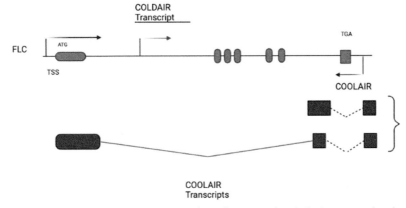

Fig. 3: FLC's noncoding RNAs. Non-protein coding transcripts in both sense and antisense are found in FLC. In contrast to FLC mRNA, COLDAIR is transcribed in the sense direction from the first intron of FLC. There are two types of antisense ncRNAs that are transcribed. The polyadenylations of COOLAIR are used to classify them (proximal pA and distal pA). Splicing variations have also been discovered in both COOLAIR classes

after vernalization and distally polyadenylate COOLAIR variations to some degree cross-over with the 5' end of FLC mRNA (Fig. 4). COOLAIR has been conjectured to possess a capability in FLC transcriptional impedance. By vernalization, in any case, insertional mutants (in which COOLAIR transcription is repressed) show lower FLC articulation (Helliwell et al., 2011). COOLAIR, then again has been proposed to play a part in 'co-transcriptional' control during the beginning phases of vernalization (De Lucia and Dean, 2010). The general transcription of antisense lncRNAs would weaken FLC transcription in this perspective. The molecular complexities of COOLAIR's job in FLC concealment interceded by vernalization are yet obscure. The second lncRNA is COLDAIR, a cold-inducible intronic lncRNA that is transcribed the other way to FLC mRNA from the principal intron of FLC (Heo and Sung, 2011). RNA polymerase II deciphers COLDAIR, which has a 5'-capped structure, yet no distinguishable polyadenylation tail. COLDAIR is created momentarily by vernalization and is directed by an obscure cold-inducible promoter inside the vernalization reaction component (Heo and Sung, 2011). COLDAIR communicates to CLF protein, a homolog of the PRC2 chromatin-changing complex's E(z) part, as demonstrated by *in vitro* transcription and pull-down assays. COLDAIR collaborates with PRC2 *in vivo*, as per RNA immunoprecipitation studies. COLDAIR is a necessary part of the vernalization reactions in *Arabidopsis*, as per a hereditary exploration utilizing the RNA-interference procedure. In contrast with parental lines, COLDAIR knockdown lines do not show a drop in FLC because of vernalization, and FLC was found to be depressed in the knockdown lines when the plants were re-established to a warm climate. COLDAIR knockdown lines show a critical decrease in CLF and H3K27me3 improvement at the FLC locus chromatin in light of vernalization, demonstrating that COLDAIR is expected for enrolling PRC2 to FLC chromatin. Therefore, COLDAIR is like hot air and Xist ncRNAs in that initiates PRC2 to chromatin targets. PRC2-intervened epigenetic gene hushing by lncRNA has all the earmarks of being a

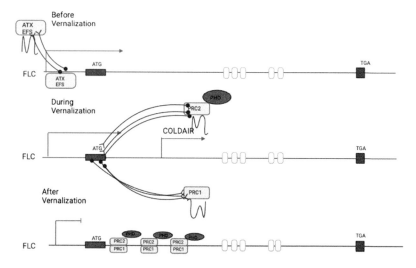

Fig. 4: Polycomb group protein complexes suppress FLC through vernalisation. FLC chromatin transitions from trithorax-dominant to polycomb-dominant chromatin in a dynamic manner. During cold exposure, COLDAIR is essential for PRC2 recruitment

developmentally monitored process in both plants and well evolved creatures, as per the discoveries.

7.3 lncRNAs Assume a Significant Role in Enlistment of Polycomb Restraint Complex 1 (PRC1)

PRC2-interceded epigenetic gene silencing by lncRNAs has been the key review centre. Mounting proof suggest that lncRNAs are engaged in the enrolment of the PRC1 complex. PRC1 distinguishes the H3K27me3 mark and subdues transcription in a consistent way (Zhang *et al.*, 2007). In plants, the PRC1 subunits have wandered. Like heterochromatin protein 1(LHP1), it ties to H3K27me3 and acts much the same way as polycomb (Mylne *et al.*, 2006). Molecular hereditary examinations uncover that non-sequence-explicit DNA-restricting proteins EMF1 and VRN1 have PRC1-like capabilities (Bastow *et al.*, 2004). In *Arabidopsis*, useful homologs of PRC1 ring finger proteins have been found. They are connected to LHP1 and work in planta to restrain PcG target genes (Xu and Shen, 2008). AtBMI1A and AtBMI1B are ring finger proteins that connect with LHP1 and EMF1 and smother PcG target genes, similar to mammalian BMI1 (a part of PRC1) (Bratzel *et al.*, 2010). It will be fascinating to check whether the *Arabidopsis* PRC1-like complex contains lncRNA parts as a feature of its gene constraint activity.

7.4 LncRNAs are Precursors to Abiotic Stress Responses

lncRNA regulatory mechanisms in response to abiotic stress is depicted (Fig. 5) revealed that lncRNAs that are embroiled in *Arabidopsis*' reaction to decreased nourishment accessibility, bringing about the disclosure of 60 differentially communicated lncRNAs. TAS3 was found to be stifled in low-nitrogen conditions and to have a high liking for the nitrate carrier. In populus (Chen *et al.*, 2016) and

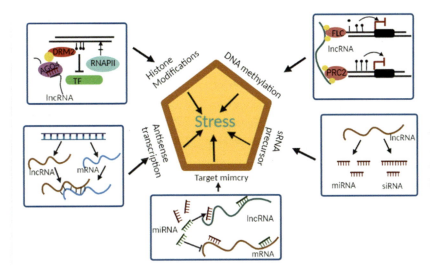

Fig. 5: Plant lncRNA regulatory mechanisms in response to abiotic stressors. The function of long noncoding RNAs in plant stress response. miRNA precursor, histone modification, target mimicry, and other modes of action are initiated by lncRNAs during abiotic stress

Arabidopsis, a broad technique was utilized to find lncRNAs differentially prompted in the light of dietary pressure (Franco-Zorrilla *et al.*, 2007). Plant lncRNA expression can also be affected by extreme temperatures. HSFB2a is a heat shock gene in *Arabidopsis* that is essential for gametophytic development and is regulated by an antisense heat-inducible lncRNA, known asHSFB2a (Wunderlich *et al.*, 2014). Surprisingly, asHSFB2a overexpression suppresses HSFB2a RNA accumulation and HSFB2a overexpression suppresses asHSFB2a expression similarly. Meanwhile, under heat and drought stress, 1,614 lncRNAs have been identified to be differently expressed in *Brassica juncea*. Grape and *Arabidopsis* have both been shown to exhibit cold-responsive lncRNAs (Calixto *et al.*, 2019). Dry spell and salt are the two most common ecological variables that diminish plant yield and both can be brought about by covering hereditary administrative pathways. Dry spell prompted lncRNA (DRIR), for instance, has been distinguished in *Arabidopsis* as a positive controller of dry season and salt pressure reactions (Qin *et al.*, 2017). The past examination has assisted in finding three up-controlled lncRNAs and 2,815 new salt-responsive lncRNAs in *Spirodela polyrhiza* (Ben Amor *et al.*, 2009). Dry season responsive lncRNAs were found to be concentrated in poplars that had been exposed to water deficiency (Shuai *et al.*, 2014). Dry season prompted long intergenic non-coding RNA expression lincRNA2752, for instance, is an objective copy of the NFYA transcription factor controller, ptc-miR169. Drought-responsive lncRNAs discovered in *Cleistogenes songoric* and *B. napus* produced similar results. Table 3 summarises abiotic stress-responsive lncRNAs.

Table 3: A summary of plant studies including abiotic stress-responsive lncRNAs

lncRNA	Stress	Plant Species	Mechanism of Regulation	Expression
COLDAIR	Cold	*Arabidopsis thaliana*	Modification of histones	Inducible
AtR8	Hypoxia	*Arabidopsis thaliana*	Trans acting	Repressive
COOLAIR	Cold	*Arabidopsis thaliana*	Histone modification	Inducible
MSTRG.6838.1	Drought	*Zea mays*	*Cis*-acting	Repressed
TAS3	Low nitrogen	*A. thaliana*	Trans-acting	Repressed
TalnRNA5	Heat	*T. aestivum*	Precursor or miRNA	Inducible
APOLO	Auxin	*A. thaliana*	Demethylation of DNA	Inducible
TCONS_00052315	Low nitrogen	*P. tomentosa*	Mimics the target	Repressive
LncRNA26929	Salt	*S. bicolour*	Mimics the target	Repressive
TCONS_00061958	Boron stress	*H. vulgare*	*Cis*-acting	Inducible
Lnc-225	High-light	*Arabidopsis thaliana*	*Cis*-acting	Inducible
IPS1	Phosphate deficiency	*Arabidopsis thaliana*	Mimics the target	Inducible
Os02g0180800-01	Drought	*O. sativa*	Nat. antisense transcript	Repressive

7.5 LncRNA in Crop Plants

Since abiotic stressors constitute a significant barrier to increasing agricultural yields, all results describing the role of lncRNAs in response to environmental challenges are especially significant in the context of crop species (Halford *et al.*, 2015). Several cases are reported for the discovery of lncRNAs during agricultural stress reactions (Pang *et al.*, 2019). In crops like wheat and rice (Jabnoune *et al.*, 2013), LncRNAs have been discovered to have a role in maintaining nutritional balance. Recent studies have shown the effects of excessive boron treatment on barley lncRNAs. Both data indicate that the interplay of miRNA-lncRNA-coding target transcript modules can cooperatively regulate boron-regulation. For specific instance, miR399's putative miRNA sponge TCONS 00043651 was favorably regulated by exposure to boron (Unver and Tombuloglu, 2020). On the other hand, miR399 expression was suppressed in this stressful situation. Heat-responsive lncRNAs were found in wheat (Xin *et al.*, 2011) and maize, yet temperature vacillations oftentimes bring about yield misfortune (Lv *et al.*, 2019). By showing Ta-miR2010 family groupings, the lncRNA TahlnRNA27 is created by heat treatment and distinguished as a potential miRNA forerunner (Xin *et al.*, 2011). Like 2,271 lncRNAs were cold-responsive in alfalfa, 182 novel cold-responsive lncRNAs are known to be differentially prompted in cassava (Zhao *et al.*, 2020). The ecological component that presently restricts farming is salt stress (Song and Wang, 2015). Notwithstanding alfalfa, barley (Karlik and Gozukirmizi, 2018), cotton and sorghum, studies have tried to get familiar with the utilitarian cycles of lncRNAs in the light of salt pressure (Sun *et al.*, 2020). For example, the overexpression of lncRNA973 further developed salt resilience and adjusted the expression of genes connected to cotton salt pressure. Studies have been performed to research the dry-spell responsive lncRNAs in crop species, including tomato, rice, wheat, maize, tomato, foxtail millet and cassava to further develop crop execution in regions restricted by water deficiency (Suksamran *et al.*, 2020). A new report utilizing maize found 124 lncRNAs that answer dry season and are named cis-acting factors (Pang *et al.*, 2019). The idea that both may be a reasonable cis-acting team is upheld by the stifled articulation association between vpp4, which encodes a vacuolar (H^+)-pumping ATPase subunit and its adjoining lncRNA MSTRG.6838.1.

7.6 Other lncRNA Classes in Plant Physiology

In plants, some lncRNAs act in chromatin modifications and also have functions in other biological processes. Leguminous plants and soil microbes engage in symbiotic relationships that are regulated by the soybean early nodulin gene, or Enod40 (Fujita *et al.*, 1993). Only two little peptides expected open-reading frames for the Enod40 transcript are present (12 and 24 amino acid residues). Because of their interactions with sucrose synthase, these peptides raise the possibility that Enod40 regulates the way sucrose is used in nodules (Rohrig *et al.*, 2002). It is interesting to note that during nodulation in *M. truncatula*, Enod40 transcripts themselves interact with MtRBP1 and cause it to move from nuclear speckles to cytoplasmic granules (Rohrig *et al.*, 2002). These findings imply that Enod40 transcripts serve as lncRNA for the re-localization of MtRBP1.

Induced by phosphate starvation 1 (IPS1), a different lncRNA, traps miRNA-399, which controls the yield of PHO2. To the phosphate (Pi) starvation-induced miRNA399, IPS1 has complementary sequences. When IPS1 and PHO2 compete

352 *Plant MicroRNAs and Stress Response*

for miRNA-399's binding, the suppression of PHO2 by miR399 is lessened (Franco Zorrilla *et al.*, 2007). Because of a misaligned loop at the cleavage site, the connection of IPS1 with miRNA399 has no effect on how the IPS1 transcript is cut (Liu *et al.*, 1997). As a result, IPS1 functions as a miRNA target mimic.

References

Alleman, M., Sidorenko, L., McGinnis, K., Seshadri, V., Dorweiler, J.E., White, J., Sikkink, K. and Chandler, V.L. (2006). An RNA-dependent RNA polymerase is required for paramutation in maize, *Nature*, 442: 295-298.

Ariel, F., Jegu, T., Latrasse, D., Romero-Barrios, N., Christ, A., Benhamed, M. and Crespi, M. (2014). Non-coding transcription by alternative RNA polymerases dynamically regulates an auxin-driven chromatin loop, *Molecular Cell*, 55(3): 383-396.

Ariel, F., Lucero, L., Christ, A., Mammarella, M.F., Jegu, T., Veluchamy, A., Mariappan, K., Latrasse, D., Blein, T., Liu, C., Benhamed, M. and Crespi, M. (2020). R-Loop-mediated trans-action of the APOLO Long Non-coding RNA, *Molecular Cell*, 77(5): 1055-1065.e4.

Ausin, I., Greenberg, M.V., Simanshu, D.K., Hale, C.J., Vashisht, A.A., Simon, S.A., Lee, T.F., Feng, S., Española, S.D., Meyers, B.C., Wohlschlegel, J.A., Patel, D.J. and Jacobsen, S.E. (2012). Involved in *de novo* involved in *de novo*2-containing complex involved in RNA-directed DNA methylation in *Arabidopsis*, *Proceedings of the National Academy of Sciences of the United States of America*, 109(22): 8374-8381.

Bao, N., Lye, K.W. and Barton, M.K. (2004). MicroRNA binding sites in *Arabidopsis* class III HD-ZIP mRNAs are required for methylation of the template chromosome, *Developmental Cell*, 7(5): 653-662.

Bastow, R., Mylne, J.S., Lister, C., Lippman, Z., Martienssen, R.A. and Dean, C. (2004). Vernalisation requires epigenetic silencing of FLC by histone methylation, Nature, 427(6970): 164-167.

Ben Amor, B., Wirth, S., Merchan, F., Laporte, P., d'Aubenton-Carafa, Y., Hirsch, J., Maizel, A., Mallory, A., Lucas, A., Deragon, J.M., Vaucheret, H., Thermes, C. and Crespi, M. (2009). Novel long non-protein coding RNAs involved in *Arabidopsis* differentiation and stress responses, *Genome Research*, 19(1): 57-69.

Bender, J. (2004). Chromatin-based silencing mechanisms, *Curr. Opin. Plant Biol.*, 7: 521-526.

Bernatavichute, Y.V., Zhang, X., Cokus, S., Pellegrini, M. and Jacobsen, S.E. (2008). Genome-wide association of histone H3 lysine nine methylation with CHG DNA methylation in *Arabidopsis thaliana*, *PLoS ONE*. 3: e3156.

Blevins, T., Podicheti, R., Mishra, V., Marasco, M., Wang, J., Rusch, D., Tang, H. and Pikaard, C.S. (2015). Identification of Pol IV and RDR2-dependent precursors of 24 nt siRNAs guiding de novo DNA methylation in *Arabidopsis*, *eLife*, 4: e09591.

Böhmdorfer, G., Sethuraman, S., Rowley, M.J., Krzyszton, M., Rothi, M.H., Bouzit, L. and Wierzbicki, A.T. (2016). Long non-coding RNA produced by RNA polymerase V determines boundaries of heterochromatin, *eLife*, 5: e19092.

Bratzel, F., López-Torrejón, G., Koch, M., Del Pozo, J.C. and Calonje, M. (2010). Keeping cell identity in *Arabidopsis* requires PRC1 RING-finger homologs that catalyse H2A monoubiquitination, *Current Biology*, CB, 20(20): 1853-1859.

Calixto, C., Tzioutziou, N.A., James, A.B., Hornyik, C., Guo, W., Zhang, R., Nimmo, H.G. and Brown, J. (2019). Cold-dependent Expression and Alternative Splicing of *Arabidopsis* Long Non-coding RNAs, *Frontiers in Plant Science*, 10: 235.

Cantu, D., Vanzetti, L.S., Sumner, A., Dubcovsky, M., Matvienko, M., Distelfeld, A., Michelmore, R.W. and Dubcovsky, J. (2010). Small RNAs, DNA methylation and transposable elements in wheat, *BMC Genomics*, 11: 408.

MicroRNAs and other Non-coding RNAs in Plant Epigenetics

Cao, X. and Jacobsen, S.E. (2002a). Locus-specific control of asymmetric and CpNpG methylation by the DRM and CMT3 methyltransferase genes, *Proc. Natl. Acad. Sci., USA*, 4: 16491-16498.

Cao, X. and Jacobsen, S.E. (2002b). Role of the *Arabidopsis* DRM methyltransferases in *de novo* DNA methylation and gene silencing, *Curr. Biol.*, 12: 1138-1144.

Cartagena, J.A., Matsunaga, S., Seki, M., Kurihara, D., Yokoyama, M., Shinozaki, K., Fujimoto, S., Azumi, Y., Uchiyama, S. and Fukui, K. (2008). The *Arabidopsis* SDG4 contributes to the regulation of pollen tube growth by methylation of histone H3 lysines 4 and 36 in mature pollen, *Dev. Biol.*, 315: 355-368.

Chan, S.W., Zhang, X., Bernatavichute, Y.V. and Jacobsen, S.E. (2006). Two-step recruitment of RNA-directed DNA methylation to tandem repeats, *PLoS Biol.*, 4: e363.

Chandler, V.L. (2007). Paramutation: From maize to mice, *Cell*, 128(4): 641-645.

Chellappan, P., Xia, J., Zhou, X., Gao, S., Zhang, X., Coutino, G., Vazquez, F., Zhang, W. and Jin, H. (2010). siRNAs from miRNA sites mediate DNA methylation of target genes, *Nucleic Acids Research*, 38(20): 6883-6894.

Chen, M., Wang, C., Bao, H., Chen, H. and Wang, Y. (2016). Genome-wide identification and characterisation of novel lncRNAs in Populus under nitrogen deficiency, *Molecular Genetics and Genomics*, MGG, 291(4): 1663-1680.

Chen, R., Li, M., Zhang, H., Duan, L., Sun, X., Jiang, Q., Zhang, H. and Hu, Z. (2019). Continuous salt stress-induced long non-coding RNAs and DNA methylation patterns in soybean roots, *BMC Genomics*, 20(1): 730.

Chen, Y., Li, X., Su, L., Chen, X., Zhang, S., Xu, X., Zhang, Z., Chen, Y., Xu Han, X., Lin, Y. and Lai, Z. (2018). Genome-wide identification and characterisation of long non-coding RNAs involved in the early somatic embryogenesis in *Dimocarpus longan* Lour, *BMC Genomics*, 19(1): 805.

Creasey, K.M., Zhai, J., Borges, F., Van Ex, F., Regulski, M., Meyers, B.C. and Martienssen, R.A. (2014). miRNAs trigger widespread epigenetically activated siRNAs from transposons in *Arabidopsis, Nature*, 508(7496): 411-415.

Csorba, T., Questa, J.I., Sun, Q. and Dean, C. (2014). Antisense COOLAIR mediates the coordinated switching of chromatin states at FLC during vernalisation, *Proceedings of the National Academy of Sciences of the United States of America*, 111(45): 16160-16165.

Das, S., Foley, N., Bryan, K., Watters, K.M., Bray, I., Murphy, D.M., Buckley, P.G. and Stallings, R.L. (2010). MicroRNA mediates DNA demethylation events triggered by retinoic acid during neuroblastoma cell differentiation, Cancer Research, 70(20): 7874-7881.

Daxinger, L., Kanno, T., Bucher, E., van der Winden, J., Naumann, U., Matzke, A.J. and Matzke, M. (2009). A stepwise pathway for biogenesis of 24- nt secondary siRNAs and spreading of DNA methylation, *EMBO J.*, 28: 48-57.

De-la-Peña, C., Nic-Can, G.I., Galaz-Ávalos, R.M., Avilez-Montalvo, R. and Loyola-Vargas, V.M. (2015). The role of chromatin modifications in somatic embryogenesis in plants, *Frontiers in Plant Science*, 6: 635.

De Lucia, F., Crevillen, P., Jones, A.M., Greb, T. and Dean, C. (2008). A PHD-polycomb repressive complex 2 triggers the epigenetic silencing of FLC during vernalidation, *Proceedings of the National Academy of Sciences of the United States of America*, 105(44): 16831-16836.

De Lucia, F. and Dean, C. (2010). Long non-coding RNAs and chromatin regulation, *Current Opinion in Plant Biology*, 14(2): 168-173.

Ding, J., Shen, J., Mao, H., Xie, W., Li, X. and Zhang, Q. (2012). RNA-directed DNA methylation is involved in regulating photoperiod-sensitive male sterility in rice, *Molecular Plant*, 5(6): 1210-1216.

Du, J., Zhong, X., Bernatavichute, Y.V., Stroud, H., Feng, S., Caro, E., Vashisht, A.A., Terragni, J., Chin, H.G., Tu, A. (2012). Dual binding of chromomethylase domains to H3K9me2-containing nucleosomes directs DNA methylation in plants, *Cell*, 151: 167-180.

Dukowic-Schulze, S., Sundararajan, A., Ramaraj, T., Kianian, S., Pawlowski, W.P., Mudge, J. and Chen, C. (2016). Novel Meiotic miRNAs and Indications for a Role of PhasiRNAs in Meiosis, *Frontiers in Plant Science*, 7: 762.

Fan, D., Dia, Y., Wang, X., Wang, Z., He, H., Yang, H., CaO, Y., Deng, X.W. and Ma, L. (2012). IBM1, a JmjC domain-containing histone demethylase, is involved in the regulation of RNA-directed DNA methylation through the epigenetic control of RDR2 and DCL3 expression in *Arabidopsis, Nucleic Acids Research*, 40(18): 8905-8916.

Fei, Q., Xia, R. and Meyers, B.C. (2013). Phased, secondary, small interfering RNAs in posttranscriptional regulatory networks, *The Plant Cell*, 25(7): 2400-2415.

Feng, S. and Jacobsen, S.E. (2011). Epigenetic modifications in plants: An evolutionary perspective, *Current Opinion in Plant Biology*, 14(2): 179-186.

Franco-Zorrilla, J.M., Valli, A., Todesco, M., Mateos, I., Puga, M.I., Rubio-Somoza, I., Leyva, A., Weigel, D., García, J.A. and Paz-Ares, J. (2007). Target mimicry provides a new mechanism for regulation of microRNA activity, *Nature Genetics*, 39(8): 1033-1037.

Fujita, T., Kouchi, H., Ichikawa, T. and Syõno, K. (1993). Isolation and characterisation of a cDNA that encodes a novel proteinase inhibitor I from a tobacco genetic tumour, *Plant & Cell Physiology*, 34(1): 137-142.

Gao, Z., Liu, H.L., Daxinger, L., Pontes, O., He, X., Qian, W., Lin, H., Xie, M., Lorkovic, Z. J., Zhang, S., Miki, D., Zhan, X., Pontier, D., Lagrange, T., Jin, H., Matzke, A.J., Matzke, M., Pikaard, C.S. and Zhu, J.K. (2010). An RNA polymerase II- and AGO4-associated protein acts in RNA-directed DNA methylation, *Nature*, 465(7294): 106-109.

Ghildiyal, M. and Zamore, P.D. (2009). Small silencing RNAs: An expanding universe, *Nature Reviews: Genetics*, 10(2): 94-108.

Goll, M.G., Kirpekar, F., Maggert, K.A., Yoder, J.A., Hsieh, C.L., Zhang, X., Golic, K.G., Jacobsen, S.E. and Bestor, T.H. (2006). Methylation of tRNAAsp by the DNA methyltransferase homolog Dnmt2, *Science* (New York, N.Y.), 311(5759): 395-398.

Halford, N.G., Curtis, T.Y., Chen, Z. and Huang, J. (2015). Effects of abiotic stress and crop management on cereal grain composition: Implications for food quality and safety, *Journal of Experimental Botany*, 66(5): 1145-1156.

Hamilton, A., Voinnet, O., Chappell, L. and Baulcombe, D. (2002). Two classes of short interfering RNA in RNA silencing, *The EMBO Journal*, 21(17): 4671-4679.

Helliwell, C.A., Robertson, M., Finnegan, E.J., Buzas, D.M. and Dennis, E.S. (2011). Vernalisation-repression of *Arabidopsis* FLC requires promoter sequences but not antisense transcripts, *PLoS ONE*, 6(6): e21513.

Henderson, I.R. and Jacobsen, S.E. (2008). Tandem repeats upstream of the *Arabidopsis* endogene SDC recruit non-CG DNA methylation and initiate siRNA spreading, *Genes Dev.*, 22: 1597-1606.

Heo, J.B. and Sung, S. (2011). Vernalisation-mediated epigenetic silencing by a long intronic noncoding RNA, *Science*, New York, N.Y., 331(6013): 76-79.

Holoch, D. and Moazed, D. (2015). RNA-mediated epigenetic regulation of gene expression, *Nature Reviews: Genetics*, 16(2): 71-84.

Houben, A., Demidov, D., Caperta, A.D., Karimi, R., Agueci, F. and Vlasenko, L. (2007). Phosphorylation of histone H3 in plants – A dynamic affair, *Biochim. Biophys. Acta.*, 1769: 308-315.

Hu, W., Wang, T., Xu, J. and Li, H. (2014). MicroRNA mediates DNA methylation of target genes, *Biochemical and Biophysical Research Communications*, 444(4): 676-681.

Inagaki, S. and Kakutani, T. (2010). Control of genic DNA methylation in *Arabidopsis, J. Plant Res.*, 123: 299-302.

Inagaki, S., Miura-Kamio, A., Nakamura, Y., Lu, F., Cui, X., Cao, X., Kimura, H., Saze, H. and Kakutani, T. (2010). Autocatalytic differentiation of epigenetic modifications within the *Arabidopsis* genome, *EMBO J.*, 29: 3496-3506.

Jabnoune, M., Secco, D., Lecampion, C., Robaglia, C., Shu, Q. and Poirier, Y. (2013). A rice cis-natural antisense RNA acts as a translational enhancer for its cognate mRNA and contributes to phosphate homeostasis and plant fitness, *The Plant Cell*, 25(10): 4166-4182.

Jha, A. and Shankar, R. (2014). miRNAting control of DNA methylation, *Journal of Biosciences*, 39(3): 365-380.

Jia, Y., Lisch, D.R., Ohtsu, K., Scanlon, M.J., Nettleton, D. and Schnable, P.S. (2009). Loss of RNA-dependent RNA polymerase 2 (RDR2) function causes widespread and unexpected changes in the expression of transposons, genes, and 24-nt small RNAs, *PLoS Genet.*, 5: e1000737. (not in text)

Jia, X., Yan, J. and Tang, G. (2011). MicroRNA-mediated DNA methylation in plants, *Front. Biol.*, 6: 133-139.

Johnson, L., Mollah, S., Garcia, B.A., Muratore, T.L., Shabanowitz, J., Hunt, D.F. and Jacobsen, S.E. (2004). Mass spectrometry analysis of *Arabidopsis* histone H3 reveals distinct combinations of post-translational modifications, *Nucleic Acids Research*, 32(22): 6511-6518.

Johnson, L., Bostick, M., Zhang, X., Kraft, E., Henderson, I., Callis, J. and Jacobsen, S.E. (2007). The SRA methyl-cytosine-binding domain links DNA and histone methylation, *Current Biology*, 17-4.

Kanno, T., Aufsatz, W., Jaligot, E., Mette, M.F., Matzke, M., and Matzke, A.J. (2005). A SNF2-like protein facilitates dynamic control of DNA methylation, *EMBO Rep.*, 6: 649-655.

Kanno, T., Bucher, E., Daxinger, L., Huettel, B., Bohmdorfer, G., Gregor, W., Kreil, D.P., Matzke, M. and Matzke, A.J. (2008). A structural-maintenance-of-chromosomes hinge domain-containing protein is required for RNA directed DNA methylation, *Nature Genetics*, 40(5): 670-675.

Karlik, E. and Gozukirmizi, N. (2018). Expression analysis of lncRNA AK370814 involved in the barley vitamin B6 salvage pathway under salinity, *Molecular Biology Reports*, 45(6): 1597-1609.

Khraiwesh, B., Arif, M.A., Seumel, G.I., Ossowski, S., Weigel, D., Reski, R. and Frank, W. (2010). Transcriptional control of gene expression by microRNAs, *Cell*, 140(1): 111-122.

Kim, D.H., Doyle, M.R., Sung, S. and Amasino, R.M. (2009). Vernalisation: Winter and the timing of flowering in plants, *Annual Review of Cell and Developmental Biology*, 25: 277-299.

Kim, D.H. and Sung, S. (2012). Environmentally coordinated epigenetic silencing of FLC by protein and long noncoding RNA components, *Current Opinion in Plant Bbiology*, 15(1): 51-56.

Kim, D.H. and Sung, S. (2013). Coordination of the vernalisation response through a VIN3 and FLC gene family regulatory network in *Arabidopsis*, *The Plant Cell*, 25(2): 454-469.

Kim, J.M., To, T.K. and Seki, M. (2012). An epigenetic integrator: New insights into genome regulation, environmental stress responses and developmental controls by histone deacetylase 6, *Plant & Cell Physiology*, 53(5): 794-800.

Köhler, C. and Springer, N. (2017). Plant epigenomics – Deciphering the mechanisms of epigenetic inheritance and plasticity in plants, *Genome Biol.*, 18: 132.

Kumar, S. and Mohapatra, T. (2021). Dynamics of DNA Methylation and Its Functions in Plant Growth and Development, *Frontiers in Plant Science*, 12: 596236.

La, H., Ding, B., Mishra, G. P., Zhou, B., Yang, H., Bellizzi, M., Chen, S., Meyers, B.C., Peng, Z., Zhu, J.K. and Wang, G.L. (2011). A 5-methylcytosine DNA glycosylase/lyase demethylates the retrotransposon Tos17 and promotes its transposition in rice, *Proceedings of the National Academy of Sciences of the United States of America*, 108(37): 15498-15503.

Lancher, M. and Jenuwein, T. (2002). The many faces of histone lysine methylation, *Current Opinion in Cell Biology*, 14(3): 286-298.

Lagana, A., Veneziano, D., Russo, F., Pulvirenti, A., Giugno, R., Croce, C.M. and Ferro, A. (2015). Computational design of artificial RNA molecules for gene regulation, *RNA Bioinformatics, Methods in Molecular Biology*, 1269: 393-412.

Law, J.A. and Jacobsen, S.E. (2010). Establishing, maintaining and modifying DNA methylation patterns in plants and animals, *Nature Reviews: Genetics*, 11(3): 204-220.

Lee, T.F., Gurazada, S.G., Zhai, J., Li, S., Simon, S.A., Matzke, M.A., Chen, X. and Meyers, B.C. (2012). RNA polymerase V-dependent small RNAs in *Arabidopsis* originate from small, intergenic loci including most SINE repeats, *Epigenetics*, 7: 781-795. (not in text)

Lelandais-Brière, C., Naya, L., Sallet, E., Calenge, F., Frugier, F., Hartmann, C., Gouzy, J. and Crespi, M. (2009). Genome-wide *Medicago truncatula* small RNA analysis revealed novel microRNAs and isoforms differentially regulated in roots and nodules, *The Plant Cell*, 21(9): 2780-2796.

Li, K.K., Luo, C., Wang, D., Jiang, H. and Zheng, Y.G. (2010). Chemical and biochemical approaches in the study of histone methylation and demethylation, *Med. Res. Rev.*, 32: 815-867.

Lindroth, A.M., Shultis, D., Jasencakova, Z., Fuchs, J., Johnson, L., Shubert, D., Patnaik, D., Pradhan, S., Goodrich, J., Schubert, I., Jenuwein, T., Khorasanizadeh, S. and Jacobsen, S.E. (2004). Dual histone H3methylation marks at lysine 9 and 27 required for interaction with CHROMOMETHYLASE3, *The EMBO Journal*, 23(21): 4286-4296.

Liu, C., Lu, F., Cui, X. and Cao, X. (2010). Histone methylation in higher plants, *Annual Review of Plant Biology*, 61: 395-420.

Liu, C., Muchhal, U.S. and Raghothama, K.G. (1997). Differential expression of TPS11, a phosphate starvation-induced gene in tomato, Plant Molecular Biology, 33(5): 867-874.

Lister, R., O'Malley, R.C., Tonti-Filippini, J., Gregory, B.D., Berry, C.C., Millar, A.H. and Ecker, J.R. (2008). Highly integrated single-base resolution maps of the epigenome in *Arabidopsis*, *Cell*, 133(3): 523-536.

Lv, Y., Hu, F., Zhou, Y., Wu, F. and Gaut, B.S. (2019). Maize transposable elements contribute to long non-coding RNAs that are regulatory hubs for abiotic stress response, *BMC Genomics*, 20(1): 864.

Matzke, M.A., Kanno, T. and Matzke, A.J. (2015). RNA-Directed DNA Methylation: The Evolution of a Complex Epigenetic Pathway in Flowering Plants, *Annual Review of Plant Biology*, 66: 243-267.

Matzke, M.A. and Mosher, R.A. (2014). RNA-directed DNA methylation: an epigenetic pathway of increasing complexity, *Nature Reviews: Genetics*, 15(6): 394-408.

Monga, I., Qureshi, A., Thakur, N., Gupta, A.K. and Kumar, M. (2017). ASPsiRNA: A Resource of ASP-siRNAs Having Therapeutic Potential for Human Genetic Disorders and Algorithm for Prediction of Their Inhibitory Efficacy, *G3* (Bethesda, Md.), 7: 2931-2943.

Mosher, R.A., Schwach, F., Studholme, D. and Baulcombe, D.C. (2008). PolIVb influences RNA-directed DNA methylation independently of its role in siRNA biogenesis, *Proceedings of the National Academy of Sciences of the United States of America*, 105(8): 3145-3150.

Mylne, J.S., Barrett, L., Tessadori, F., Mesnage, S., Johnson, L., Bernatavichute, Y.V., Jacobsen, S.E., Fransz, P. and Dean, C. (2006). LHP1, the *Arabidopsis* homologue of heterochromatin protein1, is required for epigenetic silencing of FLC, *Proceedings of the National Academy of Sciences of the United States of America*, 103(13): 5012-5017.

Nabuta, K., Lu, C., Shrivastava, R., Pillay, M., De Paoli, E., Accerbi, M., Arteaga-Vazquez, M., Sidorenko, L., Jeong, D.H., Green, P.J., Chandler, V.L. and Meyers, B.C. (2008). Distinct size distribution of endogenous siRNAs in maize: Evidence from deep sequencing in the mop1-1 mutant, *PNAS*, 105(39): 14958-14963.

Onodera, Y., Haag, J.R., Ream, T., Costa Nunes, P., Pontes, O. and Pikaard, C.S. (2005). Plant nuclear RNA polymerase IV mediates siRNA and DNA methylation-dependent heterochromatin formation, *Cell*, 120: 613-622.

Onufriev, A.V. and Schiessel, H. (2019). The nucleosome: From structure to function through physics, *Current Opinion in Structural Biology*, 56: 119-130.

Pandey, R., Müller, A., Napoli, C.A., Selinger, D.A., Pikaard, C.S., Richards, E.J., Bender, J., Mount, D.W. and Jorgensen, R.A. (2002). Analysis of histone acetyltransferase and histone deacetylase families of *Arabidopsis thaliana* suggests functional diversification of chromatin modification among multicellular eukaryotes, *Nucleic Acids Research*, 30(23): 5036-5055.

Pang, J., Zhang, X., Ma, X. and Zhao, J. (2019). Spatio-temporal Transcriptional Dynamics of Maize Long Non-Coding RNAs Responsive to Drought Stress, *Genes*, 10(2): 138.

Penterman, J., Uzawa, R. and Fischer, R.L. (2007). Genetic interactions between DNA demethylation and methylation in *Arabidopsis*, *Plant Physiol.*, 14: 1549-1557.

Pikaard, C.S. and Mittelsten Scheid, O. (2014). Epigenetic regulation in plants, *Cold Spring Harbour Perspectives in Biology*, 6(12): a019315.

MicroRNAs and other Non-coding RNAs in Plant Epigenetics **357**

Pikaard, C.S. and Tucker, S. (2009). RNA-silencing enzymes Pol IV and Pol V in maize: More than one flavor? *PLoS Genetics*, 5(11): e100736.

Pontes, O., Costa-Nunes, P., Vithayathil, P. and Pikaard, C.S. (2009). RNA polymerase V functions in *Arabidopsis* interphase heterochromatin organisation independently of the 24-nt siRNA-directed DNA methylation pathway, *Mol Plant.*, 2: 700-710.

Pontes, O., Li, C.F., Nunes, P.C., Haag, J., Ream, T., Vitins, A., Jacobsen, S.E. and Pikaard, C.S. (2006). The *Arabidopsis* chromatinmodifying nuclear siRNA pathway involves a nucleolar RNA processing centre, *Cell*, 12.

Pontier, D., Yahubyan, G., Vega, D., Bulski, A., Saez-Vasquez, J., Hakimi, M.A., Lerbs-Mache, S., Colot, V. and Lagrange, T. (2005). Reinforcement of silencing at transposons and highly repeated sequences requires the concerted action of two distinct RNA polymerases IV in *Arabidopsis*, *Genes Dev.*, 19: 2030-2040.

Qi, Y., He, X., Wang, X.J., Kohany, O., Jurka, J. and Hannon, G.J. (2006). Distinct catalytic and non-catalytic roles of ARGONAUTE4 in RNA-directed DNA methylation, *Nature*, 443(7114): 1008-1012.

Qin, T., Zhao, H., Cui, P., Albesher, N. and Xiong, L. (2017). A Nucleus-localised Long Non-Coding RNA Enhances Drought and Salt Stress Tolerance, *Plant Physiology*, 175(3): 1321-1336.

Rohrig, H., Schmidt, J., Miklashevichs, E., Schell, J. and John, M. (2002). Soybean ENOD40 encodes two peptides that bind to sucrose synthase, *Proceedings of the National Academy of Sciences of the United States of America*, 99(4): 1915-1920.

Ronemus, M. and Martienssen, R. (2005). RNA interference: Methylation mystery, *Nature*, 433(7025): 472-473.

Saleh, A., Alvarez-Venegas, R. and Avramova, Z. (2008). An efficient chromatin immunoprecipitation (ChIP) protocol for studying histone modifications in *Arabidopsis* plants. *Nature Protocols*, 3(6): 1018-1025.

Satgé, C., Moreau, S., Sallet, E., Lefort, G., Auriac, M.C., Remblière, C., Cottret, L., Gallardo, K., Noirot, C., Jardinaud, M.F. and Gamas, P. (2016). Reprogramming of DNA methylation is critical for nodule development in *Medicago truncatula*, *Nature Plants*, 2(11): 16166.

Servet, C., Conde, E., Silva, N. and Zhou, D.X. (2010). Histone acetyltransferase AtGCN5/HAG1 is a versatile regulator of developmental and inducible gene expression in *Arabidopsis*, *Molecular Plant*, 3(4): 670-677.

Shuai, P., Liang, D., Tang, S., Zhang, Z., Ye, C.Y., Su, Y., Xia, X. and Yin, W. (2014). Genome-wide identification and functional prediction of novel and drought-responsive lincRNAs in *Populus trichocarpa*, *Journal of Experimental Botany*, 65(17): 4975-4983.

Sidorenko, L., Dorweiler, J.E., Cigan, A.M., Arteaga-Vazquez, M., Vyas, M., Kermicle, J., Jurcin, D., Brzeski, J., Cai, Y. and Chandler, V.L. (2009). A dominant mutation in mediator of paramutation2, one of three second-largest subunits of a plant-specific RNA polymerase, disrupts multiple siRNA silencing processes, *PLoS Genet.*, 5: e1000725.

Sidorenka, L. and Chandler, V. (2008). RNA-dependent RNA polymerase is required for enhancer-mediated transcriptional silencing associated with paramutation at the maize p1 gene, *Genetics*, 180(4): 1983-1993.

Slotkin, R.K., Freeling, M. and Lisch, D. (2005). Heritable transposon silencing initiated by a naturally occurring transposon inverted duplication, *Nature Genetics*, 37(6): 641-644.

Soppe, W.J., Jasencakova, Z., Houben, A., Kakutani, T., Meister, A., Haung, M.S., Jacobsen, S.E., Shubert, I. and Fransz, P.F. (2002). DNA methylation controls histone H3 lysine 9 methylation and heterochromatin assembly in *Arabidopsis*, *The EMBO Journal*, 21(23): 6549-6559.

Suksamran, R., Saithong, T., Thammarongtham, C. and Kalapanulak, S. (2020). Genomic and Transcriptomic Analysis Identified Novel Putative Cassava lncRNAs Involved in Cold and Drought Stress, *Genes*, 11(4): 366.

Sun, X., Zheng, H., Li, J., Liu, L., Zhang, X. and Sui, N. (2020). Comparative Transcriptome Analysis Reveals New lncRNAs Responding to Salt Stress in Sweet Sorghum, *Frontiers in Bioengineering and Biotechnology*, 8: 331.

Sung, S. and Amasino, R.M. (2004). Vernalisation and epigenetics: How plants remember winter, *Current Opinion in Plant Biology*, 7(1): 4-10.

Swiezewski, S., Liu, F., Magusin, A. and Dean, C. (2009). Cold-induced silencing by long antisense transcripts of an *Arabidopsis* polycomb target, *Nature*, 462(7274): 799-802.

Tollefsbol, T.O. (2011). Epigenetics: The new science of genetics, pp.1-6. *In: Handbook of Epigenetics*, NY: Academic Press, New York.

Truck, F., Roudie, F., Farrona, S., Martin-Magniette, M.-L., Guillaume, E. and Buisine, N. (2007). Arabidopsis TFL2/LHP1 specifically associates with genes marked by trimethylation of histone H3 lysine 27, *PLoS Genet.*, 3(6): e86.

Unver, T. and Tombuloglu, H. (2020). Barley long non-coding RNAs (lncRNA) responsive to excess boron, *Genomics*, 112(2): 1947-1955.

Vaucheret, H. (2008). Plant Argonautes, *Trends in Plant Science*, 13(7): 350-358.

Wang, J., Meng, X., Yuan, C., Harrison, A.P. and Chen, M. (2016). The roles of cross-talk epigenetic patterns in *Arabidopsis thaliana*, *Briefings in Functional Genomics*, 15(4): 278-287.

Wierzbick, A.T., Cocklin, R., Mayampurath, A., Lister, R., Rowley, M.J., Gregory, B.D., Ecker, J.R., Tang, H. and Pikaard, C.S. (2012). Spatial and functional relationships among Pol V-associated loci, Pol IV-dependent siRNAs, and cytosine methylation in the *Arabidopsis* epigenome, *Genes Dev.*, 26: 1825-1836.

Wierzbick, A.T., Haag, J.R. and Pikaard, C.S. (2008). Non-coding transcription by RNA polymerase Pol IVb/Pol V mediates transcriptional silencing of overlapping and adjacent genes, *Cell*, 135(4): 635-648.

Wierzbicki, A.T., Ream, T.S., Haag, J.R. and Pikaard, C.S. (2009). RNA polymerase V transcription guides Argonaute4 to chromatin, *Nat. Genet.*, 41: 630-634.

Wood, C.C., Robertson, M., Tanner, G., Peacock, W.J., Dennis, E.S. and Helliwell, C.A. (2006). The *Arabidopsis thaliana* vernalisation response requires a polycomb-like protein complex that also includes vernalisation insensitive 3, *Proceedings of the National Academy of Sciences of the United States of America*, 103(39): 14631-14636.

Wu, L., Mao, L. and Qi, Y. (2012). Roles of dicer-like and Argonaute proteins in TAS-derived small interfering RNA-triggered DNA methylation, *Plant Physiology*, 160(2): 990-999.

Wu, L., Zhou, H., Zhang, Q., Zhang, J., Ni, F., Liu, C. and Qi, Y. (2010). DNA methylation mediated by a microRNA pathway, *Molecular Cell*, 38(3): 465-475.

Wunderlich, M., Gross-Hardt, R. and Schöffl, F. (2014). Heat shock factor HSFB2a involved in gametophyte development of *Arabidopsis thaliana* and its expression is controlled by a heat-inducible long non-coding antisense RNA, *Plant Molecular Biology*, 85(6): 541-550.

Xie, Z., Johansen, L.K., Gustafson, A.M., Kasschau, K.D., Lellis, A.D., Zilberman, D., Jacobsen, S.E. and Carrington, J.C. (2004). Genetic and functional diversification of small RNA pathways in plants, *PLoS Biology*, 2(5): E104.

Xin, M., Wang, Y., Yao, Y., Song, N., Hu, Z., Qin, D., Xie, C., Peng, H., Ni, Z. and Sun, Q. (2011). Identification and characterisation of wheat long non-protein coding RNAs responsive to powdery mildew infection and heat stress by using microarray analysis and SBS sequencing, *BMC Plant Biology*, 11: 61.

Xu, L. and Shen, W.H. (2008). Polycomb silencing of KNOX genes confines shoot stem cell niches in *Arabidopsis*, *Current Biology: CB*, 18(24): 1966-1971.

Xu, Y., Zhang, S., Lin, S., Guo, Y., Deng, W., Zhang, Y. and Xue, Y. (2017). WERAM: A database of writers, erasers and readers of histone acetylation and methylation in eukaryotes, *Nucleic Acids Research*, 45(D1): D264-D270.

Ye, R., Wang, W., Iki, T., Liu, C., Wu, Y., Ishikawa, M., Zhou, X. and Qi, Y. (2012). Cytoplasmic assembly and selective nuclear import of *Arabidopsis* Argonaute4/siRNA complexes, *Molecular Cell*, 46(6): 859-870.

Yoshikawa, M., Peragine, A., Park, M.Y. and Poethig, R.S. (2005). A pathway for the biogenesis of trans-acting siRNAs in *Arabidopsis*, *Genes & Development*, 19(18): 2164-2175.

MicroRNAs and other Non-coding RNAs in Plant Epigenetics

Zhang, C.J., Ning, Y.Q., Zhang, S.W., Chen, Q., Shao, C.R., Guo, Y.W., Zhou, J.X., Li, L., Chen, S. and He, X.J. (2012). IDN2 and its paralogs form a complex required for RNA-directed DNA methylation, *PLoS Genet.*, 8: e1002693.

Zhang, A., Wassarman, K.M., Rosenow, C., Tjaden, B.C., Storz, G. and Gottesmam, S. (2003). Global analysis of small RNA and mRNA targets of Hfq, *Molecular Microbiology*, 50(4): 1111-1124.

Zhang, K., Sridhar, V.V., Zhu, J., Kapoor, A. and Zhu, J.K. (2007). Distinctive core histone post-translational modification patterns in *Arabidopsis thaliana*, *PloS ONE*, 2(11): e1210.

Zhang, W., Han, Z., Guo, Q., Liu, Y., Zheng, Y., Wu, F. and Jin, W. (2014). Identification of maize long non-coding RNAs responsive to drought stress, *PLoS ONE*, 9(6): e98958.

Zhao, M., Wang, T., Sun, T., Yu, X., Tian, R. and Zhang, W.H. (2020). Identification of tissue-specific and cold-responsive lncRNAs in *Medicago truncatula* by high-throughput RNA sequencing, *BMC Plant Biology*, 20(1): 99.

Zilberman, D., Cao, X. and Jacobsen, S.E. (2003). Argonaute4 control of locus-specific siRNA accumulation and DNA and histone methylation, *Science*, 299: 716-719.

Zilberman, D., Gehring, M., Tran, R.K., Ballinger, T. and Henikoff, S. (2007). Genome-wide analysis of *Arabidopsis thaliana* DNA methylation uncovers an interdependence between methylation and transcription, *Nat Genet.*, 39: 61-69.

CHAPTER
19

MicroRNA-based Plant Genetic Engineering for Crop Improvement

Heena Tabassum[1] and Iffat Zareen Ahmad[2]*

[1] Dr. D.Y. Patil Biotechnology and Bioinformatics Institute, Dr. D.Y. Patil Vidyapeeth, Pune - 4110033, India
[2] Natural Products Laboratory, Department of Bioengineering, Integral University, Dasauli, Kursi Road, Lucknow - 226026, Uttar Pradesh, India

1. Introduction

Increasing world population as well as inadequacy of cultivable lands, water resources and constant changes in the climate have led to some major concerns regarding food crop plants and whether they will be sufficient or not to suffice the needs of the future generations (Rosegrant and Cline, 2003; Brown and Funk, 2008). In addition to water and land preservation initiatives, modern agriculture technology would be required to tackle these issues as well as maintain the world's food supply and safety. Crop advances, which give resistance to environmental stressors and soil viruses as well as a good yield, would particularly be essential (Takeda and Matsuoka, 2008). RNA interference (RNAi) is gene silencing procedure that is done by dsRNAs, which give rise to obstruction of translation or transcription. dsRNA has its cleavage products, namely short interfering RNA and microRNA (siRNA and miRNA). The utilisation of miRNA was known 10 years later after the animal miRNAs. These miRNA in plants help in crop enhancement as well as in many other factors that result in a good yield of crops. RNA interference has greatly advanced in terms of a genetically-engineered tool and functional genomics which target crop enhancement.

2. What are MicroRNAs (miRNAs)?

MicroRNAs (miRNA) are small non-coding molecules of RNA which is seen in plants, animals as well as viruses. It works like RNA silencer and regulates post-transcriptional activity in gene expression. Inside the mRNA molecules, miRNAs work through base pairing along with the complementary sequences. The main role

---

*Corresponding author: iffat@iul.ac.in

MicroRNA-based Plant Genetic Engineering for Crop Improvement **361**

played by miRNA is degeneration of mRNA as well as hampering the translation process in mRNA (Pandita, 2019; Pandita and Wani, 2019; Pandita, 2021; 2022a; 2022b; 2022c).

2.1 Origin of miRNA

- They are present endogenously.
- They originate in the nucleus. Then they come together with other proteins in the ribonucleoprotein complex, known as RISC.
- miRNA is produced following post-transcriptional alteration by RNA polymerase II and polymerase III.
- Most of the time miRNAs are processed from the introns and rarely from exons.
- miRNAs are mediated by their own promoters.
- At times they are transcribed as extended transcripts.

2.2 Properties of miRNA

The main mechanism of miRNA is mRNA cleavage while secondary mechanism is silencing of the chromatin. miRNA is encoded by the genes in the nucleus and the synthesis of miRNA is mostly done endogenously. The enzymes that take part in the synthesis of miRNA are RNA polymerase II, Drosha and Dicer. The length of miRNA is known to be around 19 to 24 base pairs.

2.3 Localisation of MicroRNA

miRNA is localised in two stages, in which one step is carried out in the nucleus and the other in cytosol. In the nucleus, its starts with primary miRNA (pri-miRNA) which is not a completely formed miRNA and is endogenous in origin. A long stretch of single-stranded RNA is formed and if they have lots self-complementary regions, then they can bind with themselves and form hairpin structures and get converted to pre-miRNA by enzyme Drosha, which cleaves from somewhere in the middle. This gives rise to short hairpin RNA, i.e., pre-miRNA, which is then delivered to the cytoplasm through export receptors that are present in the nuclear membrane called exportin 5. Then the enzyme, called Dicer, which is present in cytosol, cleaves the loop portion and leaves behind only the double-stranded miRNA (miR/miR duplex). After this, the assemblage of Argonaute protein and other proteins that give rise to RISC (RNA-induced silencing protein), remove one strand and use that single strand along with Argonaute protein to load itself on the target miRNA. Whenever they bind to the complementary miRNA, the Argonaute protein cleaves the target mRNA in a particular region. In this way, the mRNA is degraded. These are the stages of miRNA-mediated gene silencing. Figure 1 shows the schematic representation of the miRNA gene silencing process.

2.4 Impact of miRNA on Plant Structure and Growth

The structure of plants is highly determined by the extension of internode which is the amount of branching done as well as the determination of the shoots (Wang and Li, 2008). Plant structure has been discovered to correspond with agricultural qualities and the capability of plants to resist pressures of environment for ages. Therefore, its investigation has piqued interest (Peng *et al.*, 1999). Recently, our knowledge on molecular genetic foundation of plant structure has substantially improved

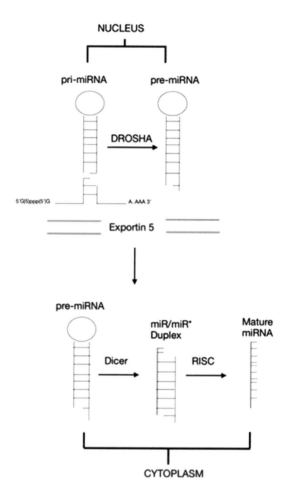

Fig. 1: Process of miRNA-mediated gene silencing

thanks to *Arabidopsis thaliana* and food crops, such as rice and maize (Reinhardt and Kuhlemeier, 2002; Babb and Muehlbauer, 2003; Wang and Li, 2008). Recent research has found that miRNAs have a function in plant structure which adds to the understanding of molecular mechanisms that seem to be responsible for plant structure. This information has also paved the way for humans to enhance crop production by enhancing the structure of crops via molecular design.

3. miRNA'S Role in Development of Shoots, Vascular Tissues and Roots

SAM (shoot apical meristem) produces plant organs which are above the ground and this goes on for the entire lifetime of higher plants. The meristem should maintain a balance between the self-renewal of a pool of core stem cells and organ commencement from peripheral cells in order to accomplish this job (Wang and Li, 2008). Till now, miR164 and miR165/miR166 are known to take part in shoot development by aiming at the NAC genes as well as Class III HD-ZIP genes (Emery *et al.*, 2003; Laufs *et al.*, 2004; Mallory *et al.*, 2004a; Baker *et al.*, 2005; Williams *et al.*, 2005; Zhou *et al.*, 2007). Furthermore, certain miRNAs have been demonstrated to control the leaf growth: By modulating the amount of ORE1 gene, miR164 inhibits leaf senescence, while miR165/miR166 conciliated HD-ZIP III works as stimulus to determine leaf polarity (Kidner and Martienssen, 2004; Mallory *et al.*, 2004b; Timmermans *et al.*, 2004; Nikovics *et al.*, 2006; Peaucelle *et al.*, 2007; Kim *et al.*, 2009). In terms of leaf maturation, it is important to note that miR319 is indeed a microRNA which regulates leaf development and senescence which modulates plant specific Class II teosinte branched 1/cycloidea/PCF transcription factors (Palatnik *et al.*, 2003; Ori *et al.*, 2007; Efroni *et al.*, 2008; Schommer *et al.*, 2008). When miR319 is expressed, it gives rise to curve to the leaves, large leaflets as well as constant maturation of leaf margins (Palatnik *et al.*, 2003; Ori *et al.*, 2007; Efroni *et al.*, 2008). Also, on the other hand, miR156 and miR172 are major regulators which promote transition of flowers by regulating SBP-box genes and genes like AP2 (Lauter *et al.*, 2005; Wu and Poethig, 2006; Schwarz *et al.*, 2008; Wang *et al.*, 2009; Wu *et al.*, 2009). miR159 also modulates the time when the blossoming of flowers will occur as well as the development of anther by aiming GAMYB (Achard *et al.*, 2004). Recently, it has been seen that miR165 and miR166 are particularly stimulated for targeting HD-ZIP-III mRNAs (Kim *et al.*, 2005; Williams *et al.*, 2005; Zhou *et al.*, 2007). HD-ZIP III homebox genes along with PHB, PHV, REV AtHB8 and ArHB15 in thale cress differentiate cells of xylem from the procambial cells (Fukuda, 2004). It is seen that miR164 plays a part in root formation as well as the miR166 which is seen modulating the branching of the shoots in many plants. It is also seen affecting the structure of the roots in legumes by post-transcription of many HD-ZIP III genes in roots as well as nodules (Boualem *et. al.*, 2008). Therefore, miR164 and miR166 emerge as very effective aids that optimise root structure for improving crops. The part played by miRNAs in shoot, vascular tissues and roots are shown in Fig. 2.

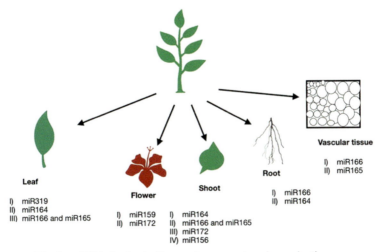

Fig. 2: miRNAs in shoots, flowers, leaves, roots and vascular tissues

4. miRNA Regulation in Plant Engineering

An immense and important role played by miRNA in plants as well as animals via RNA interference has made people adopt several techniques in manipulation of particular genes or microRNA's *in vivo* from the knowledge that already exists about the antisense technology as well as genetic engineering. Latest advancements are seen in which microRNAs modulate the function in plants as shown in Table 1. Regular miRNA interference is represented in Fig. 3a.

Table 1: Function of miRNA in different parts of plants

Plant Part	miRNA	Function	References
Shoot, leaves and flowers	miR164	Shoot morphogenesis and in a negative way modulates senescence of leaves by having a control over ORE1 transcription level.	Kim *et al.*, 2009; Peaucelle *et al.*, 2007; Nikovics *et al.*, 2006; Mallory *et al.*, 2004b
	miR165/miR166	Shoot morphogenesis and mediation of HD-ZIP III signals for specifying leaf polarity.	Kidner and Martienssen *et al.*, 2004; Timmermans *et al.*, 2004; Mallory *et al.*, 2004b
	miR319	Leads to curving of leaves, bigger leaflets as well as senescence of leaves.	Palatnik *et al.*, 2003; Ori *et al.*, 2007; Efroni *et al.*, 2008
	miR156 and miR172	Promotes transition of flowers by regulating SBP-box and AP2 genes.	Lauter *et al.*, 2005; Wu and Poethig, 2006; Schwarz *et al.*, 2008; Wang *et al.*, 2009; Wu *et al.*, 2009
	miR159	Regulates time of inflorescence and anther growth by targeting GAMYB and increased concentration of miR159 stall the inflorescence and disturb anther growth.	Achard *et al.*, 2004
Vascular tissue and root growth	miR165 and miR166	Increased expression causes sudden depletion in the transcript concentration of HD-ZIP III, which results in important phenotypes.	Zhou *et al.*, 2005
	miR164 and niR166	Plays part in root development.	Guo *et al.*, 2005; Boualem *et al.*, 2008

4.1 Effective Techniques

4.1.1 Artificial miRNAs

Artificial miRNAs (amiRNAs) are developed in such a way that they can target any gene that has good specificity and microRNAs that are good regulators towards agronomic

MicroRNA-based Plant Genetic Engineering for Crop Improvement   365

characteristics. As a result, the particular reduction of their targeting genes will be preferable so that crops can breed with the best characteristics. miRNA and siRNA are known to have effect on animals as well as plants by RNA interference interest (Elbashir *et al.*, 2001; Schwab *et al.*, 2006). Even though artificial miRNA and siRNA have some similar characteristics, they are different from each other. Various group of siRNAs can be made from an entire double-stranded RNA and therefore, siRNA not only targets eligible RNAs but also affects RNA that are not complementary (off target). One species of small RNA is made from amiRNA precursor; also an amiRNA sequence is made and it has less similarity to whichever gene of the plant could be selected to keep away from the target repercussions (Duan *et al.*, 2008; Khraiwesh *et al.*, 2008). Additionally, latest investigations show that siRNAs are either transgene derivatives or viral induced and can move from one cell to another, indicating that their functions aren't cell-independent but microRNAs aren't ambulant as well as are cell independent (Khraiwesh *et al.*, 2008; Tretter *et al.*, 2008). Thus, silencing of the gene, modulated by amiRNAs, is managed by cell independent way and this method dos not have any bio-safety problems on the environment when used in the agricultural sector. This is the reason why amiRNAs are known to be best for plant engineering. As of now, amiRNA is used in higher plants, lower plants and single-celled organisms (Alvarez *et al.*, 2006; Schwab *et al.*, 2006; Niu *et al.*, 2006; Duan *et al.*, 2008; Khraiwesh *et al.*, 2008; Warthmann *et al.*, 2008; Molnar *et al.*, 2009; Zhao *et al.*, 2009). Besides the dicotyledons, amiRNAs have also been used in down-regulating the particular genes in rice and as of now, about 61% of the anticipated loci of rice genes are not yet allotted (Alvarez *et al.*, 2006; Warthmann *et al.*, 2008). Artificial miRNas can silence more than one gene or particular alleles with good efficacy and specificity. Now that we know that amiRNA can particularly silence the genes which were not initially under the influence of miRNA and that they have been silenced in the plants is an innate mechanism of protection in response to unknown genetic components and the usage of amiRNAs in plant engineering to develop resistance against various viral vectors is colossally used (Schwab *et al.*, 2006; Niu *et al.*, 2006). Diagrammatic representation can be seen in Fig. 3b.

4.1.2 Target Mimicry

Mostly majority of miRNAs have substantial compatibility with their target genes in plants, as well as they modulate them by regulating mRNA cleavage at specific locations in coding areas. This property inspired the target mimicry technique, which uses RNA which isn't cleaved but rather sequesterate complementary miRNA. Firstly two endogenous target mimics (eTMs) have been found in 2007 in Thale cress (*Arabidopsis thaliana*). It showed that derivative transcripts from the genes IPS1 and At4 function as eTMs which controlled the inhibition effect of miR399 on gene expression of PHO2. PHO2 is a gene which takes part in P-homeostasis via its non-positive control on the expression of phosphate transporter in roots. PHR1 factor induces the miR399 where the phosphate levels are low in a shoot. Later when it is produced, miR399 transports towards the roots, where it obstructs the PHO2 expression and neutralises the non-positive effect of PHO2 on the phosphate transporter which leads to elevated amounts of P_i in the shoot and which at times, reaches toxicity levels. The amount of miR399 is managed by the IPS1 and AT4 transcripts that play the role as eTMs and offer an exact location for miR399 to bind. This, in return, sequesters miR399. The IPS1 and At4 transcripts anneal miR399 and form three nucleotide bump amongst the nucleotides 9 and 12 (important for cleaving) and therefore avoid the cleaving of the transcripts by miR399. This process is known as 'target mimicry' where IPS1 and At4 are the mimickers and PHO2 transcript is the target and miR399 is the sequester. Therefore, target mimicry is a unique method that regulates the eTMs and enhances the agronomical characteristics, giving good agricultural output. (Franco-Zorrilla *et al.*, 2007). Diagrammatic representation can be seen in Fig. 3c.

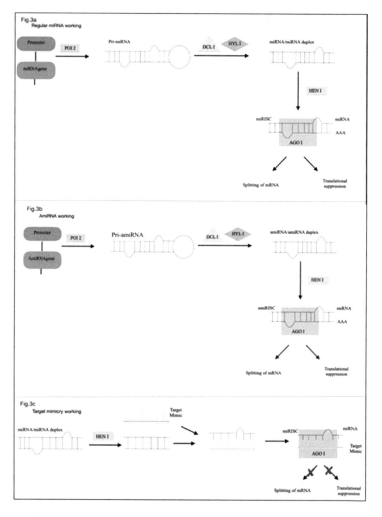

Fig. 3: (a) The pri-miRNA is further carried out by DCL1, with the help of HYL1 as well as other factors to make miRNA: miRNA duplex and the matured miRNA is assimilated in the silencing complex which has AGO1 and modulates the splitting of mRNA or translational suppression; (b) The amiRNA is engineered into miRNA precursor with the use of overlapping PCR to substitute the endogenic miRNA sequence and the pri-amiRNA carried out by DCL1 as well as functions as a regular miRNA; (c) The target mimicry uses the RNA which cannot be cleaved and which sequesters the complementary miRNA. Exact sequence complementarity permits target mimic to bind the miRNA and intervene with miRNA function

5. RNA Interference Function in Crop Enhancement

5.1 Improvement in Crop Yield

RNA interference (RNAi) could be used to improve yield of crops and fruits by manipulation of agriculture characteristics, like height, arrangement of branches, sizes and flowering as shown in Table 2. RNAi modulated knockdown of OsDWARF4 gene in rice giving rise to small plants and standing leaves structure which lead to elevated photosynthesis in lower leaves. Plants like these possess the capability of enhanced

Table 2: Role of miRNA in crop improvement

Type of Crop	miRNA	Role in Crop Improvement	Targeted Gene	References
A. thaliana	miR169	Tolerance against drought conditions.	NFYA5	Li *et al.*, 2008b
Soybean and *A. thaliana*		Tolerance against drought conditions.	GmNFYA3	Ni *et al.*, 2013
Rice	miR319	Tolerance against chill conditions.	PCF5/PCF8	Yang *et al.*, 2013
	miR160 miR1428	Sterilised pollen.	-	Zhang *et al.*, 2017
	miR172	Lowered seed setting proportion.	-	Zhang *et al.*, 2017
	miR156 miR159 miR165/ miR166 miR398	Lowered weight of grains.	-	Zhang *et al.*, 2017; Zhao *et al.*, 2017; Gao *et al.*, 2018; Zhang *et al.*, 2018b
	miR167 miR1432 miR5144-3p	Elevated weight of grains.	-	Peng *et al.*, 2018; Zhao *et al.*, 2019
	miR5144-3p	Improved tolerance towards increased and decreased temperatures.	-	Xia *et al.*, 2017
	miR159	Improved sensitivity against *Magnaporthe grisea*.	-	Chen *et al.*, 2021b
	miR156	Size, shape and superiority.	SPL14/ SPL16	Jiao *et al.*, 2010; Miura *et al.*, 2010; Wang *et al.*, 2012
	miR397	Size, shape and superiority.	OsLac	Zhang *et al.*, 2013
A. thaliana	miR393	Resistance against bacteria.	TIR1	Navarro *et al.*, 2006
A. thaliana and Tomato	miR167	Development of fruit without fertilising.	ARF8	Goetz *et al.*, 2007

368   *Plant MicroRNAs and Stress Response*

crops with the help of heavy plantation environment (Feldmann, 2006). Certain plant components are resistant to the process of breaking down complex carbohydrate in monosaccharide components, posing a significant barrier to the transformation from lignocellulosic biomass to ethanol. The issue with cell walls resistance to biomass conversion could be solved by genetically lowering lignin composition (Chen and Dixon, 2007). In transgenic alfalfa lines, down-regulating of every six lignin production enzymes produced two times quite as much glucose through stem cell walls, like wild-type plants (Reddy *et al.*, 2005). The lignin alteration even made it possible to avoid acid pretreatment (Chen and Dixon, 2007). When CG1 (Corn grass 1) microRNA belonging to miR156 family gave rise to extended vegetative stage and postponed blossoming of flowers, the results showed elevated biomass (Chuck *et al.*, 2011). Genetically engineered plants were seen to have about 250% more starch as well as enhanced digestion (Chuck *et al.*, 2011). According to a few reports, it was seen that morphology modifications and biomass yields were affiliated to the amount of exogenic rice microRNA Osa-miR156b expressed (Fu *et al.*, 2012). Transgenic switch grass crops with excessive concentrations of miR156 expression grew very slowly, whereas those with medium concentrations of miR156 expression produced 58-101% higher biomass, compared wild-type controls due to increased tiller quantities. Increased expression of rice miR156 also was found to boost the production of solubilised glucose and fodder digestion. In various plant species, such as *Arabidopsis* and rice, increased expression of miR156 resulted in higher biomass (Schwab *et al.*, 2005; Xie *et al.*, 2012). In comparison to wild poplar corn grass 1 over-expressing transgenics plants exhibited a substantial rise in axillary meristem development, smaller internode size, as well as a 30% decrease in stem lignin concentration (Rubinelli *et al.*, 2013).

5.2 Fruit Enhancement

5.2.1 Improved Nutrition Value and Edibleness

Tomato, which is commonly used as a fruit all over the world, uses RNA interference for developing with improved level of flavonoids and carotenoids that are useful for maintaining people's health. Along with some promoter particular to the fruit, RNA interference is used to repress the endogenic DET1 gene present in tomato – a photomorphogenic regulation gene which is present in suppression of many light-controlling signalling pathways (Davuluri *et al.*, 2005). In transgenic tomatoes with DET1 inhibited, DET1 was particularly destroyed, followed by a rise in flavonoid and carotenoid levels. The amino acid abscisic acid (ABA) take part in maturing of tomato fruits. RNAi repressed the SlNCED1 gene in tomato, which encodes 9-cis-epoxycarotenoid dioxygenase, an essential enzymatic component in ABA production. Fruit from RNAi lines had higher levels of upstream chemicals in the pathway, particularly lycopene and carotene. By expression down-regulation of the lycopene epsilon cyclase (-CYC) gene, RNAi have been used to boost the carotenoid concentration of rapeseed (*Brassica napus*). Seeds from transgenic Brassica plants have higher levels of zeaxanthin, violaxanthin, carotene and lutein (Yu *et al.*, 2007).

5.2.2 Improved Storage Life

Vegetable and fruit post-harvest degradation and spoiling are the main reasons of financial damage. As a result, increasing the storage period of fruits and vegetables by delaying maturing is another agriculturally important feature that RNAi/microRNA technology is addressing. The onset of maturing in climacteric fruits, such as tomatoes, is marked by a climacteric surge of ethylene, which causes expression of ripening-specific genes to be regulated (Osorio *et al.*, 2011). Therefore, in climacteric fruits, prolonged maturing could be done by gene manipulation which is implicated in ethylene synthesis and detection using antisense technology (Hamilton *et al.*, 1990; Oeller *et al.*, 1991; Theologis *et al.*, 1993; Ye *et al.*, 1996; Wilkinson *et al.*, 1997; Xiong *et al.*, 2005). An RNAi method that targets the repression of multiple homologs will be far more efficient than knocking down a one-homolog employed RNAi method to generate tomatoes along with prolonged maturing by aiming three homologs of the 1-aminocyclopropane-1-carboxylate (ACC) synthase (ACS) gene (Gupta *et al.*, 2013). miRNA also takes part in developing and maturation of tomatoes. miR156 is a gene that aims a crucial gene which helps the fruit to ripen. Fruits which do not have colour do not ripen (Moxon *et al.*, 2008; Molesini *et al.*, 2012). The colour of transgenic tomato plants overexpressing sly-miR156 was somewhat paler as compared to wild-type control, although they can ultimately ripen. In addition, excessive expression of miR156 in tomato led to small fruits, elevated number of leaves, but reduction in the height and lessened leaf size, suggesting that more the yield of the fruits and less maturing period of tomatoes can be anticipated by reduction of miR156 expression or liberating one or more miR156 targeting genes (Zhang *et al.*, 2011b).

5.2.3 Resistance Towards Biotic Stress Factors

5.2.3.1 Virus resistance

Virus-induced gene silencing (VIGS) is a technique that gives protection to plants from unknown invading genes (Beclin *et al.*, 2002; Ding, 2010). In some plants, VIGS has developed as a highly effective functional genomics method for knocking out gene expression of specific plant genes. Idea of pathogen-derived resistance (PDR) has sparked interest in genetically engineering plants that are virus-resistant (Simon-Mateo and García, 2011). PDR can be protein-modulated, including transgene-encoded proteins, or RNA-modulated, involving the transgene's transcript. Hairpin dsRNA, such as tiny hairpin RNA (shRNA), self-complementary hpRNA, and intron-spliced hpRNA, was generated *in vivo* utilising contrary repetition of sequences from viral genomes in a way to accomplish PDR. Amongst them, the approach utilising self-complementary hairpin RNAs divided by an intron evoked the most efficient PTGS. Also in the existence of inverted repeats of dsRNA-induced PTGS, plants have shown strong viral resistance (Smith *et al.*, 2000; Wesley *et al.*, 2001; Beclin *et al.*, 2002; Pandolfini *et al.*, 2003; Zrachya *et al.*, 2007). Resistance towards virus with the help of engineering is done with success in plants by aiming the coat protein gene, also known as CP via RNA interference. TMV resistance was found in transgenic tobacco, expressing the tobacco mosaic virus (TMV) CP gene, and the resilience was attributed to the expressed CP. Afterwards, this approach was expanded to produce resistance to a variety of viruses, including potato resistance to potato virus Y (Missiou *et al.*, 2004), tobacco resistance to beet necrotic yellow vein virus (Andika *et al.*, 2005), cucumis cv. melo resistance to papaya ring spot virus type W (Krubphachaya *et al.*, 2007), N (Hily *et al.*, 2007). The usage of amiRNAs, which target viral genes which take part in replication, transmission and symptom formation after viral infection, is the other potential technique for reducing virus proliferation and spread in the plant.

370   *Plant MicroRNAs and Stress Response*

Vu *et al.* (2013) employed two amiRNAs to target the centre area of the AV1 (coat protein) transcript (amiR-AV1-3) and the overlapping area of AV1 and AV2 (pre-coat protein) transcripts (amiR-AV1-1) (ToLCV). ToLCNDV tolerance was great in transgenic tomato plants expressing amiR-AV1-1, which allowed them to grow till the T2 stage.

5.2.3.2 Resistance against bacteria

To manage the diseases caused by bacteria is very challenging as the spread is fast. Repression of 2 *Agrobacterium tumefaciens* genes taking part in crown gall tumour development (iaaM and ipt) via RNA interference can drastically decrease tumour creation in *Arabidopsis* (Escobar *et al.*, 2001; Dunoyer *et al.*, 2006). This technique can be used for various plants as well. Fatty acid as well as their derive products are crucial molecules for signalling that have been known to have negative regulation in plants' resistance towards disease caused by bacteria (Li *et al.*, 2008a; Jiang *et al.*, 2009). RNAi-modulated repression of the stearoyl-acyl carrier protein desaturase gene, which encodes a fatty acid desaturase, resulting in increased tolerance to several diseases in *A. thaliana* and soybean plants (Jiang *et al.*, 2009).

In *A. thaliana*, according to some reports, miR393 suppresses auxin signals by negative regulation of F-box auxin receptors, such as TIR1 and therefore, constricting the infection caused by bacteria *P. syringae*. Overexpression of miR393 in genetically engineered *A. thaliana* plants has improved the tolerance against the bacteria along with some changes in development (Navarro *et al.*, 2006). Two microRNAs, miR398 and miR825 were known to down-regulate the infection caused by bacteria. miR398 aims the coding for CSD1 as well as CSD2 which were investigated. The CSD1 had up-regulation after infection caused by bacteria following the down-regulation of miR398 when under biotic strain (Jagadeeswaran *et al.*, 2009). The suppression of R genes by miR482 is done by viral and bacterial infection. As a result, in reaction to bacterial incursion, these pathogenic-sensitive miRNAs are possibly up- or down-regulated, affecting gene expression by repressing bad regulators and stimulating favourable regulators of immunogenicity. The detection and characterisation of these miRNA targets might aid in the discovery of novel participants in host defence pathways. If such pathogen-regulated miRNAs operate as favourable regulators of bacterial resistance, uncontrolled overexpression of miRNA or amiRNA could be used to create transgenics with improved bacterial resistance. When transgenic plants overexpress miRNAs that operate as detrimental regulators, they grow extra susceptible to bacteria. Up-regulation of their target genes accomplished by overexpression of miRNA-resistant variants of their target or utilising by amiRNA target mimics, may be an efficient technique for enhancing plant stress resistance in these circumstances (Franco-Zorrilla *et al.*, 2007).

5.2.3.3 Tolerance against fungi

Using RNAi to target genes involved in fatty acid metabolism has been an efficient technique for generating disease resistance in an assortment of agricultural plants (Jiang *et al.*, 2009). Also RNAi-mediated silencing of the rice gene OsSSI2 results in increased tolerance towards the blast fungus *Magnaporthe grisea* and the bacterium *Xanthomonas oryzae* that causes leaf blight. Furthermore, the inhibition of two genes, OsFAD7 and OsFAD8, that are 3 fatty acid desaturase, resulted in increased

MicroRNA-based Plant Genetic Engineering for Crop Improvement **371**

disease tolerance against *M. grisea* in rice (Yara *et al.*, 2007). Because of lower lignin concentration, RNAi-mediated targeting of genes participating in lignin formation improved soybean resistance to the plant pathogens *Sclerotinia sclerotiorum* (Peltier *et al.*, 2009). Twenty-four miRNAs were newly discovered to be engaged in wheat responses to the fungus *Blumeria graminis* f. sp. *tritici* (Bgt) that causes the dangerous illness, powdery mildew (Xin *et al.*, 2010). In reaction to the blast fungus *Magnaporthe oryzae*, the rice miRNA osa-miR7695 was known to adversely modulate biological resistance-associated macrophage protein 6 (OsNramp6). Increased expression of Osa-miR7696 resulted in enhanced tolerance to rice blast infection (Campo *et al.*, 2013).

5.2.3.4 Tolerance against insects and nematodes

RNAi is also used to manage insects that cause considerable amounts of crop damage (Huvenne and Smagghe, 2010). Providing dsRNA as a dietary element to insects has been found to effectively down-regulate the targeted genes in insects, which could lead to the production of a fresh breed of insect-resistant crops (Price and Gatehouse, 2008). The technique was utilised in corn plants to generate western corn rootworm (WCR)-tolerant transgenic maize plants by expressing dsRNAs for tubulin or vacuolar ATPase genes (Baum *et al.*, 2007). Plant nematode connections are also thought to be mediated by microRNAs. In *Arabidopsis*, miR161, miR164, miR167a, miR172c, miR396c, miR396a,b, and miR398a were down-regulated in reaction to infection by the worm *H. schachtii* (Hewezi *et al.*, 2008; Khraiwesh *et al.*, 2012). The soybean cyst nematode, the greatest destructive pathogen in soybean, was found to be sensitive to 101 miRNAs corresponding to 40 families in a comparable investigation of miRNA profile in soybean. It was also shown that 20 miRNAs were expressed differently in SCN resistance and susceptibility soybean cultivars (Li *et al.*, 2012).

Furthermore, nematode-induced miR-NAs and sRNAs have been reported to be important in feeding-site formation and parasitism accordingly (Hewezi *et al.*, 2008). Up-regulation of these nematode-induced potential miRNAs, as well as silencing of their respective targets, might reveal new insights into plant-nematode parasitism as well as nematode resistance in agricultural plants. Nematode resistance can even be accomplished by producing amiRNA, which combines recognised miRNA genes with seed area of a plant parasitic nematode's essential gene.

5.3 Environmental Stress Tolerance

5.3.1 Increased and Decreased Temperature Stress

In *A. thaliana* (Sunkar and Zhu, 2004; Liu *et al.*, 2008), rice (Lv *et al.*, 2010) and sugarcane (Lv *et al.*, 2010), the expression of miR319 was observed to vary in reaction to cold stressful conditions (Thiebaut *et al.*, 2011). Additionally, transgenic investigations employing wild-type plants as standards revealed that overexpression of the Osa-miR319 gene resulted in improved colder stress responses in plants after freezing acclimatisation (Yang *et al.*, 2013). MiR319 transgenic rice seedlings, on the other hand, showed a significant latency in growth. Dual RNAi lineages for the miR319 targets, OsPCF5 and OsTCP21, were created to circumvent the pleiotropic impact of miR319. These RNAi lineages had superior colder tolerance than wild-type standards after cooling acclimatisation, but have been phenotypically regular. It has been highlighted that plant cold tolerance improved further in the miR319 overexpression lines compared to the OsPCF5 and OsTCP21 RNAi lines (Yang *et al.*, 2013), mostly because of target activity redundancies in the latter instance (Yang *et al.*, 2013). It is also a problem when controlling miRNA targets rather than miRNAs for plant trait alteration. Due to certain circumstances, many targets must be down-regulated at the right time to attain the exact level of impact as overexpression of a single miRNA gene (Yang *et al.*, 2013). Guan *et al.* (2013) found a unique plant with temperature tolerance mode of action particularly for protecting the organs of reproduction. It shows inductance of miR398 for down-regulation if CSD1 and 2 and also CCS (Guan *et al.*, 2013). They discovered that CSD1 and 2 as well as CCS mutants were more resistant to heat stress than wild-type plants with increase in the levels of heat stress transcription factors and heat shock proteins and less floral harm. These findings clearly imply that altering miR398 or its targets could be a useful technique for improving heat tolerance in crop species, particularly corn, which experiences destruction of its reproductive tissues as a consequence of extended exposure to elevated summer temperatures.

5.3.2 Oxidation Stress Resistance

Plants ought to deposit ROS (reactive oxygen species) in reaction to the stimulus from the environment (Khraiwesh *et al.*, 2012). Plants have superoxide dismutases (SODs), which modify superoxide radicals to molecular oxygen and hydrogen peroxide. Increased expression of ROS (Cu/Zn-SODs) in transgenic crops, that detoxify superoxide radicals, has been used in numerous works to enhance crop stress resistance (Tepperman and Dunsmuir, 1990; Pitcher *et al.*, 1991; Gupta *et al.*, 1993; Perl *et al.*, 1993; Sunkar *et al.*, 2006). Association amongst miR398 and its targets (CSD1 and 2) also were investigated to explain why prior efforts failed. MiR398-mediated gene regulation had a deleterious effect on the expression of inserted SOD transgenes with miR398 target sites.

References

Achard, P., Herr, A., Baulcombe, D.C. and Harberd, N.P. (2004). Modulation of floral development by a gibberellin-regulated microRNA, *Development*, 131: 3357-3365.
Alvarez, J.P., Pekker, I., Goldshmidt, A., Blum, E., Amsellem, Z. and Eshed, Y. (2006). Endogenous and synthetic microRNAs stimulate simultaneous, efficient and localised regulation of multiple targets in diverse species, *Plant Cell*, 18: 1134-1151.
Andika, I.B., Kondo, H. and Tamada, T. (2005). Evidence that RNA silencing-mediated resistance to beet necrotic yellow vein virus is less effective in roots than in leaves, *Mol. Plant Microbe Interact.*, 18: 194-204.
Babb, S. and Muehlbauer, G.J. (2003). Genetic and morphological characterisation of the barley uniculm2 (cul2) mutant, *Theor. Appl. Genet.*, 106: 846-857.
Baker, C.C., Sieber, P., Wellmer, F. and Meyerowitz, E.M. (2005). The early extra petals1 mutant uncovers a role for microRNA miR164c in regulating petal number in *Arabidopsis*, *Curr Biol.*, 15: 303-315.
Baum, J.A., Bogaert, T., Clinton, W., Heck, G.R., Feldmann, P., Ilagan, O., Johnson, S., Plaetinck, G., Munyikwa, T., Pleau, M. and Vaughn, T. (2007). Control of colcopteran insect pests through RNA interference, *Nat. Biotechnol.*, 25: 1322-1326.
Beclin, C., Boutet, S., Waterhouse, P. and Vaucheret, H. (2002). A branched pathway for transgene-induced RNA silencing in plants, *Curr. Biol.*, 12: 684-688.

Boualem, A., Laporte, P., Jovanovic, M., Laffont, C., Plet, J., Combier, J.P., Niebel, A., Crespi, M. and Frugier, F. (2008). MicroRNA166 controls root and nodule development in *Medicago truncatula, Plant J.*, 54: 876-887.

Brown, M.E. and Funk, C.C. (2008). Climate: Food security under climate change, *Science*, 319: 580-581.

Campo, S., Peris-Peris, C., Sire, C., Moreno, A.B., Donaire, L., Zytnicki, Notredame, C., Llave, C. and San Segundo, B. (2013). Identification of a novel microRNA (miRNA) from rice that targets an alternatively spliced transcript of the Nramp6 (natural resistance-associated macrophage protein 6) gene involved in pathogen resistance, *New Phytol.*, 199: 212-227.

Chen, J., Zhao, Z., Li, Y., Li, T., Zhu, Y., Yang, X.M., Zhou, S.X., Wang, H., Zhao, J.Q., Pu, M. and Feng, H. (2021b). Fine-tuning roles of Osa-miR159a in rice immunity against *Magnaporthe oryzae* and development, *Rice*, 14: 26.

Chen, F. and Dixon, R.A. (2007). Lignin modification improves fermentable sugar yields for biofuel production, *Nat. Biotechnol.*, 25: 759-761.

Chuck, G.S., Tobias, C., Sun, L., Kraemer, F., Li, C., Dibble, D., Arora, R., Bragg, J.N., Vogel, J.P., Singh, S. and Simmons, B.A. (2011). Overexpression of the maize corn grass1 microRNA prevents flowering, improves digestibility, and increases starch content of switch grass, *Proc. Natl. Acad. Sci.*, USA, 108: 17550-17555.

Davuluri, G.R., Tuinen, A., Fraser, P.D., Manfredonia, A., Newman, R., Burgess, D., Brummell, D.A., King, S.R., Palys, J., Uhlig, J. and Bramley, P.M. (2005). Fruit-specific RNAi-mediated suppression of DET1 enhances carotenoid and flavonoid content in tomatoes, *Nat. Biotechnol.*, 23: 890-895.

Ding, S.W. (2010). RNA-based antiviral immunity, *Nat. Rev. Immunol.*, 10: 632-644.

Duan, C.G., Wang, C.H., Fang, R.X. and Guo, H.S. (2008). Artificial micro RNAs highly accessible to targets confer efficient virus resistance in plants, *J. Virol.*, 82: 11084-11095.

Dunoyer, P., Himber, C. and Voinnet, O. (2006). Induction, suppression and requirement of RNA silencing pathways in virulent *Agrobacterium tumefaciens* infections, *Nat. Genet.*, 38: 258-263.

Efroni, I., Blum, E., Goldshmidt, A. and Eshed, Y. (2008). A protracted and dynamic maturation schedule underlies *Arabidopsis* leaf development, *Plant Cell*, 20: 2293-2306.

Elbashir, S.M., Lendeckel, W. and Tuschl, T. (2001). RNA interference is mediated by 21- and 22-nucleotide RNAs, *Gene. Dev.*, 15: 188-200.

Escobar, M.A., Civerolo, E.L., Summerfelt, K.R. and Dandekar, A.M. (2001). RNAi-mediated oncogene silencing confers resistance to crown *Gall tumorigenesis, Proc. Natl. Acad. Sci.*, USA, 98: 13437-13442.

Feldmann, K.A. (2006). Steroid regulation improves crop yield, *Nat. Biotechnol.*, 24: 46-47.

Franco-Zorrilla, J.M., Valli, A., Todesco, M., Mateos, I., Puga, M.I., Rubio-Somoza, I., Leyva, A., Weigel, D., García, J.A. and Paz-Ares, J. (2007). Target mimicry provides a new mechanism for regulation of microRNA activity, *Nat. Genet.*, 39: 1033-1037.

Fu, C., Sunkar, R., Zhou, C., Shen, H., Zhang, J.Y., Matts, J., Wolf, J., Mann, D.G., Stewart Jr, C.N., Tang, Y. and Wang, Z.Y. (2012). Overexpression of miR156 in switch grass (*Panicum virgatum* L.) results in various morphological alterations and leads to improved biomass production, *Plant Biotechnol. J.*, 10: 443-452.

Fukuda, H. (2004). Signals that control plant vascular cell differentiation, *Nat. Rev. Mol. Cell Biol.*, 5: 379-391.

Gao, J., Chen, H., Yang, H., He, Y., Tian, Z. and Li, J. (2018). *A. brassino* steroid responsive miRNA-target module regulates gibberellin biosynthesis and plant development, *New Phytol.*, 220: 488-501.

Goetz, M., Hooper, L.C., Johnson, S.D., Rodrigues, J.C., Vivian-Smith, A. and Koltunow, A.M. (2007). Expression of aberrant forms of auxin response factor8 stimulates parthenocarpy in *Arabidopsis* and tomato, *Plant Physiol.*, 145: 351-366. Doi: 10.1104/pp.107.104174

Guan, Q., Lu, X., Zeng, H., Zhang, Y. and Zhu, J. (2013). Heat stress induction of miR398 triggers a regulatory loop that is critical for thermo tolerance in *Arabidopsis, Plant J.*, 74: 840-851.

Guo, H.S., Xie, Q., Fei, J.F. and Chua, N.H. (2005). MicroRNA directs mRNA cleavage of the transcription factor NAC1 to downregulate auxin signals for *Arabidopsis* lateral root development, *The Plant Cell*, 17: 1376-1386.

Gupta, A.S., Heinen, J.L., Holaday, A.S., Burke, J.J. and Allen, R.D. (1993). Increased resistance to oxidative stress in transgenic plants that overexpress chloroplastic Cu/Zn superoxide dismutase, *Proc. Natl. Acad. Sci.*, USA, 90: 1629-1633.

Gupta, A., Pal, R.K. and Rajama, M.V. (2013). Delayed ripening and improved fruit processing quality in tomato by RNAi-mediated silencing of three homologs of 1-aminopropane-1-carboxylate synthase gene, *J. Plant Physiol.*, 170: 987-995.

Hamilton, A.J., Lycett, G.W. and Grierson, D. (1990). Antisense gene that inhibits synthesis of the hormone ethylene in transgenic plants, *Nature*, 346: 284-287.

Hewezi, T., Howe, P., Maier, T.R. and Baum, T.J. (2008). *Arabidopsis* small RNAs and their targets during cyst nematode parasitism, *Mol. Plant Microbe Interact.*, 21: 1622-1634.

Hily, J.M., Ravelonandro, M., Damsteegt, V., Basset, C., Petri, C., Liu, Z. and Scorza, R. (2007). Plum pox virus coat protein gene intron-hair pin-RNA (ihpRNA) constructs provide resistance to Plum pox virus in *Nicotiana bethamiana* and *Prunus domestica*, *J. Am. Soc. Hort. Sci.*, 132: 850-858.

Huvenne, H. and Smagghe, G. (2010). Mechanisms of dsRNA uptake in insects and potential of RNAi for pest control: A review, *J. Insect. Physiol.*, 56: 227-235.

Jagadeeswaran, G., Saini, A. and Sunkar, R. (2009). Biotic and abiotic stress down-regulate miR398 expression in *Arabidopsis*, *Planta*, 229: 1009-1014.

Jiang, C.J., Shimono, M., Maeda, S., Inoue, H., Mori, M., Hasegawa, M., Sugano, S. and Takatsuji, H. (2009). Suppression of the rice fatty-acid desaturase gene OsSSI2 enhances resistance to blast and leaf blight diseases in rice, *Mol. Plant Microbe Int.*, 22: 820-829.

Jiao, Y., Wang, Y., Xue, D., Wang, J., Yan, M., Liu, Z., Zhu, X. and Qian, Q. (2010). Regulation of OsSPL14 by OsmiR156 defines ideal plant architecture in rice, *Nat. Genet.*, 42: 541-544.

Kang, L., Wang, Y.S., Uppalapati, S.R., Wang, K., Tang, Y., Vadapalli, V., Venables, B.J., Chapman, K.D., Blancaflor, E.B. and Mysore, K.S. (2008). Overexpression of a fatty acid amide hydrolase compromises innate immunity in *Arabidopsis*, *Plant J.*, 56: 336-349.

Khraiwesh, B., Ossowski, S., Weigel, D., Reski, R. and Frank, W. (2008). Specific gene silencing by artificial microRNAs in *Physcomitrella patens*: An alternative to targeted gene knockouts, *Plant Physiol.*, 148: 684-693.

Khraiwesh, B., Zhu, J.K. and Zhu, J. (2012). Role of miRNAs and siRNAs in biotic and abiotic stress responses of plants, *Biochim. Biophys. Acta*, 1819: 137-148.

Kidner, C.A. and Martienssen R.A. (2004). Spatially restricted microRNA directs leaf polarity through Argonaute1, *Nature*, 428: 81-84.

Kim, J., Jung, J.H., Reyes, J.L., Kim, Y.S., Kim, S.Y., Chung, K.S., Kim, J.A., Lee, M., Lee, Y., Narry Kim, V. and Chua, N.H. (2005). MicroRNA-directed cleavage of ATHB15 mRNA regulates vascular development in *Arabidopsis* inflorescence stems, *Plant J.*, 42: 84-94.

Kim, J.H., Woo, H.R., Kim, J., Lim, P.O., Lee, I.C., Choi, S.H., Hwang, D. and Nam, H.G. (2009). Trifurcate feed-forward regulation of age-dependent cell death involving miR164 in *Arabidopsis*, *Science*, 323: 1053-1057.

Krubphachaya, P., Juricek, M. and Kertbundit, S. (2007). Induction of RNA-mediated resistance to papaya ring spot virus type W, *J. Biochem. Mol. Biol.*, 40: 401-411.

Laufs, P., Peaucelle, A., Morin, H. and Traas J. (2004). MicroRNA regulation of the CUC genes is required for boundary size control in *Arabidopsis* meristems, *Development*, 131: 4311-4322.

Lauter, N., Kampani, A., Carlson, S., Goebel, M. and Moose, S.P. (2005). MicroRNA172 down-regulates glossy15 to promote vegetative phase change in maize, *Proc. Natl. Acad. Sci.*, USA, 102: 9412-9417.

Li, X., Wang, X., Zhang, S., Liu, D., Duan, Y. and Dong, W. (2012). Identification of soybean microRNAs involved in soybean cyst nematode infection by deep sequencing, *PLoS ONE*, 7: e39650.

MicroRNA-based Plant Genetic Engineering for Crop Improvement **375**

Liu, H.H., Tian, X., Li, Y.J., Wu, C.A. and Zheng, C.C. (2008). Microarray-based analysis of stress-regulated microRNAs in *Arabidopsis thaliana*, RNA, 14: 836-843.

Lv, D.K., Bai, X., Li, Y., Ding, X.D., Ge, Y., Cai, H., Ji, W., Wu, N. and Zhu, Y.M. (2010). Profiling of cold-stress-responsive miRNAs in rice by microarrays, *Gene*, 459: 39-47.

Mallory, A.C., Dugas, D.V., Bartel, D.P. and Bartel, B. (2004a). MicroRNA regulation of NAC-domain targets is required for proper formation and separation of adjacent embryonic, vegetative, and floral organs, *Curr. Biol.*, 14: 1035-1046.

Mallory, A.C., Reinhart, B.J., Jones-Rhoades, M.W., Tang, G., Zamore, P.D., Barton, M.K. and Bartel, D.P. (2004b). MicroRNA control of PHABULOSA in leaf development: Importance of pairing to the microRNA 5′ region, *EMBO J.*, 23: 3356-3364.

Missiou, A., Kalantidis, K., Boutla, A., Tzortzakaki, S., Tabler, M. and Tsagris, M. (2004). Generation of transgenic potato plants highly resistant to potato virus Y (PVY) through RNA silencing, *Mol. Breed.*, 14: 185-197.

Miura, K., Ikeda, M., Matsubara, A., Song, X.J., Ito, M., Asano, K., Matsuoka, M., Kitano, H. and Ashikari, M. (2010). OsSPL14 promotes panicle branching and higher grain productivity in rice, *Nat. Genet.*, 42: 545-549.

Molesini, B., Pii, Y. and Pandolfini, T. (2012). Fruit improvement using intragenesis and artificial microRNA, *Trends Biotechnol.*, 30: 80-88.

Molnar, A., Bassett, A., Thuenemann, E., Schwach, F., Karkare, S., Ossowski, S., Weigel, D. and Baulcombe, D. (2009). Highly specific gene silencing by artificial microRNAs in the unicellular alga *Chlamydomonas reinhardtii*, *Plant J.*, 58: 165-174.

Moxon, S., Jing, R., Szittya, G., Schwach, F., Pilcher, R.L.R., Moulton, V. and Dalmay, T. (2008). Deep sequencing of tomato short RNAs identifies microRNAs targeting genes involved in fruit ripening, *Genome Res.*, 18: 1602-1609.

Navarro, L., Dunoyer, P., Jay, F., Arnold, B., Dharmasiri, N., Estelle, M., Voinnet, O. and Jones, J.D. (2006). A plant miRNA contributes to antibacterial resistance by repressing auxin signalling, *Science*, 312: 436-439.

Ni, Z., Hu, Z., Jiang, Q. and Zhang, H. (2013). GmNFYA3, a target gene of miR169, is a positive regulator of plant tolerance to drought stress, *Plant Mol. Biol.*, 82: 113-129.

Nikovics, K., Blein, T., Peaucelle, A., Ishida, T., Morin, H., Aida, M. and Laufs, P. (2006). The balance between the miR164a and CUC2 genes controls leaf margin serration in *Arabidopsis*, *Plant Cell*, 18: 2929-2945.

Niu, Q.W., Lin, S.S., Reyes, J.L., Chen, K.C., Wu, H.W., Yeh, S.D. and Chua, N.H. (2006). Expression of artificial microRNAs in transgenic *Arabidopsis thaliana* confers virus resistance, *Nat. Biotechnol.*, 24: 1358-1359.

Oeller, P.W., Lu, M.W., Taylor, L.P., Pike, D.A. and Theologis, A. (1991). Reversible inhibition of tomato fruit senescence by antisense RNA, *Science*, 254: 437-439.

Ori, N., Cohen, A.R., Etzioni, A., Brand, A., Yanai, O., Shleizer, S., Menda, N., Amsellem, Z., Efroni, I., Pekker, I. and Alvarez, J.P. (2007). Regulation of Lanceolate by miR319 is required for compound-leaf development in tomato, *Nat. Genet.*, 39: 787-791.

Osorio, S., Alba, R., Damasceno, C.M., Lopez-Casado, G., Lohse, M., Zanor, M.I., Tohge, T., Usadel, B., Rose, J.K., Fei, Z. and Giovannoni, J.J. (2011). Systems biology of tomato fruit development: Combined transcript, protein, and metabolite analysis of tomato transcription factor (nor, rin) and ethylene receptor (Nr) mutants reveals novel regulatory interactions, *Plant Physiol.*, 157: 405-425.

Palatnik, J.F., Allen, E., Wu, X., Schommer, C., Schwab, R., Carrington, J.C. and Weigel, D. (2003). Control of leaf morphogenesis by microRNAs, *Nature*, 425: 257-263.

Pandita, D. (2019). Plant MIRnome: miRNA biogenesis and abiotic stress response. *In:* Hasanuzzaman, M., Hakeem, K., Nahar, K. and Alharby, H. (Eds.). *Plant Abiotic Stress Tolerance.* Springer, Cham, 449-474. Doi https://Doi.org/10.1007/978-3-030-06118-0_18

Pandita, D. (2021). Role of miRNAi technology and miRNAs in abiotic and biotic stress resilience. *In:* Aftab, T. and Roychoudhury, A. (Eds.). *Plant Perspectives to Global Climate Changes.* Academic Press, Elsevier, 303-330. https://Doi.org/10.1016/B978-0-323-85665-2.00015-7

Pandita, D. and Wani, S.H. (2019). MicroRNA as a tool for mitigating abiotic stress in rice (*Oryza sativa* L.). *In:* Wani, S. (Ed.). *Recent Approaches in Omics for Plant Resilience to Climate Change.* Springer, Chamz, 109-133. Doi https://Doi.org/10.1007/978-3-030-21687-0_6

Pandita, D. (2022a). How microRNAs regulate abiotic stress tolerance in wheat? A snapshot. *In:* Roychoudhury, A., Aftab, T. and Acharya, K. (Eds.). *Omics Approach to Manage Abiotic Stress in Cereals.* Springer, Singapore, 447-464. https://Doi.org/10.1007/978-981-19-0140-9_17

Pandita, D. (2022b). MicroRNAs shape the tolerance mechanisms against abiotic stress in maize. *In:* Roychoudhury, A., Aftab, T. and Acharya, K. (Eds.). *Omics Approach to Manage Abiotic Stress in Cereals,* Springer, Singapore, 479-493. https://Doi.org/10.1007/978-981-19-0140-9_19

Pandita, D. (2022c). miRNA- and RNAi-mediated metabolic engineering in plants. *In:* Aftab, T. and Hakeem, K.R. (Eds.). *Metabolic Engineering in Plants.* Springer, Singapore, 171-186. https://Doi.org/10.1007/978-981-16-7262-0_7

Pandolfini, T., Molesini, B., Avesani, L., Spena, A. and Polverari, A. (2003). Expression of self-complementary hairpin RNA under the control of the rolC promoter confers systemic disease resistance to plum pox virus without preventing local infection, *BMC Biotechnol.,* 3: 7.

Peaucelle, A., Morin, H., Traas, J. and Laufs, P. (2007). Plants expressing a miR164-resistant CUC2 gene reveal the importance of post-meristematic maintenance of phyllotaxy in *Arabidopsis, Development,* 2134: 1045-1050.

Peltier, A.J., Hatfield, R.D. and Grau, C.R. (2009). Soybean stem lignin concentration relates to resistance to *Sclerotinia sclerotiorum, Plant Dis.,* 93: 149-154.

Peng, J.R., Richards, D.E., Hartley, N.M., Murphy, G.P., Devos, K.M., Flintham, J.E. Beales, J., Fish, L.J., Worland, A.J., Pelica, F. and Sudhakar, D. (1999). 'Green revolution' genes encode mutant gibberellin response modulators, *Nature,* 400: 256-261.

Peng, T., Qiao, M., Liu, H., Teotia, S., Zhang, Z., Zhao, D., Shi, L., Zhang, C. and Le, B. (2018). A resource for inactivation of microRNAs using short tandem target mimic technology in model and crop plants, *Mol. Plant,* 11: 1400-1417.

Perl, A., Perl-Treves, R., Galili, S., Aviv, D., Shalgi, E., Malkin, S. and Galun, E. (1993). Enhanced oxidative-stress defence in transgenic potato expressing tomato Cu, Zn superoxide dismutases, *Theor. Appl. Genet.,* 85: 568-576.

Pitcher, L.H., Brennan, E., Hurley, A., Dunsmuir, P., Tepperman, J.M. and Zilinskas, B.A. (1991). Overproduction of petunia chloroplastic copper/zinc superoxide dismutase does not confer ozone tolerance in transgenic tobacco, *Plant Physiol.,* 97: 452-455.

Price, D.R.G. and Gatehouse, J.A. (2008). RNAi-mediated crop protection against insects, *Trends Biotechnol.,* 26: 393-400.

Reddy, M.S.S., Chen, F., Shadle, G., Jackson, L., Aljoe, H. and Dixon, R.A. (2005). Targeted down-regulation of cytochrome P450 enzymes for forage quality improvement in alfalfa (*Medicago sativa* L.), *Proc. Natl. Acad. Sci.,* U.S.A., 102: 16573-16578.

Reinhardt, D and Kuhlemeier, C. (2002). Plant architecture, *EMBO Reports,* 3: 846-851.

Rosegrant, M.W and Cline, S.A. (2003). Global food security: Challenges and policies, *Science,* 302: 1917-1919.

Rubinelli, P.M., Chuck, G., Li, X. and Meilan, R. (2013). Constitutive expression of the Corn-grass1 microRNA in poplar affects plant architecture and stem lignin content and composition, *Biomass Bioenerg.,* 54: 312-321.

Schommer, C., Palatnik, J.F., Aggarwal, P., Chételat, A., Cubas, P., Farmer, E.E., Nath, U. and Weigel, D. (2008). Control of jasmonate biosynthesis and senescence by miR319 targets, *PLoS Biology,* 6: e230

Schwab, R., Ossowski, S., Riester, M., Warthmann, N. and Weigel, D. (2006). Highly specific gene silencing by artificial microRNAs in *Arabidopsis, Plant Cell,* 18: 1121-1133.

Schwab, R., Palatnik, J., Riester, M., Schommer, C., Schmid, M. and Weigel, D. (2005). Specific effects of microRNAs on the plant transcriptome, *Dev. Cell,* 8: 517-527.

Schwarz, S., Grande, A.V., Bujdoso, N., Saedler, H. and Huijser, P. (2008). The microRNA regulated SBP-box genes SPL9 and SPL15 control shoot maturation in *Arabidopsis*, *Plant. Mol. Biol.*, 67: 183-195.

Simon-Mateo, C. and García, J.A. (2011). Antiviral strategies in plants based on RNA silencing, *Biochim. Biophys. Acta*, 1809: 722-731.

Smith, N.A., Singh, S.P., Wang, M.B., Stoutjesdijk, P.A. Green, A.G. and Waterhouse, P.M. (2000). Total silencing by intron-spliced hairpin RNAs, *Nature*, 407: 319-320.

Sunkar, R. and Zhu, J.K. (2004). Novel and stress-regulated micro RNAs and other small RNAs from *Arabidopsis*, *Plant Cell*, 16: 2001-2019.

Sunkar, R., Kapoor, A. and Zhu, J.K. (2006). Posttranscriptional induction of two Cu/Zn superoxide dismutase genes in *Arabidopsis* is mediated by down-regulation of miR398 and important for oxidative stress tolerance, *Plant Cell*, 18: 2051-2065.

Takeda, S. and Matsuoka, M. (2008). Genetic approaches to crop improvement: Responding to environmental and population changes, *Nat. Rev. Genet.*, 9: 444-457.

Tepperman, J.M. and Dunsmuir, P. (1990). Transformed plants with elevated levels of chloroplastic SOD are not more resistant to superoxide toxicity, *Plant Mol. Biol.*, 14: 501-511.

Theologis, A., Oeller, P.W., Wong, L.M., Rottmann, W.H. and Gantz, D.M. (1993). Use of a tomato mutant constructed with reverse genetics to study fruit ripening, a complex developmental process, *Dev. Gen.*, 14: 282-295.

Thiebaut, F., Rojas, C.A., Almeida, K.L., Grativol, C., Domiciano, G.C., Lamb, C.R.C., de Almeida Engler, J., Hemerly, A.S. and Ferreira, P.C. (2011). Regulation of miR319 during cold stress in sugarcane, *Plant Cell Environ.*, 35: 502-512.

Timmermans, M.C.P., Juarez, M.T. and Phelps-Durr, T.L. (2004). A conserved microRNA signal specifies leaf polarity, *Cold Spring Harb. Symp. Quant. Biol.*, 69: 409-417.

Tretter, E.M., Alvarez, J.P., Eshed, Y. and Bowman, J.L. (2008). Activity range of *Arabidopsis* small RNAs derived from different biogenesis pathways, *Plant Physiol.*, 147: 58-62.

Vu, T.V., Choudhury, N.R. and Mukherjee, S.K. (2013). Transgenic tomato plants expressing artificial microRNAs for silencing the pre-coat and coat proteins of a begomovirus, tomato leaf curl New Delhi virus, show tolerance to virus infection, *Virus Res.*, 172: 35-45.

Wang, J.W., Czech, B. and Weigel, D. (2009). miR156-regulated SPL transcription factors define an endogenous flowering pathway in *Arabidopsis thaliana*, *Cell*, 138: 738-749.

Wang, Y. and Li, J. (2008). Molecular basis of plant architecture, *Annu. Rev. Plant Biol.*, 59: 253-279.

Wang, S., Wu, K., Yuan, Q., Liu, X., Liu, Z., Lin, X., Zeng, R., Zhu, H., Dong, G., Qian, Q. and Zhang, G. (2012). Control of grain size, shape and quality by OsSPL16 in rice, *Nat. Genet.*, 44: 950-954.

Warthmann, N., Chen, H., Ossowski, S., Weigel, D. and Herve, P. (2008). Highly specific gene silencing by artificial miRNAs in rice, *PLoS ONE*, 3: e1829.

Wesley, S.V., Helliwell, C.A., Smith, N.A., Wang, M.B., Rouse, D.T., Liu, Q., Gooding, P.S., Singh, S.P., Abbott, D., Stoutjesdijk, P.A. and Robinson, S.P. (2001). Construct design for efficient, effective and high throughput gene silencing in plants, *Plant J.*, 27: 581-590.

Wilkinson, J.Q., Lanahan, M.B., Clark, D.G., Bleecker, A.B., Chang, C., Meyerowitz, E.M. and Klee, H.J. (1997). A dominant mutant receptor from *Arabidopsis* confers ethylene insensitivity in heterologous plants, *Nat. Biotechnol.*, 15: 444-447.

Williams, L., Grigg, S.P., Xie, M., Christensen, S. and Fletcher, J.C. (2005). Regulation of *Arabidopsis* shoot apical meristem and lateral organ formation by microRNA miR166g and its ATHD-ZIP target genes, *Development*, 132: 3657-3668.

Wu, G., Park, M.Y., Conway, S.R., Wang, J.W., Weigel, D. and Poethig, R.S. (2009). The sequential action of miR156 and miR172 regulates developmental timing in *Arabidopsis*, *Cell*, 138: 750-759.

Wu, G. and Poethig, R.S. (2006). Temporal regulation of shoot development in *Arabidopsis thaliana* by miR156 and its target SPL3, *Development*, 133: 3539-3547.

Xia, K., Zeng, X., Jiao, Z., Li, M., Xu, W., Nong, Q., Mo, H., Cheng, T. and Zhang, M. (2017). Formation of protein disulfide bonds catalysed by OsPDIL1, 1 is mediated by microRNA5144-3p in rice, *Plant Cell Physiol.*, 59: 331-342.

Xie, K., Shen, J., Hou, X., Yao, J., Li, X., Xiao, J. and Xiong, L. (2012). Gradual increase of miR156 regulates temporal expression changes of numerous genes during leaf development in rice, *Plant Physiol.*, 158: 1382-1394.

Xin, M., Wang, Y., Yao, Y., Xie, C., Peng, H., Ni, Z. and Sun, Q. (2010). Diverse set of microRNAs are responsive to powdery mildew infection and heat stress in wheat (*Triticum aestivum* L.), *BMC Plant Biol.*, 10: 123-134.

Xiong, A.S., Yao, Q.H., Peng, R.H., Li, X., Han, P.L. and Fan, H.Q. (2005). Different effects on ACC oxidase gene silencing triggered by RNA interference in transgenic tomato, *Plant Cell Rep.*, 23: 639-646.

Yang, C., Li, D., Mao, D., Liu, X., Li, C., Li, X., Zhao, X., Cheng, Z., Chen, C. and Zhu, L. (2013). Overexpression of microRNA319 impacts leaf morphogenesis and leads to enhanced cold tolerance in rice (*Oryza sativa* L.), *Plant Cell Environ.*, 36: 2207-2218.

Yara, A., Yaeno, T., Hasegawa, M., Seto, H., Montillet, J.L., Kusumi, K., Seo, S. and Iba, K. (2007). Disease resistance against *Magnaporthe grisea* is enhanced in transgenic rice with suppression of o-3 fatty acid desaturases, *Plant Cell Physiol.*, 48: 1263-1274.

Ye, Z.B., Li, H.X., Zheng, Y.L. and Liu, H.L. (1996). Inhibition of introducing antisense ACC oxidase gene into tomato genome on expression of its endogeous gene, *J. Huazhong Agric. Univ.*, 15: 305-309.

Yu, B., Lydiate, D.J., Young, L.W., Schafer, U.A. and Hannoufa, A. (2007). Enhancing the carotenoid content of *Brassica napus* seeds by down-regulating lycopene epsilon cyclase, *Transgenic Res.*, 17: 573-585.

Zhang, H., Zhang, J., Yan, J., Gou, F., Mao, Y., Tang, G., Botella, J.R. and Zhu, J.K. (2017). Short tandem target mimic rice lines uncover functions of miRNAs in regulating important agronomic traits, *Proc. Natl. Acad. Sci., U.S.A.*, 114: 5277-5282.

Zhang, J., Zhang, H., Srivastava, A.K., Pan, Y., Bai, J., Fang, J., Shi, H. and Zhu, J.K. (2018b). Knockdown of rice microRNA166 confers drought resistance by causing leaf rolling and altering stem xylem development, *Plant Physiol.*, 176: 2082-2094.

Zhang, X., Zou, Z., Zhang, J., Zhang, Y., Han, Q., Hu, T., Xu, X., Liu, H., Li, H. and Ye, Z. (2011b). Over-expression of sly-miR156a in tomato results in multiple vegetative and reproductive trait alterations and partial phenocopy of the sft mutant, *FEBS Lett.*, 585: 435-439.

Zhang, Y.C., Yu, Y., Wang, C.Y., Li, Z.Y., Liu, Q., Xu, J., Liao, J.Y., Wang, X.J., Qu, L.H., Chen, F. and Xin, P. (2013). Overexpression of microRNA OsmiR397 improves rice yield by increasing grain size and promoting panicle branching, *Nat. Biotechnol.*, 31: 848-852.

Zhang, T., Wang, W., Bai, X. and Qi, Y.J. (2009). Gene silencing by artificial microRNAs in *Chlamydomonas*, *Plant J.*, 58: 157-164.

Zhao, Y., Wen, H., Teotia, S., Du, Y., Zhang, J., Li, J., Sun, H., Tang, G., Peng, T. and Zhao, Q. (2017). Suppression of microRNA159 impacts multiple agronomic traits in rice (*Oryza sativa* L.), *BMC Plant Biol.*, 17: 215.

Zhao, Y.F., Peng, T., Sun, H.Z., Teotia, S., Wen, H.L., Du, Y.-X. *et al.* (2019). miR1432-OsACOT (acyl-CoA thioesterase) module determines grain yield via enhancing grain filling rate in rice, *Plant Biotechnol. J.*, 17: 712-723.

Zhou, G.K., Kubo, M., Zhong, R., Demura, T. and Ye, Z.H. (2007). Overexpression of miR165 affects apical meristem formation, organ polarity establishment and vascular development in *Arabidopsis*, *Plant Cell Physiol.*, 48: 391-404.

Zhou *et al.* (2005).

Zrachya, A., Glick, E., Levy, Y., Arazi, T., Citovsky, V. and Gafni, Y. (2007). Suppressor of RNA silencing encoded by tomato yellow curl virus-Israel, *Virology*, 358: 159-165.

CHAPTER
20

The miRNA-encoded Peptides

Pooja Bhadrecha[1]*, Shilpy Singh[2] and Arun Kumar[3]

[1] University Institute of Biotechnology, Chandigarh University, Punjab
[2] Department of Biotechnology, Noida International University, Uttar Pradesh
[3] Department of Agriculture Biotech, Sardar Vallabhbhai Patel University of Agriculture and Technology, Uttar Pradesh

1. Introduction

Peptides can be produced by either directly translating short open reading frames (ORFs) or assimilating precursor proteins within genomes. Non-coding RNAs (ncRNAs), like circular RNAs, long ncRNAs and major transcript RNAs, have been found to play a major role in the production of peptides (Ren *et al.*, 2021). Sequences that gave way to primary miRNA (pri-miRNA) transcripts can be found in the intergenic and intragenic sites of DNA (Jiu *et al.*, 2015). These pri-miRNAs are produced from miRNA genes and converted to mature miRNAs. The pri-miRNAs are short non-coding components involved in growth, development and stress responses (Kapusta and Feschotte, 2014). miRNAs were assumed to be derived from shattered genes or genes that had undergone inverted duplications, resulting in hair-pin-structured RNAs. Due to introns in pri-miRNA, mature miRNAs are observed to be produced more frequently (Bielewicz *et al.*, 2013; Schwab *et al.*, 2013). The regions upstream and downstream of pri-miRNAs are assumed to have no functional role following the synthesis and extraction of mature miRNAs. Recent studies, however, imply that miRNAs might have coding activities in the form of microRNA-encoded peptides (miPEPs), kicking off a new gene regulatory phase.

miPEPs are non-conventional peptides that are now being used in a number of agronomic studies for crop modification. Non-conventional peptides (NCPs), which are tiny open reading frame-encoded peptides, differ from conventional peptides in several ways. The 3' UTR, 5' UTR, intergenic regions, junctions and introns are all used to make these short peptides (Wang *et al.*, 2020). Both monocotyledons and dicotyledons might benefit from the short peptides, which could support a number of biological functions. The detection of short peptides in diverse crops, including the

---

*Corresponding author: pbhadrecha.pb@gmail.com

model plant *A. thaliana*, demonstrated that the genome may be transformed into active biomolecules for use in functional genomics research. Environmental reactions, plant growth and molecular modulation have benefited by functional NCPs incorporating miPEPs (Wang *et al.*, 2020). Due to the limitations of restricted identification platforms for such NCPs, such as peptidomic technologies and genomic annotation, knowledge of their real and fascinating properties in plants are hampered and they are frequently overlooked for future annotation (Yin *et al.*, 2019). However, in monocotyledons and dicotyledonous species, an integrative peptidogenomic pathway for facile detection of non-conventional short peptides has recently been created and utilised (Wang *et al.*, 2020). NCPs were identified in maize genome at a rate of about 70.3% among monocots. The forward strand produced more NCPs, with a majority of them coming from intergenic sequences. *Arabidopsis* was used to provide data on NCPs among dicots and research found that peptide length was shorter than typical peptides. Most of NCPs were produced from intergenic sequences on the reverse strand (Wang *et al.*, 2020). In this chapter, we look at miRNAs that actually encode peptides and discuss their roles and biological significance.

2. Biogenesis of miRNAs and miPEPS

As per the studies conducted earlier, it was reported that the microRNA genes must have evolved via processes, like induced inverted gene duplications, spontaneous evolution and derivation of transposable element. RNA PolII processes pri-miRNAs derived from microRNA genes. Dicer-like 1 (DCL1), together with some more regulatory enzymes as well as proteins, converts pri-miRNA to pre-miRNA (precursor miRNA) and then processes it into mature miRNAs. The mature miRNAs are the end product of biosynthesis, including transcription, whereas miPEPs are produced as a result of the action of tiny ORFs of pri-miRNA conclusively in the translational stage. Small ORFs in the 5′ upstream region of the coding section of these pri-miRNAs are intended for translational regulation predominantly in the cis regions. The key function of short peptides transcribed via ORFs which did not translate (uORFs) among coding section trans-regions was previously unclear (Combier *et al.*, 2008). Lauressergues *et al.* (2015) explained the relationship between short ORFs and miPEPs expressed by these ORFs. They discovered that in *A. thaliana* and *M. truncatula*, small ORFs of pri-miRNAs transcribed functional miPEPs. These miORFs, found at the 5′ end of pri-miRNA, are later translated to miPEPs that may be from 4 to 60 amino acids long. Translational control is exercised by tiny ORFs located upstream to the stem-loop segment in the pri-miRNAs. In the case of plants, a minimum of five miPEPs get translated extemporaneously, according to immunoblots utilising particular antibodies (Lauressergues *et al.*, 2015). Furthermore, regulation and expression of such miPEPs was found to be linked to miRNA expression. The study recorded that even the synthetic-kind of peptides could aid in aggregation of related miRNAs, eventually resulting in down-regulation of specific genes. Hence these peptides were found to be biologically active. Studies on *M. truncatula*, *A. thaliana*, *Glycine max* and *V. vinifera* revealed the association of miPEPs generated via matching pri-miRNAs in negative gene regulation. Almost 50 pri-miRNAs examined in *Arabidopsis* were each found to have a minimum of one putative, short ORF spanning from 3-59 amino acids (Lauressergues *et al.*, 2015).

The miRNA-encoded Peptides

3. miPEPs: The New Technology for Functional Genomics

Generally, peptides may interfere with signalling cascade and acting as mediator for cell-to-cell transmission, but the exact process that is involved remains unclear (Qu *et al.*, 2015). In another study conducted on *Arabidopsis*, it was observed that exogenously applied miPEPs, which were encoded by pri-miRNAs, like miPEP 164a, 165a and 319a, to promote the plant's growth, showed positive effects like longer stems, higher flower stalks, early inflorescence and increase in the number of flowers in comparison to untreated plants (Combier *et al.*, 2020a). Autonomous treatments of individual plant portions resulted in more favourable outcomes. Because the baseline amount of miPEPs already occur in Nature and was administered exogenously, the total amount of miPEP rose, influencing the aggregation of related miRNAs. In *Arabidopsis*, up-regulation of miPEP319a resulted in the amplification of miR319a. Later, GST tags were used in expression vector to clone the sequence which encode for the recombinant peptide, followed by testing of these recombinant peptides in *E. coli* as well. The reporter gene for GUS, immunoblotting and then the overexpression of miPEP165a and 171b were used for the experiments to determine the endogenous existence of miPEPs. Exogenous administration of miPEP165a to *Arabidopsis thaliana* improved primary root length. The administration of MiPEP165a to roots significantly boosted cell proliferation inside the meristematic tissue, showing cell elongation and division as root length increased. When miPEP165a was administered to the shoot apical meristems, the blooming day was shortened as the length of inflorescence stems increased. Ormancey *et al.* (2020), in a fascinating study, discovered that miPEPs are unable to travel across the plant. The same was demonstrated when 10 M miPEP165a was sprayed on the plant roots, which had no influence on blooming.

Exogenous administration followed by enhanced expression of miPEP171b in *M. truncatula* resulted in enhanced expression of the respective and corresponding miRNA171b as well as suppressing the related target, indicating that it plays an important role in root formation (Lv *et al.*, 2016; Lauressergues *et al.*, 2015). It was recorded that lateral roots were reduced in number and HAM genes were down-regulated as a result of overexpression. Similarly, overexpression of miPEP171b1 resulted in the aggregation of MtmiR171b, resulting in negative regulation of specific genes. In *Nicotiana benthamiana*, overexpression of MtmiR171b and MtmiPEP171b1 had comparable outcomes. The translation impact of MtmiORF171b to MtmiPEP171b1 leads to heightened accumulation of MtmiR171b, which was reliant on translational impact of MtmiORF171b to MtmiPEP171b1. MtmiPEP171b1 overexpression and also its exogenous requirements led to in increased accumulation of relevant pri-miRNAs, pre-miRNAs as well as mature miRNAs. MtmiPEP171b1 with GFP indicated that miPEPs can be found inside the nuclear bodies. Even when the ATG start codon of primiR171b's, ORF1 was replaced with an ATT and the quantity of miR171b generated was reduced, suggesting that miPEP171b works as the regulatory mediator, boosting aggregation of matching miRNA (Combier *et al.*, 2020b).

The exogenous administration of miPEP172c caused miR172c to be stimulated, resulting in the development of nodules in soybeans (Couzigou *et al.*, 2016). miR172c was overexpressed which caused increased nodulation in soybeans, whereas lower activity of miR172c caused a decrease in nodules number (Wang *et al.*, 2014). As unmodified synthetic miPEP172c was administered to *G. max* roots, nodule counts increased significantly as compared with the roots treated with scrambled miPEP172c. As an outcome of the findings, miPEP172c treatments of soybean mirrored overexpressing miR172c, followed by controlled symbiosis behaviour at both the morphological as well as biological levels (Couzigou *et al.*, 2016). When compared to untreated species, miPEP172c treated species had greater levels of ENOD40-1, NIN, NSP1 and Hb2 – all of which play critical functions in plant development and plant-microbe interactions. Exogenous administration of miPEP167c to soybeans generated by *Bradirhizobium japonicum* lead to higher nodulation, accumulation of its mediated miR167c, followed by targeting at least a single nodule-modifying gene (Combier and Andre, 2021). Root tissue profile of the small RNA of chickpea was done in conditions of water stress, with both the presence as well as absence of a plant growth-promoting rhizobacteria (PGPR) *P. putida* RA. The study recorded major changes while plotting miRNA expression profiles and the target genes, stating that miRNAs play an important role in interactions among the plant and PGPR (Jatan *et al.*, 2019). Discovering miPEPs in chickpea might be aided by analysis of upstream sequences of some significantly elevated miRNAs.

The flavonoids pathway was regulated by vvi-miPEP396a in grape berry fruit cells. This miPEP targets the MYB5 gene in *Vitis vinifera*, causing activation of the stilbene pathway and suppressing pathways which synthesise flavonoid, resulting in a decrease in overall phenolic content in fruits (Rodrigues, 2019). vvi-miPEP171d1 have been shown to increase adventitious root growth while suppressing main root growth in *V. vinifera* when given exogenously and highly expressed. The synthetic orthologs of vvi-miPEP171d1 in *A. thaliana* caused no morphological alterations in *A. thaliana* roots when administered exogenously and highly expressed (Chen *et al.*, 2020; Julkowska, 2020). miPEPs are very diverse in plant species, whereas the pri-miRNAs processed for generating mature miRNAs that are largely conserved throughout the Plantae kingdom, with the exception of miPEP165a, reported conserved across *Brassicales* (Lauressergues *et al.*, 2015). Morozov *et al.* (2019) published recent data on bioinformatics study on miPEP156a in *Brassica rapa*, demonstrating that the conserved property of miPEP156a *Brassicaceae* family is evolutionary. The functional characteristics of peptides can be changed by post-translational changes, according to the findings. MiPEPs functional analysis has also been investigated using CRISPR-based genome-editing technology. The knock-out mutants of miPEP858a in *Arabidopsis* have made it possible for researchers to work on CRISPR-based miPEPs editing in plants. Pri-miR858a, which encodes miPEP858a, controls plant growth and survival involved in phenyl-propanoids cascade. With a reduction in specific gene, miPEP858a controls the transcription of pri-miR858a (Sharma *et al.*, 2020). Higher expression levels and possibility of CRISPR-based editing of miPEP858a lines resulted in phenotypic and flavonoid levels being altered. The reduction in root length was discovered, using CRISPR-Cas9 edited plants. These findings also demonstrated that the alterations in transgenic progenies were equivalent to the changes in mature miR858a expression in plants that had been edited or overexpressed. In comparison to miR858a overexpressed lines, an optimal concentration of 0.25 molar improved the length of roots with exogenous therapy of this specific miPEP. The researchers also used a non-specific peptide (NSP) which did not have any impact on root development to improve the specificity of miPEP858a. Experiments further highlighted that miPEP was absorbed by *Arabidopsis* root cells. Furthermore, promoter/reporter research

The miRNA-encoded Peptides 383

revealed that miPEP858a is engaged in the regulation of the promoter activity of its own and also expression of GUS gene (Sharma *et al.*, 2020). This leads to the conclusion that miPEPs serve particular biological tasks and are not waste products with no defined purpose. Experiments have been accomplished by employing approaches, such as high expression, GUS reporter studies, immunoblot techniques and ribosome profiling. Even yet, some methods, such as mass spectroscopic and computational model evaluation, are difficult to master (Baerenfaller *et al.*, 2008; Waterhouse and Hellens, 2015). Combinatorial approaches, like proteomics, bioinformatics, ribosomal profiling and high-throughput sequencing might also be employed for miORFs expressing miPEPs (Morozov *et al.*, 2019). Molecular methods, like overexpression and target mimicry, and also molecular analysis and *in silico* approaches, were used to examine the role of miRNAs. Recently, *Arachis hypogaea* miPEPs were discovered by utilising ESTs from 14 related miRNAs (Ram *et al.*, 2019). Methods for editing genome, like CRISPR-based editing of miPEPs, might prove as a useful tool for modifying agronomic traits (Sharma *et al.*, 2020). miPEPs, which were endogenous, were discovered via western blotting, indicating that peptides encoded by microRNA had accumulated (Sharma *et al.*, 2020; Lauressergues *et al.*, 2015). Thus, development as well the changes in phenotype might be examined, eventually resulting in improved agronomic features of diverse crops. A recently developed integrated peptide-genomic pathway to identify NCPs of diverse plants might prove useful for studies aimed at functional genomics (Wang *et al.*, 2020).

4. Screening of Hidden Peptides

Many of the coding RNAs have a single ORF that is long and several other ORFs that are small, and in maximum cases, the long ORF is the one which codes for a functional protein; whereas the non-coding RNAs, which code mainly for hidden peptides, include the shorter ORFs with length of 300 or less nucleotides. And in a contrasting manner, the longest ORF if often not responsible for generating any functional peptides, which makes it arduous to anticipate those ORFs which are capable of encoding for hidden peptides. Evolution preserves the sequences of translated regions and cross-species genomic comparisons tend to be important for ORF prediction. PhyloCSF is an exclusive tool for ORF prediction as it can detect synonymous codon and conservative amino acid substitutions in high frequency (Lin *et al.*, 2011).

ORF identification has also been established by using experimental methods, like ribosome profile based on deep sequencing and peptidomics, based on mass spectrometry. Deep sequencing is used in ribosome profiling to detect ribosome-protected RNA snippets, which are suggestive of ongoing translation (Ingolia *et al.*, 2009; 2016). Through this method, the polysomes of bounded RNA and ribosome molecules are first isolated using centrifugation (sucrose density gradient) or via affinity pull-down of epitope-tagged ribosomes, followed by treatment with enzyme nucleases to generate ribosome-protected RNA fragments. To identify the area of active translation, the obtained fragments are next sequenced and later, mapped to the genome.

Unlike ribosome profiling, which does not give a direct evidence for translation, mass spectrometry (MS) can directly detect translation. The term 'peptidomics' refers to the variant of proteomics which blend mass spectrometry with computational prediction of collected data (Ma *et al.*, 2016). To decrease the background signal caused by bigger annotated proteins, peptides are frequently condensed by using a molecular weight cut-off filter (MWCO) or SDS-PAGE (sodium dodecyl sulfate-polyacrylamide gel electrophoresis). MS/MS spectra procured are later compared to the proprietary peptide database, which includes a list of all probable short ORFs based on the cDNA sequence databases and RNA-sequencing data sets. The remaining unique peptides are annotated as novel hidden peptides, with matched peptides relating to existing annotated proteins or peptides being removed. The whole process is well elaborated in Fig. 1 (Matsumoto and Nakayama, 2018).

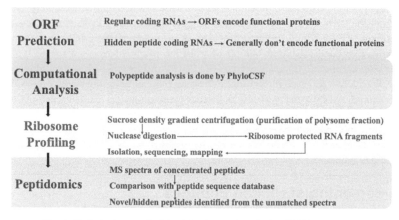

Fig. 1: Technical route for identification and analysis of hidden peptides

5. Integration and Uptake of miPEPs in Plants

Exact method of miPEPs' entrance into a plant is a crucial question for the activity and functioning of miPEPs in plants. Peptide entrance in plants involves both active and passive diffusions. One of the well-studied pathways is 'endocytosis', which is aided by clathrin-dependent as well as the membrane microdomain pathway, which are based on stimuli that are already recognised mechanisms for peptides to enter the body (Baral *et al.*, 2015). Ormancey *et al.* have lately looked into this (in 2020). In *Arabidopsis thaliana*, MiPEP165a was treated exogenously and fluorescent dye was used for labelling. Rapid penetration was seen at the root cap and meristematic zone, with a delay in the rest of the root. Internalisation of peptides in plants occurs through passive diffusion and then endocytosis. Mutants of genes implicated in the clathrin-mediated route, such as chc1-1, as well as the membrane microdomain pathway, which includes rem1-2 and rem1-3, have already been investigated in connection with miPEP165a entry. The observations stated that entry of miPEP165a was passive in the root cap, meristematic and differentiation zones, but it was inhibited in the root cells of mature zone. Rem1-miPEP 2's uptake was severely hampered, while miPEP165a had no effect. The use of pharmacologic inhibitors TyrA23 and MCD to block the clathrin-mediated and membrane microdomain pathways, respectively, was the next stage in the experiment. The favourable phenotypic effects of miPEP165a on root length were prevented by these compounds.

6. Regulatory Activities of Primary miRNA-derived Peptides

Although there is some evidence that ncRNAs in plants can code for internal peptides, the regulatory roles of peptides which are derived from pri-miRNA have been fairly understood with the help of various earlier and recent studies. Generally, a miRNA is almost 22 nucleotides long and has regulatory effects which may even suppress the transcription and post-transcriptional expression of endogenous genes (Voinnet, 2009). miRNA synthesis can easily be restricted by transcription of microRNA genes, which are pol II-dependent. Processing of mature miRNAs from considerably bigger pri-miRNAs is done by the complex of Dicer-like RNase III endonucleases (DCLs) and it is later assembled into the active RISC (RNA-induced silencing complex) by integrating them into the Argonaute1 (AGO1) protein (Yu *et al.*, 2017; Rogers and Chen, 2013). The miRNA guide strand directs RISC towards binding to the particular target gene by pairing the bases, resulting in silencing of gene via either the cleavage of target or inhibiting the process of translation (Wang *et al.*, 2019). pri-miRNAs may have brief ORFs in 5′ upstream that code for regulatory peptides and are termed as 'miPEPs', in addition to generating miRNAs. pri-miRNAs were first identified to have coding ability in *Arabidopsis* and *Medicago truncatula*, and they have been subsequently explored in cells of humans and plants, like soybean and grapes human cells (Chen *et al.*, 2020; Sharma *et al.*, 2020; Fang *et al.*, 2017; Couzigou *et al.*, 2016; Lauressergues *et al.*, 2015).

Peptide accumulation reflects presence of endogenous miPEPs, which can be easily examined and analysed with the help of western botting techniques (Sharma *et al.*, 2020; Lauressergues *et al.*, 2015). miRNAs are produced in the nucleus and their maturation takes place mostly in the nucleus only, before being exported to the cell's cytoplasm (Wang *et al.*, 2019). In starting, DCL complex in the nucleus might cut the pri-miRNAs into three parts. The first part is upstream of miRNA precursor (pre-miRNAs), the second part is pre-miRNAs and the third one is 3′ fragments having poly-adenylated tail. Later, the pre-miRNAs undergo processing to be converted to a mature miRNA, but in this process, the 3′ segments may be destroyed (Rogers and Chen, 2013). There is also a possibility of the pre-miRNA's upstream to be exported to cytoplasm to grab some guidance for translation of peptide, even though the pri-miRNA may or may not possess ORFs (Fig. 2). The coding area of pri-miRNA is probably transferred to the cytoplasm after cleaving, as evidenced by the presence of pri-miR171 in the nucleus and a little amount in the cytoplasm of grapes (Chen *et al.*, 2020; Lauressergues *et al.*, 2015). Whether re-polyadenylation protects the upstream portions of pre-miRNAs which are cleaved or not, is yet unclear and has to be investigated. Furthermore, nuclear transcription and translation coupling is thought to be an alternate possibility (Prasad *et al.*, 2021). As per some recent studies, miPEPs of plants can also function as an endogenous peptide which favourably boosts miRNA synthesis via autoregulatory feedback loop (Ormancey *et al.*, 2020). They can precisely trigger transcribing respective pri-miRNAs and may even induce the production of mature miRNAs, as well as increase accumulating themselves in such a process. Roles of miPEPs which regulate transcription of miPEPs have also been reported. Cordycepin, an RNA synthesis inhibitor, inhibits the beneficial impacts of miPEP165a on the accumulation of pri-miR165a in *Arabidopsis* (Lauressergues *et al.*, 2015).

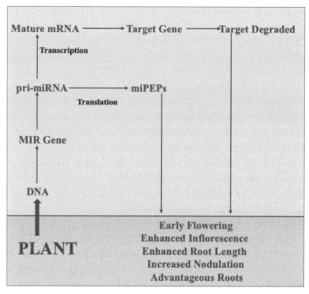

Fig. 2: miRNA biogenesis and influence on plant growth

In a recent study conducted on *Arabidopsis*, it was recorded that the promoter of miR858a typically stimulates expressing of the GUS gene in two reporter lines (PromiR858a: ATG1: GUS and PromiR858a: ORF1: GUS), where either only the start code or the complete ORF encoding miPEP858a is fused. In fact, they even recorded that supplementing the media with synthetic miPEP858a increases GUS activity, showing that miPEP858a works at the promoter region to boost transcriptional processes (Sharma *et al.*, 2020). whereas, miPEPs may operate like trans-acting factors, e.g. as transcription factor (TF) that significantly influences transcribing processes of microRNA genes, either directly or indirectly. TF expression levels in plants can be regulated by conventional peptides. As per an earlier study on *Arabidopsis*, secretion of root meristem growth factor 1, Tyr sulphated peptide, is essential in maintaining the root stem cell niche as well as for the transit amplifying cell proliferation. PLT, a root-specific transcription factor that mediates patterns of root stem cell niche, is favourably regulated by root meristem growth factor1 (Matsuzaki *et al.*, 2010). On the other hand, miPEPs can attach to a single subunit of 'Pol II transcriptional complexes' or even to Pol II itself, albeit any direct evidence for backing this up. Peptides interact with the catalysing enzyme's subunit in plants, e.g. ENOD40 peptides found in legumes, in which both peptides may exclusively bind at the nodulin100, which is a subunit of the enzyme sucrose synthase (Röhrig *et al.*, 2002). Two overlapping ORFs, situated in the 50 conserved area code for two peptides of length 12- and 24- amino acids, are referred as peptides A and B. In a nutshell, miPEPs purely regulating mechanisms remain a mystery. Those plants which do not require any extra editing can be given synthetic miPEPs externally and generate similar autoregulatory effects. It is believed that miPEPs can be internalised through passive diffusion and processes related to endocytosis, other than the precursor-derived peptides which operate like a ligand recognised through the associated receptors (Ormancey *et al.*, 2020; Oh *et al.*, 2018; Yamaguchi *et al.*, 2010). In 24 hours, miPEP165a which are fluorescently tagged, quickly invade the entire roots of plant *Arabidopsis*. At the meristematic and differentiation zones, absorption of miPEP165a is influenced when gene function is lost and this is linked to endocytosis or the use of an inhibitor of endocytosis (Ormancey *et al.*, 2020). MiPEP858a may be readily absorbed through roots and is found in the interior of the plant cell, similar to miPEP165a (Sharma *et al.*, 2020). These findings suggest that miPEPs may have regulatory effects within the cell rather than being transferred to the apoplast. The existence of even a single putative ORF, which encodes miPEPs in one of 50 *Arabidopsis* pri-miRNAs, was discovered in a study of 50 *Arabidopsis* pri-miRNAs. There aren't any shared signatures among these miPEPs, showing that each miPEP is likely unique to its miRNA. Various miPEPs constitute to be a part of a complex and specific regulatory network which regulates miRNA expression to fulfil various biological activities. The biological roles of various miPEPs are still unknown. miPEP171b and 165a are peptides of 20 and 18 amino acids, respectively, generated by *M. truncatula* and *Arabidopsis*. The accumulation of miR171b and 165a is specifically triggered by overexpressing these peptides and by their exogenous supplementation, which results in reduced development of lateral roots, as well as encouragement of the growth of main root (Lauressergues *et al.*, 2015). Soybean nodule number rises when plants are watered with synthetic miPEP172c (Couzigou *et al.*, 2016). Exogenous vvimiPEP171d1 supplementation can enhance adventitious root formation in grapes by increasing vvi-MIR171d expression (Chen *et al.*, 2020). In *Arabidopsis*, a 44 amino acid long peptide 'MiPEP858a' is transcribed by the very first ORF, which is 135 bp long and is found upstream. Molecular weight of the endogenous miPEP858a is around 6kDa, which regulates how the genes associated with signalling of auxins and phenylpropanoid pathway get expressed and this governs flavonoid production and plant growth (Sharma *et al.*, 2020). Several miPEPs' regulatory roles are reportedly verified through experimentations in many different plant species, but few crucial concerns are still unanswered. Is up-regulation of their corresponding pri-miRNAs specific to miPEPs, and if so, how is this accomplished? After all, activation specificity has only been detected by using a few miRNAs (Lauressergues *et al.*, 2015). As per another recent study, as the tissue culture plantlets of grape plant were treated with the synthesised vvi-miPEP171d1, the expression level of somemiRNAsgenessuchasvvi-MIR160c, vviMIR171a and vviMIR171i even decreases; albeit it is unclear if this drop was because of incubation time (Chen *et al.*, 2020). It has to be seen if miPEPs have a deleterious impact on the related miRNAs or even any other miRNAs.

Plants have recently been found to have peptide-encoding pri-miRNAs. Also, it has been well proved and documented that the RNAPII transcribes pri-miRNAs and is consistent with their coding potential. miR171b regulates the production of lateral roots in *M. truncatula*, and pri-miR171b has two ORFs which are putative and encode 20 or 5 amino acids long peptides. Reporter gene for A-glucuronidase linked with each of the ATG start codon is reported to be utilised to test if these ORFs of pri-miR171b are translated. The process of translation was reported only if the reporter gene got coupled with start codon of ORF which codes for 20 amino acids. Both immunoblot analysis and immunostaining with particular antibodies revealed expression of encoded 20 amino acid peptide, termed miPEP171b, endogenously, with expression visible at lateral root start locations. In *A. thaliana*, on more peptide 'miPEP165a' encoded by pri-miRNA has been discovered. *Brassicales* have a highly conserved sequence of amino acids for miPEP165a and antibodies to miPEP165a

388   *Plant MicroRNAs and Stress Response*

were used to validate its endogenous expression on immunoblots (Lauressergues *et al.*, 2015).

The functions of miPEP171b and 165a are similar and in a study when *M. truncatula* roots were treated with synthetic miPEP171b, the endogenous expression of miR171b was recorded to be enhanced along with reduced density of the lateral root, with forced production of miR171b mimicking the latter result. Specific impact of miPEP171b was noted on the density of respective corresponding miRNA and there wasn't even any influence on abundance of other miRNAs. The increase of miR165a was also seen in seedlings of miPEP165a treated *A. thaliana*. miR165a's activation was inhibited by further treatment with cordycepin, an RNA synthesis inhibitor, demonstrating that at transcription, the miPEPs trigger their pri-miRNAs (Lauressergues *et al.*, 2015).

7. Regulation of Root Growth and Yield Trait in Crops through miRNA

MicroRNA has an essential role in the regulation of plant's growth and development by exclusive target at the conserved genes' mRNA. Mostly, the majorly conserved target genes are regulated by conserved miRNAs which are responsible for hormone signalling and plant development (Zhao *et al.*, 2016). Those regulatory modules, which comprise conserved miRNAs, are targeted and the popularly studied targets are documented as 'miR156-SPLs', 'miR164-NACs' and 'miR396-GRFs'. Studies to discover miRNAs genome-wide have reported the conserved miRNA as moieties that can even regulate the targets which are not conserved widely. For example, in *Arabidopsis*, bHLH74 has been documented as a proven target of miR396, even though the bHLH74 homologs which have miR396 binding sites have been known to be available only with the species which fall under the sister families, *Cleomaceae* and *Brassicaceae* (Debernardi *et al.*, 2012). A lesser conserved target 'TaPSK5' has homologs with miR164 binding sites and restricted only to the nearly related monocot plants, like *Triticeae*, *Oryza sativa* and *Brachypodium*. In another study conducted on transgenic *Arabidopsis* and rice plants, overexpression of TaPSK5 showed lengthy roots as compared to the WT controls, agreeable to the discovery that the signalling of PSK has a particular role of promoting the growth of roots. Expected results in terms of root growth were recorded when miR164-resistant TaPSK5-D was overexpressed, which indicates that PSK hormone might affect the plants in a dose-dependent manner. Adding more to this, it was documented that GhPSK's overexpression promotes elongation of the fibres and enhances their quality (Baker *et al.*, 2017). Such reports highlight the fact that signalling of PSK hormone proves to regulate the plant's growth and development in a versatile manner, and when manipulated, the expression had exclusive applications for improving the crop quality and yield. Yield increment is a major issue for wheat breeding programmes because of increasing demand of wheat and the appropriate approach for wheat increment is marker-assisted selection for genetic improvement in crops (Ma *et al.*, 2019). TaPSK5, which is not an exclusively conserved target of miRNA 164, is well known as a novel positive regulator for the growth of roots as well as greater yields in wheat plants. Studies conducted on transgenic and haplotype proved the potential of targeting TaPSK5 to improve crop yields and growth at genetic levels. In case of *Arabidopsis* roots miPEP165a and miR165a are expressed in endodermis cells. Experimental findings proved that primary root length increased via exogenous treatment of *A. thaliana* seedlings with synthetic miPEP165a. They documented that plants which were watered with 100 µM of peptide were quite efficient for increasing the primary root length, as compared to the treatments performed with only 10 µM of peptide (Ormancey *et al.*, 2020). Figure 2 summarises the production of miRNA and their regulatory mechanisms in plants (Yadav *et al.*, 2021).

8. Agronomic Significance of miPEPs

Due to the similar tissue distribution, miPEPs boost pri-miRNA transcription, along with the expression as well as the activities of the respective corresponding mature miRNAs. As a result, from various studies conducted lately, it has become self-evident that the successful, environmentally acceptable manipulations of innovative elements of plant growth prove to be critical for long-term agricultural sustainability. miPEPs can have substantial agronomical ramifications in this respect as they indirectly influence critical plant developmental tasks, such as accelerated growth of the plant's root, blooming at an early stage along with better flower quantity and higher stem height. Alterations in rooting were recorded on external application of miPEP165a and 171b to *Arabidopsis* and *Medicago*, respectively (Ormancey *et al.*, 2020; Lauressergues *et al.*, 2015). miRNA expression is increased by spraying or watering synthetic miPEPs on the outside. In legumes, such as soybeans, miPEP172c and 167c have been reported to lead to better nodulation by increasing nodule numbers (Combier and Andre, 2021; Couzigou *et al.*, 2016). In a similar manner, vvi-miPEP171d1 controls adventitious root production in economically significant crop 'grapevine', which might be advantageous in overcoming clonal propagation constraints in it (Chen *et al.*, 2020). Eventually, the major achievement in plant growth is calculated by its yield and studies have recorded that by use of miPEPs in crop enhancement plans might replace the time-consuming task of transgenics. These findings point to previously undisclosed hidden functions of pri-miRNA sequences, reverting us to their critical roles and directing our attention to a different level of gene control. By exhausting the numerous uses, miPEPs might prove to be the latest feasible gene-modifying technique for contemporary study and it is feasible that this probably will be a useful functional tool for proteomics, for improving crops.

9. Conclusion and Future Perspectives

Various earlier and latest analyses reveal that a large variety of short peptides encoded by ORFs in lncRNAs, pose a problem for researchers looking for pri-miRNA coding functions. With the growing number of studies, a slew of new questions has arisen about interpretation of mechanisms at cellular and molecular levels which underlie the functions of miPEP. Interested facts are expected from the doubts, like, whether there is some sort of negative feedback mechanism that regulates accumulating miRNAs/miPEPs. Various methods to validate and detect these miPEPs with exclusive functions are urgently required. Exogenous applications have also prompted several unanswered problems, such as how miPEPs get over the cell membrane, the secondary cell wall, as well as into the nucleus. Still, various transporters for miPEPs may exist which might function in the penetration, long-distance communication and even for mobility.

Several extensive studies into distinct miRNAs and miPEPs followed by their respective role for environmental-efficient methods resulting in improved plant growth, productivity and agricultural development is underway. Some problems, such as pathways specific for biological activities of miPEPs, even though pri-miRNA transcription may not be the real method, whether the transcription of pri-miRNA is the real method for activation of miPEPs, and if the related miRNAs expression gets controlled negatively by miPEPs, demand answers in the coming times. Unidentified positive regulatory components mediated by miPEP have also been fascinating. Distinct invariable phenotypic alterations are generated by exogenous administration of miPEPs. As agronomic tools, miPEPs are leading gene regulators that do not require genetic alteration. miPEPs work in the form of auto-regulatory feedback. When compared to toxic chemicals, they are believed to be more efficient and less destructive when used in the environment, and they are unobjectionable to the population all over the world. Peptide molecules have proved to be regulatory and hence have gained a lot of attention in the last few years. Different kinds of peptides from various plant species are reportedly discovered as abundant in plant 'peptidome' (Olsson *et al.*, 2019; Takahashi *et al.*, 2019; Tavormina *et al.*, 2015). Additional to the typical peptides which are formed via precursor processing and the short ORFs, evidences for the occurrence and functioning of non-conventional peptides translated by transcripts' 5′ UTR or 3′ UTR, which are recently referred to as ncRNAs, is accumulating (Wang *et al.*, 2020). These peptides play a variety of roles in the development, growth, reproduction mechanisms, as well as senescence and cell death, nutritional balance, root nodulation and responses to the biotic and abiotic stress situations. Several miPEPs produced by the pri-miRNAs have been experimentally discovered in recent years, along with studies proving their function which have been proposed by overexpression experiments for several species, showing the pri-miRNA code as widespread across different species of plants. The data from numerous earlier and latest studies supports miPEPs' regulatory potential to influence the elevation in amounts of miRNA associated with them, by increasing the transcription of pri-miRNA. Still, there are a few questions which remain unanswered, like: (1) the production of miPEPs and processing of the miRNA takes place at two separate areas of pri-miRNAs, so how is translation in the cytoplasm coordinated with miRNA maturation in the nucleus? After being released from the dicing complex, followed by re-polyadenylation, the 50 upstream of pri-miRNA gets transported to the cytoplasm to direct production of miPEP. (2) How can miPEPs specifically improve miRNA gene transcription? An examination of 50 miPEPs in *Arabidopsis* revealed that the structural pattern is never common, indicating distinct regulatory role of each miPEP (Lauressergues *et al.*, 2015). It has to be further investigated that whether the miPEPs bind directly to the Pol II or to the trans-acting factors? (3) The exogenous application of miPEPs necessitates overexpression and editing by CRISPR-Cas9 in order to assess biological activities. It is unclear if modifications to the miPEPs after translation may boost their functionality. In summary, the majority of peptides that have been studied so far are thought to operate as ligands that facilitate plant intercellular communication and response. Peptides which are translated by the transcripts of non-coding RNAs enhance the plant's peptidome. The dual role of pri-miRNAs, in terms of their capabilities of coding and non-coding, can be easily revealed by identifying and analysing miPEPs. Various other miRNA-dependent gene regulation networks can be revealed by exploiting the pathways of miPEPs biosynthesis and regulation.

References

Baerenfaller, K., Grossmann, J., Grobei, M.A., Hull, R., Hirsch-Hoffmann, M., Yalovsky, S., Zimmermann, P., Grossniklaus, U., Gruissem, W. and Baginsky, S. (2008). Genome-scale proteomics reveals *Arabidopsis thaliana* gene models and proteome dynamics, *Science*, 320(5878): 938-941.

Baker, T.C., Han, J. and Borchers, C.H. (2017). Recent advancements in matrix-assisted laser desorption/ionisation mass spectrometry imaging, *Curr. Opin. Biotechnol.*, 43: 62-69.

Baral, A., Irani, N.G., Fujimoto, M., Nakano, A., Mayor, S. and Mathew, M.K. (2015). Salt-induced remodeling of spatially restricted clathrin-independent endocytic pathways in *Arabidopsis* root, *The Plant Cell*, 27(4): 1297-1315.

Bielewicz, D., Kalak, M., Kalyna, M., Windels, D., Barta, A., Vazquez, F., Szweykowska-Kulinska, Z. and Jarmolowski, A. (2013). Introns of plant pri-miRNAs enhance miRNA biogenesis, *EMBO Reports*, 14(7): 622-628.

Chen, Q.J., Deng, B.H., Gao, J., Zhao, Z.Y., Chen, Z.L., Song, S.R., Wang, L., Zhao, L.P., Xu, W.P., Zhang, C.X., Ma, C. and Wang, S.P. (2020). A miRNA-encoded small peptide, vvi-miPEP171d1, regulates adventitious root formation, *Plant Physiology*, 183(2): 656-670.

Combier, J.P. and Andre, O. (2021). Novel method for promoting nodulation in plants, *U.S. Patent Application No. 17/212:354*.

Combier, J.P., de Billy, F., Gamas, P., Niebel, A. and Rivas, S. (2008). Trans-regulation of the expression of the transcription factor MtHAP2-1 by a uORF controls root nodule development, *Genes & Development*, 22(11): 1549-1559.

Combier, J.P., Lauressergues, D. and Becard, G. (2020a).Use of micropeptides for promoting plant growth, *U.S. Patent No. 10,563,214*, Washington, DC: U.S. Patent and Trademark Office.

Combier, J.P., Lauressergues, D., Becard, G., Payre, F., Plaza, S. and Cavaille, J. (2020b). Micropeptides and use of same for modulating gene expression, *U.S. Patent Application No. 16/595,861*.

Couzigou, J.M., André, O., Guillotin, B., Alexandre, M. and Combier, J.P. (2016). Use of microRNA-encoded peptide miPEP172c to stimulate nodulation in soybean, *New Phytologist*, 211(2): 379-381.

Debernardi, J.M., Rodriguez, R.E., Mecchia, M.A. and Palatnik, J.F. (2012). Functional specialisation of the plant miR396 regulatory network through distinct microRNA-target interaction, *PLoS Genet.*, 8: e1002419.

Fang, J., Morsalin, S., Rao, V.N. and Reddy, E.S.P. (2017). Decoding of non-coding DNA and non-coding RNA: pri-micro RNA-encoded novel peptides regulate migration of cancer cells, *Journal of Pharmaceutical Sciences and Pharmacology*, 3(1): 23-27.

Ingolia, N.T. (2016). Ribosome footprint profiling of translation throughout the genome, *Cell*, 165(1): 22-33.

Ingolia, N.T., Ghaemmaghami, S., Newman, J.R. and Weissman, J.S. (2009). Genome-wide analysis *in vivo* of translation with nucleotide resolution using ribosome profiling, *Science*, 324(5924): 218-223.

Jatan, R., Tiwari, S., Asif, M.H. and Lata, C. (2019). Genome-wide profiling reveals extensive alterations in Pseudomonas putida-mediated miRNAs expression during drought stress in chickpea (*Cicer arietinum* L.), *Environmental and Experimental Botany*, 157: 217-227.

Jiu, S., Zhu, X., Wang, J., Zhang, C., Mu, Q., Wang, C. and Fang, J. (2015). Genome-wide mapping and analysis of grapevine microRNAs and their potential target genes, *The Plant Genome*, 8(2): eplantgenome2014.120091

Julkowska, M. (2020). Small but powerful: MicroRNA-derived peptides promote grape adventitious root formation, *Plant Physiology*, 183(2): 429-430.

Kapusta, A. and Feschotte, C. (2014). Volatile evolution of long noncoding RNA repertoires: Mechanisms and biological implications, *Trends in Genetics*, 30(10): 439-452.

Lauressergues, D., Couzigou, J.M., Clemente, H.S., Martinez, Y., Dunand, C., Becard, G. and Combier, J.P. (2015). Primary transcripts of microRNAs encode regulatory peptides, *Nature*, 520(7545): 90-93.

Lin, M.F., Jungreis, I. and Kellis, M. (2011). PhyloCSF: A comparative genomics method to distinguish protein coding and non-coding regions, *Bioinformatics*, 27(13): i275-i282.

Lv, S., Pan, L. and Wang, G. (2016). Commentary: Primary transcripts of microRNAs encode regulatory peptides, *Frontiers in Plant Science*, 7: 1436.

Ma, J., Diedrich, J.K., Jungreis, I., Donaldson, C., Vaughan, J., Kellis, M., Yates III, J.R. and Saghatelian, A. (2016). Improved identification and analysis of small open reading frame encoded polypeptides, *Analytical Chemistry*, 88(7): 3967-3975.

Ma, L., Hao, C.Y., Liu, H.X., Hou, J., Li, T. and Zhang, X.Y. (2019). Diversity and sub-functionalisation of TaGW8 homoeologs hold potential for genetic yield improvement in wheat, *Crop J.*, 7: 830-844.

Matsumoto, A. and Nakayama, K.I. (2018). Hidden peptides encoded by putative noncoding RNAs, *Cell Structure and Function*, 43: 75-83.

Matsuzaki, Y., Ogawa-Ohnishi, M., Mori, A. and Matsubayashi, Y. (2010). Secreted peptide signals required for maintenance of root stem cell niche in *Arabidopsis*, *Science*, 329(5995): 1065-1067.

Morozov, S.Y., Ryazantsev, D.Y. and Erokhina, T.N. (2019). Bioinformatics analysis of the novel conserved micropeptides encoded by the plants of family *Brassicaceae*, *Journal of Bioinformatics and Systems Biology*, 2(4): 66-77.

Oh, E., Seo, P.J. and Kim, J. (2018). Signalling peptides and receptors coordinating plant root development, *Trends in Plant Science*, 23(4): 337-351.

Olsson, V., Joos, L., Zhu, S., Gevaert, K., Butenko, M.A. and De Smet, I. (2019). Look closely, the beautiful may be small: Precursor-derived peptides in plants, *Annual Review of Plant Biology*, 70: 153-186.

Ormancey, M., Le Ru, A., Duboé, C., Jin, H., Thuleau, P., Plaza, S. and Combier, J.P. (2020). Internalisation of miPEP165a into *Arabidopsis* roots depends on both passive diffusion and endocytosis-associated processes, *Int. J. Mol. Sci.*, 21(7): 2266.

Prasad, A., Sharma, N. and Prasad, M. (2021). Non-coding but coding: Pri-miRNA into the action, *Trends in Plant Science*, 26(3): 204-206.

Qu, L.J., Li, L., Lan, Z. and Dresselhaus, T. (2015). Peptide signalling during the pollen tube journey and double fertilisation, *Journal of Experimental Botany*, 66(17): 5139-5150.

Ram, M.K., Mukherjee, K. and Pandey, D.M. (2019). Identification of miRNA, their targets and miPEPs in peanut (*Arachis hypogaea* L.), *Computational Biology and Chemistry*, 83: 107100.

Ren, Y., Song, Y., Zhang, L., Guo, D., He, J., Wang, L., Song, S., Xu, W., Zhang, C., Lers, A., Ma, C. and Wang, S. (2021). Coding of non-coding RNA: Insights into the regulatory functions of Pri-MicroRNA-encoded peptides in plants, *Frontiers in Plant Science*, 12: 186.

Rodrigues, J.A.S. (2019). Exploring the potential of newly-identified miRNA-encoded peptides to improve the production of bioactive secondary metabolites in grape cells (doctoral dissertation), Universidade do Minho (Portugal)).

Rogers, K. and Chen, X. (2013). Biogenesis, turnover, and mode of action of plant microRNAs, *The Plant Cell*, 25(7): 2383-2399.

Röhrig, H., Schmidt, J., Miklashevichs, E., Schell, J. and John, M. (2002). Soybean ENOD40 encodes two peptides that bind to sucrose synthase, *Proceedings of the National Academy of Sciences*, 99(4): 1915-1920.

Schwab, R., Speth, C., Laubinger, S. and Voinnet, O. (2013). Enhanced microRNA accumulation through stemloop-adjacent introns, *EMBO Reports*, 14(7): 615-621.

Sharma, A., Badola, P.K., Bhatia, C., Sharma, D. and Trivedi, P.K. (2020). Primary transcript of miR858 encodes regulatory peptide and controls flavonoid biosynthesis and development in *Arabidopsis*, *Nature Plants*, 6(10): 1262-1274.

Takahashi, F., Hanada, K., Kondo, T. and Shinozaki, K. (2019). Hormone-like peptides and small coding genes in plant stress signalling and development, *Current Opinion in Plant Biology*, 51: 88-95.

Tavormina, P., De Coninck, B., Nikonorova, N., De Smet, I. and Cammue, B.P. (2015). The plant peptidome: An expanding repertoire of structural features and biological functions, *The Plant Cell*, 27(8): 2095-2118.

Voinnet, O. (2009). Origin, biogenesis, and activity of plant microRNAs, *Cell*, 136(4): 669-687.

Wang, J., Mei, J. and Ren, G. (2019). Plant microRNAs: Biogenesis, homeostasis, and degradation, *Frontiers in Plant Science*, 10: 360.

Wang, S., Tian, L., Liu, H., Li, X., Zhang, J., Chen, X., Jia, X., Zheng, X., Wu, S., Chen, Y., Yan, J. and Wu, L. (2020). Large-scale discovery of non-conventional peptides in maize and *Arabidopsis* through an integrated peptidogenomic pipeline, *Molecular Plant*, 13(7): 1078-1093.

Wang, Y., Wang, L., Zou, Y., Chen, L., Cai, Z., Zhang, S., Zhao, F., Tian, Y., Jiang, Q., Ferguson, B.J., Gresshoff, P.M. and Li, X. (2014). Soybean miR172c targets the repressive AP2 transcription factor NNC1 to activate ENOD40 expression and regulate nodule initiation, *The Plant Cell*, 26(12): 4782-4801.

Waterhouse, P.M. and Hellens, R.P. (2015). Coding in non-coding RNAs, *Nature*, 520(7545): 41-42.

Yadav, A., Sanyal, I., Pandey, S. and Rai, C. (2021). An overview on miRNA-encoded peptides in plant biology research, *Genomics*, 113(4): 2385-2391.

Yamaguchi, Y., Huffaker, A., Bryan, A.C., Tax, F.E. and Ryan, C.A. 2010. PEPR2 is a second receptor for the Pep1 and Pep2 peptides and contributes to defense responses in *Arabidopsis*, *The Plant Cell*, 22(2): 508-522.

Yin, X., Jing, Y. and Xu, H. (2019). Mining for missed sORF-encoded peptides, *Expert Review of Proteomics*, 16(3): 257-266.

Yu, Y., Jia, T. and Chen, X. (2017). The 'how' and 'where' of plant micro RNAs, *New Phytologist.*, 216(4): 1002-1017.

Zhao, X.Y., Hong, P., Wu, J.Y.,Chen, X.B., Ye, X.G., Pan, Y.Y., Wang, J. and Zhang, X.S. (2016). The tae-miR408-mediated control of TaTOC1 genes transcription is required for the regulation of heading time in wheat, *Plant Physiol.*, 170(3): 1578-1594.

CHAPTER

21

Plant MicroRNAs: Physiological Significance in Plants and Animals

Idris Ali Dar[1], Masarat Bashir[2] and Ashraf Dar[1]*

[1] Department of Biochemistry, University of Kashmir, Srinagar - 190006, Jammu & Kashmir, India
[2] College of Temparate Sericulture, SKUAST-K, Shalimar, Srinagar, India

1. Introduction

Plant genomes typically encode hundreds of miRNAs, which are short, endogeneously expressed, small RNA molecules (19-25 nucleotides) produced by Dicer-like protein (DCL) from the stem loop regions of longer precursors RNA molecules, known as pri-miRNAs (primary microRNAs). The tissue-specific expression of different miRNAs is very important for normal physiological processes, like growth, proliferation, differentiation, development and maintenance and cell death. The miRNAs exert these functions by repressing the expression of their target genes by binding to 3′ UTR, 5′UTR or coding sequences of their corresponding mRNA transcripts. Generally, a miRNA can have more than one target and approximately 60% of mRNAs are predicted to contain miRNA-binding sites.

The common proverb 'you are what you eat' gains importance in the context of plant miRNAs regulating animal gene expression. In recent years, plant-derived miRNAs have gained increased attention not only for their roles in regulating host plant physiology, but also due to their surprising discovery in different animal systems. These plant-derived miRNAs find their way into the animal tissues via oral uptake of food, which is largely constituted of plant components. In animals, the plant-derived miRNAs regulate the expression of target genes and thereby influence the outcome of various cellular pathways. The presence of regulatory miRNAs in food or diet and their impact on human or animal physiology is an evolving area of investigation. In near future, the relevant studies will certainly help to frame new rules and guidelines for healthy nutrition.

---

*Corresponding author: ashrafdar@kashmiruniversity.ac.in

Plant MicroRNAs: Physiological Significance in Plants and Animals   **395**

2. Physiological Role of miRNAs in Plants

The development of a plant includes various processes, e.g. evolution of a zygote into an embryo, formation of flowers, fruits and seeds and root development, etc. Interplay between miRNAs and different phytohormones is attributed in plant growth and development. The miRNAs regulate gene expression through transcriptional or translational inhibition and interact with various phytohormones (Chen, 2004; Wang *et al.*, 2005).

2.1 miRNAs in Embryogensis

In the early stages of plant development, miR160 has a critical role to play. In *Arabidopsis* foc (floral organs in carpels) mutants, the 3′ regulatory sites of miR160a gene, are mutated by insertion of Ds transposon which suppresses the expression of miR160a. These mutants undergo abnormal longitudinal cell divisions in the suspensor phase of embryogenesis. The embryonic cells further fail to differentiate normally and result in asymmetric embryos, followed by asymmetric cotyledons and formation of abnormal seeds. Mechanistically, miR160a suppresses the ARF (auxin response factor) family of transcription factors, including ARF10, ARF16 and ARF17. The compromised expression of miR160a in these mutants increases levels of ARF transcription factors (Liu *et al.*, 2010). These transcription factors, when stimulated by auxin, either enhance or repress the transcription of the downstream genes of auxin signalling (Guilfoyle *et al.*, 1998).

miR164 is known to play a role in proper patterning during early embryogenesis. Targets of miR164 include members of NAC-NAM (no apical meristem); ATAF1/2 (*Arabidopsis* transcription activation factor) and CUC (cup-shaped cotyledon) transcription factors. Mutants with loss of function of CUC1 and CUC2 (cuc1 and cuc2 double mutants) develop a defective shoot apical meristem (SAM) (Aida *et al.*, 1997). miR164 overexpression phenocopies cuc1 and cuc2 double mutants having cup-shaped cotyledons (Aida *et al.*, 1997; Mallory *et al.*, 2004; Vroemen *et al.*, 2003; Laufs *et al.*, 2004). Expression of miR164-resistant CUC1 results in defective embryonic development (defects in orientation of cotyledon) (Mallory *et al.*, 2004). Furthermore, enlarged boundary domains have been witnessed in plants expressing miR164-resistant CUC2 proteins (Laufs *et al.*, 2004), suggesting that regulation of CUC1/ CUC2 by miR164 is important for normal embryo development.

2.2 miRNAs in Leaf Development

miRNAs are actively involved in leaf development. Two important regulatory circuits play a key role in leaf development process: the interplay between miR394 and LCR (leaf curling responsiveness) and between miR160 and miR165/166. miR394, which is highly abundant in protoderm, facilitates stem cell maintenance by repressing the expression of the LCR gene and the activity of miR396 is influenced by a local feedback loop established by Wuschel (WUS) and Clavata (CLV). miR160 targets three ARFs; ARF16, ARF17 and ARF10 which are critical for phyllotaxis in the rosette and maintain the local auxin peaks in the meristem (Pulido and Laufs, 2010). A number of pleiotropic phenotypes have been witnessed in *Arabidopsis* transgenic lines expressing miR160-resistant ARF17 and miR160-resistant ARF10. There is an abnormal number and position of cotyledons and excessively serrated and up-curled rosettes, suggesting the role of miR160 and ARFs in leaf initiation (Liu *et al.*, 2007; Mallory *et al.*, 2005). miR165/166 is thought to play an important regulatory role in the organisation and maintenance of the SAM (shoot apical meristem) by down-regulating transcription of class III HD-ZIP genes. Members of HD-ZIP III transcription factor family have a positive role in SAM maintenance. AGO1 mediates the loading of miR156/157 into RISC complex and AGO10 competes with AGO1 for miR165/166 and thereby prevents the degradation of HD-ZIP III transcripts. Mutant plants with gain of function of AGO10 exhibit varying degrees of leaf serration and hyponasty, whereas AGO10 loss-of-function mutants develop either totally empty or at least pin-headed apexes (Roodbarkelari *et al.*, 2015; Yu *et al.*, 2017).

During leaf development, generation of leaf polarity is one of the cardinal events and a number of miRNAs have been found to regulate this process. Generally, two polarity regulators, the adaxial (top face of leaf) determiners and the abaxial (bottom face of leaf) determiners, are the key players that maintain the opposite surfaces of a leaf. AS1 (asymmetric leaves-1), AS2 and HD-ZIPIIIs act as the adaxial determiners, whereas YABBYs and KANADIs serve as the abaxial determiners (Chitwood *et al.*, 2007; Yamaguchi *et al.*, 2012). AS1 and AS2 have been shown to positively regulate HD-ZIPIIIs whereas KANADIs have been found to up-regulate the activity of YABBYs. In addition, HD-ZIP IIs regulate the polarity identity by repressing miR165/166 (Merelo *et al.*, 2016). A number of studies have suggested the involvement of miR165/166 in the determination of leaf polarity. The targets of miR165/166 include phavoluta (PHV), phabulosa (PHB), ATHB-8, ATHB-15 and revoluta (REV) which are all the members of HD-ZIPIII transcription factor family. PHV, REV and PHB positively regulate identity of adaxial half of a leaf and share functional redundancy with one another. miR390, with four loci in *Arabidopsis* genome, has an indirect role to play in leaf polarity through the generation of ta-siRNA. This miRNA (miR390) mediates generation of ta-siRNAs through cleavage of TAS3 and these ta-siRNAs in turn, negatively regulate abaxial identity by targeting transcripts of ARF3 and ARF4. These factors are known to promote abaxial identity (Pekker *et al.*, 2005). Additionally, AS1 and AS2 stabilise adaxial-abaxial partitioning by targeting ARF3 and ARF4 by directly binding to the promoter of ARF3 and down-regulating its expression, by maintaining methylation in the coding sequence of ARF3 gene and by indirectly activating RDR6 (RNA-dependent RNA polymerase 6) pathway. miR390 also suppresses the expression of ARF3 and ARF4 (Iwasaki *et al.*, 2013). It has been found that miR390 and miR165/166 span between cells and act in concert during the formation of leaf boundaries (Marin-Gonzalez and Suarez-Lopez, 2012). ARF3/ARF4 and HD-ZIP IIIs are responsible for the fate of lower and upper surfaces of a leaf, respectively. A gradient is formed along the axis by diffusion of miR165/166 from the bottom to the top, which in turn restricts the expression of HD-ZIP III in the two uppermost layers of cells, i.e. towards the adaxial half and generates a definite boundary. The ta-siRNAs generated by miR390-directed cleavage of TAS3, also follow a similar but opposite pattern. These ta-siRNAs diffuse from their actual expression centre, i.e. the two topmost layers of a leaf and restrict the expression of ARF3 and ARF4 to the two bottom-most layers (Robinson and Roeder, 2017; Skopelitis *et al.*, 2017). Figure 2 gives an overview of regulation of leaf polarity via opposite diffusion patterns of miR166 and tasiARFs.

miR396 plays a role in leaf polarity. There is a strong correlation between expression gradients of GFR and miR396 and the divergent leaf growth polarity.

Plant MicroRNAs: Physiological Significance in Plants and Animals **397**

Basipetal growth (positive allometry) is observed when the base has a growth rate faster than that of the tip. By contrast, acropetal growth (negative allometry) is observed when the growth rate of tip exceeds that of the base. In case of basipetal leaves, miR396 is highly accumulated near the tip and at the base its levels are markedly low. In acropetal leaves, the pattern of expression of this miRNA is reversed.

The morphology of leaves differs greatly among different plant species and is dictated to a large extent by the pattern and the level of dissection in the leaf primordium margin. A number of studies have implicated the involvement of miR164 in the regulation of leaf morphology. miR164 down-regulates CUC2 (cup-shaped cotyledon-2), which is a member of CUC family of transcription factors. CUC and PIF1 (pin-formed 1) transcription factors regulate marginal leaf-serrations. In *Arabidopsis*, smooth margins and reduced serrations are observed if the activity of CUC2 is compromised. Contrary to this, the overexpression of CUC2 has been found to result in increased serration. (Nikovics *et al.*, 2006). Homodimeric CUC2 activates CUC3. CUC3 and CUC2 then interact to form a heterodimer that promotes leaf complexity. CUC3 and CUC1 share functional redundancy to promote incipient leaf-serrations (Rubio-Somoza *et al.*, 2014). Another miRNA (miR319) is involved in determining leaf complexity. miR319 targets a number of TCP transcription factors (Palatnik *et al.*, 2003). The up-regulation of miR319 has been found to result in increased serrations, whereas its down-regulation promotes opposite phenotype, i.e. reduced serrations (Rubio-Somoza *et al.*, 2014). The study of a number of TCP and miR319 mutants has attested to their importance in regulation of leaf serration, initiation of leaf senescence and defining cotyledon boundary (Koyama *et al.*, 2017; Yang *et al.*, 2013). Besides regulating the morphological attributes of leaves, some of the TCP transcription factors in collaboration with NGA (NGATHA) transcription factors have been implicated in maintenance of adaxial and abaxial identity and also the earlier events, like localisation of meristem activity (Alvarez *et al.*, 2016).

miR319 and miR164 and their respective targets establish a regulatory circuit that has an important role in regulation of leaf size. TCP3, a well-known target of miR319, directly activates miR164A. miR164 in turn down-regulates CUC2 (Koyama *et al.*, 2007). TCPs have been shown to regulate CUC2 independent of miR164 as well, resulting in reduced serrations. miR156, besides its role in regulating flowering time, regulates leaf development. SPL9, which is one of the targets of miR156, competes with CUC3 and CUC2 for interaction with TCP4. SPL9 forms a dimer with TCP4 (Chitwood and Sinha, 2014; Rubio-Somoza *et al.*, 2014). miR396 regulates leaf shape by targeting GRF. GRFs regulate leaf size by mediating cell division and expansion. Mutants with loss-of-function in GRF1, GRF2 and GRF3 show reduced leaf size, whereas plants with gain of function of these GRFs, develop comparatively bigger leaves (Kim *et al.*, 2003). The increased expression of miR396 results in down-regulation of GRFs and this serves as a cue to decide the final size of a leaf (Rodriguez *et al.*, 2010; Debernardi *et al.*, 2014; Kim and Tsukaya, 2015). By targeting bHLH74 (basic helix-loop-helix DNA-binding superfamily protein 74, miR396b regulates leaf margin and pattern of venation (Debernardi *et al.*, 2012).

miRNAs are also involved in the terminal phase of leaf development referred to as leaf senescence. miR164 regulates leaf senescence by targeting ORE1 (oresarai). The expression of ORE1 is itself up-regulated by EIN2 (ethylene insensitive2) and EIN3 in an age-dependent fashion. EIN2 and EIN3 are two important factors in ethylene signalling cascade (Li *et al.*, 2013). During the starting phases of leaf development,

398   *Plant MicroRNAs and Stress Response*

miR164 gradually suppresses the expression of ORE1. However, towards the terminal phases, its expression gets up-regulated. miR319 has also been found to regulate leaf senescence. TCP4, one of the targets of miR319, has been shown to up-regulate lipoxygenase2, which itself mediates the biosynthesis of JA (jasmonic acid). The up-regulated levels of JA are conducive to leaf senescence. Therefore, miR319-targeted TCPs regulate two separate aspects of leaf development, i.e. positively regulate leaf senescence and negatively regulate leaf growth (Schommer *et al.*, 2008).

2.3 miRNAs in Flowering

Three miRNA families, miR156/157, miR172 and miR159, are attributed to floral development in *Arabidopsis*. miR156 is highly conserved in plants and plays a role in vegetative phase transition. This miRNA also controls flowering time and floral transition in *Arabidopsis* and maize (Reinhart *et al.*, 2002; Wu and Poethig, 2006). miR156 is remarkably expressed in a large number of monocotyledons, dicotyledons, lycopods, mosses and ferns (Axtell and Bowman, 2008; Reinhart *et al.*, 2002). Overexpression of miR156/157 promotes late-flowering phenotype. miR156/157 is highly expressed in young seedlings and the expression decreases gradually as the plant develops (Wang *et al.*, 2009; Wu and Poethig, 2006; Wu *et al.*, 2009). With age, the accumulation of sugar in shoot meristem of plants acts as a signal for expression of miR156 targets. miR156 targets SPL group (~11) of transcription factors in *Arabidopsis*. These transcription factors are pivotal in the regulation of plant growth, development and stress responses (Gandikota *et al.*, 2007; Preston *et al.*, 2013; Xing *et al.*, 2010). SPL15 promotes flowering by activating the MADS-box genes, like FUL (Fruitfull), SOC1 (suppressor of overexpression of constans1), AP-1 (Apetala1) and most importantly the floral meristem identity gene LFY (leafy). AP1 and LFY are two genes whose protein products are vital components of molecular switch to flowering (Hyun *et al.*, 2017; Yamaguchi *et al.*, 2009; Wang *et al.*, 2009). SPL9 activates the expression of miR172 in leaves, which in turn promotes flowering by stimulating the expression of floral integrator, FT (flowering locust) (Wilson *et al.*, 1992; Wu *et al.*, 2009). MiR172 targets a number transcription factors, such as SNZ (schnarchzapfen), SMZ (schlafmutze), TOE1 (target of eat1), TOE2, TOE3 and AP2 which are known suppressers of photoperiod-dependent flowering process (Aukerman *et al.*, 2003; Mathieu *et al.*, 2009; Wu *et al.*, 2009; Yant *et al.*, 2010). In response to cytokinin levels of miR172 are up-regulated, since SPLs which promote miR172 expression are themselves transcriptionally up-regulated by cytokinin (Werner *et al.*, 2021).

miR159 suppresses floral transition and delays flowering by promoting the cleavage of mRNAs that code for gibberellin-specific transcriptional regulator GAMYB-related proteins, viz. MYB33, MYB65 and MYB101. These proteins act as transcription factors for GA-mediated activation of LFY. Increase in the levels of miR159 concomitantly decrease levels of LFY transcripts and therefore delay flowering (Fig. 1) (Achard *et al.*, 2004; Blazquez *et al.*, 2000; Blazquez *et al.*, 1998; Gocal *et al.*, 2001).

Flower patterning, a vital aspect of flower development, is regulated by a number of miRNAs. As per the ABC model, three class of genes, viz. Class A, Class B and Class C, are primarily involved in contributing to the specificity of different floral organs. Class A includes Apetala-1 (AP1) and Apetala-2 (AP2), Class B includes Apetala-3 (AP3) and Pistillata (PI) and Class C includes Agmaous (AG). The expression of these genes is regulated within established boundaries for precise

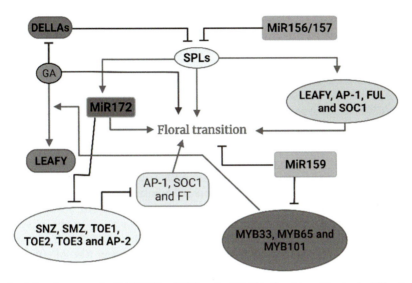

Fig. 1: Involvement of miR156/157, miR172 and miR159 in floral transition in *Arabidopsis*

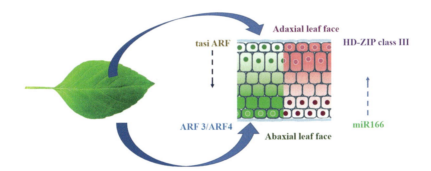

Fig. 2: Regulation of leaf polarity by opposite diffusion gradient patterns of miR166 and tasiARFs

formation of distinct whorls (Causier *et al.*, 2010). The activation of genes in the ABC model is a prerequisite for floral organ identity. AP2 mutually acts with Agmaous (AG) in an antagonistic fashion and interacts with other A class or C class genes in sepal and petal development (Jofuku *et al.*, 1994). In the centre of flower primordia, the accumulation of miR172 determines the boundaries between petals and stamens by confining the expression of AP2 to the outer two whorls of the floral meristem. *In situ* hybridisation experiments have revealed that the domains of expression of AP2 in the outer floral whorls and those of miR172 in the inner counterparts are largely complementary. However, at some points, domains of expression of AP2 show an overlap, suggesting that besides miR172, additional factors may be involved in the regulation of AP2 expression (Wollmann *et al.*, 2010; Zhou *et al.*, 2007). In presence of AP2, a transcriptional regulator, Leunig (LUG), directly represses the expression of miR172 in sepals (Grigorovo *et al.*, 2011). This indicates that for precise boundary

400   *Plant MicroRNAs and Stress Response*

formation of floral organs, a negative feedback loop, where miR172 expression is down-regulated by AP2, is crucial (Yant *et al.*, 2010). AP2 is also thought to direct the LUG transcriptional repressor complex to suppress the expression of miR172 in outer whorls to stabilise its own expression there (Grigorovo *et al.*, 2011). In cereal crops, like rice and barley, miR172 targets transcription factors, namely SNB (Supernumary bract) and cly1 (Cleistogamy1) respectively, and its overexpression has been shown to cause defective floral patterning (Lee and An, 2012; Nair *et al.*, 2010).

2.4 miRNAs in Root Development

Several studies have suggested interplay between miRNAs and phytohormones in root development. miR159, a conserved miRNA interacts with gibberellins and restricts primary root (PR) growth. The miR159 regulatory network where it targets GAMYB-like group of transcription factors like MYB65, affects size in root meristem (RM). Compared to wild-type plants, the transgenic *Arabidopsis* plants expressing rMYB65, an miR159-resistant version of MYB65 and those with loss of function of miR159 (miR159 double mutants), show longer PR and larger RM. MYB65 binds to the promoter of CYCB1(G2-M-specific CYCLINB1) and possibly activates its transcription, which is why rMYB65 plants show enhanced transcription of CYCB1. In conclusion, by down-regulating MYB65, miR159 suppresses cell cycle progression and therefore PR growth (Da Saliva *et al.*, 2017; Millar *et al.*, 2019; Xue *et al.*, 2017; Yu *et al.*, 2015). The role of regulatory network established by miR159, GAMYB genes and gibberellins has been implicated in different tissues and contexts of development (Alonso-Peral *et al.*, 2010; Da Saliva *et al.*, 2017). However, it is still unclear whether a cytokinin-auxin-miR159-GAMYB-gibberellin regulatory loop exists and whether it is has a role in regulation of PR growth.

miRNAs interact with auxin to regulate the development of lateral roots (LRs) (Peret *et al.*, 2009). For example, there exists a positive correlation between miR156 levels, auxin and number of LRs. Besides, establishment of the root meristem, the miR156 regulatory network interacts with auxins to regulate the number and emergence of LRs. The targets of miR156 include SPL3, SPL9 and SPL10 and these SPLs, especially SPL10, are potential suppressors of lateral root growth. During LR growth, SPL9 and SPL10 show a marked response to exogenous auxin treatment. Furthermore, since auxins also up-regulate the expression of miR156 in the roots, a feedback mechanism within the miR156-SPL- auxin network can be established, whereby SPL9, SPL10 and auxin induce expression of miR156, which in turn, down-regulates SPL9 and SPL10 expression in the roots (Yu *et al.*, 2015).

The role of miRNAs has been implicated in the formation of adventitious roots (ARs) as well. For example, miR156 has been found to positively regulate AR formation in a number of plant species, like *Arabidopsis, Oryza sativa, Nicotiana benthamiana, Malus xiaojinensis* and *Nicotiana tabacum* (Feng *et al.*, 2016; Xie *et al.*, 2012; Xu *et al.*, 2016; Xu *et al.*, 2017). In rice (*Oryza sativa*), for example, enhanced expression of miR156 in seedlings results in increased number of ARs with smaller size (Xie *et al.*, 2012). The *Arabidopsis* quintuple mutants with loss of function of spl2, spl9, spl11, spl13 and spl15 and those overexpressing miR156 produce almost similar number of ARs as the wild-type plants. By contrast, in plants with reduced levels of miR156, there is higher accumulation of miR156-targeted SPLs, the number of ARs significantly drops, indicating that at least some SPL genes suppress both the development of ARs and production of LRs (Xu *et al.*, 2016; Yu *et al.*, 2015). It,

Plant MicroRNAs: Physiological Significance in Plants and Animals **401**

however, remains vague whether SPLs interact with hormones during AR growth. In *M. xiaojinensis*, both miR156-overexpressing plants and the wild-type counterparts show increased number of ARs on *in vitro* treatment of their stem cuttings with IAA (indole-3-acetic acid). By contrast, in MIM156 plants (those with low levels of miR156), adventitious rooting capacity is not restored on IAA treatment (Xu *et al.*, 2017).

2.5 miRNAs in Seed Germination

The different stages of seed germination are regulated by the interplay between different phytohormones, like gibberellins (GAs), auxin, abscisic acid (ABA) and cytokinins (CKs) brassinosteroids (BRs) and ethylene (Carrera-Castano *et al.*, 2020). During seed germination in *Arabidopsis*, the transcription factors, MYB101 and MYB33, positively regulate ABA responses. miR159 down-regulates the expression of these two transcription factors by promoting cleavage of their corresponding transcripts, suggesting that miR159 has a regulatory role in seed germination. Intriguingly, the levels of miR159 increase in response to exogenous ABA treatment, suggesting that ABA suppresses its own responses through the miR159 – MYB circuit (since miR159 down-regulates MYB33, which is itself a positive regulator of ABA response). This pathway is considered to be crucial for modulating ABA responses in ensuring that reduced ABA levels are sensed by seeds (Reyes *et al.*, 2007). Since, ABA keeps seeds in dormant state, this negative regulation of ABA signalling by miR159 may serve as one of the control points in the developmental switch to seed germination. Also, ethylene interacts with ABA in many developmental events; therefore, miR159 may indirectly control ethylene-mediated seed germination.

miR160 has been implicated in seed germination. This miRNA targets auxin responsive factor10 (ARF10). Seeds overexpressing miR160a have been found to be less sensitive to ABA compared to the wild-type seeds (Liu *et al.*, 2007). There may be some sort of interplay between ABA and auxin in mature seeds undergoing germination because the ARF10 targeted by miR160 during seed germination is a component of auxin signalling (Guilfoyle *et al.*, 2007). The down-regulation of ARF10 by miR160 may serve as a regulatory step to decrease sensitivity of mature seeds towards ABA, thus making the seeds shift to the germination mode. The exact mechanism of how auxin and ABA interact during germination process is yet to be elucidated. During the germination process, ARF10 could be involved in positive regulation of ABI3 and ABI5, two important components of ABA signalling (Nonogaki, 2008).

A number of miRNAs have been profiled in studies where *Arabidopsis* seedlings were exposed to cold, salinity and dehydration stress (Sunkar and Zhu, 2004). Under stress conditions, miR402 overexpression leads to accelerated seed germination. The target of miR402 is a putative DNA glycosylase, DML3 (demeter-like protein3) which itself positively regulates DNA demethylation. Under stress conditions, mutant seeds with loss of function of DML3 (dml3 mutants) show a phenotype similar to miR402 overexpressing seeds, i.e. accelerated germination, suggesting that the miR402– DML3 circuit regulates the seed germination process. It has been proposed that up-regulation of miR402 by stress results in down-regulation of DML3 transcript, which in turn maintains DNA methylation in genes that themselves negatively regulate seed germination (Kim *et al.*, 2010; Ortega-Galisteo *et al.*, 2008).

miR395 family is comprised of six members, which regulate seed germination both positively and negatively, under abiotic stress conditions in *Arabidopsis*. SULTR,

402   *Plant MicroRNAs and Stress Response*

APS1, APS3 and APS4, which are involved in transport and assimilation of sulphate, are the targets of miR395, whereas SULTR family encodes different sulphate transporters, APS1, APS3 and APS4 code for plastid-localised ATP sulfurylases. The members of this miRNA family are quite specific for their respective targets which is clear from the fact that miR395e is unable to target APS1 and APS4, the targets of miR395c, even though there is a difference of only a single nucleotide between these two miRNAs. In *Arabidopsis*, these two miRNAs work contrary to each other during germination of seeds under dehydration or high salt stress conditions. Under dehydration or high salt stress conditions, overexpression of miR395 increases the potential of seeds to germinate, whereas miR395c overexpression leads to down-regulation of germination potential under these conditions (Kim *et al.*, 2010). miR417 negatively regulates seed germination under salt stress condition (Jung *et al.*, 2007). However, the molecular mechanism of action of this miRNA in this process is not clear.

miR408, besides its role in seed development, is also involved in regulation of light-dependent seed germination which is one of important developmental processes for many seed-bearing plants. PIF1 is a negative regulator of light-dependent germination. The degradation of PIF1 (phytochrome interacting 19 factor1) is one the prerequisites for light-dependent germination to proceed. PIF1 down-regulates the expression miR408 by binding to its promoter and miR408 itself down-regulates the expression of PLC (plantacyanin), which is another negative regulator of germination (Abdel-Ghany and Pilon, 2008; Jiang *et al.*, 2021). It is worth mentioning that the ratio between abscisic acid (ABA) and gibberellic acid (GA) needs to be tightly regulated during germination; whereas a high-ABA-low-GA state promotes seed dormancy, a low-ABA-high-GA state promotes germination. In darkness, expression of miR408 is suppressed by PIF1 accumulation because of lack of active form of phytochrome A (phyA) which could destabilise PIF1 (Jiang *et al.*, 2021; Shen *et al.*, 2008). With low levels of miR408, there is high accumulation of PLC in the storage vacuole which in turn promotes high-ABA-low-GA state and impedes germination. However, upon exposure to far-red light, phyA gets activated and destabilises PIF1 and consequently, miR408 is released from transcriptional inhibition of PIF1. Increased expression of miR408 suppresses expression of PLC which in turn promotes low-ABA- high-GA state and therefore, germination (Fig. 3) (Jiang *et al.*, 2021; Oh *et al.*, 2007; Shen *et al.*, 2008).

miR156 and miR172 regulatory network, besides having a role in flowering, is involved in seed germination. miR156 is expressed abundantly in young seedlings and by targeting SPLs (some of which promote germination), it negatively regulates germination and establishes dormancy. For correct processing of miR156 from its precursor, it requires DOG1 (delay of germination 1). By contrast, reduced expression of miR156 and therefore increased expression of SPLs results in up-regulation of miR172 (SPLs promote expression of miR172). Increased levels of miR172 have been found to promote seed germination (Huang *et al.*, 2013; Huo *et al.*, 2016; Martin *et al.*, 2010). So, the balance between the expression of these two miRNAs and the regulatory circuit formed by them decides between seed germination and dormancy.

miR9678, expressed mainly in the scutellum of developing and germinating seeds of wheat, is thought to regulate germination by directing the biogenesis of phasiRNAs (phased small interfering RNAs) from WSGAR (wheat seed germination-associated RNA) transcripts, which are long non-coding RNAs (Guanghui *et al.*, 2018; Sun *et al.*, 2014). Overexpression of miR9678 and phasiRNAs delays germination and mitigates pre-harvest sprouting. During the course of delaying germination, miR9678 is thought

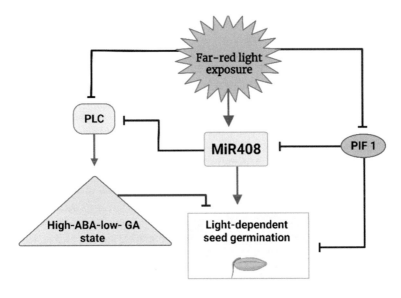

Fig. 3: Role of miR408-mediated regulation of light-dependent seed germination. This figure has been generated in Biorender

to directly target some components of GA biosynthetic pathway, irrespective of how miR9678-triggerd and WSGAR-derived phasiRNAs exert their effect. It has also been found that transcription of miR9678 relies on components ABA signalling cascade indicating a possible miR9678-mediated interaction between GA and ABA in the seed germination of wheat (Guo et al., 2018).

3. miRNAs are Differentially Expressed during Physiological Stress

Upon sensing the external stress, plants get stimulated and respond with appropriate cellular responses. This includes differential transcriptional changes to make the plant capable of surviving against the stress (Verma et al., 2013). miRNAs serve to establish an effective module in plants to respond to various biotic and abiotic stresses (Pandita, 2019; Pandita and Wani, 2019; Pandita, 2021; 2022a; 2022b; 2022c). Under stress cues, miRNAs get differentially expressed. The differential expression has a direct implication in mediating the stress response. Most miRNAs target genes whose protein products are transcriptional factors or enzymes with important roles to play in abiotic stress response (Ebrahimi et al., 2015; Kumar, R. 2014).

A number of miRNAs get differentially expressed in response to biotic stresses, like those exerted by bacteria, viruses, fungi, etc. Upon infection of *Arabidopsis* leaves with *Pseudomonas syringae pv. tomato (DC3000hrcC)* levels of two miRNAs, viz. miR160 and miR167 are up-regulated. The targets of miR167 and miR160 are TIR1 (transport inhibitor response 1) and transcription factors of ARF (auxin response factors) family respectively. When infected with the virulent strain (Pst DC3000) or even the mutant stain (Pst DC3000hrcC mutant), the levels of another miRNA, viz. miR400, decreases (Fahlgren et al., 2007; Zhang et al., 2011). Elevated expression of miR400 has been found to reduce the immunity of plants towards P. syringae pv. tomato (Pst) DC3000 infection. miR400 targets pentatricopeptide repeat (PPR) mRNA. PPR proteins concentrated in chloroplast or mitochondria have been already attributed with some indispensable roles in developmental processes like seed development and seed germination, regulation of circadian clock and post-transcriptional gene regulation (Park *et al.*, 2014). The infection of cassava (*Manihot esculenta*), grown mostly in tropical regions, by *Xanthomonas axonopodis pv. Manihotis*, which is a gram-negative bacterium leads to the development of cassava bacterial blight disease (El-Sharkawy,M.A., 2004). When infected with Xam (*Xanthomonas axonopodis pv. Manihotis*), the expression of 10 conserved miRNAs is up-regulated and that of seven miRNAs is down-regulated. Those with up-regulated expression include miR197b, miR165, miR167a, miR167b, miRE, miR171, miR160, miR390, miR393 and miR394. While miR165, miR394 and miR171 have been found to target *Arabidopsis thaliana* homeobox (ATHB) transcription factor, proteins of F-box family and a scarecrow-like transcription factor, miR160, miR167, miR390 and miR393 are involved in regulation of auxin signalling. The miRNAs that are suppressed include, miR1507, miR395, miR397, miR399, miR408, miR482 miR535. The potential targets of miR408, miR398 and miR397 are plantacyanin, copper superoxide dismutases and laccases respectively, and are therefore believed to regulate copper metabolism. miR395, miR535 and miR482 have been found to target different LRR and NB-LRR genes that are involved in disease resistance (Perez-Quintero *et al.*, 2014).

The identification and characterisation of the endogenous miRNAs having a role to play in stress response due to viral infection is somewhat limited, though the expression pattern of some miRNAs is believed to play significant regulatory roles in plants by targeting different genes. *Mungbean yellow mosaic India virus* (MYMIV), a bipartite begomovirus, causes the yellow mosaic disease (YMD) in many crop plants, especially *Vigna mungo*. Small RNA-sequencing of legume plant *V. mungo*, infected with MYMIV, has revealed the expression pattern of a number of pathogen responsive miRNAs. Upon MYMIV-infection, miR396 witnesses a sharp increase in its expression. One of the targets of miR396 is LOX, which is involved in the synthesis of jasmonic acid (JA) in plants. Elevated levels of miR396 activate SA (salicylic acid)-mediated pathway through the suppression of the JA pathway and confer resistance against MYMIV in *V. mungo*. Other miRNAs, e.g. miR166, miR159, miR1514, miR169 and miR319 have also been implicated in immune response in *V. mungo* against MYMIV. miR159 gets up-regulated whereas miR319 and miR166 are down-regulated upon infection by MYMIV. During MYMIV infection, transcript-levels of miR482 diminish with a corresponding increase in the levels of its target NB-LRR. NB-LRR is a part of disease resistance (R) protein receptors which themselves mediate the recognition of viral effector molecules. Therefore, the down-regulation of miR482 provides resistance against MYMIV. miR393 and miR160 suppress auxin perception and also target ARFs, which are the downstream effectors of auxin signalling and thereby retard normal growth and development of the plant under MYMIV infection (Kundu *et al.*, 2017).

SRBSDV (*southern rice black-streaked dwarf virus*), a potent plant-virus infects rice. Microarray profiling of miRNAs of SRBSDV-infected rice has suggested that miR396, miR164, miRNA1846 and miRNA530 might have a role in the development of various symptoms of SRBSDV-infection. During SRBSDV infection, miR164 is thought to regulate three genes: NAM, NAC1 and salicylic acid-induced protein;

Plant MicroRNAs: Physiological Significance in Plants and Animals   **405**

notably enhanced tillering, presence of aerial rootlets, SA-mediated virus resistance and branches on stem nodes could be ascribed to the regulation of these genes. Under SRBSDV infection, osa-miR396 targets three transcriptions that control transcription of AtGRF4 and AtGRF5 (orthologs of growth regulating factor) which serve crucial roles in cell proliferation in leaves (Xu *et al.*, 2014b). Using high-throughput sequencing, the miRNA-expression pattern of both healthy as well as infected rice plants suggest that a major chunk of miRNAs are expressed in leaves compared to roots. Upon RBSDV infection, the expression of miR166a-d miR166f, miR169h-m and miR156a-j increases whereas that of miR827, miR1428e and miR408 decreases both in roots and leaves. Tissue-specific differential expression also suggests that expression of miR398 and miR164abf increases in leaves but not in roots of RBSDV-infected rice plants (Sun *et al.*, 2015).

SMV (*soybean mosaic virus*), a highly pathogenic virus attacks soybean. Infection of soybean leaves by SMV has revealed the expression profiles of a number of miRNAs. During infection, miR393 and miR160 are up-regulated which in turn have been found to suppress auxin signalling factors in order to stoke resistance against the infection (Yin *et al.*, 2013).

In turmeric (*Curcuma longa L.*), infection with *Pythium aphanidermatum*, necrotrophic oomycete, causes pythium soft rot. During this infection stress, ~ 28 miRNAs, e.g. miR482, miR159, miR164 etc. are down-regulated. The decreased levels of miR164 are associated with elevated levels of NAC, a protein involved in immune response of plants. Further, decreased levels of miR160 / miR167 and miR393 are correlated with up-regulation of auxin responsive genes (Chand *et al.*, 2016).

The differential expression of some plant miRNAs has also been implicated in abiotic stress response. Under drought conditions, some conserved miRNAs, like miR169, miR164, miR171, miR408, miR2118, miR398, miR396 and miR399 have been attributed with some critical roles (Liu *et al.*, 2008). The expression patterns of these miRNAs differ from species to species. The expression of miR169 has been found to be down-regulated in *Medicago truncatula* and *Arabidopsis* (Li *et al.*, 2010; Trindale *et al.*, 2010), but up-regulated in rice and common bean upon treatment with abscisic acid. NFYA5 positively regulates drought stress response. The transgenic lines overexpressing NFYA5 show significantly higher drought tolerance. Low expression of miR169 is associated with higher levels of NFYA5 in *Arabidopsis* (Arenas-Huetero *et al.*, 2009; Li *et al.*, 2008).

The expression of miR408, miR398a and miR398b gets up-regulated in shoots and roots of *M. truncatula* under drought-stress. CSD1 (super oxide dismutase 1), plantacyanin genes and COX5b (mitochondrial cytochrome oxidase) serve as immediate targets of miR408 and miR398. One of these plantacyanin genes, namely UCL8, impacts fertility by regulating pollen tube germination and growth (Zhang *et al.*, 2018).

Up-regulation or down-regulation of miRNAs usually corresponds to potential positive or negative functional role. However, in some cases, the same miRNA exhibits an opposite expression pattern under salt-stress in different plant species, suggesting that a particular miRNA may play different roles in different plant species under salinity stress. For example, the expression of miR393, under salinity stress, is up-regulated in *Agrostis stolonifera*, *Nicotiana tabacum* and *Solanum lycopersicum* but down-regulated in *Arabidopsis thaliana*, *Spartina alterniflora* and *Oryza sativa* (Gupta *et al.*, 2014; Xia *et al.*, 2012). Osa-mR393-overexpression in *Arabidopsis* and rice is accompanied by reduced tolerance to salt and drought stress. It also increases tillers and premature flowering (Gao *et al.*, 2011; Iglesias *et al.*, 2014), and over-expresses miR393-insesitive form of TIR1 (target of miR393 in *Arabidopsis* during salt stress), increasing salt tolerance in mTIR1 transgenic plants (Chen *et al.*, 2015). However, in *Agrostis stolonifera* (creeping bentgrass), overexpression of Osa-mR393 enhances tolerance to salt-stress which is associated with elevated uptake of potassium (Zhao *et al.*, 2019). It suggests that in different plants under salt stress, the same miRNA or alternatively different miRNA belonging to the same miRNA family may have varying promotional and inhibitory effects. Overexpressing Osa-miR396c suppresses alkali and salt stress tolerance in *Arabidopsis* and rice, but in creeping bentgrass, overexpression of this miRNA increases salt tolerance. The increased salt tolerance is associated with increased chlorophyll content and cell membrane integrity, Na^+ exclusion during exposure to high salinity and enhanced water retention (Gao *et al.*, 2010). Overexpression of Osa-miR528 in creeping bentgrass, Sm-miR408 and Tae-miR408 in tobacco and miR408, miR402 and miR399f in *Arabidopsis* confers enhanced salinity tolerance (Baek *et al.*, 2016; Guo *et al.*, 2018; Kim *et al.*, 2010a; Yuan *et al.*, 2015) Contrary to this, overexpression of miRNVL5, miR414c and miR417 increases salt stress sensitivity (Gao *et al.*, 2016; Jung and Kang, 2007). In transgenic *Arabidopsis* plants, increased expression of PeNAC070, the target gene of miR164 corresponds to delayed stem elongation, increased sensitivity to salinity stress and enhanced LR development (Lu *et al.*, 2017). Overexpressing GmNFYA3 which is the target gene of miR169, leads to increased sensitivity to high salinity and exogenous ABA, improves drought tolerance and reduces leaf-specific water loss (Zhao *et al.*, 2009). In apple, overexpression of miR156a reduces resistance to salt stress, whereas overexpressed MdSPL13, the target gene of miR156a improves salt resistance (Ma *et al.*, 2020). Collectively, these studies suggest that through manipulation of miRNAs or their targets, the agronomic trait of salinity stress tolerance could be enhanced.

In salt-stressed maize roots, the expression of miR162, miR395 and miR168 is up-regulated whereas that of miR396, miR156, miR164 and miR167 is down-regulated. These miRNAs target proteins that are involved in regulation of plant's physiology and metabolism. Up-regulation of miR162 and miR168 directly targets DCL1 gene and Argonaute1 (AGO1) transcript respectively. The protein products of these genes are involved in miRNA processing (Ding *et al.*, 2009).

High or low temperatures generate either heat or cold stress in plants. In response to heat stress, plants produce heat shock proteins (HSPs) to safeguard their cellular proteins from denaturation (Iba, 2002). Several miRNAs have been found to play an important role in response to heat stress. In *Arabidopsis*, expression of miR398 is induced by two heat shock transcription factors, namely *HSFA1b and HSFA7b (*heat stress transcription factorsA1b/A7b*),* and miR398 in turn directly suppresses the expression of its target genes CSD1/CSD2 (copper/zinc superoxide dismutase 1 and 2) and copper chaperone of CSD. Up-regulation of these genes suppresses the expression of heat shock proteins. Conversely, down-regulation of CSD genes results in enhanced expression of heat shock transcription factors and HSPs, which in turn can improve heat tolerance (Song *et al.*, 2019). In heat-resistant cotton, miR160 is down-regulated on exposure to heat stress which corresponds to down-regulation of auxin signalling pathway. miR160 targets ARF10, ARF16 and ARF17 and enhances tolerance to heat. Conversely, increased expression of miR160 reduces heat tolerance (Ding *et al.*, 2017). In *Arabidopsis*, frequent heat-stress triggers the expression of miR156,

Plant MicroRNAs: Physiological Significance in Plants and Animals **407**

which is known to retain heat stress memory (Steif *et al.*, 2014). Cold stress induces miR319, a highly conserved miRNA in plants like rice and sugarcane. Overexpression of miR319 suppresses the expression of PCNA (proliferating cell nuclear antigen)-binding factor genes, teosinte branched1 and cycloidea and increases cold tolerance in both rice and sugarcane (Thiebaut *et al.*, 2012; Yang *et al.*, 2013). miR393, another cold-inducible miRNA, enhances cold tolerance by targeting the auxin receptor gene TIR1/AFB and down-regulating auxin signalling pathway (Liu *et al.*, 2017). Table 1 shows the differential expression pattern of some stress-induced miRNAs.

Table 1: The expression pattern in plant miRNAs under stress

S. No.	Type of Stress	Up-regulated miRNAs	Down-regulated miRNAs	Targets
1.	*Xanthomonas axonopodis pv.* infection in cassava (*Manihot esculenta*)	miR197b, miR165, miR167a, miR167b, miR171, miR160, miR393, miR394, miR390 and miRE	miR395, miR397, miR399, miR408, miR482, miR535 and miR1507	1. ATHB, proteins of F-box family and scarecrow-like transcription (targets of miR165, miR394, and miR171) and 2. Plantacyanin and laccases (targets of miR408 and miR397).
2.	*Mungbean yellow mosaic India virus* (MYMIV) infection in *Vigna mungo*	miR396 and miR159	miR319, miR166 and miR482	1. LOX (target of miR396) and 2. NB-LRR (target of miR482).
3.	Salt stress in maize roots	miR162, miR395 and miR168	miR396, miR156, miR164 and miR167	DCL1 gene and Argonaute 1 (targets of miR162 and miR168).
4.	Heat stress in *Arabidopsis*	miR398 and miR156	-	CSD1/CSD2 (targets of miR398).
5.	Cold stress in rice and sugarcane	miR319	-	Proliferating cell nuclear antigen binding factor genes, teosinte branched1 and cycloidea.

4. miRNAs Levels Change with Nutrient Deficiency

During their entire life cycle, plants require many mineral nutrients to support different metabolic processes and to ensure normal growth and development. Plants maintain proper levels of these mineral nutrients and any disruption in the balanced levels of the mineral nutrients may prove disastrous for their physiology and development. Nutrient deficiency activates a number of signalling molecules that convey the low nutrient-status information to the genetic material (Jung and McCouch, 2013). The expression of a number of genes is either activated or repressed to generate specific responses towards the deficiency of a particular mineral nutrient. The altered expression of a number of miRNAs has been found to play critical roles in generation of these responses. For instance, under copper (Cu) deficiency a category of conserved miRNAs including miR408, miR398 and miR397 (collectively called Cu-miRNAs) are up-regulated and they target some non-essential cuproteins for sequestration of Cu (Abdel-Ghany and Pilon, 2008). During Cu deficiency, transcription factor SPL7 up-regulates the expression of miR398. This miR398 helps in sequestration of Cu from Cu/ZnSOD by mediating the replacement of Cu/ZnSOD with Fe superoxide dismutase (FeSOD) and this sequestration increases the availability of Cu for some essential cuproteins, like plastocyanin (Yamasaki *et al.*, 2007; Yamasaki *et al.*, 2009). By contrast, the down-regulation of miR398, triggered by iron deficiency, mediates the replacement to FeSOD with Cu/ZnSOD. So, miR398 plays a role in interplay between Cu and Fe homeostasis (Waters *et al.*, 2012). Under Cu starvation, the expression of some other miRNAs, like miR857, miR397 and miR408 is also up-regulated which in turn corresponds to down-regulation of plastocyanin and laccase gene expression. miR857 down-regulates the transcripts of LAC7 and miR397 has been shown to cleave the transcripts of LAC17, LAC2 and LAC4 during Cu deficiency (Abdel-Ghany and Pilon, 2008). Both miR408 and miR398 get up-regulated in the phloem sap. Under Cu deficiency, SPL7 binds the promoter of miR408 at specific GTAC motifs and promotes its transcription. miR408 down-regulates LAC12, LAC13 and LAC3 genes. In plants, with loss in function of SPL7 (spl7 mutants), overexpression of miR408 under Cu limitation partially subsides the effects of mutation by improving the function of plastocyanin (Abdel-Ghany and Pilon, 2008).

Several miRNAs show altered expression in *Arabidopsis* during nitrogen (N) deficiency and have been divided into two groups. The first group named N-starvation induced (NSI) miRNAs, show up-regulated expression under N starvation and include members of miR846, miR839, miR829, miR826, miR319, miR171, miR169, miR160 and miR156 families. The other group of miRNAs, named N-starvation suppressed (NSS) miRNAs, show reduced expression under N starvation and include members of miR863, miR857, miR850, miR399, miR395, miR172, miR167 and miR827 families. MiR156 family shows the highest abundance among the other group members. The enhanced expression of miRNA160 under N starvation suppresses lateral root development, whereas the decreased expression of miR167 during N-limitation down-regulates the expression of ARF6/8. The down-regulation of ARF6/8 induces the development of lateral and adventitious roots (Jones-Rhoades *et al.*, 2004; Liang *et al.*, 2012). Under N limitation, the decreased abundances of miR169 in *Medicago truncatula*, results in higher expression of HAP2 (hapless) gene. By maintaining low N in the roots, HAP2 promotes the differentiation of nodule primordium (Pant *et al.*, 2009). In *Arabidopsis*, maize and soybean, some members of the miR169 family like miR169d-g, miR169a, miR169h-n and miR169bc have been found to be associated with the up-regulation of nitrate transporters (Xu *et al.*, 2011; Zhao *et al.*, 2011). The expression of miR169d-g has been found to get up-regulated during N deficiency (Liang *et al.*, 2012).

In *Arabidopsis*, miR395 shows increased expression under low sulphate limitation, whereas under sulphate deficiency, its expression decreases. So, sulphate limitation and sulphate deficiency have contrary effects on the expression of miR395. Depending on the status of sulphate (whether abundant or scarce), miR395 strongly regulates

Plant MicroRNAs: Physiological Significance in Plants and Animals **409**

different cell-specific transporters which mediate the transport of sulphate through xylem or phloem (Kawashima *et al.*, 2009). These transports include SULTR1;1, SULTR2;1 and SULTR2;2. During sulphate deficiency, miR395 is enormously expressed in phloem parenchyma, where it mediates sulphate remobilisation between leaves. In contrast, the expression of SULTR2;1 is primarily restricted to the xylem parenchyma. This restricted expression of SULTR2;1 is mediated by miR395 and is crucial for the translocation of sulphate ions from the roots to the shoots (Liang *et al.*, 2010). Under S deprivation, other miRNAs, like miR394, miR168, miR167, miR164, miR160 and miR156 also show altered expression (Kawashima *et al.*, 2011).

During phosphorous deprivation, miR399a is up-regulated in *Arabidopsis*. miR399 directs the cleavage of PHO2 (phosphate over accumulator 2) mRNA by binding to its 5'UTR. PHO2 acts as an important transporter in phosphate mobilisation and with the reduction in levels of PHO2, the levels of Pi increase in the shoots (Bari *et al.*, 2006; Pant *et al.*, 2008; Pant *et al.*, 2009). In response to phosphate limitation, some other miRNAs, e.g. miR2111, miR827 and miR778 are also up-regulated. Interestingly, there is almost a twofold reduction in the levels of miR778 and miR2111 just within three hours of phosphorus re-addition (Pant *et al.*, 2009). miR827 down-regulates the expression of NLA (nitrogen limitation adaptation) which in turn up-regulates the expression of PHO2, thereby facilitating Pi homeostasis (Kant *et al.*, 2011). In addition, expression of miR159a has been found to be up-regulated under P deficiency in soybean roots. This is in contrast to miR1507a, miR398b, miR396a and miR319a which are down-regulated under the same situation (Zeng *et al.*, 2010).

Zinc deficiency promotes altered expression of ~15 miRNAs in *B. juncea* roots. Among them two miRNAs, namely miR845a and miR399b, are down-regulated and the rest 13 miRNAs (398a, 394a, 394b, 393a, 393b, 319a, 319b, 319c, 169g, 160a, 160b, 160c and miR158b) are up-regulated (Shi *et al.*, 2013). miR319 is predicted to target some members of TCP (teosinte branched1/cincinnata/proliferating cell factor) family of transcription factors which are generally involved in biosynthesis jasmonic acid (Schommer *et al.*, 2008). miR393 targets mRNAs of auxin receptors AFB1, AFB2, AFB3 and TIR1 (Toll/interleukin-1 receptor) and of a basic helix-loop helix (bHLH) transcription factor. miR394 targets F-box transcriptional factor and miR160 is thought to target some auxin response factors (ARFs) (Shi *et al.*, 2013). A study in *sorghum bicolour* suggests that expression of 19 miRNAs belonging to eight families is altered during Zn deficiency. Among these miRNAs, four members of miR171, viz. miR171k, miR171d, miR171a and miR171b and miR528 are down-regulated in leaves. The members of other six miRNA families, namely miR408, miR399, miR398, miR319, miR168 and miR166 are up-regulated under the same scenario. The expression of the members of miR399, miR398, miR319, miR168 and miR166 gets up-regulated in roots, whereas that of miR408 gets up-regulated in leaves. Many amongst these zinc deficiency-responsive miRNAs are involved in regulation of plant growth and development. For instance, miR166 regulates root system development by targeting the Class III HD ZIP family proteins, namely ATHB15, REV and PHB (Li *et al.*, 2013).

Altered expression of miRNAs is also observed in response to potassium (K) deficiency. In a recent study on wheat, the expression profiles of a number of miRNAs were analysed under K^+ deficiency. The expression of miR171 increases approximately 5.35-fold after four days of K^+ withdrawal period. The expression of other miRNAs, like miR390, miR164, miR169, miR156 and miR166, also increases (~three-fold). Intriguingly, after extending the K^+ deficiency treatment to eight days,

410 *Plant MicroRNAs and Stress Response*

the expression of all these miRNAs, except for miR393, is inhibited (Li *et al.*, 2013; Thornburg *et al.*, 2020).

During iron (Fe) deficiency, miR2111, miR399, miR397, miR398a, miR398s, miR398b and miR398c are down-regulated. In wild type *A. thaliana,* expression of miR408 is up-regulated during iron deficiency whereas that of its target genes plantacyanin (ARPN), LAC12, LAC13 and LAC3 gets down-regulated. The up-regulated expression of miR408 is associated with enhanced sensitivity to iron. Furthermore, eight members, belonging to five miRNA families, viz. miR394, miR173, miR172, miR164 and miR159 have been found to have altered expression during Fe-deficiency (Carrió-Seguí *et al.*, 2019). Altered expression of a number of miRNAs under nutrient deficiency conditions, has been summed up in Table 2.

5. Plant-derived miRNAs Regulate Gene Expression in Animals

The animal food mostly relies on plants. Plant miRNAs find their entry into the animal or human body via oral intake of food (Table 3). These plant miRNAs are resistant to cooking heat and escape the degradation by gastric enzymes, indicating that plant miRNA can be acquired from raw and cooked meals. In human body, plant miRNAs are transported in microvesicles, which are tiny sacs loaded with lipids, RNA and proteins and play a role in intercellular communication and gene regulation. Plant-derived miRNAs have been reported in human intestinal, lung and liver tissues. The plant miRNAs isolated from human cells can be differentiated by their 2-O-mthylated 3′ end (Martinez *et al.*, 2011; Vaucheret *et al.*, 2012; Zhang *et al.*, 2012)

Recent research trends are suggesting that miRNAs derived from medicinal plants may be promoted as a new category of bioactive compounds with great therapeutic potential. Illuma Hi-sequencing of a valuable Chinese medicinal herb, *Panax ginseng,* has identified 2686 miRNAs that may potentially target approximately 50,992 human genes. These genes are linked to 296 signalling pathways, including hedgehog, TGFB, hippo and neuroactive ligand-receptor pathways. These observations suggest that miRNA component may contribute to the clinical efficacy of *Panax ginseng* in immunoregulatory function, anticancer, cardiovascular protection and cognitive improvement activities. The plant-derived miR168a has been isolated from various tissue samples and sera of animals and plays a role in intercellular communication. By feeding mice with total RNA extract prepared from rice, the level of miR168a elevates in sera and liver of these animals just after three hours of eating. By targeting open reading frame of LDLRAP1 (low-density lipoprotein receptor adapter protein 1), miR168a inhibits its expression and thereby decreases the removal of LDL from plasma (Fig. 4) (Li *et al.*, 2021; Shahid *et al.*, 2018; Zhang *et al.*, 2012).

The beebread diet for worker honey bees, made up of pollen and honey, contains three plant miRNAs including miR152a, miR162a and miR168a. These miRNAs are absent in royal jelly which is fed to the queen bee. Larvae, reared on a diet possessing 16 plant-derived miRNAs, developed traits of worker bees. Studies have further revealed that the development of larvae into worker bees and not the queen was because of, at least in part, *amTOR* (*apis mellifera* target of rapamycin) silencing by plant miR162a present in beebread. *amTOR* gene has been known to play a stimulatory role in caste development in bees (Li *et al.*, 2021; Zhu *et al.*, 2017).

Table 2: Altered expression of plant miRNAs during mineral deficiency

S. No.	Nutrient Deficiency	Up-regulated	Down-regulated	Effect on Some Concerned Genes
1.	Copper	MiR408, miR398, miR397 and 857		LAC3, LAC12 and LAC13 (targets of miR408) are down-regulated. LAC1, LAC2 and LAC 4 (targets of miR397) are down-regulated and LAC7 (target of miR857) is down-regulated.
2.	Nitrogen deficiency in *Arabidopsis*	miR846, miR839, miR829, miR826, miR319, miR171, miR169, miR160 and miR156	miR863, miR857, miR850, miR399, miR395, miR172, miR167 and miR827	ARF6 and ARF7 are down-regulated with decreased expression of miR167. HAP2 is up-regulated with decrease in miR169 levels.
3.	Phosphorus deficiency in *Arabidopsis*	miR399, miR2111, miR827 and miR778		PHO2 (target of miR399) is down-regulated.
4.	Zinc deficiency in *B. juncea* roots	398a, 394a, 394b, 393a, 393b, 319a, 319b, 319c, 169g, 160a, 160b, 160c and miR158b	miR845a and miR399b	TCP (target of miR319) is down-regulated and AFB1, AFB2, AFB3 and TIR1 (targets of miR393) are down-regulated.
5.	Zinc deficiency in *sorghum bicolor*	miR408, miR399, miR398, miR319, miR168 and miR166	miR171k, miR171d, miR171a and miR171b, and miR528	ATHB15, REV and PHB (targets of miR166) get down-regulated.

Table 3: Plant-derived miRNAs target animal genes

S. No.	miRNA	Plant Source	Target in Animals
1.	miR156a	Lettuce, spinach and cabbage	Junctional adhesion molecule A (JAMA) (human endothelial cells).
2.	miR159	Arabidopsis thaliana and Glycine max	Transcription factor 7 (breast cancer cells).
3.	miR171 variant	Tomato, A. thaliana, etc.	mTOR signalling pathway of HEK293 cells.
4.	miR168a	Rice	Low-density lipoprotein receptor adapter protein 1 (LDLRAP1) of mice/humans.
5.	miR471	Lettuce	Hepatitis B virus (HBV).
6.	miR396	Astragalus	Endothelial transcription factor 3 of mice Th2 cells.
7.	miR156	Wheat, maize and soybean	Wnt10b in mice.
8.	miR2911	Lonicera caprifolia (honeysuckle)	NS1 and PB2 genes of influenza A viruses (H1N1, H5N1, H7N9).
9.	Gas-miR01 and Gas-miR02	Gastrodia elata	A20 gene of mice and 293T cells.
10.	mdo-miR7267-3p	Ginger	LGG monooxygenase ycnE of microbiota (mouse gut).
11.	gma-miR159a-3p	Soybean	Transcription factor 7 (human colonic Caco-2 cancer cell).
12.	miR2911	Honeysuckle	Viral genome of SARS -CoV-2 (novel coronavirus).
13.	miR519	Lettuce	Hepatitis B virus (HBV).
14.	miR167e-5p	Maize and Moringa oleifera	β-catenin of Caco-2 and IPEC-J2 cells.
15.	aba miRNA 9497	Atropa belladonna	ZNF-691 (zinc-finger transcription factor) of HNG cells.
16.	FvmiR168	Strawberry fruits	Toll-like receptor 3 in mice.
17.	miR5338	Rape bee pollen	Mitochondrial fusion protein 1 (Mfn1) in rat.

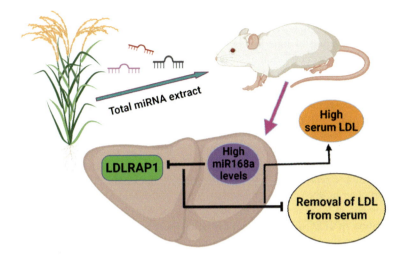

Fig. 4: miR168a from rice increases serum LDL by inhibiting LDLRAP1.
This figure has been generated in Biorender

Four common miRNAs from edible plants may perfectly target 22 human transcripts. The protein products of these transcripts play a role in muscle contraction and tumour suppression. miR168a from *Moringa oleifera* significantly decreases SIRT1 protein in HepG2 cells, suggesting a key role in regulating cell cycle, apoptosis and inflammation in humans (Stefano *et al.*, 2016).

Presence of low levels of plant miR159 is correlated with incidences of breast cancer and its progression. Feeding mice with miR159 at a dose of 25 mg/kg of body weight for 16 days dramatically reduces growth in tumour xenografts. miR159a from soybean induces apoptosis in Caco-2 cells by inhibiting expression of TCF7 (transcription factor 7) and MYC oncoprotein. Transfecting cancer cell lines with broccoli miR160 lower the expression of CDC6, POL2RF, PYCR1, MTHFD2, NCBP2, SHMT2; CELSR3 and FKBP4 genes, suggesting that miR160 may contribute to cancer prevention properties of broccoli (Chin *et al.*, 2016; Pastrello *et al.*, 2016; Liu *et al.*, 2020).

Junctional adhesion molecule A (JAMA) is a transmembrane protein. It is involved in cell invasion, lymphocyte adhesion and platelet aggregation. miR156a, from broccoli, inhibits JAMA expression and promotes up-regulation of E-cadherin and down-regulation of vimentin in human nasopharyngeal carcinoma cell lines, HONE1 and CNE2. As a consequence, these cells significantly lose their invasive capacity. miRNA171 is a ubiquitous and conserved plant miRNA that modulates mTOR pathway in 293 human kidney embryonic cells, suggesting that miRNA171 may have a greater importance in the treatment of various cancers. miR14 from Curcuma longa targets 20 human genes which pay a role in inflammation (Gismondi *et al.*, 2021; Tian *et al.*, 2016).

miR156a is known to have beneficial effects in cardiovascular disease (CVD). After eating lettuce, miR156a was found in human serum and its levels peaked within three hours after intake of a meal. Low levels of miR156a have been found in serum and blood vessel of CVD patients. Plant miR156a targets 3'-UTR of JAMA mRNA in human cells. By inhibiting JAMA expression, miR156a blocks inflammation-induced monocytes adhesion in CVD patients (Hou *et al.*, 2018; Wang *et al.*, 2018).

Gavage feeding of rape bee pollen to BPH (benign prostatic hyperplasia) rat models results in accumulation of seven different plant miRNAs in the prostate gland of rats and correlated with a decreased size of posterior lobes. These miRNAs include miR894, miR5338, miR3440-5p, miR2878- 5p, miR7754-5p, miR5015 and miR7731-3p. Among these, miR5338 is highly enriched miRNA in rat prostate from rape bee pollen treatment. miR5338 prevents BPH by decreasing Mfn1 expression and inhibits mitochondrial fusion and apoptosis (via Bcl-2 pathway) (Xu *et al.*, 2012).

miR167e-5p is a conserved plant miRNA. It is highly expressed in maize and *Moringa oleifera* Lam. This miRNA significantly reduces proliferation of intestinal IPEC-J2 and Caco-2 cells by down-regulating β-catenin. Further, miR156, a common plant miRNA found in soybean, maize and wheat plays a role in intestinal development. miR156 is thought to bind 3'-UTR of Wnt10b, thereby suppressing the proliferation of IPEC-J2 cells. Administration of maize diet containing miR156 to mice has been shown to significantly reduce the villus height:crypt depth ratio in the duodenum. This is further associated with down-regulation of Wnt10b and β-catenin. In conclusion, it is inferred from these studies that plant miRNA156 and miRNA167e-5p are involved in regulation of Wnt/β-catenin pathway to maintain homeostasis in intestinal epithelium and also offer protection against colitis (Li *et al.*, 2019a; Li *et al.*, 2019b).

References

Abdel-Ghany, S.E. and Pilon, M. (2008). MicroRNA-mediated systemic down-regulation of copper protein expression in response to low copper availability in *Arabidopsis*, *J. Biol. Chem.*, 283: 15932-15945.

Achard, P., Herr, A., Baulcombe, D.C. and Harberd, N.P. (2004). Modulation of floral development by a gibberellin-regulated microRNA, *Development*, 131: 3357-3365.

Aida, M., Ishida, T., Fukaki, H., Fujisawa, H. and Tasaka, M. (1997). Genes involved in organ separation in *Arabidopsis*: An analysis of the cup-shaped cotyledon mutant, *Plant Cell*, 9: 841-857.

Alonso-Peral, M.M., Li, J., Li, Y., Allen, R.S., Schnippenkoetter, W., Ohms, S., White, R.G., and Millar, A.A. (2010). The microRNA159-regulated GAMYB-like genes inhibit growth and promote programmed cell death in *Arabidopsis*, *Plant Physiology*, 154(2): 757-771.

Alvarez, J.P., Furumizu, C., Efroni, I., Eshed, Y. and Bowman, J.L. (2016). Active suppression of a leaf meristem orchestrates determinate leaf growth, *eLife*, 5: e15023.

Arenas-Huertero, C., Pérez, B., Rabanal, F., Blanco-Melo, D., De la Rosa, C., Estrada-Navarrete, G., Sanchez, F., Covarrubias, A. and Reyes, J. (2009). Conserved and novel miRNAs in the legume *Phaseolus vulgaris* in response to stress, *Plant Mol. Biol.*, 70: 385-401.

Aukerman, M.J. and Sakai, H. (2003). Regulation of flowering time and floral organ identity by a microRNA and its Apetala2-like target genes, *Plant Cell*, 15: 2730-2741.

Axtell, M.J. and Bowman, J.L. (2008). Evolution of plant microRNAs and their targets, *Trends Plant Sci.*, 13: 343-349.

Baek, D., Chun, H.J., Kang, S., Shin, G., Park, S.J., Hong, H., Kim, C., Kim, D.H., Lee, S.Y., Kim, M.C. and Yun, D.J. (2016). A role for *Arabidopsis* miR399f in salt, drought, and ABA signalling, *Molecules and Cells*, 39(2): 111-118.

Bari, R., Pant, B.D., Stitt, M. and Scheible, W.R. (2006). PHO2, microRNA399 and PHR1 define a phosphate-signalling pathway in plants, *Plant Physiol.*, 141: 988-999.

Blázquez, M.A., Green, R., Nilsson, O., Sussman, M.R. and Weigel, D. (1998). Gibberellins promote flowering of *Arabidopsis* by activating the leafy promoter, *Plant Cell*, 10: 791-800.

Blázquez, M.A. and Weigel, D. (2000). Integration of floral inductive signals in *Arabidopsis*, *Nature*, 404: 889-892.

Carrera-Castano, G., Calleja-Cabrera, J., Pernas, M., Gomez, L. and Onate-Sanchez, L. (2020). An updated overview on the regulation of seed germination, *Plants*, 9: 703.

Carrió-Seguí, À., Ruiz-Rivero, O., Villamayor-Belinchón, L., Puig, S., Perea-García, A. and Peñarrubia, L. (2019). The altered expression of microRNA408 influences the *Arabidopsis* response to iron deficiency, *Frontiers in Plant Science*, 10: 324.

Causier, B., Schwarz-Sommer, Z. and Davies, B. (2010). Floral organ identity: 20 years of ABCs, *Semin. Cell Dev. Biol.*, 21: 73-79.

Chand, S.K., Nanda, S., Rout, E., Mohanty, J., Mishra, R. and Joshi, R.K. (2016). Identification and characterisation of microRNAs in turmeric (*Curcuma longa* L.) responsive to infection with the pathogenic fungus *Pythium aphanidermatum*, *Physiol. Mol. Plant Pathol.*, 93: 119-128.

Chen, X. (2004). A microRNA as a translational repressor of Apetala2 in *Arabidopsis* flower development, *Science*, 303: 2022-2025.

Chen, Z., Hu, L., Han, N., Hu, J., Yang, Y., Xiang, T., Zhang, X. and Wang, L. (2015). Overexpression of a miR393-resistant form of transport inhibitor response protein 1 (mTIR1) enhances salt tolerance by increased osmoregulation and Na$^+$ exclusion in *Arabidopsis thaliana*, *Plant and Cell Physiology*, 56(1): 73-83.

Chin, A.R., Fong, M.Y., Somlo, G., Wu, J., Swiderski, P., Wu, X. and Wang, S.E. (2016). Cross-kingdom inhibition of breast cancer growth by plant MIR159, *Cell Res.*, 26: 217-228.

Chitwood, D.H. and Sinha, N.R. (2014). Plant development: Small RNAs and the metamorphosis of leaves, *Curr. Biol.*, 24: R1087-R1089.

Chitwood, D.H., Guo, M., Nogueira, F.T. and Timmermans, M.C.P. (2007). Establishing leaf polarity: The role of small RNAs and positional signals in the shoot apex, *Development*, 134: 813-823.

Da Silva, E., Ferreira, E., Silva, G.F., Bidoia, D., Azevedo, M., De Jesus, F., Pino, L., Peres, L., Carrera, E., Lopez-Diaz, I. and Nogueira, F.T. (2017). MicroRNA159-targeted SlGAMYB transcription factors are required for fruit set in tomato, *The Plant Journal*, 92: 95-109.

Debernardi, J.M., Mecchia, M.A., Vercruyssen, L., Smaczniak, C., Kaufmann, K. and Inze, D. (2014). Post-transcriptional control of GRF transcription factors by microRNA miR396 and GIF co-activator affects leaf size and longevity, *Plant J.*, 79: 413-426.

Debernardi, J.M., Rodriguez, R.E., Mecchia, M.A. and Palatnik, J.F. (2012). Functional specialisation of the plant miR396 regulatory network through distinct microRNA target interactions, *PLoS Genet.*, 8: e1002419.

Ding, D., Zhang, L., Wang, H., Liu, Z., Zhang, Z. and Zheng, Y. (2009). Differential expression of miRNAs in response to salt stress in maize roots, *Ann Bot.*, 103(1): 29-38.

Ding, Y., Ma, Y., Liu, N., Xu, J. and Hu, Q. (2017). MicroRNAs involved in auxin signalling modulate male sterility under high-temperature stress in cotton (*Gossypium hirsutum*), *Plant J.*, 91: 977-994.

Ebrahimi Khaksefidi, R., Mirlohi, S., Khalaji, F., Fakhari, Z., Shiran, B., Fallahi, H., Rafiei, F., Budak, H. and Ebrahimie, E. (2015). Differential expression of seven conserved microRNAs in response to abiotic stress and their regulatory network in *Helianthus annuus*, *Frontiers in Plant Science*, 6: 741.

El-Sharkawy, M.A. (2004). Cassava biology and physiology, *Plant Mol. Biol.*, 56(4): 481-501.

Fahlgren, N., Howell, M.D., Kasschau, K.D., Chapman, E.J., Sullivan, C.M., Cumbie, J.S., Givan, S.A., Law, T.F., Grant, S.R., Dangl, J.L. and Carrington, J.C. (2007). High-throughput sequencing of *Arabidopsis* microRNAs: Evidence for frequent birth and death of miRNA genes, *PLoS ONE*, 2: e219.

Feng, S., Xu, Y. and Guo, C. (2016). Modulation of miR156 to identify traits associated with vegetative phase change in tobacco (*Nicotiana tabacum*), *Journal of Experimental Botany*, 67(5): 1493-1504.

Gandikota, M., Birkenbihl, R.P., Höhmann, S., Cardon, G.H., Saedler, H. and Huijser, P. (2007). The miRNA156/157 recognition element in the 3' UTR of the *Arabidopsis* SBP box gene SPL3 prevents early flowering by translational inhibition in seedlings, *The Plant Journal: For Cell and Molecular Biology*, 49(4): 683-693.

Gao, P., Bai, X., Yang, L., Lv, D., Li, Y., Cai, H., Ji, W., Guo, D. and Zhu, Y. (2010). Overexpression of osa-miR396c decreases salt and alkali stress tolerance, *Planta*, 231(5): 991-1001.

Gao, P., Bai, X., Yang, L., Lv, D., Pan, X. and Li, Y. (2011). Osa-MIR393: A salinity- and alkaline stress-related microRNA gene, *Mol. Biol. Rep.*, 38: 237-242.

Gao, S., Yang, L., Zeng, H.Q., Zhou, Z.S., Yang, Z.M., Li, H., Sun, D., Xie, F. and Zhang, B. (2016). A cotton miRNA is involved in regulation of plant response to salt stress, *Scientific Reports*, 6: 19736.

Gismondi, A., Nanni, V., Monteleone, V., Colao, C., Di Marco, G. and Canini, A. (2021). Plant miR171 modulates mTOR pathway in HEK293 cells by targeting GNA12, *Mol. Biol. Rep.*, 48: 435-449.

Gocal, G.F.W., Sheldon, C.C., Gubler, F., Moritz, T., Bagnall, D.J., MacMillan, C.P., Li, S.F., Parish, R.W., Dennis, E.S., Weigel, D. and King, R.W. (2001). GAMYB-like genes, flowering and gibberellin signalling in *Arabidopsis*, *Plant Physiol.*, 127: 1682-1693.

Grigorovo, B., Mara, C., Hollender, C., Sijacic, P., Chen, X. and Liu, Z. (2011). Leunig and SEUSS co-repressors regulate miR172 expression in *Arabidopsis* flowers, *Development*, 138: 2451-2456.

Guilfoyle, T.J. and Hagen, G. (2007). Auxin response factors, *Curr. Opin. Plant Biol.*, 10: 453-460.

Guilfoyle, T.J., Ulmasov, T. and Hagen, G. (1998). The ARF family of transcription factors and their role in plant hormone-responsive transcription, *Cell Mol. Life Sci.*, 54: 619-627.

Guo, G., Liu, X., Sun, F., Cao, J., Huo, N., Wuda, B., Xin, M., Hu, Z., Du, J., Xia, R., Rossi, V., Peng, H., Ni, Z., Sun, Q. and Yao, Y. (2018). Wheat miR9678 affects seed germination by generating phased siRNAs and modulating abscisic acid/gibberellin signalling, *The Plant Cell*, 30(4): 796-814.

Guo, X., Niu, J. and Cao, X. (2018). Heterologous expression of *Salvia militiorrhiza* MicroRNA408 enhances tolerance to salt stress in *Nicotiana benthamiana*, *International Journal of Molecular Sciences*, 19(12): 3985.

Gupta, O.P., Meena, N.L., Sharma, I. and Sharma, P. (2014). Differential regulation of microRNAs in response to osmotic, salt and cold stresses in wheat, *Mol. Biol. Rep.*, 41: 4623-4629.

Hou, D., He, F., Ma, L., Cao, M., Zhou, Z., Wei, Z., Xue, Y., Sang, X., Chong, H., Tian, C., Zheng, S., Li, J., Zen, K., Chen, X., Hong, Z., Zhang, C.Y. and Jiang, J. (2018). The potential atheroprotective role of plant MIR156a as a repressor of monocyte recruitment on inflamed human endothelial cells, *The Journal of Nutritional Biochemistry*, 57: 197-205.

Huang, D., Koh, C., Feurtado, J.A., Tsang, E.W. and Cutler, A.J. (2013). MicroRNAs and their putative targets in *Brassica napus* seed maturation, *BMC Genomics*, 14: 140.

Huo, H. (2016). Delay of germination (DOG1) regulates both seed dormancy and flowering time through microRNA pathways. *In:* Wei, S. and Bradford, K.J. (Eds.). *Proceedings of the National Academy of Sciences of the United States of America*, 113(15): E2199-E2206.

Hyun, Y., Richter, R. and Coupland, G. (2017). Competence to flower: Age-controlled sensitivity to environmental cues, *Plant Physiol.*, 173: 36-46.

Iba, K. (2002). Acclimative response to temperature stress in higher plants: Approaches of gene engineering for temperature tolerance, *Annu. Rev. Plant Biol.*, 53: 225-245.

Iglesias, M.J., Terrile, M.C., Windels, D., Lombardo, M.C., Bartoli, C.G. and Vazquez, F. (2014). MiR393 regulation of auxin signalling and redox-related components during acclimation to salinity in *Arabidopsis*, *PLoS ONE*, 9: e107678.

Iwasaki, M., Takahashi, H., Iwakawa, H., Nakagawa, A., Ishikawa, T., Tanaka, H., Matsumura, Y., Pekker, I., Eshed, Y., Vial-Pradel, S., Ito, T., Watanabe, Y., Ueno, Y., Fukazawa, H., Kojima, S., Machida, Y. and Machida, C. (2013). Dual regulation of ETTIN (ARF3) gene expression by AS1-AS2, which maintains the DNA methylation level, is involved in

Plant MicroRNAs: Physiological Significance in Plants and Animals **417**

stabilisation of leaf adaxial-abaxial partitioning in *Arabidopsis, Development* (Cambridge, England), 140(9): 1958-1969.

Jiang, A., Guo, Z., Pan, J., Yang, Y., Zhuang, Y., Zuo, D., Hao, C., Gao, Z., Xin, P., Chu, J., Zhong, S. and Li, L. (2021). The PIF1-miR408-PLANTACYANIN repression cascade regulates light-dependent seed germination, *The Plant Cell*, 33(5): 1506-1529.

Jofuku, K.D., Den Boer, B.G.W., Van Montagu, M. and Okamuro, J.K. (1994). Control of *Arabidopsis* flower and seed development by the homeotic gene apetala2, *Plant Cell*, 6: 1211-1225.

Jones-Rhoades, M.W. and Bartel, D.P. (2004). Computational identification of plant microRNAs and their targets, including a stress-induced miRNA, *Mol. Cell*, 14: 787-799.

Jung, H.J. and Kang, H. (2007). Expression and functional analyses of microRNA417 in *Arabidopsis thaliana* under stress conditions, *Plant Physiol. Biochem.*, 45: 805-811.

Jung, J.H., Seo, Y.H., Seo, P.J., Reyes, J.L., Yun, J. and Chua, N.H. (2007). The Gigantea-regulated microRNA172 mediates photoperiodic flowering independent of Constans in *Arabidopsis, Plant Cell*, 19: 2736-2748.

Jung, J.K.H. and McCouch, S. (2013). Getting to the roots of it: Genetic and hormonal control of root architecture, *Front. Plant Sci.*, 4: 186.

Kant, S., Peng, M. and Rothstein, S.J. (2011). Genetic regulation by NLA and microRNA827 for maintaining nitrate-dependent phosphate homeostasis in *Arabidopsis, PLoS Genet.*, 7: e1002021.

Kawashima, C.G., Matthewman, C.A., Huang, S., Lee, B.R., Yoshimoto, N. and Koprivova, A. (2011). Interplay of SLIM1 and miR395 in the regulation of sulphate assimilation in *Arabidopsis, Plant J.*, 66: 863-876.

Kawashima, C.G., Yoshimoto, N., Maruyama-Nakashita, A., Tsuchiya, Y.N., Saito, K. and Takahashi, H. (2009). Sulphur starvation induces the expression of microRNA-395 and one of its target genes but in different cell types, *Plant J.*, 57: 313-321.

Kim, J.H. and Tsukaya, H. (2015). Regulation of plant growth and development by the growth-regulating factor and GRF-interacting factor duo, *J. Exp. Bot.*, 66: 6093-6107.

Kim, J.H., Choi, D. and Kende, H. (2003). The AtGRF family of putative transcription factors is involved in leaf and cotyledon growth in *Arabidopsis, Plant J.*, 36: 94-104.

Kim, J.Y., Kwak, K.J., Jung, H.J., Lee, H.J. and Kang, H. (2010). MicroRNA402 affects seed germination of *Arabidopsis thaliana* under stress conditions via targeting demeter-like protein3 mRNA, *Plant Cell Physiol.*, 51: 1079-1083.

Kim, J.Y., Lee, H.J., Jung, H.J., Maruyama, K., Suzuki, N. and Kang, H. (2010). Overexpression of microRNA395c or 395e affects differently the seed germination of *Arabidopsis thaliana* under stress conditions, *Planta*, 232: 1447-1454.

Koyama, T., Sato, F. and Ohme-Takagi, M. (2017). Roles of miR319 and TCP transcription factors in leaf development, *Plant Physiol.*, 175: 874-885.

Koyama, T., Furutani, M., Tasaka, M. and Ohme-Takagi, M. (2007). TCP transcription factors control the morphology of shoot lateral organs via negative regulation of the expression of boundary-specific genes in *Arabidopsis, Plant Cell*, 19: 473-484.

Kumar, R. (2014). Role of microRNAs in biotic and abiotic stress responses in crop plants, *Applied Biochemistry and Biotechnology*, 174(1): 93-115.

Kundu, A., Paul, S., Dey, A. and Pal, A. (2017). High throughput sequencing reveals modulation of microRNAs in *Vigna mungo* upon mungbean yellow mosaic India virus inoculation highlighting stress regulation, *Plant Sci.*, 257: 96-105.

Laufs, P., Peaucelle, A., Morin, H. and Traas, J. (2004). MicroRNA regulation of the CUC genes is required for boundary size control in *Arabidopsis* meristems, *Development*, 131: 4311-4322.

Lee, D.Y. and An, G. (2012). Two AP2 family genes, supernumerary bract (SNB) and OsIndeterminate spikelet 1 (OsIDS1), synergistically control inflorescence architecture and floral meristem establishment in rice, *Plant J.*, 69: 445-461.

Li, D., Yang, J., Yang, Y., Liu, J., Li, H., Li, R., Cao, C., Shi, L., Wu, W. and He, K. (2021).

A timely review of cross-kingdom regulation of plant-derived microRNAs, *Frontiers in Genetics*, 12: 613197.

Li, H., Deng, Y., Wu, T., Subramanian, S. and Yu, O. (2010). Mis-expression of miR482, miR1512, and miR1515 increases soybean nodulation, *Plant Physiol.*, 153: 1759-1770.

Liu, H.H., Tian, X., Li, Y.J., Wu, C.A. and Zheng, C.C. (2008). Microarray-based analysis of stress-regulated microRNAs in *Arabidopsis thaliana*, *RNA*, 14: 836-843.

Li, M., Chen, T., He, J.J., Wu, J.H., Luo, J.Y., Ye, R.S., Xie, M.Y., Zhang, H.J., Zeng, B., Liu, J., Xi, Q.Y., Jiang, Q.Y., Sun, J.J. and Zhang, Y.L. (2019a). Plant MIR167e-5p inhibits enterocyte proliferation by targeting β-catenin, *Cells*, 8: 1385-1391.

Li, M., Chen, T., Wang, R., Luo, J.Y., He, J.J., Ye, R.S., Xie, M.Y., Zhang, H.J., Zeng, B., Liu, J., Xi, Q.Y., Jiang, Q.Y., Sun, J.J. and Zhang, Y.L. (2019b). Plant MIR156 regulates intestinal growth in mammals by targeting the Wnt/β-catenin pathway, *Am. J. Physiol-Cell Ph.*, 317: 434-448.

Li, W.X., Oono, Y., Zhu, J., He, X.J., Wu, J.M., Iida, K., Lu, X.Y., Cui, X., Jin, H. and Zhu, J.K. (2008). The *Arabidopsis* NFYA5 transcription factor is regulated transcriptionally and post-transcriptionally to promote drought resistance, *The Plant Cell*, 20(8): 2238-2251.

Li, Y., Zhang, Y., Shi, D., Liu, X., Qin, J., Ge, Q. and Xu, J. (2013). Spatial-temporal analysis of zinc homeostasis reveals the response mechanisms to acute zinc deficiency in Sorghum bicolor, *New Phytologist.*, 200(4): 1102-1115.

Li, Z., Peng, J., Wen, X. and Guo, H. (2013). Ethylene-insensitive3 is a senescence associated gene that accelerates age-dependent leaf senescence by directly repressing miR164 transcription in *Arabidopsis*, *Plant Cell*, 25: 3311-3328.

Liang, G. and Yu, D. (2010). Reciprocal regulation among and miR395 post-transcriptionally APS and SULTR2; 1 *Arabidopsis thaliana*, *Plant Signal. Behav.*, 5: 1257-1259.

Liang, G., He, H. and Yu, D. (2012). Identification of nitrogen starvation responsive miRNAs in *Arabidopsis thaliana*, *PLoS ONE*, 7: e48951.

Liang, G., Yang, F. and Yu, D. (2010). MicroRNA395 mediates regulation of sulphate accumulation and allocation in *Arabidopsis thaliana*, *Plant J.*, 62: 1046-1105.

Liu, J., Wang, F., Weng, Z., Sui, X., Fang, Y., Tang, X. and Shen, X. (2020). Soybean-derived miRNAs specifically inhibit proliferation and stimulate apoptosis of human colonic Caco-2 cancer cells but not normal mucosal cells in culture, *Genomics*, 112(5): 2949-2958.

Liu, P.P., Montgomery, T.A., Fahlgren, N., Kasschau, K.D., Nonogaki, H. and Carrington, J.C. (2007). Repression of auxin responsive factor10 by microRNA160 is critical for seed germination and post-germination stages, *The Plant Journal: For Cell and Molecular Biology*, 52(1): 133-146.

Liu, X., Huang, J., Wang, Y., Khanna, K., Xie, Z., Owen, H.A. and Zhao, D. (2010). The role of floral organs in carpels, an *Arabidopsis* loss-of-function mutation in microRNA160a, in organogenesis and the mechanism regulating its expression, *Plant J.*, 62: 416-428.

Liu, Y.-C., Chen, W.L., Kung, W.-H. and Huang, H.-D. (2017). Plant miRNAs found in human circulating system provide evidences of cross kingdom RNAi, *BMC Genom.*, 18: 112.

Liu, Y., Wang, K., Li, D., Yan, J. and Zhang, W. (2017). Enhanced cold tolerance and tillering in switchgrass (*Panicum virgatum* L.) by heterologous expression of Osa-miR393a, *Plant Cell Physiol.*, 58: 2226-2240.

Lu, X., Dun, H., Lian, C., Zhang, X., Yin, W. and Xia, X. (2017). The role of peu-miR164 and its target PeNAC genes in response to abiotic stress in *Populas euphratica*, *Plant Physiology and Biochemistry: PBB*, 115: 418-438.

Ma, Y., Xue, H., Zhang, F., Jiang, Q., Yang, S., Yue, P., Wang, F., Zhang, Y., Li, L., He, P. and Zhang, Z. (2021). The miR156/SPL module regulates apple salt stress tolerance by activating MdWRKY100 expression, *Plant Biotechnology Journal*, 19(2): 311-323.

Mallory, A.C., Bartel, D.P. and Bartel, B. (2005). MicroRNA-directed regulation of *Arabidopsis* auxin response factor17 is essential for proper development and modulates expression of early auxin response genes, *Plant Cell*, 17: 1360-1375.

Mallory, A.C., Dugasm, D.V., Bartel, D.P. and Bartel, B. (2004). MicroRNA regulation of NAC-domain targets is required for proper formation and separation of adjacent embryonic, vegetative, and floral organs, *Curr. Biol.*, 14: 1035-1046.

Marin-Gonzalez, E. and Suarez-Lopez, P. (2012). And yet it moves: Cell-to cell and long-distance signalling by plant microRNAs. *Plant Sci.*, 196: 18-30.

Martin, R.C., Liu, P.P., Goloviznina, N.A. and Nonogaki, H. (2010). MicroRNA, seeds, and Darwin? Diverse function of miRNA in seed biology and plant responses to stress, *J. Exp. Bot.*, 61: 2229-2234.

Martinez de Alba, A.E., Jauvion, V., Mallory, A.C., Bouteiller N. and Vaucheret, H. (2011). The miRNA pathway limits AGO1 availability during siRNA-mediated PTGS defence against exogenous RNA, *Nucleic Acids Res.*, 39: 9339-9344.

Mathieu, J., Yant, L.J., Murdter, F., Kuttner, F. and Schmid, M. (2009). Repression of flowering by the miR172 target SMZ, *PLoS Biol.*, 7: e1000148.

Merelo, P., Ram H., Pia Caggianoa, M., Ohnoa, C., Ottb, F., Straubc, D., Graeffc, M., Cho, S.K., Yangc, S.W., Wenkelc, S. and Heisler, M.G. (2016). Regulation of MIR165/166 by class II and class III homeodomain leucine zipper proteins establishes leaf polarity, *Proc. Natl. Acad Sci. USA*, 113(42): 11973-11978. doi:10.1073/pnas.1516110113

Millar, A.A., Lohe, A. and Wong, G. (2019). Biology and function of miR159 in plants, *Plants* (Basel), 8: 255

Nair, S.K., Wang, N., Turuspekov, Y., Pourkheirandish, M., Sinsuvongwat, S., Chen, G., Sameri, M., Tagiri, A., Honda, I., Watanabe, Y., Kanamori, H., Wicker, T., Stein, N., Nagamura, Y., Matsumoto, T. and Komatsuda, T. (2010). Cleistogamous flowering in barley arises from the suppression of microRNA-guided HvAP2 mRNA cleavage, *Proc. Natl. Acad Sci. USA*, 107(1): 490-495. doi:10.1073/pnas.0909097107

Nath, U., Crawford, B.C., Carpenter, R. and Coen, E. (2003). Genetic control of surface curvature, *Science*, 299: 1404-1407.

Nikovics, K., Blein, T., Peaucelle, A., Ishida, T., Morin, H. and Aida, M. (2006). The balance between the MIR164A and CUC2 genes controls leaf margin serration in *Arabidopsis*, *Plant Cell*, 18: 2929-2945.

Nonogaki, H. (2008). Repression of transcription factors by microRNA during seed germination and post-germination: Another level of molecular repression in seeds? *Plant Signal. Behav.*, 3: 65-67.

Oh, E., Yamaguchi, S., Hu, J.H., Yusuke, J., Jung, B., Paik, I., Lee, H.S., Sun, T.P., Kamiya, Y. and Choi, G. (2007). PIL5, a phytochrome interacting bHLH protein, regulates gibberellin responsiveness by binding directly to the GAI and RGA promoters in *Arabidopsis* seeds, *Plant Cell*, 19: 1192-1208.

Ortega-Galisteo, A., Morales-Ruiz, T., Ariza, R. and Roldán-Arjona, T. (2008). *Arabidopsis* Demeter-like proteins DML2 and DML3 are required for appropriate distribution of DNA methylation marks, *Plant Mol. Biol.*, 67: 671-681.

Palatnik, J.F., Allen, E., Wu, X., Schommer, C., Schwab, R. and Carrington, J.C. (2003). Control of leaf morphogenesis by microRNAs, *Nature*, 425: 257-263.

Pandita, D. (2019). Plant MIRnome: miRNA biogenesis and abiotic stress response. *In:* Hasanuzzaman, M., Hakeem, K., Nahar, K. and Alharby, H. (Eds.). *Plant Abiotic Stress Tolerance*. Springer, Cham, 449-474. Doi https://Doi.org/10.1007/978-3-030-06118-0_18

Pandita, D. and Wani, S.H. (2019). MicroRNA as a tool for mitigating abiotic stress in rice (*Oryza sativa* L.). *In:* Wani, S. (Ed.). *Recent Approaches in Omics for Plant Resilience to Climate Change*, Springer, Chamz, 109-133. Doi https://Doi.org/10.1007/978-3-030-21687-0_6

Pandita, D. (2021). Role of miRNAi technology and miRNAs in abiotic and biotic stress resilience. *In:* Aftab, T. and Roychoudhury, A. (Eds.) *Plant Perspectives to Global Climate Changes*. Academic Press, Elsevier, 303-330. https://Doi.org/10.1016/B978-0-323-85665-2.00015-7

Pandita, D. (2022a). How microRNAs regulate abiotic stress tolerance in wheat? A snapshot. *In:* Roychoudhury, A., Aftab, T. and Acharya, K. (Eds.). *Omics Approach to Manage Abiotic Stress in Cereals*. Springer, Singapore, 447-464. https://Doi.org/10.1007/978-981-19-0140-9_17

Pandita, D. (2022b). MicroRNAs shape the tolerance mechanisms against abiotic stress in maize. *In:* Roychoudhury, A., Aftab, T. and Acharya, K. (Eds.). *Omics Approach to Manage Abiotic Stress in Cereals.* Springer, Singapore, 479-493. https://Doi.org/10.1007/978-981-19-0140-9_19

Pandita, D. (2022c). miRNA- and RNAi-mediated metabolic engineering in plants. *In:* Aftab, T. and Hakeem, K.R. (Eds.). *Metabolic Engineering in Plants.* Springer, Singapore, 171-186. https://Doi.org/10.1007/978-981-16-7262-0_7

Pant, B.D., Buhtz, A., Kehr, J. and Scheible, W.R. (2008). MicroRNA399 is a long distance signal for the regulation of plant phosphate homeostasis, *Plant J.*, 53: 731-738.

Pant, B.D., Musialak-lange, M., Nuc, P., May, P., Buhtz, A. and Kehr, J. (2009). Identification of nutrient-responsive *Arabidopsis* and rapeseed microRNAs by comprehensive real-time polymerase chain reaction profiling and small RNA sequencing, *Plant Physiol.*, 150: 1541-1555.

Park, Y.J., Lee, H.J., Kwak, K.J., Lee, K., Hong, S.W. and Kang, H. (2014). MicroRNA400-guided cleavage of pentatricopeptide repeat protein mRNAs renders *Arabidopsis thaliana* more susceptible to pathogenic bacteria and fungi, *Plant Cell Physiol.*, 55(9): 1660-1668.

Pastrello, C., Tsay, M., McQuaid, R., Abovsky, M., Pasini, E. and Shirdel, E. (2016). Circulating plant miRNAs can regulate human gene expression *in vitro*, *Sci. Rep.*, 6: 32773.

Pekker, I., Alvarez, J.P. and Eshed, Y. (2005). Auxin response factors mediate *Arabidopsis* organ asymmetry via modulation of Kanadi activity, *Plant Cell*, 17: 2899-2910.

Péret, B., De Rybel, B., Casimiro, I., Benková, E., Swarup, R., Laplaze, L., Beeckman, T. and Bennett, M.J. (2009). *Arabidopsis* lateral root development: An emerging story, *Trends in Plant Science*, 14: 399-408.

Pérez-Quintero, Á.L., Quintero, A., Urrego, O., Vanegas, P. and López, C. (2012). Bioinformatic identification of cassava miRNAs differentially expressed in response to infection by *Xanthomonas axonopodis* pv. Manihotis, *BMC Plant Biol.*, 12(1): 29.

Preston, J.C. and Hileman, L.C. (2013). Functional evolution in the plant squamosa-promoter binding protein-like (SPL) gene family, *Front. Plant Sci.*, 4: 80.

Pulido, A. and Laufs, P. (2010). Co-ordination of developmental processes by small RNAs during leaf development, *J. Exp. Bot.*, 61: 1277-1291.

Reinhart, B.J., Weinstein, E.G., Rhoades, M.W., Bartel, B. and Bartel, D.P. (2002). MicroRNAs in plants, *Genes Dev.*, 16: 1616-1626.

Reyes, J.L. and Chua, N.-H. (2007). ABA induction of miR159 controls transcript levels of two MYB factors during *Arabidopsis* seed germination, *Plant J.*, 49: 592-606.

Robinson, D.O. and Roeder, A.H.K. (2017). Small RNAs turn over a new leaf as morphogens, *Dev. Cell*, 43: 253-254.

Rodriguez, R.E., Mecchia, M.A., Debernardi, J.M., Schommer, C., Weigel, D. and Palatnik, J.F. (2010). Control of cell proliferation in *Arabidopsis thaliana* by microRNA miR396, *Development*, 137: 103-112.

Roodbarkelari, F., Du, F., Truernit, E. and Laux, T. (2015). ZLL/AGO10 maintains shoot meristem stem cells during *Arabidopsis* embryogenesis by down-regulating ARF2-mediated auxin response, *BMC Biol.*, 13: 74.

Rubio-Somoza, I., Zhou, C.M., Confraria, A., Martinho, C., von Born, P. and Baena-Gonzalez, E. (2014). Temporal control of leaf complexity by miRNA-regulated licensing of protein complexes, *Curr. Biol.*, 24: 2714-2719.

Schommer, C., Palatnik, J.F., Aggarwal, P., Chételat, A., Cubas, P., Farmer, E.E. and Weigel, D. (2008). Control of jasmonate biosynthesis and senescence by miR319 targets, *PLoS Biology*, 6(9): e230.

Shahid, S., Kim, G., Johnson, N.R., Wafula, E., Wang, F. and Coruh, C. (2018). MicroRNAs from the parasitic plant *Cuscuta campestris* target host messenger RNAs, *Nature*, 553: 82-85.

Shen, H., Zhu, L., Castillon, A., Majee, M., Downie, B. and Huq, E. (2008). Light-induced phosphorylation and degradation of the negative regulator phytochrome-interacting factor1 from *Arabidopsis* depend upon its direct physical interactions with photo-activated phytochromes, *Plant Cell*, 20: 1586-1602.

Shi, D.Q., Zhang, Y., Li, Y.L. and Jin, X.U. (2013). Identification of zinc deficiency-responsive microRNAs in *Brassica juncea* roots by small RNA sequencing, *Journal of Integrative Agriculture*, 12(11): 2036-2044.

Skopelitis, D.S., Benkovics, A.H., Husbands, A.Y. and Timmermans, M.C.P. (2017). Boundary formation through a direct threshold-based readout of mobile small RNA gradients, *Dev. Cell*, 43: 265-273.

Song, X., Li, Y., Cao, X. and Qi, Y. (2019). MicroRNAs and their regulatory roles in plant-environment interactions, *Ann. Rev. Plant Biol.*, 70: 489-525.

Stefano, P., Letizia, Z., Maurice, K., Carla, M., Antonella, M. and Marina, P. (2016). MicroRNA from *Moringa oleifera*: Identification by high throughput sequencing and their potential contribution to plant medicinal value, *PLoS ONE*, 11: e0149495.

Steif, A., Altmann, S., Hoffmann, K., Pant, B.D., Scheible, W.R. and Baurle, I. (2014). *Arabidopsis* miR156 regulates tolerance to recurring environmental stress through SPL transcription factors, *Plant Cell*, 26: 1792-1807.

Sun, F., Guo, G., Du, J., Guo, W., Peng, H., Ni, Z., Sun, Q. and Yao, Y. (2014). Whole-genome discovery of miRNAs and their targets in wheat (*Triticum aestivum* L.), *BMC Plant Biol.*, 14: 142.

Sun, Z., He, Y., Li, J., Wang, X. and Chen, J. (2015). Genome-wide characterisation of rice black streaked dwarf virus-responsive microRNAs in rice leaves and roots by small RNA and degradome sequencing, *Plant Cell Physiol.*, 56(4): 688-699.

Sunkar, R. and Zhu, J.K. (2004). Novel and stress-regulated microRNAs and other small RNAs from *Arabidopsis*, *Plant Cell*, 16: 2001-2019.

Thiebaut, F., Rojas, C.A., Almeida, K.L., Grativol, C. and Domiciano, G.C. (2012). Regulation of miR319 during cold stress in sugarcane, *Plant Cell Environ.*, 35: 502-512.

Thornburg, T.E., Liu, J., Li, Q., Xue, H., Wang, G., Li, L., Fontana, J.E., Davis, K.E., Liu, W., Zhang, B., Zhang, Z., Liu, M. and Pan, X. (2020). Potassium deficiency significantly affected plant growth and development as well as microRNA-mediated mechanism in wheat (*Triticum aestivum* L.), *Frontiers in Plant Science*, 11: 1219.

Tian, Y., Cai, L., Tian, Y., Tu, Y., Qiu, H., Xie, G., Huang, D., Zheng, R. and Zhang, W. (2016). miR156a mimic represses the epithelial-mesenchymal transition of human nasopharyngeal cancer cells by targeting junctional adhesion molecule A, *PLoS ONE*, 11(6): e0157686.

Trindade, I., Capitao, C., Dalmay, T., Fevereiro, M.P. and Santos, D.M. (2010). miR398 and miR408 are up-regulated in response to water deficit in *Medicago truncatula*, *Planta*, 231: 705-716.

Vaucheret, H. and Chupeau, Y. (2012). Ingested plant miRNAs regulate gene expression in animals, *Cell Research*, 22(1): 3-5.

Verma, S., Nizam, S. and Verma, P.K. (2013). Biotic and abiotic stress signalling in plants. *In: Stress Signalling in Plants: Genomics and Proteomics Perspective*, vol. 1, 25-49.

Vroemen, C.W., Mordhorst, A.P., Albrecht, C. and Kwaaitaal, de Vries, S.C. (2003). The cup-shaped cotyledon3 gene is required for boundary and shoot meristem formation in *Arabidopsis*, *Plant Cell*, 15: 1563-1577.

Wang, J.W., Czech, B. and Weigel, D. (2009). miR156-regulated SPL transcription factors define an endogenous flowering pathway in *Arabidopsis thaliana*, *Cell*, 138: 738-749.

Wang, J.W., Wang, L.J., Mao, Y.B., Cai, W.J., Xue, H.W. and Chen, X.Y. (2005). Control of root cap formation by microRNA-targeted auxin response factors in *Arabidopsis*, *Plant Cell*, 17(8): 2204-2216.

Wang, W., Liu, D., Zhang, X., Chen, D., Cheng, Y. and Shen, F. (2018). Plant MicroRNAs in cross-kingdom regulation of gene expression, *International Journal of Molecular Sciences*, 19(7): 2007.

Waters, B.M., McInturf, S.A. and Stein, R.J. (2012). Rossette iron deficiency transcript and microRNA profiling reveals links between copper and iron homeostasis in *Arabidopsis thaliana*, *J. Exp. Bot.*, 63: 5903-5918.

Werner, S., Bartrina, I. and Schmülling, T. (2021). Cytokinin regulates vegetative phase change in *Arabidopsis thaliana* through the miR172/TOE1-TOE2 module, *Nature Communications*, 12(1): 5816.

Wilson, R.N., Heckman, J.W. and Somerville, C.R. (1992). Gibberellin is required for flowering in *Arabidopsis thaliana* under short days, *Plant Physiol.*, 100: 403-408.

Wollmann, H., Mica, E., Todesco, M., Long, J.A. and Weigel, D. (2010). On reconciling the interactions between APETALA2, miR172 and AGAMOUS with the ABC model of flower development, *Development*, 137: 3633-3642.

Wu, G., Park, M.Y., Conway, S.R., Wang, J.W., Weigel, D. and Poethig, R.S. (2009). The sequential action of miR156 and miR172 regulates developmental timing in *Arabidopsis*, *Cell*, 138: 75.

Wu, G. and Poethig, R.S. (2006). Temporal regulation of shoot development in *Arabidopsis thaliana* by miR156 and its target SPL3, *Development*, 133: 3539-3547.

Xia, K., Wang, R., Ou, X., Fang, Z., Tian, C. and Duan, J. (2012). OsTIR1 and OsAFB2 down-regulation via OsmiR393 overexpression leads to more tillers, early flowering and less tolerance to salt and drought in rice, *PLoS ONE*, 7(1): e30039.

Xie, K., Shen, J., Hou, X., Yao, J., Li, X., Xiao, J. and Xiong, L. (2012). Gradual increase of miR156 regulates temporal expression changes of numerous genes during leaf development in rice, *Plant Physiology*, 158: 1382-1394.

Xing, S., Salinas, M., Hohmann, S., Berndtgen, R. and Huijser, P. (2010). miR156-targeted and non-targeted SBP-box transcription factors act in concert to secure male fertility in *Arabidopsis*, *Plant Cell*, 22: 3935-3950.

Xu, D., Mou, G., Wang, K. and Zhou, G. (2014b). MicroRNAs responding to southern rice black-streaked dwarf virus infection and their target genes associated with symptom development in rice, *Virus Res.*, 190: 60-68.

Xu, F., Liu, Q., Chen, L., Kuang, J., Walk, T. and Wang, J. (2013). Genome-wide identification of soybean microRNAs and their targets reveals their organs specificity and responses to phosphate starvation, *BMC Genomics*, 14: 66.

Xu, J., Wang, Y., Tan, X. and Jing, H. (2012). MicroRNAs in autophagy and their emerging roles in crosstalk with apoptosis, *Autophagy*, 8: 873-882.

Xu, M., Hu, T., Zhao, J., Park, M.Y., Earley, K.W., Wu, G., Yang, L. and Poethig, R.S. (2016). Developmental Functions of miR156-regulated Squamosa Promoter binding protein-like (SPL) genes in *Arabidopsis thaliana*, *PLoS Genetics*, 12(8): e1006263.

Xu, X., Li, X., Hu, X., Wu, T., Wang, Y., Xu, X., Zhang, X. and Han, Z. (2017). High miR156 Expression is required for auxin-induced adventitious root formation via MxSPL26 independent of PINs and ARFs in *Malus xiaojinensis*, *Frontiers in Plant Science*, 8: 1059.

Xu, Z., Zhong, S., Li, X., Li, W., Rothstein, S.J. and Zhang, S. (2011). Genomewide identification of microRNAs in response to low nitrate availability in maize leaves and roots, *PLoS ONE*, 6: e28009.

Xue, Tao, Liu, Zhenhua, Dai, Xuehuan and Xiang, Fengning (2017). Primary root growth in *Arabidopsis thaliana* is inhibited by the miR159 mediated repression of MYB33, MYB65 and MYB101, *Plant Science*, 262: 182-189.

Yamaguchi, A., Wu, M.-F., Yang, L., Wu, G., Poethig, R.S. and Wagner, D. (2009). The microRNA-regulated SBP-box transcription factor SPL3 is a direct upstream activator of leafy, fruitful and Apetala1, *Dev. Cell*, 17: 268-278.

Yamaguchi, T., Nukazuka, A. and Tsukaya, H. (2012). Leaf adaxial-abaxial polarity specification and lamina outgrowth: Evolution and development, *Plant Cell Physiol.*, 53: 1180-1194.

Yamasaki, H., Abdel-Ghany, S.E., Cohu, C.M., Kobayashi, Y., Shikanai, T. and Pilon, M. (2007). Regulation of copper homeostasis by micro-RNA in *Arabidopsis*, *J. Biolumin. Chemilumin.*, 282: 16369-16378.

Yamasaki, H., Hayashi, M., Fukazawa, M., Kobayashi, Y. and Shikanai, T. (2009). Squamosa promoter binding protein-like7 is a central regulator for copper homeostasis in *Arabidopsis*, *Plant Cell*, 21: 347-361.

Yang, C., Li, D., Mao, D., Liu, X.U.E., Ji, C., Li, X., Zhao, X., Cheng, Z., Chen, C. and Zhu, L. (2013). Overexpression of microRNA319 impacts leaf morphogenesis and leads to enhanced cold tolerance in rice (*Oryza sativa* L.), *Plant, Cell Environ.*, 36: 2207-2218.

Yant, L., Mathieu, J., Dinh, T.T., Ott, F., Lanz, C. and Wollmann, H. (2010). Orchestration of the floral transition and floral development in *Arabidopsis* by the bifunctional transcription factor Apetala2, *Plant Cell*, 22: 2156-2170.

Yin, X., Wang, J., Cheng, H., Wang, X. and Yu, D. (2013). Detection and evolutionary analysis of soybean miRNAs responsive to soybean mosaic virus, *Planta*, 237(5): 1213-1225.

Yuan, S., Li, Z., Li, D., Yuan, N., Hu, Q. and Luo, H. (2015). Constitutive expression of rice MicroRNA528 alters plant development and enhances tolerance to salinity stress and nitrogen starvation in creeping bentgrass, *Plant Physiology*, 169(1): 576-593.

Yu, N., Niu, Q.W., Ng, K.H. and Chua, N.H. (2015). The role of miR156/SPLs modules in *Arabidopsis* lateral root development, *The Plant Journal*, 83: 673-685.

Yu, Y., Ji, L., Le, B.H., Zhai, J., Chen, J., Luscher, E., Gao, L., Liu, C., Cao, X., Mo, B., Ma, J., Meyers, B.C. and Chen X. (2017). Argonaute10 promotes the degradation of miR165/6 through the SDN1 and SDN2 exonucleases in *Arabidopsis*, *PLoS Biol.*, 15: e2001272.

Zeng, H.Q., Zhu, Y.Y., Huang, S.Q. and Yang, Z.M. (2010). Analysis of phosphorus-deficient responsive miRNAs and cis elements from soybean (*Glycine max* L.), *J. Plant Physiol.*, 167: 1289-1297.

Zhang, F., Zhang, Y. and Zhang, J. (2018). Rice UCL8, a plantacyanin gene targeted by miR408, regulates fertility by controlling pollen tube germination and growth, *Rice*, 11(60): 1-6.

Zhang, L., Hou, D., Chen, X., Li, D., Zhu, L., Zhang, Y., Li, J., Bian, Z., Liang, X., Cai, X., Yin, Y., Wang, C., Zhang, T., Zhu, D., Zhang, D., Xu, J., Chen, Q., Ba, Y., Liu, J., Wang, Q, … Zhang, C.Y. (2012). Exogenous plant MIR168a specifically targets mammalian LDLRAP1: Evidence of cross-kingdom regulation by microRNA, *Cell Research*, 22(1): 107-126.

Zhang, W., Gao, S., Zhou, X., Chellappan, P., Chen, Z., Zhou, X., Zhang, X., Fromuth, N., Coutino, G., Coffey, M. and Jin, H. (2011). Bacteria-responsive microRNAs regulate plant innate immunity by modulating plant hormone networks, *Plant Mol. Biol.*, 75(1-2): 93-105.

Zhao, B., Ge, L., Liang, R., Li, W., Ruan, K., Lin, H. and Jin, Y. (2009). Members of miR-169 family are induced by high salinity and transiently inhibit the NF-YA transcription factor, *BMC Mol Biol.*, 10: 29.

Zhao, J., Yuan, S., Zhou, M., Yuan, N., Li, Z. and Hu, Q. (2019). Transgenic creeping bentgrass overexpressing Osa-miR393a exhibits altered plant development and improved multiple stress tolerance, *Plant Biotechnology, J.*, 17: 233-251.

Zhao, M., Ding, H., Zhu, J.K., Zhang, F. and Li, W.X. (2011). Involvement of miR169 in the nitrogen-starvation responses in *Arabidopsis*, *New Phytol.*, 190: 906-915.

Zhou, G.K., Kubo, M., Zhong, R., Demura, T. and Ye, Z.H. (2007). Overexpression of miR165 affects apical meristem formation, organ polarity establishment and vascular development in *Arabidopsis*, *Plant Cell Physiol.*, 48: 391-404.

Zhu, K., Liu, M., Fu, Z., Zhou, Z., Kong, Y., Liang, H., Lin, Z., Luo, J., Zheng, H., Wan, P., Zhang, J., Zen, K., Chen, J., Hu, F., Zhang, C.Y., Ren, J. and Chen, X. (2017). Plant microRNAs in larval food regulate honeybee caste development, *PLoS Genet.*, 13: e1006946.

Index

A

Abiotic stress, 52, 54, 55, 183-185, 190, 192, 198, 201, 202, 204, 212, 303, 306-309, 313-317, 321, 322
Adaptation, 91, 93, 103, 104
Animal miRNA, 16-19, 21, 23, 31, 33, 36, 37
Antiviral defence, 247, 248, 250
Artificial miRNAs, 364, 365

B

Biogenesis of miRNA, 17, 18, 27, 36
Biotechnology, 360
Biotic and abiotic stress, 269
Biotic stress, 56, 58, 207, 209, 305, 315, 318
Breeding, 121, 123, 132, 175, 180

C

Chelating and signalling molecules, 184
Chilling, 175-179
Cold stress, 175-179
Crop improvement system, 360, 367

D

Development, 394, 395-402, 404, 406-410, 414
Disease resistance, 243, 250
Drosha, 16, 18, 27, 31-33, 36

E

Embryogenesis, 66, 67, 69, 71
Epigenetic regulatory mechanisms, 341

F

Flowering, 66, 70, 71, 397, 398, 402, 406
Food security, 187

G

Gene expression, 50-52, 54, 56-58
Gene regulation, 199
Gene silencing, 17, 19, 22, 25, 50, 51, 53, 57, 361, 369
Global warming, 121, 122

H

Heat stress responses, 91
Heavy metals, 183-185, 187, 188, 192
Herbicide resistance, 220, 221, 231-233
Herbicides, 221, 230-234
High temperatures, 91-94, 100, 101, 103
High throughput sequencing, 95, 100, 104, 106

L

LDL, 410, 413
Light signalling, 156
lmiRNAs, 332, 333, 335-337
lncRNAs, 342, 345-351

M

Microprocessor complex, 18, 31
MicroRNA, 1, 50, 51, 53, 54, 57, 90, 91, 94-96, 98, 220, 221, 223, 227, 228, 231, 303, 305, 306, 308, 309, 314-322
microRNA, 257, 258, 261, 262
miPEP, 379-382, 384, 387, 389, 390
miRNA evolution, 16, 33
miRNA turnover, 2, 10, 13
miRNA, 65-72, 74-83, 120-135, 144-151, 153-159, 184-193, 198, 199, 201-212, 243-251, 277-279, 281, 282, 285-290, 292-295, 329, 332-340, 379-383, 385-390, 394-398, 400-414
miRNA-based marker, 250

Index **425**

Molecular markers, 302-304, 308, 309, 315-318, 322

N

Non-coding RNAs, 303, 305-309, 379, 383, 390

O

Open reading frames, 379
Osmolyte stress, 167
Oxidative stress, 149, 150

P

Phytohormones, 257-259, 261, 264, 265, 269, 270
Plant defence, 278, 283, 285, 289, 292
Plant miRNA, 16, 17, 20, 21, 25, 26, 29, 31, 33, 34, 38
Plant virus, 242, 243, 247-251
Plant-derived miRNAs, 394, 410-412
Plant-microbe interaction, 277, 279, 295
Plants, 74-77, 79, 81-82, 220-233
Post-transcriptional response, 168
Post-transcriptional silencing, 22
Pri-miRNA, 379, 380, 385, 387, 389, 390

R

Regulatory network, 123, 127, 176, 177
RISC assembly, 6-8
RNA silencing, 243, 247-251

RNA-induced silencing complex, 258, 259

S

Salinity, 167-173
Seed development, 66, 67, 71
Seed germination, 401-404
Signalling pathway, 168-170
Signalling pathways in plants, 257
Silencing, 1, 4, 8, 74, 76-80
siRNAs, 145, 147, 148, 150, 159, 331-346
Stresses, 77, 79-81, 83, 398, 401-407

T

Target gene, 281, 282, 284-286, 290, 294, 295
Target mRNAs, 146
Transcription factors, 175-177, 179, 180
Translational inhibition, 21, 23, 26
Transporters, 184, 187, 188

U

UV radiation, 144, 145, 149, 151-153

V

Vegetative phase, 70

W

Water deficit, 121, 125, 126, 128, 132-135
Weed control, 220, 221, 231-233